Endorsements for *Communication Networking: An Analytical Approach*

The authors show excellent taste in their selection of topics. The book should be both a treat for those who are analytically minded and an appetizer for those who are not.

−Vivek Borkar, School of Technology and Computer Science,
Tata Institute of Fundamental Research

Communication Networking: An Analytical Approach provides a timely, complete discussion of current networking paradigms and state-of-the-art analysis and modeling techniques.

−Professor Gustavo de Veciana, The University of Texas at Austin

This book combines an innovative and uniform representation of a variety of communication networks, simple to understand motivations for real design problems for these networks, intuitive approaches to solutions, and rigorous mathematical analysis where appropriate. It will be very valuable both as a textbook and as a reference for practitioners.

−Bharat Doshi, Director of Transformational Communication
Johns Hopkins University, Applied Physics Laboratory

This book provides a quantitative approach to networking and is a much needed addition to the field. Not only is the coverage up to date in a rapidly evolving field, but it also covers the appropriate analytical methodologies for design and evaluation of communication networks.

−P. R. Kumar, Professor, Dept. of Electrical and Computer Engineering, and
Research Professor, Coordinated Science Lab, University of Illinois

It has been very difficult to write a textbook on networking that is relevant and rigorous because the field is diverse and fast changing. This book stands out in both providing the readers with the essential domain knowledge and equipping them with fundamental tools to analyze and design new systems as the networking field evolves.

−Steven Low, California Institute of Technology

This book is a well-researched compendium of theoretical modeling applied to a number of practical networking problems. Some interesting topics of note are important insights in the design of packet switches, performance of TCP under various conditions, and the design of packet address prefix lookups.

Although much of the material is mathematically advanced, the book contains a comprehensive set of appendices useful as a reference for the researcher or advanced practitioner.

–Dr. David E. McDysan, Fellow, MCI Internet Architecture and Technology

This book fills an important gap since good advanced books in networking are rare. It nicely combines in-depth technology aspects with analysis. The feedback from students who took my graduate class on "advanced topics in networking" here at Purdue was very positive. I recommend it!

–Catherine Rosenberg, Professor, School of ECE and Director, Center for Wireless Systems and Applications, Purdue University

I found the book *Communication Networking* very illuminating. The book lucidly describes important research results in networks and makes these accessible for first and second year graduate students.

–Saswati Sarkar, Department of Electrical Engineering, University of Pennsylvania

This is one of the first textbooks of its kind that gives excellent motivation combined with detailed explanation of key concepts in communication networks. The main strength of the book is the analytical and engineering approach taken by the authors. The book is highly recommended as a senior level undergraduate text or a graduate text in universities. With its unique approach, the book should also be extremely useful to networking researchers in industry.

–Rajeev Shorey, Computer Science Department, National University of Singapore & Research Staff Member, IBM India Research Laboratory

A truly great book that provides an in depth and detailed treatment of both fundamental and advanced networking concepts. The book is very well written and will be extremely useful in the classroom as well as for practitioners and researchers.

–Biplab Sikdar, Department of ECSE, Rensselaer Polytechnic Institute

Functionally organized under multiplexing, switching, and routing, this book provides clear, precise, and substantive coverage of contemporary issues at a level that is readily accessible to graduate and advanced undergraduate students. It splendidly fills a void in an area that has seen remarkable progress over the past decade.

–Joseph Thomas, University of Maryland, Baltimore

COMMUNICATION NETWORKING

An Analytical Approach

The Morgan Kaufmann Series in Networking

Series Editor, David Clark, M.I.T.

For further information on these books and for a list of forthcoming titles, please visit our website at http://www.mkp.com

COMMUNICATION NETWORKING

An Analytical Approach

Anurag Kumar
Department of Electrical Communication Engineering
Indian Institute of Science

D. Manjunath
Department of Electrical Engineering
Indian Institute of Technology, Bombay

Joy Kuri
Centre for Electronics Design and Technology
Indian Institute of Science

ELSEVIER

Amsterdam Boston Heidelberg
London New York Oxford
Paris San Diego San Francisco
Singapore Sydney Tokyo

MORGAN KAUFMANN PUBLISHERS

Morgan Kaufmann Publishers is an imprint of Elsevier

Senior Editor	Rick Adams
Publishing Services Manager	Simon Crump
Associate Editor	Karyn Johnson
Cover Design	Ross Carron
Composition	Cepha Imaging Pvt. Ltd
Copyeditor	Betsy Hardinger
Proofreader	Deborah Prato
Indexer	Robert Swanson
Interior printer	Maple-Vail Book Manufacturing Group
Cover printer	Phoenix Color

Morgan Kaufmann Publishers is an imprint of Elsevier.
500 Sansome Street, Suite 400, San Francisco, CA 94111

This book is printed on acid-free paper.

Library of Congress Cataloging-in-Publication Data
Application submitted

ISBN: 0-12-428751-4
For information on all Morgan Kaufmann publications,
visit our Web site at *www.mkp.com*.

Printed in the United States of America
04 05 06 07 08   5 4 3 2 1

To our teachers, our colleagues, and our students, from all of whom we have learnt.

Contents

Preface

Raison d' être

The field of study popularly known as "networking" is concerned with the science and technology for building communication networks that provide reliable and timely information transport services to distributed applications over a physical network of communication links. Thus communication networks are engineering systems, and, as with other engineering systems, an analytical approach can be brought to bear on their study and design. Although qualitative heuristics can be used for simple networks, for high-speed integrated services networks the analytical approach can yield more efficient solutions. With rapidly expanding link bandwidths, increasing memory densities, frequent releases of ever more powerful microprocessors, and shrinking time to market, there is often little time to determine optimal implementations. Nevertheless the analytical approach yields important insights that can drive good heuristics. Furthermore, in the design of high-speed wide area networks, we are faced with physical limits (such as the speed of light and geographical distances) and technological limits (such as the lag between transmission speeds and memory access speeds), and an approach based on quantitative evaluation of alternatives is indispensable in handling these limits.

The analytical approach has played an important role in the engineering of circuit-multiplexed telephone networks (e.g., for network sizing and for improving adaptive routing strategies) and is making a significant impact on the design of high-speed packet networks that carry a variety of traffic, each with its own service requirements. Some examples of such impact are the design of high-performance routers, the routing of bandwidth guaranteed paths, and traffic controls for Web transfers. In networks that offer only store-and-forward services, the technology has been driven more by an effort to develop a best-effort data transport that would work under diverse conditions (links with widely different bit rates and bit error rates and nonhomogeneous hosts and operating systems). Among the best-effort transport technologies, the Internet protocols have by far the largest installed base. With the Internet rapidly becoming the vehicle of choice for transporting all kinds of information, modeling and analysis are proving invaluable in understanding the limitations of the Internet protocols and for improving them so that the Internet evolves into an integrated services packet network that can provide some level of quality of service assurance to each type of traffic it carries. Indeed this effort is strongly influenced by the analytical research that has gone into the development of ATM (Asynchronous Transfer Mode) networking technology.

About This Book

In this book we take a fresh view to the teaching of the quantitative aspects of communication networking. We view networking in terms of the elements required to build a communication network out of the basic "plumbing" of communication links. We identify these elements as multiplexing, switching, routing, and network management. This book is about the first three elements and has one part devoted to each of these elements. Thus, from the point of view of communications engineering, the material presented in this book can be seen as a natural progression from physical layer engineering, which yields communication links, to the engineering of communication networks built out of these links.

The approach in this book is to raise various engineering issues in multiplexing, switching, and routing and to provide analysis and design techniques to address them via mathematical models. Yet this is *not* a book on mathematical tools for networking (specialized monographs are available on such topics), but one on network engineering via mathematical models and their analyses. The mathematical analyses are developed when they are needed as the discussion unfolds. This helps in emphasizing and consequently developing the structure of the right performance questions to be asked during the engineering of network subsystems. In some cases the models attempt to capture the details of particular implementations (e.g., adaptive windowing in TCP); in many cases the models abstract one or more important issues arising in a system architecture, thus permitting the study of various alternatives (e.g., input queueing in packet switches). In each case we show how the models and their analyses can also yield important insights into the underlying phenomena. To the extent possible, the presentation of the material is generic and not tied to specific networking technologies. The latter are discussed as examples or applications. The Internet's Transmission Control Protocol (TCP) is an important exception; this class of adaptive window-based congestion control algorithms is so important that several pages have been devoted to its analysis.

The coverage in this book includes material from recent literature on the modeling, analysis, control, and optimization of communication networks so as to systematically address the issues in multiplexing, switching, and routing. For example, recent advances in the following topics have been covered:

- Deterministic and stochastic network calculus

- Congestion control for elastic traffic: TCP modeling and performance and utility optimization models
- Wireless access networks and multihop ad hoc networks
- Queueing in packet switches
- Input processing in packet switches, such as route lookups and classification
- Routing for quality of service

The book draws heavily on research literature, and each chapter is accompanied by extensive notes on the literature. Inline exercises (that may require providing simple proofs, analyzing simple cases, or carrying out a calculation) help the reader to better appreciate the material. A large number of supplementary problems are provided at the end of all but the introductory chapters. There are about two hundred exercises and problems in the book. Extensive use of schematic figures helps to illustrate the discussions.

Target Audience for This Book

This book has come out of courses taught by us at the Indian Institute of Science and the Indian Institute of Technology, primarily to final-year undergraduate and first-year graduate students. In reading this book, it will be useful to have a familiarity with the "big picture" of networking through at least a descriptive-level knowledge of networking concepts and technologies. To make this book as self-contained as possible, we provide a descriptive overview of the principles and practice of networking before delving into the analytical aspects. Also, a first course in probability and random processes would be useful background for reading this book. We invoke some results from probability theory, linear and nonlinear optimization, discrete event random processes, and queueing theory. An appendix that summarizes the results we use is also provided.

This book is first of all intended as a text to accompany an analytical course in networking for final-year undergraduate and first-year graduate students (in electrical engineering or computer science). The book is also targeted at networking professionals whose work is primarily architecture definition and implementation and who would like to obtain a quantitative understanding of the comparisons between design choices. This book will also be useful to performance modeling and analysis experts who would like to get a better understanding of how models arise from the physical problems, so that the models they analyze can be better tuned to represent more realistic situations.

Outlines of the Chapters

The book is organized into four parts along with five appendices. Some chapters have their own appendices, containing proofs that are technical in nature and that need not be read inline.

The preamble part contains Chapters 1 and 2. Chapter 1 (Introduction: Two Examples) motivates the role of analytical models via examples of a packet voice multiplexing system and a fast packet switch. Chapter 2 (Networking: Functional Elements and Current Practice) first describes networking as being concerned with the problem of sharing the resources of a physical network of communication links. The basic functional elements of networking are then identified and discussed. A description of current network technologies provides the "big picture" understanding of communication networks.

Part I is on multiplexing and contains Chapters 3–8. The discussion on multiplexing begins in Chapter 3 (Multiplexing: Performance Measures and Engineering Issues) with classification of traffic into two types—stream and elastic traffic—and a discussion on the performance and engineering issues associated with each of these traffic types.

Chapter 4 (Stream Sessions: Deterministic Network Analysis) develops the end-to-end network calculus based on worst-case traffic bounds and the characterization of network elements by service curves, and then discusses packet scheduling. The theory developed in the chapter is tied to practice by showing how the RSVP protocol is used to set up a guaranteed delay connection in the Internet's IntServ architecture. In Chapter 5 (Stream Sessions: Stochastic Analysis), we make a natural progression to develop packet multiplexer capacity designs based on probabilistic models and asymptotic results such as the central limit theorem, Chernoff's bound, and Cramer's theorem. The technique of effective bandwidths is covered in detail in this chapter. The chapter ends with a glimpse of why these techniques are not useful for long-range-dependent traffic. While Chapters 4 and 5 are concerned with in-call performance of stream traffic, models for call-level performance are considered in Chapter 6 (Circuit-Multiplexed Networks). Here we first present the analysis of circuit multiplexed links in a multiclass network and then discuss blocking analysis in a network using the Erlang fixed point approach. Applications of the techniques from the chapter to a cellular system and to a WDM optical network are also discussed.

Elastic traffic is discussed in detail in Chapter 7 (Adaptive Bandwidth Sharing for Elastic Traffic) from the point-of-view of dynamic sharing of links among the elastic flows. We begin by using o.d.e. techniques to analyze generic rate-controlled (as in ATM/ABR) and window-controlled (as in TCP) flows through a single bottleneck link. TCP is described and analyzed under various situations: a wide area connection with random loss, multiple connections with random drop, and randomly arriving finite volume connections. Fairness for elastic flows in a network is studied in detail using utility optimization techniques. We also study the long range dependence of TCP controlled traffic.

Chapter 8 (Multiple Access: Wireless Networks) is the last chapter in Part I and details the multiplexing issues in the emerging area of wireless networks. A taxonomy of wireless networks is followed by a discussion on the physical layer issues. There is a brief discussion of cross-layer techniques. The performance of TCP over lossy wireless links is discussed. An extensive discussion on Aloha and IEEE 802.11 random access networks follows. Multihop ad hoc wireless networks are also discussed; the topics covered include connectivity, link scheduling, and capacity scaling laws for dense ad hoc networks.

Chapters 9–12 constitute Part II on switching and are primarily concerned with packet switches. Chapter 9 (Performance and Architectural Issues) is an introduction to switching with a discussion on performance measures for switches and the architectural choices for packet switches. In Chapter 10 (Queueing in Packet Switches), we treat the nonblocking switch fabric as a blackbox and first discuss saturation throughput and queueing delays. Virtual output queueing, switch scheduling for increased throughput, and output queue emulation are also discussed. In Chapter 11 (Switching Fabrics), we progress to designing the switch fabric. We first describe elementary switching structures and then develop the design of switching networks including Clos networks and self-routing networks. The design of a self-routing broadcast network and the multicast switch are also presented. Finally, in Chapter 12 (Packet Processing), the design of fast packet processing (IP address lookup and packet classification) and general issues in packet switch design are discussed. A brief overview of network processors is also provided.

Chapters 13–16 form Part III, which focuses on routing. Chapter 13 (Routing: Engineering Issues) is an introduction to this part. In this chapter, we discuss general issues and terminology that appear in the study of routing. Chapter 14 (Shortest Path Routing of Elastic Aggregates) is about shortest path routing in networks. We formulate and analyze an optimal routing problem for routing a given set of elastic traffic demands that can be arbitrarily split across multiple routes. Then we explore the idea of realizing optimal routes by setting appropriate link weights and then using shortest paths according to these link weights. Finally, we discuss well-known shortest path algorithms and their generalizations, as well as two standard routing protocols. Chapter 15 (Virtual-Path Routing of Elastic Aggregates) is motivated by the issue of routing MPLS paths. We consider the generalized problem of on-demand routing, where prior information about the demands is not known. We also assume the constraint that demands cannot be split across multiple routes and study minimum interference routing formulations and heuristics. Chapter 16 (Routing of Stream-Type Sessions) considers the problem of routing stream traffic sessions so as to satisfy their end-to-end performance requirements. We consider nonadditive and additive performance metrics, as well as networks with rate-based and non-rate-based multiplexers, and study algorithms to obtain feasible routes.

Appendix A is a glossary of the acronyms and the mathematical notation used in the book. Appendices B–E provide a review of some of the mathematical techniques used in the book.

Using the Book

We have made the book self-contained by providing all the advanced concepts used in the book in the appendices. There are some "starred" sections (numbered or unnumbered and indicated by the symbol (⋆) in the heading) that deal with specialized or more advanced material and can be skipped on a first reading without losing the continuity of the presentation. The dependencies between the different parts of the book are shown in Figure 1. Thus, for example, Chapters 3, 4, and 5 must be read in sequence, and these can be followed by Chapter 7, but it is also possible to develop a course that uses Chapters 3, 7, and 8, with the instructor filling in some material from Chapter 5 as needed in Chapter 7.

There are many ways in which the book can be used to teach an analytical course in networking. The material in Chapters 1 and 2 can be given as a reading assignment at the beginning of the course or covered in 3–5 lecture hours. One lecture hour each on the introductory chapters to the three parts— Chapters 3, 9, and 13—may be useful to explain the issues in each of the components. The ideal way to use the book would be in a two-semester course with Part I being covered in the first semester and Parts II and III in the second semester. A one-semester course based on parts of Chapters 4–8, 10, and 14 has been taught by one of the authors. A course based on Chapter 4, Part II (except Chapter 10), and Part III, along with a summary of the results from Chapters 5–8

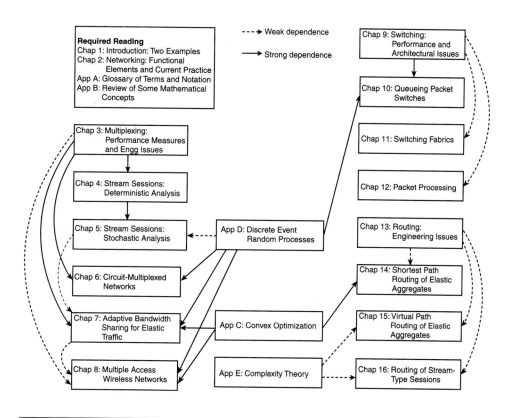

Figure 1 **Dependencies between the material covered in the various chapters of the book.**

and Chapter 10, will be accessible to graduate students who do not have a strong background in stochastic analysis.

Our experience in conveying some of the seemingly difficult probability concepts to students with possibly limited exposure to such material has been very positive and we recommend that an instructor bravely try it.

The web site for the book is at www.books.elsevier.com\mk\?ISBN= 0124287514. This web site contains errata, additional problems, PostScript files of the figures used in the book, and other instructional material. An instructor's manual containing detailed course plan alternatives and solutions to all the exercises and problems is also available via password protected access (see web site for details).

Acknowledgments

We are grateful to Vivek S. Borkar, Bharat T. Doshi, P. R. Kumar, Steven Low, David McDysan, Galen Sasaki, and Don Towsley for the patience with which they read the manuscript and for the detailed comments and critique they provided. These comments have helped to smooth many rough edges in the presentation.

Sanjay Bose, A. Chockalingam, Supratim Deb, Madhav Desai, Munish Goyal, N. Hemachandra, Koushik Kar, Akshay Kashyap, Nilesh Khude, P. Mathivanan, Harish Pillai, and Amol Sahasrabudhe read many of the draft chapters and provided valuable feedback. We are thankful to them. We thank the many students who pointed out errors and omissions when parts of the books were used in our courses.

In the planning and preparation of this book the authors have had useful and encouraging discussions with Gustavo De Veciana, Sanjay Shakkottai, and Pramod Viswanath. We are grateful to them for sparing their time for these discussions.

Parts of this book were class tested by Alhoussein Abouzeid (Rensselaer Polytechnic Institute), Steven Low (California Institute of Technology), Armand Makowski (University of Maryland, College Park), Utpal Mukherji (Indian Institute of Science), Catherine Rosenberg (Purdue University), Saswati Sarkar (University of Pennsylvania), Rajeev Shorey (National University of Singapore), Biplab Sikdar (Rensselaer Polytechnic Institute), and Joseph Thomas (University of Maryland, Baltimore County). We thank them for their confidence in the book and the valuable feedback they have given us.

The team at Morgan Kaufmann and Elsevier were a pleasure to work with. Our thanks are due to Rick Adams and Karyn Johnson for providing excellent support during the development of the book and for quick responses to all our questions and concerns, to Betsy Hardinger for the pain staking copy editing, to Marcy Barnes-Henrie and Simon Crump for efficiently managing the production, to Gopinath Iyengar for a superb job of compositing the material, to Eric DeCicco for patiently discussing the cover design with us, and to Brent dela Cruz for managing the publicity.

Finally we express our gratitude to our families for their love, support, and patience during the preparation of the book. Anurag Kumar thanks his wife, Pamela, for her constant encouragement, his mother for her ever present support, and his children, Anupama and Siddhartha, for their patience during his long periods of preoccupation with this project. D. Manjunath thanks ManjulaRani for maintaining sanity at home during the many times the timelines of her thesis and this book were in phase and AmruthaVarshini (more than half her life has been spent in the shadow of this book) for bearing his absence and supporting KMK. Joy Kuri thanks his parents, Nanda Kumar and Bela, his wife, Manjula, and his daughter, Ushnaa, for their love, support, and encouragement.

Anurag Kumar   D. Manjunath   Joy Kuri
I.I.Sc., Bangalore   I.I.T., Bombay   I.I.Sc., Bangalore

CHAPTER 1

Introduction: Two Examples

To help illustrate the emphasis and tenor of this book, we begin with two examples. Although they are very different from each other, they show how quantitative models and analyses provide crucial insights into the behavior and performance of a packet network subsystem. In both cases, they helped to determine the direction of the research and development efforts on these subsystems.

1.1 Efficient Transport of Packet Voice Calls

A packet voice communication system is one in which the voice signal is digitized, the bits are grouped into packets, and the packets are transported from the source to the destination. A call made using such a system is called a packet voice call.

Figure 1.1 depicts a situation in which a communication link between two locations must carry several packet voice calls. The speed of the link is C (bits per second). Associated with each telephone instrument is a voice digitizer, coder, and packetizer. A random stream of voice packets from each call enters a communication switch (or *packet router*), at which point the packets are *multiplexed* into the link. At the other end of the link, a *router* extracts (or *demultiplexes*) the packets from the link and routes them to devices that *depacketize* and decode the voice. A mirror image of this path is followed by voice in the reverse direction. The objective of the system designer is that, given a link of speed C, the number of calls that can be simultaneously carried should be maximized subject to a constraint on voice quality.

There are two aspects to voice quality: distortion and delay. The voice coder is assumed to perform *embedded* (or *layered*) coding. Thus, if the peak coding rate is, say, 4 bits per sample, and if every packet is delivered intact, then essentially distortion-free (telephone-quality) speech is heard by the listener. On the other hand, embedded coding permits, for example, 1 bit per sample to be dropped (in an appropriate way) from each voice packet to obtain a coded voice stream with

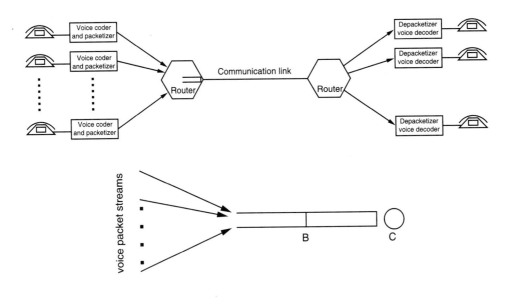

Figure 1.1   An interlocation link is used to carry packet voice calls (top). The LAN-WAN packet router on the left multiplexes several digitized, coded, and packet voice sources into the link, and the router on the right extracts packets from the link. On the bottom left is a queuing model for studying the performance of packet voice multiplexing.

3 bits per sample. This facility can be used to reduce the traffic, thus permitting more calls to be carried simultaneously. The quality of the received speech is acceptable, however, only if at least a fraction α of the coded bits are delivered (for example, with $\alpha = 0.95$ and 4 bits peak rate per sample, at least 3.8 bits per sample would need to be delivered). Because the arrivals of packets into the link are random, queuing of packets will necessarily occur, leading to delays. It is very difficult to carry out an interactive conversation when there is substantial delay. In practice, there are several components of the end-to-end delay (see Chapter 3), but here we focus only on the delay in the multiplexer. We can specify the multiplexer delay requirement in terms of a level B (bits) that should be exceeded only rarely in the multiplexer buffer (see Figure 1.1); this says that we desire the delay bound of $\frac{B}{C}$ to be rarely exceeded.

In this problem setting we have two design alternatives.

1. *Bit-dropping at the multiplexer:* The speech is coded at the full rate. When a packet arrives in the multiplexer buffer, if accepting this packet would cause the buffer level to exceed B, some bits are dropped from the arriving and

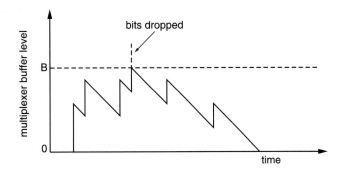

Figure 1.2 Evolution of the multiplexer buffer level with bit dropping control. At a packet arrival instant the buffer increases by the number of bits from the packet that are allowed into the buffer. The buffer decreases at the link rate C in between the packet arrival instants.

existing packets so that after accepting the new packet, the buffer level is exactly B (see Figure 1.2). To smooth out the effect of lower-rate coding, the bit-dropping can be evenly distributed over all the packets involved (to the extent possible, given the discrete nature of bits). We can then ask the question, what is the maximum number of voice calls that can be simultaneously handled by the link so that the system delivers at least a fraction α of the coded bits? Another way to implement such queue-length-dependent bit-dropping is to provide a signaling mechanism between the multiplexer and the coder; it is the coder that drops the bits based on the buffer level signaled by the multiplexer. If the signaling delay is zero and if only bits from the arriving packet are dropped, then this *buffer-adaptive coding* would be equivalent to our description of dropping bits at the multiplexer. Note that because the state of the system (namely, the queue length) is used to determine the instants and the amounts of bit dropping, this is *closed-loop* control.

2. *Lower-bit-rate coding at the source coder:* It can be argued that, because the application can tolerate a lower bit rate, the embedded coder can itself just send a fraction α of the peak number of bits per sample. The link must then carry all the bits of all the packets that it receives, and hence bit-dropping control is not possible in the multiplexer. Owing to randomness in the arrival process, however, the buffer can build up to beyond B. The question in this case is, what is the maximum number of voice calls that can

be simultaneously handled by the link so that the probability (fraction of time) of exceeding the buffer level B in the multiplexer is less than a small number ϵ—say, 0.001? We can view this as *open-loop* control because the coders reduce the bit rate without any consideration of the multiplexer queue length.

To analyze these two alternatives, we will make the simplifying assumption that the random process of arrival instants of packets into the multiplexer is Poisson (with rate λ) and that the packet lengths are continuous and exponentially distributed, with (*full-rate-coded*) mean $\frac{1}{\mu}$. Such a simplified model is sufficient for our present purpose of providing some insights into the problem. Analysis, albeit more complex, can also be carried out with more general assumptions (see Chapter 5).

We will assume that the mean packet size is fixed (a characteristic of the speech and the coder), whereas the packet arrival rate λ relates to the number of voice calls. The total, uncontrolled arrival rate of bits is $\frac{\lambda}{\mu}$. Define $\rho := \frac{\lambda}{C\mu}$, which is the total arrival rate of (full-rate) bits from the sources, normalized to the link speed. It is clear that it is sufficient to determine the maximum value of ρ that can be supported in each of the two alternatives.

The model can be analyzed to obtain the distribution of the steady state multiplexer buffer level for the bit-dropping multiplexer, and for lower-bit-rate coding at the source. In each case the probability distribution has a continuous part with a density and also has a point mass at 0. Let us denote by $f(\cdot)$ the density function, and by z the point mass at 0. The following results can be obtained quite easily by the technique of *level crossing analysis* (see Appendix D, Section D.1.9, where the following equations are derived).

The stationary distribution of the buffer level with bit-dropping is supported on the interval $[0, B]$. For $0 < x < B$, we have

$$f_{bit\text{-}dropping}(x) = \frac{\mu(1-\rho)\rho}{1 - \rho e^{-\mu(1-\rho)B}}\, e^{-\mu(1-\rho)x} \tag{1.1}$$

and

$$z_{bit\text{-}dropping} = \frac{(1-\rho)}{1 - \rho e^{-\mu(1-\rho)B}} \tag{1.2}$$

Because the link is sending bits whenever the buffer is not empty, the rate at which the multiplexer is putting bits through is $(1-z)C$. Hence the fraction of the arriving bits that are carried is given by $\frac{(1-z)C}{(\lambda/\mu)} = \frac{1-z}{\rho}$. With bit-dropping at the multiplexer, the fraction of offered bits that are carried by the multiplexer should be at least α (e.g., 0.95). Using the formula for z, this requirement translates to the following requirement:

$$\frac{1 - e^{-\mu(1-\rho)B}}{1 - \rho e^{-\mu(1-\rho)B}} \geq \alpha \tag{1.3}$$

Note that, by definition, a bit-dropping multiplexer bounds the delay to the time taken to serve B bits.

Let us define the maximum load that can be offered to a bit-dropping multiplexer, so that the average bits per sample objective is met, by

$$\rho_{bit\text{-}dropping}^{max} := \sup\left\{\rho: 0 \leq \rho \leq 1, \frac{1 - e^{-\mu(1-\rho)B}}{1 - \rho e^{-\mu(1-\rho)B}} \geq \alpha\right\}$$

The stationary distribution of the buffer level with lower-rate-source coding is supported on $[0, \infty)$. For $x \geq 0$, we have

$$f_{low\text{-}rate\text{-}coding}(x) = \mu(1 - \alpha\rho)\rho e^{-\frac{\mu}{\alpha}(1-\alpha\rho)x} \tag{1.4}$$

and

$$z_{low\text{-}rate\text{-}coding} = 1 - \alpha\rho \tag{1.5}$$

This result requires that $\alpha\rho < 1$, and the distribution can be obtained from the one for bit-dropping as follows. Because a fraction α of the bits are retained by the source, the mean packet length is $\frac{\alpha}{\mu}$. Furthermore, we cannot drop any more bits in the multiplexer; this is equivalent to letting $B \to \infty$. Now consider a bit-dropping multiplexer with packet arrival rate λ and mean packet length $\frac{\alpha}{\mu}$. The preceding distribution will be obtained from Equations 1.1 and 1.2 by letting $B \to \infty$.

We, however, still need to meet the multiplexer delay objective. The requirement that the probability of the buffer level exceeding B be less than ϵ leads to the following inequality.

$$\alpha\rho e^{-\frac{\mu}{\alpha}(1-\alpha\rho)B} \leq \epsilon \tag{1.6}$$

Define the maximum voice load that can be offered to the multiplexer (after lowering the coding rate by a fraction α) by

$$\rho_{low\text{-}rate\text{-}coding}^{max} := \sup\left\{\rho\colon 0 \leq \rho \leq 1, \alpha\rho e^{-\frac{\mu}{\alpha}(1-\alpha\rho)B} \leq \epsilon\right\}$$

In Figure 1.3 we plot $\rho_{bit\text{-}dropping}^{max}$ and $\rho_{low\text{-}rate\text{-}coding}^{max}$ as a function of B normalized to the mean packet size $\frac{1}{\mu}$ (i.e., we plot the maximum load versus μB). Note that these plots depend on only two parameters—namely, α and ϵ—both specifying the performance objectives.

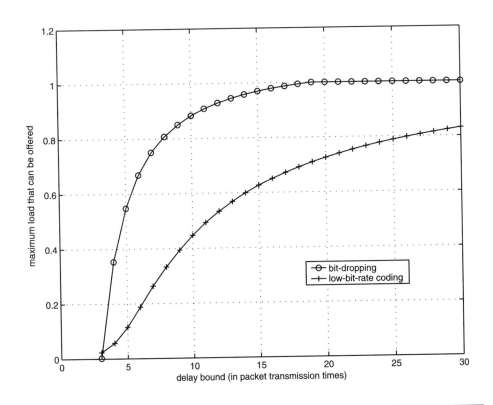

Figure 1.3   Maximum offered load that meets the performance objectives for bit-dropping and low-bit-rate coding, plotted versus the delay objective expressed as μB. The parameters are $\alpha = 0.95$ and $\epsilon = 0.001$.

The plots in Figure 1.3 show that bit-dropping (or closed-loop control) can permit the system to carry significantly more load than simply coding the voice at just enough bits per sample (open-loop control). For example, for $\mu B = 10$ (i.e., a delay bound of 10 packet service times), the bit-dropping approach will carry twice as many voice calls as the open-loop approach. The large difference is significant. The advantage of closed-loop control can be intuitively explained as follows. The closed-loop control intelligently drops bits only when there is need to—that is, when the delay threatens to exceed the target delay. On the other hand, the open-loop control naively drops all the bits it possibly can during coding itself. The variability in the buffer process remains, and there is no more leverage in the multiplexer to take any action to control congestion.

This simplified model has glossed over several practical issues (for example, the perceptual effect of variability in the coding rate as a result of bit-dropping), but the insights gained are remarkable. A simple model, along with quick analysis, has taught us a lot. In the engineering process, more detailed modeling and simulation experiments then take our understanding further, all the way to a complete design. (See [269] for an implementation example.) Such a complete design would yield a multiplexing and control strategy that would maximize the rate at which calls could be offered to the system, while meeting the quality of service objectives (bits per sample, packet delay, and call blocking).

1.2   Achievable Throughput in an Input-Queueing Packet Switch

As a second example, consider a packet switch that is at the heart of most high-speed routers, such as the Cisco 12000 series of Internet routers. Such a high-speed packet switch is typically designed to handle fixed-length packets, also called *cells*. Cell switches are usually operated in a time-slotted manner; time is divided into *slots*, and the activities at all the ports—such as cell arrivals from the input links, switching of the cells to the respective outputs, and departures on the output links—are synchronized to the slot boundaries. Each slot duration is equal to the cell transmission time. Consider such an $N \times N$ cell switch having N input ports and N output ports. Let the transmission rates on the input and output links be equal. Let d_i denote the destination of the cell in input i at the beginning of a slot. Assume that the switch is *nonblocking*, meaning that if $[d_1, d_2, \ldots, d_N]$ is a permutation of $[1, 2, \ldots, N]$, then the cell at input i can be sent to output d_i in that slot, for all i. The design of nonblocking switches is well understood and was originally considered in the context of telephone switches and in interconnection networks used for processor and memory subsystem communication in parallel computing systems.

Now consider the operation of a cell switch. In each slot, up to one cell can be received on each of the N input links, and the cell can have any of the N output links as its destination. Thus it is possible that in a slot, more than one cell with the same destination can arrive at the inputs, causing *destination conflicts*. Because the output link rates are equal to the input link rates, in each slot at most one cell can be transmitted on each of the output links. So how do we handle the excess (more than one) cells destined for the same output that may have arrived in the slot? Obviously, dropping them is not a good idea; instead, they need to be buffered or queued. But where? We can see that the queue can be placed either at the output ports or at the input ports, and accordingly we will have an *output-queued* (OQ) or an *input-queued* (IQ) switch.

In the OQ switch, all the cells destined to a port are switched to the output in the same slot, and one of them is transmitted on the output link if no other packet from a previous slot is waiting. Clearly, the OQ switch can provide 100% throughput. In other words, if the average rate at which each output receives cells is less than the output link rate, then the queue will not grow in an unbounded manner. However, by definition, the OQ switch should be capable of switching up to N cells to an output in a slot. More importantly, the memory that is used for the output queue should be capable of supporting such a read-write speed. Improvements in memory access speed have not kept pace with that of processor technologies or even with the size of memory. Thus the OQ switch does not scale very well with N. This means that it might be feasible for small N, but as N becomes large, especially if the transmission rates are high, it can become technologically infeasible.

The IQ switch works as follows. There is a *first-in-first-out* (FIFO) queue maintained at each input and an arriving cell is placed at the tail of this queue. In each slot, at most one cell is transferred to each output, which is immediately transmitted on the output link. Thus the switch should be capable of switching one cell to each output and one cell from each input. The speed of the memory used to maintain the cell queues should allow one read and one write in each slot. This is obviously a more scalable alternative to the OQ switch. However, as is shown in the example in Figure 1.4, it suffers from *head-of-line* (HOL) blocking, which severely affects its achievable throughput. The question then is, how low is this reduction in the throughput? To answer this question, we proceed as follows.

Consider an $N \times N$ IQ switch operating under the extreme condition of input saturation: Every time a cell is switched from the input to its output, it is immediately replaced by another cell. Any of the N output ports is equally

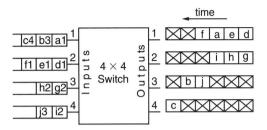

Figure 1.4 Example of a head-of-line (HOL) blocking situation in an IQ switch. At the input, the cells are shown with a label denoted by a letter and the destination port number. At the output, the cell departure sequence in six slots is shown. It is assumed that no more packets arrive at the inputs during these slots. Crosses indicate that there was no transmission from the output in that slot. Observe that although output 4 is free, packet c is blocked by its HOL packets, a and b, which are waiting for their outputs to become free.

likely to be the destination of this cell and is independent of everything else (e.g., destinations of other cells). Consider the case for $N = 2$. We will assume that in the case of destination conflict, one of the cells is chosen randomly to be switched and transmitted on the output link. Now consider the system at the beginning of every slot. Let $d_n^{(i)}$ be the destination of the cell in input i during slot n. $d_n := [d_n^{(1)}, d_n^{(2)}]$ can be called the state of the switch at the beginning of slot n. There are four possible states that the switch can be in: $[1, 1]$, $[1, 2]$, $[2, 1]$, and $[2, 2]$. Observe that to determine the probabilities for the state of the switch at the beginning of slot $n + 1$, it is sufficient to know the state at the beginning of slot n. Thus the switch can be modeled as a four-state discrete time Markov chain, with transition probabilities as shown in Figure 1.5.

Solving the Markov chain of Figure 1.5, we get the stationary probability of the switch being in each of the four states to be 0.25. To obtain the throughput of the system we use the following observation. If the system is in state $[1, 1]$ or $[2, 2]$, only one of the cells can be switched to the output, and the throughput of the switch will be 0.5 cell per port. If the switch is in either $[1, 2]$ or $[2, 1]$, both cells can be switched to their output and the throughput is 1 cell per port. Thus the stationary saturation throughput of this switch will be $(0.25 \times 0.5 + 0.25 \times 0.5 + 0.25 \times 1.0 + 0.25 \times 1.0) = 0.75$ cells per port. Although this analysis can be extended to larger values of N, we immediately see a scalability issue in solving the model. The number of states in the Markov chain will be N^N, and even for small values of N it becomes impossible to enumerate the states, describe them, and solve

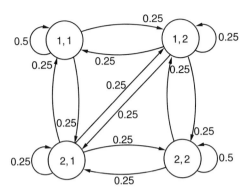

Figure 1.5 Markov chain representation of the state of the 2 × 2 input queued switch at the beginning of every slot. The numbers next to the arrows are the transition probabilities. The state of the Markov chain is $[d^{(1)}, d^{(2)}]$, where $d^{(i)}$ is the destination port of the packet at input i.

the Markov chain. The saturation throughputs for eight values of N are shown in Table 1.1 after obtaining them using this technique. As shown in the table, the throughput decreases as N increases. The question then is, does the saturation throughput converge to some value sufficiently greater than zero, or would it continue to decrease to zero with increasing N, thereby negating the technological advantages of the input-queued switch through a significant performance penalty for large N? The answer to this question is in the asymptotic analysis, discussed

N	Saturation throughput
1	1.0000
2	0.7500
3	0.6825
4	0.6553
5	0.6399
6	0.6302
7	0.6234
8	0.6184

Table 1.1 Saturation throughput of an $N \times N$ input-queued switch for $N = 1, 2, \cdots, 8$.

in Chapter 10, where we will show that the saturation throughput converges to $2 - \sqrt{2}$ (≈ 0.586) cells per port as N becomes large.

The saturation throughput of the switch is an indication of the *capacity* of the switch: the maximum packet arrival rate that can be applied to the inputs and yet keep the queue lengths bounded. In fact, it is widely believed, but not proved except for a special case, that the capacity is equal to the saturation throughput. From the foregoing discussion, the maximum per port packet arrival rate sustainable by an IQ switch is asymptotically 0.586, whereas it is 1.0 for the OQ switch.

Thus the capacity of the FIFO IQ switch—the input arrival rate that it can sustain at each input—is significantly greater than 0 even for large N but significantly worse than the ideal of 1.0 of the OQ switch. The natural interest then is in improving its capacity, possibly by doing some computing to smartly schedule the cells that will be offered to the switch fabric from each input, to make the capacity approach the ideal of 1.0. This will allow us to trade off switching speed for computing power. As has been shown in the example of Figure 1.4, the source of throughput degradation is the FIFO nature of the input queues and the consequent head-of-line blocking. To minimize this effect, many ad hoc schemes were tried, and it was not clear until 1996 whether IQ switches could ever achieve the ideal capacity. In that year, it was conclusively shown that an IQ switch can be made to yield a throughput of 1.0 per port by devising a scheme that takes an extreme step to break the HOL effect. In this scheme, called *virtual output queueing* (VOQ), each input, denoted by i, maintains a separate queue of cells for each output, denoted by j, called the virtual output queue at input i for output j and denoted by VOQ_{ij}. A weighted *bipartite* graph with a vertex for each input and each output is formed with edges defined only between an input and an output vertex. The weight of the edge (i,j), is Q_{ij}, the number of cells in VOQ_{ij}. A 2×2 VOQ switch and the weighted bipartite graph are shown in Figure 1.6. A *maximum-weight matching* algorithm is executed on the weighted bipartite graph at the beginning of each slot, and the matching determines from which of the N virtual output queues at each input a cell will be switched to the respective outputs. We will study this and other algorithms in more detail in Chapter 10.

Note that an identical problem was solved in the context of multihop *packet radio networks* (PRNs). A multihop PRN is represented by a graph $G = (V, E)$, where V is the set of nodes in the network and E is a set of directed edges. A directed edge $e = (i, j) \in E$, where $i, j \in V$, indicates that node j is in the transmission range of node i—that is, node i can receive the transmissions from node j. Figure 1.7 shows an example. At any time, if link (i, j) is *active*, it means that node i is transmitting to node j. Radio networks differ from wired networks in the important aspect that not all links can be active at the same time. This is

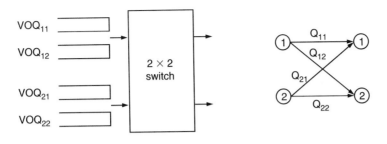

Figure 1.6   A 2 x 2 virtual output queue switch. The left panel shows the four queues—$Q_{11}, Q_{12}, Q_{21},$ and Q_{22}—at the inputs, and the right panel shows the corresponding bipartite graph.

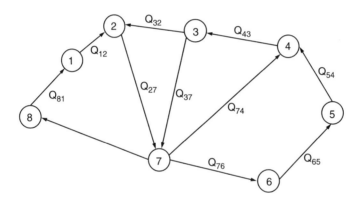

Figure 1.7   An example of a graph representing a packet radio network. The link weights are Q_{ij}, the number of packets in i with destination j.

for two reasons. First, in such networks, the radio transceiver can either send or receive at any point in time but cannot do both simultaneously. Second, a radio receiver can decode only one transmission at a time; multiple signals received at a node can destructively interfere so that neither is received correctly. On the other hand, if a radio receives one strong signal and another weak one, it may be able to detect the data in the strong signal. Thus, for example, in Figure 1.7, links $(1, 2)$ and $(3, 2)$ cannot be active at the same time because node 2 cannot receive two simultaneous transmissions from such nearby nodes. On the other hand, we can specify that $(6, 5)$ and $(3, 2)$ can be active at the same time because the interfering signals are weak enough at this distance. The set of edges that can be simultaneously active is called a *feasible schedule*. Let S be the set of all feasible schedules.

Consider a time-slotted PRN with slot length equal to a packet transmission time. Let the packet arrivals form a random process, and, at the beginning of slot n, let $Q_n^{(ij)}$ be the number of packets that are at node i with destination to a neighbor j. Let us also assume that all transmissions need to go only one hop; when node i sends to node j, the packet exits the network at j. It is obvious that the sequence of schedules determines the capacity of the network: the maximum arrival rates that can be allowed between the source destination pairs without the queues becoming unbounded. Now consider the problem of scheduling the links to maximize the capacity of the packet radio network. The following algorithm is shown to provide the maximum capacity. At the beginning of every slot n, obtain a maximum-weight-matching on the weighted graph G with weight of edge (i, j) equal to $Q_n^{(ij)}$, and activate the links in the match. Observe that this problem, and the solution, is identical to the VOQ switch scheduling problem discussed earlier except that the graph in the PRN is not a bipartite graph. We discuss multihop PRNs in Chapter 8.

From this discussion, we can see that two important practical problems in networking, seemingly disparate, in the abstract have an almost identical solution technique. This underscores the importance of an analytical approach to networking, which is the focus of this book.

1.3   The Importance of Quantitative Modeling in the Engineering of Telecommunication Networks

In engineering, mathematical modeling is typically used for two purposes:

1. Simple "toy" models are developed and analyzed to obtain insights into phenomena. These models abstract away a lot of detail, and the results obtained from them are usually not directly useful in the design of actual systems. Such models are amenable to mathematical analysis, yielding provable mathematical relationships and sometimes even yielding closed-form expressions for the way performance varies with system parameters. These models can be used to derive control algorithms and scheduling policies and to motivate heuristics. The examples given earlier fall into this category and are only two among numerous instances in which mathematical

models and their analysis have provided important insights into phenomena in telecommunication networks.

2. More complicated models that capture more system detail are developed and analyzed to yield numerical results that can be used in actual system design. Often these models are too complex for obtaining closed-form expressions, and we must take recourse to numerical computation or even computer simulation. In such situations, simplified instances of the complete model, but amenable to mathematical analysis, are useful in validating the computer simulations. Algorithms and heuristics derived from the simpler models can be experimented with on these more detailed models.

Perhaps the earliest use of engineering models in telecommunication networks was in the *teletraffic engineering* of circuit-multiplexed telephone networks. In the early days of telephony, the deployment of long distance telephone circuits was expensive, and hence it was important to understand the call attempt rate that they could be offered for a given probability of call blocking. The most basic model used in the engineering of telephone circuits is the celebrated *Erlang-B* model. This model comprises several identical resources, along with a random arrival process of "customers," each of which requires exactly one (any one suffices) of the resources for a random duration of time, after which the customer departs. A customer that finds all resources occupied is sent away and is not heard from again. The mathematical assumptions are that the request arrival process is Poisson and that the resource holding times are independent and identically distributed; the model yields the probability of turning a customer away, or *call blocking*. Beginning from this model, there was hectic research activity in the area of trunk engineering with alternate-routed traffic (an available route is chosen from a set of routes using a predefined algorithm), time-varying traffic, and with different classes of calls requiring different performance objectives. The Erlang-B model, when extended to a circuit-multiplexed network with *fixed routes*, gave rise to the *network Erlang* model. A natural approximation was extensively studied for obtaining route-blocking probabilities from the network Erlang model. The area of *dynamic routing* was driven by the insights obtained from such engineering models. Although packet voice was explored as recently as the 1970s and 1980s, the idea of exploiting silences in speech to efficiently manage transoceanic telephone channels was studied via mathematical models as early as in the 1950s.

With the emergence of packet networking, attention was paid to the development of models of buffered-congestion systems. In the early days of this technology, packet networks were viewed as *computer networks*: networks for interconnecting computers. Interestingly, the development of the theory of *queueing network* models, and of optimal scheduling in queueing systems, was given considerable impetus by the problems that arose in modeling interconnected computers that were executing distributed computing tasks. In the 1980s it was realized that there could be efficiencies if all kinds of telecommunications traffic (data, voice, video, etc.) could be integrated into packet networks. Because packet networks mix all traffic into the same links, it became necessary to develop local transmission scheduling at links and distributed traffic controls across the network in order for the different flows to obtain the desired performance. The need to understand the performance of *integrated packet networks* gave rise to the need to develop models of buffered-congestion systems with a variety of complex arrival processes (modeling, for example, superposed traffic from packet voice calls), complex service disciplines, and congestion-dependent traffic controls.

The use of analytical models for the end-to-end engineering of integrated packet networks is still in its infancy, but significant engineering insights have been obtained. In addition to the two examples given at the beginning of this chapter, the following are more such striking results:

- A "network calculus" for the end-to-end deterministic analysis of the performance of flows in networks, and for design with worst case performance guarantees.

- A stochastic calculus based on the notion of "effective bandwidths" for link sizing with stochastic performance guarantees.

- Distributed asynchronous algorithms for rate allocation to elastic flows sharing a packet network.

- The observation that packet flows in the Internet are not Poisson-like but instead have slowly decaying correlations (i.e., have long-range dependence), and an understanding of the origins of this phenomenon.

- The proof that the maximum-weight-matching algorithm achieves 100% throughput in an IQ packet switch, and the development of a large class of "maximal" scheduling algorithms that are significantly simpler to implement but perform as well as the maximum-weight-matching algorithm.

- The development of nonhierarchical routing algorithms for telephone networks. This led to the discovery of the problem of the network being in a "bad state" (high call-blocking probabilities) for extended periods of time, and its solution by the method of trunk reservation and state protection.

- The insight that many optimal routing problems are equivalent to shortest path problems on the network topology when appropriate link weights are used.

One of the characteristic aspects of networking technologies (generally not seen in physical layer technologies) is that designs are typically not "derived" from engineering models but, in fact, precede any such modeling exercise. Simple, back-of-the-envelope calculations are taken to provide sufficient support for such designs. Simple, but not naive, technologies such as Aloha, Ethernet, and Transmission Control Protocol (TCP) are among the many important examples. Because such technologies adequately serve their initial purposes, they are quickly deployed. They work well to provide simple services, but to extend them beyond their original brief requires understanding their behavior. Analytical modeling is indispensable for this purpose because the generality of the conclusions is not possible with experimental investigations alone. In the context of TCP, for example, such analytical models have provided very important results on TCP performance problems and their improvement over wireless channels, and on TCP performance optimization and differentiation using Active Queue Management.

The approach of abstracting out the essentials of a problem and formulating it mathematically also permits the discovery of recurrent themes, as we saw in the packet switching and the packet radio network examples in Section 1.2. The problems of resource sharing in networks are usually very complex, and even to develop heuristics we often need a conceptual starting point. The analytical approach permits the development of optimal solutions from which better heuristics can be inferred.

As we have hinted at, the complete design of packet networks using stochastic models is a hard problem, and satisfactory solutions remain elusive. Into this void have stepped deterministic flow-based formulations and related constrained optimization problems. Practical problems—such as capacitated network topology design, setting of link weights for shortest path routing, and network overprovisioning to take care of failure situations—have been addressed using such approaches. These techniques at least provide a network operator with reasonable alternatives that can be starting points for manual optimizations.

This discussion has traced the role that mathematical models have played in the engineering of telecommunication networks. Quantitative modeling has been, and continues to be, indispensable for the proper understanding of and design of telecommunication networks. Such modeling can lead to major gains even in the *qualitative* understanding of systems. For example, in the bit-dropping example of Section 1.1, we found that congestion-dependent bit-dropping could lead to a doubling of the load that could be carried for the same application-level performance. Ultimately, to the network operator the call-handling rate is what matters, because each accepted call provides revenue. Such insights are extremely valuable because they can help us to make judicious choices between various proposals and conjectures, and they can be obtained only via models and their analysis. In this book we cover most of the mathematical models and analysis techniques that have been mentioned.

1.4 Summary

This chapter aims to motivate our analytical approach to the topic of communication networking by presenting two examples that show how important insights are obtained by adopting the approach of mathematical modeling and analysis. One example is in the context of transporting voice over packet networks, and the other is on the problem of queueing and scheduling in high-speed switches. Using these examples as a springboard, we then recall the role of quantitative modeling in telecommunication networks, from the design of telephone networks to the analysis of myriad issues in integrated packet networks.

1.5 Notes on the Literature

The comparison of the performance of packet voice multiplexers with and without bit-dropping was performed by Sriram and Lucantoni [268], and the bit-dropping technique was implemented in a system described in [269]. We provide the results for a simple version of the authors' more detailed model. This analysis can be done using level crossing analysis, for which Brill and Posner [43] is the standard reference.

The saturation throughput of the input queued switch was derived by Karol, Hluchyj, and Morgan [158]. The maximum-weight-matching algorithm for the virtual output queued switch and the fact that this can lead to maximum capacity was shown by McKeown, Anantharam, and Walrand [210]. The packet radio network problem and its solution were provided by Tassiulas and Ephremides [282], where they consider a more general system in which packets can traverse multiple wireless hops.

> One of the first books to provide an analytical treatment of communication networks is the widely used textbook by Bertsekas and Gallager [33]. More modern treatments are provided by Schwartz [258] and by Walrand and Varaiya [295].

Problems

The following problems are in the spirit of this chapter, which is to illustrate the theme and tenor of the book. The solution of these problems is not critical to understanding this chapter or the rest of the book. However, at least thinking about them helps you to get into the mood.

1.1 In analyzing the buffer level in the multiplexer in the two schemes discussed in Section 1.1, we claimed that there will be a point mass at 0. Why will this be true?

1.2 If $\rho = 1.0$, show that $f(\cdot)$, as given in Equation 1.1, is independent of x.

1.3 With reference to Figure 1.3, would the difference in the number of calls carried by the two schemes approach a constant as the delay bound increases, or would it eventually become zero? Explain.

1.4 Consider an $N \times N$ cell switch that operates in a time-slotted manner. Assume that in a slot all the N inputs get a cell. Also assume that a cell has output j as its destination, with probability $\frac{1}{N}$ for $j = 1, \ldots, N$, and that the destination of a cell is independent of the other cells. What is the probability that there will be no destination conflict—that is, that each cell will choose a different destination?

1.5 Consider the same case as in 1.4 except that a cell arrives at an input with probability λ. Among the cells that arrived in the same slot, what is the probability that there is no destination conflict?

CHAPTER 2

Networking: Functional Elements and Current Practice

Corporations, businesses, banks, government agencies, medical establishments, and educational and research institutions have begun to rely heavily on distributed information applications for storing, transporting, and accessing data, for distributed computing, for telemetry and remote control, and for communicating by audio and visual media. Communication networks provide the transport infrastructure that carries the information flows between such distributed information applications.

In this chapter we begin by developing a three-layered view of communication networks. We delineate the subject matter of this book—networking—as dealing with the problem of sharing the resources of a bit-carrier infrastructure. We then identify the basic functional elements of networking: multiplexing, switching, routing, and network management. Then, in the first part of the chapter, we discuss the functions that are carried out by each of these elements. This is followed, in the second part of the chapter, by a discussion of current practice in communication networks. In this part, we first describe the dominant bit-carrier technologies: optical fiber backbones and cellular wireless access networks. Next, we describe the seven-layer Open Systems Interconnection (OSI) architecture, which has traditionally been used to conceptualize communication networks, although it is not always strictly followed. We then provide an overview of telephone networks (the most widely deployed communication networks), followed by the two dominant packet networking technologies: the Internet and Asynchronous Transfer Mode (ATM) networking. For completeness, a brief discussion of X.25 networks, Frame Relay networks, and Integrated Services Digital Network (ISDN) is also provided. These are precursors to today's networks, and many of the concepts that have now matured had their inception in these technologies.

The remainder of the book is organized into three parts: Part I on multiplexing, Part II on switching, and Part III on routing. This chapter deals with functional and architectural issues, and Chapters 3, 9, and 13 discuss the performance and engineering issues associated with multiplexing, switching, and routing. Chapters 3, 9, and 13 are the first chapters in their respective parts.

2.1   Networking as Resource Sharing

The points at which distributed information applications connect to the generators and absorbers of information flows can be viewed as *sources* and *sinks* of traffic (see Figure 2.1). Examples of traffic sources are telephone instruments, video cameras, or a file on a computer disk that is being transmitted to another location. Examples of traffic sinks are telephone receivers, television monitors, or computer storage devices.

To focus on the emphasis of this book, we explode Figure 2.1 into the layered view shown in Figure 2.2. In this view, the sources and sinks of information and the distributed applications connect to the communication network via common *information services*. The information services layer comprises all the hardware and software required to facilitate the necessary transport services and to attach the sources or sinks to the communication network—for example, voice coding, packet buffering and playout, and voice decoding (for packet telephony), or mail preparation and forwarding software (for electronic mail), or a browser (for the World Wide Web).

The communication network is built from the raw material of *communication links*. Information covers geographical distance by being carried over such links. The design and fabrication of such links involve the consideration of electromagnetic propagation in various media, transducers, modulation schemes, error-control coding, and physical interfaces. Modern communication networks

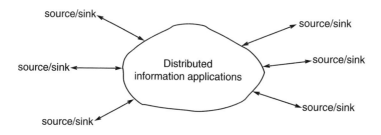

Figure 2.1   Sources and sinks of information attached to distributed information applications; *source/sink* **is a device that can be a source or a sink or both.**

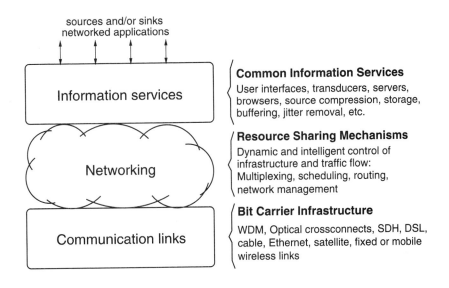

Figure 2.2 A three-layered view of a communication network. Networking is concerned with resource-sharing mechanisms that efficiently share the bit-carrier infrastructure and control the quality of service provided to the various applications using the network.

are largely digital, and the information they carry is transported as digital data—hence the term *data communication networks*. Thus, from the viewpoint of communication network engineering, or *networking*, we will view the communication links simply as *imperfect bit-pipes*, the imperfection being that the bit-pipes can delay, lose, or modify the information they carry. (Although this abstraction suffices for most of the book, when we turn to wireless networks in Chapter 8 we provide an understanding of digital communication over mobile wireless channels.) In the right side of Figure 2.2 is a list of several bit-carrier technologies over copper, fiber, and wireless media. Given the basic raw material of communication links, the next task in developing a communication network is to interconnect several links to form a bit-carrier network. In this book we do not consider the problem of the design of physical network topologies (i.e., the placement of network nodes and the choice of which nodes to interconnect and with what link speeds); we implicitly assume that a network topology is given.

Residing between the information services and the bit-carrier network is the set of functions that constitute the subject of *networking*, as depicted in Figure 2.2.

Ideally, each instance of an information flow from the information services layer can expect from the network a certain *quality of service* (QoS) (e.g., in terms of guaranteed or statistical bounds on service denial, information loss, throughput, and delay). The bit-carrier network is an expensive resource and must be efficiently shared among the arriving information transport demands so that each demand obtains its required QoS. Thus, as shown in Figure 2.2, the broad functionality that networking is concerned with is *resource sharing*. The resources of the bit-carrier infrastructure must be shared between the contending information flows so that the flows obtain their required QoS and the resources are utilized efficiently. Even if a networking technology does not offer per-flow QoS, there would still be the requirement of efficient sharing of the bit carrier resources so as to maximize the traffic carried, subject to some aggregate performance objectives.

We note that when a communications engineer designs a communication link, the primary quality objective is to obtain the highest possible bit transmission rate with an acceptable bit error rate, given the various physical and resource constraints. In designing a network out of the raw material of communication links, however, we need to become aware of the QoS requirements of the flows that will be carried in the network. Some flows are sensitive to delay but can tolerate loss, whereas others may require reliable transfer and high throughput. Thus, in this sense, networking is concerned with information flows rather than only bit flows. High-quality bit carriers, if integrated into a poorly designed communication network (as seen by an information flow), can lead to poor overall performance. On the other hand, awareness of the limitations of link quality in the process of network design can lead to better utilization of the bit carrier infrastructure while satisfying the QoS requirements of the various flows to be carried in the network. The latter is especially true when wireless links are involved.

Analogy with the Operating System of a Computer

To readers familiar with computer systems, the analogy depicted in Figure 2.3 provides another perspective on the point of view presented here. A computer system has several hardware resources (such as the processor(s), the memory, the secondary storage, and the various peripherals such as displays and printers). In a multitasking computer, several applications could be running simultaneously, thus needing to share these resources. The operating system contains the various resource-sharing algorithms that permit the applications to efficiently share the system resources. Networking occupies much the same position between the bit carriers and the networked applications. This is shown on the left side of Figure 2.3. Distributed network algorithms (often called *protocols* in the networking jargon) implement the resource-sharing mechanisms. Just as operating systems have

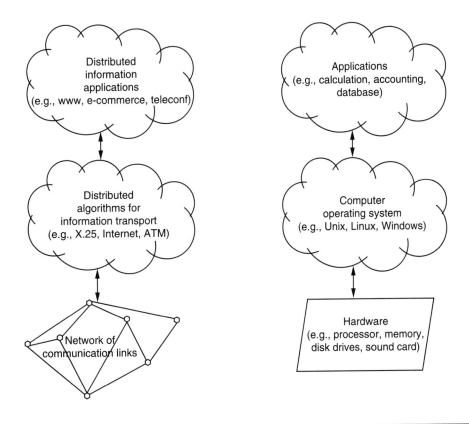

Figure 2.3 Networking is concerned with distributed algorithms for efficient sharing of the resources of a bit-carrier network, in much the same way as a computer's operating system helps computer applications to use and share the hardware resources of a computer.

evolved over time, and just as there could be several competing computer operating systems, so are there various networking technologies.

2.2 The Functional Elements

When a demand arrives for the transport of a bit stream between points in a communication network, we must decide on which set of links of the physical bit-carrier network this demand should be *routed*, and on those links how the bit stream of this demand should be *multiplexed* with the existing bit streams. At the nodes where links meet, the bit stream must be *switched* from one link to the other.

All this must be done so that the QoS of the existing demands is not disturbed and the QoS of the new demand is met. In principle, it is possible for the system to determine the best way to accept the demand by evaluating all possible routes in the network and a variety of possible service strategies at each hop on each route. For example, this could be done so as to maximize the amount of traffic carried in the network in the long run. In practice, however, not all these decisions are made for each demand arrival; this would be impracticable. Instead, a *divide-and-conquer* approach is followed. For example, routing can be taken to be fixed between each pair of points. Even the multiplexing strategy can be specified at each link; for example, the strategy could be to give strict nonpreemptive priority to interactive voice calls and to put all other packets into a single queue. When a new demand arrives along a route, the decision can simply be whether to accept or reject the demand; in some strategies it may be necessary to reparameterize the multiplexers along the path of the demand. The routing can then be optimized so as to maximize the load-carrying capacity of the network. One can also study the effect of different multiplexing strategies on the network capacity.

Apart from the three activities of multiplexing, switching, and routing, networks invariably have *network management*, which can be viewed as procedures for monitoring the network and reporting actual performance (in comparison with performance objectives), and for taking care of situations "not engineered for."

In the remainder of this section we elaborate on the issues involved in these activities, and in the next section we provide an overview of how current networks address these issues. We conclude this chapter with some perspectives. For completeness, we discuss network management briefly in this chapter, and this topic is not covered further in the book.

2.2.1   Multiplexing

At any instant in time, several information flows must be carried by a network. Given the routing, the task of systematically merging several flows into a network is called multiplexing. We obtain the multiplexing effect at the network level by using multiplexing techniques at the individual links in the network. After the multiplexing at an individual link has been defined and analyzed, we can put together a network of links and study the overall effect of the multiplexing techniques implemented at each of the links. Multiplexing is the subject of Part I of this book.

There are broadly two classes of multiplexing techniques: *circuit multiplexing* and *packet multiplexing*.

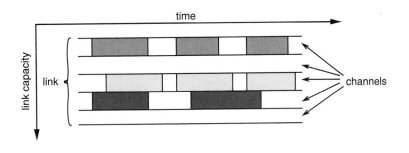

Figure 2.4 Circuit multiplexing: static partitioning of a bit-pipe.

In circuit multiplexing, there is a statically partitioned allocation of the capacity (i.e., bit transmission rate) of the bit-pipe to the various flows. This is depicted in Figure 2.4, where a link's transmission rate (depicted as the space between the two thick lines) is partitioned into several *channels* (five in the figure). In practice, the link rates and channel rates are not arbitrary but rather conform to standards that specify the possible set of link rates and the channel rates into which a link can be partitioned. For example, the International Telecommunications Union (ITU) specifies the E-1 digital link, which has a raw bit rate of 2.048 Mbps, of which 1.920 Mbps can be used to carry user data; this rate can be divided into 30 channels, each one 64 Kbps. If a link is circuit multiplexed, each *conversation* (or *flow*) is allocated to a channel and holds the channel for its entire duration.

Most traffic flows do not generate data at a sustained constant rate but instead typically send data in bursts. As shown in Figure 2.5, all the common sources of traffic have such behavior. This point is also reflected in Figure 2.4, where it is shown that the channels are occupied by bursts of activity, with idle periods in between. A circuit-multiplexed link transmits the data from a source at a fixed rate allocated to the source at *connection setup* time. If all the channels in the link are busy and another request arrives, such a request is *blocked*. Thus the main performance objective when engineering a circuit multiplexed link is the *probability of blocking* requests for channels. If the link handles different *classes* of flows, each with its own blocking objective, then the design of a channel allocation strategy becomes an important problem.

In contrast, in a packet-multiplexed link the entire bit rate of the link is applied to the service of any data that arrive from the sources. We see from Figure 2.6 that this results in the bursts of data from the various sources being interleaved into the link. Comparing this figure with Figure 2.4, notice that the

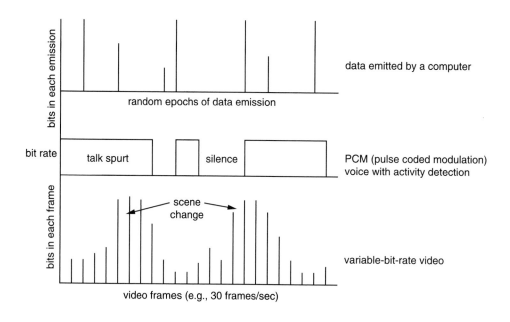

Figure 2.5   Traffic flow from sources is typically not smooth, but _bursty_.

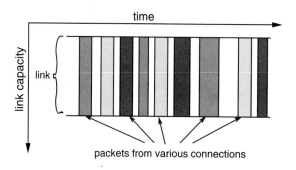

Figure 2.6   Packet multiplexing: no partitioning of the bit-pipe.

bursts occupying the link are shown to be shorter, because they are transmitted at the full link rate. The sources emit information in chunks of bits (*packets*). Intuitively, it is clear that the average rate at which the sources, being multiplexed into the link, emit data must be less than the bit rate of the link. On the other hand, the total *peak* rate at which the sources can emit data typically exceeds the link's bit rate. At such times, the excess data arriving to the link must be *queued*, and such queueing causes data to be *delayed*. If the space available to buffer the arriving data is inadequate, then there could also be data *loss*. From a modeling perspective it is common to view the link as a bit "server" that serves the "customers" waiting at the packet queue. It follows that, when we design a packet multiplexer, the performance objective is usually a delay or loss objective; for example, the mean delay should be less than some given value, or the probability of the delay exceeding a required bound should be small, or the probability of packet loss should be no more than a small value.

In a circuit-multiplexed network, a flow in the network is identified at setup time by the sequence of channels it is assigned to. With packet multiplexing, however, there is additional *overhead*, because each packet must carry a *header* (and perhaps a *trailer*) that identifies the flow to which the packet belongs.

Whereas circuit multiplexing is essentially only the one concept, packet multiplexing has many variations that arise depending on the networking context. Figure 2.7 shows several computers and terminals interlinked by a

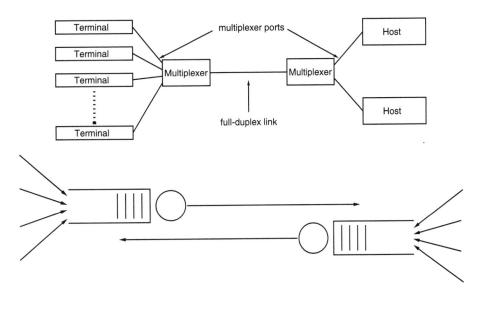

Figure 2.7   In centralized packet multiplexing, the multiplexers have full control over the link's transmission rate. The bottom part of the figure shows a queueing schematic of the traffic being multiplexed into each end of the link.

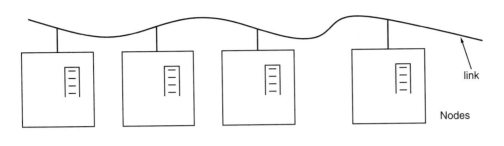

Figure 2.8   In distributed packet multiplexing, the various sources share the link in a distributed fashion with no central control.

point-to-point bit-pipe. A multiplexer at each end "stuffs" packets (destined for the other end) into the link. Notice that each multiplexer has full control of the link in the direction in which it sends packets. The packet *scheduler* at the multiplexer can make decisions about which packet to send next into the link, depending on the various QoS objectives. We can call such an arrangement *centralized* packet multiplexing.

In contrast, consider the situation depicted in Figure 2.8. Several computers, or other computerized devices (which we will call nodes), are attached to a *multipoint* link at various points along its length. This is physically possible for a wired link by means of a device called a *tap*, at which a transmitter of an attached computer can inject electrical energy that propagates over the wire. The same figure can also be a depiction of a wireless communication setup. The common feature is that if a station transmits, all other stations can potentially receive variously attenuated versions of the transmitted signal. Obviously, if the sharing of the link is completely uncoordinated, then, owing to *collisions*, even under moderate load most of the transmissions will be wasted, leading to very poor effective transmission rates, as seen by the sources and sinks of traffic.

The simplest "coordination" mechanism is called *random access*, in which nodes attempt to transmit and try to determine whether their transmissions succeeded; if not, they wait for random amounts of time before reattempting. Such random *backoff* simply reduces the probability of a collision, in case the failure of transmission was the result of a collision. In a wired network, a collision can be sensed because the transmitter can measure the power on the medium as it transmits and thus can determine whether the power level exceeds a certain threshold (indicating that someone else is transmitting as well). In a wireless network, the idea is to very carefully avoid collisions by sensing the channel for

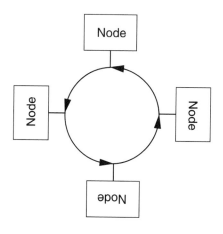

Figure 2.9   Polling-based distributed packet multiplexing; the token ring architecture.

activity before transmitting, and attempting to reserve the channel for the duration of a transmission.

A more coordinated distributed packet multiplexing approach derives from the basic concept of *polling*. Figure 2.9 depicts a *token ring network*, in which a token is passed around and possession of the token permits a node to transmit. Yet another approach is for a master station to poll each of the other stations and thus give turns for data transmission between each station and the master or between the stations themselves.

Figure 2.10 shows a satellite network in which there are several (geographically widespread) small terminals (attached to individual computers or the local networks of small organizations), and these share a satellite link to a large hub. The channel inbound from the terminals to the hub needs to be shared. An important difference between this distributed multiplexing situation and the ones discussed earlier is that there is a large propagation delay across the link, owing simply to the fact that electromagnetic waves travel at a finite speed (a rough value is 0.25 second each way, for a link using a single geostationary satellite hop). Contention- or polling-based access in such a situation can waste a lot of link transmission rate, because the overheads of these strategies grow with the propagation delay. This is because in a contention mechanism, when a collision occurs it takes time of the order of the propagation delay to detect and resolve the collision, and in polling a whole propagation delay's worth of time is wasted each

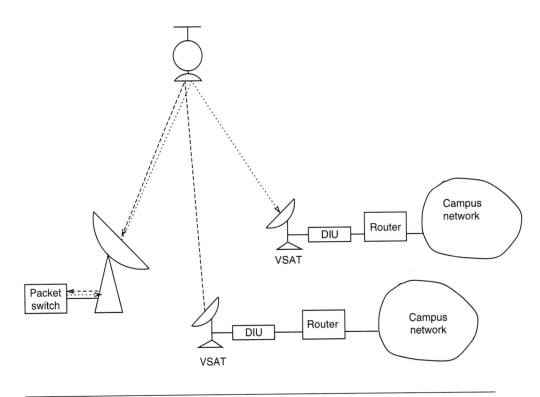

Figure 2.10 Reservation-based distributed packet multiplexing; a very small aperture terminal (VSAT) satellite network. A DIU (digital interface unit) connects the radio frequency equipment on the satellite side to digital equipment on the customer's network.

time an idle terminal is polled. Instead, the terminals request reservations of time on the inbound channel. The reservations are granted on the outbound channel (which, as we can see, is centrally packet multiplexed at the hub). When the reservations are granted, the terminals take turns accessing the inbound direction of the link. Reservations are periodically renegotiated over durations referred to as *frames*.

The situations depicted in Figures 2.7–2.10 are the simplest possible; in general, there could be a network with multiple links, each of which could have its own multiplexing strategy. Devices interconnecting the links would be capable of performing the appropriate multiplexing on their respective interfaces.

Figure 2.11 provides a summary depiction of the taxonomy of link multiplexing that we have presented. It is also possible to create hybrid multiplexing schemes that combine circuit multiplexing and packet multiplexing. The approach is depicted in Figure 2.12. The link's transmission rate is partitioned

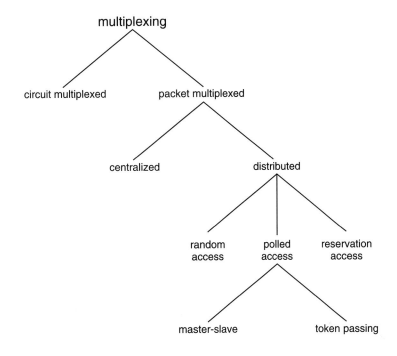

Figure 2.11   A summary view of a link multiplexing taxonomy.

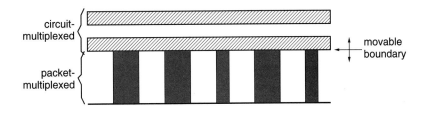

Figure 2.12   Hybrid link multiplexing; combining circuit multiplexing and packet multiplexing on a link.

into several digital channels; some of them can be used in the circuit-multiplexed mode, and the others in the packet-multiplexed mode. As Figure 2.12 shows, in general, the boundary between the circuit-multiplexed part and the packet-multiplexed part can be dynamically varied. An important example is the Integrated Services Digital Network (ISDN) service, in which a telephone network subscriber can use the twisted pair telephone wire to access circuit-switched telephony as well as packet-multiplexed network services. For example, let us consider the 2B+D ISDN interface between a user and an ISDN access node; here, on a telephone wire, one carries two 64 Kbps B (or bearer) channels and one 16 Kbps packet D (or "data") channel. The D channel is used for packet signaling between the user and the network. The B channels can be used in various ways. For example, the user can set up a circuit-multiplexed phone call on one channel and a fax call on the other channel. Alternatively, the user can use one channel for a phone call while simultaneously using the other channel for accessing a packet-multiplexed network. For a more detailed discussion of ISDN technology, see Section 2.3.3.

Many packet-multiplexed networks also use *virtual circuit multiplexing*, in which the path for the packet flow between the source and destination is decided by a call setup procedure, very much like that in circuit-multiplexed networks.

2.2.2 Switching

Typically, a data flow will need to traverse more than one link. To see why, consider a network of N sources or sinks. Assume only *unicast* flows—that is, each flow has only one source and one destination. A maximum of $N(N-1)$ flows are possible, and in the general case, the network must be capable of supporting each of these flows by providing a possible path for each of them. To support all possible flows, a brute force approach would be to construct a fully connected network with $N(N-1)/2$ duplex transmission links, as shown in Figure 2.13a. In such a network, the number of links increases in proportion to N^2 and is obviously economically infeasible. Furthermore, a large number of these links would be very long, and many of them would also probably never be used. A more economical arrangement would be to have a centralized network in which a central node collects the data flows from the sources and then distributes them to the appropriate sinks. Such a network uses only N transmission links and is shown in Figure 2.13b. This can also become infeasible for large operational areas because of the long transmission links that will be required. However, this procedure can be repeated recursively using subsets of N to construct a network in which the path from a source to a sink involves multiple links or hops. Such a network is

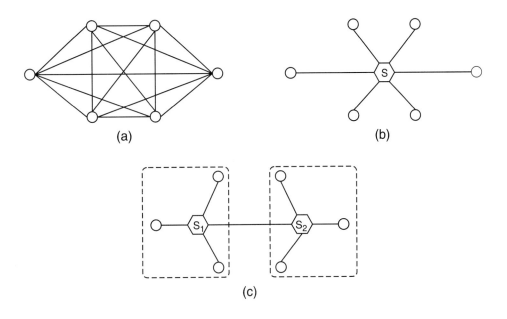

Figure 2.13   A six-node network constructed in three ways: (a) a fully connected network; (b) a centralized network with every node connected to a central node that selectively establishes paths between the communicating nodes; (c) a hierarchical network with two networks connecting nodes in a smaller area.

shown in Figure 2.13c. Observe that the link between S_1 and S_2 in Figure 2.13c is now a shared resource among the nodes of the network and will be made available to the nodes by the switches on demand. The capacity of this link can be designed so as to satisfy a large fraction of the demands rather than every demand. With such a design criterion, the capacity of the links can be significantly lower than that required to satisfy every demand. Also, because the links interconnecting the subnetworks will be shared among a large number of users, the sizing of the capacity of these links is a multiplexing problem.

For simplicity, the preceding discussion makes the assumption that the link cost is proportional to the distance. In practice, the link costs depend on many factors, such as the terrain, the medium, the bandwidth, and so on, and is not strictly proportional to the length of the link nor the transmission rate on it. A generally accepted notion is that longer links are more cost-efficient when the transmission rates are high.

The hierarchical network of Figure 2.13c achieves many objectives. Building smaller networks around a central node makes them more manageable. The links between different subnetworks will be shared by a larger number of users, thereby increasing its usage efficiency. At the higher levels of the hierarchy, the links will require higher bandwidths because they interconnect larger groups of users and may also traverse longer distances. This also works out well because, as we have said before, the cost of a transmission link is a sublinear function of its capacity and distance, and high-capacity links are more economical over longer distances. By appropriately designing the interconnections, we can improve network reliability and avoid single points of failure, as in the networks of Figure 2.13b and 2.13c.

From this discussion, we can define a switch as a device that sits at the junction of two or more links and moves the flow units between them to allow the sharing of these links among a large number of users. Figure 2.14 shows such a block diagram view of a switch. A switch allows us to replace transmission links with a device that can switch flows between the links. The construction of the switch can be quite complex. However, this is a fair trade (between transmission and switching) because, traditionally, transmission capacity is significantly more expensive than switching capability.

The switch functionality is achieved by first *demultiplexing* the flows on each incoming link, *identifying* the output link for the flow, *forwarding* or *switching* the element of each flow to its destination output link, and then *multiplexing* the outgoing flows onto each output link. These are very fast timescale functions and must be performed on every packet or slot of a *time-division-multiplexed* (TDM) frame. In addition, in circuit-multiplexed networks, the switches must process call or flow requests. These are slower timescale functions and include performing the necessary signaling and, if available, reserving resources to maintain the call. Although it is rarely done, resource reservation is also possible in packet-multiplexed networks, with a resource reservation phase preceding the actual flow of data. In both circuit- and packet-multiplexed networks, the switches

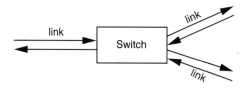

Figure 2.14   A switch moves data between links.

also must exchange information about the network and switch conditions and must calculate routes to different destinations in the network. These functions are typically performed on even slower timescales of a few minutes to hours, the timescales at which the traffic characteristics are expected to change.

We detail these functions with reference to packet and circuit switches separately. First, consider a packet switch, such as a router in the Internet or a cell switch in an ATM (Asynchronous Transfer Mode; see Section 2.3.6) network. The packet lengths could be fixed as in the ATM network (where they are 53 bytes long), in which case the input links can be assumed to be slotted in time, with the length of the time slot equal to the packet transmission time. The packet lengths could also be variable, as in IP networks, where we must consider the links to be unslotted. Typically, in this case there will be upper and lower bounds on the packet lengths. A *store-and-forward* packet switch waits until the entire packet is received before processing it, whereas a *cut-through* packet switch starts processing the packet header as soon as the header is received. In the latter case, the transmission on the output port could start before the entire packet is received on the input port. The advantages of cut-through switching are limited, and it is rarely implemented in practice.

The components of a packet switch are shown in Figure 2.15. A line interface extracts the packet from the input link by appropriately identifying the

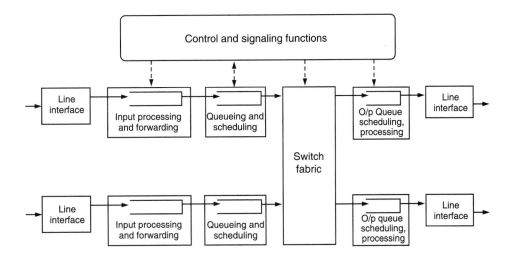

Figure 2.15   The components of a packet switch.

boundaries of the bits and of the packet. An input processor then extracts the header, and an associated forwarding engine performs a *route lookup* using this header information by consulting a *routing table* to determine the output link. This is essentially the demultiplexing function mentioned earlier. In a multiservice network, the service type of the packet is also determined at this stage to identify the kind of service that the packet is to be provided. The type of service determines the scheduling of the packet's transmission on the output link and drop priorities during periods of congestion. If it is not possible to send the packet to the output port immediately, it is queued and scheduled to be moved to the output port according to its service type. The switch fabric next moves the packet to the output queue. An output processor determines the queue position and *schedules* its transmission on the output link. Finally the packet is transmitted on the physical interface by the output line interface.

In addition to processing packets, other functions are performed by the switch. In virtual-circuit based packet multiplexed networks, such as ATM, Frame Relay, and X.25 networks, the actual information flow is preceded by a call setup phase. In this phase, the path for the packets corresponding to the information flow is set up much as in the case of circuit multiplexing. During this phase other parameters of the call, such as the bandwidth to be reserved for the call and the permissible delays and loss rate on the packets of the flow, can also be determined. This is very similar to call setup in a telephone network except that many more parameters for the call are negotiated, and information flow is in the form of packets and perhaps of variable rate. In datagram packet switching, the path for the packets from the flow is not explicitly set up. The destination port is decided separately for each packet by consulting a routing table, and, in principle, it is possible that different packets from the same flow follow different paths through the network. This kind of stateless, next-hop packet switching is used in the Internet. In packet-multiplexed networks, routing and other signaling protocols and the routing algorithms must be executed so that the network can construct the routing table that determines the output port for a packet. All these functions are shown as control and signaling functions in Figure 2.15. Note that the output of the control and signaling block interacts with all the other packet processing blocks in the switch.

In Chapters 9–12 we discuss the architectures and design choices for packet switches and also the details of some of the functional blocks in the switch.

Next, let us consider a switch in a circuit-multiplexed network—specifically, that in a telephone network. The telephone network has been using TDM-based digital communication channels beyond the local exchange for quite some time now, and we will assume that is the case. To understand what a switch in this

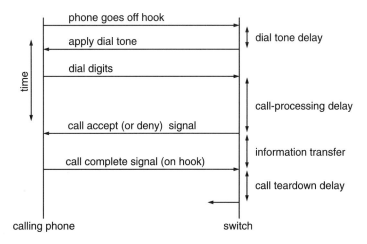

Figure 2.16 Call setup procedure and the delays in a switch in a circuit-multiplexed network. Call-processing delay is also called postdial delay or ringing delay.

network does, consider the call setup with the sequence of events shown in Figure 2.16. When the calling phone goes off hook, the condition is recognized by the switch in the local exchange and a dial tone is presented to the calling phone, indicating readiness of the switch to accept digits. The dialed digits are read by the exchange and analyzed to determine the destination of the call. The routing algorithm is used to determine the path to be assigned to this call in the network. This could involve an exchange of signaling messages with other nodes in the network. If a path is available, it is reserved for this call; if it is not, the call is dropped and assumed lost forever. From the point of view of the switch, a path is determined by the two pairs "input port:TDM slot" and "output port:TDM slot." After the path is established in this manner, the switching fabric is programmed to move the contents of the "input port:TDM slot" to "output port:TDM slot" in every frame. This is shown in Figure 2.17, where the contents of slot x of input i are copied into slot y of output link o. On completion of the call, the reserved path is released for use by other calls, and the billing functions are executed. In addition to these functions associated with a call, the switch must also perform background functions for executing the routing protocols and algorithms as well as management and maintenance of the switch.

It is possible that in a circuit switch the input and output links are free but the switch cannot set up a path between them *through* the switch. This can happen

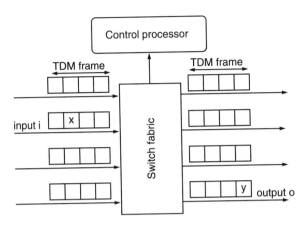

Figure 2.17   The operation of a circuit switch.

because of the paths through the switch that the other active calls may be taking, a condition called *switch blocking*. Switches in circuit-multiplexed networks should be designed to minimize switch-blocking probabilities and to handle a specified call attempt rate. In Chapter 9 we discuss the issues in such designs in more detail.

2.2.3   Routing

Routing is concerned with establishing end-to-end paths between sources and sinks of information. A route can be thought of as a concatenation of communication links and switches that connects a source and a sink. Both routing and multiplexing are activities in which a network switch, also called a node, is involved. The difference is that multiplexing several flows onto a link is the concern of a single network node, whereas the function of routing necessarily involves multiple network nodes.

It is useful to distinguish between the terms *routing* and *forwarding*. Forwarding refers to looking up the outgoing link on which a data packet is to be transmitted (for packet-multiplexed networks) or the time slot in the outgoing TDM frame to which a time slot in the incoming TDM frame must be mapped (for circuit-multiplexed networks). As discussed in Section 2.2.2, this is done by consulting a routing table, which maintains the necessary information. Forwarding is a *data* or *user plane* activity because it deals with data. In contrast, routing can be viewed as the collection of actions, involving multiple nodes, that has the goal of building the routing table. In this sense, routing is a *control plane* activity. The data and control planes are discussed further in Section 2.3.2; see also Figure 2.22. Several issues that are intimately connected with the routing problem are discussed next.

Information related to network topology and destination reachability is of crucial importance. Depending on how routing is planned in a network, two cases arise. In *centralized routing*, all the topology information is stored at a central point, where routes are computed. Topology information is collected by a distributed protocol, but the route computation is done at the central point. These routes are then conveyed to the nodes in the network where they are needed. In *distributed routing*, the nodes in a network participate in a distributed protocol that not only collects topology-related information but also computes routes in a distributed manner. The distributed protocol may involve exchange of summarized information, as in, for example, the distance to a set of destinations. It may also involve exchange of local topology information, as in, for example, a node advertising information on the status of its attached links.

Related to the topology database is the issue of *aggregation* of topology information. This is important for the scalability of a routing protocol. When a network becomes large, it is possible for the regular topological updates to become so frequent that significant processing and bandwidth resources are consumed. The solution is to design hierarchies of topology-related information, with only summarized or aggregated information about one level in the hierarchy being available to others.

Although establishing end-to-end paths is a basic requirement, a routing algorithm is expected to do more: The routes assigned to flows must be such that their quality of service (QoS) demands are satisfied and, at the same time, network resources are efficiently utilized. Achieving both objectives simultaneously is a challenging task because they are conflicting: It is easy to satisfy QoS objectives by keeping the network lightly loaded, whereas we can easily achieve high resource utilization by not caring about QoS objectives.

There are two ways routes are found across a network. When *hop-by-hop* routing is chosen, each node is interested in computing only the next hop toward a destination. The full path to the destination is not sought. Thus, the topology database is used merely to decide the best next hop toward the destination. The other alternative is to find the complete path to the destination. Evidently, the topology database must contain sufficiently detailed information for this to be possible. The second alternative is referred to as *source routing* because the first node on the path of the traffic (the source node) can compute the complete route to the destination.

The control plane activities related to routing have several timescales associated with them, as elaborated on in Section 2.2.5. To begin with, we have the regular exchange of topology information that is necessary to build the topology database. However, route computation need not necessarily occur at this timescale. If a network is static in the sense that no node or link is going down or coming up, then the topology is static. Furthermore, if the traffic pattern is stationary, then routes need not be recomputed often. But the distributed protocol that disseminates topology information must continue its exchanges at regular and frequent intervals. Route recomputation occurs at slower timescales, driven by significant changes in the network topology, such as link or node failure, or by major changes in the traffic pattern. Route recomputation may (or may not) also be driven by performance objectives. For example, when the average transfer delay on a path begins to exceed some acceptable value, indicating congestion somewhere on the path, a route recomputation may be triggered. We note, however, that even though route recomputation can, in theory, be triggered by the onset of congestion, many practical networks may not do this at all. For example, in the Internet, route computation does not consider congestion levels and traffic patterns directly but instead utilizes *weights* defined by network administrators. It is expected that the weight definitions will take into account projected traffic demands and link capacities.

In general, we would need performance measures to gauge the quality of routing. Good routing should mean that under low offered load, information transfer delays are relatively small. As the load offered to a network increases, eventually a part of the offered load must be rejected if QoS commitments are to be met. This is the point at which *connection blocking* begins. A consequence of good routing is that the onset of blocking is delayed. That is, the network can accommodate substantially more traffic before blocking begins.

Again, we hasten to add that, in practice, some networks may not use the notion of connection blocking at all. Indeed, in the Internet, the IP layer is connectionless, and therefore *at the IP layer*, there is no concept of a connection and hence no question of connection blocking. With increasing load, performance measures (for example, average packet delay) indicate higher levels of congestion, and consequently performance degrades, but the IP layer does nothing about it.

In a circuit-multiplexed network, the route is determined when the call is set up. Subsequently, forwarding simply reduces to switching of the appropriate TDM time slot on the input side to the corresponding slot on the output side. Section 2.2.2 discusses this in greater detail. Similarly, virtual circuit-based packet-multiplexed networks have a connection establishment phase (or call setup phase) prior to data transfer, and, again, routes are determined during this phase.

Every packet associated with the call is tagged with a label at every switch; the label identifies the outgoing port to which the packet should be switched. The labels have local significance only, because they merely indicate an input-port-to-output-port map at a particular switch that was established when the call was set up. In contrast, in a datagram-based packet-multiplexed network, there is no connection establishment phase before data transfer begins. Each packet is routed as a separate entity, without reference to the connection or session to which it belongs. Indeed, in a datagram-based network, the very notion of a connection (at the network layer) does not exist. Therefore, each packet carries the complete source and destination addresses that may possibly be used to find a route for that packet through the network. This mode of operation can be contrasted with that in virtual circuit-based packet-multiplexed networks, where each packet need not carry the complete source and destination addresses; it is sufficient for it to carry the label assigned at the time the connection was established, because this label is used to obtain the route.

The recent development of *reactive* routing has been driven by the advent of wireless ad hoc networking. In the preceding discussion, we have tacitly assumed that routing-related control-path activity is carried on irrespective of the presence of data traffic. For example, topology updates are periodically sent and received by a network node even when it has no data packets to forward. The idea is to have routes available for use whenever needed. This approach is referred to as *proactive* routing. However, in wireless ad hoc networks, such continued background control traffic processing can be expensive in terms of excessive power consumption, leading to reduced battery life. This has motivated the approach of acquiring topology information only when it is needed—that is, only when a node needs to forward a data packet and finds that it has no appropriate route. This approach to routing is called reactive to contrast it with the traditional proactive approach.

Next, we take a brief look at routing in packet-multiplexed and circuit-multiplexed networks. We will see that even though the core issues are common to both types of networks, the mechanisms that have evolved to address the issues are sometimes different. We first consider packet-multiplexed networks, followed by circuit-multiplexed networks.

Both virtual circuit and datagram-based networks usually run specific protocols aimed at acquiring information related to neighbor reachability. For example, the Hello protocol is used in both OSPF (Open Shortest Path First)—a routing protocol in the Internet, a datagram-based packet network—and PNNI (Private Network-to-Network Interface), a routing protocol used in ATM networks, which are virtual circuit-based packet networks.

In "small" virtual circuit networks, source routing is normally used. Thus, the first ATM switch on a connection's path would provide a full path, listing all intermediate switch identifiers, to the connection setup packet. As the network grows larger, this approach may not scale because too many nodes may have to be specified. As mentioned before, hierarchical routing schemes would be required in such cases. In datagram networks, hop-by-hop routing is commonly implemented. However, some routing protocols, such as OSPF, do have enough information to obtain complete paths to a destination rather than only the next hop.

Route selection in virtual circuit networks is often accompanied by resource reservation at the same time. Thus, as the connection setup packet travels down the chosen path, the resources necessary to provide the requested QoS are also reserved. This leads naturally to the *crankback* feature found in virtual circuit networks. Here, a source node has a set of alternative routes available to it, and, to begin with, it tries the first of these. If call setup is not successful because some switch on the path has inadequate resources, then a failure message is sent back to the source node. This causes the source to crank back and try other routes in its set, avoiding those routes in the set that pass through the switch where resources are currently limited. In datagram networks, the crankback feature is not usually found. Traditionally, datagram networks carried best-effort traffic, and there was no particular concern about finding adequate resources on a path. In case of drastic events such as link failure, the routing protocol would detect the change in topology, and new routes, bypassing the failed link, would be computed.

Routing in packet-multiplexed networks has the dual objectives of meeting QoS requirements at both call and packet levels and ensuring efficient resource utilization. This is discussed in greater detail in Section 3.1. Thus, minimizing call-blocking probability is an important goal, but so is ensuring that packet-level QoS measures—for example, packet transfer delay and packet loss probability— are also adequate. It is worth noting that the multiplexing policies implemented at switches have a significant impact on packet-level QoS measures. This indicates again that routing and multiplexing need to be considered jointly if we are to achieve overall QoS objectives. However, as mentioned in Section 2.2, we follow a divide-and-conquer approach and consider multiplexing and routing separately.

In circuit-multiplexed networks, a notable point is that even though the network that carries user-generated information—for example, voice—is circuit-multiplexed, the network used for exchange of control information is packet-multiplexed. This is the case, for example, in the telephone network (see Section 2.3.3), where the common channel signaling (CCS) network carries the control information that needs to be exchanged by the switches.

In packet-multiplexed networks, however, both user-generated information and control information are carried as packets and coexist on the same network.

The network topology in circuit-multiplexed networks is considered to be relatively much more static. Thus, there is little need for an automated process such as the Hello protocol, which keeps checking at frequent intervals whether neighboring nodes are "alive." Normally, network topology is administratively configured, and manual reconfiguration is done as necessary. Circuit-multiplexed networks follow the source-routing philosophy.

The absence of a protocol that generates regular topology update messages indicates that the control traffic in a circuit-multiplexed network flows on a slower timescale. The intervals at which routing tables are updated depend on the routing method chosen. For example, many operational telephone networks use fixed traffic routing, in which the switches are equipped with a set of routes corresponding to each destination; these routes are computed offline during the network design phase and are programmed into the switches. This is discussed in greater detail in Section 3.3. When a call is to be set up, a switch tries the stored routes one by one, and if adequate resources are available on a route, the call succeeds. Dynamic traffic routing allows routing tables to be changed dynamically, either in a planned manner or in response to changes in network state. For example, telephone networks may update routing tables at the beginning of the "busy hour" to cope with the increase in traffic; this is an example of planned dynamic routing.

In circuit-multiplexed networks, the objective of routing is to minimize call-blocking probability and utilize links efficiently. After a call is set up, the in-call QoS is automatically ensured, because an end-to-end path is reserved specifically for this call. Thus, in contrast to packet-multiplexed networks, only call-level QoS objectives are of concern here.

Topics related to routing are discussed in Chapters 13, 14, 15, and 16.

2.2.4 Network Management

When a network is established, the topology, the bandwidths, and the multiplexing, switching, routing, and traffic controls are chosen so as to provide a certain grade of service, however that is defined, to handle a specified engineered load and certain kinds of faults. An operational network is a complex system and must be continuously monitored and measured to ensure that the achieved grade of service is as per the specification and design. Often the load may exceed the engineered load because of excessive failures, delay in capacity upgrading, or any other event that was not accounted for during planning. These situations need to be managed by external means, which depend on aggregate measurements and slower timescale controls. This is in the domain of *network management*. Note that accounting for every eventuality during the network design stage would lead to a very expensive network design; hence, in practice, only a certain range of operational conditions are designed for, and excursions outside this range are carefully monitored and managed when they occur.

All operational networks define a network management architecture that provides them with a mechanism to collect performance data from the various network elements and to remotely configure and hence control these elements in a centralized manner. Examples of performance data include bit and packet error rates on links, the traffic volume delivered by a link, and the volume dropped by it. Mechanisms are also included to identify the specifications of the network elements and their configuration details. Network management includes the ability to isolate faults and reconfigure the network during failures so as to restore service. Service may be restored by using spare resources and does not imply repair of the fault.

Securing the network against threats is also becoming an important aspect of network management. Protection against unauthorized access is provided by firewalls that allow very specific traffic into or out of a network. Managing the access through these devices is a network management function. *Denial-of-service* attacks make a target network handle large volumes of "irrelevant traffic" and thereby choke it. This in turn prevents legitimate users from gaining access to the network. Security management functions should be able to recognize that such attacks are developing and minimize their effect. Many times it is not possible to recognize a security breach when it is in progress. An important aspect of network security management is to examine network logs and do a post facto analysis to recognize security breaches.

Network management protocols decide the kind of data to be collected by the different *managed* nodes and specify how these data are made available to the manager. These protocols also decide which configuration parameters of the managed devices can be read by the network manager. TCP/IP networks (see Section 2.3.5) use the Simple Network Management Protocol (SNMP) and the Remote Monitoring (RMON) protocol to collect device and network statistics. These protocols also permit alarms to be triggered based on preset thresholds, and elementary control actions to be exercised by the network manager. The ITU has developed a network management architecture called Telecommunication Management Network (TMN) that is useful in managing large, complex, multivendor networks. Most telecommunications vendors conform to the TMN specifications, making it possible for an operator to easily integrate equipment purchased from various vendors into a network with unified network management.

Although network management protocols may be standardized, network management practice is usually a collection of heuristics that are developed from operational experience. Little systematic engineering appears to have been brought to bear on this area, and hence we will not discuss network management further in this book.

2.2.5 Traffic Controls and Timescales

Network traffic can be described across many timescales, from microseconds to days or even years. Thus the control mechanisms for the resource-sharing activities also need to be viewed as operating over various timescales. In fact, these resource-sharing and control mechanisms can also be understood in terms of the timescales over which they act. Rather than consider absolute time we identify the following four relative timescales.

- Packet timescale (packet transmission time; microseconds or milliseconds)
- Session, call, or flow timescale (typically minutes)
- Busy hour or traffic variation timescale (typically hours)
- Provisioning timescale (usually hours to days or weeks)

Let us now discuss the relation between these timescales and the four basic activities that we have outlined.

Of course, the packet timescale applies only to packet-multiplexed networks. Packet timescale controls discriminate between the treatment that the network provides to individual packets. These controls include mechanisms such as priority queueing, adaptive window-based congestion control (as in TCP in the Internet), and random packet dropping (for TCP performance optimization). For example, packets of a real-time service, such as interactive telephony, need to be expedited, and hence these can be given priority when competing for transmission over a congested link. When a link is shared between two sets of flows, then a scheduling policy can be set up that allocates half the link's capacity to each set of flows; this requires packet-by-packet decision making at the point where the two flows are being multiplexed into the link. File transfers in an internet (see Section 2.3.5) are controlled by a set of packet-level mechanisms in the end-to-end TCP protocol. A TCP transmitter works with individual packets, deciding when to transmit packets based on acknowledgments or lack of them. We can control the performance of these mechanisms by deliberately and randomly dropping individual packets from TCP flows.

When a session or call (e.g., a phone call in the phone network, or a video streaming session in a packet network) arrives at a network, several decisions can be made. Should this call be accepted or not? A circuit-multiplexed network would block a call if there are no channels remaining to carry a new call of the arriving class (e.g., a nonemergency call). In a packet network, a call may not be accepted if the network cannot offer it the required QoS. If the call is accepted, how should it be routed, and how much transmission rate will the call use up on that route? The route should be such that it has enough resources to handle the call. Such decisions are made only when flows or calls arrive and hence can be viewed as operating over flow arrival and completion timescales, which are certainly slower than packet timescales.

In telecommunication networks the traffic processes are not stationary but instead vary continually over the hours of the day. Different traffic patterns exist during different times of the day. An obvious example is traffic between organizations, which would be heavy during business hours, whereas the residential networks would carry traffic during other times. With networks becoming international, time zones and time differences also govern the traffic pattern. Most networks are engineered for a *busy hour* during which the traffic is at its peak.

The controls at the packet and flow timescales are typically parameterized to optimize network performance for a specified traffic pattern. If the traffic pattern changes, however, a different set of configurations for these controls may be more efficient. Hence at traffic variation timescales it becomes necessary to adjust the packet- and flow-level controls in order to optimize network performance. For example, the network routing may need to be adjusted to accommodate varying patterns of traffic at different times of the day. Thus it is at these timescales that the network operator would adjust routing policies in order to optimize the network performance. In the context of telephone networks, the routing plan could be different for different times of the day or on different days of the week. This is also called the *traffic engineering* timescale.

Notice that the controls at the packet timescale and session timescale are concerned with the control of the way traffic is multiplexed and switched. Similarly, the flow and traffic engineering timescale controls are applied through routing functions.

Resource provisioning, also known as capacity planning, is an important part of network management. In the hierarchy of control timescales listed earlier, provisioning is the slowest timescale control. It entails increasing the transmission rate on some links or adding new links and switches. If the fastest timescale controls (based on controlled multiplexing and switching, or adjusting the routes) do not provide the desired service quality to the offered traffic, clearly it is time to add capacity to the network. For example, reprovisioning could be triggered when attempts to reroute traffic or to reconfigure the other traffic controls fail to achieve the desired results, and this happens frequently over several traffic engineering timescales. Actions at the provisioning timescale could be to increase the capacity of some links or to establish new links, which might in turn entail upgrading switching equipment. Modern networks are being built on optical fiber links, with intelligent optical crossconnects. Such an infrastructure can be used to rapidly provision additional bandwidth with little or no manual intervention.

2.3 Current Practice

Sections 2.1 and 2.2 provide a generic view of the issues involved in developing a communication network. Over the past 100 years, several communication network technologies have been developed and deployed. In this section we provide an overview of the current practice. Our aim here is to provide the "big picture" rather than deal with the myriad details. In keeping with the view presented in Figure 2.2, we present this material in two parts. First we discuss the network infrastructure, where we describe the way communication links are derived from physical communication technologies. Second, we discuss networking architectures, describing the way information transport services are developed on top of the bit-carrier infrastructure. We follow this discussion with examples of the major networking technologies: the circuit-switched phone network, IP packet networks (internets), and ATM packet networks. We also provide a brief overview of X.25 networks, Frame Relay networks, and ISDN.

2.3.1 Network Infrastructure

By *network infrastructure* we basically mean the network of communication links (see Figure 2.2). In modern networks, the interswitch links are typically derived from an optical fiber infrastructure. This is becoming so even within large buildings and campuses, where *local area network* (LAN) switches are increasingly being connected by optical fiber links. What remains is the connection of terminals (sources and sinks) to the switches; such access continues to be over copper media, and increasingly over multiple-access wireless links. In the following discussion we provide an overview of optical fiber backbones and wireless access technologies.

Optical Fiber Backbones

Figure 2.18 depicts how an optical fiber infrastructure is used in a packet network; some of the terminology in this diagram is from the Internet, the currently

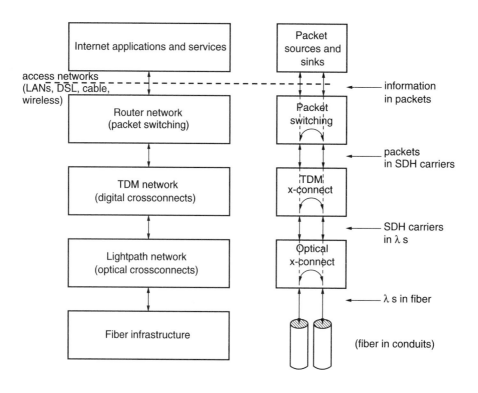

Figure 2.18 From fiber to packets: how links between packet switches are derived from an optical fiber infrastructure. The left side shows the various physical networks that are derived from the fiber infrastructure, and the right side shows the network components that are needed to achieve these layers of networks.

most popular packet network technology, which we describe in more detail in Section 2.3.5.

First, notice the *layering* in the architecture. By layering we mean that one level of the network provides a service to the next upper layer without the upper layer being aware of the functional details of the layer below it. Let us begin with the bottommost layer, which comprises strands of optical fiber sheathed in protective cladding and laid inside conduits. Normally such lengths of fiber terminate in junction boxes in buildings at each end and lie unused until the need arises to provide additional transmission capacity between the end points of the fiber cable. Then the *dark fiber* is *lit up* by terminating each end with connectors, which are then attached to optical interfaces in transmission equipment. Each fiber can carry several tens of *wavelengths* (or colors), also called λs owing to the symbol often used to denote wavelength. Each wavelength is *modulated* to carry a bit stream; for example, a common modulation scheme is to simply turn the light on and off to signal 1's and 0's. Current technology permits each wavelength to carry a bit rate of about 10 Gbps.

Figure 2.18 shows that the first level of transmission equipment creates a network of *lightpaths* from the *fiber plant*. The simplest case of a lightpath is one in which on every fiber traversed by the path, the same color is used. The electronic device that modulates (or demodulates) the bits into (or from) the lightpath needs to exist only at the ends of the lightpath. In between we require only wavelength switches, or, as they are called in the industry, *optical crossconnects*. An optical crossconnect simply modulates a color on an outgoing fiber according to the modulating light signal it receives on an incoming fiber. If the incoming and outgoing wavelengths are different, then the crossconnect also needs to do *wavelength conversion*. The technology of carrying information over several wavelengths in a fiber, and then optically switching this information between wavelengths in different fibers, is called *wavelength division multiplexing* (WDM).

At the end points of the lightpaths, the bit stream carried by the light must be extracted. This is shown as the next layer in Figure 2.18. The bit stream actually has a periodic, hierarchical format corresponding to a digital transmission standard, such as SDH (Synchronous Digital Hierarchy) or SONET (Synchronous Optical Network). Thus, a bit stream with a 9.95328 Gbps information rate can actually be viewed as comprising 64 separate TDM bit streams, each with an information rate of 155.52 Mbps. A *digital crossconnect* can interconnect these 155.52 Mbps TDM *subcarriers* between the various lightpaths terminating at this crossconnect. Notice that by doing this we have obtained end-to-end digital links at, say, 155.52 Mbps. An *add-drop multiplexer* is used to extract a subcarrier that needs to be terminated at a location, or to insert a subcarrier into a TDM stream.

These digital links, in turn, carry multiplexed information as circuits or packets. In Figure 2.18 we assume the example of a packet network, such as an internet. At places where packet switches are deployed, a digital crossconnect extracts a digital subcarrier, which is then connected to a port of the packet switch. The packet switch then extracts packets from the digital bit stream. Some of these packets are passed on to terminal equipment (shown as packet sources and sinks in Figure 2.18—e.g., computers or packet phones) at the location of the packet switch; as shown in the figure, the terminal equipment is connected to the *backbone* network, by an access network, which could be based on wired technologies (such as Ethernet, television cable, or digital subscriber loops (DSL)) or any of the several emerging wireless access technologies. At the packet switch, packets not meant for local terminals are switched into output ports of the switch. These are then inserted into a digital subcarrier by the output interface of the packet switch. This subcarrier is then multiplexed into a higher-speed digital carrier, which is then modulated into a lightpath, which is in turn multiplexed into an optical fiber.

In practice, of course, the same optical transmission facilities are used to carry both circuit-multiplexed networks and packet-multiplexed networks. If it is a circuit-switched network that is being built on top of the optical fiber infrastructure, then the digital carriers terminate into circuit switches. Essentially, one can think of a circuit switch as a digital crossconnect that rapidly sets up and tears down end-to-end digital bit-pipes on demand from end users.

With this discussion in mind, it is clear that there are engineering issues at all levels in the architecture shown in Figure 2.18—from the laying of the fiber cables to the design of the lightpath and digital carrier network to the placement of the routers and the choice of link speeds to interconnect them. The physical laying of fiber is, of course, a long-term decision that one cannot revisit or modify frequently. On the other hand, current technology permits the rapid reconfiguration of optical crossconnects and digital crossconnects. Yet current practice does not actually exploit this possibility except over fairly slow timescales: typically days or weeks, at least. The main reason is that standard network management procedures that can effectively exploit this flexibility are not yet available. We can expect that such procedures will be developed as the demand for fast timescale bandwidth reprovisioning grows. Hence, in this book we confine ourselves to the design problem at the highest of the four levels of the infrastructure shown in Figure 2.18. At this level we see only switches (circuit or packet), the links between them, access networks, and terminals. If a circuit-multiplexed network and a packet-multiplexed network are carried on the same transmission infrastructure, then at this level they become logically separate networks. Thus, Figure 2.19 shows a packet network with the details of the transmission facilities abstracted away. This is an example of a network operated by an Internet service provider (ISP). The backbone comprises packet switches interconnected by digital links (e.g., 34 Mbps or 155 Mbps). This network attaches to other networks by *peering links*. Users access the network over high-speed LANs or over lower-speed serial links (e.g., 2 Mbps or 34 Mbps).

As we have stated before, in this book we do not discuss the problems of physical network design—fiber routing, node placement, topology design, and link sizing. We assume that a network infrastructure such as that shown in Figure 2.19 has been realized, and we concern ourselves with the problems of sharing the resources of such a network infrastructure.

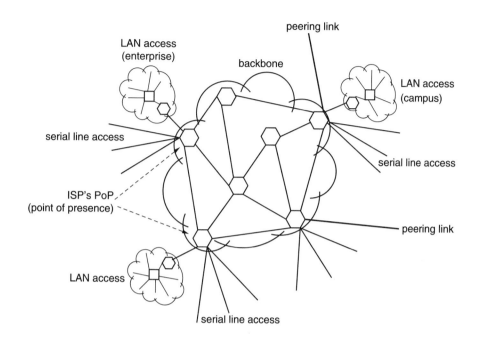

Figure 2.19   A view of a packet network with the optical fiber infrastructure abstracted away. An Internet service provider (ISP) connects to its customers at points of presence (PoPs).

Mobile Wireless Access Networks

So where does wireless transmission fit into the overall picture shown in Figure 2.18? There are two places. Until a few years ago, long distance digital transmission links were carried mainly on microwave wireless facilities. Owing to decreasing costs of optical transmission technology and to the much higher bandwidths available, digital microwave terrestrial networks are rapidly being replaced by optical transmission facilities. On the other hand, over the past 20 years, mobile wireless communication technology has made major advances, spurred by new modulation, coding, and signal processing techniques and the availability of high-performance digital signal processing chips. With hundreds of millions of cellular telephony subscribers, this technology is already playing a major role in the access to circuit-switched networks. In the architecture of new-generation *cellular wireless networks*, the availability of bandwidth for access to packet-switched networks is a major consideration. Furthermore, *wireless local area networks* are now providing transmission speeds in the tens of megabits per second. Given the major attraction of mobility, wireless networks can be expected to become the access networks of choice in office buildings and homes. Here we provide a brief overview of the communication infrastructure in cellular wireless networks; we discuss wireless local area networks in Chapter 8.

Figure 2.20 shows the components of a typical cellular wireless network infrastructure. The figure is modeled after the popular GSM (Global System for Mobile communications) system for mobile telephony, which is deployed in many regions throughout the world. The wireless links are only between the *mobile stations*, or MSs (shown as cellular phone handsets in the figure), and the *base transceiver stations* (BTSs). An MS can be in the vicinity of several BTSs, but at any point in time, an active MS is associated with one BTS: the one with which it is determined to have the highest probability of reliable communication. The region around a BTS in which MSs would most probably associate themselves with this BTS is called a *cell*. Thus the deployment of several BTSs in a geographical region essentially partitions the region into cells. If during a call, an MS moves between cells, a connection *handover* must be done between two BTSs. To facilitate handovers, the BTSs are laid out so that the cells overlap at their peripheries.

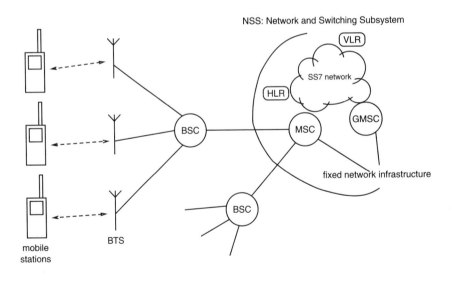

Figure 2.20   A typical cellular wireless network infrastructure.

Several BTSs are linked to *base station controllers* (BSCs) by wired links. Together, the BTSs and the associated BSC are called a BSS (*base station subsystem*). The BTSs provide the fixed ends of the radio links to the MSs; it is the BSC that has the intelligence to participate in the signaling involved in connection handovers. In turn, the BSCs are connected to the *mobile switching center* (MSC), which connects to the fixed network infrastructure.

A cellular operator leases spectrum (say, in the 900 MHz band) from the spectrum management authorities in the region. The spectrum is assigned in two equal-size segments: one to be used for communication from the MSs to the BSSs (this is called the *uplink* direction), and the other to be used from the BSSs to the MSs (the *downlink* direction). This spectrum is then partitioned, and each cell is assigned one of the parts. There is a fixed association between uplink and downlink frequencies, so that if an uplink frequency is assigned to a BTS, then the corresponding downlink frequency is also assigned to the same BTS; hence we can think in terms of the assignment of, say, only the downlink frequencies. The assignment of parts of the spectrum to cells must be done so that the same parts of the spectrum are reused in cells far enough apart so that they do not interfere with each other. Obviously, the less we need to partition the spectrum and the more we can reuse the frequencies over cells, the more communication traffic can be handled by the cellular network.

A digital modulation scheme is used to transmit digital data over the wireless spectrum allocated to a BTS. The approach taken in GSM is to first partition the spectrum into several *carriers*. Each carrier (there is a pairing of uplink and downlink carriers) is digitally modulated. On each of these digitally modulated carriers, several subchannels are defined; these are used for carrying user traffic and for signaling. Note that after an MS is connected to the network, it not only must send and receive data bits but also must exchange connection control messages to, for example, facilitate handovers.

Because the resources (i.e., the spectrum) of a cellular wireless network are limited, an MS cannot have permanent access to the network but instead must make a request for connection. Thus, because an MS is not always connected to the network, there are two problems that must be addressed.

1. Between the time that an MS last accessed the network and the time that it next needs to access, the MS may have moved; hence, it is first necessary to locate the MS and associate it with one of the cells of the network.

2. Because the MS initially does not have any access bandwidth assigned to it, some mechanism is needed for it to initiate a call or to respond to an incoming call.

Call setup and location management are the major activities that need to be overlaid on the basic cellular wireless infrastructure to address the first problem. In Figure 2.20 we show the additional components that are needed. Together, these are called the *network and switching subsystem* (NSS) and comprise the MSC, the GMSC (*gateway MSC*), the HLR (*home location register*), the VLR (*visitor location register*), and the signaling network (standardized as Signaling System 7 by the ITU (see also Section 2.3.3). The SS7 signaling network already exists where there is a modern circuit-multiplexed phone network. As their names suggest, the HLR carries the registration of an MS at its home location, and a VLR in an area enters the picture when the MS is *roaming* in that area. Each operator has a GMSC at which all calls to MSs that are handled by the operator must first arrive. The GMSC, HLR, and VLR exchange signaling messages over the SS7 network, and together they help in setting up a call to a roaming user.

Let us now turn to the second of the two problems just enumerated. In the GSM system there are several permanent channels defined in each cell. Whenever an MS enters a cell it "locks into" these channels. One of these channels is called the *paging and access grant channel* (PAGCH). If a call arrives for an MS and it is determined that the MS may be in a cell or in a group of cells, then the MS is "paged" in all these cells. Another such common channel is basically a slotted Aloha *random access channel* (RACH) (see Chapter 8) and is shared by all the MSs in the cell. When an MS must respond to an incoming call (i.e., it is paged on the PAGCH) or must initiate a call, it contends on the RACH in the cell and conveys a short message to the network. Subsequently, the network allocates a channel to the MS, and call setup signaling starts.

Several other standards exist for using the wireless spectrum to create digital carriers between the MSs and BTSs. Note that GSM divides the spectrum into frequency bands, digitally modulates each band, and then shares these digital bit-pipes in time; thus it creates bit carriers from the spectrum by using an FDM/TDM (frequency-division-multiplexing/time-division-multiplexing) approach. A major contender to GSM is CDMA (*code division multiple access*) technology, which does not partition the spectrum but separates the users by making their signals appear to be orthogonal to each other. We do not provide details of this approach in this book. For the purpose of this chapter, suffice it to observe that the spectrum-sharing technology, along with the call setup signaling and mobility management, provides a basic access infrastructure to mobile terminals. It is important to note that while an MS is connected to the network the mobility management is transparent to the communication, and the MS can be viewed simply as having a link into the network for the duration of the connection (see the dashed line labeled "access networks" in Figure 2.18). The link would be error-prone and subject to fading, and it could temporarily break during handovers.

Cellular wireless networks facilitate wide area mobile access to circuit-multiplexed phone networks and packet-multiplexed data networks. Because they must support highly mobile users and long distances (hundreds of meters to kilometers) between the mobile and the fixed infrastructure, the bit rates supported are quite small (Kbps in the second-generation cellular networks). On the other hand, wireless local area networks provide mobile access within buildings. These networks provide a communication range of a few tens of meters, tens of Mbps of bit rate, and mobility within the building. Unlike cellular networks, wireless LANs use some form of random access for sharing the wireless spectrum between the users. We discuss these topics at some length in Chapter 8. To provide continuous network access as a person moves from indoors to outdoors and back, access devices are being developed that can use the wireless LAN access in a building and cellular access outdoors, while seamlessly handing over between these as necessary.

2.3.2 Networking Architectures

We have seen in Sections 2.1 and 2.2 that several mechanisms and functions must be built on top of the basic bit-carrier infrastructure (described in Section 2.3.1) to provide information transport services. In Figure 2.2 we further partition these mechanisms and functions into two parts: the networking functions and information services. The former refers to the mechanisms for sharing the resources of the communication infrastructure among the various contending information flows, and the latter refers to certain common services that facilitate the utilization of the network. Many of these mechanisms and functions require the implementation of distributed procedures, which involve two or more remote entities executing their respective parts of the procedure to achieve some overall objective. In networking terminology, such a procedure is called a *protocol*. Because protocols are implemented in a distributed fashion, it is important that they be designed so that correct results are obtained in spite of asynchronous execution and delayed and unreliable information exchange.

Recall that a circuit-multiplexed network can basically be viewed as an extension of the TDM bit-carrier infrastructure (see Section 2.3.1). The circuit switches that interconnect individual user demands are crossconnects that operate rapidly to set up and tear down end-to-end circuits. Thus at the multiplexing and switching level there are no further architectural issues. However, as you have seen, associated with these circuit-multiplexed networks are signaling networks that are themselves packet-switched. The discussion in this section also applies to such packet-multiplexed signaling networks.

There are many protocols involved in the implementation of a complete communication network. When commercial networks were first developed, there was no common architecture that was followed. Architectures were vendor-specific, and network systems from different vendors would not interoperate. To systematize the implementation of network protocols, a seven-layer architecture, known as the *Open Systems Interconnection* (OSI) model, was developed; it is shown in Figure 2.21. Each function required in a network is mapped into one of the layers. A layer is viewed as providing *services* to the layers above, and as obtaining services from the layers below. Each layer can carry out its function satisfactorily only if it is provided certain requisite services from the layers below. For example, a distributed algorithm may need reliable and sequential delivery of interentity messages in order to operate correctly. Such a distributed algorithm must therefore obtain such a service from the layers below it.

In addition to codifying the places of the various functions in the overall network architecture, an important objective of the OSI model is to facilitate *open systems* as opposed to proprietary systems. The idea is that if the layer interfaces are well defined, then different vendors can supply various layers of the architecture. If each vendor correctly implements the layer it supplies, the whole system should work correctly.

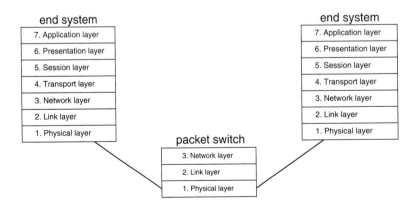

Figure 2.21   The Open Systems Interconnection (OSI) model.

The OSI model is an architecture for packet communications. The peer layers exchange packets in carrying out their distributed procedures. Some of these packets may carry user data, whereas others may carry messages only between the peer entities; these messages (e.g., acknowledgment packets) are used in carrying out the procedures. Peer layers exchange packets by utilizing the services of the layer below. Conceptually, the packets generated by a layer are queued at the interface between it and the layer below. When it is ready, the layer below takes each packet from the layer above, attaches some bits of its own (for carrying out its own functions in conjunction with its peer), and then queues the packet in turn at the next layer boundary. Similarly, when a packet is received it moves up the layers. Each layer removes the bits attached by its peer and carries out its functions, passing the remaining packet to the layer above.

The Layers of the OSI Model

The function of the *physical layer* (Layer 1) is to send and receive bits over the communication infrastructure. All the issues involved in digital modulation, demodulation, and channel coding, and even the details of the physical connectors, are subsumed in this layer. Note that the channel coding in the physical layer may not always be able to eliminate bit errors; hence there will be a residual bit error rate at the physical layer. Furthermore, because the bits are transmitted over electromagnetic media, there is a finite speed at which signals travel, and this introduces propagation delay between the bits entering the physical layer at one end of a bit-pipe and emerging at the physical layer at the other end. Delays may also be introduced due to the modulation and demodulation mechanisms. With reference to Figure 2.2, the physical layer provides the communication links. Because every device, whether it is a switch or a terminal, must send and receive bits, the physical layer exists in all devices.

Given the raw bit-pipe provided by the physical layer, the *link layer* (Layer 2) makes for reliable communication between the two ends of the pipe. The link layer offers *frames* to the physical layer for transmission. We achieve error control by attaching to the data in a frame certain redundant bits that are derived from the data bits. At the receiving link layer (at the other end of the communication link), a mathematical check is performed over the received bits in each frame, and it can be determined with a very high probability whether or not an error has occurred. In case of an error the frame can be retransmitted by the sending link layer. This can be attempted a few times, after which the link layer gives up and silently discard the packet(s) it could not send. In networks with highly reliable links (e.g., optical fiber links), this retransmission function of the link layer may not be implemented; in such cases the link layer receiver would discard a received packet upon error detection, leaving the application to handle the loss of the packet.

When the layers above the link layer offer it a packet to be sent, the link layer performs its framing function and then queues the packet at the boundary between itself and the physical layer. It is here that the issues of multiplexing of several flows into the physical link are handled (see Section 2.2.1). If the physical link is shared by many transmitters and receivers, then the sharing of the distributed medium is also viewed as belonging to the link layer. This is called *medium access control* (MAC). In the OSI model, the so-called MAC layer is positioned between the link layer and the physical layer. The link layer also exists in every device in the network, whether it is a switch or a terminal.

The *network layer* (Layer 3) ties together the links (point-to-point or multipoint) into a network. At this layer, devices in the network are provided with globally unique *network addresses*. The primary function of the network layer is that of routing packets between these network addresses. Figure 2.21 shows that in a packet switch, only the bottom three layers need to be implemented; these are sufficient for the switch to carry out its function of packet forwarding (but see Section 2.3.5 to see what actually goes into an Internet packet switch). Between the packet switches in a network, a distributed routing protocol is implemented in the network layer to learn good routes. The outcome of this route computation is that a routing table is obtained at each switch; this table determines how packets that need to go out of the switch should be forwarded. When a packet arrives into the network layer from the link layer and needs to be forwarded, the table is looked up and then the packet is again given back to the link layer, where it is queued at the appropriate output port. The network layer also exists in terminals, because they need to know how to forward packets to other terminals on their local network (e.g., a switched Ethernet) or to a router on the network if the packet needs to leave the local network.

The transport layer (Layer 4) does not exist in packet switches in the network but instead is an end-to-end layer that resides in the end terminals. Important functions of protocols at this layer are to associate communicating entities in the end terminals and to provide each such instance of communication with additional services. These services bring up the level of services provided by the network layer to a level suitable to be utilized by the communicating entities. If several logical end points in a terminal with *one network address* need to communicate separately with other such end points in other terminals, the transport layer takes care of this *logical* multiplexing and demultiplexing of data from and to the communicating entities. This should be distinguished from the physical multiplexing of several flows into a bit carrier, a function that in this architecture is carried out by the link layer.

The following are two examples of how a transport layer helps to adapt the service provided by the network layer to the needs of an application. The packets of a packet telephony application are carried asynchronously by the packet network. At the receiver these packets are "played out" according to their original timing relationships, and missing packets are interpolated. Timing and sequence information that can be used by the receiver to carry out such playout can be provided by additional protocol fields added by the transport layer. A file transfer application requires reliable, sequential transfer of its data packets even though the network layer may not provide such service; the transport layer can also provide this service. In the Internet, the procedures of dynamic bandwidth sharing between competing file transfers are also implemented in the transport layer.

The session layer (Layer 5) provides services for dialog management between users. For example, if a long file transfer is interrupted, the task of carrying on from that point is a function of this layer.

Functions such as file format conversion, data encryption, and data compression are possible services at the *presentation layer* (Layer 6).

The application layer (Layer 7) provides commonly used services such as file transfer, e-mail handling interfaces, terminal emulation, and directory services.

Note that although the bottom four layers are clearly distinguishable as separate entities, common practice typically embeds the functions of the upper three layers into applications themselves. Thus, session recovery and data compression may be embedded within a file transfer application rather than being implemented as separate common services. With reference to Figure 2.2, we can also discern that networking, which is the focus of this book, corresponds to the functions in layers 2, 3, and 4 of the OSI model, whereas the information services layer corresponds to the top three layers.

In addition to transporting user data, communication networks carry *control* and *management* messages. The control functions in networks include connection control, or *signaling*, and adaptive routing. Signaling procedures are used to set up associations between entities that want to use the network to communicate (e.g., packet voice call setup) and to control these associations during communication (e.g., putting an ongoing call on hold). Control also includes routing protocols that communicate between network elements to discover good routes in the network and to relearn new routes in case of network element failures. Network management is concerned with monitoring the flow of traffic and the status of the facilities in the network, and also with actions that may need to be

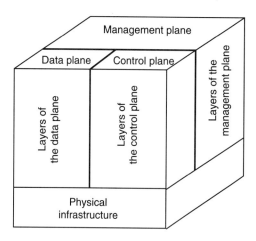

Figure 2.22   Network functions can be viewed in terms of three planes: data, control, and management. They share the same physical infrastructure for carrying out their functions. This model has been adapted from ITU standards.

taken care of—for example, failures and overloads. It is therefore customary to think of networks as having three *planes:* (1) the user or data plane, (2) the control plane, and (3) the management plane (see Figure 2.22). Each of these planes can, in principle, have its own protocols at each of the OSI layers, but they planes would share the physical communication infrastructure. In the Internet, for example, these three planes share the protocols up to the network layer.

Cross-Layer Interactions

As we have mentioned, what we call the networking functions in Section 2.1 correspond to the link/MAC, network, and transport layers of the OSI model. Two important features of this layered model for communication are that each communication function is provided at only one layer, and that layer boundaries are strict, with adjacent layers communicating over well-defined interfaces. Although the seven-layer model has been extremely useful in the conceptualization and design of communication network protocols, the technology has reached a point where the strict layering of functions is routinely violated. A simple example is the use (in the Internet context) of TCP port numbers to identify flows that require a certain QoS, and then the use of this information to manage queues at lower layers in routers and Ethernet switches (which are, respectively, layer 3 and layer 2 packet-switching devices). Functions are often duplicated: Important examples arise in *IP over ATM*, in which case the ATM layer is often considered to be akin to a link layer. In this case, there is a routing protocol at the ATM level, and another one at the IP level; if the ATM connections use the *Available Bit Rate* (ABR) service (see Section 2.3.6), then there is an adaptive window congestion control at the TCP level and a rate-based congestion control at the ATM level (see Chapter 7). In the latter context, there are research proposals to feed rate information from the ABR control into the adaptive windowing protocol at the TCP level. Of course, all this is happening because of the evolutionary nature of communication network technology and an attempt to create new services from a combination of new technology and an embedded base of old protocols. In view of this we have chosen to deal with the core issues in networking. Wherever it proves useful, we attempt to relate our discussion to the way current technology has addressed the issues.

The interactions between layers take place because of the evolutionary nature of network designs and the need to interwork existing (legacy) systems with new technologies. The traditional layered approach to network design simplifies the conceptualization, design, and implementation of networks. However, recently it has begun to be realized that there are compelling reasons to retain the coupling between the various layers in the process of network design. We can recognize an instance of this in the first example in Chapter 1. We saw that if the voice coder could adapt its coding rate to the occupancy of the link buffer, then there could be a substantial gain in network efficiency. Note that a buffer is a link-level entity, whereas voice coding is an application-level activity. Thus we have an example of beneficial cross-layer interaction.

This issue of possibly beneficial cross-layer interactions is particularly important in mobile communication networks built over wireless communication links. The links vary with time because of multipath fading. These links are shared by a number of traffic flows having various QoS requirements. In addition the end points are mobile and hence resource-limited, a significant limitation being the batteries; mobility also constrains the bit rate that can be supported on the wireless link. In such situations, it becomes very important to make the access and utilization of the network highly adaptive to the channel and traffic conditions, and hence it necessarily leads one to seek approaches that attempt to jointly solve problems at several levels in the layered model. In fact, channel condition, battery utilization, multiple access, multiplexing, and routing may all need to be interdependent for efficient operation that meets QoS objectives. See Section 8.4 for a more detailed discussion of this topic.

2.3.3 Telephone and ISDN Networks

Until a few years ago, the *public switched telephone network* (PSTN) was used primarily for carrying telephone-quality voice calls but recently has been used for carrying all kinds of other traffic such as data and compressed video. The links in a telephone network are logically partitioned into channels, each of which can carry a voice call; in telephony jargon these channels are called *trunks*, and the group of channels on the link is called a *trunk group*. Each traffic flow on a link is assigned one or more trunks for the entire duration of the flow. Thus the PSTN links are circuit-multiplexed. The switches in the PSTN (also called *telephone exchanges*) establish paths between the channels on the links to which they are attached. Every *call* (an instance of traffic flow) goes through a call setup phase during which a path is chosen for the call. In the telephone network, the predominant calls are requests for "plain old telephone service" (POTS). The exchanges determine a route for each call and the availability of free channels on the links along the route. If possible, channels are allocated on the links along the chosen path, and the intermediate switches are set up to interconnect these channels. If the required number of channels cannot be allocated along a route, alternative routes may be attempted, and if these also fail, the call is blocked. After the circuit for a call is set up, the switches are transparent to the traffic flow of the call. When the end points of the call (the source/sink pair) signal the end of the call, the resources reserved for the call (i.e., channels and switch paths) are torn down.

The telephone network provides exactly one service to any flow that it carries; it can establish bidirectional circuits that can carry high-quality voice signals that have fixed end-to-end data transport delay (the sum of the propagation delay and the switching delay). The sources/sinks need to adapt their needs to the constraints imposed by this service. The QoS offered by a telephone network is specified in terms of the probability that a circuit cannot be allocated to a call (*the call-blocking probability*) and the time it takes to set up the call (*the call setup delay*). The call-blocking probability depends on the network topology and the number of channels on each link. With most of the telephone network being digital and computer-controlled, the call setup delay depends on the computing power of the call-processing computers at the switches, and on the interswitch signaling.

Routing and traffic control in the telephone network facilitate achieving an acceptable call-blocking probability and call setup delay, while using the network resources efficiently. When all the channels on each link on a path are busy, in a richly connected network calls can be *alternate-routed*. If routes or switches are congested, traffic controls are used to reduce new call attempts and to reroute call attempts that have already made progress in call setup.

The telephone network consists of a *local network* serving a *calling area* and the *long distance network*. The phone at the customer premises connects to a *local exchange* (LE), also called *end office* or *central office*, through a local loop. The local loop may also be provided by a cellular telephony infrastructure. In this case the mobile switching center (MSC; see Section 2.3.1) has a function similar to that of the local exchange. In the wireline network, data transmission on the local loop typically is in analog form, and every other part of the network is predominantly digital, with each channel (or trunk) on a link (trunk group) having a fixed capacity of 64 Kbps. The end office location is chosen to minimize the wiring costs to connect the phones distributed in the area served by it. A local network may be served by many end offices. The exchanges in a calling area are interconnected through a *tandem exchange*, and if there is significant traffic between two exchanges, they may be directly connected by *high-usage* (HU) trunks. The use of HU trunks between two exchanges in a calling area depends on the traffic between them and the distances. A call between two LEs is first offered to the HU trunks, and if all the circuits on that are busy, it is offered to the *final route* through the tandem exchange.

An important factor in the dimensioning of the links is its efficiency. The number of circuits on a link depends on the traffic volume expected to flow on that link and the allowable blocking probability. The time average of the occupancy of a link can be defined to be its efficiency. It can be shown that, for a given call-blocking probability objective, higher-capacity links handling larger volumes are more efficient than smaller-capacity links.

The design of the long distance network is similar to that of the local network except that there are more levels in the hierarchy, and trunk efficiency is a major design issue. The traffic volume between geographically distant LEs is typically small, and direct trunks may be inefficient. Efficiency becomes important with increasing distance because a more expensive resource must be used more efficiently. Thus a switching hierarchy is constructed for the telephone network. However, if there is significant traffic between any two exchanges, they can be connected directly by high-usage trunks, much as in the local network. The telephone network switching hierarchy is shown in Figure 2.23. A number of LEs are served by a primary center, or class 4, exchange, and the hierarchy is built up to quaternary centers, or class 1 exchanges. The exchanges at the top of the hierarchy are connected in a mesh much as in Figure 2.13a. Every exchange, except the class 1 exchanges, *homes* into a higher-class exchange. The first-choice route for a call between phones connected to two different exchanges is through the high-usage trunk toward the destination exchange. If such a path is not available, then the call is attempted through the next level exchange in the hierarchy. Obviously, the

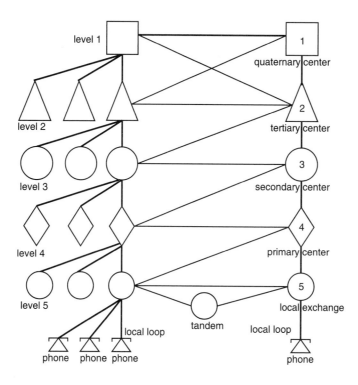

Figure 2.23 Hierarchical arrangement of the telephone network exchanges. The thick lines are links (trunk groups) to be used on the final routes. The notation used in the North American network is shown on the left side, and the ITU-T nomenclature on the right.

highest-level exchange that might be involved is the lowest level that is common to the two phones. Note that a route is always available through the hierarchy, and this route is considered the final route for the call. The final route is shown using thick lines in Figure 2.23. In recent times, nonhierarchical routing schemes have also been devised.

An important aspect of circuit-multiplexed networks such as the telephone networks is call setup. The local exchange communicates status information with the terminal device (such as a phone). All the exchanges communicate control and status information, collectively called *signaling*, among themselves to facilitate path setup and maintenance for the duration of the call. In the early networks, this was accomplished through the exchange of tones of different frequencies on the same physical channel as that used to carry the voice. Very little information can be reliably exchanged in a reasonable time in this manner to achieve an acceptable call setup delay. This limits the services that can be offered both to the network and to the user. To provide advanced *intelligent network* (IN) features and also to enable the use of more efficient resource (link) utilization capabilities, a fairly sophisticated signaling technology called common channel signaling (CCS) has been developed. CCS uses a logically separate channel for signaling between nodes. The latest version is SS7, and the signaling messages are exchanged over a logically separate network called the signaling network. This logical separation is shown in Figure 2.24. The signaling network corresponds to the control plane of the telephone network and controls the resources in the PSTN.

The network elements of the SS7 network are shown in Figure 2.25. There are three types of nodes in the SS7 network. The *signal switching points* (SSPs) are exchanges that have SS7 capabilities and can hence receive or initiate signaling messages. The *signal transfer points* (STPs) are packet switches in the SS7 network that route the signaling messages, and the *signal control points* (SCPs) are databases that contain the status information that can be used to process a call setup request or provide other services. Normally, each SSP and SCP is connected to more than one STP to achieve signaling reliability. This is not shown in Figure 2.25.

The following is an example call setup procedure between phones connected to exchanges, say *A* and *B*. For the following we assume that the exchanges are directly connected.

1. After the digits are dialed by the calling phone, exchange *A* recognizes that the target exchange is *B*, decides which trunk group to use for the call, and sends an *initial address message* (IAM) to exchange *B* with that information

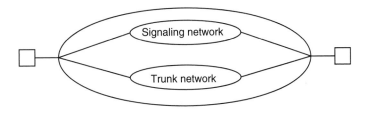

Figure 2.24 The trunk network and the signaling network are logically separate networks. Nodes (exchanges) exchange status and control information over the SS7 signaling network to allocate and control the resources of the trunk network.

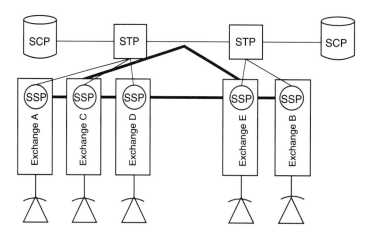

Figure 2.25   The SS7 network. The thick lines are the trunks, and the thin lines are the logical paths for the signaling messages.

over the SS7 network. The IAM passes through one or more STPs to reach the SSP in exchange *B*.

2. On receiving the IAM from *A*, exchange *B* recognizes that the called phone is connected to itself and acknowledges the message with an *address complete message* (ACM) to *A*. Again, STPs are involved in this transfer. Simultaneously, exchange *B* sends a ringing tone over the trunk that *A* has decided to use for the call.

3. On receiving the ACM, exchange *A* connects the local loop of the calling phone to the trunk that it has reserved for the call, and the ringing tone is now available to the calling phone.

4. When the called phone is picked up, *B* sends an *answer message* (ANM) to *A* to indicate that the call is successful.

Call release is achieved in a similar manner.

The SS7 network is built as a packet-switched network. The SSPs are the sources of the signaling packets, and the other SSPs or SCPs can be the destination. The STPs essentially act as routing devices. In fact the protocols used to exchange the signaling messages have a design that follows the layered approach of the seven-layer OSI model discussed in Section 2.3.2.

The SS7 network enables exchange of detailed control and status information. Adding some computation power to this has enabled many new services such as the toll-free service. In this service the toll-free numbers are mapped to different real telephone numbers for different calling areas. The SSP obtains the map by querying the databases at an SCP.

The SS7 network has also been used to provide entirely new access methods. For example, from our discussion of the cellular system in Section 2.3.1, recall the significant amount of signaling that is necessary to provide roaming services to a cellular network user. Also recall that much of this signaling is carried over the SS7 network. Another service that the SS7 has enabled is the Integrated Services Digital Network (ISDN). We provide a brief description of this service.

As mentioned earlier, in the telephone network, digital links are deployed in the backbone of the network at the higher levels of the hierarchy of Figure 2.23, whereas the local loop remains analog. This means that the access interface—the point where the phone instrument connects to the local exchange—is analog, and the benefits of end-to-end digital communication do not reach the end user. In the 1980s ISDN, a new digital access interface to the PSTN, was developed. This interface permits applications to use end-to-end circuits with bit rates in multiples of 64 Kbps. (In ISDN terminology, a 64-Kbps channel is called a bearer (B) channel.) A 16-Kbps data (D) channel is also available, and it can be used for signaling and data transfer. Typically two types of ISDN services are available for an end user. The *basic rate interface* (BRI) is at 144 Kbps and consists of "2B+D" channels. In North America, the *primary rate interface* (PRI) is at 1.536 Mbps and consists of "23B+1D," where the D channel is 64 Kbps, whereas in Europe the PRI is at 1.984 Mbps and consists of "30B+1D" channels, with the D channel being at 64 Kbps. The H channels provide a way to aggregate B channels, and the following aggregations are defined:

- 6 B channels for a 384 Kbps H0 channel

- 23 B channels for a 1472 Kbps H10 channel

- 24 B channels for a 1536 Kbps H11 channel

This ability to let the users demand ($n \times 64$ Kbps) channels for different n makes ISDN a multiservice network. Another important capability of ISDN is the simultaneous establishment of multiple calls, of possibly different types, across an access interface. For example, a 64 Kbps telephone call and a 384 Kbps video connection can be established simultaneously across the same access interface.

An ISDN-compatible end device is called a TE1, and a non-ISDN device is called a TE2 device. TE2 devices connect to the ISDN network through *terminal adapters* (TAs). The two-wire interface from the local exchange is called the *U interface* and is converted to a four-wire interface (called the *T interface*) by network termination equipment called NT1. In North American networks, the NT1 is a customer premises equipment (CPE) device, whereas in many other countries, NT1 termination may be provided by the service provider. This means that the customer premises will have only the T interface. ISDN devices connecting to the ISDN network over the T interface treat the interface as one end of an end-to-end circuit, and they use a communication stack that includes all the layers above the physical layer. A second type of termination, called NT2, is typically used by ISDN-capable *private branch exchanges* (PBXs) to connect to the public ISDN network. An NT2 has the link layer and network layer functions of the OSI model of Figure 2.21 and connects a TE1 over the S interface.

The link layer protocol used in ISDN is called the Link Access Protocol, type D (LAPD) and is specified by the ITU-T Q.920 and Q.921 standards. The network layer protocol is similar to that used by the X.25 network (discussed next) and is specified by the ITU-T Q.930 and Q.931 standards.

The various terminals, interfaces, and protocols are shown in Figure 2.26. Note that Figure 2.26 is an abstract model, and in practice many of the interfaces will be combined into one device.

2.3.4 X.25 and Frame Relay Networks

Early attempts by the telephone companies to introduce public data networks were based on the X.25 protocol suite. This effort gave rise to *packet-switched public data networks* (PSPDNs) in many countries. In this technology, the network links are packet-multiplexed. The links are interconnected by packet switches. A fixed bidirectional path is chosen for all packets of a call; because the path is fixed, even though bandwidth is not reserved along the path, such a call is called a *virtual circuit*. Thus there is a call setup and tear-down phase for X.25 calls. Even though each terminal point of the network has a unique (and long) address, a path for all packets of a call is preestablished, so the switching of packets of a call is done on the basis of short virtual-circuit *labels* carried by the data packets. Traffic control is exercised by means of an end-to-end *window flow control*. Because X.25 was originally designed for poor-quality links, an elaborate error control protocol, LAPB (Link Access Protocol, type B), was implemented at each hop.

X.25-based PSPDNs have been used to carry computer–computer or terminal–computer traffic. The terminal speeds were usually a few kilobits per

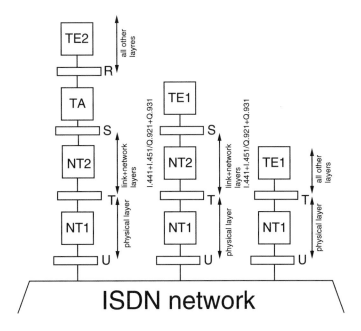

Figure 2.26 Terminals, interfaces, and protocols in an ISDN network.

second (e.g., 19.6 Kbps), and the host attachment speeds were rarely more than 64 Kbps. The links were typically 64 Kbps or 1.5 Mbps. Thus these were low-performance networks, meant primarily for store-and-forward applications that had no QoS requirement from the transport network.

In conjunction with the digitalization of transmission facilities and the development of ISDN, the need was felt for a more efficient packet transport service than X.25. This gave rise to the frame relay protocols and services. Like X.25, Frame Relay is based on virtual-circuit based packet-multiplexing. These networks were designed to operate on high-quality digital links with low error rates, and hence hop-by-hop error control was eliminated. The link layer in Frame Relay networks is a derivative of ISDN's layer 2 protocol, namely LAPD. An LAPD receiver checks an incoming frame for errors; if an error is detected the frame is dropped, to be recovered by the end applications. Frames are switched at the link layer rather than at layer 3, as in X.25. Longer frames can be used because the error rates on the links are small. In addition, a Frame Relay network can provide a rate assurance to each connection that is set up through it. Such rate assurances can be enforced by edge switches, which mark frames that violate their rate assurances, and by network switches, which drop such violating frames during periods of congestion. The most popular application of Frame Relay networks is for the interconnection of corporate local area networks via assured-rate permanent virtual circuits in commercial Frame Relay networks.

2.3.5 The Internet

The Internet refers to the worldwide interconnection of packet networks that all use a suite of protocols that originated in the famous ARPANET project of the 1970s. In this protocol suite, IP (Internet Protocol) is the network layer protocol, and TCP (Transmission Control Protocol) and UDP (User Datagram Protocol) are the most commonly used protocols at the transport layer. The common noun "internet" is often used to connote a network that uses the Internet protocol suite. The IP protocol can operate over any link layer (and, by implication, any physical layer) that can transport IP packets. Because it simply requires the implementation of a packet *driver* to carry packets over any bit carrier, an internet can be operated over essentially any bit-carrier infrastructure. The Internet protocol suite also does not define the layers above the transport layer. Thus, for the Internet, Figure 2.21 simplifies to the depiction in Figure 2.27 (we show many more packet switches, and the physical layer is implicit in the links). We will see later that this is a simplified representation; for example, an application may run directly over IP (thus taking care of its transport needs itself).

The most widely deployed version of IP is version 4, which uses 32-bit addresses. The network address of an entity is also called its *IP address*. As in most communication networks, the addresses in the Internet are hierarchically assigned. The address of each device comprises some contiguous high-order bits that identify the subnetwork in which the device resides; this is also called the *network prefix*. The remaining bits identify the device uniquely in the subnetwork. So, for example, all the addresses in a campus may be of the form 10010000.00010000.010xxxxx.xxxxxxxx, where each x can be 0 or 1. In such a case the network prefix is 10010000.00010000.010, and it is of length 19 bits.

Unlike circuit-multiplexed networks or the packet-multiplexed X.25 and ATM networks, internets do not fix a path for the packet flow on a connection. The network simply provides connectivity between end points. Every packet carries the full network address of its destination end point. Each packet switch looks at the arriving packets, consults a routing table (which actually deals with network prefixes), and forwards the packet to an outgoing link that, hopefully, carries it closer to its destination. By rejecting the virtual-circuit approach in favor of

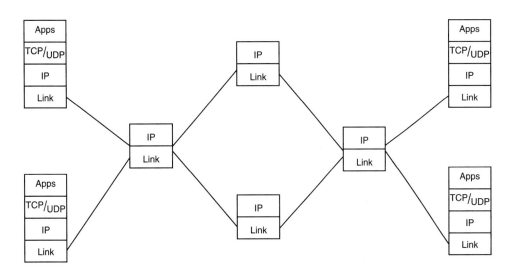

Figure 2.27   The Internet protocol architecture; "Apps" means applications. In this simplified depiction, the packet switches are shown to have only IP and the link layer. The end nodes have the transport protocol in addition to IP and the link layer.

per-packet, hop-by-hop, best-effort routing, the Internet gains the advantages of (1) quick delivery of small amounts of data, (2) automatic resilience to link failures, and (3) ease of *multicast* (i.e., the transmission of a packet to multiple destinations by replicating it at appropriate points in the network, rather than by the source sending multiple copies of the packet).

Notice that because IP routes each packet as a separate entity, it is possible for consecutive packets of the same session to follow different routes and then, owing to different delays on the routes, arrive out of order. The IP layer at the destination simply delivers the packets out of order! Measurements showing that there can be significant packet reordering in the Internet have been reported in [29]. In addition, a link layer may discard a packet after unsuccessfully attempting it a few times. A packet may also be discarded at the queue of a physical link because of exhaustion of buffer space or some packet scheduling decision. Hence the packet delivery service that the IP layer provides is *unreliable* and *nonsequential*. This is known as a datagram delivery service.

Figure 2.28 shows a fragment of an internet's topology. Each router is attached to physical links by its interfaces. Note that a multipoint link can have many routers attached to it. All the devices attached to a link are each

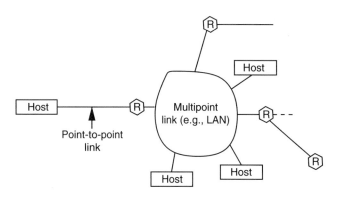

Figure 2.28   A fragment of an internet, showing routers (labeled R), hosts, and links (including multipoint "links," such as LANs).

other's neighbors. The Internet is equipped with *routing protocols* that permit routers to identify good paths on which to forward packets. These protocols work on the basis of metrics assigned to the network links and a distributed algorithm for determining shortest paths under these metrics. Note that a routing protocol is also an application running on the network! Hence it needs to use the packet transport services of the network. Fortunately there are distributed algorithms that can learn shortest paths through the network, and, in the execution of these algorithms, nodes need only exchange packets with their neighbors. Hence routing in an internet can bootstrap itself. A simple protocol (aptly called the Hello protocol) is used by routers to discover neighbors. After neighbors are discovered they begin to exchange routing protocol packets, which are used in computations that gradually lead to all the routers to learn shortest paths to network prefixes. One such distributed algorithm works by each router informing its neighbors about the status of the links to which the router is attached. This information is *flooded* through the network. Eventually every router obtains these *link state advertisements* (LSAs), which can be put together to obtain a full topology view in each router. A shortest path computation can then be locally performed and routing tables built up in each router. This algorithm is implemented by the currently most popular routing protocol, OSPF (Open Shortest Path First). The OSPF protocol is a routing application protocol, but it does not utilize the services of a transport protocol, instead running directly on the IP layer in routers.

The basic Internet packet transport does not promise any QoS to the traffic streams it carries. The network provides connectivity and routing; any user attached to the Internet can initiate multiple flows. There is no *connection admission control* (CAC); all the flows end up sharing the network resources. Thus the Internet transport provides a highly variable quality of service that can vary widely depending on geographical location and time of day. To bring some sanity to the situation and to enforce some fairness in bandwidth sharing, the end-to-end Transmission Control Protocol (TCP) implements an adaptive window protocol. This protocol reacts to packet losses by reducing its window size and then slowly building it up again. This function of *congestion control* and *bandwidth sharing* is in addition to the two other functions that TCP performs: (1) reliable and sequential packet transport over IP's unreliable and nonsequential packet transport service, and (2) sender–receiver flow control (which prevents, for example, a fast computer from flooding a slow network printer). Chapter 7 discusses TCP's congestion control and bandwidth-sharing function at length.

UDP (User Datagram Protocol) is another popular layer 4 protocol. This protocol simply permits a user above IP to utilize the basic datagram delivery service, thereby multiplexing several flows into one IP address. Note that this is the *logical* multiplexing of several flows originating and terminating at a common IP address. It should be distinguished from the physical multiplexing of flows into a bit carrier, something that is a link layer function. We distinguish the different flows by assigning them different UDP *port numbers*. UDP is used by applications, such as packet voice telephony, that must receive guaranteed service rates in the network and cannot deal with packet loss by retransmission, and hence cannot use the services of TCP. Other mechanisms, above the transport layer, are used to facilitate such applications.

We have stated that the Internet's packet transport is not designed to provide specific QoS to the flows it carries. The service model is quite simple. Nodes that have valid IP addresses can attach themselves to the network and send IP packets back and forth between themselves. The network provides an unreliable, nonsequential packet transport service, with no guarantee of transfer rate or delay, often called best-effort service. The network does not distinguish between the various traffic flows; it does the best it can, treating everyone alike. The idea is that the applications best know what they need and should adapt to the transport that the network provides by using end-to-end mechanisms. TCP's mechanisms for achieving a reliable and sequential packet transport service over the Internet, and some sort of fair sharing of network bandwidth, are a prototypical example of the end-to-end approach. Over the past decade, however, the Internet has become *the* packet transport network of choice for all kinds of store-and-forward data transfer and increasingly is being used by applications that require some minimal quality of packet transport. Broadly there are two approaches that can be followed to provide some level of QoS over the Internet: new QoS architectures and traffic engineering.

Two QoS architectures have been proposed and extensively studied: the *Integrated Services Architecture*, abbreviated as IntServ, and the *Differentiated Services Architecture*, abbreviated as DiffServ. The proposals in the IntServ architecture essentially allow each session arriving to the network to request QoS guarantees; it must declare its traffic characteristics to the network, and the network has the choice of rejecting the request or accepting it at some lower level of QoS. This architecture requires signaling protocols to be put in place, and packet-scheduling mechanisms are needed at the packet-multiplexed links. Evidently, these protocols and scheduling mechanisms need to be implemented in every router in parts of the network over which such QoS guarantees are needed.

In the high-speed core of the network, the session arrival rates and the packet rates are too high to permit session-by-session analysis and packet-by-packet scheduling. Hence, as in other transport systems (e.g., airlines and railways), it has been argued that only a few levels of differentiation may suffice (e.g., first class, business class, and economy class). The classes are, of course, priced differently. At the simplest level, there may be one priority class of traffic (reserved, for example, for interactive packet telephony), and another class for the remaining traffic. There could be one additional class for the premium store-and-forward traffic, and the remaining traffic could be handled by the default best-effort packet transport. This idea has led to the DiffServ proposals. The scheduling at the links distinguishes a few classes of packets (e.g., 2, 3, or 8, depending on the choice of the network operator). The class of a packet is identified by the contents of its header; such classification can be based on six special bits in the IP header, called the *DS code* (DiffServ code), and could also be based on source and destination addresses and even the source and destination transport protocol port numbers. The schedulers are aware only of such packet *aggregates*; there is no awareness of individual so-called microflows. A DiffServ core network may put a limit on the amount of traffic of each class it is willing to accept from each customer network that transports traffic through it. If a network violates such restrictions, the DiffServ core can reject the excess traffic or handle it at lower levels of service. It is up to the edge nodes of the customer network to police the traffic that it offers to the DiffServ core.

It has been argued that in conjunction with an IntServ architecture in the lower-speed edges of the network, a DiffServ architecture would suffice in the core to provide an overall end-to-end QoS to applications.

The other approach for providing QoS over the Internet is traffic engineering. Clearly, if there is sufficient bandwidth the best-effort packet transport suffices.

With the rapid deployment of optical networks, even to within a few hundred meters of network access points, it is becoming easier and less expensive to quickly deploy additional bandwidth where there are bottlenecks. Hence, it has been argued that it suffices to manage QoS by network and traffic engineering. The network topology (node placement, node interconnection, and link capacities) should be properly designed with the expected traffic in mind, and it should be tolerant of link failures. It also should be possible to deploy new bandwidth when needed. Furthermore, in the operating network the traffic should be carefully routed to prevent the formation of bottlenecks, and the routing should be monitored and revised from time to time. Traffic engineering alone, of course, cannot address the problem of congestion in access networks.

Finally, to end this section, we turn to the components of Internet hosts and routers. Figure 2.29 shows the protocols implemented in a typical Internet host. This host can handle email because it has SMTP (Simple Message Transfer Protocol) and can browse the Web because it has HTTP (Hypertext Transfer Protocol); both of these protocols run over TCP. The host can also participate in a packet voice call, something that requires the RTP (Real Time Transport) protocol, which in turn runs on UDP's simple datagram service.

Now let us look at the components of an Internet router (see Figure 2.30). A router is a packet switch and hence moves packets between several links. In the figure, the router shown has three links: a link to a *wide area network* (WAN), a link into a LAN, and a *dial-up link* (into which a connection can be made over the telephone network). There is a link protocol for each physical link: HDLC (High-Level Data Link Control) for the WAN link, the IEEE standard link protocol

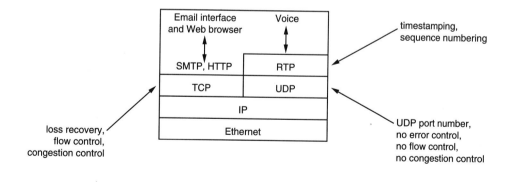

Figure 2.29   The typical protocols in an end system (or host) attached to the Internet.

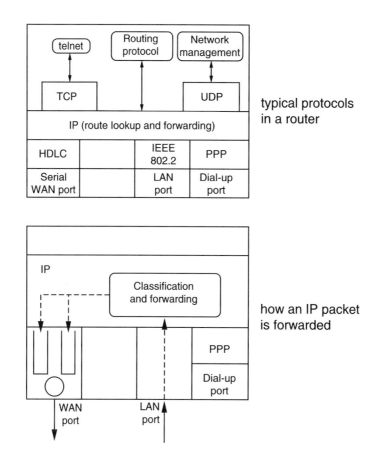

typical protocols
in a router

how an IP packet
is forwarded

Figure 2.30   The protocols in a router (top) and the way an IP packet is forwarded from one interface of a router to another (bottom).

(IEEE 802.2) for the LAN link, and PPP (Point to Point Protocol) for the dial-up link. The IP layer runs across all these link layers and forwards packets between them. The packet switching could simply be done by reading into and copying out of the processor memory, or there could be a hardware switching fabric in a high-capacity router; the switch is not shown. In addition to these components (which were also depicted in Figure 2.27), there is a routing protocol (shown here directly over IP, as is the case for OSPF). There is *telnet*, a protocol for permitting an administrator to log in to the router to configure its parameters; because telnet requires TCP, TCP must also be implemented on a typical router. Furthermore, there is a network management protocol that is used to monitor the traffic in the router and the status of its links. As mentioned earlier, SNMP is the commonly used protocol for network monitoring and management in the Internet, and, as shown in Figure 2.30, it operates with UDP as the transport protocol. The bottom part of Figure 2.30 shows what is involved in routing a packet from one port to another. Not only must IP do forwarding, but there is also the need to classify the packets into QoS classes to appropriately queue them at the output port; two queues are shown in the figure, perhaps a high-priority queue for voice packets and a low-priority one for data packets.

2.3.6 Asynchronous Transfer Mode (ATM) Networks

The development of ATM technology was motivated by the desire to implement a Broadband Integrated Services Digital Network (B-ISDN), which is a generalization of ISDN (see Section 2.3.3). Instead of access rates in multiples of 64 Kbps as in ISDN, B-ISDN supports access at almost any rate that the user desires. ATM offers a specific technique of packet switching and multiplexing onto a communication link. The packets generated by sources of information are, in general, of variable size. The most prominent feature of ATM is that it is designed to multiplex, switch, and transport small, fixed-size packets called *cells*. Each cell in an ATM network is 53 octets long and consists of a 5-octet header and a 48-octet payload.

The small size of the ATM cell can offer useful advantages. The fixed size helps in *cell delineation* at very high speeds. That is, identifying the boundaries of a cell becomes simpler because of the fixed size. Also, cell switching at very high speeds becomes possible because the units to be switched (i.e., cells) are always of the same size.

To see another advantage, suppose that packetized voice is being carried on the network along with bulk file transfer traffic and that the two information flows are to be multiplexed onto the same link. In such a case, it is natural to give priority to the voice flow so that the cell carrying digitized voice is transmitted as soon as possible. If the first cell generated from the packetized voice application arrives just after the first cell from the file transfer application, then the voice cell will have to wait at most one cell transmission time, while the file transfer cell is being transmitted. Because cell sizes are small and fixed, it is clear that the waiting time for a delay-sensitive application such as packetized voice is kept small. This would not have been possible for arbitrary packet sizes. However, it is also worth noting that the small cell size in ATM helps more at lower link speeds than at higher link speeds. As link speed increases, the time taken to transmit large packets reduces, and the nonpreemption delay seen by a voice cell may not be significant.

On the downside, a 5-octet header in a 53-octet cell implies a fixed overhead of 9.4%. Owing to the fixed cell size, it is not possible to amortize the overhead over large packet sizes.

The word "Asynchronous" in ATM distinguishes it from Synchronous Transfer Mode (STM) techniques. You can see the distinction by considering the ways in which STM and ATM utilize the raw transmission capacity provided by the physical layer below (see Figure 2.31). As mentioned in Section 2.3.1, the capacity is provided by a digital transmission standard, in which frames are defined. When the STM is used, a particular connection's data can be found at the same relative positions in successive frames, as shown in Figure 2.31. However, when ATM is used, the data is inserted into the next available free slot, and specific positions are not maintained. It is possible that some frames do not contain any data from a particular connection because the connection had no data to send. In this case, the slots can be used to carry traffic from other connections. On the other hand, it is possible that a connection generates a burst of data so that the full burst cannot be

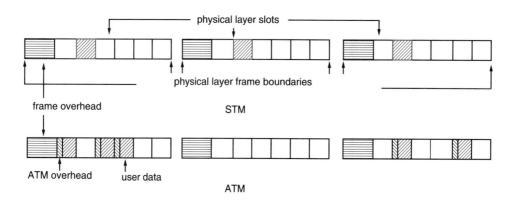

Figure 2.31 The top panel shows a particular connection's traffic carried using STM on a digital bit carrier. The traffic occupies the same relative positions in successive frames. The corresponding slots are shaded. The bottom panel shows a connection's traffic carried using ATM on the same carrier. The traffic is pushed into the frame without maintaining the same relative position in every frame. When ATM is used, part of the bandwidth resources are wasted because each ATM cell has an overhead.

carried in one frame; this causes queueing at the interface between the ATM and physical layers. The excess data would then be carried in free slots in the ensuing frames.

Of course, we cannot expect that the applications that are the sources of information will generate fixed-size packets. This is particularly true because many existing applications that were designed before the advent of ATM made no attempt to generate fixed-size packets. But if ATM technology is to be used, all units to be transported must be 48 octets long. We achieve this by using an ATM *adaptation layer* (AAL), whose purpose is to segment large packets into cells at the transmitting end and then reassemble the original packets from the corresponding cells at the receiving end. Thus, the AAL makes it possible for legacy applications to utilize ATM transport without requiring any change in the applications.

From the point of view of the OSI model (see Section 2.3.2), ATM technology is a way to implement *both* the network layer (layer 3) and as the link layer (layer 2). First, ATM does not require any particular link layer technology (for example, Ethernet) to carry the *protocol data units* generated by it. In fact, ATM defines its own link layer. In contrast, in the Internet, the IP layer is strictly at layer 3 and requires an underlying link layer to carry the packets generated by it. Moreover, ATM provides full routing capabilities through a mesh of ATM switches. For this purpose, the Private Network-to-Network Interface (PNNI) protocol has been defined in the ATM Forum (the standardization body for ATM technology). Hence, ATM technology also implements layer 3 functionality, of which routing is a typical example.

As in any layered communication architecture, a layer may offer connection-oriented or connectionless services to the layer above it. The ATM layer was designed to offer connection-oriented services to the AAL above. *Connection-oriented* service implies that an association between the two communicating end points must be set up before information transfer begins. Also, as part of the connection setup procedure, a *traffic contract* is negotiated between the ATM layer and its user. This contract characterizes the cell traffic that the ATM connection is supposed to carry and also specifies the QoS expected by compliant traffic. Because a commitment to QoS is being made, the ATM layer must reserve adequate resources in the network. If sufficient resources are not available at all ATM switches along the path, then a connection setup request can be blocked.

Going up the protocol stack, the service offered to the user of the AAL can be connection-oriented or connectionless, as shown in Figure 2.32. Thus, applications that expect connection-oriented service as well as those that expect connectionless service can be seamlessly carried over the ATM network.

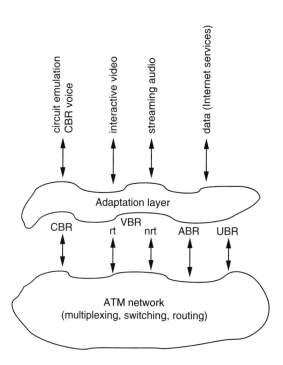

Figure 2.32 The ATM layer offers connection-oriented services to the AAL above it. But the service offered to the user of the AAL can be connection-oriented or connectionless. Different AALs provide different services. The AALs that have been standardized are AAL 1, AAL 2, AAL 3/4, and AAL 5. Some applications that utilize the services of the AAL are also shown.

Unlike the Internet, ATM networking technology was designed with QoS in mind from its conception. As we have explained, the small cell size was chosen to reduce packetization and nonpreemption delays when carrying packetized voice over an integrated services packet network. In addition, the ATM Forum has defined five types of standardized service categories that can be offered by the ATM layer (see Figure 2.32). The constant bit rate (CBR) service is a vehicle for carrying traffic that has a regular periodic structure. Traffic that is less regular can utilize the *variable bit rate* (VBR) service, of which there are two types: *non-real-time* and *real-time*. If the application generating the variable-bit-rate traffic can characterize it in terms of the maximum burst size and the peak and average rates, then an appropriate VBR connection can be set up. For example, a packet voice application can often specify its maximum burst size by considering the peak rate at which the voice coder emits bytes and the duration of a talkspurt. Delay-sensitive applications such as interactive video would choose a real-time VBR connection, whereas delay-tolerant applications such as streaming audio and streaming video can be carried by a non-real-time VBR connection.

Whenever the ATM network accepts a request for setting up a CBR or VBR connection, it must to commit bandwidth and buffer resources for carrying the corresponding traffic. Because a given ATM switch normally handles many connection requests, groups of connections must be multiplexed onto outgoing links. Each CBR source must be allocated a bandwidth equal to its declared rate. For VBR sources, however, the resource allocation algorithms running on the switch can take advantage of the statistical nature of traffic generation and can multiplex more connections than simple peak-rate allocation would suggest. Note that, because resources are limited, an ATM switch cannot keep accepting CBR and VBR connection requests indefinitely. This implies that the resource allocation algorithm must also have connection admission control (CAC) functionality so that it can refuse requests when resources are overallocated.

Typical data applications, however, do not have a simple way of characterizing the traffic they generated. Moreover, data traffic can tolerate variable delays, and applications can actually modify their data transfer rates according to network conditions. These factors indicate that it is hard to carry data traffic efficiently using the CBR and VBR services. The Available Bit Rate (ABR) and Unspecified Bit Rate (UBR) services are designed to carry data traffic.

The ABR service indicates to the application the amount of the bandwidth available in the network for carrying its traffic. The source of the traffic is then expected to modify its rate of data generation accordingly, in response to feedback from the network regarding resource availability. In practice, resource management (RM) cells, generated at regular intervals by traffic sources, are used by the network to indicate the amount of available resources. A network switch can provide information indirectly, by indicating that it is getting congested; alternatively, it can explicitly indicate the data rate that it can handle. Together, the traffic source using the ABR service and the network switches form a closed-loop feedback system (see Chapter 7). Users of the ABR service can expect some level of performance guarantees from the network; for example, a source can ask for and obtain a *minimum cell rate* (MCR) guarantee.

From this discussion, it appears that applications using the VBR service generate variable-bit-rate traffic, and so do applications using the ABR service. What, then, is the difference between the two? Consider a long file being transferred across the network. The significance of the long file is that the source always has data to send. Now, variability in ABR traffic occurs because the source adapts its rate of traffic generation in response to network feedback. On the other hand, variability in VBR traffic is completely independent of network conditions. It is an intrinsic property of the source, determined by its own characteristics, such as distributions of talkspurts and silence periods.

The UBR service is also meant for carrying data traffic. It is a typical best-effort service without any guarantees. Essentially, UBR traffic is treated as low-priority traffic that is carried using network resources that are not demanded by other categories of higher priority. The network makes no effort to indicate resource availability to a UBR source. Because no resources are committed, the bit rate is left unspecified.

As mentioned before, the ATM layer offers connection-oriented services to layers above it, and hence, before data transfer can begin, a connection between the communicating end points *at the ATM layer* must be established. This ATM-level connection must extend right through the ATM network, from one end point to the other. This is called a virtual circuit (VC).

Because the VC extends across the complete ATM network, it crosses several ATM switches. Hence, a VC consists of several *legs*, with a leg being the segment of the connection between a pair of neighboring switches. To identify a VC, therefore, an ATM switch must maintain an association between the incoming leg of the VC on an input port of the switch and the outgoing leg of the VC on an output port. At the time the connection is established, each ATM switch assigns *local identifiers* to the VC legs and stores the labels in a table in its memory. The idea is shown in Figure 2.33. For example, the first row of the table shows that for a VC passing through this switch, the incoming leg on input port 1 has been assigned

Input port	Input label	Output port	Output label
1	22	4	16
6	49	9	27
4	13	7	45

Figure 2.33   An ATM switch maintains a map between labels identifying the incoming and outgoing legs of a VC passing through it. This table shows that the label 22 on input port 1 is associated with the label 16 on output port 4.

the label 22, and it is associated with the outgoing leg label 16 on output port 4. Note that the labels are meaningful only to the switch that assigns them. Thus, labels have local significance only. A user of an ATM VC does not need to know the complete sequence of labels to identify the VC. It is sufficient for the user to insert into the cell the correct label on the *first* leg only. The tables maintained in the ATM switches ensure that the cell is forwarded correctly along the VC that has already been set up. In fact, as the cell proceeds along the VC, the ATM switches remove and insert labels by consulting the forwarding table. This is called *label swapping*.

ATM uses a hierarchical labeling scheme having two levels. A label consists of the *virtual path identifier* (VPI) at the "higher" level and the *virtual circuit identifier* (VCI) at the "lower" level. The label associated with a VC at a particular switch port is specified by the VPI–VCI combination. However, the hierarchy indicates that a group of VCs that share the same VPI, but differ in the VCI, can be treated as a unit for switching. For example, if there are 10 VCs, with the VPI of each being 22 and the VCIs being $5, 6, \dots, 14$, then this group can be treated as a single unit by virtue of the common VPI. Such a bundle of VCs is called a *virtual path* (VP). The idea is shown in Figure 2.34.

When a cell arrives at a port, the switch searches its table to identify the outgoing port and the new label that must be inserted. As the number of VCs passing through the switch increases, the time taken to search the table goes up. This indicates a possible scalability issue: As the number of VCs increases, a switch may not be able to cope with cells arriving rapidly, because it does not have sufficient time to search the large table in its memory. In this situation, the VPI serves as a handy tool for aggregation. If only the VPI is considered as a label instead of the VPI–VCI combination, there will be fewer entries in the

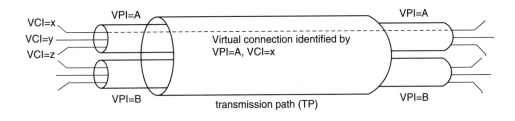

Figure 2.34   A schematic showing the relationship between VCs, VPs, and the transmission path (TP). The transmission path is a concatenation of one or more physical links.

table and searching will take less time. We can think of VPI-based switching as switching at coarse granuarity, whereas VPI–VCI-based switching occurs at finer granularity. Switching devices that consider only VPIs instead of the VPI–VCI combination are called crossconnects. Because the label associated with a connection at a crossconnect is the VPI only, label swapping means swapping of the VPIs alone; the VCI carried in the cell header is not modified by a crossconnect.

VCs can be classified according to whether they are *permanent virtual circuits* or *switched virtual circuits* (PVCs or SVCs). (A corresponding classification can be made for VPs.) A PVC between two end points is set up manually by a network operator. VPI and VCI values are defined for the permanent connection, and the values are entered in the table of each switch in the path. PVCs normally last for long periods: days or months. In contrast, an SVC is set up for a specific call by means of signaling between the user and the network and lasts for the duration of the call: minutes or hours. Thus, the timescales associated with setting up and clearing PVCs and SVCs are very different.

Standard interfaces have been defined in the ATM architecture so that devices from different manufacturers can interoperate. Figure 2.35 shows the main interfaces.

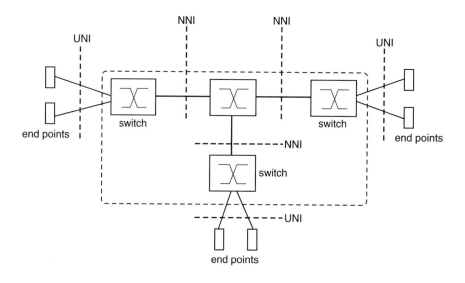

Figure 2.35   The user network interfaces and network node interfaces in an ATM network.

The *user network interface* (UNI), as the name implies, is the standard interface between the end-user equipment and the ATM devices inside the network. Communication across the UNI is structured in several planes (see Figure 2.22). The user plane is one in which user information is transmitted. Signaling information between the user and the network, used in establishing and releasing VCs, is exchanged in the control plane. Finally, the management plane is used for conveying various types of information related to management of the end-user equipment and ATM devices in the network.

The end result of standardization is the specification of protocols according to which peer entities at a specific layer interact. For example, it has been decided that when the end user wishes to initiate a switched virtual circuit across the UNI on the control plane, the relevant protocol will be DSS2 (Digital Subscriber Signaling System No. 2, described in ITU-T Q.2931) at the application layer. This is a part of the SS7 suite of signaling protocols, which is used in the telephone and ISDN networks (see Section 2.3.3).

The ATM *network node interface* (NNI) is used between nodes (ATM crossconnects and ATM switches) in the network. As with the UNI, the purpose of standardization here is to fix protocols at different layers so that compliant devices from different manufacturers can interoperate. For example, on the control plane, the protocol that two ATM switches should talk in order to set up a switched virtual circuit is B-ISUP MTP 3 (Broadband Integrated Services User Part, Message Transfer Protocol 3). As discussed in Section 2.3.3, the SSP functionality in the ATM switch is responsible for creating and processing the MTP3 signaling packet as well as transmitting it over the signaling network.

2.4 Summary and Our Way Forward

In Section 2.1 we argue that networking is basically about resource sharing, with the basic functional elements being multiplexing, switching, routing, and management. Section 2.3 provides a fairly detailed overview of the manner in which many of the telecommunication technologies carry out the functions of networking.

From this discussion we argue that the layered model will not serve us well in organizing our understanding of telecommunication network design and analysis. Rather, we will be well served by studying the various functions, the way these functions can be performed, and the way they interact or are isolated. As an example of isolation of the design of the functional blocks, recall the discussion of switches for circuit-multiplexed networks. The call-blocking probability depends on the blocking probabilities in the switches and the links on the path.

The switch-blocking probability is made negligible compared with the link-blocking probability and is ignored in the link capacity design. The problem of designing switches with low blocking is taken up as a separate problem. In packet networks, in principle, routing and multiplexing can be considered together, but in practice they are treated as being separated by operating over different timescales. On the other hand, in wireless networks over a limited radio spectrum and time-varying channels, the functions at several layers may need to be studied together. Thus, we adopt the functional approach, and the rest of this book is divided into three parts: one each to discuss multiplexing (Part I), switching (Part II), and routing (Part III). Our approach to the presentation of the material is that instead of just formally developing a battery of models and the associated results, we use engineering issues to motivate the presentation of the theory. The results obtained from the theory are then used to suggest solutions and to explain how these engineering issues are addressed in current practice. We expect that this approach will provide a more useful presentation for the network practitioner and the practical performance analyst.

2.5 Notes on the Literature

In this chapter we provide a point of view about communication networking. We explain the "big picture" of networking as it is practiced today. Finally, we provide additional perspectives to support our approach in adopting a functional (rather than layered) approach in organizing this book. The following is a representative survey of books that helps to supplement the overview of networking that we have provided.

Descriptive, textbook treatments of the vast area of computer networks are provided in the books by Tanenbaum [281], Keshav [168], Peterson and Davie [237], and Kurose and Ross [186]. The current most authoritative book on optical networking is the one by Ramaswami and Sivarajan [243]. A comprehensive book on the basics of wireless mobile communications is the one by Goodman [124]. Detailed coverage of the OSI model is provided in Jain's and Agarwala's book [152].

One of the best resources for understanding the telephone network is the Bell Systems operations handbook [250]. The book by Bellamy [24] is a good source book for all aspects of current practice in digital telephone networks, including access network design, backbone transport systems, and capacity design. It also covers some of the newer access technologies such as xDSL, as well as wireless and transport technologies such as SONET and SDH. The handbook by Freeman [109] is another source that also has a discussion of the engineering of the analog networks. SS7 is very well covered by van Bosse [289].

The book by Stallings [270] provides an overview of ISDN, Frame Relay and ATM networks. For ATM networking alone the books by de Prycker [78] and by Händel et al. [135] provide very detailed coverage of the concepts and standards. For Internet technology the book by Keshav [168] provides an insightful treatment. The most authoritative coverage of Internet protocols is found in the books by Comer and Stevens [66] and by Stevens [273].

The book by Rose [254] is the first on SNMP. The books by Subramanian [279] and Stallings [271] describe the most recent versions of the SNMP and RMON standards in detail. Of course, the Requests for Comments (RFCs) from the Internet Engineering Task Force (IETF) are the original source for all the SNMP standards.

The books by Kershenbaum [167] and Cahn [46] are good references on the topic of network topology design, which is not covered in this book.

Problems

The following can be called "toy" problems. They require the reader to carry out elementary calculations for some simple models that arise in the context of the functional elements discussed in this chapter. Solving these problems will help illustrate some engineering issues and will also review some of the basic analytical prerequisites required for reading this book.

2.1 Consider a channel in which bit errors occur independently from bit to bit, and the bit error probability (or bit error rate) is p. This is called the Bernoulli packet error model. Show that the probability that a K-bit packet is received in error is approximated by Kp. Examine the validity of the approximation for $K = 10,000$ bits, for $p = 10^{-7}$, and for $p = 10^{-4}$, and discuss. If the packet length is a random variable, K, with $f(k)$ being the probability that the packet has k bits, show that the packet error probability is approximated by $p\mathrm{E}(K)$.

2.2 A simple model to introduce memory in the error process is as follows: A bit error follows another bit error with probability p, and a correct bit follows another correct bit with probability q. The channel state can be represented by a two-state Markov chain, as shown in Figure 2.36. If bits

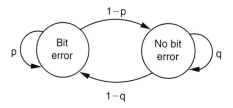

Figure 2.36   Markov error model for problem 2.2.

are being continuously transmitted on the channel, find the probability, π_e, that a random bit is in error. Find the probability that a K-bit packet, randomly located in time, is in error. For what values of p and q do we recover the Bernoulli packet error model with bit error probability p?

2.3   A P bit packet is to be transported over L store-and-forward links; d_l and C_l are the propagation delay and bit rate, respectively, on link l. Assuming zero queueing and processing delays on the links, argue that the end-to-end delay would be $\sum_{l=1}^{L} \left(\frac{P+H}{C_l} + d_l \right)$, where H is the number of header bits in a packet. If an M bit file is to be transmitted as K packets and if $C_l = C$ for all l, show that the end-to-end file transfer completion time is $\frac{1}{C} \left[\left(\frac{M}{K} + H \right)(L-1) + M + KH \right] + \sum_{l=1}^{L} d_l$. Discuss the trade-offs of choosing K. Show that the K that minimizes the delay is $\sqrt{\frac{M}{H}(L-1)}$.

2.4   If the links in problem 2.3 used cut-through switching rather than store-and-forward, obtain the end-to-end delays for the packet and the file with K packets. Assume $C_l = C$ for all l.

2.5   A circuit-multiplexed point-to-point link can accommodate one call at any time. The arrival times of calls X and Y are uniformly distributed in the time interval $[0, 5]$. The holding time of call X is one unit, and that of Y is two units. An arriving call that finds a busy trunk is lost. Find the blocking probability for calls X and Y.

2.6   A source generates bits according to a Poisson process of rate λ. The bits from the source are to be transmitted as K bit packets. The time to accumulate a full packet will be called the packetization delay. Find the distribution of the packetization delay.

2.7 A video source generates K fixed-length packets per frame, which are then transmitted over a link. The frame rate is 25 frames per second; that is every 40 ms the source outputs K packets into the link's buffer. The transmission time of each packet on the link is 1 ms. If $K \leq 40$, find the average packet delay and the time average occupancy of the link buffer. Assume that a packet is stored in the buffer until its last bit has been transmitted. Now assume that K is a random variable with probability mass function $\Pr(K = k) = \alpha^{k-1}(1 - \alpha)$ for $k = 1, 2, \ldots$, with $0 < \alpha < 1$. In each frame, packets in excess of 40 are lost. Find the probability of losing packets in a frame and the average number of packets lost per frame. Also, find the average packet delay and the average buffer occupancy.

2.8 In a network with distributed multiplexing using random access, assume that the distance between all node pairs is a. Can you think of a physical network in which this could be possible?

2.9 In a slotted multiaccess system, K nodes are each attempting transmission with probability p in each slot independent of the other nodes. If N is the random number of slots consumed before the first success, find the probability mass function of N.

2.10 Three kinds of polling can be identified: token ring network, in which N nodes are arranged in a physical ring and a token is passed around the nodes to control access to the ring; *hub polling* on a *token bus* network, in which the token is passed over the broadcast bus to the "next node" in the sequence; and centralized roll call polling, where a central node sends an explicit query to each node in the network and receives an ACK/NACK. Identify the minimum delays between consecutive opportunities to transmit (called the *walk time*) in each of the three types of polling.

2.11 $N + 1$ nodes, numbered $0, \ldots, N$, are located on a straight line starting at the origin, and node i is at distance i from the origin. Each node is connected to a switch nearest to it. Assuming $N = 2K$, what would be the total length of the transmission links with only one switch at location K? Repeat the calculation for a network with $N = 4K + 1$, switches at K and $3K + 1$, and a transmission link connecting these switches.

2.12 To get a perspective on packet and call timescales, consider a switch with 10 100 Mbps links. If the packet lengths can vary from 500 bits to 12,000 bits, with an average packet length of 1000 bits, determine the

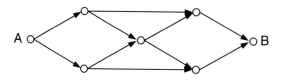

Figure 2.37   Network for problem 2.14.

maximum and average packet arrival rate to the switch. If stream traffic requires 50 Kbps per call, with a call being active for 200 seconds, what is the maximum rate at which accepted calls arrive? Compare the mean time between packets and between call arrivals.

2.13 Consider a hierarchical construction of a network. At the lowest level, the nodes are formed into groups of K_1 nodes and are connected to one level-1 switch; K_2 level-1 switches are grouped together to be connected to one level-2 switch. Let there be L levels of this hierarchical construction. For $K_i = K$ and $i = 1, \ldots, L$, find the number of switches in the network and the number of links. Note that the links interconnecting the switches at the higher levels will be longer, but because they aggregate more traffic, they will also be of increasing capacities.

2.14 How many possible routes are there between a node pair in (a) a linear network where the nodes are arranged in a straight line, (b) a ring topology network, (c) a star network, and (d) a fully connected network, and (e) between nodes A and B in the network shown in Figure 2.37?

2.15 One link in a network is failure-prone, and the other links are stable. The network topology is sampled every second, and, if the topology has changed, the routes are recalculated. Each second, the failure-prone link changes from up to down with probability p, and from down to up with probability q. What is the rate at which the routes in the network are calculated?

2.16 On a link the mean delay can be approximated by $\frac{1}{1-\rho}$, where ρ is the utilization of the link. If the network monitoring and management traffic increases the utilization by α, find the percentage increase in the mean delay and plot the increase as a function of ρ. What do you observe?

Part I
Multiplexing

CHAPTER 3

Multiplexing: Performance Measures and Engineering Issues

W henever several flows need to share a communication network, a multiplexing mechanism is required. In Section 2.2.1, we introduce and discuss a taxonomy for multiplexing schemes. In this chapter we provide an overview of the performance and engineering issues that arise when traffic of various types is carried in packet- or circuit-multiplexed networks. This discussion motivates the kinds of engineering models that arise in such networks. A discussion of these models forms the content of the remaining chapters in this part of the book.

As presented in Chapter 2, the basic alternatives are packet multiplexing and circuit multiplexing. Although there have been some efforts to design entire networks whose links use hybrid multiplexing, actual implementations use a combination of packet and circuit multiplexing only over the network access links. Important examples are ISDN access and satellite access, in which the frames in the digital access link are dynamically partitioned into a packet-multiplexed part and a circuit-multiplexed part. When the traffic reaches the first network switch (at the local exchange in the case of ISDN, or at the satellite ground-station in the case of satellite access), the packet traffic is switched to a packet-multiplexed network, and the circuit traffic is switched to a circuit-multiplexed network. For simplicity, in this book we consider networks that are either only packet-multiplexed or only circuit-multiplexed.

3.1 Network Performance and Source Characterization

From the discussion in Chapter 2, we can see that after a call is set up in a circuit-multiplexed network, the flow of traffic within the call is transparent to the network and is not affected by the flows from other sources.

In a packet-multiplexed network, however, transport procedures in the network continue to affect the flow of traffic within a "call" even after the call is set up. Hence, to engineer a packet network it is essential to model not only call (or session) arrivals but also the nature of the traffic that the source of traffic emits during a session. In a circuit-multiplexed network, when a call arrives, the network need only check whether or not the required bandwidth is available on any of the routes that the call can take to its destination. If the bandwidth is available, then the call is accepted; otherwise, it is blocked. Such blocking, or *admission control*, is nontrivial in an integrated-services packet network. Each arriving call can, potentially, have its own traffic characterization and requirement for quality of service. Hence, when a call arrives the resources required to handle the call need to be evaluated *online*; based on this evaluation, a decision as to acceptance or rejection must be made. Thus, in the case of packet-multiplexed networks, not only do we need to concern ourselves with source characterization, but we also need quick and efficient techniques for analyzing resource requirements for the purpose of admission control. Much of the material in Chapters 4 and 5 deals with various ways to characterize sources and with the development of models and analyses for using these source characterizations, along with QoS objectives, to determine resource requirements. Such models can be used in network engineering and connection admission control.

3.1.1 Two Types of Traffic

We will develop our discussion in relation to Figure 3.1, which depicts a typical packet-multiplexed network. Because the terminals must at least process packet headers (for transmission or reception), they must be smarter than simple phone instruments. Typically, packet network terminals (also called *hosts*) are processor-based systems of varying levels of complexity. The figure shows a generic computer system comprising a processor, memory, a disk, and peripherals, all connected by a system bus. Audiovisual communication with such a terminal is facilitated by end equipment (microphone, speakers, camera, and video display) connected to audio- or video-processing hardware cards, which in turn are connected to the system processor via the system bus. The terminal is connected to the packet transport network by a *network interface card* (NIC).

We can broadly classify traffic sources into two classes: *elastic* and *stream*.

Elastic Traffic

Consider a data file, residing on the disk of the terminal shown in Figure 3.1, that needs to be transferred to the disk of another terminal across the packet network.

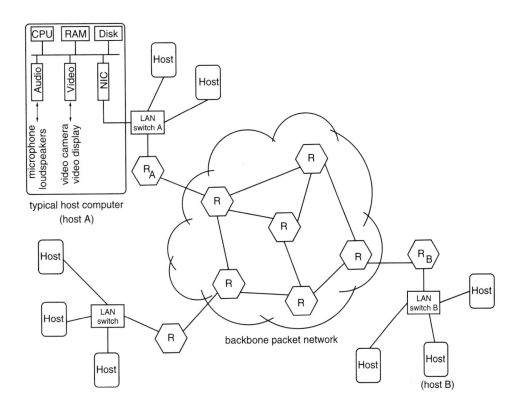

Figure 3.1   A schematic view of a packet network, showing one terminal in detail. R denotes a router.

Although the human (or application) who wishes to make this file transfer would like to have the transfer completed in, say, a second or two, the source of data itself does not demand any specific transfer rate. If the data transfer does not lose data, no matter how fast or slow it is (but as long as the rate is positive), the file will sooner or later get transferred to the destination disk. We say that, from the point of view of the network, this source of traffic is *elastic*. Many store-and-forward services (with the exception of media streaming services) are elastic; examples are file transfer, WWW downloads, and electronic mail (e-mail). In this list, the first two are distinguished by the fact that they are *nondeferrable* (i.e., the network should initiate the transfer immediately), whereas e-mail is *deferrable*.

Elastic traffic does not have an intrinsic temporal behavior and can be transported at arbitrary transfer rates. Thus the following are the QoS requirements of elastic traffic.

- Transfer delay and delay variability can be tolerated. An elastic transfer can be performed over a wide range of transfer rates, and the rate can even vary over the duration of the transfer.

- The application cannot tolerate data loss. This does not mean, however, that the network cannot lose any data. Packets can be lost in the network—owing to uncorrectable transmission errors or buffer overflows—provided that the lost packets are recovered by an automatic retransmission procedure. Thus, effectively the application would see a lossless transport service. Because elastic sources do not require delay guarantees, the delay involved in recovering lost packets can be tolerated.

Users, of course, will not tolerate arbitrarily poor throughput, high throughput variability, and long delays. Hence a network carrying elastic traffic will need to manage its resource-sharing mechanisms so that some minimum level of service is provided.

Elastic traffic can also be carried over circuit-multiplexed networks. In this case the shaping of the traffic to fit into the allocated bandwidth should be carried out by the source. Obvious examples would be Internet service over a dial-up line or the interconnection of two computers over a phone line using modems.

Stream Traffic

Consider a user, at the terminal shown in Figure 3.1, speaking into the microphone and wishing to converse with another user across the packet network. The audio card samples and digitizes the voice; the digitized voice can then be compressed, and *silence periods* can be removed to further economize on bandwidth. An adaptation protocol (running in the system processor or in the NIC) then packetizes the speech and transmits it over the packet network. Obviously, the pattern of packet arrivals closely follows the pattern of speech generation; that is, this source of traffic has an *intrinsic temporal behavior*, and this pattern must be preserved for faithful reproduction of the speech at the receiver. The packet network will introduce delay: fixed propagation delay as well as queueing delay that can vary from packet to packet. Playout delay will need to be introduced at the receiver to mitigate the effect of random packet delay variation. The more variable the packet delay the longer will be the required playout delay. Hence the network cannot serve such a source at arbitrary rates, as it can in the case of elastic traffic. In fact, depending on the adaptability of a stream source, the network may need to reserve bandwidth and buffers in order to provide an adequate transport service to a stream source. Applications such as real-time interactive speech or video telephony are examples of stream sources.

The following are typical QoS requirements of stream sources.

- Delay (average and variation) must be controlled. Real-time interactive traffic such as that from packet telephony would require tight control of end-to-end delay; for example, for packet telephony the end-to-end delay may need to be controlled to less than 200 ms, with a probability more than 0.99.

- There is tolerance for data loss. Because of the high levels of redundancy in speech and images, a certain amount of data loss is imperceptible. As an example, for packet voice in which each packet carries 20 ms of speech and the receiver does lost packet interpolation, 5 to 10% of the packets can be lost without significant degradation of the speech quality ([179], [136]). Because of the delay constraints, the acceptable data loss target cannot be achieved by first losing and then recovering the lost packets; in other words, stream traffic expects a specific *intrinsic loss rate* from the packet transport service.

Observe that calls in a circuit-multiplexed network can be treated as stream traffic, and, as discussed at the beginning of this section, the issue of modeling the in-call performance does not arise.

Remarks 3.1

It is important to distinguish what we described here as stream traffic from the kind of traffic that is generated by applications such as streaming audio and video. Such applications basically involve a one-way transfer of an audio or video file stored on the disk of a media server. For the received video to be useful, however, the playout device should be continuously "fed" with video frames so that it can reproduce a smooth video output. One way to achieve this is to provide a guaranteed rate to the transfer. Alternatively, because the transfer is one-way, a more economical way is to treat the transfer as elastic and buffer the video frames as they are received. Playout is initiated only after a sufficient number of video frames have been buffered so that a smooth video playout can be achieved in spite of a variable transfer rate across the network. Thus, the problem of transporting streaming audio or video becomes just another case of transferring elastic traffic, with appropriate receiver adaptation. We can also support simple interactivity, such as the ability to rewind, by having the receiver store frames that have already been played out. This, of course, puts a burden on the amount of storage that the playout device needs to have. An alternative is to trade off sophistication at the receiver with the possibility of interactivity across the network; the press of the rewind button stops the video playout, frames stored in the playout device are used to create a rewind effect, and meanwhile additional past frames are fetched from the server. But this approach would need some delay and throughput guarantees from the network, requiring a service model somewhere in between the pure elastic and pure stream model that we have described.

3.1.2   A Historical Note

Circuit-switched networks have a history of about 100 years (most countries have public switched telephone networks, which are used to carry mainly telephone calls). In contrast, packet-switched networks began to be deployed only in the late 1960s and early 1970s (the ARPANET, along with public switched packet data networks based on X.25, were the earliest such networks). Until today, the predominant use of packet networks has been to carry store-and-forward elastic traffic using best-effort bandwidth sharing. Much of the popularity of the Internet is because of the ease, speed, and efficiency of correspondence using e-mail and of sharing information using the World Wide Web (WWW, or just "the Web"). It is important to note, therefore, that whereas quantitative engineering of circuit-switched phone networks dates back to Erlang (whose famous Erlang-B formula was published in 1917), even after 20 years of operational experience with carrying elastic traffic there are few effective engineering models for designing large packet networks for carrying elastic traffic. On the other hand, in conjunction with the development of ATM technology, much work has been done on traffic models and network engineering models for stream traffic. This research can now be applied to the Internet.

One of the reasons for this situation is that, at least in principle, open-loop queueing theory can be directly used in models for packet multiplexing of stream traffic. One characterizes the traffic from the stream sources, using deterministic or stochastic models, and then a queue, or a network of queues, with these inputs is used as the engineering model. New queue analysis techniques, exact or approximate, may need to be developed. One of the engineering techniques that has been explored for stream traffic is based on the idea of *effective bandwidth* (see Chapter 5). This idea can be used in conjunction with a generalization of the theory of circuit-multiplexed networks to engineer an integrated packet network for stream traffic. Note that although the predominant circuit-switched network is the uniservice PSTN network, the theory for multiservice networks supporting different types of traffic has been developed and is quite mature. We will study some of this generalized theory in this book.

In the case of elastic traffic, however, the performance of the individual streams (e.g., the average or minimum throughput obtained by a Web session) depends crucially on the closed-loop congestion control and bandwidth-sharing controls that are used in the network. This makes the performance analysis and traffic engineering of packet networks for carrying elastic traffic a significantly harder problem. In view of this, we first study engineering models for packet-multiplexed stream connections. Later, in Chapter 7, we examine some approaches for quantitative performance evaluation of the transport of elastic traffic in packet networks.

Remarks 3.2

Recall from Chapter 2 that packet switches can introduce delays in many stages. Given that, it might seem that multiplexing and switching in packet-multiplexed networks would be interrelated and one would need to perform a joint analysis. In this book, we take a divide-and-conquer approach, and adopt the following viewpoint: The switch can move data from the input link to the queue of the output link with a *switching delay* that has a variance that is significantly lower than the delay variance in the output queue. This viewpoint enables us to discuss multiplexing and switching in packet networks separately and allows us to treat each link as a centralized packet multiplexer. Note that the fixed switching delay adds to the other two constant delays experienced by the packet in traversing the preceding link—the propagation delay (determined by the distance) and the transmission delay (determined by the packet length and link speed)—and may have to be taken into account.

3.2 Stream Sessions in a Packet Network: Delay Guarantees

To help in the development of our exposition of the models for stream traffic in packet networks, we will use an example based on the network shown in Figure 3.1. Consider a voice call between host A (shown in the upper left of the figure) and host B (shown in the lower right of the figure). Electrical signals from the microphone are digitized and coded by a *speech coder* on the audio card. A typical approach is to sample the analog signal from the microphone at

8000 samples per second, quantize the resulting continuous amplitude samples into 256 predetermined levels, and then encode each of these levels into 8 bits (one byte). The output of such a speech coder is called *pulse code modulation* (PCM)-coded speech. The simplest sound cards in most PCs do basically just this; the bytes that are output by the coder are then read by the PC's processor. Three functions are then performed:

- *Voice activity detection (VAD)*: Inactive periods are identified; these are low-energy periods that correspond to silences while the speaker listens, or to gaps between words, sentences, and utterances. The coder output corresponding to these *inactive periods* is discarded. Segments of the *active periods* are then *packetized* for transport over a packet network. Note that although the inactive periods do not contain speech information, the duration of the gaps is indeed information that needs to be conveyed to the receiver. One of the difficulties in packetized transport of speech is in the retention of such timing information. Because packets are transmitted only during active periods, the inactive periods can be replicated only approximately at the receiver. It has been found that the resulting errors are not noticeable if the inactive periods are long. Thus the VAD function does not discard bytes from short inactive periods. For example, any inactive period less than 200 ms in duration can be considered to be a part of a continuous talkspurt, and packets are sent even during such periods. In such a case it is said that the VAD has a *hangover* of 200 ms. The speech that results after VAD is said to have talkspurts and silences; during a talkspurt the data rate is 64 Kbps. This is an example of a variable-bit-rate source; in particular, this is called an on–off VBR source. The original 64 Kbps PCM output of the speech coder is called a *constant-bit-rate* (CBR) source.

- *Shaping*: The VBR speech source obtained after VAD can be *shaped*. To understand the need for shaping, let us consider the problem of transmitting the VBR source over a link of rate C (see Figure 3.2). Let us denote by R the peak rate of the VBR source, and by $\bar{r}$ the average rate. Thus, for example, if the on–off VBR source has an average *on* duration of 400 ms and average *off* duration of 600 ms, then with $R = 64$ Kbps, we will have $\bar{r} = \frac{400}{400+600} \times R = 25.6$ Kbps. It is clear that it is a waste of bandwidth to make $C > R$, and it is necessary that $C \geq \bar{r}$ (we discuss this latter point more fully in Chapter 5). Now suppose we take $C < R$. Notice that when the voice source is emitting data at rate R, the link buffer builds

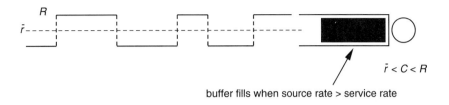

$\bar{r} < C < R$

buffer fills when source rate > service rate

Figure 3.2   An on–off VBR source, of peak rate *R* Kbps and average rate *r̄* Kbps, being carried by a link of capacity *C* Kbps, *r̄* < *C* < *R*.

up at rate $(R - C)$ Kbps. Any byte that arrives when the buffer level is, say, B bits will be delayed by $\frac{B}{C}$ ms. A priori, we do not know for how long this *rate mismatch* will last (the average rate $\bar{r} = 25.6$ Kbps could have been obtained with a 4-sec *on* time and a 6-sec *off* time, too!). Hence, if we want to bound the delay of the voice bytes in the link buffer, in the absence of any other information about the source, our only recourse is to use $C = R$. On the other hand, however, if the source were to provide the network with a more detailed characterization of its output (for example, a bound on the *on* times), then the network could more efficiently reserve bandwidth for the source, and the source would end up paying less for its call. *A source can shape itself and then declare its shape parameters to the network.* We discuss this in detail in Chapter 4. Note here that shaping basically entails occasionally delaying the source output just enough so that it conforms to an output profile. Thus it is important to note that a shaper will, in general, introduce *shaping delay*.

- *Packetization*: The next step is to take contiguous segments of talkspurts and packetize them for transmission over the network. One approach is to take a certain number of bytes from the source (e.g., 160 bytes) and generate a packet from these. It may happen that a talkspurt finishes before 160 bytes have been collected; in such a case a short packet will be generated. In technologies in which the packet length is fixed (e.g., ATM cells), when there are insufficient data bytes to fill a packet, *padding* bytes are added in the packet, and a length field is used to inform the receiver as to the actual number of data bytes in the packet. The maximum number of bytes in a packet (160 bytes in our example) is an important parameter. The packetizer must wait to accumulate a packet; thus bytes that arrive early in the packet must wait for those that arrive later until the packet is formed. This results in a *packetization delay*. Packets cannot be very short, because there could be a significant amount of header overhead in each packet (e.g., in the context of the Internet there would be at least 12 bytes for RTP, 8 bytes for UDP, and 20 bytes for IP; see Figure 2.29).

3.2.1 Network Delay, Delay Jitter, and Playout Delay

Ultimately, the procedures just discussed result in the generation of a stream of packets, which then must be transported over the packet network from host A to host B (refer again to Figure 3.1). We use the Internet terminology in the following discussion. Host A has a link to an Ethernet switch, which in turn has a link to the router R_A, which connects A's campus network to the Internet (this router is often called a LAN–WAN router because one of its interfaces is on a LAN).

We assume that the LANs in A's and B's campuses are switched Ethernets, with each link to each computer providing *full duplex* (i.e., simultaneous transfer in both directions) access. This implies that the shared medium (Carrier Sense Multiple Access with Collision Detection, or CSMA/CD) aspect of Ethernet does not play any role; such an assumption is quite appropriate for Ethernet-based LANs in which there is a strong trend away from shared medium installations (i.e., the use of Ethernet hubs) in favor of switched Ethernet installations. We can then think of Ethernet as playing only the role of framing and packet switching using Ethernet addresses. Typical link speeds for this technology are now 100 Mbps (and even 1000 Mbps (1 Gbps) for switched Ethernet campus backbones). In contrast to these LAN speeds, typical WAN link speeds (between the routers in Figure 3.1) could range from as low as 64 Kbps to 2.4 Gbps. Although Gbps speeds could possibly prevail in the core network in some places, typical *edge link speeds* (i.e., the speeds of links connecting campuses or enterprises to the Internet backbone) are in the range 64 Kbps to 2 Mbps. Thus, relative to the LAN speed between host A and the LAN–WAN router R_A, the WAN access link speed would be quite low. Also, switched Ethernet LANs typically tend to operate at a low utilization, because it is less expensive to upgrade a LAN than to upgrade a WAN link (a high-speed WAN link would entail a substantial recurring expense).

In any case, for our discussion we assume that the utilizations of the campus networks at host A and host B are low. Hence the voice packets generated at host A would not experience any significant delay until they reach router R_A. There, these packets may need to queue up for transmission over the egress link out of A's campus. Furthermore, the packets from the voice call may need to queue up at each link between the routers on the path to router R_B. Router R_B has an

Ethernet link into the campus LAN to which host B is connected. Thus, after the packets arrive into router R_B they move rapidly to host B, and, as before, we can ignore the queueing delays after the voice packets enter into router R_B.

At host B there is the additional problem of playing out the individual packets in a way that the original voice patterns are reproduced in spite of the random network delays introduced by the packet network. To understand this problem, look at Figure 3.3. If each packet that left host A arrived at host B instantly, then the packet send times and packet receive times would lie along the "$y = x$" line in the figure. However, as the voice packets are transported across the WAN, they will encounter transmission, queueing, and propagation delays. In addition, the queueing delays will be random, leading to delay variation, also called *delay jitter*. These random network delays are depicted in Figure 3.3 by the vertical bars standing on the slanting line. Each bar corresponds to one packet and is positioned on the x-axis at the send time of the packet. The bursts of packets from a talkspurt can be identified as consecutive, periodically occurring vertical bars. The gaps between these bursts are the silence periods. The height of each bar represents the network delay experienced by the packet. Suppose the receiver adopted the policy of playing out each packet as soon as it was received. We look

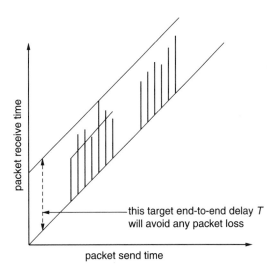

Figure 3.3   Jitter in the delay of voice packets results in the need for playout delay at the receiver.

at the first talkspurt and notice that if the first packet were played out beginning from the time that it was received, its bits would be played out along the slanting line (parallel to the $y = x$ line) starting from the tip of the first vertical bar. The intersection of this slant line and the second bar is where the playout of the first packet would complete, and the speech decoder would be ready for the next packet. This packet, however, would be too late. It follows that if an *immediate playout* policy is adopted at the receiver, then in the first talkspurt four out of the seven packets would arrive too late for playout. The coder could attempt to *interpolate* these lost packets, but such a high frequency of interpolation would lead to very poor speech quality.

An obvious alternative is to adopt the policy of *deferred playout*. A *playout delay* is applied to each packet to allow trailing packets to "catch up." From Figure 3.3 it is clear that if all packets are played out starting from the uppermost slant line, then (for the fragment of the packet process shown) no packet would be late. Obviously, this naive approach presents two practical problems. We do not know in advance the maximum delay that any packet in the connection will encounter, and in any case this worst case delay could be very large. Notice that playout delay adds to the end-to-end voice delay, the so-called *mouth-to-ear (MtoE) delay*. Suppose, however, that we are able to determine a value T such that the packet delay rarely exceeds T. Then, as shown in Figure 3.4, the receiver can stretch out the delay of each arriving packet to T; packets that are delayed more than T are lost and can possibly be interpolated.

We are still left with the problem of determining a value for T. There are two alternatives. The network may have the ability to provide a delay guarantee at call setup (e.g., $\Pr(X > T) < \epsilon$ where X is the delay of packets in the network; see Figure 3.4). In such a case the end points specify their traffic characteristics and

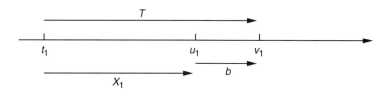

Figure 3.4   Arriving voice packet delays are stretched to a target delay T. A voice packet that left the source at time t_1 arrives at the receiver at time u_1, thus incurring a network delay of X_1. It is buffered for a time b so that its end-to-end delay is T. If $X_1 > T$, then the packet is discarded.

the values of T and ϵ. The network evaluates whether the call can be accepted, and if the call is accepted the network sets up the appropriate mechanisms along the path of the call so that the delay objective is met. Now T is known to the receiver at call setup time, and the procedure shown in Figure 3.4 can be performed. If the network cannot provide delay guarantees, then the receiver would need to estimate T as the call progresses. Timestamps carried by the voice packets in their headers would be used to obtain a statistical estimate of T. This estimate could then be used to set the playout delays of arriving packets. Because there is no guarantee, the value of T could be larger than desired and could vary over time as congestion in the network varies.

3.2.2   QoS Objectives

The preceding discussion is summarized in Figure 3.5. We gather that the MtoE delay for the packet voice call between host A and host B is the sum of several terms as shown in the following equation:

$$\text{MtoE Delay} = \text{coding delay} + \text{shaping delay}$$

$$+ \text{packetization delay} + \text{WAN propagation delay}$$

$$+ \text{WAN transmission and queueing delay}$$

$$+ \text{receiver playout delay}$$

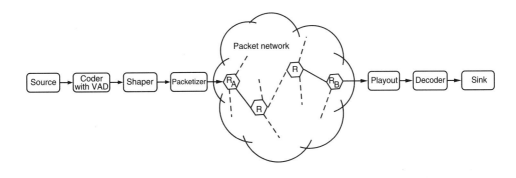

Figure 3.5   The various components involved in voice transport over a packet network.

In this equation, two terms appear that we have not discussed. In the simple PCM scheme the coding delay is simply the time taken by the voice coder to sample and digitize one sample; this would be 125 μsec for PCM. More sophisticated voice compression schemes need longer segments of voice to determine dependencies and redundancies and to achieve *source compression*, and hence these schemes would introduce longer coding delays (typically, 10 ms to 30 ms). For the present example, we consider only PCM speech.

The other term that we have not yet discussed is the WAN propagation delay. This is the signal propagation delay over the various media interconnecting the routers in the path of the call. A rule of thumb is to compute this fixed delay as 5 ms per 1000 Km of cabled transmission. Thus, for example, between points in the continental United States and India separated by a distance of about 20,000 Km, the one-way WAN propagation delay would be about 100 ms. For a geostationary satellite link, the one-way propagation delay is computed as the time taken for radio waves to travel from the transmitter up to the satellite and then down to the receiving ground station, or about 250 ms.

In addition to the MtoE delay, some voice packets can be lost, either because of buffer overflows in routers or because they arrive after their scheduled playout time at the receiver. Thus an example of the QoS expected by a voice call could be

$$\text{Probability(MtoE Delay} > 200 \text{ ms}) < 0.02$$

and

$$\text{Probability(Packet Loss}) < 0.05$$

Notice that the MtoE delay has some fixed parts (coding delay, packetization delay, and propagation delay) and some variable parts (shaping delay, transmission and queueing delay, and playout delay). It is these variable delays that are governed by the characteristics of the traffic emitted by the source and by the way the traffic is handled in the network (i.e., the other traffic that it is multiplexed with, and the queueing, scheduling, and bandwidth allocation at the network links). The material in Chapters 4 and 5 is devoted to traffic models, along with models for delays in networks. In Chapter 4 we use deterministic worst case bounds to model the traffic, and we present a network calculus based on such bounds and deterministic characterization of network elements. Such analysis can be used for providing worst case performance guarantees to stream connections. Although such an approach is fairly complete, it leads to conservative designs. In Chapter 5 we deal with stochastic models for traffic, and network analysis with stochastic performance objectives. By permitting a small probability of violation of the delay objectives, we obtain a substantial gain in resource efficiency. In fact, as we will see in Section 5.6, this objective can be expressed as a bandwidth requirement that can be reserved by the network for exclusive use of the stream. With this view, network engineering can use the generalized theory of circuit-multiplexed networks, where the bandwidth requirements of calls is not a fixed number but can depend on the call. Chapter 6 discusses such circuit-multiplexed networks.

3.3 Circuit-Multiplexed Networks

Circuit multiplexing is best studied through the PSTN model, albeit with a small generalization that calls can have different bandwidth requirements. Figure 3.6 is a schematic of the ubiquitous circuit-switched PSTN, which is used primarily for making telephone calls. (Observe that this figure is similar to Figure 2.25 except that now we look at the complete PSTN network rather than only the signaling

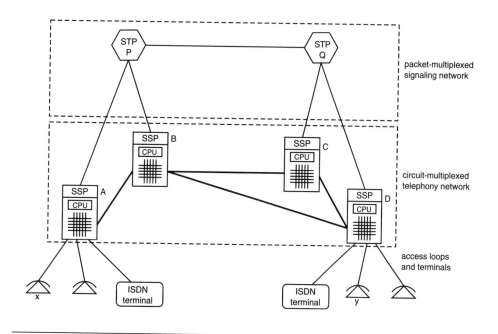

Figure 3.6 The facilities in a circuit-switched network, showing the terminals ("plain old telephones" or ISDN terminals), the network trunks, the switch fabrics, and the signaling network.

network part of it.) Such networks have evolved over the past 100 years, from analog trunks with human-operated call switching to digital trunking, digital switching, and overlaid packet-switched signaling networks (as shown in the figure). We assume the current technology in our discussion here.

Recall the operation of the circuit-multiplexed PSTN or the ISDN network from Chapter 2. A terminal—say, x—initiates a call by signaling to its local exchange—say, A—and providing it the number of the called terminal (y) and the amount of bandwidth required for the call. The call requires that an end-to-end circuit of the required bandwidth be set up, and the LE sets up this call over the signaling network as follows. From the number provided and the routing plan, the LE determines that the next exchange on the path to y is B and reserves the required bandwidth on the AB trunk. B reserves the corresponding bandwidth on the same trunk AB at its end and determines that the target exchange D is connected to it directly; the process of reserving bandwidth on the link AB is repeated to reserve bandwidth on link BD. The call thus progresses hop-by-hop to the called terminal. If the required bandwidth is not available on BD, then the call is tried along BCD. If the bandwidth is not available on that path either, then the call is not admitted and is lost to the network unless the caller makes another attempt after some random delay.

From this discussion the following issues emerge in the engineering of circuit-multiplexed networks. Associated with every possible source destination pair in the network, there is a routing plan that is essentially an ordered sequence of paths that are to be tried to complete a call. The call is completed on the first path of this sequence over which the required bandwidth is available on all the links of the path. Although the routing plan may be dynamic, the changes are infrequent compared with call arrivals, and to model the multiplexing, it can be assumed to be static. An important feature of circuit-multiplexed networks is that if the bandwidth required by the call is not available on any of the paths in the routing plan for that call, the call request is *not queued* but rather is blocked and deemed lost to the network. Thus a circuit-multiplexed network is a *loss system*, and the main design objective in the engineering of these networks is to efficiently minimize the call-blocking probability, the probability that call requests are not accommodated because bandwidth could not be found on the paths that are permitted for the call. The study of a circuit-multiplexing network concentrates primarily on the blocking probability analysis.

Recall from Chapter 2 that switches can themselves block. Thus, call-blocking probability is determined by the switch blocking (a path cannot be established inside the switch between the free input and output links) and link blocking (capacity not available on the output link) probabilities on the path. Thus it again seems that switching and multiplexing are interrelated in a circuit-multiplexed network, and the two must be considered jointly. We resolve this as follows. For wide area networks, the transmission capacity is far more expensive than switching capacity, and network designs should be limited by the transmission capacity rather than by switching capacity. This means that the switch design objective should be to efficiently minimize the path-blocking probability inside the switch and make it insignificant compared with the link-blocking probability. This in turn will help us ignore the effect of switch blocking in analyzing network performance in circuit-multiplexed networks and continue with the divide-and-conquer approach mentioned in the context of packet-switched networks. We take this view in this book and discuss the design of circuit-multiplexed networks in Chapter 6. We also briefly discuss in Chapter 11 the switch architectures that are used in the construction of circuit switches.

The existence of alternate routes between a source–destination pair and the ordered manner in which they are tried raises the issue of the nature of the *overflow* traffic: that portion of the traffic that could not be accommodated on the first-choice route and had to be offered to the second-choice route. It is easy to see that the overflow process will be more bursty than the original traffic; that is, it will have a higher variance. This calls for a different model for the overflow traffic than that of the original, or *first-offered*, traffic. In our example, the xy traffic overflowing from the BD link is offered to BC and CD.

Another aspect of circuit-multiplexed networks is the possibility of admission control. In the example, consider the case when the bandwidth is not available on the path ABD but is available on $ABCD$. Clearly, routing the call along $ABCD$ consumes more resources than routing it along ABD. If the occupancy of the link BC is already high, then accommodating this new call might make the network block calls that would have required fewer resources and hence cause revenue reduction. (Typically, the revenue from routing the call along $ABCD$ would not be higher than that of routing it along ABD.) Hence, depending on the occupancy of the links on the alternate routes, to possibly accommodate future calls having possibly higher revenue potential, the network can choose not to admit a call even though resources are available for it.

We discuss these issues in Chapter 6.

3.4   Elastic Transfers in a Packet Network: Feedback Control

The predominant use of packet networks is by elastic traffic, which is generated by applications whose basic objective is to move chunks of data between the disks of two computers connected to the network. Elastic flows can be speeded up or slowed down depending on the availability of bandwidth in the network. Various studies have indicated that 85 to 95% of the traffic carried by packet networks falls into this class. Of the volume of data carried, Web browsing, e-mail, and file transfers are the main elastic applications. With increasing standardization and improvements in technology, it can be expected that real-time interactive applications, such as packet voice telephony, will see increasing deployment. Given the limitations of the human telephony interface and the increasing use of high-quality voice coders, however, packet telephony will remain a low-bandwidth application. The growth of telephony is small (figures for the United States suggest a less than 10% annual growth), and we can expect the market to be saturated soon. On the other hand, indications are that the growth of elastic traffic in the Internet is very rapid (some reports have shown growth rates of at least 100% per year, and others have even suggested doubling of such traffic every six months).

Furthermore, it can be expected that it is from novel elastic applications that Internet users will have increasing expectations. The following (elastic) applications, already available on local area networks (LANs), would be high on Internet users' wish lists for use over wide area networks (WANs).

- *Diskless nomadic computing*: Anyone who has taken work on a trip on a laptop disk knows about the headaches involved in synchronizing the work files between the office workstation and the laptop. Travelers would like to carry diskless laptops, which, when hooked into the network *anywhere*, would give the users the same performance as over a local storage area network. In this context it is interesting to note that an IETF group has been defining a transport protocol for the popular SCSI disk interface (used in small computers) that will fetch blocks of files from remote disks to diskless clients.

- *Remote medical consultation*: Expert medical consultants could make their expertise available even to the remotest areas, if only they could easily and rapidly browse through a number of high-quality X-ray or scan images over the network.

- *Movie editing over a wide area network*: A movie editor could browse through movie clips stored on various servers over the wide area network and piece together a movie.

Hence a good understanding of the performance of elastic sessions in packet networks becomes very important. Chapter 7 is devoted to developing such an understanding via analytical models.

3.4.1   The File Transfer Abstraction

Figure 3.7 shows that at the most basic level, an elastic session simply involves the transfer of some files from one host attached to the packet network to another host. For example, the two hosts could be e-mail relays; each file transfer would then correspond to an e-mail being forwarded toward its destination mail server. Alternatively, the source host could be a file archive; at the destination host a user downloads several files during a File Transfer Protocol (FTP) session. Similarly, the source might be a Web server, and the destination might be a client with a Web browser using HTTP to browse the files at the server. In the Internet, for example, when a user requests a Web page (using an HTTP *get* request), a *base* file is downloaded, which in turn may trigger the transfer of several *embedded objects*, such as images. When there are embedded objects, the exact mechanism for downloading the objects depends on the version of HTTP in use. In HTTP 1.0, for the transfer of the base file and for each embedded object file, a separate TCP connection would be set up between the client and the server. In HTTP 1.1, to reduce connection setup overheads, a TCP connection would be reused for several file transfers between the same client and server.

In all these cases the basic problem is to transfer each file in its entirety from the source machine to the destination machine. This is the primary objective. There is no intrinsic rate at which the files *must* be transferred. In fact the transfer rate can vary as a file is transferred. A user downloading files would want to receive the files quickly, but this requirement is not really a part of the service definition of an elastic session.

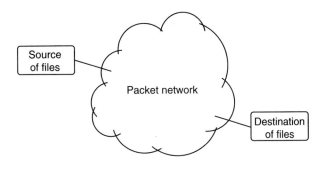

Figure 3.7   An elastic session simply involves the transfer of one or more files from one host to another.

Furthermore, there is no intrinsic packet size that the files need to be segmented into during their transfer. The transfer protocol can view each file simply as a byte stream and can transfer varying amounts of it in each packet.

We are dealing primarily with point-to-point elastic sessions, and this is what we mean when we use the phrase "elastic session." Thus an elastic session involves an association between two end points. The network determines a route between the end points in each direction; if the session lasts long enough there is a possibility that the route, in either direction, may change during the session. During a session, data transfers may take place in either direction, with possible gaps between the successive transfers. For example, if a user at a computer logs on to a file server, then FTP's *get* or *put* commands can be used to download or upload files. The user may need to do some other activities in between the file transfers (e.g., read what is downloaded and make notes); in user models these gaps are often referred to as *think times*. Similarly, a user browsing a Web server might download a Web page, and spend some time looking at it, before downloading another Web page from the same site. If the user shifts to browsing another Web server, we view this as another session starting, typically over a different pair of network routes.

3.4.2   Congestion Control, Feedback, and Bandwidth Sharing

Figure 3.8 shows a very simple "network" comprising a single link over which several users, on their respective hosts, are downloading files from some Web servers. Let us take the link capacity to be C bps, and assume that the local network attaching the users and the Web servers to this link is infinitely fast. We use this simple scenario to illustrate and discuss some basic issues that arise when several elastic sessions share the network bandwidth.

Suppose, to begin with, a single user initiates a download from a Web server over the link. It is reasonable to expect that an *ideal* data transfer protocol will (and should) provide this file transfer with a throughput of C bps. This much bandwidth is available, and if all of it is provided to the transfer, the session will be out of the system as early as possible. Now suppose another user starts a session while the first file transfer is still progressing. When the corresponding Web server starts transferring data to the user, the total input rate into the link (from the Web server's direction) will exceed C bps. If the first file transfer is proceeding at C bps, then the link's service rate will be exceeded no matter how slowly the second server sends its data. This will lead to link *congestion*. The network device that interconnects the server's LAN to the backbone link will have buffers "behind" this link. These buffers can absorb excess data that accumulate during this overload, provided that the situation does not continue for long. In addition, this situation

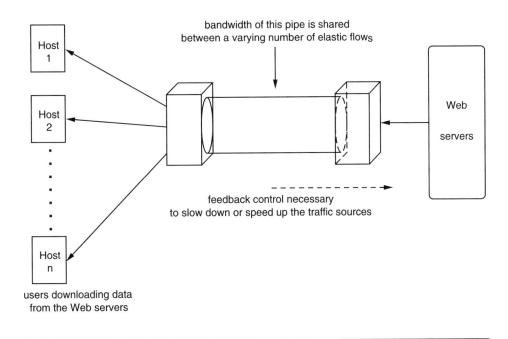

Figure 3.8   Several users dynamically share a link to download files from Web servers.

is clearly *unfair*, with one user getting the full link rate and the other user getting almost no throughput; this situation should not be allowed to persist.

Both of these issues (congestion and unfairness) require that there be some kind of *feedback* (explicit or implicit) to the data sources so that the rate of the first transfer is reduced and that of the second transfer is increased so that ultimately each transfer obtains a rate of $\frac{C}{2}$ bps. Now suppose that the first file transfer completes; if the second transfer continues to proceed at $\frac{C}{2}$ bps, the link's bandwidth is wasted, and the second session is unnecessarily prolonged. Hence, when the first session departs, the source of the second session should increase its transfer rate so that a throughput of C bps is obtained.

In summary, we conclude that an explicit or implicit feedback control mechanism needs to be in place so that as the number of sessions varies, the transfer rate provided to each session varies accordingly. *Explicit feedback* means that control packets flow between the traffic sources, sinks, and the network, and these packets carry information (e.g., an explicit rate or a rate reduction signal) that is used by the sources to adapt their sending rates. On the other hand, *implicit feedback* can be provided by packet loss or an increase in network delay; that is, a source can reduce its rate on sensing that one of the packets it sent may not have reached the destination. For example, in ATM networks the Available Bit Rate (ABR) service has a congestion management mechanism that is based on explicit feedback, whereas in much of the current Internet, TCP uses a congestion control mechanism based on implicit feedback. In an ATM network, associated with each ABR session is a flow of control cells (called *resource management* (RM) *cells*) generated by the source. As the RM cells of a session flow through the network, the ATM cell switches in their path can set an explicit rate value in these cells. The sink returns the RM cells to the source, and the source can then use the explicit rate in the returned RM cells to adjust its cell emission rate. On the other hand, TCP uses a window-based transmission protocol. A TCP source detects lost packets, takes these as indications of rate mismatch and network congestion, and voluntarily reduces its transmission rate by reducing the transmission window. We discuss these two control approaches at length in Sections 7.4 and 7.5, respectively. The Internet's window-based protocol, TCP, is discussed extensively in Section 7.6. In Section 7.7, we discuss the problem of fair bandwidth sharing in an arbitrary network of links, along with associated distributed algorithms.

3.5 Packet Multiplexing over Wireless Networks

An implicit assumption in the discussions of the previous sections is that the links interconnecting the nodes are *wireline*, or *guided*, media. Such links have bit error rates that are low enough to be ignored in optimizing the protocols and analyzing their performance. (The low error rate cannot, however, be assumed away in the design for correctness of the protocols.) Also, in such networks, the link characteristics can be assumed to be time-independent. These assumptions need to be revised when we consider wireless networks.

An important aspect of wireless networks is that the frequency band that is available for its exclusive use is limited. Also, the channel is a *broadcast medium*: A transmission is heard by everybody within the range of the transmitter. This constraint is addressed in the design of wireless networks in two ways.

- The capacity in the spectrum is shared among multiple users over the geographical area. This requires that the transmissions by the different users be made orthogonal to each other so that two simultaneous but unrelated transmissions with overlapping *footprints* can coexist and that the respective receivers can distinguish them. Orthogonality of transmissions can be achieved by either time, frequency, or code division multiplexing techniques or combinations thereof. Rather than use strict time division multiplexing, we can use multiple access protocols such as the distributed multiplexing schemes of Figure 2.11 to share the capacity of the spectrum among a number of users.

• The spectrum is reused at a geographically distant location. Obviously, a smaller *reuse distance*—the minimum distance at which the spectrum can be reused without the transmissions interfering significantly—increases the network capacity. This requires that the transmission power be as low as possible.

Channel reuse causes *cochannel* interference from transmissions that use the same frequency band. In analyzing the channel, interference is added to the receiver noise and the *signal-to-noise plus interference ratio* (SNIR) at the receiver determines the bit error rates (BERs) on the channel. Typical BERs on wireless channels can be six to eight orders of magnitude higher than that over wireline channels. Because the packet error rate (PER) can be approximated by the product of the packet length and the BER, the PER can become significant even for moderate channel BERs and cannot be ignored in protocol design.

Most wireless networks have nodes that are mobile. Mobility of the transmitter and the receiver and also of the elements that make up the radio environment causes the wireless channel to have time-varying characteristics. This means that for constant transmission power, the strength of the received signal at the receiver is time-varying. The interference characteristics are also time-varying because channel reuse and the method of providing orthogonality among the users contribute to the interference at the receiver. These in turn are traffic-dependent, and traffic is, of course, random. Thus in a wireless channel both the received signal power and the interference at the receiver (and hence the SNIR) are time-varying. This means that the channel will have time-varying BERs and PERs.

Because the channel characteristics are time-varying, the channel can be said to be in a "bad state" when the PER is high. Transmitting during these times will not only deplete precious battery energy but also will cause cochannel interference at the receivers that are reusing the spectrum. This is because although the channel condition to the intended receiver may be bad, the channel condition to the unintended receiver may be good. Thus, using channel state information to schedule transmissions not only helps conserve energy but also increases capacity by limiting cochannel interference. This in turn means that we need techniques to measure and estimate channel conditions, as well as algorithms that can take this into account.

To summarize, multiplexing techniques over wireless channels should be able to share precious capacity in the allocated spectrum among a large number of nodes in a power- and energy-efficient manner while taking into account the time-varying channel propagation characteristics. Let us recall the view presented in Figure 2.2, where we depict networking as being concerned with problems of sharing the resources of a bit-carrier infrastructure. The bit carriers are assumed to be given, and the problem is to develop mechanisms for sharing the given bit-carrier infrastructure. In fact, this viewpoint is more appropriate for wired networks. In the case of wireless networks, however, the problem can be viewed more generally as that of sharing a radio spectrum, and we may not want to treat the bit carriers as being given. This yields the view shown in Figure 3.9. In fact, the entire concept of cross-layer interactions and design (discussed in Section 2.3.2) becomes very relevant in wireless networks.

Different forms of wireless networks are emerging. Wireless LANs (WLANs) are deployed over small geographical regions and are connected to an internet through *access points*, which are similar to the base stations of the cellular

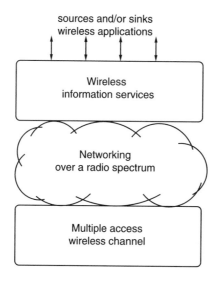

Figure 3.9   The modified view of networking over a shared radio spectrum.

networks discussed in Section 2.3.1. In these networks, there is typically one wireless hop, which uses a multiple access protocol between a node and the access point or directly between two nodes. The important issue here is the performance of the multiple access protocol.

A wireless network could be deployed over a larger area, with the nodes being peers and the path between two nodes possibly involving multiple wireless hops. Nodes in such a network could cooperate with each other to provide end-to-end paths using routing protocols similar to those used in the Internet. Of course, these protocols must be adapted for use in wireless networks. The topology of these networks could be arbitrary, and because they can also be deployed arbitrarily, they are called wireless ad hoc networks (WANETs). Among the issues of interest in such networks are the connectedness and the capacity of the network. When WANETs are used for Internet applications, there are issues of the performance of Internet protocols such as TCP over such networks.

Instead of providing a full set of information services, as do the nodes in a wireless ad hoc internet, a node could be a specialized device such as a *sensor*. A large number of sensors can be randomly deployed over a relatively small region. The nodes cooperate among themselves to perform a global computation function, and the communication is adapted to enabling this. In addition to overall connectivity, the capability of a set of randomly placed nodes to organize themselves to provide the global function, and the expected rate at which such functions can be performed, is of interest.

Although many local area network protocols, such as Ethernet and token ring, were designed as multiple access protocols over a wired broadcast medium, a large fraction of wired local area networks deployed today are switched networks with point-to-point links from the nodes to a central switch. There are, of course, many protocols, such as the one used in digital data transmission over cable TV networks, that in reality continue to be multiple access, too. A brief overview of the multiple access protocols used in such networks is presented in Section 2.2.1, and we do not discuss such networks in detail in this book. In our discussion of multiple access protocols, we focus only on wireless network subsystems. Specifically, with reference to Figure 2.18, we discuss only wireless access networks. Note that wireless links and multiple access protocols were used in one of the first computer networks ever deployed: the Aloha network of the University of Hawaii. The Aloha protocol is the precursor to the CSMA/CD mechanism of the Ethernet and the Carrier Sense Multiple Access with Collision Avoidance (CSMA/CA) mechanism of the IEEE 802.11 networks.

CHAPTER 4

Stream Sessions: Deterministic Network Analysis

When developing network engineering techniques, we are interested in using the simplest mathematical models with the widest applicability. Although network traffic in general, and the traffic emanating from stream sources in particular, is intrinsically stochastic, the past decade has seen the development of a *network calculus* that works with deterministic bounds on the traffic. The calculus can be used to engineer networks with worst case performance guarantees—for example, that the delay of packets in a packet voice call will never exceed T (as opposed to probabilistic performance objectives, such as that the delay will be less than T with a probability of 0.99). The models are quite simple (one often uses linear bounds), and, remarkably, end-to-end performance can be analyzed even when a variety of network elements (e.g., shapers, multiplexers, and propagation delay) are involved in the transport of the traffic.

In this chapter we first set down the notation that is used throughout our presentation of the material on packet multiplexing. We then present some very general observations, followed by a study of multiplexing via deterministic network calculus. The discussion is self-contained, and a knowledge of basic undergraduate mathematics suffices.

4.1 Events and Processes in Packet Multiplexer Models: Universal Concepts

Let us consider a packet moving from host A to host B (see Figure 3.1). The packet will follow some route through the backbone network. There is a certain minimum delay that the packet must experience: This is the sum of the propagation and bit transmission delays along the links that constitute the route taken by the packet. In addition to this fixed end-to-end delay, at the LAN interface card in host A (or at the switches or routers along its path), the packet may have to await the transmission of other packets before gaining access to the link it must go out on. Because the number of packets it must wait for is random and varies from hop to hop, the resulting additional delay is also random. The delay experienced by packets of a given packet stream at a link depends on the pattern of arrivals in that stream (arrival instants and the number of bits in the arriving packets) and the way the link transmits packets from the stream (the link may be shared in some way between two or more packet streams). To analyze such situations we use mathematical models that are variously called traffic models, congestion models, or queueing models.

In the following discussion, we assume that when packets must be transmitted onto a link, they wait in a buffer that is associated with that link and contains packets that are destined for transmission only on that link. In the context of packet-switching architectures, this amounts to assuming that the output contention resolution and the queueing in the switch are equivalent to *output queueing* (see Figure 4.1; Section 1.2; and Chapter 10). Such an output buffer, along with a packet-scheduling policy, is called a *packet multiplexer*.

In practice the link buffer may be implemented in the memory of the main control processor, with the control processor itself doing the scheduling; alternatively, the buffering and scheduling may be implemented in intelligent link-interface hardware. In high-performance packet switches the latter will invariably be the case. In the former case (scheduling being done by the host processor), it is important to ensure that after the scheduler offers a packet to the link-interface hardware to transmit, there should be negligible queueing in the interface hardware itself, because such substantial queueing delay at this stage would nullify the calculations done by the scheduler.

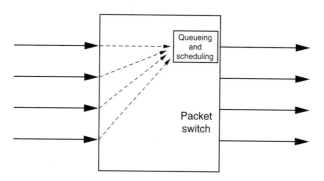

Figure 4.1   **Packets from various input links are switched to a buffer at an output link.**

4.1.1 Notation and General Results

When packets arrive into a queue associated with an output link of a switch, they arrive over an input link and hence arrive at a finite speed. A packet of length L bits arriving over a link of bit rate C will start arriving at some time t and will finish arriving at time $t + \frac{L}{C}$. Notice from this that the next packet on the same input link cannot start arriving before $t + \frac{L}{C}$. Thus, in practice, the interarrival times between packets on a link are bounded below by the transmission times of the packets on the link. Because the output link of a switch will receive packets from several input links, packets on different input links could be arriving simultaneously. This discussion is illustrated in the top part of Figure 4.2. If input and output links are of equal speeds and if only one input link feeds an output link, then a packet arriving at the link buffer will never find another packet in service and hence will not experience queueing delay. Queueing delay can, however, occur if several input links feed packets to an output link; this is typically the case.

Each packet carries several bytes of header, which contains information that the switch uses to determine the output link to which the packet must be switched. Assuming that the header is at the beginning of each packet, it should be possible to determine the output port of the packet before the entire packet arrives on the input link. Note that such a feature would require a *header error check* feature in each packet. Clearly there are two possibilities.

- The entire packet can be allowed to accumulate in the input processor of the switch before it is moved (switched) to the output port. This adds an additional store-and-forward delay at each hop.

- After reception of the header and determination of the output port, the header is (after some necessary modification) passed to the output port, followed immediately by the data part of the packet, even as the bits arrive on the input link. This cut-through switching has the obvious advantage of reducing store-and-forward delay.

In the following discussion we limit our scope to store-and-forward packet switching. This is what actually happens when packet switching takes place above layer 2; it is in fact the practice in Internet packet transport and also is usually the case for packet switching at layer 2. With this assumption, the middle part of Figure 4.2 shows the *packet arrival instants* on each input link as seen by the output buffer. The lower part of the same figure shows the superimposed packet arrival instants at the output link buffer, with the heights of the lines indicating the lengths of the corresponding packets.

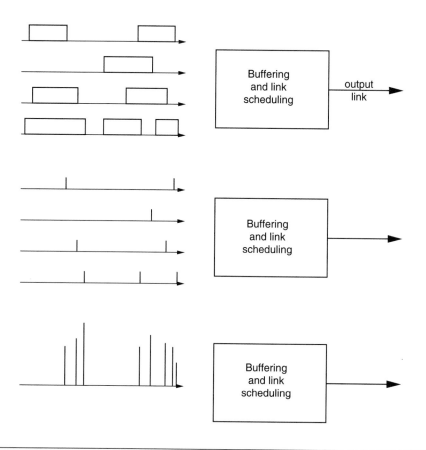

Figure 4.2 Packets from various input links arrive, at the input link speed, and are switched to the buffering and scheduling module of an output link. Time progresses to the right. The top display shows the packets arriving over the links. The middle display shows packet arrival instants assuming store-and-forward packet switching. In the bottom part of the figure, the heights of the vertical lines indicate the lengths of the arriving packets.

We now set down some notation that is used repeatedly in the book.

a_k: the time instant of the kth packet arrival into the buffer.

L_k: the number of bits in the kth packet arrival; even though bits are discrete, we make the simplification that L_k is a nonnegative real number—that is, the packets bring in *fluid* bits.

d_k: the time instant at which the kth arriving packet departs the buffer.

We say that the link scheduler is *non-idling* or *work-conserving* if the link is not allowed to idle when there is data (or "work to be done") in the buffer. Thus, if the scheduler is work-conserving, then whenever there are bits in the buffer the buffer level is reducing at the bit rate of the link, C bits per second (bps). Figure 4.3 shows the pattern of bit arrivals into the buffer and, assuming a work-conserving scheduler and infinite buffer space, shows the number of bits in the buffer at any time. When the buffer is nonempty it is depleted at rate C; a new packet arrival causes the contents of the buffer to increase by the number of bits in the arriving packet. The buffer becomes nonempty at a_1, and we see that between the times a_3 and a_4 there is a period during which the buffer again becomes empty. The time periods during which the buffer is nonempty are called *busy periods*, and the periods during which the buffer is empty are called *idle periods*.

We now introduce additional notation:

$A(t)$: cumulative data arrivals (i.e., the total number of bits) into the multiplexer until time t. We take $A(t) = 0$ for $t < 0$. Thus $A(t)$ is the amount of data that arrives in the interval $[0, t]$. Note that we include in $A(t)$ a packet arrival, if any, at time 0.

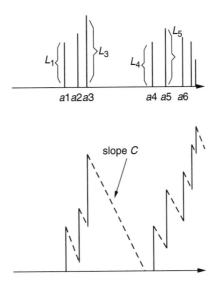

Figure 4.3 The bit arrival process and the number of bits in the buffer process. A busy period is followed by an idle period, and a part of the following busy period is also seen.

$D(t)$: cumulative data departures until t. $D(t)$ is the amount of data that departs in the interval $[0, t]$. Since, obviously, $D(t) \leq A(t)$, $D(t) = 0$ for $t < 0$.

$X(t)$: queue length at the multiplexer (amount of bits queued) at time t (including the partial packet currently being transferred). Clearly, $X(t) = A(t) - D(t)$.

We assume that $A(t), X(t)$, and $D(t)$ are right-continuous and have left-hand limits. Thus a packet arrival at t is included in the cumulative arrivals counted by $A(t)$, and the total volume of data that arrives *at* an instant t is given by $A(t) - A(t_-)$.[1] Also, because $A(t)$ and $D(t)$ are cumulative processes they are nondecreasing in t.

We observe now, from Figure 4.3, that for a work-conserving scheduler, the evolution of $X(t), t \geq 0$, depends only on the arrival instants $(a_k, k \geq 1)$ and the packet lengths $(L_k, k \geq 1)$, and not at all on the order in which the packets are scheduled for transmission on the link. In turn this implies that the busy and idle periods are also invariant with the scheduling policy, as long as the policy is work-conserving.

From an examination of Figure 4.3 we cannot determine the departure instants $d_k, k \geq 1$. These depend on the actual scheduling policy involved. However, *it is clear that the same packets depart in each busy period under any scheduling policy, and the difference between scheduling policies lies in the order in which the packets depart in each busy period*. For example, in Figure 4.3, packets $1, 2$, and 3 depart in the first busy period, and (at least) packets $4, 5$, and 6 depart in the second busy period.

A simple measure of performance at the multiplexer is the average number of bits in the buffer. If this average is large, it can be inferred that the link is congested and that the packets passing through this link will experience large delays. Define, for $t_2 > t_1$, $\overline{X}(t_1, t_2) = $ the time average of the function $X(t)$ over the time interval (t_1, t_2); that is,

$$\overline{X}(t_1, t_2) = \frac{1}{t_2 - t_1} \int_{t_1}^{t_2} X(u)\,du$$

Obviously, because $X(t), t \geq 0$, is invariant with the scheduling policy, so is $\overline{X}(t_1, t_2)$.

---

[1]For an instant t, t_- can be read as "just before" the instant t; more formally $A(t_-) = \lim_{s \uparrow t} A(s)$.

The packets coming over the various links into the output queue at a link would actually be from several sessions. For example, during an interval (t_1, t_2) the link may be carrying a few packet voice sessions and some file transfer sessions. Letting $X_s(t)$ denote the number of bits in the buffer from a session s, we see that

$$X(t) = \sum_{\text{sessions } s} X_s(t)$$

Defining $\overline{X}_s$ analogously to $\overline{X}$, we then easily see that

$$\overline{X}(t_1, t_2) = \sum_{\text{sessions } s} \overline{X}_s(t_1, t_2) \tag{4.1}$$

This simple relationship, derived from first principles, is called a *conservation law*. If the scheduler attempts to favor a particular session s by giving it better service and thereby reducing its $\overline{X}_s$, the average buffer occupancy (and hence delay) of another session must increase, because the total average buffer occupancy $\overline{X}$ must be "conserved."

Recall that the earlier discussion assumes that the buffer size is infinite. In practice, of course, the amount of memory space available to buffer the data at a link is finite. The infinite buffer idealization is a good model only if the arrival pattern and link service rate are such that $X(t)$ rarely reaches the available buffer capacity. When $X(t)$ does reach close to the buffer capacity, any new packet arrivals that would cause the buffer capacity to be exceeded would result in data loss. At such arrival instants, packets can be discarded, or lost, in two ways:

- The arriving packet can be discarded.

- The arriving packet can be accepted, but one or more "less important" buffered packets could be *pushed out*, or *dropped*.

In either case there would be packet loss. The effect of packet losses depends on the application. A stream application, such as packet telephony, will suffer because of distortion in the played-out speech. An elastic application, such as file transfer, will need to recover lost packets and hence will experience additional delay in the completion of the transfer. It is not a good idea, however, to reduce packet loss by simply increasing the buffer capacity. In the example of packet telephony, if the buffer builds up to large values, then even though a packet joining the tail of the buffer may not be lost, it will experience a large delay and may arrive at the receiver too late to be played out. In the example of file transfers, the effect of large buffers is more subtle, and we discuss this at length in Chapter 7.

What happens to the conservation law if the buffer capacity is finite? Let the buffer capacity be B bytes. Figure 4.3 will now change. In particular, $X(t)$ will never exceed B. If a packet of length L bytes arrives at a and if $B - X(a_-) \geq L$, then this packet can be accepted; otherwise, this or other (buffered) packets must be discarded. Denote the buffer occupancy processes for the finite buffer model with a superscript (B), as in $X^{(B)}(t)$. A version of the conservation law can then be developed when the arriving packet is the one to be dropped.

Exercise 4.1
Consider the buffered link model with finite buffer B. Assume that when a packet arrival will cause the buffer to overflow, *the arriving packet is discarded*. Show that the sample paths of the process $X^{(B)}(t)$ are invariant to the scheduling discipline, and hence that

$$\overline{X}^{(B)}(t_1, t_2) = \sum_{\text{sessions } s} \overline{X}_s^{(B)}(t_1, t_2) \tag{4.2}$$

4.2 Deterministic Traffic Models and Network Calculus

In general, the outputs of traffic sources (e.g., the speech coder in Figure 3.1) are a priori unknown and nondeterministic. Hence, one modeling methodology is to use a stochastic process model for the traffic emitted by sources. Such models permit us to make statistical statements about network performance (e.g., the probability of the end-to-end packet delay exceeding 50 ms is less than 0.01). In each instance of network operation, however, the network actually sees only a *sample path* of the random process; that is, the cumulative arrival process would be $A(t, \omega)$, where ω is the usual notation for a particular sample path. For a fixed ω, the arrival process evolves in time and hence is a function of t. The deterministic methodology can be seen as working with such individual sample paths. Thus, in the deterministic methodology, when we say that the arrival process from a source has a certain property (e.g., the peak arrival rate is R), then under the stochastic process model it would mean that *every* sample path of the arrival process would have that property (i.e., a peak rate of R would imply that, for every sample path ω, $A(t, \omega) \leq Rt$). In this sense, we can say that the deterministic methodology considers the worst case performance of the network and therefore yields conservative results. We illustrate this by an example in Chapter 5.

4.2.1 Reich's Equation and the Convolution Operator

We begin by developing relationships between the various processes we have defined. The following is a very general result for the work-conserving case; some authors refer to it as Reich's equation. The proof is provided in the Appendix at the end of this chapter. The concepts of sup and inf are explained in Appendix B.

Assume that just before time 0 (i.e., at 0_-), the buffer is empty: $X(0_-) = 0$. For all $t \geq 0$,

$$X(t) = \sup_{0_- \leq s \leq t} (A(t) - A(s) - C \cdot (t - s)) \tag{4.3}$$

It is important to observe the structure of this expression. Each term in the sup is the difference between the cumulative arrivals over the interval (s, t) and the total number of bits that can be transmitted from the buffer in that interval. Notice also that the right edge of the intervals is anchored at t and that the left edge varies over $(0_-, t)$. Write Equation 4.3 as

$$X(t) = \sup_{0_- \leq s \leq t} (A(t) - (A(s) + C \cdot (t - s)))$$

Rewriting the equation in this way helps in a geometric interpretation (see Figure 4.4).[2] The stepped curve is the cumulative arrival process $A(t)$; each packet arrival causes the cumulative arrivals to increase by the packet length. For any t, we obtain $X(t)$ by considering, for each $s \leq t$, straight lines with equations $A(s) + C \cdot (t - s)$. The value of $X(t)$ is the maximum gap between $A(t)$ and any of these lines. Notice that this maximum is obtained for the slant line corresponding to the instant b, which is the start of the busy period that includes the instant t. The busy periods of the queue are now evident from Figure 4.4. In between the busy periods, where the queue lengths are zero, we have the idle periods. Clearly, when the queue is busy the departure rate is C, and when idle the departure rate is 0. Thus the thick

---

[2]Strictly speaking, in Figure 4.4 we are viewing bits as fluid, or as being infinitely divisible.

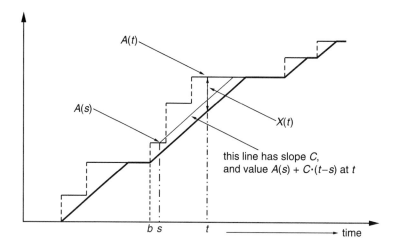

Figure 4.4   The geometry of Reich's equation. The piecewise constant (stepped) curve is the arrival process $A(t)$; the slope of the thick slant lines is C, the service rate; the vertical gap between $A(t)$ and the thick solid curve is $X(t)$.

solid line in Figure 4.4—comprising the slant lines of slope C during busy periods, and the flat lines—is the cumulative departure process $D(t)$.

We now provide an equation for the departure process. For all $t \geq 0$,

$$D(t) = \inf_{0_- \leq s \leq t} (A(s) + C \cdot (t - s)) \tag{4.4}$$

A direct proof of this equation is given in the chapter Appendix. We can also obtain this expression from the relationship between $A(t)$, $X(t)$, and $D(t)$ as follows:

$$D(t) = A(t) - X(t)$$

$$= A(t) - \sup_{0_- \leq s \leq t} ((A(t) - A(s)) - C \cdot (t - s))$$

$$= \inf_{0_- \leq s \leq t} (A(s) + C \cdot (t - s))$$

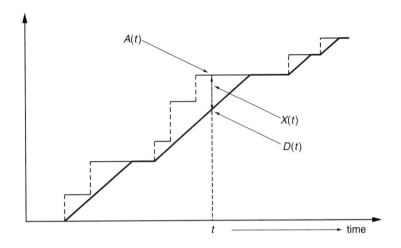

Figure 4.5 The arrival, backlog, and departure processes.

For clarity, the three processes $A(t), X(t)$, and $D(t)$ are shown separately in Figure 4.5. Observe here that $D(t)$ depicts the cumulative bit departures.

If the multiplexer's service is *first-come-first-served* (FCFS), the instants of packet departures can be inferred from $D(t)$; a packet that arrives at the instant a departs at the instant d such that $d = \inf\{u : D(u) \geq A(a)\}$—that is, at the earliest instant at which the number of bits served is just equal to the cumulative bit arrivals in $[0, a]$.

Observe that if in Equation 4.4 we replace the term $C \cdot (t - s)$ with $C \cdot (t - s)^+$, then the value of the right side of the equation remains unchanged. Furthermore, because $A(t) = 0$ for $t < 0$ and because $A(t)$ is nondecreasing with t, notice that

$$D(t) = \inf_{s \in \mathbb{R}} \; (A(s) + C \cdot (t - s)^+)$$

Writing $B(t) = C \cdot t^+$, we can then write

$$D(t) = \inf_{s \in \mathbb{R}} \; (A(s) + B(t - s))$$

This expression motivates the following general definition.

Definition 4.1

If $A(t)$ and $B(t), t \in \mathbb{R}$ are nonnegative, nondecreasing functions (possibly also $+\infty$ valued), then the *convolution* of these functions is defined as

$$(A * B)(t) := \inf_{\tau \in \mathbb{R}} \ (A(\tau) + B(t - \tau))$$

∎

The term *causal* is used to denote an $A(t), t \in \mathbb{R}$, which has the property that $A(t) = 0$ for $t < 0$.

Remarks 4.1

The operator $*$, introduced in Definition 4.1, is very useful in deterministic network calculus. The following comments relate to its interpretation and computation.

- From Definition 4.1, it is clear that for causal $A(t)$ and $B(t)$,

$$(A * B)(t) = \inf_{0_- \leq \tau \leq t_+} \ (A(\tau) + B(t - \tau))$$

  Here, we need the limits 0_- and t_+ in the inf to take care of jumps in $A(t)$ and $B(t)$ at $t = 0$. In Equation 4.4 we do not have the upper limit in the inf as t_+ because the function Ct^+ does not have a jump at 0.

- We can view the definition of $B * A$ as follows. For each $\tau \in \mathbb{R}$, define a function of t by $f_\tau(t) = B(\tau) + A(t - \tau)$. Then

$$(B * A)(t) = \inf_{\tau \in \mathbb{R}} f_\tau(t)$$

  To illustrate this way of computing $B * A$, consider $B(t) = Ct^+$, and $A(t) = (\sigma + \rho t)I_{\{t \geq 0\}}$. These functions are shown at the top of Figure 4.6. The process of obtaining the convolution is depicted in the lower graph in the figure. $(B * A)(t)$ is the lower "envelope" of all the functions $B(\tau) + A(t-\tau)$, indexed by $\tau \in \mathbb{R}$. Notice that in this example, $(B * A)(t) = \min\{B(t), A(t)\}$.

- It is interesting to compare the convolution operation defined here with the familiar convolution operation used in linear systems theory: $(A * B)(t) = \int_{\tau \in \mathbb{R}} A(\tau) \cdot B(t - \tau) d\tau$. In comparison with this convolution, the convolution in Definition 4.1 replaces the multiplication between

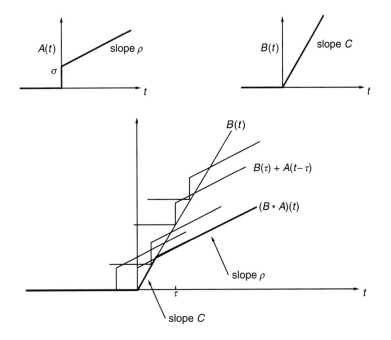

Figure 4.6   A visualization of the convolution operation $(B * A)(t)$. Note that $B(\tau) +$ $A(t - \tau)$ is shown for several values of τ, but only one such τ is shown plotted on the t axis. The lower graph, drawn with the thick line, is the plot of $(B * A)(t)$.

$A(\tau)$ and $B(t - \tau)$ with a summation, and replaces the integration over $\{\tau \in \mathbb{R}\}$ with an infimum over $\{\tau \in \mathbb{R}\}$. Thus this new operation can be called a $(\min, +)$ convolution, in contrast with the familiar $(+, \times)$ convolution. In this context, notice that $+$ distributes over min (i.e., $a + \min(b, c) = \min(a + b, a + c)$), just as $\times$ distributes over $+$. ∎

Define

$$\delta(t) = \begin{cases} 0 & \text{for} \quad t < 0 \\ \infty & \text{for} \quad t \geq 0 \end{cases}$$

By the constructive way of viewing the $*$ operator, it is easy to see that $A * \delta = A$.

Notice further that $\delta_d(t) = \delta(t-d)$ represents a delay element because

$$(A * \delta_d)(t) = A(t-d)$$

For B causal, $B \leq \delta$, and hence

$$A * B \leq A * \delta = A$$

And, in particular, for A causal, $A * A \leq A$.

Exercise 4.2   (Properties of $*$)
Establish the following properties of the binary operator $*$.

   (i) $*$ is commutative: $A * B = B * A$

  (ii) $*$ is associative: $(A * B) * C = A * (B * C)$

 (iii) $*$ is distributive over min: $A * \min\{B, C\} = \min\{(A * B), (A * C)\}$

4.2.2 Service Curves for Network Elements

Figure 4.7 shows an arrival process $A(t)$ entering a network element. We also show the corresponding departure process, denoted by $D(t)$. Note that there could be other arrival and departure processes in the system, but we are focusing on this particular pair; in a sense, the other arrivals and departures are viewed as being a part of the environment that provides service to the arrivals we are focusing on.

In the discussion in the beginning of Section 4.2.1, the network element was only a link with service rate C. Over a time interval of length t, this network element simply removes up to Ct bits from its buffer. As we will see,

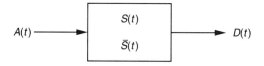

Figure 4.7 Lower and upper service curves characterize a relation between the input and output processes of a network element.

to provide service differentiation and performance guarantees to traffic flowing through a network, other service elements are required. We thus need a more general characterization of the way a network element "serves" the traffic passing through it.

Let $S(t)$ and $\bar{S}(t)$ be nonnegative, nondecreasing, causal functions. If $D \geq A * S$, then we say that S is a *lower service curve*, whereas if $D \leq A * \bar{S}$, then $\bar{S}$ will be called an *upper service curve*. If S is such that it is both a lower and an upper service curve, it is called the *service curve*, and then $D = A * S$.

Example 4.1

(i) With reference to the packet voice example, Figure 4.8 shows the output of a voice coder being fed into a packetizer. The coder emits a stream of bytes. The packetizer takes blocks of bytes (at most $L_{\max}$ bytes to a block), puts headers around these blocks, and emits them as packets. The coder would typically emit bursts of data, and the bursts may not be an integral multiple of $L_{\max}$. Hence, at the end of a burst of data from the encoder, the packetizer packetizes the accumulated data, even if it is less than $L_{\max}$ bytes. Assume that when the coder is emitting data, the minimum byte arrival rate from the coder is r. Notice that any byte is delayed by at most $\frac{L_{\max}}{r}$. It follows that $D_p(t) \geq A\left(t - \frac{L_{\max}}{r}\right)$. Hence a lower service curve for the packetizer is $\delta_{\frac{L_{\max}}{r}}$.

(ii) Figure 4.9 shows an arrival process $A(t)$ feeding a constant-rate fluid server with an infinite buffer space. By *fluid server* we mean that the server can serve and emit arbitrarily small amounts of data; that is, any amount of data that is output is considered a valid (useful) departure. This idealization may be useful for networks where the packet sizes are very small, such as cells in an ATM network.

It is then clear from our earlier discussions (see Equation 4.4) that $S(t) = Ct^+$ is the service curve for the constant-rate (fluid output) server,

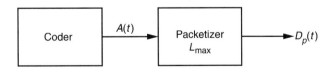

Figure 4.8   A source coder followed by a packetizer.

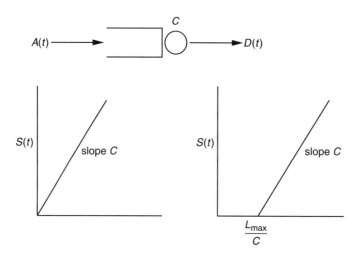

Figure 4.9 Lower service curves for a constant-rate server with infinite buffer. The curve on the left is for fluid output, and the one on the right is for packet output, with L_{max} being the maximum packet length.

and $D = A * S$. Suppose, however, that the departures are meaningful only at packet boundaries (the receiving entity can make use of them only as full packets); then at most L_{max} bits must emerge from the network element before we have a usable packet. Hence a lower service curve is $(Ct^+ - L_{max})^+$, which is the same as $C(t - \frac{L_{max}}{C})^+$. This is shown in the lower-right plot in Figure 4.9. Note that the curve Ct^+ is an upper service curve for the packet output case. Finally, observe that $C(t - \frac{L_{max}}{C})^+ = Ct^+ * \delta_{\frac{L_{max}}{C}}$, which again provides an interpretation of reception at packet boundaries as a delay element.

(iii) Figure 4.10 shows a coder followed by a packetizer. The output of the packetizer is fed into a constant-rate link server. Write $S(t) = Ct^+$. Then, using the notation in Figure 4.10, we have

$$D = (D_p * S)$$
$$\geq (A * \delta_{\frac{L_{max}}{r}}) * S$$
$$= A * (\delta_{\frac{L_{max}}{r}} * S)$$

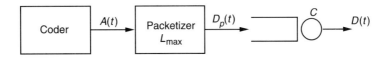

Figure 4.10   A source coder followed by a packetizer and a constant-rate server.

It follows that an effective lower service curve for the tandem configuration of the packetizer and the constant-rate server is $\left(\delta_{\frac{L_{max}}{r}} * S \right)$. The form of this lower service curve is the same as that shown in the lower-right plot of Figure 4.9.   ∎

Latency Rate Service Elements

In the preceding example we have twice encountered service elements (or combinations of service elements) with lower service curves of the general form $\delta_d * rt^+$, or equivalently $r(t - d)^+$. Such service elements are called *latency rate servers*; they are characterized by two parameters: a rate r, and a delay (or latency) d.

Service Elements in Tandem

In being transported through a network, the data from a source typically pass through several network elements. Suppose we can characterize each such network element by its upper and lower service curve, as shown in Figure 4.11. It is then of interest to determine how the eventual departure process $(D_n(t)$ in Figure 4.11) is related to the arrival process $A(t)$. From the foregoing development it is easy to see that the service curves of the network elements in tandem can be composed.

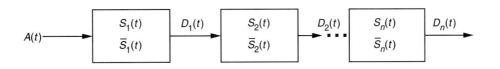

Figure 4.11   An arrival process A(t) passing through several network elements in tandem. The kth element has lower service curve S_k and upper service curve $\bar{S}_k$.

We can recurse backward from the last hop, as follows. By definition of a lower service curve, we get

$$D_n \geq D_{n-1} * S_n$$

$$\geq (D_{n-2} * S_{n-1}) * S_n$$

$$\geq A * (S_1 * S_2 * \cdots * S_n)$$

Hence a lower service curve for the tandem of network elements is $S = S_1 * S_2 * \cdots * S_n$. Note that in this calculation we have used the associativity property of $*$ (see Exercise 4.2). Similarly, it is easily seen that an upper service curve for the tandem of network elements is $\overline{S} = \overline{S}_1 * \overline{S}_2 * \cdots * \overline{S}_n$.

Example 4.2

We saw earlier that latency rate servers arise in certain situations, and we will encounter them again. The lower service curve of a tandem of latency rate servers takes a particularly simple form. Consider n such servers in tandem, with the lower service curve of the kth server being $S_k(t) = r_k(t - d_k)^+$. Then a lower service curve of the tandem of elements is given by the following:

$$S(t) = (S_1 * S_2 * \cdots * S_n)(t)$$

$$= r_1(t - d_1)^+ * r_2(t - d_2)^+ * \cdots * r_n(t - d_n)^+$$

$$= (r_1 t^+ * \delta_{d_1}(t)) * (r_2 t^+ * \delta_{d_2}(t)) * \cdots * (r_n t^+ * \delta_{d_n}(t))$$

$$= (r_1 t^+ * r_2 t^+ * \cdots * r_n t^+) * (\delta_{d_1}(t) * \delta_{d_2}(t) * \cdots * \delta_{d_n}(t))$$

$$= (\min_{1 \leq k \leq n} r_k) t^+ * \delta_{\sum_{1 \leq k \leq n} d_k}(t)$$

$$= r(t - d)^+$$

where $r = \min_{1 \leq k \leq n} r_k$, and $d = \sum_{1 \leq k \leq n} d_k$. Hence the tandem of latency rate servers is also effectively a latency rate server.   ∎

4.2.3 Delay in a Service Element

A stream session would normally expect the network to deliver its packets to the sink in the same order in which they were injected into the network by

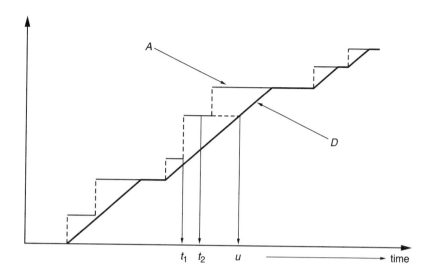

Figure 4.12   The horizontal distance between the arrival and departure curves yields the delay through the network element.

the source. Thus network elements always handle packets *from a particular session* so that packets of this session pass through the element in a first-in-first-out (FIFO) fashion. Note that this restriction applies only to the delivery of packets of each session. Across sessions, the packet movement through a network element does not need to be FIFO.[3]

Figure 4.12 shows the arrival and departure processes of a session passing through a network element. Notice that all the data arriving in the interval $[0, t_1]$, or the interval $[0, t_2]$, departs the network element by u. More formally, all the data arriving in $[0, t]$ has departed by $u = \inf\{s : D(s) \geq A(t)\}$. Hence the maximum delay through the network element is

$$\Delta_{\max} = \sup_{t \geq 0_-} \{u - t: u = \inf\{s: D(s) \geq A(t)\}\}$$

---

[3]FIFO delivery within a session does not preclude packet loss in the network element; packets (in a session) that *successfully* pass through the network element will emerge in the same order as they entered the element.

In other words, the max delay Δ_{max} is the maximum horizontal distance between A and D. It follows that $D \geq A * \delta_{\Delta_{max}}$, or, in words, if A is shifted to the right by Δ_{max} the curve A will lie below the curve D.

4.2.4 Envelopes and Regulators

As discussed in Section 3.2, unless a stream source provides more information to the network than only its peak rate, the best the network can do is to attempt to reserve the peak rate for the source. In this section we discuss an approach for deterministic characterization of arrivals from a source into the network.

Let $A(t)$ be the cumulative arrival process from a source. Let $E(t)$ be a nondecreasing and nonnegative function. We say that A has the *envelope E* if, for all t and $\tau, 0 \leq \tau \leq t, A(t) - A(\tau) \leq E(t - \tau)$. That is, over any interval of time of length u, the amount of data brought by the arrival process $A(t)$ is bounded by $E(u)$, irrespective of where in time we take this interval.

From the definition of $*$ it is easily inferred that A has envelope E if and only if

$$A \leq A * E$$

An envelope is *causal* if $E \leq \delta$. It follows, then, that if A is causal and has a causal envelope E,

$$A * E \leq A * \delta$$

$$= A \leq A * E$$

It follows that for a causal envelope, $A = A * E$.

An envelope is said to be *subadditive* if, for all $\tau \leq t, E(t) \leq E(\tau) + E(t - \tau)$. It follows that for a subadditive envelope $E \leq E * E$.

Exercise 4.3

Show that if $E(t)$ is causal, nondecreasing, and concave, then $E(t)$ is subadditive.

A little later we examine the advantages of knowing that an arrival process A has an envelope E. Typically, the output of a source is characterizable only

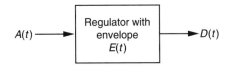

Figure 4.13 A regulator element with arrival process A, departure process D, and envelope E.

statistically. How then does it come about that A can have a given envelope? This is achieved by passing the source output through a regulator.

A network element is a *regulator* with envelope E, if for *any* input arrival process A, the departure process D has envelope E (see Figure 4.13). Thus, for all t and $\tau, 0 \leq \tau \leq t$, $D(t) - D(\tau) \leq E(t - \tau)$, or

$$D \leq D * E$$

Example 4.3

Figure 4.14 shows a network element called a *buffered leaky bucket* (LB) regulator (or *LB shaper*). There is a token bucket into which tokens are fed at rate ρ. The bucket can hold up to σ tokens. Data is allowed to depart from the regulator only if there are matching tokens available. If data arrives and there are no matching tokens, then the data is buffered in the source buffer.

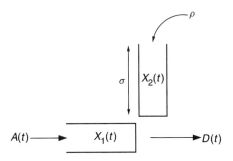

Figure 4.14 A buffered leaky bucket regulator, with token bucket depth σ and token arrival rate ρ. Also shown is a buffer in which arrivals can wait if no tokens are available.

Notice that no matter what the input process A is, the output process D satisfies (for $0 \leq \tau \leq t$)

$$D(t) - D(\tau) \leq \sigma + \rho(t - \tau)$$

This is because in the time interval $[\tau, t]$ the maximum amount of data that can depart is the number of tokens at τ plus the number of tokens that arrive in $[\tau, t]$. A process that can be bounded as above is said to be (σ, ρ) regulated, or a (σ, ρ) process.

It follows that $D \leq D * E$, where $E(t) = \sigma + \rho t$ for $t \geq 0$, and $E(t) = 0$ for $t < 0$. Thus the leaky bucket regulator has an envelope $E(t)$. Notice also that this envelope of a leaky bucket regulator is subadditive (see Exercise 4.3).

An LB regulator with parameters σ and ρ is also called a (σ, ρ) regulator. Thus a (σ, ρ) regulator's output process is (σ, ρ) regulated. ∎

We see from the foregoing description that a leaky bucket (LB) can be used to regulate or smooth out the traffic it receives. If data arrives at some peak rate $R > \rho$, then a maximum amount of $\frac{\sigma R}{R - \rho}$ can emerge from the LB at rate R. After a burst of this size, additional data arriving at the peak rate must be buffered in the source buffer of the LB, and the LB output will be rate-limited to ρ. In addition, the LB output rate averaged over long time intervals can be no more than ρ, because this is the rate at which tokens arrive into the LB.

Figure 4.15 shows a sample evolution of the various processes associated with a leaky bucket regulator. The thin solid line is the cumulative amount of data entering the LB regulator (i.e., $A(t)$), the thin dashed line is the cumulative amount of tokens entering the token bucket, and the thick solid line is the cumulative departure process from the regulator. At time 0 the bucket is full, with σ tokens. In this example, the arrival process has two states: an *on* state when the arrival rate is $R > \rho$, and an *off* state when there are no arrivals. The token bucket stays at level σ until the first arrival burst starts; during this time, in a sense, the arriving tokens are being lost. When the arrival burst starts, the token bucket begins to be drained at rate $R - \rho$. During this period the rate of the output process is R. This continues until the token bucket becomes empty, and the source buffer begins to fill up. Note that for fluid arrivals and fluid tokens, *the token bucket and the source buffer* cannot both be nonempty, and exactly one of the two is always nonempty. When the source buffer is nonempty the output rate is ρ. When the arrival burst ends, the source buffer depletes at rate ρ; eventually the source buffer empties out,

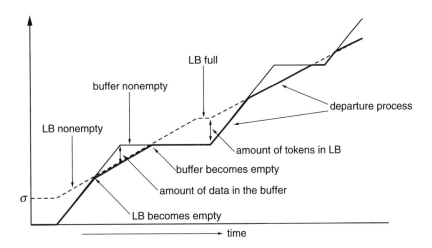

Figure 4.15 A sample evolution of processes associated with an LB regulator. The thin solid line is the cumulative amount of data entering the LB regulator (i.e., $A(t)$), the thin dashed line is the cumulative amount of tokens entering the token bucket, and the thick solid line is the cumulative departure process from the regulator.

and the token bucket begins to fill up. If the source stays in the *off* state for enough time, the bucket again fills up to level σ. This is as shown in the figure.

We now argue that the leaky bucket regulator has the following important property. We assume fluid arrivals and fluid tokens; if the downstream element can work only with packets, then a packetization element must follow the leaky bucket regulator.

Theorem 4.1

If A is an arrival process into a leaky bucket regulator and if E is the envelope of the regulator (i.e., $E(t) = \sigma + \rho t$ for $t \geq 0$, and $E(t) = 0$ for $t < 0$), then the departure process D of the regulator satisfies

$$D = A * E$$

Remark: It follows that the envelope $E(t)$ is the service curve of the leaky bucket regulator service element. The following simple proof is easier to understand after a study of Figure 4.15.

Proof: For any $t \geq 0$, and $0 \leq \tau \leq t$, it is clearly true that

$$D(t) = D(\tau) + D(t) - D(\tau)$$

$$\leq A(\tau) + \sigma + \rho(t - \tau)$$

It follows that

$$D(t) \leq \inf_{\tau \in \mathbb{R}} (A(\tau) + E(t - \tau))$$

Now, with reference to Figure 4.15, consider any time instant $t \geq 0$. If the source buffer at the leaky bucket regulator is empty, then we have

$$D(t) = A(t)$$

$$\geq \inf_{\tau \in \mathbb{R}} (A(\tau) + E(t - \tau))$$

To see this, note that $A(t) = \inf_{\tau > t}(A(\tau) + E(t - \tau))$. On the other hand, if the source buffer is nonempty (i.e., the leaky bucket is empty), then define

$$v = \sup\{s \leq t : \text{leaky bucket is full at } s\}$$

Thus, up to v_- all the arriving data have passed through. Because at t the source buffer is nonempty, it follows that all tokens that were added since v have been matched with data departures. We therefore see that

$$D(t) = A(v_-) + \sigma + \rho(t - v)$$

$$\geq \inf_{\tau \in \mathbb{R}} (A(\tau) + E(t - \tau))$$

Putting these observations together, we have

$$D(t) = \inf_{\tau \in \mathbb{R}} (A(\tau) + E(t - \tau))$$

$$= A * E \qquad \blacksquare$$

Consider a causal and subadditive envelope E. By subadditivity $E \le E * E$ and by causality $E * E \le E * \delta = E$, and thus we have that $E = E * E$. A leaky bucket regulator has a causal and subadditive envelope $E(t) = (\sigma + \rho t)I_{\{t \ge 0\}}$. Now if an arrival process A is passed through a leaky bucket controller, and if the output $D_1 (= A * E)$ is passed through a second leaky bucket with envelope E, then we see that the final output $D_2 = D_1 * E = (A * E) * E = A * (E * E) = A * E = D_1$. As expected, the second leaky bucket regulator with the same parameters does not smooth process A any further.

Exercise 4.4

A process A with envelope E_1 is passed through a leaky bucket regulator with envelope E_2. Show that the resulting output has envelope $E_1 * E_2$.

As an application of this exercise, consider the output of a voice coder with peak rate R (e.g., 64 Kbps). It is easily seen that this process A has envelope $E_1 = Rt^+$. If this process is regulated by a leaky bucket shaper with parameters σ and ρ, the result in Exercise 4.4 says that the resulting process will have envelope $E(t) = \min(Rt^+, (\sigma + \rho t)I_{\{t \ge 0\}})$ (recall the example in Figure 4.6). This envelope is shown as the solid line in Figure 4.16. Suppose the voice coder emits frames; for example, a G.729 coder encodes 10 ms of speech, emitting 10 bytes per frame, with an overall peak rate of 1 byte per ms, or 8 Kbps. If the frame length is L bits, then observe that the coder output process has envelope $E_1 = (L + Rt)I_{\{t \ge 0\}}$. If this source is regulated by a (σ, ρ) LB, the resulting process will have envelope $E(t) = \min((L + Rt)I_{\{t \ge 0\}}, (\sigma + \rho t)I_{\{t \ge 0\}})$.

4.2.5   Network Performance and Design

Let us now examine the usefulness of knowing that an arrival process A has an envelope E. Consider passing the arrival process A through a network element with a lower service curve $S(t)$. Let D be the departure process. Define

$$d_{\max} = \inf \{d: \ E * \delta_d \le S\}$$

That is, $d_{\max}$ is the least amount we should shift the envelope $E(t)$ to the right so that it falls below the curve $S(t)$.

In Figure 4.16 the service curve $S(t) = Ct^+$ is also shown. It is easily checked that for this service curve and for $E(t) = \min(Rt^+, (\sigma + \rho t)I_{\{t \ge 0\}})$,

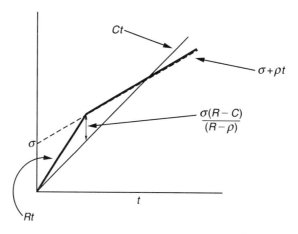

Figure 4.16 The envelope of a peak rate controlled process shaped by a (σ, ρ) regulator. Shown also is the maximum buffer that accumulates if the source is served by a link of rate C.

we get $d_{\max} = \frac{\sigma(R-C)}{C(R-\rho)}$. With $R > \rho$, we see that as C reduces from R to ρ, $d_{\max}$ increases from 0 to $\frac{\sigma}{C}$. Notice that if $C < \rho$, then the delay cannot be bounded; this is to be expected because the output of the LB can have an average rate as large as ρ, and hence a link service rate less than ρ can lead to an infinitely large backlog building up in the link buffer.

In general, if $E(t)$ is an envelope, then for the (lower) service curve $S(t) = Ct^+$, $d_{\max}$ must satisfy, for all $t \geq 0$,

$$Ct + Cd_{\max} > E(t)$$

Think about this as shifting the service curve to the *left* by $d_{\max}$ so that it will become greater than $E(t)$ for all $t \geq 0$. Hence, for the (lower) service curve $S(t) = Ct^+$,

$$d_{\max} = \sup_{t \geq 0} \left(\frac{E(t) - Ct}{C} \right) \tag{4.5}$$

The delay bound for a source $A(t)$ with envelope $E(t)$ serviced by an element with lower service curve $S(t)$ is provided by the following theorem.

Theorem 4.2

$$D \geq A * \delta_{d_{\max}}$$

That is, the delay of the process A through this service element is upper bounded by $d_{\max}$.

Proof: Because S is a lower service curve, we have

$$D \geq A * S$$

Furthermore, by the definition of $d_{\max}$, it follows that

$$A * S \geq A * \left(E * \delta_{d_{\max}}\right)$$

Finally, using the associativity of $*$ and the fact that E is an envelope of A, we obtain

$$D \geq (A * E) * \delta_{d_{\max}}$$

$$\geq A * \delta_{d_{\max}} \qquad \blacksquare$$

It follows that, given an arrival process with envelope $E(t)$ and a multiplexer with service curve $S(t) = Ct^+$, we know that the delay through the multiplexer is bounded above by $\sup_{t \geq 0} \left(\frac{E(t) - Ct}{C}\right)$. Suppose we want to choose the service rate C so that the delay is guaranteed to be less than some required delay bound T. Then we can ask for the minimum value of C such that

$$\sup_{t \geq 0} \left(\frac{E(t) - Ct}{C}\right) \leq T$$

Let $C_{\min}$ be this minimum service rate. Then we have

$$\sup_{t \geq 0} \left(\frac{E(t) - C_{\min} t}{C_{\min}}\right) = T$$

It follows that, for all $t \geq 0$,

$$E(t) - C_{min}t \leq C_{min}T$$

That is,

$$C_{min} \geq \sup_{t \geq 0} \left(\frac{E(t)}{T + t} \right)$$

It can be checked that strict inequality cannot hold in this inequality. For, suppose $\sup_{t \geq 0} \left(\frac{E(t)}{T+t} \right) = C_{min} - \epsilon$, for some $\epsilon > 0$, then we find that $\sup_{t \geq 0} \left(\frac{E(t) - (C_{min} - \epsilon)t}{C_{min} - \epsilon} \right) \leq T$, a contradiction to C_{min} being the minimum required service rate. It thus follows that

$$C_{min} = \sup_{t \geq 0} \left(\frac{E(t)}{T + t} \right) \tag{4.6}$$

We have thus obtained a solution to the following design problem: What is the minimum link capacity required so that an arrival process with envelope $E(t)$ suffers a delay no more than T?

Remark: For the (σ, ρ, R) controlled source (see the thick curve in Figure 4.16) served by a fixed rate server of rate C, the value of d_{max} can be obtained directly by the following intuitive argument. We argue that the buffer will never exceed $b_{max} = \frac{\sigma(R-C)}{(R-\rho)}$. Suppose that at some time t, the buffer does exceed this level (see Figure 4.17). Consider the most recent instant $t - \tau$ at which the buffer was last empty—that is, the start of the current busy period. During the interval $[t - \tau, t]$, the server was continuously busy, and hence the amount of data that has arrived from the source is greater than $Ct + \frac{\sigma(R-C)}{(R-\rho)}$. This contradicts the fact that the source has a (σ, ρ, R) envelope; see Figure 4.16.

Exercise 4.5

Consider a source with peak rate R and packet size L—that is, the packets are spaced by no less than $\frac{L}{R}$ seconds. For example, a voice coder and packetizer emit 200-byte packets (160 bytes of PCM voice plus 40 bytes of RTP, UDP, or IP headers) every 20 ms, yielding $L = 200$ bytes and $R = 10$ KBps. This source is shaped with an LB having parameters (σ, ρ),

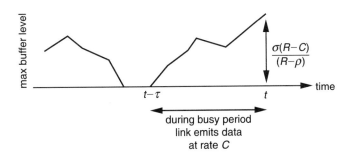

Figure 4.17   **A source with a (σ, ρ, R) envelope is served by a server of rate C. A contradiction occurs if we assume that at some time t, the buffer level exceeds $\frac{\sigma(R-C)}{(R-\rho)}$.**

---

with $\sigma \geq L$ and $\rho \leq R$. The source is served by a latency rate server with a lower service curve that has rate r and latency d (or a tandem of service elements that has this lower service curve). Show that

   a. The source has an envelope $(L + Rt)I_{\{t \geq 0\}}$.

   b. The output of the LB has an envelope $\min((L + Rt)I_{\{t \geq 0\}}, (\sigma + \rho t)I_{\{t \geq 0\}})$.

   c. For $r < R$, the delay experienced by the source is bounded by

$$d + \left(\frac{\sigma - L}{r}\right)\left(\frac{R - r}{R - \rho}\right) + \frac{L}{r}$$

This formula is used in the Internet's IntServ architecture proposals. In this architecture each arriving call (e.g., a packet voice call) declares its LB parameters and an end-to-end delay bound. Notice that if we know d, L, R, σ, and ρ, then to guarantee an end-to-end delay, we can use the formula (in Exercise 4.5) to calculate the required value of r, the rate of the latency rate server. If the network cannot set up a path with the required properties, then the call is blocked. The signaling between the end points and the network is proposed to be carried out by the Resource reSerVation Protocol (RSVP) (see Section 4.5). If the source is served by a tandem of latency rate servers, then the minimum rate of the servers, over the hops of the connection, should be r. One possibility is to choose to make the rate of the latency rate service at every hop the same. It is important to point out that the latency d may also be a function of r; this will become clear in Section 4.3.

4.3 Scheduling

We have seen in Section 4.2.5 that, given an envelope for a source, the service given to the source can be designed to provide some delay guarantee. Scheduling techniques provide a systematic way to design the service provided to individual flows through a network element.

At an output port of a packet switch, the data that are switched into this output port may belong to one of N classes. For example, there could be one class for each packet voice call being carried, and a second one for all the elastic traffic being carried. Alternatively all voice calls could be aggregated into one class, and the elastic traffic into another class.

Each class requires some guaranteed service from this output link. The buffer at this output port is organized into N buffers, one for each class. Data from each class is queued in a FIFO fashion in the corresponding buffer. In this section we discuss how the service offered by the link to these queues can be scheduled so as to achieve specific objectives.

At a packet-multiplexed link, at any point in time there would be several packets in the link buffer, and one packet would be in transmission by the link. Because the receiver, at the other end of the link, must know where each packet starts and ends, in practice a packet is transmitted to completion once its transmission is initiated. In situations where the packet length needs to be reduced for some reason (a high bit error rate on the link or the desire to limit the delay of higher-priority packets), the link (or layer 2) protocol can *fragment* long packets and *reassemble* them at the other end of the link. In this case, it is these shorter packets that are queued at the physical link, and our discussion here applies to these packets. Hence a scheduling policy at a packet-multiplexed link comes into play at packet transmission boundaries. Upon the completion of each packet transmission, the policy determines which packet should be transmitted next, given the current set of queued packets. A popular scheduling technique at a packet-multiplexed link is WFQ (weighted fair queueing), and we start with its study.

4.3.1 Weighted Fair Queueing (WFQ)

To understand WFQ, a scheduling technique for packet-multiplexed links, we must first study a fluid scheduling policy called *generalized processor*

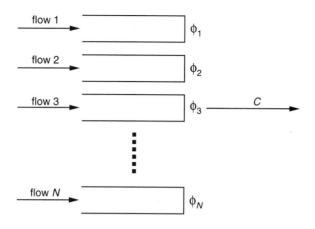

Figure 4.18 *N* **flows, each with its own queue, being serviced at a GPS scheduler with weights** $\phi_1, \phi_2, \phi_3, \ldots, \phi_N$.

sharing (GPS). WFQ is simply the packetized version of GPS and has also been called packet GPS (PGPS).

As shown in Figure 4.18, associated with class $i, 1 \leq i \leq N$, is a weight ϕ_i, which is a positive real number. If these weights are integers, they can be roughly interpreted as follows: While queues i and j are nonempty, the link serves ϕ_i bits from queue i for each ϕ_j bits served from queue j. The server can be viewed as visiting the queues in a *round robin* manner, serving up to ϕ_i bits from queue i in each round. GPS is a fluid version of the bit-wise *weighted round robin* (WRR) scheduling.

Why the name "generalized" processor sharing? What is processor sharing (PS) that GPS is a generalization of? PS is a scheduling model in which if there are $n(t)$ customers (each packet could be a customer) in the queue at time t, then the server applies $\frac{1}{n(t)}$ of its service effort to each customer (see Appendix D for more details). In this service model, various classes of traffic do not have their own queues, and a FIFO order of packet service within a class will not be guaranteed. When packets of different classes queue up in separate queues, then a variation of PS is head-of-the-line PS (HOLPS), in which the server applies an equal share of its service effort to each of the HOL packets in the nonempty queues. GPS is a generalization of HOLPS in which the service effort applied to each of the HOL packets is in proportion to the weights associated with the queues.

Assuming fluid arrivals (i.e., the arriving data can be served in infinitesimally small quantities), we now formally define GPS scheduling. We first establish some notation. Let, for time $t \geq 0$,

$$
Z_j(t) = \begin{cases} 1 & \text{if queue } j \text{ is nonempty} \\ 0 & \text{if queue } j \text{ is empty} \end{cases}
$$

$$
Z(t) = \begin{cases} 1 & \text{if at least one queue is nonempty} \\ 0 & \text{if all queues are empty} \end{cases}
$$

Let the link service rate be C (bits per second, but because we have a fluid model, fractions of bits can also be served). In GPS, the service rate applied to queue j at time u (or the rate at which fluid is drained from queue j) is

$$
\frac{\phi_j \, Z_j(u)}{\sum_{i=1}^{N} \phi_i Z_i(u)} \, C
$$

where we interpret this expression to be zero if $Z(u) = 0$. Thus if there are 10 queues and if at time u queues 3 and 7 are nonempty, then (at time u) the fluid in queue 3 is being served at the rate $\frac{\phi_3}{\phi_3 + \phi_7} C$, and queue 7 is being served at the rate $\frac{\phi_7}{\phi_3 + \phi_7} C$. Hence, the cumulative service provided to queue j over the interval $[0, t]$ is given by

$$
\int_0^t \frac{\phi_j \, Z_j(u)}{\sum_{i=1}^{N} \phi_i Z_i(u)} \, C \, du
$$

The average service rate applied to a queue over the interval $[0, t]$ is the ratio of the cumulative service over the interval to the duration in that interval during which the queue was nonempty. Thus the average service rate applied to queue j over the interval $[0, t]$ is given by

$$
\frac{\int_0^t \frac{\phi_j \, Z_j(u)}{\sum_{i=1}^{N} \phi_i Z_i(u)} \, C \, du}{\int_0^t Z_j(u) \, du}
$$

Again, this is interpreted as 0 if $\int_0^t Z_j(u)du = 0$—that is, queue j is empty in the interval $[0, t]$. Now because $Z_i(u) \leq 1, 1 \leq i \leq N$, the previous expression is

$$\geq \phi_j \, \frac{\int_0^t \frac{Z_j(u)}{\sum_{i=1}^N \phi_i} \, C du}{\int_0^t Z_j(u) du}$$

$$= \frac{\phi_j}{\sum_{i=1}^N \phi_i} \, C$$

We have thus shown that under GPS scheduling, the average service rate provided to queue j over *any* interval $[0, t]$ is at least $\frac{\phi_j}{\sum_{i=1}^N \phi_i} C$. Thus over *any* interval that queue j is continuously nonempty, the service rate applied to queue j is at least $\frac{\phi_j}{\sum_{i=1}^N \phi_i} C$.

Remarks 4.2

From our discussions in Section 4.2.2 it then follows that for class j fluid arrivals the GPS server offers a lower service curve:

$$S_{GPS}(t) = \frac{\phi_j}{\sum_{i=1}^N \phi_i} \, C t^+$$

∎

Virtual Time

To proceed further with the analysis it is useful to define

$$V(t) = \int_0^t \frac{C}{\sum_{i=1}^N \phi_i Z_i(u)} \, Z(u) \, du \qquad (4.7)$$

Notice that $V(t)$ *is nondecreasing with t.* That is, $t_1 \geq t_2 \Rightarrow V(t_1) \geq V(t_2)$, and hence is sometimes called *virtual time*.

A nice way to think about the function $V(t)$ is to view the server as making the rounds of the queues, serving ϕ_i "bits" from queue i if it is nonempty. $V(t)$ can then be roughly interpreted as the server's *round number* at time t. For example, if $N = 2$, $\phi_1 = 10$ bits, and $\phi_2 = 20$ bits, then during a period in which both queues are nonempty, the server makes $\frac{C}{30}$ rounds per second. Note that $\{V(t), t \geq 0\}$ is a stochastic process, because it depends on the sequence of packet arrivals and the sequence of packet lengths.

The function $V(t)$ should be viewed as a global function that applies to the entire multiplexer (i.e., switch output port) and not to any particular subqueue in the multiplexer. Thus we say that $V(t)$ is the virtual time of the system at "wall-clock" time t. It is clear that the function $V(t)$ has the following properties.

- $V(0) = 0$, and $V(t) \geq 0$ for $t \geq 0$

- $V(t)$ is a piecewise linear function whose slope increases whenever a queue becomes empty (the round number increases faster when there is one less queue to serve in a round) and decreases when an empty queue receives some new data.

- The slope of $V(t)$ is 0 whenever the system has no data to serve.

For the GPS system we define, for $1 \leq j \leq N$,

$a_k^{(j)}$: the arrival instant of the kth packet in queue j

$L_k^{(j)}$: the amount of data in the kth packet arriving into queue j

$s_k^{(j)}$: the service initiation instant of the kth arriving packet into queue j

$d_k^{(j)}$: the departure instant of the kth arriving packet into queue j

It can be seen that if queue j is being served by a dedicated, fixed-capacity, work-conserving server with service rate c_j, then

$$d_{k+1}^{(j)} = \max\left\{d_k^{(j)}, a_{k+1}^{(j)}\right\} + \frac{L_{k+1}^{(j)}}{c_j} \qquad (4.8)$$

To see this, note that the first term on the right side of this equation is the time at which packet $k+1$ in stream j starts service (the later of the service completion instant of packet k and the arrival instant of packet $k+1$), and the second term is the service time. (When a packet is being served, it is served at rate c_j, and service is taken to completion; there is no interruption in service once service is started.)

Notice that, using this recursive relation, the departure instant of a packet arriving into a fixed-service-rate queue can be computed at the instant the packet arrives into the queue. Such an equation does not hold for GPS service for the following reason. Within a class, the service order is FCFS, so the first term in Equation 4.8 holds. With GPS service, however, the service rate keeps varying during the HOL packet's service at queue j, and hence the second term in Equation 4.8 no longer holds. On the other hand, if we think in terms of virtual time, then a similar equation can be obtained as follows. Notice that

$$V\left(d_{k+1}^{(j)}\right) = \int_0^{d_{k+1}^{(j)}} \frac{C}{\sum_{i=1}^N \phi_i \, Z_i(u)} \, Z(u) \, du$$

$$= \int_0^{s_{k+1}^{(j)}} \frac{C}{\sum_{i=1}^N \phi_i \, Z_i(u)} \, Z(u) \, du + \int_{s_{k+1}^{(j)}}^{d_{k+1}^{(j)}} \frac{C}{\sum_{i=1}^N \phi_i \, Z_i(u)} \, Z(u) \, du$$

Observe that because $Z_j(u) = 1$ during $\left(s_{k+1}^{(j)}, d_{k+1}^{(j)}\right)$, the increase in $V(t)$ during this interval will be the number of rounds required to serve the packet of length $L_{k+1}^{(j)}$, which is $\frac{L_{k+1}^{(j)}}{\phi_j}$. Formally, we have

$$\int_{s_{k+1}^{(j)}}^{d_{k+1}^{(j)}} \frac{\phi_j \cdot C \, du}{\sum_{i=1}^N \phi_i \, Z_i(u)} = L_{k+1}^{(j)}$$

Hence,

$$\int_{s_{k+1}^{(j)}}^{d_{k+1}^{(j)}} \frac{C}{\sum_{i=1}^N \phi_i \, Z_i(u)} \, du = \frac{L_{k+1}^{(j)}}{\phi_j}$$

Owing to FCFS service, $s_{k+1}^{(j)} = \max\left(d_k^{(j)}, a_{k+1}^{(j)}\right)$. Hence, because $V(t)$ is nondecreasing, we get

$$V\left(d_{k+1}^{(j)}\right) = \max\left(V\left(d_k^{(j)}\right), V\left(a_{k+1}^{(j)}\right)\right) + \frac{L_{k+1}^{(j)}}{\phi_j} \tag{4.9}$$

Viewed in the light of the interpretation of $V(t)$ as the total round number, the recursion in Equation 4.9 parallels the recursion in Equation 4.8. With this recursion, in the fluid context, the GPS scheduler can be implemented as follows. The function $V(t)$ is simulated by using Equation 4.7; at each arrival or departure the slope of the function changes, and the function is linear in between. Upon each packet arrival, the virtual time at this packet's departure instant (the packet's *virtual finish time*) is computed using Equation 4.9; the packet is marked with this value and stored. The marked packets are sorted by their virtual finish times. In this fluid GPS system, when the simulated function $V(t)$ reaches a packet's virtual finish time, then that packet is ejected from the scheduler and departs the system. Note that scaling all the ϕs by the same factor will just result in scaling $V(t)$ and will not change the ordering of the virtual finish times of the packets.

Packetized GPS

In actual networks, packets must be served in their entirety, and hence the fluid model on which GPS is based is not practical. We can, however, use GPS to define scheduling policies that have bounded performance as compared with GPS. Broadly these policies are referred to as packet GPS (PGPS) schedulers. We begin by discussing the most well known among these, weighted fair queueing, which works as follows:

1. The virtual time process $V(t)$ is simulated, as if there were a GPS scheduler operating.

2. When a packet arrives into the scheduler it is marked with its virtual finish time in the GPS scheduler.

3. When a packet is selected for transmission, it is transmitted completely.

4. After completion of service of a packet, the next packet to be transmitted is the one that has the smallest virtual finish time among all the packets in the multiplexer. (Ties can be broken in various ways—for example, by smallest queue index if the tie is between packets of several queues.)

To provide a simple illustration of WFQ, let us consider the situation in Figure 4.19. There are 6 queues with weights 50, 10, 10, 10, 10, and 10. There are 5 packets in queue 1, and 1 each of the other 5 queues. All the packets are of the same length. Thus, when all the queues are nonempty, 5 packets will be served in queue 1 during the time that 1 packet is served in either of the other 5 queues. In GPS the packet services are interleaved, and, in Figure 4.19, the diagram marked "fluid service" shows the way the 10 packets are served. The fifth packet in queue 1 departs at the same time as each of the packets in the other 5 queues. To implement WFQ, we simulate the GPS system, and hence we know the order of departure of

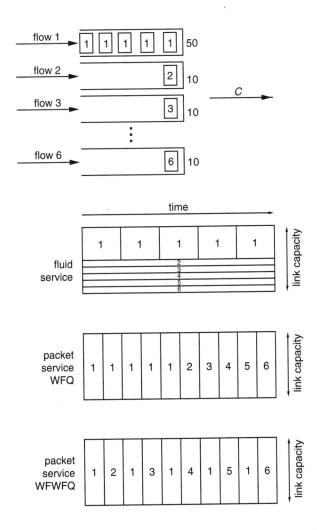

Figure 4.19 An example illustrating how a fluid simulation of GPS is used to determine the service order for PGPS (WFQ). An alternative, WFWFQ (worst case fair WFQ), is also shown. In GPS, the link is simultaneously shared among the various queues because fluid data is assumed; in WFQ and WFWFQ, only one packet from a queue is served at a time, at the full link rate.

the packets. Now the packetized implementation starts by serving the first packet to depart in the GPS system, which is the first packet of queue 1. The next packet to be served is also from queue 1. This goes on until 4 packets from queue 1 depart (see, in Figure 4.19, the diagram labeled "packet service WFQ"). The remaining 6 packets, one from each queue, have the same finish time in the GPS system, and in our example we break ties by smallest queue index.

Some points are worth noting from this example:

- Recall from Section 4.1 that if the link is non-idling, for a given process of arrival instants and packet sizes, for all packet scheduling policies, the busy periods and the packets served in each busy period are invariant. We see this in the preceding example; under GPS and WFQ we have the same busy period. The 10 packets depart in this busy period, but their order of departure is different for the two policies.

- In this example, looking at the way WFQ serves packets, the service seems to be quite unfair, with all the packets of flow 1 departing before any of the packets of the other flows depart. Shown in Figure 4.19 (in the diagram labeled "packet service WFWFQ") is another service schedule that also requires the simulation of GPS. It differs from WFQ in that the rule for selecting the next packet to serve is the following: Select packets that would *already have started service in GPS*, and among all such packets select the ones that have the least finish time, and then break ties if this leaves more than one packet. In GPS, at time 0, the first packet in queue 1 starts service, as do all the packets in the other 5 queues. However, the first packet in queue 1 has the earlier finish time in GPS and hence is served. Taking the packet length to be L, this packet finishes service at $\frac{L}{C}$ (whereas in GPS it would finish service at $\frac{2L}{C}$). At this time the second packet in queue 1 would not have started service in GPS, and hence only the 5 packets in the other 5 queues remain as candidates, and among them the packet from queue 2 is selected. It can now easily be seen that the order of service shown in Figure 4.19 is obtained. This would certainly appear to be a fairer policy than WFQ. WFWFQ is the abbrevation for worst case fair weighted fair queueing and is discussed later in this section.

- Comparing the departure order of the packets under GPS and WFQ, we notice that, in this example, the order of packet departures is the same under the two policies (where we order tied departures in GPS by their queue indices). The reason for this can be seen from the simple Lemma 4.1.

Lemma 4.1

If two packets—say, i_1 and i_2—are present in the multiplexer at a scheduling instant, then packet i_1 is selected to be served in WFQ (and hence departs earlier than packet i_2) if and only if the same order of departure exists in GPS.

Proof: Because packets i_1 and i_2 are present together in WFQ, they must belong to the same busy period in both systems (GPS and WFQ). Let d_{i_1} and d_{i_2} be the departure instants of the two packets under GPS. By the WFQ scheduling rule, packet i_1 is selected if and only if $V(d_{i_1}) \le V(d_{i_2})$ (and, in case of equality, i_1 satisfies the tie-breaking criterion), which is true if and only if $d_{i_1} \le d_{i_2}$. Here we have used the fact that d_{i_1} and d_{i_2} are in the same busy period in GPS, and within a busy period the function $V(t)$ is strictly increasing. Note here that if two packets are present in the same busy period and if in GPS they have the same virtual finish times, then they must have the same departure times in GPS. ■

We can explain the observation that the order of departures in the example in Figure 4.19 is the same under GPS and WFQ by noting that all the packets were present together at the start of the busy period.

> **Exercise 4.6**
> Provide an example of packet arrivals to a number of queues so that the GPS service order is different from the WFQ order.

Thus, in general, packets can depart later in WFQ than they depart in GPS. It is possible to bound how much later they can be, and this will then lead to a latency-rate lower service curve for WFQ. We now proceed to obtain this bound.

Let $L_{\max}$ denote a bound on the lengths of the packets being handled by the multiplexer. Consider some indexing, $k = 1, 2, \ldots$, of the packets passing through the multiplexer. Let d_k denote the GPS departure instant of the kth packet, and define $\hat{d}_k$ as the WFQ departure instant of the same packet. L_k will denote the length of packet k. The following theorem relates the delay performance under WFQ to that under GPS.

Theorem 4.3

For all packets

$$\hat{d}_k \le d_k + \frac{L_{\max}}{C}$$

where $L_{\max}$ is the maximum packet length, and C is the full rate of the link.

Proof: Both GPS and WFQ are work-conserving (non-idling) disciplines. Hence it suffices to consider a busy period, which will, of course, be the same under both. During the busy period the same packets will be served by both disciplines except that the departure order of the packets may be different. Index the packets by their order of departure under WFQ; packet p_1 is the first to depart under WFQ in this busy period, and packet p_k is the kth to depart. If the busy period starts at t_0, because WFQ serves packets to completion, notice that $\hat{d}_1 = t_0 + \frac{L_1}{C}$, $\hat{d}_2 = t_0 + \frac{L_1 + L_2}{C}, \ldots, \hat{d}_k = t_0 + \frac{\sum_{i=1}^{i=k} L_i}{C}$ (see Figure 4.20). Without loss of generality, in the following discussion we take $t_0 = 0$.

Now consider d_k, the departure instant of packet p_k under GPS. There are two cases.

Case 1: Under GPS none of the packets $p_1, p_2, \ldots, p_{k-1}$ departs (strictly) after d_k. In this case it is clear that

$$d_k \geq \frac{\sum_{i=1}^{i=k} L_i}{C} = \hat{d}_k$$

and we are finished.

Case 2: There is an index $m, 1 \leq m \leq k-1$, such that $d_m > d_k$, and m is the largest such index. Thus $d_{m+1} \leq d_k, d_{m+2} \leq d_k, \ldots, d_{k-1} \leq d_k$. Now consider $\hat{s}_m$, the instant at which the service of packet p_m began under WFQ. By Lemma 4.1, in the WFQ system packets $p_{m+1}, \ldots, p_k$ cannot be present at time $\hat{s}_m$,

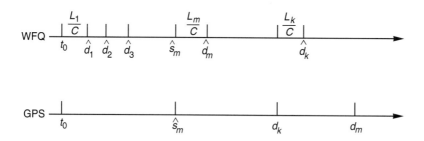

Figure 4.20   Scheduling events for GPS and WFQ during a busy period.

for if they were, one of them would have been chosen for service instead of packet p_m, because in GPS p_m departs after packets $p_{m+1}, \ldots, p_k$. Hence $p_{m+1}, \ldots, p_k$ arrived after $\hat{s}_m$, and in GPS they all depart before (or at) d_k. This key observation leads to the following (see Figure 4.20):

$$d_k \geq \hat{s}_m + \frac{\sum_{i=m+1}^{i=k} L_i}{C}$$

$$= \hat{d}_k - \frac{L_m}{C}$$

from which the result follows, because $L_m \leq L_{\max}$. ∎

Let $U^{(j)}(t)$ be the amount of work done on queue j by the GPS scheduler in the interval $[0, t]$, and let $\hat{U}^{(j)}(t)$ be the corresponding quantity for WFQ. The following result states that the work done on queue j by the WFQ scheduler can be behind that done by the GPS scheduler by at most $L_{\max}$, the maximum packet size in the traffic being carried by the multiplexer.

Theorem 4.4

$$U^{(j)}(t) - \hat{U}^{(j)}(t) \leq L_{\max}$$

Proof: See the Appendix of this chapter. ∎

Remarks 4.3

Consider a WFQ scheduler, and suppose that the weights $\phi_1, \phi_2, \ldots, \phi_N$ are such that the minimum service rate applied to queue j is $c_j = \frac{\phi_j}{\sum_{i=1}^{N} \phi_i} C$. It follows from Theorem 4.3 and the observation in Remarks 4.2 that a minimum service curve at queue j is $S_{WFQ}(t) = c_j (t - \frac{L_{\max}}{C})^+$. Thus the service at each queue in a WFQ scheduler can be modeled as a latency rate server, with latency equal to $\frac{L_{\max}}{C}$ and rate equal to the minimum service rate as determined by the weights.

For a peak-rate-controlled and leaky-bucket-controlled source having parameters (σ_j, ρ_j, R_j), the delay through a WFQ scheduler will be bounded by $\frac{\sigma_j(R_j - c_j)}{c_j(R_j - \rho_j)} + \frac{L_{\max}}{C}$. (See Figure 4.21; compare with Figure 4.16, and notice that the service curve has simply been shifted to the right by $\frac{L_{\max}}{C}$.) ∎

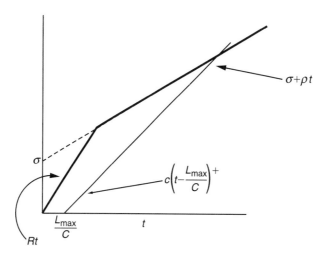

Figure 4.21   The envelope of a peak-rate-controlled process shaped by a (σ, ρ) regulator. Also shown is the lower service curve given to this source at a WFQ server with minimum rate c, total link rate C, and maximum packet length L_{max}.

The output link of a router handles a variety of traffic, including TCP-controlled traffic (Web, FTP, and e-mail) and packet voice traffic. The maximum packet size in the multiplexer is typically the maximum packet size of a TCP file or Web transfer session. A common value is $L_{max} = 1500B$. The following are some values of the additional delays owing to packetized service in WFQ.

$C = 384$ Kbps; $\frac{L_{max}}{C} = 31.25$ ms

$C = 2$ Mbps; $\frac{L_{max}}{C} = 6$ ms

$C = 34$ Mbps; $\frac{L_{max}}{C} = 350$ μs

$C = 150$ Mbps; $\frac{L_{max}}{C} = 80$ μs

Notice that at lower link speeds the additional delay can be a significant fraction of the desired end-to-end packet voice delay bound of about 200 ms.

Worst Case Fair WFQ (WFWFQ)

Theorem 4.3 states that under WFQ, packets will depart no later than $\frac{L_{max}}{C}$ after their departure times in GPS, but it says nothing about how much earlier they can depart. In the example of Figure 4.19 WFQ causes all the packets from queue 1 to depart before any of the packets from the other queues. In that discussion we also show the result from WFWFQ, which results in nicely interleaving the 5 packets from queue 1 and the packets from each of the other queues. WFWFQ adds the following rule to the WFQ policy: After completing the service of a packet, select packets that would *already have started service in GPS*, and among all such packets select the ones that have the least finish time, and then break ties if this leaves more than one packet. It has been shown that Theorem 4.3 and Theorem 4.4 apply to WFWFQ as well, but in addition the following result also holds. For each queue j,

$$U^{(j)}_{WFWFQ}(t) - U^{(j)}_{GPS}(t) \leq (1 - \frac{c_j}{C})L^{(j)}_{\max}$$

where c_j is the service rate allocated to queue j. Hence at each queue the WFWFQ scheduler can get ahead of the GPS scheduler by only a bounded amount of work done.

4.4 Application to a Packet Voice Example

Before we continue with our discussion, we now apply much of what we have discuss in Sections 4.2 and 4.3 to the example of a packet voice session that we introduce in Section 3.2. Figure 4.22 shows a model for the transport of this session up to the LAN–WAN router R_A.

The coder emits bytes at the peak rate of R bytes per second. The shaper has a source buffer. The shaping is typically done in the source system and involves the handling of low-speed data. In the source there is plenty of low-speed memory, and hence memory for the source buffer is not an expensive resource. Furthermore, the source system will not drop its own data. Hence, essentially we can take the shaper buffer to be of infinite size. The greater the buffer occupancy, however, the greater the delay that the bytes from the coder experience in the shaper. Let X_1 denote the amount of data in the shaper source buffer. If a byte from the coder arrives to find X_1 bytes in the buffer (including itself), then the delay of this byte is bounded by $\frac{X_1}{\rho}$.

The shaper is followed by a packetizer that emits packets whose maximum length is $L_{\max}$. Finally, Figure 4.22 shows that the voice packets are buffered in a queue of a WFQ scheduler at the output port of the router. The total service rate of the link outgoing from the router is C bps. Let r bps be the service rate allocated by the WFQ scheduler to the packet voice session; the weight ϕ is chosen such that the minimum service rate applied to the packet voice session is r bps.

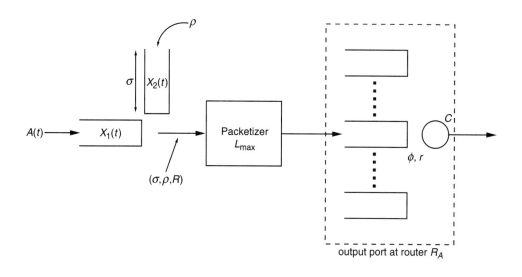

Figure 4.22 The output of the voice coder is shaped, packetized, and then multiplexed onto the outgoing wide area link from router R_A.

Let $V_{\max}$ be the maximum packet size (over all flows) in the output port multiplexer at the router. It is then easily seen from the foregoing discussions that the delay is bounded by

$$D := \frac{X_1}{\rho} + \frac{\sigma(R-r)}{r(R-\rho)} + \frac{L_{\max}}{\rho} + \frac{V_{\max}}{C}$$

Note that the last three terms on the right side of this equation are deterministic, whereas the first term is random. Define

$$T' := \frac{\sigma(R-r)}{r(R-\rho)} + \frac{L_{\max}}{\rho} + \frac{V_{\max}}{C}$$

Now suppose that the session's QoS requirement is that

$$\Pr(D > T) < \epsilon$$

where ϵ is a small number, such as 0.05. Hence, substituting the expression for D, we get

$$\mathrm{Pr}(D > T) = \mathrm{Pr}\big(X_1 > (T - T')\rho\big)$$

which yields the performance requirement

$$\mathrm{Pr}\big(X_1 > (T - T')\rho\big) < \epsilon$$

Remarks 4.4

Why is this QoS requirement meaningful or useful? Let us discuss this at some length. Consider a buffer into which bytes arrive at rate $a(t)$ at time t. The time average arrival rate is $\bar{r}$, and the peak arrival rate is R (e.g., 64 Kbps for a PCM voice source). For a discussion of the notion of time averages, see Sections 5.4 and 5.5. What we would really like to ensure is that the fraction of bytes from the encoder that experience a delay exceeding T (e.g., $T = 200$ ms) should be less than a small number—say, $\delta, 0 < \delta < 1$. What we are asking for is that the *fraction of time* the buffer exceeds some level is bounded by a small number ϵ.

The bytes in the buffer are served in FIFO order. We want to ensure that the fraction of bytes that arrive to find more than B bytes in the buffer is less than δ. If the rate at which the buffer is drained is lower bounded, then this requirement bounds the delay of bytes in the buffer, with a certain probability (note that in the case of the LB in Figure 4.22, the rate at which the source buffer is drained is ρ). Let $X(t)$ be the occupancy of the buffer at the instant t. The QoS requirement then translates to the following; with probability 1, we require that

$$\lim_{t \to \infty} \frac{\int_0^t I_{\{X(u)>B\}}\, a(u)\, du}{\int_0^t a(u)\, du} < \delta$$

In the fraction on the left side of this expression, the numerator is the number of bytes in $[0, t]$ that arrive to find more than B bytes in the buffer (bytes are taken as being fluid). The denominator is the number of bytes that arrive in $[0, t]$. Thus the left side is the long-run fraction of bytes that arrive to find more than B bytes in the buffer. This is a random variable, and we require that this be less than δ with probability 1. If the system has a stationary limit, let X denote the stationary queue length. By the definition of arrival rate, with probability 1, we have

$$\lim_{t \to \infty} \frac{1}{t} \int_0^t a(u) \, du = \bar{r}$$

Furthermore,

$$\lim_{t \to \infty} \frac{1}{t} \int_0^t I_{\{X(u)>B\}} \, a(u) \, du \leq \lim_{t \to \infty} \frac{1}{t} \int_0^t I_{\{X(u)>B\}} R \, du$$

$$= R \, \Pr(X > B)$$

where the last equality is with probability 1. It follows that, with probability 1,

$$\lim_{t \to \infty} \frac{\frac{1}{t} \int_0^t I_{\{X(u)>B\}} \, a(u) \, du}{\frac{1}{t} \int_0^t a(u) \, du} \leq \frac{R}{\bar{r}} \, \Pr(X > B)$$

Hence to ensure $\lim_{t \to \infty} \frac{\int_0^t I_{\{X(u)>B\}} \, a(u) \, du}{\int_0^t a(u) \, du} < \delta$, it is sufficient to require that

$$\Pr(X > B) < \epsilon$$

where $\epsilon = \frac{\bar{r}}{R} \delta$. Note that this is a conservative QoS measure but is computable via good approximations and hence more useful than the exact measure. A considerable part of the effort in Chapter 5 is to show how such "tail" probabilities can be approximated. ∎

Exercise 4.7
Figure 4.22 also shows the contents of the leaky bucket as the process $X_2(t)$. Consider the queueing model shown in Figure 4.23. Let X denote the stationary random variable for the content of the queue in Figure 4.23. Let X_1 denote the stationary random variable for the process $X_1(t)$. Then show that, for some given buffer level B_s, $\Pr(X_1 > B_s) = \Pr(X > B_s + \sigma)$.

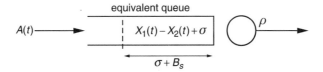

Figure 4.23 A single-server queue with service rate ρ and infinite buffer, which is equivalent to the (σ, ρ) leaky bucket regulator for the computation of the probability of exceeding a level in the leaky bucket regulator source buffer.

From this exercise it follows that to satisfy the QoS $\Pr(D > T) < \epsilon$ it suffices to satisfy

$$\Pr\big(X > (T - T')\rho + \sigma\big) < \epsilon$$

where X is the stationary buffer occupancy of the simple system in Figure 4.23.

Thus we find from this example that, because we have only a statistical characterization of the output of the coder, we finally do need to solve a stochastic queueing model. The approach we have outlined here shows how deterministic network calculus can be combined with a stochastic model to obtain an overall analysis and design approach.

We turn our attention to stochastic models in Chapter 5.

4.5 Connection Set Up: The RSVP Approach

We now turn to the problem depicted in Figure 4.24. An LB-shaped and peak-rate-limited source having parameters (σ, ρ, R) and packetization size L must be carried across K hops of a packet-switched network. It is assumed that the path the connection takes is given by a routing protocol, and that there is no interaction between the routing and the process of guaranteeing a service to the connection. An end-to-end network delay bound of T is required. The network uses WFQ at each hop to provide service guarantees. We assume that at each hop the same service rate c is reserved for the connection. The objective is to determine c.

Before we proceed, note that such an approach may not lead to a successful connection setup, because the path assigned by the routing protocol may not have the required resources, whereas other available paths may have the resources. However, the interaction between routing and the provision of guaranteed services is a topic in the area of QoS routing, which is dealt with in Part III of the book.

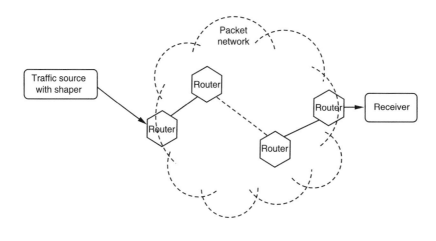

Figure 4.24   A leaky-bucket-shaped source to be transported over several router hops.

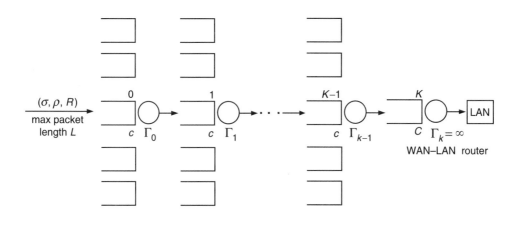

Figure 4.25   A leaky-bucket-shaped source being transported over a tandem of links with WFQ scheduling at each hop.

Figure 4.25 depicts the connection being carried over K WFQ hops. The first queue, indexed 0, is at the LAN–WAN interface, and we denote this WAN link speed by Γ_0. There follow $K - 1$ WAN links, with speeds Γ_j, $1 \leq j \leq K - 1$. Each of these uses WFQ scheduling at its respective queue. The last queue is at the WAN–LAN interface. We assume that the LAN speed is much greater than the WAN speeds, and we denote the service rate at this queue by $\Gamma_K = \infty$. Let the maximum packet lengths at each of the WAN links be denoted by $L_{\max j}$.

We proceed by representing the service received by the connection at each hop by a lower service curve. Recalling Remarks 4.3, we see that a lower service curve at queue 0 is given by $c(t - \frac{L_{\max 0}}{\Gamma_0})^+$. Because the next hop can work only with complete packets of the connection, however, an additional latency of $\frac{L}{c}$ must be added, yielding the latency-rate lower service curve $c(t - (\frac{L_{\max 0}}{\Gamma_0} + \frac{L}{c}))^+$. Proceeding in this manner, the lower service curves at each of the K hops can be written down as $c(t - (\frac{L_{\max j}}{\Gamma_j} + \frac{L}{c}))^+$, $0 \le j \le K - 1$.

It then follows (see Example 4.2) that an end-to-end lower service curve for the connection is given by $c(t - (\sum_{j=0}^{K-1} \frac{L_{\max j}}{\Gamma_j} + K\frac{L}{c}))^+$. Then using the result derived in Exercise 4.5, we obtain the following queueing delay bound for the connection, for the chosen value of c.

$$\left(\frac{\sigma - L}{c}\right)\left(\frac{R - c}{R - \rho}\right) + \frac{L}{c} + \sum_{j=0}^{K-1} \frac{L_{\max j}}{\Gamma_j} + K\frac{L}{c}$$

Let the propagation delay at the jth WAN link, $0 \le j \le K - 1$, be denoted by T_{propj}. Then for an end-to-end delay bound of T, we need the following inequality to hold.

$$\left(\frac{\sigma - L}{c}\right)\left(\frac{R - c}{R - \rho}\right) + \frac{L}{c} + \sum_{j=0}^{K-1} \left(\frac{L_{\max j}}{\Gamma_j} + T_{propj}\right) + K\frac{L}{c} < T$$

If all the terms in the left side of this inequality are known, then it is possible to determine the smallest value of c (if any) to satisfy the inequality, and this much bandwidth can be reserved for the connection at every hop. However, a priori, we do not know the value of K, the propagation delays T_{propj}, the maximum packet lengths at each hop, $L_{\max j}$, and the link speeds, Γ_j.

In the Integrated Services framework for QoS in the Internet, the protocol RSVP (Resource reSerVation Protocol) learns the required parameters of the path and determines the value of c as depicted in Figure 4.26. The source initiates the connection by sending a PATH message bearing its LB parameters (σ, ρ, R) and its packet length, L. As this PATH message traverses the network, the parameters

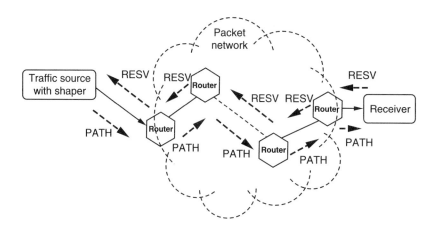

Figure 4.26   Guaranteed bandwidth connection setup with the RSVP protocol.

$\sum_{j=0}^{K-1}\left(\frac{L_{\max j}}{\Gamma_j}+T_{propj}\right)$ and K are accumulated in the PATH message, and a connection state is established in the routers. At each hop, this state contains the previous hop from which the PATH message arrived. On reaching the receiver, all the information to carry out the bandwidth calculation is available (the receiver would know the delay bound). The value of c is calculated and sent back in a RESV (RESerVe) message. This message can retrace the path of the PATH message by virtue of the previous hop entries in the connection state at each hop. (Note that normal IP routing from the receiver to the source may not have sent the RESV message on the same reverse path.) If at each hop the bandwidth c is available, the reservation succeeds; otherwise, a reservation failure packet is generated.

4.6   Scheduling (Continued)

Returning to the implementation of WFQ, recall that we need to compute the virtual finish time for each arriving packet via Equation 4.9. This computation requires the virtual time at the arrival instant. Recall that virtual time is computed using Equation 4.7. $V(t)$ is a nondecreasing, piecewise linear function whose slope changes each time the number of busy queues changes. Thus $V(t)$ can be computed with the knowledge of the instants at which the number of nonempty queues changes, and the indices of the queues that are nonempty during the intervals between successive such instants. If the number of queues is large, then the computational effort to keep an updated value of $V(t)$ can become significant. One way of viewing this complexity is to observe that if the scheduler is handling N queues, then between two packet arrivals there can be up to N changes in the slope of $V(t)$ (by virtue of up to N queues changing from backlogged to empty). We say that the computational complexity of WFQ is $O(N)$. Problem 4.10 develops the calculation of virtual time only at arrival instants.

In the packet voice example provided earlier, we assumed that the single voice stream had its own queue in the WFQ scheduler. If there is one queue for every connection passing through the scheduler, then the scheduler could be handling thousands of individual connections, and clearly the computational complexity will lead to a serious performance bottleneck. Note that in Section 4.4 when computing a delay bound we did not account for any processing overheads in the scheduler; it was assumed that it took zero time to perform the tasks associated with identifying which queue an arriving packet belonged to[4] and computing its virtual finish time. Because it is extremely difficult to incorporate such overheads in the delay analysis, the effort in router design is to make these activities so fast (or so simple) that the packet delay caused by them is approximately zero as compared with the other delays involved (propagation delay, and queueing delay behind the transmission link). There are two ways in which the complexity of packet scheduling can be reduced:

- By reducing the number of queues. Several connections can be aggregated, with each aggregate being assigned to one queue. Analysis can be done to account for the interactions between the connections assigned to a queue so that the delay requirements for the individual flows are still met. For example, Figure 4.27 shows several sources shaped by the same LB parameters being carried over a path in a packet network; at each hop WFQ is used, but all these sources are mapped into the same queue at each hop. Such a situation could arise from two packet voice PBXs at two offices of a company that are connected by a packet network. The PBXs use the same LB parameters for all the voice calls, and when there are n calls active, the network reserves the bandwidth nc, as shown in the figure. An envelope for the superposition of the LB-controlled sources can be obtained, and then a calculation just like the one shown in Section 4.5 will yield a value of nc to be reserved. Such an approach should be efficient for connections having similar traffic characteristics and delay constraints.

---

[4]This is also called *packet classification*. Algorithms and implementations for fast packet classification has been an important research area. We present this topic in Chapter 12.

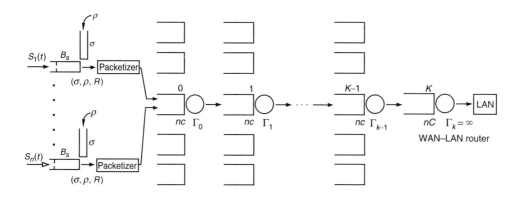

Figure 4.27   Several leaky-bucket-shaped sources being transported over a tandem of links with WFQ scheduling at each hop.

- By simplifying the scheduling algorithm itself. This can be done by using an approximation of the virtual time $V(t)$ such that the resulting scheduling decisions still have good properties. We briefly present an approximation without its analysis. Recall that WFQ requires the maintenance of the virtual time function so that it can compute the virtual finish time of any arrival via Equation 4.9, which we reproduce here:

$$V\left(d_{k+1}^{(j)}\right) = \max\left\{V\left(d_k^{(j)}\right),\ V\left(a_{k+1}^{(j)}\right)\right\} + \frac{L_{k+1}^{(j)}}{\phi_j}$$

This need to continuously keep $V(t)$ updated is the reason for the high computational complexity of WFQ. Note that WFQ can be viewed as assigning to the $(k+1)$th arrival in queue j the finish tag $F_{k+1}^{(j)}$ computed as

$$F_0^{(j)} = 0$$

$$F_{k+1}^{(j)} = \max\left\{F_k^{(j)},\ V\left(a_{k+1}^{(j)}\right)\right\} + \frac{L_{k+1}^{(j)}}{\phi_j}$$

WFQ serves the packets in order of their finish tags. Thus we need the virtual time at arrival instants in order to compute the finish numbers. *Self-clocked fair queueing* (SCFQ) approximates $V\left(a_{k+1}^{(j)}\right)$ by taking the following approach. At the time the kth packet in queue j arrives, there would be some packet k' in service from queue j'. This packet in service would have been given a finish tag $F_{k'}^{(j')}$ at the time of its own arrival. SCFQ takes the virtual time for any arrival during the service of packet k' from queue j' as $F_{k'}^{(j')}$—that is, the finish tag of the packet in service at the arrival instant. Using this approximation of the virtual time, each arrival is assigned a finish tag and packets are served in order of their finish tags. It can be shown that the delay bound is now looser than that for WFQ, and, unlike WFQ, the delay bound depends on the sum of the packet lengths of each of the other flows in the multiplexer, and hence on the number of such flows.

4.7 Summary

In this chapter we introduce some general notations that are used repeatedly in the context of traffic flows through packet networks. Then we develop a deterministic network calculus for packet networks. This calculus takes as inputs deterministic bounds on the incoming traffic and a characterization of network elements via service curves. As outputs, this calculus produces characterizations of departure processes and provides end-to-end delay bounds. A key input-output relationship is provided by the (min, +) convolution (Definition 4.1), much like the (+, ×) convolution of linear systems theory. We learned that given traffic processes can be made to conform to deterministic bounds (or envelopes) by passing them through regulators. The most popular traffic regulator is the leaky bucket, which we analyze in detail. Scheduling is the mechanism by which desired service curves are obtained for the various traffic flows sharing a link. We discuss weighted fair queueing (WFQ) in detail and prove a performance bound that relates the packet delay in WFQ to that in generalized processor sharing (GPS) (Theorem 4.3). Finally, we show how some of the calculus developed in this chapter is proposed to be used in the IntServ proposals for the provision of QoS in the Internet.

4.8 Notes on the Literature

The foundations of deterministic network calculus were laid down in the seminal papers [70] and [71] by Rene Cruz. Subsequently several researchers observed that Cruz's calculus could be formalized in a manner similar to the input-output theory of linear systems, and this gave rise to (min, +) systems theory, which is the approach we have presented in this chapter. Our treatment is along the lines of the development in the compact but comprehensive paper [5] by Agrawal et al. Two recent monographs ([191] by Le Boudec and [53] by Chang) provide an extensive and deeper treatment of (min, +) network calculus and its applications.

The WFQ scheduling algorithm was formalized via the virtual time technique and studied in the paper [74] by Demers, Keshav, and Shenker. The delay bound for WFQ in relation to GPS was proved by Parekh and Gallager in [233]; the proofs of Theorems 4.3 and 4.4 are from this paper. The WFWFQ algorithm was proposed and analyzed by Bennett and Zhang in [28]. Motivated by the implementation complexity of scheduling algorithms (see also [290]) that require the simulation of GPS (such as WFQ and WFWFQ), several researchers have proposed alternatives that basically hinge on some approximation to the virtual time function: self-clocked fair queueing was proposed by Golestani in [123], start time fair queueing was proposed and analyzed by Goyal et al. in [126]; in [275], Stiliadis and Varma used their idea of rate proportional servers [274] to develop another class of efficient fair scheduling algorithms. In [115] Georgiadis et al. show that more efficient network provisioning can be done if sources are reshaped inside the network, that is—between successive hops.

A comprehensive survey of scheduling algorithms and the deterministic calculus-based approach to providing delay guarantees is provided in [308] and [114].

Appendix

Proof of Equation 4.3: Clearly, for all s, $s \leq t$

$$X(t) \geq (A(t) - A(s) - C(t - s))$$

To understand this inequality, note that $A(t) - A(s)$ is the amount of data that arrives in the interval $[s, t]$, and $C \cdot (t - s)$ is the maximum amount of data that can depart in the same interval. The inequality occurs because there may already have been a positive backlog at instant s, and also the server could be idle for some time in $[s, t]$, and hence not all of the possible service $C \cdot (t - s)$ need have been utilized. Hence

$$X(t) \geq \sup_{s \leq t}(A(t) - A(s) - C \cdot (t - s))$$

Because $A(t) = 0$ for $t < 0$, it is clear that

$$X(t) \geq \sup_{0_- \leq s \leq t} (A(t) - A(s) - C \cdot (t - s))$$

because making s negative does not change $A(t) - A(s)$ but increases $C \cdot (t - s)$. We must take the lower limit of s in the sup as 0_- to allow for a jump in the arrival process at time 0. Now let us define

$$v = \sup \{0_- \leq s \leq t \ : \ X(s) = 0\}$$

Notice that if a packet arrival occurs at $t = 0$ and if the buffer is nonempty over the interval $[0, t]$, then the preceding definition of v will yield $v = 0$. In general, we have $X(v_-) = 0$, and the queue is nonempty over $[v, t]$, and hence over this interval the entire service effort of the link is utilized. Thus we have

$$X(t) = (A(t) - A(v_-) - C \cdot (t - v))$$

Observe that if $v = t$, then $X(t) = A(t) - A(t_-)$; this means that the queue was empty until t_-, and at t it is equal to the size of the packet (if any) that arrived at t. This situation also applies to the case in which we have fluid arrivals that over $[0, t]$ had a rate less than C so that the queue never accumulated; in this case $X(t) = 0$, because with fluid arrivals $A(t) = A(t_-)$. Now

$$(A(t) - A(v_-) - C \cdot (t - v)) \leq \sup_{0_- \leq s \leq t} (A(t) - A(s) - C \cdot (t - s))$$

Hence, it follows that

$$X(t) \leq \sup_{0_- \leq s \leq t} (A(t) - A(s) - C \cdot (t - s))$$

Thus

$$X(t) = \sup_{0_- \le s \le t} (A(t) - A(s) - C \cdot (t - s))$$

This completes the proof.

Proof of Equation 4.4: For all s, $0 \le s \le t$, observe that

$$D(t) = D(s) + (D(t) - D(s)) \le A(s) + C \cdot (t - s)$$

This inequality follows because the total amount of departure in $[0, s]$ can be no more than the total amount that arrived in $(0, s]$; furthermore, the total amount of data that departs in $(s, t]$ can be no more than the maximum amount of service that can be applied during $(s, t]$. The latter is just $C \cdot (t - s)$. Because the preceding inequality is true for all $s, 0_- \le s \le t$, it follows that

$$D(t) \le \inf_{0_- \le s \le t} (A(s) + C \cdot (t - s))$$

Now consider the instant $v = \sup\{s : 0_- \le s \le t, \ X(s) = 0\}$. Notice that

$$D(t) = A(v_-) + C \cdot (t - v)$$

$$\ge \inf_{0_- \le s \le t} (A(s) + C \cdot (t - s))$$

That is,

$$D(t) = \inf_{0_- \le s \le t} (A(s) + C \cdot (t - s))$$

Proof of Theorem 4.4: For packets passing through queue j, let $d_k^{(j)}$ denote the packet departure instants and $s_k^{(j)}$ the service initiation instants under GPS, and let $\hat{d}_k^{(j)}$ and $\hat{s}_k^{(j)}$ denote the corresponding quantities under WFQ. Within queue j

the service by both the schedulers is FCFS, and hence the amount of work done in both systems until the kth departure from queue j is the same:

$$U^{(j)}(d_k^{(j)}) = \hat{U}^{(j)}(\hat{d}_k^{(j)})$$

Obviously, because WFQ dedicates the link to a packet during its service, $\hat{U}^{(j)}(\hat{d}_k^{(j)}) = \hat{U}^{(j)}(\hat{s}_k^{(j)}) + L_k^{(j)}$, where $L_k^{(j)}$ is the length of the kth packet served at queue j. By Theorem 4.3 we have

$$d_k^{(j)} \geq \hat{d}_k^{(j)} - \frac{L_{\max}}{C}$$

where C is the total link capacity. Hence

$$\hat{s}_k^{(j)} - d_k^{(j)} \leq \hat{s}_k^{(j)} - \hat{d}_k^{(j)} + \frac{L_{\max}}{C}$$

or, noting that $\hat{d}_k^{(j)} - \hat{s}_k^{(j)} = \frac{L_k^{(j)}}{C}$,

$$\hat{s}_k^{(j)} \leq d_k^{(j)} + \frac{L_{\max} - L_k^{(j)}}{C}$$

It follows that, because $D(t)$ is nondecreasing and because the maximum service rate applied under GPS to queue j is C,

$$U^{(j)}(\hat{s}_k^{(j)}) \leq U^{(j)}(d_k^{(j)}) + L_{\max} - L_k^{(j)}$$

Hence, using $\hat{U}^{(j)}(\hat{d}_k^{(j)}) = \hat{U}^{(j)}(\hat{s}_k^{(j)}) + L_k^{(j)}$, we obtain

$$U^{(j)}(\hat{s}_k^{(j)}) - \hat{U}^{(j)}(\hat{s}_k^{(j)}) \leq U^{(j)}(d_k^{(j)}) + L_{\max} - L_k^{(j)} - (\hat{U}^{(j)}(\hat{d}_k^{(j)}) - L_k^{(j)})$$

But $U^{(j)}(d_k^{(j)}) = \hat{U}^{(j)}(\hat{d}_k^{(j)})$, and we have shown that at all service initiation instants in the WFQ system, $U^{(j)}(\hat{s}_k^{(j)}) - \hat{U}^{(j)}(\hat{s}_k^{(j)}) \leq L_{\max}$. This suffices to establish the result because when WFQ is serving packets at queue j, it serves at rate C, and the difference $U^{(j)}(t) - \hat{U}^{(j)}(t)$ can only decrease in these periods; and when WFQ is not serving queue j, then $\hat{U}^{(j)}(t)$ is constant, and the difference $U^{(j)}(t) - \hat{U}^{(j)}(t)$ only increases until the next service initiation instant in WFQ.

Problems

4.1   A (σ, ρ) regulated source is served at a network link with transmission rate $C > \rho$. Show, using Reich's equation, that the queue length in the link at any time t is bounded by σ.

4.2   Consider a (deterministic) $(b+\sigma, \rho)$ fluid source applied to a leaky bucket controller with buffer size b, bucket size σ, and token addition rate ρ.

   a. Show that the fluid source suffers no loss in this controller.

   b. Show that the output is (σ, ρ) controlled.

4.3   Consider a network element with service curve S and an input process A with envelope E. Show that E is also an envelope for the departure process D.

4.4

   a. An arrival process A has envelope E_1 and is passed through a regulator with causal subadditive envelope E_2. Show that $D = A * E_2$ has envelope $E_1 * E_2$.

   b. If E_1 is also causal and subadditive, then show that $E_1 * E_2$ is causal and subadditive.

4.5   Obtain the overall envelope of two LBs in tandem with parameters (σ_1, ρ_1) and (σ_2, ρ_2), as depicted in Figure 4.28.

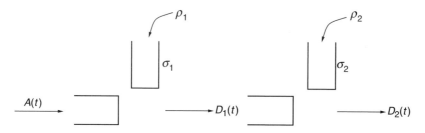

Figure 4.28   Leaky buckets in tandem for Problem 4.5.

4.6

a. An arrival process A with envelope E_1 is passed through a leaky bucket regulator with envelope E_2. Show that the resulting output has envelope $E_1 * E_2$.

b. Use (a) to show that if A has peak rate R, then the output of a (σ, ρ) shaped A is (σ, ρ, R) controlled.

4.7 A source A is put through an LB regulator with parameters (σ, ρ), and the output is carried by a link of rate R. Show that the departure process on the link has envelope $E(t) = \min(Rt^+, (\sigma + \rho t)I_{\{t \geq 0\}})$.

4.8

a. What should be the parameters of a bufferless leaky bucket so that the superposition of (σ_i, ρ_i) controlled sources, $1 \leq i \leq m$, does not suffer any loss?

b. A deterministic periodic on–off source with peak rate R, average rate ρ, and burst length $\frac{\sigma}{R-\rho}$ is fed to a buffer with service rate ρ. Determine the buffer depth required for no fluid loss, and sketch the output fluid rate function for this buffer size.

4.9

a. $A(t)$ is a (σ, ρ) controlled fluid source (see Figure 4.29(a)) that passes through a system with maximum delay $d_{\max}$ (i.e., any fluid entering before t will be drained out before $t + d_{\max}$). Show that $B(t)$ is $(\sigma + \rho d_{\max}, \rho)$ constrained.

Figure 4.29   The two alternatives in Problem 4.9.

b. $A(t)$ is (σ, ρ) controlled, and the server has an infinite buffer (see Figure 4.29(b)). First apply (a) to get a constraint for $B(t)$. Then obtain another constraint from direct arguments on this model. Finally, write down the more precise of these two constraints.

4.10 Consider a GPS scheduler at a link of rate C with fluid packet arrivals at $(t_0, t_1, t_2, \ldots)$. There are N queues with weights $\phi_1, \phi_2, \ldots, \phi_N$. $V(t)$ denotes the virtual time for the scheduler.

a. If $V(t)$ is updated at each arrival and each departure event, show the update equations for $V(t)$. (Hint: Define $\Phi(t) = \sum_{i=1}^{N} \phi_i Z_i(t)$, where $Z_i(t) = 1$ if queue i is nonempty at t, and 0 otherwise.)

b. With $V(t_0) = 0$, show how $V(t_1), V(t_2), \ldots$ can be calculated without any intermediate calculations. (Hint: You will need to keep the largest virtual finish time for each queue, and observe that between arrivals the queues can only deplete.)

CHAPTER 5

Stream Sessions: Stochastic Analysis

A t the end of Section 4.4 we note that out of necessity (because a useful deterministic bound may not be known for the output of the voice coder), we need to use a stochastic model of the coder output to analyze the behavior of the leaky bucket. We also mention in the beginning of Section 4.2 that deterministic traffic models and worst case analysis can lead to very conservative design, because deterministic calculus designs the network for the worst case behavior of the input. We begin this chapter by illustrating this point by a simple example. This will motivate the study of stochastic models of stream traffic flows in packet networks.

5.1 Deterministic Calculus Can Yield Loose Bounds

Consider a multiplexer of service rate C whose input is the superposition of n statistically independent sources $A_1(t), A_2(t), \ldots, A_n(t)$. Each source has the envelope $\min(Rt^+, (\sigma + \rho t)I_{\{t \geq 0\}})$. It is easy to see that the worst case deterministic envelope of the superposition of the sources will be $E(t) = \min(nRt^+, n(\sigma + \rho t)I_{\{t \geq 0\}})$. If a worst case delay bound is $d_{\max}$, then it will follow that if capacity $C = c$ is required for $n = 1$, then $C = nc$ will be required for the superposition of n sources. The problem is that the worst case analysis designs for the worst case behavior over *any* time interval. Denote the superposition of the sources by $A(t)$, a random process. For $0 < t < \frac{\sigma}{R - \rho}$, we can infer from the envelope $E(t)$ that the cumulative data from the n sources will not exceed nRt. We will see later, however, that $A(t)$ is very likely to be significantly smaller than $E(t)$.

According to our notation, the process $A_i(t)$ is the cumulative bit arrival process. Let $B_i(t)$ denote the bit rate process from source i; that is, $B_i(t)$ is the instantaneous bit rate emanating from source i at time t. Let $B(t)$ denote the aggregate bit rate process:

$$B(t) = \sum_{i=1}^{n} B_i(t)$$

Hence $A(t)$ is the cumulative bit arrival process derived from $B(t)$:

$$A(t) = \int_0^t B(u)\,du$$

Example 5.1

Suppose each bit rate process $B_i(t)$ alternates between R bps for T_{on} seconds, and 0 bps for T_{off} seconds. The average bit rate is r bps, where $r \leq \rho$. Assume that $T_{on} = \frac{\sigma}{R-\rho}$, and $T_{off} = \frac{\sigma}{R-\rho}\frac{R-r}{r}$; check that these are consistent with the average rate r. Also, it can be verified that for this source the cumulative process $A_i(t)$ conforms to the envelope $E(t) = \min(Rt^+, (\sigma + \rho t)I_{\{t \geq 0\}})$. In particular, let us take $r = \frac{R}{2}$ so that $T_{on} = T_{off}$; define T to be this common value. Assume that the bit rate processes $B_1(t), B_2(t), \ldots, B_n(t)$ are individually stationary. Typical sample paths of these processes are shown in Figure 5.1. At time $t = 0$, each process is *on* or *off* with probability $\frac{1}{2}$. The worst case behavior of $A(t)$ is obtained if at time $t = 0$ all the processes $B_1(t), B_2(t), \ldots, B_n(t)$ are just starting their *on* phase. This, of course, has a very small probability for large n.

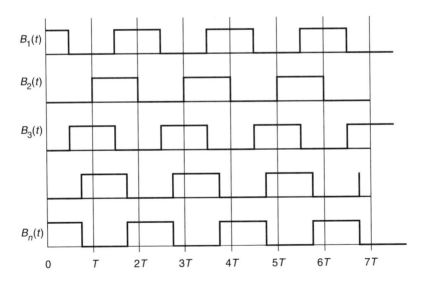

Figure 5.1 Sample paths of *n* independent on–off bit rate processes.

Let us therefore analyze the superposition process when the processes $B_i(t)$, $1 \leq i \leq n$, are stationary and independent.

Stationarity of each $B_i(t)$ process implies the following: Given that it is on at time 0, the length of the remaining *on* time is uniformly distributed over $[0, T]$; similarly, given that the source is off at time 0, the time until the next *on* period starts is uniformly distributed over $[0, T]$. Clearly, the average bit rate at any instant $\mathsf{E}(B_i(t)) = \frac{R}{2}$, and $\mathsf{E}(B(t)) = n\frac{R}{2}$. Because $A(t)$ is the cumulative process of $B(t)$, it follows that

$$\mathsf{E}(A(t)) = \mathsf{E}\left(\int_0^t B(t)\, dt \right) = \int_0^t \mathsf{E}(B(t))\, dt = n\frac{R}{2}t$$

The relationship between $\mathsf{E}(A(t))$ and the worst case envelope of $A(t)$ (i.e., $\min(nRt^+, n(\sigma + \rho t)I_{\{t \geq 0\}})$) is shown in Figure 5.2. It can also be shown that for each of the processes $A_i(t)$ the standard deviation at any t is bounded by a finite

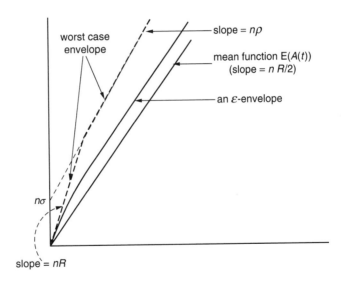

Figure 5.2   Relationship between a worst case envelope, the mean rate function, and an ϵ stochastic envelope for the superposition cumulative bit arrival process $A(t)$.

number $s_{\max} = \frac{RT}{2\sqrt{3}}$. By the central limit theorem (Theorem B.4 in Appendix B) it then follows that for large n, for all t,

$$\Pr\left(A(t) > n\left(\frac{R}{2}t + a_\epsilon \frac{1}{\sqrt{n}}s_{\max}\right)\right) < \epsilon$$

where a_ϵ is defined by $1 - \Phi(a_\epsilon) = \epsilon$, $\Phi(\cdot)$ being the cumulative distribution function of the standard normal distribution. With this observation in mind, Figure 5.2 also shows an ϵ-*envelope* (say, $A_\epsilon(t)$) such that, for all t, $\Pr(A(t) > A_\epsilon(t)) < \epsilon$. Notice that for large n the worst case deterministic envelope $E(t)$ can be much larger than the ϵ-envelope.

Now observe that a worst case design using $E(t)$ will ensure that the delay bound is never violated but will require a large value of multiplexer bandwidth C. On the other hand if the ϵ-envelope is used for the design, then a much smaller value of C will suffice, but the delay requirement will be violated with a small probability. For most applications such a probabilistic "guarantee" would suffice.   ∎

This example raises several questions:

- What are good stochastic models for traffic from stream sources?

- What are the appropriate QoS measures or objectives?

- How do we design to achieve such objectives using stochastic models for the traffic and the network elements? For this we will need to understand certain techniques for stochastic analysis.

In this chapter we discuss the concepts and techniques to use in addressing these questions in the context of stream traffic in packet communications networks. The standard Markovian queueing models that are traditionally encountered in queueing theory (such as M/M/1, M/M/K, M/G/1, etc.) are generally not directly useful. This is mainly because traffic processes in packet networks are not well modeled by the Poisson process. However, the stochastic process analysis that one encounters in queueing theory provides a set of very useful tools and concepts. Furthermore, some of these traditional models may themselves be useful for gaining quick insight and for providing examples and counterexamples when performance issues are being studied and design alternatives are being evaluated. In our presentation, we proceed in a self-contained manner without taking recourse to formal queueing theory; some of the results needed for the analysis are provided in Appendix D.

5.2 Stochastic Traffic Models

In experimentation with network equipment we can use the outputs of actual sources, and in computer simulations we can use *traces*, or very detailed models of the sources. For the purposes of analytical modeling, however, we need stochastic models that model the sources well while being analytically tractable. In this section we discuss some commonly used stochastic models for stream sources.

There is a natural periodicity in the instants at which a speech or video source sends packets into the network. A PCM speech encoder emits a byte every 125 μs, or 8 bytes per ms; if the packetization interval is 20 ms and if the header overhead per packet is 40 bytes, then every 20 ms the network will receive a 200 byte packet. If the packetizer suppresses packets during speech inactivity periods, then the packetizer output will comprise bursts of packets during active speech periods. During these bursts, 200 byte packets will arrive periodically every 20 ms. During inactive periods, the packetizer will not send packets. More complex voice coders emit frames of coded speech every 10 ms. For example, the ITU standard G.729 encoder produces a 10 byte coded voice frame every 10 ms (thus yielding 8 Kbps coding). If the packetizer packs two such voice frames into one packet, then, together with the header, the network receives a 60 byte packet every 20 ms during speech activity periods.

Video is basically a sequence of picture frames—for example, at 30 frames per second. The video coder produces code bits for every frame that it is presented with. To achieve a high compression ratio, *interframe coding* is done; rather than recode the entire information in each frame, only the new information in each frame is encoded. Thus at each frame, unlike speech coders, a video coder emits a random number of bits of coded video.

5.2.1 Models for a Single Stream Source

In either of the cases just described, we can model the packet process from the packetizer as follows. There is a basic time interval T; at instants $t_k = kT, k \geq 0$, the packetizer sends B_k packets into the network. We assume that all the packets are of a fixed length, L bytes; this is sometimes the case in practice (as, for example, in ATM networks). In case there are insufficient bytes to make up a "full" packet, the packetizer pads the packet with some fill bytes, indicating this in the header. The receiving packetizer can then remove the fill bytes before passing the code bytes to the decoder.

Thus for the case of G.729 speech, $T = 20$ ms, $B_k = 1$ during speech activity periods and $B_k = 0$ during inactivity periods, and $L = 60$ bytes. The randomness comes from the duration of the active periods, called talkspurts. The intervening periods during which no packets are sent are called silence periods. A commonly used model, which has been supported by measurements from actual speech, is that the number of intervals in a talkspurt is *geometrically* distributed, as is the number of intervals in a silence period, and that the talkspurt and silence period durations are independent. Denoting the active periods by $U_j, j \geq 1$ and the silence periods by $V_j, j \geq 1$, we can write, for $m \geq 1$,

$$\Pr(U_j = m) = (1 - \tau)^{m-1}\tau$$

and

$$\Pr(V_j = m) = (1 - \sigma)^{m-1}\sigma$$

where $0 < \tau < 1$ and $0 < \sigma < 1$. It can be checked that the mean talkspurt duration is $\frac{1}{\tau}T$ ms and the silence period duration is $\frac{1}{\sigma}T$ ms. Thus, for example, if $T = 20$ ms, and if the mean talkspurt duration is 400 ms, and if the mean silence period duration is 600 ms, then $\tau = \frac{1}{20}$ and $\sigma = \frac{1}{30}$, and the source is active 40% of the time.

Exercise 5.1

With this model for the talkspurt and silence periods, show that the process $B_k, k \geq 0$, is a discrete time Markov chain (DTMC; see Appendix D) on the state space $\{0, 1\}$, with transition probabilities $p_{01} = \sigma$ and $p_{10} = \tau$. The talkspurt and silence periods are geometrically distributed with the distributions shown earlier. Furthermore, the mean byte rate from the source is $\frac{\sigma}{\sigma + \tau} \frac{L}{T}$ bytes per second.

In the case of packetized video, there is no notion of a "silence" period during which the packet rate is zero. Thus the process $B_k, k \geq 0$ can take any nonnegative integer value. Various stochastic models have been proposed for this process, the simplest being that B_k is a discrete time Markov chain on $\{0, 1, 2, \ldots, M\}$, where M is the maximum number of packets that can correspond to a single coded video frame.

5.2.2   Superposition of Several Streams

A packet multiplexer at the network edge (e.g., at the first-hop WAN link, exiting from the LAN–WAN router R_A in Figure 3.1) would typically receive a number of stream sources to multiplex into the outgoing link. Let us suppose that packets from all the streams are placed into a common buffer, and the link transmits the packets from this buffer. It therefore becomes necessary to characterize the aggregate process of packet arrivals into this buffer.

First suppose that we are multiplexing several packet voice sources (as characterized earlier) with the same coding and framing rate and the same packetization interval T. Because the various sources are not synchronized, the packet arrivals at the multiplexer will not arrive at multiples of the period T. In fact, if there are n sources, every T ms up to n packets will arrive into the multiplexer. Depending on the parameters of the problem and the kind of results we seek, we can model the superposition in one of three ways.

Discrete Time Markovian Batch Arrival Model

Consider the time instants $t_k = kT$, $k \in 0, 1, 2, \ldots$. For $k \geq 1$, we gather all the packet arrivals from the n sources during the interval $(t_{k-1}, t_k]$ and take these packets to arrive at t_k. Let the resulting number of packets arriving at t_k be denoted by B_k. It can then be seen that, for the packet voice model discussed earlier, B_k is a Markov chain on the state space $\{0, 1, \ldots, n\}$. This can be seen as follows. When there are j voice sources in the talk spurt state, j packets arrive into the multiplexer; in the next time interval, some of the active sources can become inactive, and some of the inactive sources can become active.

Exercise 5.2

Determine the transition probabilities of the superposition Markov chain B_k described here.

Notice that this model of the superposition process yields larger queueing delays in the multiplexer than does the actual superposition process in which packets probably will not arrive in batches in multiples of T. Suppose each packet is 200 bytes, $T = 20$ ms, and the link rate is 240 bytes per ms (i.e., 1.920 Mbps, which is the effective payload rate of an E-1 digital carrier). The link can transmit 4800 bytes in each interarrival interval of each voice source; that is, the link can transmit 24 packets in each interval of length T. If there are 20 sources, then all the packets that arrive in each interval can be served in less than one interval, and if the arrivals are uniformly spread out over the interval of length T, there should be no queueing; hence, arriving voice packets would rarely have to wait for other previously arrived packets to be transmitted. In the superposition model, however, a batch arrival at each $t_k = kT$ will cause a queue to accumulate, thus resulting in delays. Hence this model will overestimate the delay in the multiplexer.

Observe, however, that if each source is only 60% active (as in the packet voice example here), then the link can carry the average packet rate from 35 sources (these will generate an *average* of 21 packets per interval, as against the 24 packets per interval that the link can carry). With this load, because of the randomness in the talkspurt and silence periods, significant queueing can occur in any case (e.g., all the 35 sources may be active at the same time, but the link can serve only 24 packets per interval of length T), and the batch arrival superposition model may be adequate.

The same superposition approach can also be applied to the packet video source model. Because the number of packets coming from the video packetizer at each frame can be more than 1, the characterization of the superposition Markov chain will not be as simple as in the case of packet speech. For n packet video sources, denote by $B_k^{(j)}, k \geq 0$, the number of packets from the jth source in the kth interval. Then given the vector $(B_k^{(1)}, B_k^{(2)}, \ldots, B_k^{(n)})$, the number of packets in the superposition process is $B_k = \sum_{j=1}^{n} B_k^{(j)}$, and the transition probability from $(B_k^{(1)}, B_k^{(2)}, \ldots, B_k^{(n)})$ to $(B_{k+1}^{(1)}, B_{k+1}^{(2)}, \ldots, B_{k+1}^{(n)})$ can be obtained as the Kronecker product (see Appendix D) of the transition probabilities of the individual sources.

Markov Modulated Poisson Process

In the model described earlier, batches of packets arrive at periodic instants. The actual packet arrival process is smoother than this. An alternative is to model the aggregate packet arrival process from n sources as a Poisson process whose rate changes at certain transition instants according to a continuous time Markov chain, a so-called Markov modulated Poisson process (MMPP).

In this approach, for the superposition of n on–off packet voice sources, we could proceed as follows. The modulating Markov chain has $n + 1$ states, $i \in \{0, 1, \ldots, n\}$, corresponding to the number of active sources. In state i, packets arrive in a Poisson process of rate $\frac{i}{T}$ (T as defined earlier). When in state i, the modulating Markov chain's transition rate to state $i - 1$ is $i\frac{\tau}{T}$, and to the state $i + 1$ is $(n - i)\frac{\sigma}{T}$ (with τ and σ, as before).

Sum of Time-Varying Number of Deterministic Processes

Yet another modeling approach arises from the observation that when n sources are active (i.e., sending packets), then one packet will arrive from each active source in every interval $(t_{k-1}, t_k]$, and, because each active source generates packets with

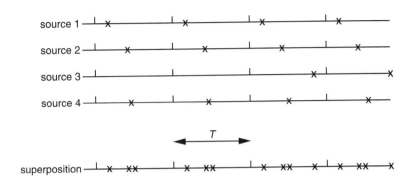

Figure 5.3 The superposition of a time-varying number of periodic packet voice sources. Source 3 becomes active in the third time interval.

a period T, the pattern of arrivals of packets over the successive intervals will repeat. This is illustrated in Figure 5.3, where we show the superposition of four sources. During the first two intervals sources 1, 2, and 4 are active. In this time, exactly three packets arrive in each interval, and the arrival pattern repeats. When source 3 becomes active, four packets arrive in each interval, and again the arrival pattern repeats. Thus when the number of sources is fixed, the arrival pattern is periodic with period T. When there are n sources active, the distribution of the n arrival instants over the interval T can be taken to be that obtained by throwing n points independently and uniformly over an interval of length T. Note that this random sampling is done only in the first interval, and subsequently the pattern repeats.

This periodic arrival model is most appropriate when packets generated during silence periods are not suppressed; that is, during the entire duration of a voice call, packets are sent into the network at the rate of one per T. Such a voice source is called a constant bit rate (CBR) source (as opposed to a variable bit rate (VBR) source obtained by silence suppression). With CBR calls we can expect that over relatively long periods of time (relative to the VBR case), the number of active sources remains constant. As the CBR packet voice calls arrive and depart, the number of superimposed sources varies, but when there are n sources active we can model the input into the multiplexer as superimposed periodic sources, as discussed earlier. Two analyses can now be combined to obtain the multiplexer performance: a multiplexer analysis with n superimposed periodic sources, for every n, and an analysis that yields the distribution of the number of active calls.

The latter analysis is the same as that done for circuit-multiplexed voice trunks and is covered in Chapter 6.

5.3 Additional Notation

We recall the notation introduced in Section 4.1 and define some additional notation here.

$\mathcal{A}(t)$: the number of packet arrivals into the multiplexer in the time interval $[0, t]$; this is the number of arrival instants, a_k, that fall in the interval $[0, t]$.

$\mathcal{D}(t)$: the number of packet departures from the multiplexer in the time interval $[0, t]$; this is the number of departure instants, d_k, that fall in the interval $[0, t]$.

W_k: the total time spent in the multiplexer by the kth packet.

$N(t)$: the number of packets in the multiplexer at time t; clearly $N(t) = \mathcal{A}(t) - \mathcal{D}(t)$.

Note that the processes $\mathcal{A}(t)$ and $\mathcal{D}(t)$ must be distinguished from the processes $A(t)$ and $D(t)$ defined earlier. The former count the cumulative number of arrivals or departures up to t (they are *point-counting processes*), whereas the latter denote the *cumulative amount of data* arrived or departed up to t.

When there are a number of sessions being handled (say, m sessions) by a multiplexer, as before, we distinguish the processes of the various sessions by the superscript $^{(j)}$.

5.4 Performance Measures

Let us consider several packet sessions or connections being carried by a network subsystem, as shown in Figure 5.4. The subsystem could be one output port of a packet switch, or an entire packet switch, or an entire network of links and switches. In this section we formally describe some of the common performance measures associated with such a situation.

We start by assuming that every packet that enters the system eventually leaves (after a finite amount of time). Without further comment, in this section we take the limits in the following discussion to exist.

- *Average number of packets in the subsystem:* If we do not wish to distinguish between various sessions, then the total number of packets

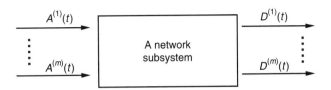

Figure 5.4   Arrival and departure processes of several sessions being transported by a network subsystem.

in the system, at time t, is $N(t) = \sum_{j=1}^{m}(\mathcal{A}^{(j)}(t) - \mathcal{D}^{(j)}(t))$. Then the *time average* number of packets in the system, N, is given by

$$N = \lim_{t \to \infty} \frac{1}{t} \int_{0}^{t} N(u)\, du$$

If we wish to focus on individual sessions, then the number of packets in the system from the jth session is $N^{(j)}(t) = \mathcal{A}^{(j)}(t) - \mathcal{D}^{(j)}(t)$, and the time average number of packets in the system from session j is

$$N^{(j)} = \lim_{t \to \infty} \frac{1}{t} \int_{0}^{t} N^{(j)}(u)\, du$$

- *Average time spent in the subsystem:* If we do not distinguish between the sessions, then let k index all the packets in their order of arrival, and let a_k and d_k be the arrival and departure instants of the kth packet in this ordering. Then the system sojourn time of packet k is $W_k = d_k - a_k$. The average system time of the packets is defined as

$$W = \lim_{n \to \infty} \frac{1}{n} \sum_{k=1}^{n} W_k$$

Because a system time is associated with each packet and because we are averaging over packets to obtain W, we call this a *customer average* (or, in this context, a *packet average*) as compared with the time average (number in the system) defined earlier.

Focusing on individual sessions, we have $W_k^{(j)} = d_k^{(j)} - a_k^{(j)}$, and the average system time for packets of session j is defined by

$$W^{(j)} = \lim_{n \to \infty} \frac{1}{n} \sum_{k=1}^{n} W_k^{(j)}$$

In making measurements, or estimates from simulations, it is important not to confuse time average and customer average performance measures. For example, suppose that there is a single session that brings two packets of length L bits every T seconds, to a link of capacity C bps. Suppose that $\frac{2L}{C} < T$, which implies that the link's buffer becomes empty between consecutive arrival instants. Suppose that we index each of the arriving packets by $k \geq 1$ and denote the number of packets found by the kth packet (including itself) by N_k. Note that in each "batch" of packet arrivals one packet finds the buffer empty, and the other finds one packet (the one it comes together with).

Exercise 5.3

In the preceding example show that the time average number in the system $N = \frac{3L}{CT}$ (which depends on L, C, and T and can be small if CT is large), whereas the average of N_k over $k \geq 1$ is $\frac{3}{2}$.

Thus an attempt to estimate N by "sampling" the number in the system at packet arrival instants can lead to serious errors.

In Section 5.5, we provide theorems that relate certain time average and customer average quantities.

5.5   Little's Theorem, Brumelle's Theorem, and Applications

A very powerful and important result that must be learned and understood by anyone dealing with stochastic models of congestion systems is Little's law, or Little's formula. We provide a simple (incomplete) derivation of this result; details can be found in many textbooks on queueing theory (a rigorous proof is available in [300]). Brumelle's theorem is a generalization of Little's result. These results can be applied very fruitfully when we deal with average performance measures. We provide some examples illustrating the insights that can be derived from these simple but powerful results.

5.5.1 Little's Theorem

Consider a system into which packets arrive at the instants $a_k, k \geq 1$. The system may be an output queue of a packet switch, may be the entire switch, or may be an entire network. We are not specifying any order in which the packets are served; for example, some of the packets may belong to voice streams and may therefore be given priority over the others. In fact, the system could be handling other packet streams that we are not concerned about. In view of this, the packets of the stream that we are concerned with do not necessarily depart in the order in which they arrive. Each packet spends some time in the system, and this time comprises some waiting time and some "service" time. Note that we are not even requiring that the packet be served in its entirety in a single service attempt. For example, in the fabric of a packet switch a packet may get fragmented into small, fixed-length cells, which are then switched and then reassembled at the switch output. Variable-length packets may enter a network and may get fragmented into smaller packets, which are then reassembled before leaving the network.

Let d_k denote the instant at which the kth arriving packet leaves the system. Figure 5.5 shows horizontal bars of thickness 1 for the entire duration that each packet is in the system. Note that the d_k sequence is not in order, indicating that packets do not depart in their arrival order. Thus, associated with packet k is a

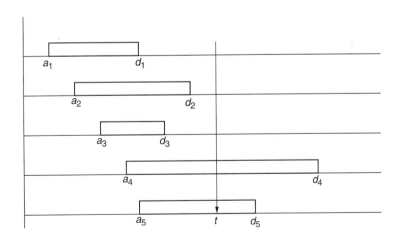

Figure 5.5 Packet indicator functions for several packets passing through a multiplexer; note that FIFO service is not assumed.

function $I_k(t), t \geq 0$ such that

$$I_k(t) = \begin{cases} 1 & \text{if packet } k \text{ is present at time } t \\ 0 & \text{otherwise} \end{cases}$$

That is, $I_k(t) = 1$ for $a_k \leq t \leq a_k + W_k$, and $I_k(t) = 0$ otherwise. Packet k is in the system at time t if and only if $I_k(t) = 1$. It follows that, for all $t \geq 0$, the number in the system is given by

$$N(t) = \sum_{j=1}^{\infty} I_j(t)$$

because the right side of the equation counts all those packets that are in the system at time t.

Little's theorem relates the long-run averages of $N(t)$, W_k, and $A(t)$. Consider a time instant t (see, for example, Figure 5.5). Because of the preceding observation we can write the integral of $N(\cdot)$ over $[0, t]$ as follows:

$$\int_0^t N(u)\, du = \int_0^t \sum_{j=1}^{\infty} I_j(u)\, du$$

That is, the area under the $N(u)$ sample path over $[0, t]$ can be written as the sum of the areas under the functions $I_k(u)$. Dividing both sides by t, we have time averages over $[0, t]$:

$$\frac{1}{t} \int_0^t N(u)\, du = \frac{1}{t} \int_0^t \sum_{j=1}^{\infty} I_j(u)\, du \tag{5.1}$$

The right side of this equation is affected by only those $I_k(\cdot), k \geq 1$, for which $a_k \leq t$. Consider the time instant t in Figure 5.5, and notice that some of the packets that arrived in $[0, t]$ have departed, whereas others are still present in the system. The integral on the right side of Equation 5.1 is the total area under the $I_k(\cdot)$ functions up to time t. Noticing that the area under the curve $I_k(\cdot)$ is W_k, we can then see that the following inequalities hold:

$$\frac{1}{t}\sum_{j=1}^{\mathcal{D}(t)} W_j \leq \frac{1}{t}\int_0^t \sum_{j=1}^{\infty} I_j(u)\,du \leq \frac{1}{t}\sum_{j=1}^{\mathcal{A}(t)} W_j \tag{5.2}$$

The lower bound accounts only for those packets that have arrived and departed by t, whereas the upper bound adds up the total areas under $I_k(\cdot)$ for all the packets that arrive in $[0, t]$.

Consider a particular sample function (evolution) of the process for which the following limits exist:

$$\lim_{n\to\infty} \frac{1}{n}\sum_{j=1}^{n} W_j = W \tag{5.3}$$

$$\lim_{t\to\infty} \frac{\mathcal{A}(t)}{t} = \lambda > 0 \tag{5.4}$$

We are asserting that for this sample path the packet average system time, W, exists, and the time average rate of arrivals, λ, exists and is positive. Observe that because $\lambda > 0$ it follows that $\mathcal{A}(t) \to \infty$; that is, for increasing time the number of arrivals increases to infinity (as λt). Rewriting the upper bound in Equation 5.2,

$$\frac{\mathcal{A}(t)}{t}\,\frac{1}{\mathcal{A}(t)}\sum_{j=1}^{\mathcal{A}(t)} W_j$$

and taking limit $t \to \infty$, we see from the hypotheses 5.3 and 5.4 that this bound converges to λW. Rewrite the lower bound in Equation 5.2 as

$$\frac{\mathcal{D}(t)}{t}\,\frac{1}{\mathcal{D}(t)}\sum_{j=1}^{\mathcal{D}(t)} W_j$$

We would like to show that the hypotheses in Equations 5.3 and 5.4 imply that $\frac{\mathcal{D}(t)}{t} \to \lambda > 0$ and $\mathcal{D}(t) \to \infty$. We skip these technical details here. Basically the existence of the packet average system time and the time average arrival rate can be used to show that all customers that arrive eventually depart, and the time average departure rate is the same as the arrival rate, λ. It then follows, using an argument exactly as used for the upper bound, that the lower bound in Equation 5.2 converges to λW, as $t \to \infty$.

We have proved the first part of the following result:

Theorem 5.1 (Little)

(i) If, for a sample path (i.e., realization of the process) denoted by ω, $\frac{A(t)}{t} \to \lambda(\omega)$, and $\frac{1}{n} \sum_{j=1}^{n} W_j \to W(\omega)$, then $\lim_{t \to \infty} \frac{1}{t} \int_0^t N(u)\, du = N(\omega)$ exists, and $N(\omega) = \lambda(\omega) W(\omega)$.

(ii) If the limits, for the various sample paths ω, are equal to the constants λ and W with probability 1, then $\lim_{t \to \infty} \frac{1}{t} \int_0^t N(u)\, du = N(\omega) = \lambda W$ with probability 1; we denote by N the average number in the system, and we obtain $N = \lambda W$ with probability 1.   ∎

Part (i) of Little's theorem holds for any sample path for which the required time averages exist. The second part (which is the one that is usually used) states that if the packet average system time and the packet arrival rate are constant over a set of sample paths of probability 1, taking the values W and λ on this set, then, with probability 1, the time average number of packets in the system is $N = \lambda W$ with probability 1. In applications, we would know of the *existence* of N, λ, and W by some other means, and Little's theorem would be used to relate these quantities—for example, to determine one quantity when the other two are known.

It is important to emphasize that Little's theorem is a relationship between certain average system performance measures. N is the *time* average number of packets in the system, W is the *packet* average time that packets spend in the system, and λ is the *time* average number of arrivals into the system.

The preceding derivation of Little's theorem was for a continuous time model; the process $N(\cdot)$ evolves over continuous time. In a discrete time multiplexer model, we will need to work with N_k, the number of packets in the buffer at discrete time instants $t_k = kT$, for some given interval T. In this situation, the exact meaning of N_k becomes important: Is N_k the queue length *at* t_k, just *before* t_k, or just *after* t_k? Care must be taken in writing down a Little's theorem for such problems, because the exact interpretation of N_k could result in slightly different answers.

Example 5.2

Consider a multiplexer with link rate C. The multiplexing discipline is work-conserving (or non-idling). Packets arrive at the rate λ. Let $N_l(t) = 1$ if the link is serving packets, and $N_l = 0$ if the multiplexer is empty. $N_l(t)$ is the number of packets in transmission from the multiplexer at time t. Assuming that all the arriving packets are served, the rate of packets passing through the link is λ. If the length of the kth packet to be served at the link is L_k, with an average length $L \ (= \lim_{n\to\infty} \frac{1}{n} \sum_{k=1}^{n} L_k)$, the average time spent by the link in transmitting a packet is $\frac{L}{C}$. Little's theorem (applied only to the packet transmitter) then says that $\lim_{t\to\infty} \frac{1}{t} \int_0^t N_l(u)\, du = \lambda \frac{L}{C}$. In other words the fraction of time that the link is carrying data is the arriving packet bit rate normalized by the link's bit rate.   ∎

5.5.2 Invariance of Mean System Time

In spite of its simplicity, Little's theorem can be a powerful analytical tool. We now demonstrate its use for establishing a basic result for a class of multiplexing problems. It will be useful to review the material in Section 4.1.1.

Packets enter the multiplexer at arrival instants $a_k, k \geq 1$; the packet arriving at a_k has the length L_k. The packet arrival rate is λ (packets per second). We make the following assumptions about the system:

- The packet lengths $(L_k, k \geq 1)$ are independently and identically distributed, with some distribution $L(l)$; that is, $\Pr$ (packet length $\leq l) = L(l)$.

- After the completion of the transmission of a packet, the scheduler chooses the next packet for transmission without any regard to the service times of the waiting packets. Thus, for example, a scheduling policy that transmits shorter packets first is not under consideration in this discussion.

- When the transmission of a packet is initiated, the link is dedicated to the packet and the packet is transmitted completely.

The second and third assumptions ensure that when the link completes the transmission of a packet and the scheduler looks for a new packet to transmit, the service times of the waiting packets are as if they were "freshly" sampled (in an independent and identically distributed (i.i.d.) fashion) from the distribution $L(l)$. Let $N_{FCFS}(t), \ t \geq 0$, denote the number of packets in the system when the scheduling policy is FCFS. $N_{FCFS}(t)$ is a random process; observe that it is completely determined by the random sequences a_k and L_k. Let P denote another policy that satisfies the preceding assumptions, and let $N_P(t)$ denote the corresponding packet queue length process. Let us now modify the system under policy P in the following way. The packet arrival instants are unchanged, but the kth packet *to be transmitted* is assigned the length L_k *when it is scheduled for transmission*. Note that for a non-FIFO policy the kth packet to be transmitted need not be the kth packet to arrive. Denote the multiplexer queue length (for this alternative way of sampling packet lengths) by $\tilde{N}_P(t)$. Because of the assumptions about the policies, a little thought shows that $N_P(t)$ and $\tilde{N}_P(t)$ are statistically indistinguishable; a probability question about either of these processes yields the same answer. By the preceding construction, it can also be easily seen that $\tilde{N}_P(t) = N_{FCFS}(t)$ for all t. Further assume that with the scheduling policies under consideration, the system is "stable" (formally discussed later in this chapter; see Section 5.6.2), and hence the following time averages exist with probability 1: $N_{FCFS} = \tilde{N}_P = N_P$, W_{FCFS} and W_P. By Little's theorem we have

$$N_{FCFS} = \lambda W_{FCFS}$$

$$N_P = \lambda W_P$$

We conclude that $W_{FCFS} = W_P$; that is, the mean time that a packet spends in the system is invariant with the choice of policy in this class. It can be shown that the higher moments of the system time do depend on the policy and that the FCFS policy is optimal in a certain sense.

5.5.3　Generalization of Little's Theorem: Brumelle's Theorem

When establishing Little's result, we associate with each packet the indicator function $I_k(t)$. Suppose that, in general, we can associate with customer k a function $f_k(t)$, $t \geq 0$. For example, $f_k(t)$ could be the untransmitted number of bits of packet k when the packet is in the system, and $f_k(t) = 0$ otherwise. Define, for $j \geq 1$, and for $t \geq 0$,

$$G_j = \int_0^\infty f_j(u) \, du$$

$$H(t) = \sum_{j=1}^\infty f_j(t)$$

Analogously to Little's theorem, we define the packet average of the G_j sequence and the time average of $H(t)$.

$$G = \lim_{n \to \infty} \frac{1}{n} \sum_{j=1}^{n} G_j$$

$$H = \lim_{t \to \infty} \frac{1}{t} \int_0^t H(u) \, du$$

The following is a version of Brumelle's theorem with simplified hypotheses. The more general result can be found in [300].

Theorem 5.2   (Brumelle)

(i) If for a sample path ω, for all k, $f_k(t) = 0$ for $t \notin [t_k, t_k + W_k]$, $\lim_{n \to \infty} \frac{1}{n} \sum_{j=0}^{n} W_j = W(\omega)$, $\lim_{t \to \infty} \frac{A(t)}{t} = \lambda(\omega) > 0$, and $G(\omega)$ exists, then for ω, $H(\omega)$ exists and

$$H(\omega) = \lambda(\omega) G(\omega)$$

(ii) If, in addition to the hypotheses in (i), with probability 1, $\lambda(\omega) = \lambda$, and $G(\omega) = G$, then, with probability 1,

$$H(\omega) = H = \lambda G \qquad \blacksquare$$

The technical condition in part (i) of Theorem 5.2 basically says that the function $f_k(t)$ is nonzero only when packet k is in the system. This is a less general condition than in the full theorem but is adequate for most applications. In the next section we use Brumelle's theorem to derive the mean sojourn time formula for the M/G/1 queue.

5.5.4 The M/G/1 Model: Formula for Mean Number of Bits in a Buffer

Packets arrive to a buffered link, of transmission rate C bits per sec and are transmitted FCFS. No packets are lost from the buffer, which we idealize and assume has infinite capacity. The packet lengths $L_k, k \geq 1$, are i.i.d., with expectation $\mathsf{E}(L)$ (note that this i.i.d. assumption includes the important case in which all packets are of the same length). The delay until the start of service of packet k is U_k.

Associate with packet k the function $f_k(\cdot)$ such that $f_k(t) = 0$ when the packet is not in the system, and $f_k(t) = $ the residual number of bits in the packet when the packet is in the system. Thus $f_k(t) = L_k$ for $t \in [a_k, a_k + U_k]$, and $f_k(t)$ decreases linearly from L_k to 0, at rate C, over the interval $[a_k + U_k, d_k]$, during which time the packet is being transmitted. Clearly, the "area" under $f_k(t)$ is $(L_k U_k + \frac{1}{2} L_k^2 / C)$ bit seconds.

Observe that $\sum_{k=1}^{\infty} f_k(t)$ is the total number of bits in the buffer (i.e., $X(t)$). Assuming that this system is stable, Brumelle's theorem says

$$\lim_{t \to \infty} \frac{1}{t} \int_0^t X(u)\, du = \lambda \lim_{n \to \infty} \frac{1}{n} \sum_{k=1}^{n} \left(L_k U_k + \frac{1}{2} \frac{L_k^2}{C} \right)$$

Because the buffer is drained FCFS, it follows that $U_k = \frac{X(a_k)}{C}$, where $X(a_k)$ is the number of bits in the buffer in front of the kth arrival. Let X_a denote the random variable of the number of bits that arrivals find in the buffer. It follows that

$$E(X) = \lambda \left(\frac{E(L X_a)}{C} + \frac{1}{2} \frac{E(L^2)}{C} \right)$$

where the strong law of large numbers has been invoked to conclude that $\lim_{n \to \infty} \frac{1}{n} \sum_{k=1}^{n} L_k^2 = E(L^2)$. Because the number of bits brought by an arrival is independent of the number it finds in the buffer, we have $E(L X_a) = E(L)\, E(X_a)$, thus yielding

$$E(X) = \lambda \left(\frac{E(L)\, E(X_a)}{C} + \frac{1}{2} \frac{E(L^2)}{C} \right)$$

If packet arrival instants can be modeled as a Poisson process (as may be the case for the superposition of arrival processes), then we can invoke the PASTA result (Poisson arrivals see time averages; see Appendix D) to write $E(X_a) = E(X)$. It finally follows that

$$E(X) = \frac{\rho}{1 - \rho} \frac{1}{2} \frac{E(L^2)}{E(L)} \tag{5.5}$$

where $\rho := \frac{\lambda E(L)}{C}$.

Thus Brumelle's theorem has permitted the derivation of a formula for the time average number of bits in the link's buffer in a particular situation. A single-station queueing system with Poisson arrivals and i.i.d., generally distributed service requirements is called an *M/G/1 queue*. The derived parameter $\rho := \frac{\lambda E(L)}{C}$ is the rate of bit arrivals normalized to the link's bit rate, and hence is the normalized *load* on the link. Intuitively, we expect the buffer not to "blow up" if $\rho < 1$; we formally examine this a little later in the chapter. From Example 5.2 we also infer that when $\rho < 1$ then ρ is the fraction of time the link is occupied, or the *link occupancy*.

Remarks 5.1

Notice that we derive the formula for the mean number of bits in the buffer for a multiplexer modeled as an M/G/1 queue under the assumption of FCFS service in the buffer. However, from our discussion in Section 4.1.1, it follows that, for a non-idling link, the same formula for the average number of bits in the buffer holds under any order of service of the packets.

We observe from this formula that for a fixed mean packet length, increasing the variance of the packet length will lead to an increase in the average number of bits queued. Also, for a fixed mean and variance of the packet length, as the normalized load on the link increases to 1 (i.e., $\rho \to 1$), the mean number of bits in the buffer increases to ∞.   ∎

5.6 Multiplexer Analysis with Stationary and Ergodic Traffic

We now turn to techniques for analyzing multiplexer performance with fairly general models for arrival processes. We limit the detailed derivations to discrete time models. Analyses of the continuous time models yield very similar results, but the derivations are technically much more involved without yielding any new engineering insights.

We consider m sources, indexed by $i = 1, 2, \ldots, m$, bringing amounts of data $\{B_k^{(i)}, \ k \geq 0\}$, at times $t_k = kT, k \geq 0$, where T is some appropriate time interval in the model. For example, recalling the packet voice model, with a voice coder generating voice frames, T could be taken to be the coder's frame time—for example, 10 ms, 20 ms, or 30 ms depending on the coding algorithm. In this example, $B_k^{(i)}$ is the number of bits in each coded voice frame, or 0, depending on whether or not source i is active. Thus the total amount of data arriving in the interval $(t_{k-1}, t_k]$ is given by $\sum_{i=1}^{m} B_k^{(i)}$. In each interval of length T, the multiplexer can emit up to C units of data. Let $X_k, k = 0, 1, 2, \ldots$, denote the amount of the data in the multiplexer buffer at times t_k. We now show how to develop an analysis for the steady state behavior of the X_k process under fairly loose assumptions on the data arrival processes $\{B_k^{(i)}, \ k \geq 0\}$.

5.6.1   Analysis with Marginal Buffering

We consider first the situation in which the multiplexer has no buffer to store data that arrives in a slot but cannot be served in that slot. We call this *marginal* buffering as we will see that the performance depends only on the marginal distribution of the arrival process, and not on the correlations between the arrivals from slot to slot. This is interesting for the following three reasons:

- In the case of stream sources, the arriving packets would have a deadline after which they are useless even if delivered. For example, consider a superposition of packet voice sources whose packet generation period is 10 ms. If we take this as the value of T in our arrival model, and if packet delay must be bounded by 20 ms, then it is reasonable to require that packets arriving in the interval $(kT, (k + 1)T]$ be transmitted in the interval $((k + 1)T, (k + 2)T]$, and any residual packets be dropped.

- For a large class of stochastic models for data sources, the distribution of the stationary occupancy of a buffered multiplexer can be approximated in terms of the marginal buffer behavior and the large buffer behavior. We analyze the small buffer behavior in this section, and the large buffer behavior in Section 5.6.2.

- In some situations buffers may be very expensive, and hence it becomes necessary to understand the limits of performance with little or no buffering. Such a situation will arise in all-optical packet-switching networks, in which packet buffering is also optical. In the near future, optical buffers will be realized with optical fiber delay lines, and a large number of packet buffers will not be feasible.

Practically, it should be clear that packet switching is unachievable with zero buffering. We need some storage space to store at least the header of a packet entering a switch. Some packet switches can cut through the data part from the input link to the output link without needing to store it. Most packet switches, however, are store-and-forward. An arriving packet is entirely copied into the switch from the link on which it arrives, and then is copied out into the output link.

Thus in the small-buffering models that we will discuss we are permitting enough buffering to store the packets that arrive in a slot, but no packets that are unserved in a slot can be carried over to the next.

For this case, we can ask several questions. What is the long-run fraction of slots in which data is lost? Or symbolically:

$$\lim_{n \to \infty} \frac{1}{n} \sum_{k=0}^{n-1} I_{\left\{ \sum_{i=1}^{m} B_k^{(i)} > C \right\}} \tag{5.6}$$

This is because there is data loss in slot $k, k \geq 0$, if $I_{\left\{ \sum_{i=1}^{m} B_k^{(i)} > C \right\}} = 1$. Recall that, as per the notation, the bits or packets that are taken to arrive at t_k (i.e., $\sum_{i=1}^{m} B_k^{(i)}$) are served in slot k—that is, $(t_k, t_{k+1}]$. We could also ask for the long-run fraction of data that is lost:

$$\lim_{n \to \infty} \frac{\sum_{k=0}^{n-1} \left(\sum_{i=1}^{m} B_k^{(i)} - C \right)^+}{\sum_{k=0}^{n-1} \left(\sum_{i=1}^{m} B_k^{(i)} \right)} \tag{5.7}$$

where the numerator inside the limit is the amount of data lost up to the nth slot, and the denominator is the amount of data that has arrived up to the nth slot.

Exercise 5.4

In the foregoing framework, assume that each source has a peak rate (i.e., for each i, $B_k^{(i)} < R^{(i)}$). Use the bounding idea of Exercise 4.4 to show that an objective based on the performance measure in Equation 5.7 can be achieved by imposing an objective on the performance measure in Equation 5.6. (Hint: Notice that we can write $\sum_{k=0}^{n-1} \left(\sum_{i=1}^{m} B_k^{(i)} - C \right)^+ = \sum_{k=0}^{n-1} I_{\left\{ \sum_{i=1}^{m} B_k^{(i)} > C \right\}} \left(\sum_{i=1}^{m} B_k^{(i)} - C \right)^+ .$)

We assume that for each i, the data arrival process $\left\{ B_k^{(i)}, k \geq 0 \right\}$ is stationary and ergodic (see Appendix B), and further that the arrival processes from the various sources $i = 1, 2, \ldots, m$ are mutually independent. It then follows that the vector process $\{ \underline{B}_k := (B_k^{(1)}, B_k^{(2)}, \ldots, B_k^{(m)}), \ k \geq 0 \}$ is stationary and ergodic.

By Birkhoff's strong ergodic theorem (see Theorem B.5 in Appendix B) we have, with probability 1,

$$\lim_{n\to\infty} \frac{1}{n} \sum_{k=0}^{n-1} I_{\left\{\sum_{i=1}^{m} B_k^{(i)} > C\right\}} = \mathsf{E}\left(I_{\left\{\sum_{i=1}^{m} B_0^{(i)} > C\right\}} \right)$$

$$= P\left(\sum_{i=1}^{m} B_0^{(i)} > C \right)$$

where the first equality is as a result of the ergodic theorem. Similarly, for the second question, the ergodic theorem yields, with probability 1,

$$\lim_{n\to\infty} \frac{\sum_{k=0}^{n-1} \left(\sum_{i=1}^{m} B_k^{(i)} - C \right)^+}{\sum_{k=0}^{n-1} \left(\sum_{i=1}^{m} B_k^{(i)} \right)} = \lim_{n\to\infty} \frac{\frac{1}{n}\sum_{k=0}^{n-1} \left(\sum_{i=1}^{m} B_k^{(i)} - C \right)^+}{\frac{1}{n}\sum_{k=0}^{n-1} \left(\sum_{i=1}^{m} B_k^{(i)} \right)}$$

$$= \frac{\mathsf{E}\left(\sum_{i=1}^{m} B_0^{(i)} - C \right)^+}{\mathsf{E}\left(\sum_{i=1}^{m} B_0^{(i)} \right)}$$

provided that the expectations are finite.

Two observations are in order from these results. First, note that the assumptions are very general; the only thing that has been assumed about the data arrival processes is that they are stationary and ergodic. A common model for such arrival processes is a stationary finite state, irreducible Markov chain, for which the assumptions hold (see Theorem D.3 in Appendix D). Second, we observe that for the marginal buffering case the long-run performance measures depend only on the *marginal distribution* of the data arrival process, and not on the correlations between the number of arrivals over the various slots. If the data arrival process from each source is a Markov modulated process, then the superposition of the arrival processes is again a Markov modulated process (see Section 5.2.2), and we conclude that the preceding performance measures for a marginal buffering multiplexer depend only on the stationary probability distribution of this Markov process.

We emphasize that for independent sources the marginal random variables $B_0^{(i)}, 1 \leq i \leq m$, are independent random variables.

Remarks 5.2 *The Continuous Time Case*

Similar results hold in continuous time. Let $\{B^{(i)}(t),\ t \geq 0\}$ denote the rate of arrival of bits from the ith source at time t. Assume that the process for each source is stationary and ergodic and that the sources are statistically independent. C is the rate of the link serving the multiplexer. It can then be shown that, as in the discrete time case, with probability 1,

$$\lim_{t \to \infty} \frac{1}{t} \int_0^t I_{\{\sum_{i=1}^m B^{(i)}(u) > C\}}\, du = \mathsf{Pr}\left(\sum_{i=1}^m B^{(i)}(0) > C\right)$$

If the data is modeled as a fluid (arriving at rate $B^{(i)}(t)$ at time t from source i), then we can consider a truly zero buffer model. The link emits fluid at rate C. If the arrival rate is greater than C, then fluid will be continuously lost at the rate $\sum_{i=1}^m B^{(i)}(t) - C$. Nothing needs to be buffered. ∎

Example 5.3 *A Two-State Markov Source*

Consider m independent, discrete time Markov modulated sources with two states 0 and 1. In state 1 a source emits one (fixed-length) packet in the slot, and when in state 0 the source emits nothing. Such a source model would apply to a voice coder with voice activity detection, if we model the speech talkspurts and silent periods as geometrically distributed random variables. Denote the mean talkspurt length by $\frac{1}{\tau}$ slots, and the mean silent period length by $\frac{1}{\sigma}$ slot. It follows that the Markov chain's transition probabilities are $p_{10} = \tau = 1 - p_{11}$, and $p_{01} = \sigma = 1 - p_{00}$. For example, if the mean talkspurt length is 400 ms, the mean silent period length is 600 ms, and the coder's frame time is 20 ms, then $\frac{1}{\tau} = 20$, and $\frac{1}{\sigma} = 30$.

The stationary distribution for this Markov chain is $\pi_1 = \frac{\sigma}{\sigma + \tau} = 1 - \pi_0$. In the stationary regime, this is the marginal distribution of the number of packets emitted by each source in a slot—that is, the distribution of $B_0^{(i)}$. Because the sources are independent, it is easily seen that $\sum_{i=1}^m B_0^{(i)}$ is Binomial $(m, \frac{\sigma}{\sigma + \tau})$ distributed. If the number of packets that the link can serve is C (say, a positive integer less than m), then we find from the results discussed earlier that, in the marginal buffering case, the long-run fraction of time that packets are lost is given by

$$\sum_{k=C+1}^m \binom{m}{k} \left(\frac{\sigma}{\sigma + \tau}\right)^k \left(1 - \frac{\sigma}{\sigma + \tau}\right)^{m-k}$$

Notice that this distribution depends only on the ratio $\frac{\sigma}{\tau}$. Thus if the mean talkspurt length and silence period length are changed from 400 ms and 600 ms, to 40 ms and 60 ms, or to 4000 ms and 6000 ms, in either case the marginal buffering performance remains unchanged. This emphasizes the fact that this performance measure depends only on the marginal distribution of the arrival process. In Section 5.6.2, we contrast this with the situation when the buffer is allowed to build up from slot to slot. ∎

If we are actually interested in designing a multiplexer with marginal buffering—that is, with no buffers for carrying over packets that could not be served in a slot—then we would like to make the probability of this event small. Given the sources that must be multiplexed, the design problem can be stated as one of choosing the link rate C so as to make $\Pr\left(\sum_{i=1}^{m} B_0^{(i)} > C\right)$ small; we have seen in Exercise 5.4 that this objective suffices even if we are interested in bounding the fraction of data lost. Here $B_0^{(i)}, i = 1, 2, \ldots, m$, are the independent marginals of the sources. Consider the case where the sources are identically distributed, and, dropping the subscript 0, denote the i.i.d. marginal random variables by $B^{(1)}, B^{(2)}, \ldots, B^{(m)}$. So we need C such that $\Pr\left(\sum_{i=1}^{m} B^{(i)} > C\right) < \epsilon$ for some suitable small number $1 > \epsilon > 0$—for example, $\epsilon = 10^{-K}$ for $K = 3$, or $K = 5$, or $K = 8$, etc.

Suppose that $B^{(i)} \leq R$ (the maximum amount of data that can arrive in an interval); $\mathsf{E}(B)$ is the mean. Take each source to be a stationary on–off source with *on* probability $\frac{\sigma}{\sigma+\tau}$ (as in the earlier example); hence $\mathsf{E}(B) = R\frac{\sigma}{\sigma+\tau}$. For example, the source could emit a 200 byte packet in each slot (i.e., $R = 1600$ bits) or emit nothing, and $\frac{\sigma}{\sigma+\tau} = 0.4$. Consider $m = 1$, and take $C < R$, and notice that $\Pr(B^{(1)} > C) = \frac{\sigma}{\sigma+\tau}$, which would be larger than any useful ϵ. Hence, clearly, for a single source we need $C = R$. Let us denote by c the capacity of the link per source; that is, $C = mc$. Now for $m > 1$, we first note that the law of large numbers yields

$$\lim_{m \to \infty} \Pr\left(\sum_{i=1}^{m} B^{(i)} > m(\mathsf{E}(B) - \delta)\right) = 1$$

Hence for large m we must take $c \geq \mathsf{E}(B)$. Thus we need to choose c such that $\mathsf{E}(B) \leq c \leq R$.

We now examine two approaches for studying $\Pr(\sum_{i=1}^{m} B^{(i)} > mc)$ to determine a suitable c. One is based on the central limit theorem, and the other is based on Cramer's theorem.

Analysis Using the Central Limit Theorem

We are basically concerned with the distribution of the sum of i.i.d. random variables $\sum_{i=1}^{m} B^{(i)}$ or equivalently $\sum_{i=1}^{m}(B^{(i)} - \mathsf{E}(B))$. The density function or the probability mass function of this (centered) sum is the m fold convolution of that of the individual one. We know that this distribution will spread out as m increases. If the sum is normalized by m, then the law of large numbers (Theorems B.2 and B.3 in Appendix B) states that, as m increases, the probability distribution will not spread but will accumulate around 0, converging to a point mass (degenerate distribution) at 0 as $m \to \infty$. On the other hand if the sum is normalized by $\sqrt{m}$, then it converges to the Normal distribution. This is called the central limit theorem (Theorem B.4, Appendix B). Formally (writing $\sigma(B)$) as the common standard deviation of the marginal random variables,

$$\lim_{m\to\infty} \mathsf{Pr}\left(\frac{1}{\sigma(B)\sqrt{m}} \sum_{i=1}^{m}(B^{(i)} - \mathsf{E}(B)) > z \right) = 1 - \Phi(z)$$

where $\Phi(z)$ is the cumulative distribution function of the Normal $(0, 1)$ distribution.

We are, however, interested in computing

$$\mathsf{Pr}\left(\sum_{i=1}^{m} B^{(i)} > mc \right)$$

where c is the per-source capacity of the link, and $c > \mathsf{E}(B)$. This is equivalent to

$$\mathsf{Pr}\left(\frac{1}{\sigma(B)\sqrt{m}} \sum_{i=1}^{m}(B^{(i)} - \mathsf{E}(B)) > \frac{\sqrt{m}(c - \mathsf{E}(B))}{\sigma(B)} \right)$$

By the central limit theorem, for large m, this probability can be expected to be well approximated by $1 - \Phi\left(\frac{\sqrt{m}(c - \mathsf{E}(B))}{\sigma(B)} \right)$. Note that in the statement of the central limit theorem, z is fixed, whereas to apply the central limit theorem here amounts to asking whether the following result holds:

$$\lim_{m\to\infty} \mathsf{Pr}\left(\frac{1}{\sigma(B)\sqrt{m}} \sum_{i=1}^{m}(B^{(i)} - \mathsf{E}(B)) > z_m \right) \sim 1 - \Phi(z_m)$$

where z_m is a sequence of numbers that increases with m, and $\sim$ denotes that the ratio of the left expression to the right expression goes to 1 as $m \to \infty$. The existence of such a result depends on the specific distribution of B and on the rate of increase of z_m.[1] In any case, for a given fixed and large m we could try to use the central limit theorem to approximate the desired probability. It is found that this works well for small m and for c close to $\mathsf{E}(B)$. A more precise result is available that provides a direct approximation to $P(\sum_{i=1}^{m} B^{(i)} > mc)$. This and related *large deviations* techniques form the topic of our next discussion. Later in this chapter we discuss a technique that combines the approximation based on the central limit theorem with a large deviations-based approximation.

Analysis Using Chernoff's Bound and Cramer's Theorem

We first derive an upper bound to the desired probability of exceeding the capacity by using Chernoff's bound, and then we state Cramer's theorem. We proceed generically. To this end, consider generic i.i.d. random variables $\{X_1, X_2, \ldots, \}$. Let $a > \mathsf{E}(X_1)$. Motivated by the need to determine $\Pr(\sum_{i=1}^{m} B^{(i)} > mc)$, we are interested in bounding or approximating

$$\Pr\left(\sum_{i=1}^{n} X_i \geq na \right) = \Pr\left(\sum_{i=1}^{n} (X_i - a) \geq 0 \right)$$

Observe that, for $\theta \geq 0$, and for any random variable Y,

$$I_{\{Y \geq 0\}} \leq e^{\theta Y}$$

Taking expectations on both sides, the following Chernoff's bound is obtained

$$\Pr(Y \geq 0) \leq \mathsf{E}\left(e^{\theta Y} \right) \tag{5.8}$$

---

[1]For example, it can be shown that if the $B^{(i)}$'s are $\{0, 1\}$ random variables, so that $\sum_{i=1}^{m} B^{(i)}$ is Binomial, then for $z_m \to \infty$ such that $\frac{z_m^3}{\sqrt{m}} \to 0$ (see [99, page 193])

$$\lim_{m \to \infty} \Pr\left(\frac{1}{\sigma(B)\sqrt{m}} \sum_{i=1}^{m} (B^{(i)} - \mathsf{E}(B)) > z_m \right) \sim 1 - \Phi(z_m)$$

where the case $E(e^{\theta Y}) = \infty$ is allowed. Applying this to the random variable $\sum_{i=1}^{n}(X_i - a)$,

$$\Pr\left(\sum_{i=1}^{n}(X_i - a) \geq 0\right) \leq E\left(e^{\theta(\sum_{i=1}^{n}(X_i-a))}\right)$$

$$= \left(e^{-\theta a}\, E\left(e^{\theta X_1}\right)\right)^n$$

$$= e^{-n(\theta a - \ln M(\theta))}$$

where in the first equality we have used the fact that the X_i are i.i.d., and in the second equality we have defined

$$M(\theta) = E\left(e^{\theta X_1}\right) \tag{5.9}$$

Because this is true for all $\theta \geq 0$, we can write

$$\Pr\left(\sum_{i=1}^{n}(X_i - a) \geq 0\right) \leq \inf_{\theta \geq 0} e^{(-n(\theta a - \ln M(\theta)))}$$

$$= e^{\inf_{\theta \geq 0}(-n(\theta a - \ln M(\theta)))}$$

$$= e^{-n \sup_{\theta \geq 0}(\theta a - \ln M(\theta))} \tag{5.10}$$

where the first equality follows from the fact that e^x is increasing in x. We know that $e^{\theta x}$ is convex in x for any θ, and hence, for all θ, by Jensen's inequality (see Appendix B) we obtain $E(e^{\theta X_1}) \geq e^{\theta E(X_1)}$, where equality holds for $\theta = 0$. It follows that

$$\theta a - \ln M(\theta) = \theta a - \ln E\left(e^{\theta X_1}\right)$$

$$\leq \theta a - \ln e^{\theta E(X_1)}$$

$$= \theta a - \theta E(X_1)$$

For $a > E(X_1)$ and $\theta \leq 0$, we therefore obtain

$$\theta a - \ln M(\theta) \leq 0$$

with equality for $\theta = 0$. Thus, for $\theta \leq 0$, $\theta a - \ln M(\theta)$ is nonpositive and achieves its maximum value, 0, at $\theta = 0$. We have thus shown that

$$\sup_{\theta \in \mathbb{R}} (\theta a - \ln M(\theta)) = \sup_{\theta \geq 0} (\theta a - \ln M(\theta))$$

Hence, using Equation 5.10, we obtain

$$\Pr\left(\sum_{i=1}^{n} X_i \geq na\right) \leq e^{-nl(a)} \tag{5.11}$$

where we have defined

$$l(a) = \sup_{\theta} (\theta a - \ln M(\theta)) \tag{5.12}$$

Notice that Equation 5.11 can be rewritten as

$$\frac{1}{n} \ln \Pr\left(\sum_{i=1}^{n} X_i \geq na\right) \leq -l(a)$$

Remarks 5.3   *Some Properties of the Exponent $l(a)$*

(i) $l(a) \geq 0$. This is true because for $\theta = 0$, $\theta a - \ln M(\theta) = 0$.

(ii) $l(a)$ is convex. See Exercise 5.5.

(iii) $l(E(X_1)) = 0$; that is, $\sup_{\theta} \left(\theta E(X_1) - \ln E\left(e^{\theta X_1}\right)\right) = 0$. This follows because $\theta E(X_1) - \ln E\left(e^{\theta X_1}\right) \leq \theta E(X_1) - \theta E(X_1) = 0$, where we have used Jensen's inequality (see Appendix B), in which equality holds for $\theta = 0$.   ∎

Exercise 5.5

   a. Show that the function $l(a)$ is convex; that is, for $0 \le \alpha \le 1$, $\alpha l(a_1) + (1 - \alpha) l(a_2) \ge l(\alpha a_1 + (1 - \alpha) a_2)$ (see Appendix C). (Hint: We have, by definition, $l(a) = \sup_\theta (\theta a - \ln M(\theta))$; use a property of sup to complete the proof.)

   b. Combine (ii) and (iii) in Remarks 5.3 to show that for $a > \mathsf{E}(X_1)$, $l(a)$ is nondecreasing. (Hint: use Theorem C.1 in Appendix C.)

Before we proceed further, we will look at an example.

Example 5.4   *The Two-State Markov Source*

Recall that the marginal random variables of the m independent on–off sources are denoted by $B^{(i)}$. Let p denote the probability that a source is in the *on* state:

$$B^{(i)} = \begin{cases} R & \text{with probability } p \\ 0 & \text{with probability } 1 - p \end{cases}$$

Hence, we need a link with per-source capacity c such that $pR \le c \le R$. We wish to determine what c should be so that $\Pr(\sum_{i=1}^{n} B^{(i)} \ge nc) \le e^{-\delta}$, say. Note that $e^{-9} \approx 10^{-4}$ and $e^{-12} \approx 10^{-6}$. Let us compute $l(a)$ as defined earlier for the marginal distribution of a source.

We have, for any real number θ,

$$M(\theta) = \left((1 - p) + p e^{R\theta} \right)$$

Figure 5.6 shows a plot of $\ln M(\theta)$ versus θ for the parameters $p = 0.4$ and $R = 1$. As expected, the slope of this curve at $\theta = 0$ is equal to $\mathsf{E}(B^{(i)})$ or $0.4R$. Now let us consider the exponent in the upper bound obtained from Chernoff's bound. As defined earlier, we have

$$l(a) = \sup_\theta \left(\theta a - \ln \left((1 - p) + p e^{R\theta} \right) \right)$$

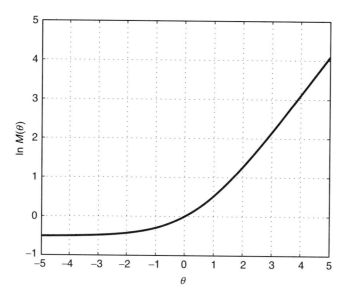

Figure 5.6   Plot of the log of the moment generating function ln M(θ) vs θ for the marginal distribution of an on–off source, with p = 0.4 and R = 1.

A simple calculation then yields the following expression for $l(a)$:

$$
l(a) = \begin{cases} \infty & \text{for } a < 0 \\ \frac{a}{R} \ln \frac{(a/R)}{p} + \left(1 - \frac{a}{R}\right) \ln\left(\frac{1-(a/R)}{1-p}\right) & \text{for } 0 \le a \le R \\ \infty & \text{for } a > R \end{cases} \tag{5.13}
$$

This function is plotted in Figure 5.7. Several important features are noticeable here:

- As mentioned in Remarks 5.3, $l(pR) = 0$.

- In this particular example, $l(a) = \infty$ for $a < 0$ and for $a > R$, because the source is bounded between these two values.

- The function $l(a)$ is also seen to be nonnegative and convex.

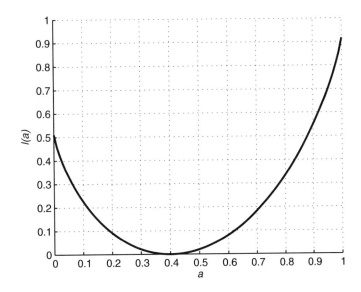

Figure 5.7 **Plot of the rate function $I(a)$ vs a for the marginal distribution of an on–off source, with $p = 0.4$ and $R = 1$.**

Now returning to the problem of obtaining a link capacity $C = nc$ such that $\Pr\left(\sum_{i=1}^{n} B^{(i)} \geq nc\right) \leq e^{-\delta}$, we can proceed by determining a c, $pR < c < R$, such that $I(c) = \frac{\delta}{n}$. The bound obtained earlier would then assert that

$$\Pr\left(\sum_{i=1}^{n} B^{(i)} \geq nc\right) \leq e^{-nI(c)} = e^{-\delta}$$

For example, with $\delta = 12$ and $n = 100$, the plot in Figure 5.7 yields, approximately, $c = 0.643$. Because we have taken $R = 1$, the bandwidth is normalized to the source peak rate. So if $R = 64$ Kbps, then we get $c = 41.15$ Kbps as the capacity required per source so that the fraction of slots in which data is lost is less than 1 in a million, on the average. ∎

A natural question to ask at this point is whether the bound on $\Pr\left(\sum_{i=1}^{n} B^{(i)} \geq nc\right)$ we have obtained is loose, or how conservative is the multiplexer design that this approach yields. This question is answered by the following theorem.

Theorem 5.3   (Cramer)
If $X_1, X_2, X_3, \ldots$ are i.i.d. random variables with mean $E(X)$, then for $a > E(X)$,

$$\lim_{n \to \infty} \frac{1}{n} \ln \Pr\left(\sum_{i=1}^{n} X_i > na \right) = -l(a) \qquad \blacksquare$$

In other words, for large n, $e^{-nl(a)}$ is a good approximation of $\Pr(\sum_{i=1}^{n} X_i > na)$, and not just an upper bound. It is an approximation nevertheless, and for any finite n the error would depend on terms that are not captured by the theorem. More precisely, we can write for large n,

$$\Pr\left(\sum_{i=1}^{n} X_i > na \right) = e^{-nl(a) + o(n)}$$

where $o(n)$ denotes a term that when divided by n goes to 0 as $n \to \infty$ (for example, $\ln(K)$ where K is a constant, or $\ln n$; see also Appendix A). On taking logs on both sides, dividing by n, and letting $n \to \infty$, we obtain the statement of Cramer's theorem.[2]

Finally, to conclude this discussion of Cramer's theorem, we point out a relationship with the law of large numbers. From the weak law of large numbers (see Theorem B.2 in Appendix B) we have, for all $\delta > 0$,

$$\lim_{n \to \infty} \Pr\left(\sum_{i=1}^{n} B^{(i)} > n(E(B) + \delta) \right) = 0$$

Clearly, Cramer's theorem leads to the same conclusion; just take $a = (E(B) + \delta)$ in Theorem 5.3. In addition we learn that the convergence to 0 takes place as $e^{-nl(E(B)+\delta)}$—that is, exponentially fast.

---

[2] The Bahadur–Rao theorem provides a refinement of Cramer's theorem and states (roughly) that

$$\Pr\left(\sum_{i=1}^{n} X_i > na \right) \approx \frac{1}{\theta(a)\sqrt{2\pi n}\sqrt{M''(\theta(a))}} e^{-nl(a)}$$

where $\theta(a)$ achieves the sup in $\sup_\theta (\theta a - \ln M(\theta))$.

Multiplexing Gain, Link Engineering, and Admission Control

From the point of view of network engineering, an important observation must be made from Example 5.4. Notice that as n is increased (i.e., more sources are multiplexed) then $\frac{\delta}{n}$ decreases, and the per-source link capacity computed by the preceding approach decreases toward the mean of 0.4. This is called *statistical multiplexing gain*; an increasing number of sources requires a decreasing amount of bandwidth requirement per source, for the same quality of service objective. A measure of multiplexing gain is the following:

$$\frac{\text{the number of sources that can be handled with statistical QoS}}{\text{the number of sources that can be handled if each is provided its peak rate}}$$

Of course, this gain is obtained at the cost of a nondeterministic QoS. Ultimately, for very large n the capacity required can be made arbitrarily close to (but must be strictly larger than) the mean pR. Hence, for the on–off source example, the statistical multiplexing gain can be no more than $\frac{C/(pR)}{C/R} = \frac{1}{p}$.

Now suppose we have a link of capacity C. For a given type of source and for a given probability of capacity overflow (characterized by δ, as before), how many sources can be handled by this link? Observe that for this we must solve the equation

$$\frac{\delta}{C/c} = l(c)$$

This yields a solution c^*, $EX_1 < c^* \le R$. Hence no more than $\lfloor \frac{C}{c^*} \rfloor =: N$ sources can be admitted; for if $n > N$ sources are admitted, then the per-source capacity $\frac{C}{n} < c^*$, which implies that $\frac{\delta}{n} > l\left(\frac{C}{n}\right)$, a violation of our engineering rule. There is thus a need for *admission control*. Newly arriving calls (or sessions) are admitted as long as the number of ongoing calls is less than N; any call that arrives when N calls are in progress is blocked. Observe now that at the *call level* this yields a system that is exactly the same as we would get in circuit switching over a trunk with a capacity to hold N calls (see Chapter 6). Hence, if it is assumed that the call arrival instants constitute a Poisson process, then the classical Erlang-B model (see Chapter 6) can be used to analyze the system at the call level and obtain the call-blocking probability.

Thus we arrive at an important point in the discussion of traffic engineering for stream traffic in a packet network. There are two QoS specifications that must be made:

- *In-call QoS:* This is the quality of service that governs the performance within a call. For a packet voice call, for example, in-call QoS specifications would require that the end-to-end packet delay be below some upper bound with a probability close to 1 and that the fraction of lost or late packets be small. In the earlier discussion on marginal buffering we guaranteed a delay bound by ensuring that no transmitted packet is in the multiplexer for more than two slots (one in which it arrives, and one in which it may possibly be served), and we placed a statistical requirement on the packet loss. These specifications will determine the number of calls that can be handled by a packet-multiplexed link of a given bandwidth C.

- *Call-level QoS:* This is the call-blocking probability; a typical value is 1%. For a specified call-blocking probability, there will be a maximum call arrival rate.

Thus, together the in-call and call-level QoS specifications determine the maximum call arrival rate that the network can handle. In the engineering of a packet-multiplexed voice trunk, the operator can program the admission control rule (or table) into the call controller software. Such a call controller is usually involved in the signaling required for call set up and can accept or block new calls depending on the number that are currently active.

5.6.2 Analysis with Arbitrary Buffering

With marginal buffering, just discussed, the link capacity also needs to be large because there should not be any accumulation of data from slot to slot. Clearly, more efficient use of the link bandwidth can be made if more than marginal buffering is provided. Data that are not transmitted in a slot, owing to insufficient capacity, can then be stored in the buffer. Consider, for example, a number of on–off sources feeding a multiplexer. When the number of active sources is such that the total input rate exceeds the link rate, then data will accumulate in the buffer. When the number of active sources decreases such that the total input rate drops below the link capacity, then the buffer will deplete. With proper choice of link capacity, the buffer will remain stable (i.e., will not blow up), and the delay of packets in the buffer will be within tolerable limits. Thus when there is a burst of data at the input with an input rate exceeding the link rate, then the buffer builds up. For this reason it is also called *burst scale* buffering.

Characterization of the Stationary Queue Length (Continuous Time)

Consider a multiplexer with cumulative arrival process $A(t)$, $A(t) = 0$, $t < 0$. At time 0_- the buffer is empty. The link transmits at rate C bps, and the scheduler is non-idling. The cumulative arrival process $A(t)$ has stationary increments. This implies (at least) the following: The distribution of $A(t + s) - A(t)$ (the amount of data that arrives in the interval $(t, t+s]$) does not depend on t, but depends only on the length of the interval s; that is, in any interval of length s the distribution of the amount of data that arrives is the same as in any other such interval. $A(t+s) - A(t)$ is called an *increment* of the process $A(t)$. In fact, more holds; if we take the joint distribution of several increments of $A(t)$ (at different points in time and over intervals of possibly different lengths), then this joint distribution is invariant to a shift of all these intervals by the same amount (see also Appendix B).

Now recall Reich's equation (Equation 4.3), which says that for all $t \geq 0$, the amount of data in the buffer $X(t)$ is given by

$$X(t) = \sup_{0_- \leq s \leq t} \left(A(t) - A(s) - C \cdot (t - s) \right)$$

Our objective here is to characterize the probability distribution of $X(t)$ as $t \to \infty$. Note that in this equation we cannot invoke the stationarity of the increments of $A(t)$ to obtain the simplification $X(t) \overset{dist}{=} \sup_{0 \leq \tau \leq t_+} \left(A(\tau) - C \cdot \tau \right)$. This would be wrong because in the expression for $X(t)$ all the increments of the arrival process have their right end-points at t, whereas in the proposed simplified expression all the increments have their *left* end-points at 0. Hence equality in distribution will not hold.

If we simply let t go to ∞ in Reich's equation, we do not get anything useful. Notice that as t increases in Reich's equation, the right edge of the interval over which the sup is being taken increases, and hence the increments of $A(t)$ that are involved in the expression are not a superset of those for smaller t's. Hence we can say nothing about how the sup varies as we increase t.

Instead, let us assume that the arrival process extends over the entire time line; we need only work with the stationary increments of the process. "Start" the multiplexer with an empty buffer at $-t_-$ (i.e., just before the instant $-t$, the buffer is empty). With this "initial" condition denote the amount of data in the buffer by $U_{-t}(\cdot)$; then consider $U_{-t}(0)$. Rewriting Reich's equation we get

$$U_{-t}(0) = \sup_{-t_- \leq s \leq 0} (A(0) - A(s) - C \cdot (-s))$$

This is equivalent to writing

$$U_{-t}(0) = \sup_{0 \leq s \leq t_+} (A(0) - A(-s) - Cs)$$

Now because of the stationarity of the increments of $A(t)$, we see that $U_{-t}(0)$ and $X(t)$ have the same distributions. This follows because both of these random variables are the queue lengths after an amount of time t starting with an empty buffer, and the arrival process in each case is identically distributed (this is because of the stationary increments of $A(t)$).

Now we fix a sample path of the arrival process and let t go to ∞. Notice that as we increase t, the sup in the right side of the expression for $U_{-t}(0)$ is over an increasing set (for a fixed sample path of $A(t)$). Thus, as t increases, $U_{-t}(0)$ is nondecreasing and hence converges, possibly to ∞. Call the limit U_∞:

$$U_\infty = \sup_{s \geq 0} (A(0) - A(-s) - Cs)$$

By the observation that $U_{-t}(0)$ and $X(t)$ have the same distributions, it follows that the distribution of $X(t)$ converges to that of U_∞. We can say that the multiplexer is *stable* if the limiting distribution of $X(t)$ is finite with probability 1. Suppose we know that, with probability 1,

$$\lim_{t \to \infty} \frac{A(0) - A(-t)}{t} = A < C \tag{5.14}$$

This will happen if the time average data arrival rate is less than C. It then follows that as t goes to ∞, $A(0) - A(-t) - Ct$ goes to $-\infty$ with probability 1, and hence $\sup_{s \geq 0}(A(0) - A(-s) - Cs)$ stays bounded with probability 1. We conclude that

$$X(t) \overset{dist}{\to} \sup_{s \geq 0} (A(0) - A(-s) - Cs) \tag{5.15}$$

and this limit is finite with probability 1, if $\lim_{t\to\infty} \frac{A(t)}{t} = A < C$ (see Appendix B for a discussion of the notion of convergence in distribution).

Armed with this result, we can now study the long-run fraction of time that the buffer exceeds some level—say, B. This is given by

$$\Pr\left(\sup_{t\geq 0}(A(0) - A(-t) - Ct) > B\right)$$

If we want the delay in this multiplexer to exceed T with a small probability, then such a requirement could be expressed as

$$\Pr\left(\sup_{t\geq 0}(A(0) - A(-t) - Ct) > CT\right) < \epsilon$$

or equivalently,

$$\Pr\left(\sup_{t\geq 0}\frac{(A(0) - A(-t) - Ct)}{C} > T\right) < \epsilon \qquad (5.16)$$

We have thus expressed the stochastic delay objective purely in terms of the arrival process and the multiplexer capacity. From this we can expect to determine the smallest value of the link bandwidth that would satisfy the QoS requirement. The solution that we get will depend on the stochastic characterization of the $A(t)$ that we have.

Characterization of the Stationary Queue Length (Discrete Time)

Although we will have occasion to use the continuous time analysis, the detailed study of the stationary queue length is easier to perform in discrete time. Hence we now obtain, in discrete time, expressions that are parallel to those obtained earlier. Using notation introduced earlier, the amount of data in the buffer at t_k, denoted by X_k, can be written as

$$X_k = \max\left\{(X_{k-1} + B_k - C), 0\right\} \qquad (5.17)$$

where B_k is the amount of data that arrives in the kth interval, C is the amount of data that the link can emit in each interval, and $X_0 = 0$. It is then easy to see that (Reich's equation in discrete time)

$$X_n = \max_{0 \le k \le n} (A_n - A_k - (n - k)C) \qquad (5.18)$$

where, for $n \ge 1$, $A_n := \sum_{k=1}^{n} B_k$, and $A_0 = 0$.

Exercise 5.6

Following an argument exactly as in the continuous time case, show that, as $n \to \infty$,

$$X_n \overset{dist}{\to} \sup_{k \ge 0} (A_0 - A_{-k} - kC) \qquad (5.19)$$

where, as in the continuous time case, we have taken the arrival process for all $k \in \{\ldots, -3, -2, -1, 0, 1, 2, 3, \ldots\}$.

The limit in Equation 5.19 is finite with probability 1, if $\lim_{n \to \infty} \frac{A_n}{n} < C$. That is, the rate of arrivals per interval is less than the amount of data that can be served in that interval.

Analysis Using Chernoff's Bound: Effective Bandwidths

First, let us consider the discrete time case. Let us denote by X the limiting random variable obtained in Equation 5.19 (assuming that the random variable is finite with probability 1). X is the random amount of data in the multiplexer buffer, in steady state. We are interested in studying $\Pr(X > x)$, the probability that in steady state there are more than x bytes of data in the buffer. We know that X is the limit, in distribution, of the random variables X_n, where the equation

$$X_n \overset{dist}{=} \max_{0 \le k \le n} (A_0 - A_{-k} - kC)$$

relates the distribution of X to the random process A_k and the link rate C. Let us write, for $k \geq 0$, $S_{-k} = A_0 - A_{-k} - kC$. Then we can write

$$
\Pr(X_n > x) = \Pr\left(\left(\max_{0 \leq k \leq n} S_{-k}\right) > x\right)
$$

$$
= \Pr\left(\bigcup_{0 \leq k \leq n} \{S_{-k} > x\}\right)
$$

$$
\leq \sum_{k=0}^{n} \Pr(S_{-k} > x) \tag{5.20}
$$

where the second equality is obtained because $(\max_{0 \leq k \leq n} S_{-k}) > x$ if and only if there is some k such that $S_{-k} > x$, and the last inequality is just the union bound (the probability of the union of sets is bounded above by the sum of the probabilities of the sets). For any $\theta > 0$, using Chernoff's bound (see Equation 5.8), we then get

$$
\Pr(X_n > x) \leq \sum_{k=0}^{n} \mathsf{E}\left(e^{\theta(S_{-k} - x)}\right)
$$

$$
= e^{-\theta x} \sum_{k=0}^{n} \mathsf{E}\left(e^{\theta S_{-k}}\right) \tag{5.21}
$$

Let us assume that the data arrival process A_k is such that the following *log-moment generating function* exists at θ:

$$
\Gamma(\theta) = \lim_{n \to \infty} \frac{1}{n} \ln \mathsf{E}\left(e^{\theta(A_n - A_0)}\right) \tag{5.22}
$$

It follows that, for large n,

$$
\mathsf{E}\left(e^{\theta(A_n - A_0)}\right) \approx e^{n\Gamma(\theta)}
$$

where $\approx$ is read as "is approximately equal to." Because A_k has stationary increments, we have $\mathsf{E}(e^{\theta(A_0 - A_{-k})}) = \mathsf{E}(e^{\theta(A_k - A_0)})$. Hence, we can write, for large k,

$$\mathsf{E}\left(e^{\theta S_{-k}}\right) \approx e^{k(\Gamma(\theta) - \theta C)}$$

Now suppose the following condition holds:

$$\frac{\Gamma(\theta)}{\theta} < C \qquad\qquad (5.23)$$

That is, for some $\epsilon > 0$, $\frac{\Gamma(\theta)}{\theta} + \epsilon = C$. Then we observe that the terms in the sum on the right side of Equation 5.21 are geometrically decreasing (with k), and we can conclude that the sum is bounded as n goes to infinity; let us write a bound on that term as K. Hence we can write, for large n,

$$\Pr(X_n > x) \leq Ke^{-\theta x}$$

Because $\Pr(X_n > x) \to_{n \to \infty} \Pr(X > x)$, we can also conclude that

$$\Pr(X > x) \leq Ke^{-\theta x}$$

This says that the decay of the probability of the steady state buffer exceeding x is at least exponential with parameter θ if the link rate is greater than $\frac{\Gamma(\theta)}{\theta}$. A more precise statement can be made, but before we work on that we make some observations.

If we can express the QoS requirement at the multiplexer as a requirement on the tail probability decay rate, then to ensure a particular decay rate θ, it is sufficient to choose a link capacity that is more than $\frac{\Gamma(\theta)}{\theta}$ by an arbitrarily small, positive amount. In view of this, $\frac{\Gamma(\theta)}{\theta}$ can be called the *effective bandwidth* of the source for the QoS requirement θ; we denote this by $e(\theta)$.

Now let us turn to the continuous time case. Using the characterization of the stationary queue length for the continuous time case provided earlier, we can also develop an effective bandwidth result. Let us consider a fluid source, with $A(t)$ denoting the aggregate data that arrives over the interval $[0, t]$. The multiplexer has a fluid server of rate C. Parallel to the definition of $\Gamma(\theta)$ for a discrete time source, we define

$$\Gamma(\theta) = \lim_{t \to \infty} \frac{1}{t} \ln \mathsf{E}\left(e^{\theta A(t)}\right) \tag{5.24}$$

It can then be shown that to satisfy the requirement that the stationary queue length distribution decay faster than the exponential rate $\theta > 0$,

$$\Pr(X > x) \le K e^{-\theta x}$$

it suffices to have

$$\frac{\Gamma(\theta)}{\theta} < C$$

Hence, even for a continuous time fluid model, $\frac{\Gamma(\theta)}{\theta}$ can be called the effective bandwidth of the source for the QoS requirement θ; we denote this as well by $e(\theta)$.

Example 5.5 *The Two-State Markov Source*

Consider a two-state stationary source B_k with $B_k \in \{0, R\}$, with $\Pr(B_k = R) = p = 1 - \Pr(B_k = 0)$, as in Example 5.4. When the B_k are i.i.d., then clearly, as in Example 5.4,

$$\Gamma(\theta) = \ln\left((1 - p) + p e^{R\theta}\right)$$

In Figure 5.8 we plot $\Gamma(\theta)$ and $C\theta$ on the same set of axes; here $C > pR$, the mean rate of the source. The slope of $\Gamma(\theta)$ at $\theta = 0$ is $\mathsf{E}(B) = pR$, and hence C must be greater than pR for there to be a $\theta > 0$ such that $C\theta > \Gamma(\theta)$. This just confirms what will be proved in the remarks to follow: that the effective bandwidth must be greater than the mean rate of the source. In this example $R = 1$, $pR = 0.4$, and we have chosen $C = 0.75$. This could correspond to the practical situation of a packet voice source with peak rate 64 Kbps, activity factor 0.4, and the assigned service rate $C = 48$ Kbps. Notice that there is a value $\theta_{\max}$ (approximately 3.4 in Figure 5.8) such that $C\theta > \Gamma(\theta)$ only for $0 < \theta < \theta_{\max}$. Thus with this value of C, we can achieve tail probability decay rates only in this range. We see that as C increases $\theta_{\max}$ increases. ∎

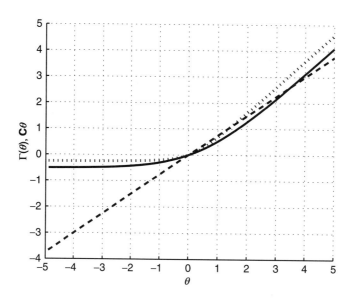

Figure 5.8   Plot of the log moment generating functions Γ(θ) for an i.i.d. Bernoulli source with p = 0.4 and R = 1 (solid curve), and a Markov source with the same mean but with a burst length of 3 (dotted curve), and the function Cθ, where C = 0.75 (dashed line).

Exercise 5.7

Observe that the asymptote of $\Gamma(\theta)$ as $\theta \to \infty$ has slope R, and hence for $C \to R$, $\theta \to \infty$. What is the interpretation of this?

Now suppose the B_k is modulated by a two-state Markov chain, as in Example 5.3 ($B_k = R$ when the Markov chain is in state 1; the transition probabilities are $p_{10} = \tau = 1 - p_{11}$, and $p_{01} = \sigma = 1 - p_{00}$). We will then see later that for this source,

$$\Gamma(\theta) = \ln \frac{1}{2} \Big[((1 - \tau)e^{\theta R} + (1 - \sigma)) \tag{5.25}$$

$$+ \sqrt{((1 - \tau)e^{\theta R} + (1 - \sigma))^2 - 4e^{\theta R}(1 - \tau - \sigma)} \Big]$$

In Figure 5.8, we have plotted $\Gamma(\theta)$ for a source with *the same mean as the i.i.d. source* earlier, but with a mean *on* period equal to 3 (the mean *on* period for the i.i.d. (Bernoulli) source is $\frac{5}{3}$). Notice that now with the same link rate (i.e., 0.75), θ_{max} is only about 1.3. As would be expected, a source with longer bursts (higher correlations over time) would result in longer periods during which the source arrival rate and the link rate would be mismatched, thus requiring more buffering. In other words, if the same θ_{max} must be ensured for the Markov source, then more capacity would be required. It can be verified from the curves that a capacity of about 0.88 is required to achieve $\theta_{max} = 3.4$ with the burstier Markov source.

Remarks 5.4   *Some Properties of $e(\theta)$*

a. If $A_n \leq nR$ (or R is the peak rate of the source), then, for $\theta \geq 0$, $e(\theta) \leq R$. We can see this as follows.

$$\Gamma(\theta) = \lim_{n\to\infty} \frac{1}{n} \ln \mathsf{E}\left(e^{\theta(A_n - A_0)}\right)$$

Hence, because $\theta \geq 0$,

$$\Gamma(\theta) \leq \lim_{n\to\infty} \frac{1}{n} \ln \mathsf{E}\left(e^{\theta nR}\right)$$

$$= \theta R$$

It follows that

$$e(\theta) = \frac{\Gamma(\theta)}{\theta} \leq R$$

Thus, as expected, the effective bandwidth is less than or equal to the peak rate of the source.

b. If $\lim_{n\to\infty} \frac{\mathsf{E}(A_n)}{n} = r$ (i.e., r is the average source rate), then $e(\theta) \geq r$. To prove this, we use Jensen's inequality (see Appendix B) to write

$$\Gamma(\theta) \geq \lim_{n\to\infty} \frac{1}{n} \ln e^{\theta \mathsf{E}(A_n - A_0)}$$

$$= \theta r$$

Hence

$$e(\theta) = \frac{\Gamma(\theta)}{\theta} \geq r$$

as expected. The effective bandwidth is no less than the average rate of the source. This is fine, because if the link rate is taken to be less than the average source rate, then the buffer process X_n is itself not stable.

c. *Effective bandwidths add up.* If the multiplexer input is the superposition of several independent sources, then we would be interested in the effective bandwidth of this composite source. Exercise 5.8 concludes that the effective bandwidth of the superposition of independent sources (for a given value of θ) is the sum of their effective bandwidths.

d. *Units:* In numerical calculations the question comes up as to the units of $\Gamma(\theta)$ and θ. From the previous discussion, it is easy to see that $\Gamma(\theta)$ has units of time^{-1} (e.g., sec^{-1}), and θ has units of the unit of data^{-1} (e.g., bits^{-1}). From this we see that, as required, the units of $\frac{\Gamma(\theta)}{\theta}$ are data per unit time (e.g., bits/sec).
∎

Exercise 5.8

If $A_n^{(i)}, 1 \leq i \leq m$, are independent sources with log-moment generating functions $\Gamma^{(i)}(\theta)$, then the log-moment generating function of their superposition $A_n = \sum_{i=1}^{m} A_n^{(i)}$ is $\Gamma(\theta) = \sum_{i=1}^{m} \Gamma^{(i)}(\theta)$. Hence conclude that effective bandwidth $e(\theta)$, of the process A_n, is given by $e(\theta) = \sum_{i=1}^{m} e_i(\theta)$, where $e_i(\theta)$ is the effective bandwidth of the source $A_n^{(i)}$.

Calculating $\Gamma(\theta)$ for a Discrete Time Markov Source

Consider a discrete time model for which the amount of data arriving in the kth interval $B_k, k \geq 1$, is a stationary and ergodic Markov chain on the state space $\{0, 1, 2, \ldots, R\}$. Let $\mathbf{P}$ denote the transition probability matrix of the Markov chain. Let π_b be the stationary probability that the Markov chain is in the state $b \in \{0, 1, 2, \ldots, R\}$, and let π denote the probability vector. We would like to calculate

$$\Gamma(\theta) = \lim_{n \to \infty} \frac{1}{n} \ln \mathsf{E}_\pi e^{\theta \sum_{k=1}^{n} B_k}$$

where E_π denotes that the expectation is with respect to the stationary probability law of the process B_k. This means that $\Pr(B_1 = b) = \pi_b, b \in \{0, 1, 2, \ldots, R\}$. With a slight abuse of notation, we will use E_b to denote the expectation when $B_1 = b$. We proceed by conditioning on the state at time 1 (this is done in the second equality in the following calculation).

$$E_\pi e^{\theta \sum_{k=1}^n B_k} = E_\pi e^{\theta B_1} e^{\theta \sum_{k=2}^n B_k}$$

$$= \sum_{b_1=0}^R \pi_{b_1} e^{\theta b_1} \sum_{b_2=0}^R p_{b_1 b_2} E_{b_2} e^{\theta \sum_{k=1}^{n-1} B_k}$$

$$= \pi Q P (f(n-1))^T$$

where Q is the diagonal matrix with diagonal terms $[e^0, e^\theta, e^{2\theta}, \ldots, e^{R\theta}]$, and we define the row vectors $f(n-1)$ as follows: for $b \in \{0, 1, 2, \ldots, R\}$,

$$f_b(n) = E_b e^{\theta \sum_{k=1}^n B_k}$$

$$f_b(1) = e^{\theta b}$$

It follows that

$$E_\pi \, e^{\theta \sum_{k=1}^n B_k} = \pi (QP)^{(n-1)} (f(1))^T$$

$$= e^{\theta R n} \pi (\tilde{Q} P)^{(n-1)} (\tilde{f}(1))^T$$

where $\tilde{Q} = e^{-R\theta} Q$, and $\tilde{f}(1) = e^{-R\theta} f(1)$. Define

$$\tilde{P} = \tilde{Q} P$$

That is,

$$\tilde{P} = \begin{bmatrix} e^{-\theta R} & 0 & \cdots & 0 \\ 0 & e^{-\theta(R-1)} & \cdots & 0 \\ \vdots & 0 & \ddots & \vdots \\ 0 & \cdots & 0 & 1 \end{bmatrix} P$$

Hence we can write

$$\mathsf{E}_{\boldsymbol{\pi}}\, e^{\theta\, \sum_{k=1}^{n} B_k} = e^{\theta R n} \boldsymbol{\pi}\, \tilde{\mathbf{P}}^{(n-1)}(\tilde{\mathbf{f}}(1))^T$$

Let us assume that the transition probability matrix $\mathbf{P}$ is aperiodic and irreducible (see Section D.1). It can then be shown that $\tilde{\mathbf{P}}$ has an eigenvalue $\eta(\theta)$ that (1) is real, positive, and simple, and (2) has strictly positive left and right eigenvectors, $\mathbf{w}$ and $\mathbf{v}$, and (3) $\eta(\theta) > |\lambda|$ for any other eigenvalue λ. $\eta(\theta)$ is called the Perron–Frobenius eigenvalue of $\tilde{\mathbf{P}}$. We can also see that, for $\theta > 0$, $\tilde{\mathbf{P}}$ is a substochastic matrix (i.e., its row sums are less than 1), and hence $\eta(\theta) < 1$. Also, the following can be stated, for $n \to \infty$,

$$\tilde{\mathbf{P}}^n \sim \eta(\theta)^n \,(\mathbf{v}^T \circ \mathbf{w})$$

where $\circ$ denotes the outer product (see Appendix D), and $\sim$ denotes asymptotic equivalence (see Appendix A). It follows that as $n \to \infty$,

$$\mathsf{E}_{\boldsymbol{\pi}} e^{\theta\, \sum_{k=1}^{n} B_k} \sim e^{n\theta R}\, \eta(\theta)^{n-1} \boldsymbol{\pi}(\mathbf{v}^T \circ \mathbf{w})\, (\tilde{\mathbf{f}}(1))^T$$

Because $\boldsymbol{\pi}(\mathbf{v}^T \circ \mathbf{w})(\tilde{\mathbf{f}}(1))^T$ is a constant scalar term, not depending on n, we obtain the following by taking logarithms, dividing by n, and then letting $n \to \infty$:

$$\Gamma(\theta) = \theta R + \ln \eta(\theta) \qquad (5.26)$$

Observe that, for $\theta > 0$, because $\eta(\theta) < 1$,

$$\frac{\Gamma(\theta)}{\theta} < R$$

as expected (see Remarks 5.4).

Exercise 5.9

For the discrete time, two-state Markov modulated source, obtain Equation 5.25 by applying the formula in Equation 5.26.

Calculating $\Gamma(\theta)$ for a Continuous Time Markov Fluid Source

Consider a continuous time fluid source model, for which the data arrive at time t at rate $B(t)$. $B(t)$ is a continous time Markov process taking values among the set of rates $r_1 < r_2 < \cdots < r_{n-1} < R$. The transition rate matrix of $B(t)$ is denoted by $\mathbf{Q}$. Take $\mathbf{Q}$ to be irreducible, and hence there exists a stationary probability distribution π on $\{r_1, r_2, \ldots, r_{n-1}, R\}$ (see Theorem D.3). We would like to calculate

$$\Gamma(\theta) = \lim_{t \to \infty} \frac{1}{t} \ln \mathsf{E}_\pi e^{\theta \int_0^t B(t)\, dt}$$

where E_π denotes that the expectation is with respect to the stationary probability law of the process $B(t)$. Let $\mathbf{D} = diag(r_1, r_2, \ldots, r_{n-1}, R)$. It can then be shown that $\Gamma(\theta) = \eta(\theta)$, where $\eta(\theta)$ is the Perron–Frobenius eigenvalue of the matrix $\theta \mathbf{D} + \mathbf{Q}$.

The "Necessity" of Effective Bandwidths (⋆)

Let us again look at the analysis that led to the idea of an effective bandwith for a source, and observe that we proceeded by upper bounding $\Pr(X_n > x)$. We showed that it is sufficient to have $C > \frac{\Gamma(\theta)}{\theta}$ to ensure that $\ln \Pr(X > x) \leq -\theta x$ for large x. Because we proceeded by using an upper bound on the desired probability, a natural question to ask is whether it is *necessary* to have $C > \frac{\Gamma(\theta)}{\theta}$ to ensure this QoS objective. Or is this an overestimate of the required capacity? With reference to our discussion on marginal buffering, where also we initially use the Chernoff bound, the question is similar to asking whether a result like Cramer's theorem holds for the sequence of correlated random variables, $B_k, k \geq 0$ (the amounts of data arriving in successive intervals). Such a result is the Gärtner–Ellis theorem, of which we will state a simple case that is adequate for our purposes.

Let the maximum value of B_k (the amount of data that arrives in the kth interval) be R, and let the average data rate in the source be r; that is, $r = \lim_{n \to \infty} \frac{1}{n} \sum_{k=0}^{n-1} B_k$.

Recall Equation 5.22 and assume that the $\Gamma(\theta)$ defined there is finite for all real numbers θ and is also differentiable. Notice that the $\Gamma(\theta)$ in Equation 5.25, for the discrete time two-state Markov source, has this property. Similar to

$l(a)$ in Section 5.6.1., define

$$\gamma(a) = \sup_{\theta} (\theta a - \Gamma(\theta)) \tag{5.27}$$

In fact, observe that if the B_k sequence is i.i.d. then $\gamma(a)$ defined here is exactly the same as $l(a)$.

> **Exercise 5.10**
> Show that $\gamma(a)$ is nonnegative, convex, and differentiable, $\gamma(r) = 0$, and, for $a > r$, $\gamma(a)$ is nondecreasing with a.

The Gärtner–Ellis theorem then states that, for $a > r$,

$$\lim_{n \to \infty} \frac{1}{n} \ln \Pr\left(\sum_{k=0}^{n-1} B_k > na \right) = -\gamma(a) \tag{5.28}$$

Notice the similarity between this result and Cramer's theorem. If the B_k sequence is stationary and ergodic, then we know that $\Pr\left(\sum_{k=0}^{n-1} B_k > na \right) \to 0$, as $n \to \infty$. This follows from Birkhoff's ergodic theorem (Theorem B.5 in Appendix B). We see that the Gärtner–Ellis theorem shows that this convergence is exponential and provides us with the exponent.

Remarks 5.5

We note here that the technical conditions (one being that $\Gamma(\theta)$ exists) for the Gärtner–Ellis theorem to hold do not themselves hold if the input process is *long-range-dependent*; we return to this point in Section 5.11. The conditions do, however, hold for finite state Markov traffic models for stream traffic. ∎

It can be argued from the Gärtner–Ellis theorem and the convexity of $\gamma(a)$ that, for $a > r$, the event $\{\sum_{k=0}^{n-1} B_k = na\}$ occurs most probably by $\sum_{k=0}^{m} B_k$ growing along the *straight line* path of slope a, and this path has probability approximately $e^{-n\gamma(a)}$. Let us accept this statement and proceed. Note that for $a > r$, if $\sum_{k=0}^{m} B_k$ grows along the path $a \cdot m$ it is a deviation from the mean path $r \cdot m$, and hence the probability is smaller the larger the value of a.

With these results we can now proceed to lower bound $\Pr(X_n > x)$ as follows. We begin as we did in getting the upper bound to $\Pr(X_n > x)$ in Equation 5.20.

$$\Pr(X_n > x) = \Pr\left(\left(\max_{0 \le k \le n} S_{-k}\right) > x\right)$$

$$= \Pr(\cup_{0 \le k \le n}\{S_{-k} > x\})$$

$$\ge \Pr(S_{-n} > x) \qquad (5.29)$$

where, we recall, $S_{-k} = A(0) - A(-k) - kC$. Because the arrival process has been assumed to have stationary increments, we see that $\Pr(S_{-n} > x) = \Pr(A(0) - A(-n) - nC > x) = \Pr(A(n) - A(0) - nC > x)$. We thus have the lower bound

$$\Pr(X_n > x) \ge \Pr\left(\sum_{k=1}^{n} B_k > nC + x\right)$$

where the capacity C is greater than the mean data arrival rate r (which, in any case, is a requirement for stability of the buffer).

Let us now approximate this lower bounding probability by using the Gärtner–Ellis theorem. For the event $\sum_{k=1}^{n} B_k > nC + x$ to occur, the cumulative data arrivals, $A(j) - A(0) = \sum_{k=1}^{j} B_k$, will most probably grow along a straight line of slope $s > C$ (as discussed earlier; and recall that $C > r$). If the slope is s, the cumulative data will exceed the line $kC + x$ at $k = \frac{x}{s-C}$ (see Figure 5.9). The probability of this path is approximately $e^{-\frac{x}{s-C}\gamma(s)}$. The possible paths are those with slopes such that the crossing occurs before time n; that is, $\frac{x}{s-C} < n$, or $s > C + \frac{x}{n}$, where we can ignore the term $\frac{x}{n}$ for large n. Putting all this together, we approximate

$$\Pr\left(\sum_{k=1}^{n} B_k > nC + x\right) \approx \sum_{s>C} e^{-\frac{x}{s-C}\gamma(s)}$$

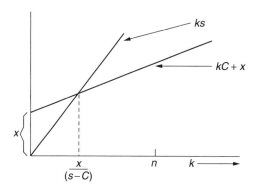

Figure 5.9   A straight line growth path with slope s exceeds the buffer level x at time $\frac{x}{(s - C)}$.

For large x, we can then approximate the sum in the right side by the dominant exponential term, to get

$$\Pr\left(\sum_{k=1}^{n} B_k > nC + x \right) \approx e^{-x \inf_{s > C} \frac{\gamma(s)}{s - C}}$$

Now we wish to determine a requirement on C so that the right hand side decays no slower than $e^{-\theta x}$, our QoS objective.

Exercise 5.11

Show that to obtain $\inf_{s > C} \frac{\gamma(s)}{s - C} = \theta$ we need $C = \frac{\Gamma(\theta)}{\theta}$. Hint: Write $\gamma(a) = \theta(a)a - \Gamma(\theta(a))$, where $\theta(a)$ is the maximizer in the definition of $\gamma(a)$ (i.e., $\Gamma'(\theta(a)) = a$); then examine the minimizer of $\inf_{s > C} \frac{\gamma(s)}{s - C}$ by differentiating and setting equal to 0.

We thus find that a capacity of $C = \frac{\Gamma(\theta)}{\theta}$ is not only sufficient but also necessary if we are to achieve the desired QoS objective θ.

The Stationary Buffer Distribution Has an Exponential Tail

In the foregoing development we sought a capacity C such that the QoS requirement

$$\lim_{x \to \infty} \frac{1}{x} \ln \Pr(X > x) = -\theta$$

was satisfied; that is, the dominant behavior of $\Pr(X > x)$ for large x is $e^{-\theta x}$. Here X denotes the stationary amount of data in the buffer. Suppose, on the other hand, that we are given a link of capacity C and that an arrival process $A(k)$ feeds a buffer that is served by this link; C is larger than the average data rate, r, and C is less than the peak rate R. What is the tail behavior of the stationary buffer distribution in this case? Under the technical conditions mentioned earlier, we simply need to solve the following equation for η:

$$C\eta = \Gamma(\eta) \tag{5.30}$$

The tail of the buffer distribution will then be exponential with exponent η. Writing this as $\eta(C)$ to show the dependence on the link capacity, we thus have

$$\lim_{x \to \infty} \frac{1}{x} \ln \Pr(X > x) = -\eta(C)$$

The exponential decay of $\Pr(X > x)$ for large x does not say very much for small x. A more precise approximation has been obtained in [92] for Markovian sources. It is argued that the following is a good approximation of the stationary queue length distribution $\Pr(X > x)$.

$$\Pr(X > x) \approx a_0 e^{-\eta x} \tag{5.31}$$

where a_0 *is the loss probability with a bufferless multiplexer* (for example, for the continuous time case, for a multiplexer service rate C, $a_0 = P(\sum_{i=1}^{m} B^{(i)}(0) > C))$, and η is the parameter of the exponential tail. Equivalently,

$$\ln \Pr(X > x) \approx \ln a_0 - \eta x$$

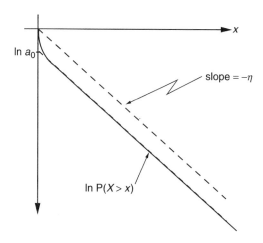

Figure 5.10 Typical plot of ln Pr($X > x$) for Markovian sources, and an affine approximation.

Thus, we have an *affine approximation* to $\ln \Pr(X > x)$. This approximation is depicted in Figure 5.10, where the shape of $\ln \Pr(X > x)$ near the origin is also shown. We see that the curve near $x = 0$ in $\ln \Pr(X > x)$ is approximated by a jump of size $\ln a_0$, and from here the approximation is taken to be the linear asymptote. In other words, the approximation asserts that the vertical distance between the linear asymptote of $\ln \Pr(X > x)$ and the straight line of slope $-\eta$ passing through the origin is just $\ln a_0$.

5.7 The Effective Bandwidth Approach for Admission Control

In Figure 5.11, we provide a simple illustration of an admission control approach based on effective bandwidths. We assume that the approximation depicted in Figure 5.10 holds. A single stream source is being handled by a multiplexer, and the following is the requirement on the stationary queue length, X, for this source in the multiplexer.

$$\Pr(X > x_{\max}) \leq \epsilon$$

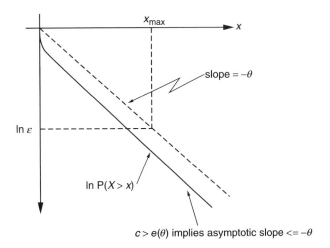

Figure 5.11   Relation between ln Pr($X > x$) and the QoS requirement Pr($X > x_{max}$) > ϵ when $c > e(\theta)$.

Define $\theta = -\frac{\ln \epsilon}{x_{max}}$. Figure 5.11 shows a line with slope $-\theta$, passing through the origin and the point $(x_{max}, \ln \epsilon)$. This line summarizes the QoS requirement. We also show a sketch of the natural logarithm of the complementary c.d.f. of X ($\ln \Pr(X > x)$) when the multiplexer capacity $c > e(\theta)$. Notice that if the link rate at the multiplexer is larger than the source's effective bandwidth (corresponding to the QoS requirement), then we can meet the QoS target.

Suppose several sources, $A^{(i)}(t), 1 \leq i \leq m$, are multiplexed into a link with a single, first-come-first-served buffer. The QoS requirement is

$$\Pr(X > x_{max}) \leq \epsilon$$

where X is the stationary queue length in the buffer. Following the effective bandwidth (EBW) approach discussed earlier, it is sufficient to provide a service capacity $c > e(\theta) = \frac{\Gamma(\theta)}{\theta}$, where $\Gamma(\theta)$ is the log-moment generating function of the superposition arrival process. Furthermore, because $\Gamma(\theta) = \sum_{i=1}^{m} M^{(i)}(\theta)$, we see that $e(\theta) = \sum_{i=1}^{m} e^{(i)}(\theta)$; that is, the EBW of the superposition arrival process is the sum of the EBWs of its component arrival processes. Notice how the additivity of

EBW can be very useful in the admission control process. One simply calculates the EBW of the arriving source for the QoS objective θ and checks whether the total EBW is less than the link capacity.

Although this additivity of effective bandwidths is an attractive feature, we will now argue that directly using this approach can lead to a very conservative design; that is, it may lead to an allocation of capacity much higher than what is actually required to meet the QoS objective.

Let us first examine what happens if we superimpose a number of independent but statistically identical sources and use the EBW approach. Under the EBW approach, the capacity required for m sources is required to scale as $m\,e^{(1)}(\theta)$, where $e^{(1)}(\theta)$ is the EBW of one of the sources. As shown in Figure 5.12, this scaling with the number of sources is necessary to maintain the asymptotic slope of $\ln \Pr(X > x)$ to be less than $-\theta$. However, near $x = 0$ the form of $\ln \Pr(X > x)$ is governed by small-buffer behavior, which improves when capacity is increased linearly with the number of sources. Notice, in Figure 5.12, that the EBW approach yields an increasingly overengineered system as we increase the number of independent sources in the superposition.

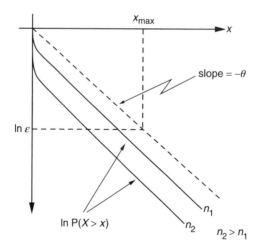

Figure 5.12   Relation between ln Pr($X > x$) and the QoS requirement Pr($X > x_{max}$) < ϵ when $c > e(\theta)$, for increasing number of sources.

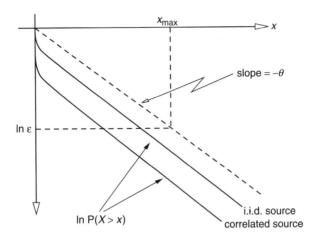

Figure 5.13   Relation between ln Pr($X > x$) and the QoS requirement Pr($X > x_{\max}$) $< \epsilon$ when $c > e(\theta)$, for increasing time correlation in a fixed number of sources, keeping the same marginal distribution of the arrival process.

Figure 5.13 shows a sketch of how $\ln \Pr(X > x)$ changes as we increase the time correlations of the source while keeping the marginal distribution fixed. For an example of this, see Example 5.5, where we have two two-state sources with the same marginal distributions, but the Markov source has positive time correlations. Notice that increasing the time correlations entails an increase in the EBW, and hence the required capacity from the EBW approach. However, because the marginal distribution does not change, increasing the capacity leads to an improvement in the small-buffer behavior, thus leading to a more conservative design (see Figure 5.13).

The Guerin, Ahmadi, Nagshineh (GAN) Approach

Let $B^{(i)}, 1 \leq i \leq m$, denote the marginal rates of m independent arrival processes. The QoS requirement is, as before,

$$\Pr(X > x_{\max}) \leq \epsilon$$

The GAN approach computes two capacities and then uses the smaller of the two. One of these, C_0, is taken to be the minimum capacity required to achieve the following:

$$\Pr\left(\sum_{i=1}^{m} B^{(i)} > C_0\right) \le \epsilon$$

The other, C_{EBW}, is taken to be equal to $\sum_{i=1}^{m} e^{(i)}(\theta)$, with $\theta = -\frac{\ln \epsilon}{x_{\max}}$. Note that C_0 can be obtained either by using a Gaussian approximation (via the central limit theorem) or by using Cramer's theorem with the Bahadur–Rao correction; see Theorem 5.3 and associated discussion in Section 5.6.1. Finally, the buffer is allocated the capacity $C = \min\{C_0, C_{EBW}\}$.

We now discuss why the GAN approach might be able to provide a value of capacity that meets the QoS objective while not being overly conservative? Suppose it turns out that $C = C_0$; that is, $C < C_{EBW}$. Figure 5.14 illustrates this situation. Because $C < C_{EBW}$, the tail of $\ln \Pr(X > x)$ crosses over the line of slope $-\theta$. However, the small-buffer behavior is sufficient to satisfy the requirement $\Pr(X > x_{\max}) \le \epsilon$. Notice that essentially the approach ignores the available buffers. On the other hand, suppose it turns out that $C = C_{EBW}$ (i.e., $C < C_0$). In Figure 5.15 we show this situation. Because $C < C_0$, marginal buffering does not suffice, and the design must utilize the buffers. Because $C = C_{EBW}$, the tail

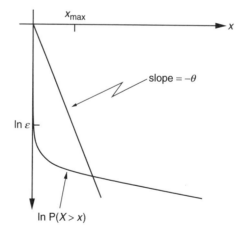

Figure 5.14 Illustrating the GAN approach, when $C = C_0$. Relation between $\ln \Pr(X > x)$ and the QoS requirement $\Pr(X > x_{\max}) < \epsilon$.

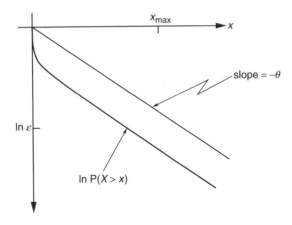

Figure 5.15 Illustrating the GAN approach, when $C = C_{EBW}$. Relation between ln Pr($X > x$) and the QoS requirement Pr($X > x_{max}$) < ϵ.

of $\ln \text{Pr}(X > x)$ does not cross over the line of slope $-\theta$, and the requirement $\text{Pr}(X > x_{\text{max}}) \leq \epsilon$ is again satisfied.

Exercise 5.12

In practice, the QoS requirement for stream sources at a multiplexer is to bound the delay experienced by the data. If the bandwidth assigned to the source (or superposition of sources) is C and if the delay bound is T, then the QoS requirement is $\text{Pr}(X > TC) \leq \epsilon$. Develop an approach for solving this problem using the affine approximation for $\ln \text{Pr}(X > x)$.

5.8 Application to the Packet Voice Example

We now continue at the point where we left off at the end of Section 4.4. We used a deterministic bound for the delay in the WFQ multiplexer. In doing this, we saw that to bound the shaping plus multiplexing delay for the voice source by T with probability $1 - \epsilon$, we needed to satisfy the following inequality:

$$\text{Pr}\big(X > (T - T')\rho + \sigma\big) < \epsilon$$

where X is the stationary queue length in an equivalent model for the leaky bucket shaper. This equivalent model is simply an infinite buffer served by a constant-rate fluid server of rate ρ. The voice source is modeled as a two-state Markov modulated fluid source. For the given voice source and fixed ρ in the equivalent queue for the leaky bucket, let us define $g_\epsilon(\rho)$ as follows:

$$g_\epsilon(\rho) = \inf \{x : \Pr(X > x) \le \epsilon\}$$

That is, $g_\epsilon(\rho)$ is the smallest buffer level that, in steady state, is exceeded with probability less than ϵ. Observe that we can now aim to choose $\sigma, \rho,$ and r such that

$$g_\epsilon(\rho) \le (T - T')\rho + \sigma$$

This will imply the inequality $\Pr(X > (T - T')\rho + \sigma) < \epsilon$.

Let us simplify the design and take $r = \rho$. This results in $T' = \frac{\sigma}{\rho} + \frac{L_{\max}}{\rho} + \frac{V_{\max}}{C}$. Hence, substituting for T', we find that we need ρ such that

$$g_\epsilon(\rho) \le \left(T - \frac{V_{\max}}{C}\right)\rho - L_{\max}$$

The problem is now to determine $g_\epsilon(\rho)$. Let $\Gamma(\theta)$ be the log-moment generating function for the two-state fluid voice source feeding the leaky bucket. We know from the theory developed earlier that for a given ρ, for $\theta > 0$, if we define

$$\delta(\rho) = \sup \{\theta : \rho\theta > \Gamma(\theta)\} \tag{5.32}$$

then, for a given ρ, $\delta(\rho)$ is the maximum value of θ for which it holds that $\lim_{x \to \infty} \frac{1}{x} \ln \Pr(X > x) \le -\theta$.

Then invoking the affine approximation in Equation 5.31 and ignoring the intercept at 0, we can take $g_\epsilon(\rho) = \frac{-\ln \epsilon}{\delta(\rho)}$. This will yield

$$\Pr(X > g_\epsilon(\rho)) \le \epsilon$$

We are being conservative by using only the linear bound.

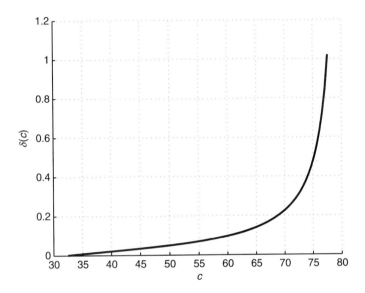

Figure 5.16 Plot of the function $\delta(\cdot)$. Two-state Markov source with rate $R = 80$ Kbps in the *on* state, and zero in the *off* state; average rate 32 Kbps; mean *on* time 400 ms.

Given $g_\epsilon(\rho)$, we can then choose ρ such that

$$\frac{-\ln \epsilon}{\delta(\rho)} \le \left(T - \frac{V_{\max}}{C}\right)\rho - L_{\max}$$

If we can determine $\delta(\cdot)$, then this inequality provides a way to determine ρ. Because we know how to compute $\Gamma(\theta)$ for Markov sources, the defining Equation 5.32 can be used to determine $\delta(\rho)$. In Figure 5.16 we show a sample plot of $\delta(\cdot)$ for a two-state (on–off) Markov source.

In Figure 5.17 we plot the left and right sides of the preceding inequality. We have considered a two-state Markov fluid model for the output of the voice coder. The peak rate is 80 Kbps (obtained from a 64 Kbps peak rate, 20 ms packetization, and 40 B overhead per packet), average rate 32 Kbps, mean burst length 400 ms, $C = 2.048$ Mbps, $V_{\max} = 1500$ B, $L_{\max} = 200$ B, and the delay bound $T = 100$ ms. We have taken $\epsilon = 0.05$. Notice in Figure 5.17 that the inequality is satisfied for $\rho > 76$ Kbps, and hence (as per this approach) the minimum bandwidth in the multiplexer required to handle this source is 76 Kbps. We note that

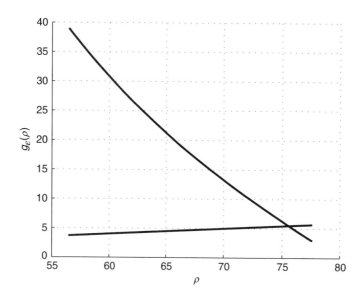

Figure 5.17   Plot of the approximation to $g_\epsilon(\rho)$ vs ρ, with $\epsilon = 0.05$. Two-state Markov source with rate R = 80 Kbps in the *on* state, and zero in the *off* state; average rate 32 Kbps; mean *on* time 400 ms. On the axes is also plotted the line $\left(T - \frac{V_{max}}{C}\right)\rho - L_{max}$, for T = 100 ms, V_{max} = 1500 B, C = 2.048 Mbps, and L_{max} = 200 B.

the approximation for $g_\epsilon(\rho)$ discussed earlier is conservative; the approximation overestimates $g_\epsilon(\rho)$ because it uses the linear upper bound for $\ln \Pr(X > x)$. Hence the value of ρ obtained from this calculation is an overestimate.

5.9   Stochastic Analysis with Shaped Traffic

If the analysis in Section 5.6 is used for online admission of calls into a network, then the arriving calls must specify to the network their detailed statistical characterization, or their effective bandwidths for the desired QoS objectives. Alternatively, the network can admit a call by allocating it its peak rate and could then make statistical measurements; from these measurements, the required bandwidth could be estimated, and the extra bandwidth released. This latter approach is called *measurement-based admission control*.

Yet another alternative is for the sources to shape themselves by using the leaky bucket (LB) regulator introduced in Chapter 4. Given these LB parameters,

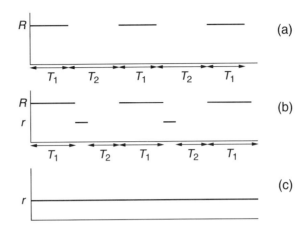

Figure 5.18 Examples of source outputs compatible with the LB parameters (σ, ρ, R).
(a) An on–off source with peak rate R, burst length $T_1 \leq \frac{\sigma}{R-\rho}$, and average rate $r \leq \rho$
(so that $T_2 = \frac{T_1(R-r)}{r}$). (b) A two-level source with peak rate R, $T_1 \leq \frac{\sigma}{R-\rho}$, and average
rate r (so that $T_2 = \frac{T_1(R-r)}{r}$). (c) A constant-rate source of rate r.

one alternative for the network is to perform deterministic, worst case traffic engineering. Recall that at the beginning of this chapter we showed that this approach can lead to a huge overestimation of the resources required, and hence we had motivated stochastic analysis with stochastic QoS objectives. Another alternative is to carry out a stochastic analysis with the given LB parameters. In this section we provide some results of the stochastic analysis approach given LB shaped sources.

The main difficulty in carrying out traffic engineering with LB parameters alone is that these parameters specify only the worst case behavior and do not uniquely specify a statistical characterization of the source. A source $A(t)$ with LB parameters (σ, ρ, R) has peak rate R and average rate $r \leq \rho$, and furthermore, it cannot emit data at the peak rate for a duration exceeding $\frac{\sigma}{R-\rho}$. Figure 5.18 shows three examples of source outputs that are compatible with these LB parameters. In the case of the on–off model or the two-level model, randomness comes from a random phasing of the source with respect to other sources being handled by the multiplexer.

One approach for statistical traffic engineering with LB shaped sources is to assume the source output that is the *worst* for the desired network performance.

Given m statistically independent sources $A_1(t), A_2(t), \dots, A_m(t)$, with LB parameters $(\sigma_i, \rho_i, R_i), 1 \leq i \leq m$, the approach is to assume for each source a model compatible with the LB parameters but one that leads to the worst performance. Network resources are allocated (and, if needed, some calls may be blocked) so as to achieve the desired performance under the worst case source output assumption. This ensures that no matter what the actual source output, the performance will not be any worse than that desired. Hence the problem reduces to one of determining the worst case stochastic models for a set of independent LB shaped sources.

Again we look separately at the marginal buffering case and the arbitrary buffering case.

5.9.1   Analysis with Marginal Buffering

We begin by looking at multiplexer design with marginal buffering. There are m statistically independent sources $A_1(t), A_2(t), \dots, A_m(t)$, with LB parameters (σ_i, ρ_i, R_i), being handled by a link of rate C. We are interested in designing for a target value of packet loss rate. Recalling our notation for the data arriving in each interval, this performance measure is

$$\lim_{n \to \infty} \frac{1}{n} \sum_{k=0}^{n-1} \left(\sum_{i=1}^{m} B_k^{(i)} - C \right)^+$$

Assuming that the sources are stationary and ergodic, this performance objective reduces to (see Theorem B.5 in Appendix B)

$$\mathsf{E} \left(\sum_{i=1}^{m} B_0^{(i)} - C \right)^+$$

which depends only on the marginal distributions of $B_0^{(i)}, 1 \leq i \leq m$. The ith marginal random variable is bounded between 0 and R_i and has a mean value no more than ρ_i. Note that these marginal random variables are mutually independent.

For ease of notation, we now drop the subscript 0. Suppose we assume some distributions for $B^{(i)}, 2 \leq i \leq m$. Which distribution for $B^{(1)}$ will make the packet loss rate the worst? The argument proceeds as follows. For given C, define the function

$$f(u) = (u - C)^+$$

for all real numbers u; notice the $f(u)$ is convex and increasing in u. The performance measure thus becomes

$$\mathsf{E}\left(f\left(\sum_{i=1}^{m} B^{(i)}\right)\right)$$

Fixing the distributions of $B^{(i)}, 2 \leq i \leq m$, we wish to choose the distribution of $B^{(1)}$ (independent of $B^{(i)}, 2 \leq i \leq m$) so as to maximize

$$\mathsf{E}\left(f\left(B^{(1)} + \sum_{i=2}^{m} B^{(i)}\right)\right)$$

under the constraint that $0 \leq B^{(1)} \leq R_1$, and $\mathsf{E}(B^{(1)}) = r_1 \leq \rho_1$. Observe that for $0 \leq x \leq R_1$ and y some real number

$$f(x + y) \leq \frac{x}{R_1} f(R_1 + y) + \left(1 - \frac{x}{R_1}\right) f(0 + y)$$

by the convexity of $f(\cdot)$. Hence, for the random variable $B^{(1)}$, constrained as required earlier, we have

$$f(B^{(1)} + y) \leq \frac{B^{(1)}}{R_1} f(R_1 + y) + \left(1 - \frac{B^{(1)}}{R_1}\right) f(0 + y) \tag{5.33}$$

Now by using the fact that the random variables $B^{(i)}, 2 \leq i \leq m$, and $B^{(1)}$ are independent, it follows that

$$\mathsf{E}\left(f\left(B^{(1)} + \sum_{i=2}^{m} B^{(i)}\right)\right) = \mathsf{E}_{\{B^{(i)}, 2 \leq i \leq m\}} \mathsf{E}_{B^{(1)}} f\left(B^{(1)} + \sum_{i=2}^{m} B^{(i)}\right)$$

where the notation E_X denotes expectation with respect to the distribution of X, and we have used the fact that if X and Y are independent random variables, then for a function $g(\cdot, \cdot)$, $E(g(X, Y)) = E_X E_Y g(X, Y)$. It then follows by applying Equation 5.33 that

$$E_{\{B^{(i)}, 2 \leq i \leq m\}} E_{B^{(1)}} f \left(B^{(1)} + \sum_{i=2}^{m} B^{(i)} \right) \leq E_{\{B^{(i)}, 2 \leq i \leq m\}} E_{B^{(1)}} \left(\frac{B^{(1)}}{R_1} f \left(R_1 + \sum_{i=2}^{m} B^{(i)} \right) \right.$$
$$\left. + \left(1 - \frac{B^{(1)}}{R_1} \right) f \left(0 + \sum_{i=2}^{m} B^{(i)} \right) \right)$$

But $E_{B^{(1)}} B^{(1)} = r_1$, and hence we can continue the calculation as follows:

$$= E_{\{B^{(i)}, 2 \leq i \leq m\}} \left(\frac{r_1}{R_1} f \left(R_1 + \sum_{i=2}^{m} B^{(i)} \right) + \left(1 - \frac{r_1}{R_1} \right) f \left(0 + \sum_{i=2}^{m} B^{(i)} \right) \right)$$
$$= E \left(f \left(B^{(1)*} + \sum_{i=2}^{m} B^{(i)} \right) \right)$$

where $B^{(1)*}$ is a random variable such that $\Pr(B^{(1)*} = R_1) = \frac{r_1}{R_1} = 1 - \Pr(B^{(1)*} = 0)$.

We conclude that, for a given value of $r_1 (\leq \rho_1)$, and given distributions of $B^{(i)}, 2 \leq i \leq m$, the packet loss rate $E(\sum_{i=1}^{m} B^{(i)} - C)^+$ is maximized when $B^{(1)}$ is an on–off source (taking values R_1 and 0) with mean rate r_1.

Exercise 5.13

Show that the packet loss rate is further maximized if we take the largest possible value of $E(B^{(1)})$ (i.e., $r_1 = \rho_1$). For this, use the fact that $f(\cdot)$ is nondecreasing.

Fixing the means of $B^{(i)}, 1 \leq i \leq m$, to $r_i (\leq \rho_i), 1 \leq i \leq m$, the preceding result can be applied recursively to obtain

$$\mathsf{E}\left(f\left(\sum_{i=1}^{m} B^{(i)}\right)\right) \leq \mathsf{E}\left(f\left(\sum_{i=1}^{m} B^{(i)*}\right)\right)$$

where, for each i, $B^{(i)*}$ is an on–off source (taking values R_i and 0) with mean rate r_i. The packet loss rate is further maximized if source i has average rate ρ_i.

Earlier we dealt with packet loss rate. Let us now examine what happens if we consider the *fraction* of packets lost as the performance measure. In terms of time averages this measure is the total amount of data lost over an interval divided by the total amount of data that arrived over that interval:

$$\lim_{n \to \infty} \frac{\sum_{k=0}^{n-1} \left(\sum_{i=1}^{m} B_k^{(i)} - C\right)^+}{\left(\sum_{k=0}^{n-1} \sum_{i=1}^{m} B_k^{(i)}\right)}$$

Using the ergodicity of the data arrival process and the function $f(\cdot)$ defined earlier, this can be written as

$$\frac{\mathsf{E}\left(f\left(\sum_{i=1}^{m} B^{(i)}\right)\right)}{\sum_{i=1}^{m} \mathsf{E}\left(B^{(i)}\right)}$$

where we have used the ergodic theorem (Theorem B.5 in Appendix B).

If we choose the mean rates of the sources as $\mathsf{E}\left(B^{(i)}\right) = r_i \leq \rho_i, 1 \leq i \leq m$, then it follows from the preceding results that the fraction of packets lost is maximized if each source is taken to be an on–off source. For this performance measure, however, it is not clear whether choosing the mean rates $r_i = \rho_i, 1 \leq i \leq m$, is the worst case. In fact, it is known that this is not the worst case in general. If the sources are statistically identical and have the same LB parameters (σ, ρ, R), then it can be shown that taking $r_i = \rho, 1 \leq i \leq m$, yields the worst case fraction of lost packets (see Problem 5.8).

In this context, it is important to note the difference between the loss rate for individual sources and that for the aggregate. In general, the aggregate loss rate could be very unfairly distributed over the various sources. Only in the statistically identical sources case, discussed here, can we say, by symmetry, that the individual loss rates will be the same as the aggregate loss rate. For this case, we see that the design problem (for either performance measure: packet loss rate or fraction of packets lost) has reduced to the problem with the superposition of identically distributed two-state sources discussed in Example 5.4.

5.9.2 Analysis with Arbitrary Buffering

We turn now to the case of connection admission control with burst scale buffering, with LB shaped sources and a statistical QoS objective. Again the approach is to design the multiplexer for the worst case output of the leaky bucket for each source. For an LB controlled source with parameters (σ, ρ, R), one might expect that the worst case source output is one that switches between the peak rate R and 0, with the maximum possible burst length $\frac{\sigma}{R-\rho}$; this can be called the *extremal on–off source* for the LB parameters (σ, ρ, R). In fact, it has been shown that, in general, this source is not the one that gives the worst case performance. A two-level source (see Figure 5.18) can yield worse performance; the intuition is that, by being active longer, it can sustain congestion longer, thus causing loss for other sources in the multiplexer.

Nevertheless, some researchers have developed design methodologies by assuming the extremal on–off source as the output from each LB. The randomness, and hence the statistical multiplexing gain, comes from the random phase relationships between the sources being multiplexed. The techniques discussed in Section 5.6.2 can be applied to this superposition of on–off sources. Additional approximations have been provided in the literature to simplify the design problem (e.g., [93], [200]).

5.10 Multihop Networks

All the models and analysis presented so far in this chapter are for a single multiplexer served by a single link. Figure 5.19 depicts a situation where the design of a single-link "network" is all that is needed. The situation depicted could represent an enterprise that leases a link from a long distance bandwidth provider, using this link for its interlocation voice and data traffic. The enterprise would deploy packet PBXs (e.g., IP PBXs) at each location, and interlocation voice would be digitized, coded, and packetized to be sent over the packet-multiplexed leased link. Because traffic enters and leaves the backbone link over high-speed LANs, the interlocation link is the bottleneck because performance is limited by it alone. One question is to determine the link speed needed to support a certain voice call rate, given some QoS requirements for the packetized voice and for

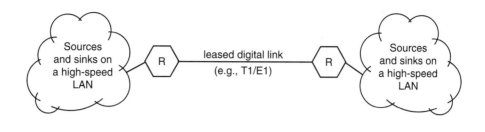

Figure 5.19 Two locations of an enterprise connected by a single link. Traffic enters and leaves the link over high-speed local area networks.

the data transfers. Another question is to determine whether an arriving stream connection can be admitted, given the traffic that is already being carried by the link. Observe that these problems reduce to one of analyzing a single multiplexer (one in each direction) served by a link.

Another important issue, evident from the example in Figure 5.19, is that a link would need to carry both elastic and stream traffic. Consider the single-hop case first. A WFQ scheduler can be implemented in the LAN–WAN router interface at which packets are multiplexed into the wide area link. The nonpreemption delay (see Section 4.3.1) can be subtracted from the delay target for the stream traffic, and then the techniques in this chapter can be used to obtain the capacity required to achieve the desired stochastic QoS (for the stream traffic). The WFQ weights would then be set up so as to apply the required service rate to the stream traffic. The remaining capacity would apply to the elastic traffic. Of course, elastic traffic cannot be arbitrarily squeezed out, because eventually that would affect the transfer rates seen by users of elastic services (file transfers and Web browsing). We will study the performance of such traffic in Chapter 7.

In general, of course, we need to design multihop networks with arbitrary topologies. The important new aspect is that in a multihop network, a flow may need to traverse multiple links, in tandem, in going from its source to its destination. Figure 5.20 depicts such a situation. Let us focus on stream traffic (e.g., packet voice streams between packet PBXs). Voice originating at location 1 and destined for location 2 enters the network at router 1 and leaves the network at router 2. On the other hand, voice originating at location 1 and destined for location 3 leaves the link from router 1 to router 2 and enters the link from router 2 to router 3. At this hop, the two-hop traffic will be multiplexed, with voice emanating from location 2 and terminating at location 3. A similar situation exists in the reverse direction. Each direction can be analyzed as a separate network.

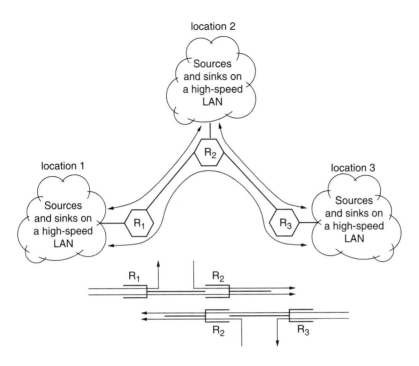

Figure 5.20   A three-location enterprise connected by a packet-multiplexed network with two links. Traffic enters and leaves the backbone network over high-speed local area networks. The schematic at the bottom of the figure shows the bottleneck multiplexers and links for each direction of traffic.

The stochastic analysis of multihop packet-multiplexed networks becomes difficult because of the following three aspects:

- Whereas the traffic from a source may be well characterized at the point where it enters the network (by a statistical or envelope characterization), after multiplexing with other traffic, we need a stochastic characterization of the *departure* process of each flow from each hop in order to carry out analysis of subsequent hops.

- Whereas at the first hop the various flows can be assumed to be statistically independent, after multiplexing at the first hop, the flows become dependent; this dependence is very difficult to characterize. If flows that

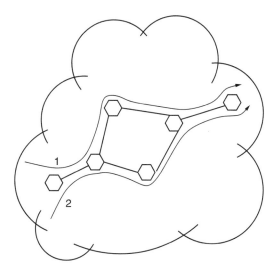

Figure 5.21 Flows 1 and 2 are multiplexed at one hop and are again multiplexed together at a subsequent hop.

are multiplexed at a link need to be multiplexed again at subsequent hops, then this dependence must be taken into account. This aspect of the problem does not exist in the situation depicted in Figure 5.20; see, however, Figure 5.21. One possibility is to design the network using only *end-to-end virtual paths*, into which traffic enters only at the ingress point and leaves at the egress point; no new traffic enters anywhere else along the path. If the links along the path can apply a *fixed* service rate to this virtual path all the way along the path, then the design techniques presented in this section can be directly used to design each such virtual path. Of course, this simplification will not arise if traffic from virtual paths is dropped and new traffic is added at various points along the paths.

- As mentioned earlier in this section, the network must carry both elastic and stream traffic. One possibility is to design a separate network for each type of traffic. The ideal situation, however, would be to design fully integrated packet networks, carrying all types of services, each getting its own QoS. Evidently, we need to implement packet schedulers at each link (e.g., WFQ schedulers), which can apply the requisite service rates to each type of flow through the link. This, however, leads to interactions between the flows, and these interactions are difficult to account for in the design. For example, consider only two kinds of flows being multiplexed into a buffered link with WFQ scheduling: one a superposition of stream flows, and the other a superposition of elastic flows. When the elastic buffer is empty, the entire service effort is applied to the stream traffic. This *burstiness* in the service must be accounted for in designing the multiplexer at the next hop. One solution is to *reshape* the traffic departing from the WFQ scheduler.

Because of these difficulties, the end-to-end stochastic analysis of multihop packet networks has not yet yielded a complete solution. In the remainder of this section we provide two approaches that offer solutions in certain limited situations. But first we discuss the important issue of the allocation of the end-to-end QoS objective to objectives at each hop.

5.10.1   Splitting the End-to-End Quality of Service Objective

For a stream session in a packet network, a typical stochastic QoS objective is that the end-to-end delay is bounded by T with probability $1 - \epsilon$. As explained in Section 3.2, in the context of packet telephony, such a QoS objective permits the receiver to apply a playout delay so that the probability of packets arriving later than their playout times is no more than ϵ. In principle, we can formulate a model for the entire network, with the traffic arriving at the network egress points, and then proceed to analyze this model to obtain, for each stream connection, the probability of violating its delay objective. In general, such an approach is intractable. Instead, one breaks down the end-to-end QoS objective into per-hop objectives. Thus, for an n-hop session, one could break up the end-to-end delay bound T as $T = T_1 + T_2 + \cdots + T_n$, and the probability of violating the delay bound can be split up as $\epsilon = \epsilon_1 + \epsilon_2 + \cdots + \epsilon_n$. Let the stationary random delay at the ith hop be denoted by $D_i, 1 \leq i \leq n$. Suppose we ensure that, for each $i, 1 \leq i \leq n$,

$$\Pr(D_i > T_i) < \epsilon_i$$

It then follows that

$$\Pr(D > T) = \Pr\left(\sum_{i=1}^{n} D_i > \sum_{i=1}^{n} T_i\right)$$

$$\leq \Pr\left(\bigcup_{i=1}^{n}\{D_i > T_i\}\right)$$

$$< \sum_{i=1}^{n}\epsilon_i = \epsilon$$

where the first inequality follows from the simple observation that if $\sum_{i=1}^{n} D_i > \sum_{i=1}^{n} T_i$, then it cannot be that for every i, $D_i \leq T_i$, and the last inequality is just the union bound.

Hence it suffices to design for the per-hop QoS objective $\Pr(D_i > T_i) < \epsilon_i$. Note that it is not clear how we should split up the QoS in this way. In addition, recall that a stream source can be shaped before being offered to the network. Shaping entails packet delay, which must be added to the network delay. Thus we actually need to split the end-to-end delay bound T, and the delay bound violation probability ϵ, into QoS objectives for the shaper and for the network, and then further split the delay and delay violation probability allocated to the network into per-hop targets. One could conceive of an *optimal split* that minimizes the bandwidth requirement in the network for given load. Consider one shaper in tandem with one link. If the entire delay violation probability is assigned to the shaper, then the multiplexer requirement becomes very stringent, and we would require a large link bandwidth. On the other hand if the entire delay violation probability is assigned to the multiplexer, then the source would not be smooth, and again the link bandwidth requirement would be large. This suggests that there is an optimal split that assigns positive delay violation probability to the shaper and to the multiplexer.

We proceed in the remainder of this section under the assumption that some split of the QoS objective has been chosen. A simple approach is to split the delay bound and the delay violation probability equally over the hops of a connection.

5.10.2 Marginal Buffering at Every Hop

To illustrate this approach let us consider the simple example depicted earlier in Figure 5.20. Statistically independent and smoothed stream sources arrive at the network. Let us discuss the analysis of the traffic flow from location 1 to locations 2 and 3, and from location 2 to location 3 (i.e., the flow from left to right; see the schematic in that figure). Each source has a peak rate and a maximum average rate, which are known at the time of call admission (these parameters are obtained from the leaky bucket parameters). The marginal buffering approach is applied at each hop.

Consider the first hop, and recall from Section 5.9 that each source can be taken to be an on–off source that is either at its peak or at 0, because this will yield the worst possible packet loss rate. If the sources are homogeneous (e.g., all packet voice sources with the same coding rate and statistical behavior), then we also can take the average rate of the on–off sources to be the average rate bound provided to the network (the LB token rate for the source); this will yield the worst possible packet loss ratio (which in this marginal buffering case is the same as the delay violation probability). We now have a superposition of on–off sources with marginal buffered multiplexing, and the analysis in Section 5.6.1 applies.

At the next hop the two-hop sessions are multiplexed with the freshly arriving sessions from location 2 to location 3 (see Figure 5.20). Now there is a difficulty: The two-hop sessions are no longer independent! If they were, the same analysis could be applied at this hop as well. Note that, because of marginal buffering, at the input of the second hop the same peak rate holds for each of the two-hop sources multiplexed at the first hop. However, the sources are not independent. One alternative is to take a conservative approach by viewing all the two-hop sessions at the first hop as only one session. Consider the homogeneous case, and let the calls have peak rate R and maximum average rate ρ. If there are $n_{1,3}$ calls from location 1 to location 3, then the aggregate session will have peak rate $n_{1,3}R$ and maximum average rate $n_{1,3}\rho$. The worst case arrival process for this aggregate session will now be on–off with rate $n_{1,3}R$, or 0. Now at the second hop we can again treat this as one source and carry out an analysis of it being multiplexed with the freshly arriving traffic from the location 2 to location 3 calls. Evidently, this approach (of aggregating the sessions along a path) will lose much of the stochastic multiplexing gain that is possible.

5.10.3   Arbitrary Buffering at Every Hop: Effective Envelopes

We saw in Chapter 4 that by characterizing input processes via envelopes and network elements via service curves, we can develop a complete end-to-end deterministic network calculus. This calculus is suitable if worst case guarantees are required. However, we also saw, at the beginning of this chapter, that designing for worst case guarantees can lead to very conservative designs. Permitting even a small probability of violating a delay bound, for example, can substantially increase the load capacity of a network. Here we discuss an approach via a stochastic version of traffic envelopes, which its originators call *effective envelopes*. We will see one way in which effective envelopes can be utilized for network design.

Consider a cumulative arrival process $A(t)$, which has stationary increments (see Appendix B).

Definition 5.1

For given $\epsilon > 0$, a function $E_\epsilon(t)$ is an *effective envelope* (or, more precisely, an *ϵ-effective envelope*) for $A(t)$ if, for every t and $\tau \geq 0$, $\Pr(A(t+\tau) - A(t) > E_\epsilon(\tau)) \leq \epsilon$. ■

If the input process is a superposition of several statistically homogeneous processes, each with a deterministic envelope, then an ϵ-effective envelope can be obtained via the central limit theorem or the Chernoff bound. We do not provide these derivations here but refer to the relevant literature in Section 5.13.

First we develop a one-hop design for a buffered multiplexer using an effective envelope. As before, let $A(t)$ denote the cumulative arrival process into the multiplexer. The process has stationary increments and is defined over $t \in (-\infty, \infty)$. The arrival process enters a multiplexer, where it is given a service rate C. Assuming that the arrival rate is less than the service rate and that a stationary regime exists, let X denote the stationary queue length random variable. We have seen, in Section 5.6.2, that the distribution of X is the same as that of

$$\sup_{\tau \geq 0} ((A(0) - A(-\tau)) - C\tau)$$

As before (see Section 5.6.2), let us write $S(-\tau) = (A(0) - A(-\tau)) - C\tau$. Suppose the QoS requirement is

$$\Pr\left(\frac{X}{C} > T\right) \leq \epsilon$$

where $\frac{X}{C}$ is the stationary delay in the multiplexer. This then translates to

$$\Pr\left(\sup_{\tau \geq 0} \left(\frac{S(-\tau)}{C}\right) > T\right) \leq \epsilon$$

Now, for every $t \geq 0$, we have

$$\Pr\left(\sup_{\tau \geq 0} \left(\frac{S(-\tau)}{C}\right) > T\right) \geq \Pr\left(\left(\frac{S(-t)}{C}\right) > T\right)$$

because the event on the right side implies the event on the left. Because this is true for every $t \geq 0$, it follows that

$$\Pr\left(\sup_{\tau \geq 0}\left(\frac{S(-\tau)}{C}\right) > T\right) \geq \sup_{\tau \geq 0}\Pr\left(\left(\frac{S(-\tau)}{C}\right) > T\right)$$

Note that we can also upper bound $\Pr\left(\sup_{\tau \geq 0}\left(\frac{S(-\tau)}{C}\right) > T\right)$ as follows

$$\Pr\left(\sup_{\tau \geq 0}\left(\frac{S(-\tau)}{C}\right) > T\right) = \Pr\left(\cup_{\tau \geq 0}\left\{\frac{S(-\tau)}{C} > T\right\}\right)$$

$$\leq \sum_{\tau \geq 0}\Pr\left(\frac{S(-\tau)}{C} > T\right)$$

where the first equality occurs because $\sup_{\tau \geq 0}\left(\frac{S(-\tau)}{C}\right) > T$ if and only if there is some τ such that $\frac{S(-\tau)}{C} > T$, and the inequality is just the union bound. It can be argued that the largest term in the sum in the upper bound is a good approximation of this sum, and because this term is also a lower bound to $\Pr\left(\sup_{\tau \geq 0}\left(\frac{S(-\tau)}{C}\right) > T\right)$, we obtain the following commonly used approximation (e.g., [89]):

$$\Pr\left(\sup_{\tau \geq 0}\left(\frac{S(-\tau)}{C}\right) > T\right) \approx \sup_{\tau \geq 0}\Pr\left(\left(\frac{S(-\tau)}{C}\right) > T\right)$$

Now, owing to the assumption of stationary increments of the input process $A(t)$, we have $\Pr\left(\left(\frac{(A(0)-A(-\tau))-C\tau}{C}\right) > T\right) = \Pr\left(\left(\frac{(A(\tau)-A(0))-C\tau}{C}\right) > T\right)$, and hence, finally the approximation is

$$\Pr\left(\frac{X}{C} > T\right) \approx \sup_{\tau \geq 0}\Pr\left(\left(\frac{(A(\tau)-A(0))-C\tau}{C}\right) > T\right)$$

We would like to determine C such that $\Pr\left(\frac{X}{C} > T\right) \leq \epsilon$.

Now recall Equation 4.6, which relates the (deterministic) envelope of an input process to the capacity C required to ensure that the delay does not exceed T. Consider the same expression with the envelope $E(t)$ replaced with the ϵ-effective envelope $E_\epsilon(t)$:

$$C_\epsilon = \sup_{t \geq 0} \left(\frac{E_\epsilon(t)}{T + t} \right) \tag{5.34}$$

With C_ϵ defined in this way, we have, for all $t \geq 0$,

$$E_\epsilon(t) \leq C_\epsilon T + C_\epsilon t$$

Exercise 5.14

Show that C_ϵ as defined earlier satisfies the QoS requirement

$$\sup_{\tau \geq 0} \mathsf{Pr}\left(\left(\frac{(A(\tau) - A(0)) - C_\epsilon \tau}{C_\epsilon} \right) > T \right) \leq \epsilon$$

Thus we easily get the required capacity at the first hop by using an effective envelope. Consider now a two-hop problem. The end-to-end delay target is T, and the delay violation probability is ϵ, and these are split up as $T = T_1 + T_2$ and $\epsilon = \epsilon_1 + \epsilon_2$. An effective envelope E_{ϵ_1} is used for the first hop design; this yields a capacity C_1. The next task is to see whether we can obtain an effective envelope for the output process of $A(t)$ served by a fixed rate of C_1. Take the first hop to have achieved stationarity, and denote the stationary buffer size at the first hop by X_1. Because the arrival process has stationary increments, the increments of the stationary output process also have stationary increments. Then (see the derivations in Chapter 4) we can write the increments of the output process $D(t) - D(0)$ as

$$D(t) - D(0) = \inf_{0_- \leq \tau \leq t} (\overline{A}(\tau) + C_1(t - \tau)^+)$$

where we define $\overline{A}(0_-) = 0$ and, for $\tau \geq 0$, $\overline{A}(\tau) = A(\tau) - A(0_-) + X_1$. A way to think about this expression is to view the buffer occupancy in hop 1 at time 0 as an additional "arrival" at time 0. Now let us define a candidate envelope for the first hop output process as follows. First, as an interim definition, for $\epsilon_1 > 0$, let

$$\overline{E}_{\epsilon_1}(t) = \begin{cases} 0 & \text{for } t = 0_- \\ C_1 T_1 + E_{\epsilon_1}(t) & \text{for } t > 0 \end{cases}$$

where $E_{\epsilon_1}(t)$ is the ϵ_1-effective envelope at the first hop, as developed earlier. Finally, define

$$Z_{\epsilon_1}(t) = \inf_{0_- \leq \tau \leq t} (\overline{E}_{\epsilon_1}(\tau) + C_1(t - \tau)^+) \tag{5.35}$$

or, equivalently, $Z_{\epsilon_1} = \overline{E}_{\epsilon_1} * C_1 t^+$.

Exercise 5.15

Show that $\Pr(D(t) - D(0) > Z_{\epsilon_1}(t)) \leq 2\epsilon_1$. Hint: The union bound is used in the argument.

Hence, if $\epsilon_2 = 2\epsilon_1$, then $Z_{\epsilon_1}(t)$ can be used to design the next hop multiplexer. Sharper effective envelopes for the output process can also be sought.

Note that the discussion of the envelope of the output process of the first hop assumes that the service at the first hop is at the rate C_1. The same link could, however, also be serving other traffic—for example, elastic traffic. In this case a GPS scheduler could be used, with a minimum rate of C_1 applied to the stream traffic under consideration. With GPS, however, the service rate actually applied to this stream traffic could exceed C_1 during times when the other traffic is small; the point is that $C_1 t^+$ is just a lower service curve. This would upset the output envelope calculations that we have shown. In such a situation, a peak rate controller will need to be used to reshape the stream traffic after service from the GPS scheduler. Basically, the GPS scheduler followed by the shaper should appear as an element with service curve $C_1 t^+$.

5.11   Long-Range-Dependent Traffic

In this section we provide a brief introduction to *long-range-dependent* (LRD) traffic sources, primarily for the purpose of pointing out that design approaches based on exponential asymptotics of the multiplexer buffer occupancy do not apply in this case. In addition, this material provides background for our discussion of heavy-tailed file transfers and long-range dependence of traffic on links carrying elastic traffic, topics that are discussed in Section 7.6.8.

Let B_k denote the output in interval k from a stationary discrete time source. Then the autocorrelation function of this process is defined as

$$r_k = \mathsf{E}\left(\left(B_j - \mathsf{E}\left(B_j\right)\right)\left(B_{j+k} - \mathsf{E}\left(B_{j+k}\right)\right)\right)$$

where the left side depends only on k owing to the stationarity of the process. Let $\sigma^2 = r_0$; that is; $\mathsf{Var}\left(B_j\right) = \sigma^2$, for all j.

The examples of source traffic models that we have given earlier in this chapter can all be classified as stationary *finite state* Markov models. A characteristic property of such models is that the autocovariance function of the traffic between different points of time decays geometrically (or exponentially, for the continuous time case) in the difference in time; that is, for $k \geq 1$,

$$r_k \sim_{k \to \infty} a^{-k}$$

for some constant a.

Measurements have shown, however, that the autocovariance function behaves very differently for certain kinds of traffic, and this can have profound consequences for the design of multiplexers that carry such traffic. The typical behavior that we are concerned with is

$$r_k \sim_{k \to \infty} k^{-\beta} \tag{5.36}$$

where $0 < \beta < 1$.

Exercise 5.16
Show that $\mathsf{Var}\left(\sum_{i=1}^n B_i\right) = n\sigma^2\left(1 + 2\sum_{k=1}^{(n-1)}\left(1 - \frac{k}{n}\right)r_k\right)$. Hence show that if $r_k \sim_{k \to \infty} a^{-k}$, for $k \geq 1$, then $\lim_{n \to \infty} \frac{\mathsf{Var}\left(\sum_{i=1}^n B_i\right)}{n}$ exists.

In fact, for the conclusion in Exercise 5.16 to hold, it suffices that $\sum_{k=0}^\infty r_k < \infty$ (see [30]). We can also state this conclusion as $\mathsf{Var}\left(\sum_{i=1}^n B_i\right) = O(n)$.

On the other hand, it can be shown that, when Equation 5.36 holds, then the following is true:

$$\lim_{n \to \infty} \frac{\mathsf{Var}\left(\sum_{i=1}^{n} B_i\right)}{n^{(2-\beta)}} \quad \text{converges to a constant} \tag{5.37}$$

That is, $\mathsf{Var}\left(\sum_{i=1}^{n} B_i\right)$ grows at a rate faster than n, because $0 < \beta < 1$.

Let us now suppose that B_k is a stationary Gaussian process. Thus each of the B_k is Gaussian with a common mean and variance, and any finite set of random variables taken from this process is jointly Gaussian. We point out that Gaussian processes have, in fact, been used to model the superposition of a large number of traffic processes. Here we are using this model only to make a point about the inapplicability of the earlier effective bandwidth results. To compute an effective bandwidth, we must compute the log-moment generating function for the B_k process; that is, we need

$$\Gamma(\theta) = \lim_{n \to \infty} \frac{1}{n} \ln \mathsf{E}_{\pi} e^{\theta \sum_{k=1}^{n} B_k}$$

where the subscript π indicates that the expectation is with respect to the stationary distribution of the process. With B_k being a Gaussian process, $\sum_{k=1}^{n} B_k$ is a Gaussian random variable. It is well known that for a Gaussian random variable Z, with mean μ and variance σ^2, we have $\ln \mathsf{E} e^{\theta Z} = \theta \mu + \frac{1}{2}\theta^2 \sigma^2$. It follows that

$$\frac{1}{n} \ln \mathsf{E}_{\pi} e^{\theta \sum_{k=1}^{n} B_k} = \frac{1}{n}\left(n\theta \, \mathsf{E}_{\pi} B_1 + \frac{1}{2}\theta^2 \, \mathsf{Var}_{\pi}\left(\sum_{k=1}^{n} B_k\right)\right)$$

Notice, however (from the fact shown in Equation 5.37), that the term $\frac{1}{n}\mathsf{Var}_{\pi}\left(\sum_{k=1}^{n} B_k\right)$ diverges (for $0 < \beta < 1$), and hence $\Gamma(\theta)$ does not exist, as defined. It can be shown that when such traffic is fed to a buffer served by a constant-rate link, the tail of the stationary buffer distribution decays slower than exponentially. We refer to appropriate literature on this topic in Section 5.13.

5.12   Summary

We begin this chapter by showing, via an example, that, in general, deterministic network calculus can yield very conservative network designs. We then show how we can develop stochastic models for the outputs of packetized voice and video sources and for the superposition of such sources. Next, we provide two very useful results from queueing theory—namely, Little's theorem (5.1) and Brumelle's theorem (5.2)—and show a couple of their applications.

We then turn our attention to developing packet multiplexer design criteria based on asymptotic results such as the central limit theorem, Chernoff's bound, and Cramer's theorem. We discuss small-buffer designs (which we call marginal buffering) and burst scale designs (which we call arbitrary buffering). For the latter situation we develop the idea of effective bandwidths and discuss its usefulness and refinements.

We then discuss stochastic analysis with leaky bucket shaped sources. Unlike deterministic network calculus, there is no general end-to-end stochastic network calculus. We provide a brief discussion of stochastic analysis for multihop networks. Much of the asymptotic analysis discussed thus far yields exponential bounds on the tails of multiplexer buffers. In Section 5.11 we introduce long-range-dependent traffic and show why such exponential bounds would not be expected for such traffic.

5.13   Notes on the Literature

Although much of the modeling and analysis presented in this chapter are concerned with the analysis of queueing systems, we have deliberately avoided most of the traditional queueing theory. Among many others, the books by Kleinrock ([173, 174]), and by Wolff ([300]) provide excellent coverage of this vast topic, with the latter also covering the underlying stochastic processes theory. Instead, in this chapter, we provide more recent material on approximate analysis via bounds and asymptotics. What we call "marginal buffering" has been referred to in the literature as "bufferless" multiplexing, and what we call "arbitrary buffering" is called "burst scale" buffering in the literature.

The seminal paper [13] by Anick, Mitra, and Sondhi, dating back to 1982, provided the asymptotics of the tail distribution of a buffer fed by a superposition of on–off Markov fluid sources. The approach was via the differential equation characterizing the stationary distribution of buffer occupancy. The advent of ATM technology resulted in renewed interest in such problems, leading to the development of the notion of effective bandwidth and related admission control techniques during the early 1990s. Some of the significant papers were [164, 118, 91, 92, 79, 169, 52]. Our derivation of the effective bandwidth result for a buffered multiplexer follows the heuristic treatment by Kesidis, Walrand, and Chang [169]. A rigorous proof was provided by Glynn and Whitt in [120]. We have presented the approach in which the number of sources being multiplexed is kept fixed, the buffer is assumed to be infinite, and large-buffer asymptotics are characterized. It is also possible to develop an analysis in which the buffer is finite and the number of sources is scaled; see, for example, Likhanov and Mazumdar [197]. That the use of additive effective bandwidths can lead to very conservative designs was observed by many researchers; two reports on this topic were [64] and [246]. The technique of combining the burst scale effective bandwidth along with marginal buffer analysis was developed in [130]. A related technique, which also uses the marginal buffer analysis to correct for the conservatism in additive effective bandwidths, was developed by Elwalid et al. in [92]. A comprehensive recent survey of the approaches for single-link analysis was provided by Knightly and Shroff in [176]. In this survey is also reported a Gaussian approximation from the paper [63] by Choe and Shroff.

The idea of using worst case traffic compatible with leaky bucket shaped sources was developed by Doshi ([86]). For buffered multiplexers receiving leaky bucket shaped traffic, the approach of using extremal on–off sources was introduced in [93] by Elwalid, Mitra, and Wentworth, and then extended by LoPresti et al. in [200]. Of the limited literature on the analysis of multihop networks, the paper [247] reported the work on bufferless multiplexers at each hop, whereas the papers [40, 196] developed the idea of effective envelopes.

The observation that traffic in packet networks can have long-range dependence was made in the landmark paper [195], by Leland, Taqqu, Willinger, and Wilson. These measurement results were further analyzed statistically by the same research group in [299]. The impact of such traffic on the performance of multiplexers has been studied by several authors; some important references are [287, 137, 94, 226]. Buffer analysis with a Gaussian LRD process was performed by Norros in [226]. In [137], Heath, Resnick, and Samorodnitsky analyze an LRD on–off process; we will come across this model in Section 7.6.8. We have pointed out that for an LRD input the tail of the buffer does not decay as fast as exponential; the approach based on the Gärtner–Ellis theorem has been extended to handle the LRD case by Duffield and O'Connell [89]. Cox's paper [68] and the book [30] by Beran are the main references for the general theory of LRD processes.

Problems

5.1   Consider n independent, on–off Markov modulated fluid sources multiplexed into an unbuffered channel with service rate C fluid units/sec. Each source has a peak rate $\hat{a}$, average rate a, and exponentially distributed burst length with mean β.

    a. What is the mean *off* period of each source?

    b. For $n = 1000$, $\beta = 0.4 = a$, $\hat{a} = 1$, estimate the value of C required for the fraction of time that fluid is lost to be less than 10^{-6}.

    c. For the capacity you found in (b), find a bound on the fluid loss ratio.

5.2   Consider a stationary and ergodic cell arrival process $\{B_k, k \geq 0\}$ with $B_k \in \{0,1,2,\ldots\hat{a}\}$. Letting X denote the stationary queue length distribution, and given $\Gamma(\cdot)$, the log-moment generating function of the process B_k, what value of C will you use to ensure $\Pr(X > 100) < 10^{-8}$?

5.3 $\{X_k, k \geq 0\}$ is a discrete time Markov source that brings 0 or 1 cell at each slot boundary (i.e., $X_k \in \{0,1\}$). The transition probabilities are $p_{00} = \beta$, $p_{11} = \alpha$. The packets arrive into a buffer of size B, served by a link of service rate C packets per slot. If X is the stationary buffer size, we wish to accept only so many such sources so that $\Pr(X > B) \leq \epsilon$.

    a. Find the effective bandwidth of each source.

    b. Check your answer by considering $\epsilon \to 0$ and $\epsilon \to 1$.

    c. For $\epsilon = 10^{-10}$, $B = 100$, $C = 100$, $\alpha = 0.9$, and $\beta = 0.5$, estimate the number of sources that can be accepted.

    d. What is the statistical multiplexing gain for your answer in part (c)?

5.4 X is a random variable, $0 \leq X \leq R$, where R is a constant.

    a. For $\theta \geq 0$, show that

$$E\left(e^{\theta X}\right) \leq E\left(e^{\theta \hat{X}}\right)$$

    where

$$\hat{X} = \begin{cases} R & \text{with prob. } \frac{E(X)}{R} \\ 0 & \text{with prob. } \left(1 - \frac{E(X)}{R}\right) \end{cases}$$

    (Hint: Use the series expansion of $e^{\theta X}$ and observe that $X^k \leq XR^{k-1}$.)

b. Show that, for $\theta \geq 0$ and $\{X_i, \ i \geq 1\}$ i.i.d., defining $\hat{\mu}(\theta) = \ln \mathsf{E}\!\left(e^{\theta \hat{X}_1}\right)$,

$$\mathsf{Pr}\left(\sum_{i=1}^{n} X_i \geq nc\right) \leq e^{-n\sup_{\theta \geq 0}(\theta c - \hat{\mu}(\theta))}$$

c. What is the practical implication of (b)?

5.5 Consider a discrete time stationary and ergodic arrival process $\{B_k, \ k \geq 0\}$. Let $\Gamma(\theta)$ denote the log-moment generating function for the process $\{B_k\}$.

a. Show that if $\{B_k\}$ is scaled by n (i.e., a bit arriving in $\{B_k\}$ results in n bits arriving in the scaled process), then the log-moment generating function of the resulting scaled process $\Gamma_n(\theta)$ is given by $\Gamma(n\theta)$.

b. The scaled and unscaled sources are each fed into separate multiplexers with infinite buffers. Show that the effective bandwidth for the scaled source, $e_n(\theta)$, is related to that for the unscaled source, $e(\theta)$, by $e_n(\theta/n) = ne(\theta)$.

c.

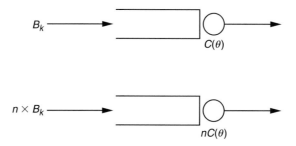

Figure 5.22 Scaling of sources and service rates in Problem 5.5.

Pictorially depict the relationship obtained in (b) by sketching the logarithm of the stationary c.d.f. of the buffer occupancy in each case shown in Figure 5.22 (on the same set of axes).

d. Is there a contradiction between the observation in (c) and the additivity of effective bandwidth? Explain.

5.6 Consider a discrete time multiplexer that can serve up to C_k packets in slot k, $k \geq 1$; that is, the service rate of the multiplexer is itself a random process. If B_k, $k \geq 1$, packets arrive in slot k, we can write the queue length X_k at the beginning of slot $k + 1$, $k \geq 1$, as

$$X_k = \left(X_{k-1} - C_k\right)^+ + B_k$$

Let $X_0 = B_0$.

a. Show that for $n \geq 0$,

$$X_n = \max_{0 \leq k \leq n} \left(\sum_{i=k}^{n} B_i - \sum_{i=k+1}^{n} C_i \right)$$

b. Let $\{B_k\}$ and $\{C_k\}$ be stationary and ergodic processes. Write down your guess for the stability condition for the process $\{X_n\}$.

c. Now consider $B_k = B$, for all $k \geq 0$, where B is consistent with your answer in (b). Let X denote the stationary queue length random variable. We would like to guarantee that, for some $\theta > 0$,

$$\lim_{x \to \infty} \frac{1}{x} \ln \Pr(X > x) \leq -\theta$$

Define

$$\Gamma(\theta) = \lim_{n \to \infty} \frac{1}{n} \ln \mathsf{E}\left(e^{-\theta \sum_{i=1}^{n} C_i}\right)$$

Show that a sufficient condition for this objective to hold is

$$B < \frac{-\Gamma(\theta)}{\theta}$$

d. Show that $\frac{-\Gamma(\theta)}{\theta} \leq \lim_{n \to \infty} \frac{1}{n} \sum_{i=1}^{n} C_i$ with probability one.

5.7 An alternating on–off variable bit rate source has *on* times $(U_1, U_2, U_3 \dots)$, where U_i, $i \geq 1$, are i.i.d. exponential with mean α^{-1}.

The *off* times $(V_1, V_2, \dots)$ are such that $V_i = U_i$. The peak rate is $2c$, and the source is served by a link of rate c with infinite buffer. Let $X(t)$ denote the buffer level process. Obtain and sketch $\ln(\lim_{t \to \infty} \frac{1}{t} \int_0^t I_{\{X(u) > B\}} \, du)$— that is, the fraction of time that the buffer is above B vs B. (Hint: Sketch a sample path of the buffer process and notice that the buffer empties out at the end of each cycle of the source; use the renewal reward theorem (see Appendix D))

5.8 N statistically identical sources are shaped by leaky buckets with the same parameters (σ, ρ, R). They are then multiplexed into a link of rate C. Assuming marginal buffering, show that the fraction of packets lost is maximized if the LB outputs are extremal on–off sources (i.e., the sources take values R and 0, and their average rates are ρ). Hint: By symmetry, the average rates of all sources must be the same value $r \leq \rho$. For any such rate r, we know that on–off sources with this rate maximize the packet loss rate (and hence the fraction of packets lost). The fraction of lost packets is given by $\frac{1}{Nr} \sum_{k=1}^{N} (kR - C)^+ \binom{N}{k} (\frac{r}{R})^k (1 - \frac{r}{R})^{(N-k)}$. Show that, over $0 \leq r \leq \rho$, this expression is maximized for $r = \rho$.

CHAPTER 6

Circuit-Multiplexed Networks

The theory of circuit-multiplexed networks was originally developed in the context of telephone networks. In fact the origins of queueing theory can be traced mainly to the analysis and dimensioning of telephone network resources. In this chapter, wherever possible, we consider a generalized circuit-multiplexed network that supports many classes of traffic, with the classes distinguished by varying resource requirements. At times, when the discussion of the generalized network becomes difficult or if analytically tractable models are not available, we consider single-service networks such as the telephone network. Much of the theory that we develop for circuit-multiplexed networks can be extended fairly simply to the design of cellular networks and a class of optical networks. We provide a sample analysis for each of these.

6.1 Introduction and Sample Applications

Traffic is offered to circuit-multiplexed networks as calls. A call is essentially a request from an end point to reserve a fixed set of resources in the network for exclusive use by the information flow generated by the call. The offered traffic to a circuit-multiplexed network can be described by a call arrival process (from which can be derived a call arrival rate) and a call holding time distribution. Typically, the call arrival times and the call holding times are assumed to be mutually independent. The call arrival rate, λ, is defined by $\lambda := \lim_{t \to \infty}$ (Number of Arrivals in $[0, t]$)$/t$. Because the calls are allocated a fixed amount of bandwidth for the duration of the call, the call holding time depends only on the call, and the network has no effect on it. The traffic intensity is a measure of the *offered load* in a circuit-multiplexed network. It is defined as the ratio of the average call holding time to the average call interarrival time. Alternatively, it is also the product of the call arrival rate and the mean call holding time. Although traffic intensity is a ratio and is dimensionless, it is sometimes expressed as Erlangs. For example, a call arrival rate of 10 calls per hour with an average holding time of 3 minutes per call will correspond to a load of 0.5 Erlangs. Another way to look at the traffic intensity is to say that it is a normalization of the call arrival rate to the mean call holding time; that is, it expresses the call arrival rate as the average number of call arrivals during a call. Yet another interpretation of traffic intensity is that it is the average number of active calls when the capacity is infinite.

The generalized circuit-multiplexed network that we consider in this chapter supports many classes of calls. It is also called an integrated network. A class of calls is distinguished by the call arrival process, call holding times, and resource (bandwidth) requirements from the network. To make our analysis simple, we make some reasonable assumptions: There is a finite number of call classes, the bandwidth requirement of each class is an integer multiple of a *basic rate*, and the link capacities are integer multiples of this basic rate. The last two assumptions allow us to specify the link capacities and the bandwidth requirements as integers.

The discussion that we develop in this chapter is applicable to a wide class of networks in addition to the traditional telephone network. As a first example, recall our discussion on ISDN networks in Chapter 2, where the terminals can make calls that require $n \times 64$ Kbps. Clearly, the backbone of this network will be offered different classes of calls. As a second example, consider an integrated packet network supporting multiple classes of stream traffic, with many of the classes requiring a quality of service commitment from the network. Recall from our discussion in Section 5.6.2 that one of the ways to provide this service is for the source to translate its in-call packet arrival statistics and QoS specification into an effective bandwidth requirement from the network. The network must reserve that much bandwidth on all the links that will be used by the call. The call is admitted if and only if the network can reserve the specified bandwidth on all the links on the route without affecting the active calls. Otherwise, the call is lost.

The analytical techniques that we discuss in this chapter can be applied to analyzing networks in which fixed amounts of resources are reserved for the call while it is active and the admissibility of the call itself is subject to a set of constraints on resource usage. Here is an example of a resource usage constraint: In cellular networks, when a call arrives, only a channel that is not in use in a set of neighboring cells is available for allocation to the call. Another such example: In optical networks, calls arrive as lightpath requests that require a wavelength on every link of the route. A typical constraint is that the same wavelength must be allocated to the lightpath on every link of the route.

A circuit-multiplexed system is modeled as a loss system; an arriving call will be accommodated and its service will begin "immediately" if the resources required for it are available. Otherwise, it is assumed that the call is dropped, or lost, forever. The call is not made to wait in a queue until resources become available.

The models that we consider for circuit-multiplexed networks are thus called *loss models*. Thus the most important performance issue is the loss probability, or blocking probability. This in turn can be used to obtain the *carried load*, the part of offered traffic that is carried by the link. Thus in all the discussions in this chapter, our focus is on obtaining these performance measures.

6.2   Multiclass Traffic on a Single Link

Consider a single link in a multiclass, circuit-multiplexed network. Let C, an integer, be the capacity of the link. As discussed earlier, in practice, C will be a multiple of a basic rate that is also used to express the bandwidth requirement of the different classes of calls. For example, as we saw in Chapter 2 the basic rate in ISDN networks is 64 Kbps; in time or frequency division multiplexed cellular networks, it is the bandwidth of a channel required to carry a call; and in optical networks, it is a wavelength. We assume that there are K classes of calls, and a call of class k, $k \in \{1, 2, \ldots, K\}$, arrives according to a time-homogeneous, or stationary, Poisson process of rate λ_k. The holding times of the calls of class k are i.i.d. exponentially distributed with mean μ_k^{-1} and require b_k units of bandwidth when active. For now, we assume that a new call of class k is admitted on the link if there are b_k units of unused bandwidth available at that time.

An analogy of the multiclass link with the well-known knapsack resource-allocation problem can be made: Objects of different sizes (calls requiring different capacities) are to be fit into a box (link) of a fixed capacity. Because the arrival times and sizes of the objects (calls) are random, a multiclass link is often called a *stochastic knapsack*. Figure 6.1 shows a snapshot in time of the occupancy of a multiclass link. In time division multiplexing, allocating circuits to a call involves allocating slots in every frame, with each slot corresponding to a specific bandwidth. If the slots are numbered from the beginning of the frame, many practical systems will require that a call be allocated a block of contiguous slots in each frame. To simplify the analysis, we do not consider this case here. Note that in the allocation shown in Figure 6.1 we have not made any attempt to provide each call with contiguous channels.

Let us now introduce some notation. Let n_k denote the number of class k calls active on the link. Define two K-dimensional column vectors **b** and **n** as follows:

$$\mathbf{b} := [b_1, \ldots, b_K]^T$$

$$\mathbf{n} := [n_1, \ldots, n_K]^T$$

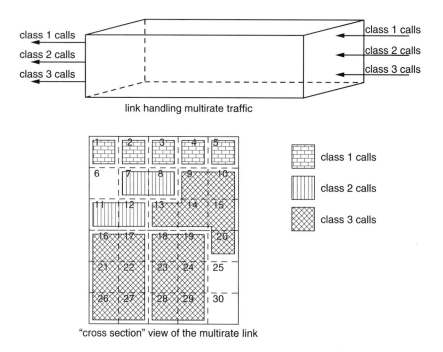

link handling multirate traffic

"cross section" view of the multirate link

Figure 6.1 Sample snapshot of the occupancy of a multiclass link with $C = 30$, $K = 3$, $b_1 = 1$, $b_2 = 2$, and $b_3 = 6$. In the example, $n_1 = 5$, $n_2 = 2$, and $n_3 = 3$. When the link is in this state, a new call of class 1 or 2 can be accepted, but a call of class 3 cannot be admitted. The numbers on the top right corners indicate the possible slot numbers in a TDM frame. Observe that a call has not always been allocated contiguous channels.

Using this notation, the admission rule is to admit a call of class k' if the following condition is satisfied:

$$b_{k'} \leq C - \sum_{k=1}^{K} n_k b_k \tag{6.1}$$

If this condition is not satisfied, the call is blocked and lost forever. Allowing the blocked calls to retry after random delay makes the models complex and often analytically intractable. (We discuss such systems briefly in Section 6.7.)

Equation 6.1, in matrix notation, can be written as

$$b_{k'} \leq C - \mathbf{n}^T \cdot \mathbf{b}$$

The state of the link seen by an arriving call determines whether it is blocked or accepted. The assumption of Poisson call arrivals and the Poisson arrivals see time averages (PASTA) property (see Section D.1.8 in Appendix D) implies that the fraction of arriving calls that will see the link in state $\mathbf{n}$ is the same as the fraction of time for which the link is in state $\mathbf{n}$; that is, the probability that a call is blocked is exactly the same as the probability that the system is in the corresponding blocking state. Thus we need to find the stationary probabilities for the system state.

Now consider the operation of a link of capacity C. Let $X_k(t)$ be an integer denoting the number of active calls of class k at time t, and let the vector

$$\{\mathbf{X}(t) := [X_1(t), \dots, X_k(t), \dots, X_K(t)]^T, \quad t \geq 0\}$$

denote the random process of the number of calls active on the link at time t. We will call $\mathbf{X}(t)$ the state of the link at time t. $\mathbf{X}(t)$ evolves over a state space S defined by $S = \{\mathbf{n} : \mathbf{n}^T \cdot \mathbf{b} \leq C\}$. Figure 6.2 shows a sample state space for a multiclass link handling two classes of calls. Note that S is finite but can be very large.

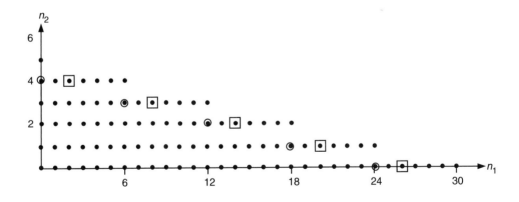

Figure 6.2 **State space for the multiclass link with** $C=30$, $K=2$, $b_1=1$, **and** $b_2=6$. **The dots represent the elements of the state space. The states marked by a circle form the set** $S_{\mathbf{busy}}(24)$, **and those marked by a square form the set** $S_{\mathbf{busy}}(26)$.

The following exercise will help in explaining the growth in the cardinality of $\mathcal{S}$ with increasing C and K.

Exercise 6.1

If $b_k = 1$ for $k = 1, \ldots, K$, using a combinatorial argument show that the cardinality of $\mathcal{S}$, the number of elements in S, is given by

$$|\mathcal{S}| = \sum_{j=0}^{C} \binom{K+j-1}{K-1}$$

Evaluate $|\mathcal{S}|$, the number of elements in $\mathcal{S}$, for $K = 5$ and $K = 10$. Hint: For each j, you can treat the problem as the number of ways of putting j indistinguishable balls in K urns.

The state of the link can change by a call departure (service completion) or by a call arrival. Because the call holding times are i.i.d. exponential and the call arrivals are from a stationary Poisson process, $\{X(t),\ t \geq 0\}$ is a *continuous time Markov chain* (CTMC). The state change due to an arrival of a class k customer occurs with rate λ_k for all $\mathbf{n}$ that can admit a class k call. Because we are considering a loss system and because all admitted customers begin service immediately, the departure rate of a class k call when $X(t) = \mathbf{n} = [n_1, \ldots, n_k, \ldots, n_K]$ is $n_k \mu_k$. The Markov chain for $X(t)$ is thus completely defined. Figure 6.3 shows a part of the Markov chain and the transition rates from interior states, for a link that is offered two classes of traffic.

The Markov chain representation of $\{X(t)\}$ is irreducible. Furthermore, because it is defined over a finite state space, it is positive recurrent, which implies that a stationary distribution exists. To help us obtain the stationary probabilities for the number of active calls on the link, we first show that the CTMC $X(t)$ is reversible, which in turn is used to show that the stationary probability has a simple *product form*.

Let $C = \infty$. Then, unlike when C is finite, there is no interaction among the different classes, and the number of active calls in the system of one class does not affect the admission of arriving calls of other classes. The individual components of $\{X(t)\}$ (i.e., $X_k(t)$), evolve independently of each other and are a CTMC, as shown in Figure 6.4. Let $\pi_\infty(n_k)$ be the stationary probability of $X_k(t) = n_k$. From Figure 6.4 we see that the graph of the CTMC of $X_k(t)$ is a tree,

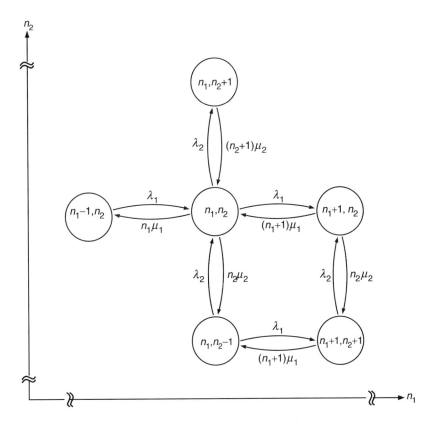

Figure 6.3 The transition rates in the continuous time Markov chain representing the multiclass link that is offered two classes of calls (i.e., $K=2$). The transition rates from an arbitrary interior state are shown.

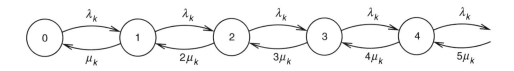

Figure 6.4 The transition rates for the CTMC of $X_k(t)$ when $C=\infty$.

and the Kolmogorov criterion for reversibility holds. (See Section D.3.1 for more details.) Thus $X_k(t)$ is reversible, and we can write the detailed balance equations $\lambda_k \pi_\infty(n_k) = (n_k+1)\mu_k \pi_\infty(n_k+1)$ for $n_k \geq 0$. Solving the detailed balance equations we get

$$\pi_\infty(n_k) = e^{-\rho_k} \frac{\rho_k^{n_k}}{n_k!} \tag{6.2}$$

where $\rho_k := \lambda_k/\mu_k$. The stationary probability that $\mathbf{X}(t) = \mathbf{n} = [n_1,\ldots,n_K]$, $\pi_\infty(\mathbf{n})$, will be a product of the probability of $X_k(t) = n_k$:

$$\pi_\infty(\mathbf{n}) := \pi_\infty(n_1,\ldots,n_K) = \prod_{k=1}^{K} \pi_\infty(n_k) = \prod_{k=1}^{K} e^{-\rho_k} \frac{\rho_k^{n_k}}{n_k!}$$

Let $\mathbf{e}_k$ denote a vector, with the kth element being 1 and the other elements being 0. State $\mathbf{n} - \mathbf{e}_k$ will have one class k call less than state $\mathbf{n}$, and $\mathbf{n} + \mathbf{e}_k$ will have one more class k call than state $\mathbf{n}$. For $k = 1,\ldots,K$ and considering states $\mathbf{n}$ and $\mathbf{n} + \mathbf{e}_k$, we can write

$$\pi_\infty(\mathbf{n} + \mathbf{e}_k) = \pi_\infty(n_k + 1) \prod_{\substack{k'=1 \\ k' \neq k}}^{K} \pi_\infty(n_{k'})$$

$$= \frac{\lambda_k}{(n_k + 1)\mu_k} \pi_\infty(n_k) \prod_{\substack{k'=1 \\ k' \neq k}}^{K} \pi_\infty(n_{k'})$$

$$\lambda_k \pi_\infty(\mathbf{n}) = (n_k + 1)\mu_k \pi_\infty(\mathbf{n} + \mathbf{e}_k) \tag{6.3}$$

The second equality above is obtained from the detailed balance equation of $X_k(t)$. For $\mathbf{X}(t)$, the only transitions possible from state $\mathbf{n}$ are to either $\mathbf{n} + \mathbf{e}_k$ or $\mathbf{n} - \mathbf{e}_k$ (if $n_k > 0$) at rates λ_k and $n_k\mu_k$, respectively, for $k = 1,\ldots,K$. Thus Equation 6.3 verifies the detailed balance equations for $\mathbf{X}(t)$, and hence $\mathbf{X}(t)$ is also reversible. The Kolmogorov criterion for $\mathbf{X}(t)$ can also be verified. For $K = 2$, in Figure 6.3

consider the loop from state $[n_1, n_2]$: $[n_1, n_2] \to [n_1 + 1, n_2] \to [n_1 + 1, n_2 - 1] \to$ $[n_1, n_2 - 1] \to [n_1, n_2]$. The following can be verified for this loop.

$$\lambda_1 (n_2 \mu_2)(n_1 + 1)\mu_1 \lambda_2 = (n_1 + 1)\mu_1 (n_2\ \mu_2)\lambda_1 \lambda_2$$

When C is finite, the state space is a truncation of that when $C = \infty$. Hence, from Theorem D.22 in Section D.3.1, the stationary probability of the truncated CTMC, $\pi_C(\mathbf{n})$, is the same as $\pi_\infty(\mathbf{n})$ restricted to the state space $\mathcal{S}$:

$$\pi_C(n_1, n_2, \ldots, n_K) = \frac{\pi_\infty(n_1, n_2, \ldots, n_K)}{\sum_{\mathbf{n} \in \mathcal{S}} \pi_\infty(n_1, n_2, \ldots, n_K)}$$

We have thus proved the following theorem.

Theorem 6.1

The stationary probability for $\mathbf{X}(t)$

$$\pi_C(n_1, n_2, \ldots, n_K) = \frac{1}{G} \prod_{k=1}^{K} \frac{\rho_k^{n_k}}{n_k!} \tag{6.4}$$

where G is the normalizing constant that is necessary to make the stationary probabilities sum to 1 and is obtained as

$$G := \sum_{\mathbf{n} \in \mathcal{S}} \prod_{k=1}^{K} \frac{\rho_k^{n_k}}{n_k!} \qquad \blacksquare$$

It turns out that Theorem 6.1 holds even if the holding time distribution is not exponential. We do not discuss this generalization.

Note that the detailed balance equations hold for $\mathbf{X}(t)$ even when C is finite, and it is therefore reversible.

For any numerical study of the circuit-multiplexed link, the numerical value of the normalizing constant G is essential. In addition we may also need to sum the right side of Equation 6.4 over a subset of $\mathcal{S}$. For example, consider the evaluation of the blocking probability of class k calls, $P_B(C, k)$. Let the set of states that will block a class k call be denoted by $\mathcal{S}_B(k)$.

$$\mathcal{S}_B(k) := \{\mathbf{n}: C - \mathbf{n}^T \cdot \mathbf{b} < b_k\}$$

From this we can write

$$P_B(C, k) \;=\; \sum_{\mathbf{n} \in \mathcal{S}_B(k)} \pi_C(\mathbf{n}) \;=\; \frac{1}{G} \sum_{\mathbf{n} \in \mathcal{S}_B(k)} \prod_{k'=1}^{K} \frac{\rho_{k'}^{n_{k'}}}{n_{k'}!} \tag{6.5}$$

As another example, consider the probability that at any time exactly c circuits are busy. Denote this probability by $P_{\text{busy}}(C, c)$, and the states for which c circuits are busy by $\mathcal{S}_{\text{busy}}(c)$. We can write the following.

$$\mathcal{S}_{\text{busy}}(c) = \{\mathbf{n}: \mathbf{n}^T \cdot \mathbf{b} = c\}$$

$$P_{\text{busy}}(C, c) = \sum_{\mathbf{n} \in \mathcal{S}_{\text{busy}}(c)} \pi_C(\mathbf{n}) \;=\; \frac{1}{G} \sum_{\mathbf{n} \in \mathcal{S}_{\text{busy}}(c)} \prod_{k=1}^{K} \frac{\rho_k^{n_k}}{n_k!}$$

Recall from Exercise 6.1 that the cardinality of $\mathcal{S}$ can be very large. The same will be true for the subsets of $\mathcal{S}$ of interest. Hence, enumerating the terms in these summations may not be feasible. Also, the terms in the summation involve evaluation of ratios of very large integers, and this may cause significant errors when we evaluate using finite-precision real number arithmetic. Thus any technique to efficiently evaluate these expressions should also be numerically stable. We next describe a recursive algorithm for such summations that is also numerically stable.

6.2.1   The Kaufman–Roberts Recursion

From the detailed balance equations for $\mathbf{X}(t)$, for $\mathbf{n} \in \mathcal{S}$ and $n_k > 0$,

$$n_k \, \pi(\mathbf{n}) = \rho_k \, \pi(\mathbf{n} - \mathbf{e}_k)$$

Summing the left and the right sides of this equation over the set of elements of $S_{\text{busy}}(c)$ for which $n_k > 0$, we can write the following for $c \geq b_k$:

$$\sum_{\substack{\mathbf{n} \in S_{\text{busy}}(c) \\ n_k > 0}} n_k \, \pi(\mathbf{n}) = \sum_{\mathbf{n} \in S_{\text{busy}}(c)} n_k \, \pi(\mathbf{n}) = \sum_{\substack{\mathbf{n} \in S_{\text{busy}}(c) \\ n_k > 0}} \rho_k \, \pi(\mathbf{n} - \mathbf{e}_k) \qquad (6.6)$$

The first equality is obtained by seeing that the summand for $n_k = 0$ is 0. See Figure 6.5 for an illustration of the sets over which the summation is being carried out.

Consider the set of states defined by $S' := \{\mathbf{n} : \mathbf{n} \in S_{\text{busy}}(c) \text{ and } n_k > 0\}$. Now consider the set of states S'', obtained from S' by removing one class k call from the states in S'. The number of channels occupied by the states in S'' is $c - b_k$ because there is one less active class k call. Thus $S'' \subset S_{\text{busy}}(c - b_k)$, and, from the

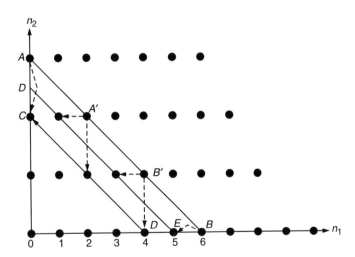

Figure 6.5 **Illustrating the recursion arguments in the computation of G with $b_1 = 1$ and $b_2 = 2$. The states of $S_{\text{busy}}(6)$ are on AB, and those of $S' = \{S_{\text{busy}}(6), n_2 > 0\}$ are on AB'. The vertical arrows from the states in S' are to states with one less class 2 call. Notice that these states form the set $S_{\text{busy}}(4) = S_{\text{busy}}(6 - b_2)$. Similarly, the states in $S' = \{S_{\text{busy}}(6), n_1 > 0\}$ lie on $A'B$. The horizontal arrows are to states with one less class 1 call, and these states form the set $S_{\text{busy}}(5) = S_{busy}(6 - b_1)$.**

foregoing discussion, $\mathbf{n} \in \mathcal{S}'$ implies $(\mathbf{n} - \mathbf{e}_k) \in \mathcal{S}_{\text{busy}}(c - b_k)$. Summarizing,

$$\mathbf{n} \in \mathcal{S}_{\text{busy}}(c) \text{ and } n_k > 0 \;\Rightarrow\; (\mathbf{n} - \mathbf{e}_k) \in \mathcal{S}_{\text{busy}}(c - b_k)$$

Using similar arguments, we can show that

$$(\mathbf{n} - \mathbf{e}_k) \in \mathcal{S}_{\text{busy}}(c - b_k) \;\Rightarrow\; \mathbf{n} \in \mathcal{S}_{\text{busy}}(c) \text{ and } n_k > 0$$

This means that the sets $\{\mathbf{n} : \mathbf{n} \in \mathcal{S}_{\text{busy}}(c) \text{ and } n_k > 0\}$ and $\{\mathbf{n} : (\mathbf{n} - \mathbf{e}_k) \in \mathcal{S}_{\text{busy}}(c - b_k)\}$ are the same. Saying it another way, the set of states defined by $\{\mathbf{n} - \mathbf{e}_k : \mathbf{n} \in \mathcal{S}_{\text{busy}}(c) \text{ and } n_k > 0\}$ and $\{\mathbf{n}' : \mathbf{n}' \in \mathcal{S}_{\text{busy}}(c - b_k)\}$ are the same. Figure 6.5 illustrates this equivalence for a link with two classes of traffic.

Continuing with Equation 6.6, for $1 \le k \le K$ and $b_k < c \le C$,

$$\sum_{\mathbf{n} \in \mathcal{S}_{\text{busy}}(c)} n_k \, \pi(\mathbf{n}) = \sum_{\substack{\mathbf{n} \in \mathcal{S}_{\text{busy}}(c) \\ n_k > 0}} \rho_k \, \pi(\mathbf{n} - \mathbf{e}_k)$$

$$= \rho_k \sum_{\mathbf{n} \in \mathcal{S}_{\text{busy}}(c - b_k)} \pi(\mathbf{n})$$

$$= \rho_k \, P_{\text{busy}}(C, c - b_k) \tag{6.7}$$

The last equality follows from the definition of $P_{\text{busy}}(\cdot)$. Multiplying both sides of Equation 6.7 by b_k and summing from $k = 1, \ldots, K$, we get

$$\sum_{k=1}^{K} b_k \, \rho_k \, P_{\text{busy}}(C, c - b_k) = \sum_{k=1}^{K} b_k \sum_{\mathbf{n} \in \mathcal{S}_{\text{busy}}(c)} n_k \pi(\mathbf{n})$$

$$= \sum_{\mathbf{n} \in \mathcal{S}_{\text{busy}}(c)} \left(\sum_{k=1}^{K} b_k n_k \right) \pi(\mathbf{n})$$

Input: $\{\lambda_k, \mu_k\}$ for $k = 1, \ldots, K$ and C
Initialize: $g(c) = 0$ for $c < 0$, $g(0) = 1$
for $c = 1, \ldots, C$ **do**
$g(c) = \frac{1}{c} \sum_{k=1}^{K} b_k \rho_k g(c - b_k)$
$G = \sum_{c=0}^{C} g(c)$
$P_{\text{busy}}(C, c) = \frac{g(c)}{G}$
$P_B(C, k) = \sum_{i=0}^{b_k - 1} P_{\text{busy}}(C, C - i)$

Algorithm 6.1 **Compute_P_B**

For $n \in S_{\text{busy}}(c)$, the index of the outer summation, $\sum_{k=1}^{K} b_k n_k = c$, and we can write the right side of the preceding equation as $cP_{\text{busy}}(C, c)$. Thus we have

$$cP_{\text{busy}}(C, c) = \sum_{k=1}^{K} b_k \rho_k \, P_{\text{busy}}(C, c - b_k) \qquad (6.8)$$

Equation 6.8 is called the Kaufman–Roberts recursion. The discussion can now be summarized into a computationally efficient algorithm, **Compute_P_B** (see Algorithm 6.1). It is easy to verify that $P_{\text{busy}}(C, 0) = \frac{1}{G}$. Define $g(0) := 1, g(c) := 0$ for $c < 0$ and

$$g(c) := \frac{1}{c} \sum_{k=1}^{K} b_k \rho_k \, g(c - b_k)$$

From Equation 6.8, we see that $P_{\text{busy}}(C, c) = \frac{g(c)}{G}$ and

$$\sum_{c=0}^{C} g(c) = \sum_{c=0}^{C} G P_{\text{busy}}(C, c) = G \sum_{c=0}^{C} P_{\text{busy}}(C, c) = G$$

The Erlang Blocking Model

The case of $K = 1$ and $b_1 = 1$ is an important special case. This is the model for a link with C circuits in the traditional circuit-multiplexed telephone network.

The blocking probability for calls arriving on the link is given by

$$E_B(\rho, C) := P_B(C, 1) = \frac{1}{\sum_{c=0}^{C} \rho^c/c!} \frac{\rho^C}{C!} \tag{6.9}$$

This is the celebrated Erlang-B formula for the blocking probability in a circuit-multiplexed link. The Erlang-B formula has been used extensively in the design of telephone networks. Most early telecommunication network designers used an Erlang table, in which the combination of the traffic intensity, the blocking probability, and the number of circuits was arranged in many ways for easy reference. Table 6.1 shows a sample Erlang table, in which the number of circuits that would be required to obtain a specified blocking probability for a given load can be obtained.

Consider an area in which long distance or trunk calls are made according to a Poisson process of rate 120 calls per hour. Let the call durations be exponentially distributed, with the average duration of the call being five minutes. This corresponds to a traffic intensity of 10 Erlangs. Assume that all the long distance calls are to be carried to a nearby trunk exchange. For a blocking probability of at most 5% on the link to the trunk exchange, 15 circuits would be required. A blocking probability of 1% would require 18 circuits, 0.1% blocking would require 21, and a blocking probability of 0.01% would require 24 circuits.

Exercise 6.2

For the special case of $K = 1$ and $b_k = 1$, show that $E_B(\rho, C)$ is given by the recursive relation

$$E_B(\rho, C) = \left(1 + \frac{C}{\rho} \frac{1}{E_B(\rho, C - 1)}\right)^{-1}$$

Starting with $E_B(\rho, 0) = 1$, this is an efficient recursion to compute the Erlang blocking formula.

Consider a telephone exchange providing phone services and dial-up Internet access. The bandwidth for both classes of calls is the same 64 Kbps, but the call arrival rates and the holding times will be different for each class of calls. For example, the mean call holding time of Internet access calls will be significantly longer than that of voice calls, whereas the arrival rate may be lower.

$C \downarrow$	Loss Probability									$C \downarrow$
	0.0001	0.001	0.002	0.005	0.01	0.02	0.05	0.1	0.2	
15	4.7812	6.0772	6.5822	7.3755	8.1080	9.0096	10.633	12.484	15.608	15
16	5.3390	6.7215	7.2582	8.0995	8.8750	9.8284	11.544	13.500	16.807	16
17	5.9110	7.3781	7.9457	8.8340	9.6516	10.656	12.461	14.522	18.010	17
18	6.4959	8.0459	8.6437	9.5780	10.437	11.491	13.385	15.548	19.216	18
19	7.0927	8.7239	9.3515	10.331	11.230	12.333	14.315	16.579	20.424	19
20	7.7005	9.4115	10.068	11.092	12.031	13.182	15.249	17.613	21.635	20
21	8.3186	10.108	10.793	11.860	12.838	14.036	16.189	18.651	22.848	21
22	8.9462	10.812	11.525	12.635	13.651	14.896	17.132	19.692	24.064	22
23	9.5826	11.524	12.265	13.416	14.470	15.761	18.080	20.737	25.281	23
24	10.227	12.243	13.011	14.204	15.295	16.631	19.031	21.784	26.499	24
25	10.880	12.969	13.763	14.997	16.125	17.505	19.985	22.833	27.720	25

Table 6.1 Part of the Erlang table showing the traffic intensity that can be offered to a link of capacity C (rows) circuits for specified blocking probabilities (columns).

Generalizing this scenario, consider $b_k = 1$ for $k = 1, \ldots, K$. From Equation 6.8, we can write

$$cP_{\text{busy}}(C, c) = P_{\text{busy}}(C, c - 1) \left(\sum_{k=1}^{K} \rho_k \right)$$

Writing $\sum_{k=1}^{K} \rho_k := \rho$, and recognizing that $P_{\text{busy}}(C, c)$ is the probability that c circuits are busy, we can rewrite the equation as

$$c\pi(c) = \pi(c - 1)\rho \quad \text{for } c = 1, \ldots, C$$

$$\pi(c) = \pi(0) \frac{\rho^c}{c!} = \frac{1}{\sum_{i=0}^{C} \rho^i / i!} \frac{\rho^c}{c!}$$

This is identical to the Erlang-B formula. Because the bandwidth requirement of all the classes is the same, the blocking condition for all classes is also the same and occurs when all the C circuits are busy. The blocking probability for all classes is therefore given by $E_B(\rho, C)$. This means that when the bandwidth requirements are the same, with respect to blocking probability, the system is insensitive to the presence of classes of calls with different arrival rates and mean holding times; we can group their loads and treat all the classes as one class, with traffic intensity equal to the sum of the traffic intensities of the individual classes.

Carried Load

In a blocking system such as the ones we consider in this chapter, the notion of *carried load* is important. This is the traffic that is not blocked. The carried load of class k is given by $\lambda_k(1 - P_B(C, k))$. Because the call holding times are independent, the carried load in Erlangs is given by $\rho_k(1 - P_B(C, k))$. In Problem 6.2 we will use Little's law to show that

$$\rho_k(1 - P_B(C, k)) = \sum_{n=1}^{C} n\pi(n_k)$$

where $\pi(n_k)$ is the probability that there are n_k calls of class k active. Thus the right side of the equation is the mean number of active calls of class k on the link.

Trunking Efficiency

Table 6.1 demonstrates an important phenomenon in statistical multiplexing. Let $\rho(C, \epsilon)$ be the maximum load that can be offered to a link with C circuits and achieve a blocking probability of less than ϵ. For a given grade of service ϵ, the trunking efficiency, $TE(\epsilon)$, is defined as $TE(\epsilon) := \rho(C, \epsilon)/C$. For $C = 1$, $P_B = \rho/(1 + \rho)$, and we can obtain $TE(\epsilon) = \epsilon/(1 - \epsilon)$. With 1% blocking and $C = 1$, $TE(0.01) = 0.01/0.99 \approx 0.01$. For $C = 15$ and 1% blocking, we get $TE(0.01) = 8.1080/15 \approx 0.54$, and for $C = 25$, $TE(0.01) = 0.645$. Thus trunking efficiency increases as the number of circuits increases.

> **Exercise 6.3**
> Obtain the trunking efficiencies for $C = 1, 15, 20$, and 25 when $\epsilon = 0.05$ and $\epsilon = 0.1$. Make a qualitative argument to explain why $TE(\epsilon)$ increases with ϵ.

6.3   Overflow and Non-Poisson Traffic

From our discussions in Section 2.3.3 of PSTN, ISDN, and other predominantly circuit-multiplexed networks, recall that for every class of calls there will be a routing plan that is a sequence of routes, or paths, to be tried in an effort to service calls of that class. If the first-choice route is blocking the call, then the second-choice route is tried and so on. If we want to obtain the overall call-blocking probability, we need to know the probability of the call being blocked on all the route options.

Consider a single-service network such as PSTN with a Poisson arrival process for calls between a node pair A and B. Let the routing plan for these calls have a primary path and an overflow path, as shown in Figure 6.6. For Poisson arrivals, the probability that the call is blocked on the primary path is obtained using the Erlang-B formula derived in Section 6.2. Let C_P be the capacity of link AB (the primary path), and C_S the capacity on link AC of the overflow path. Assume that link CB has sufficient bandwidth and does not contribute to the blocking of AB calls. Consider the following numerical example. Let $C_P = 4$ and $C_S = 2$, and let the offered load be 3.5 Erlangs. As just mentioned, assume that blocking on the overflow path occurs only due to nonavailability of circuits on link AC. We obtain the overall call-blocking probability by observing that a call is blocked if there is no capacity available on *both* the primary and the overflow paths—that is, if more than 6 calls are in progress between A and B. The overall blocking probability is $E_B(3.5, 6) = 0.0825$, and the overall

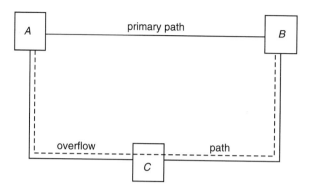

Figure 6.6   Two alternative paths for calls between *A* and *B*. The direct primary path has capacity *C*_P, whereas the secondary path through *C* has maximum capacity *C*_S.

blocked traffic is 0.2887 Erlangs. The blocking probability on the primary path is $E_B(3.5, 4) = 0.2603$, and hence the offered load to the overflow path is $\rho \times E_B(3.5, 4) = 3.5 \times 0.2603 = 0.9109$ Erlangs. If this were a Poisson process, the blocking probability on the overflow path would be $E_B(0.911, 2) = 0.1784$. The blocked traffic on the overflow route (and hence the overall blocked traffic) would be 0.1625 Erlangs. This is significantly lower than that calculated earlier. So obviously, the blocking probability in the overflow path is significantly higher (0.311) than the 0.1784 obtained from the Erlang-B formula. We can therefore argue that the arrival process to the overflow path is not Poisson. Arriving calls are not independently blocked; that is, the overflow process is not a Bernoulli splitting with probability P_B of the original arrival process. For $C > 1$, the probability that an overflow follows another is higher than P_B (see Problem 6.3); that is, overflows tend to happen together and are bursty. In fact, we show in the following that the overflow process has a higher variance than the Poisson process, and this is reflected in the higher blocking probability on the overflow path. To understand the behavior of a network in which blocked calls may be routed on an overflow path, we need to understand the behavior of the overflow process. This is then used to develop models in which many levels of overflow may occur and many overflows may be superposed.

6.3.1   Characterizing the Overflow Process

Consider a link for which the call interarrival times are i.i.d. random variables with an arbitrary distribution and mean $1/\lambda$, and the call holding times are i.i.d. exponential with mean $1/\mu$. We will obtain a recursive relation for the interarrival times of the overflow traffic from a link with $k + 1$ circuits in terms of those with k circuits.

Let the circuits on the link be numbered $1, 2, \ldots$. An arriving call is allocated to the lowest-numbered circuit that is free. A call arriving when circuits $1, 2, \ldots, k$ are busy will be called a k-overflow, meaning that it could not be accommodated in circuits $1, 2, \ldots, k$. Let $f_k(y)$ be the probability density function of the time between successive k-overflows, and $\tilde{f}_k(s)$ the Laplace transform of $f_k(y)$. In the following discussion we use the example of the overflow process shown in Figure 6.7 for $k = 2$.

A $(k + 1)$-overflow is also a k-overflow. In the example, arrivals at a, b, c, and g are 3-overflows and also 2-overflows. Consider a $k + 1$-overflow instant. Let t_k denote the time until the next k-overflow, and t_{k+1} the time until the next $k + 1$-overflow. For example, with reference to the 3-overflow (and 2-overflow) at time a, t_3, the time until the next 3-overflow is the interval ab. This is also the time until the next 2-overflow. For the 3-overflow at c, the time until the next 2-overflow is cd, and the time until the next 3-overflow is cg.

Let s_{k+1} denote the service time remaining on the call being carried by circuit $k + 1$ at the $k + 1$-overflow instant. Consider the instant of a k-overflow following a $k + 1$-overflow. We consider two cases: Case 1 occurs when this k-overflow is also a $k + 1$-overflow, and case 2 occurs when it is not. Case 1 implies $t_k = t_{k+1}$

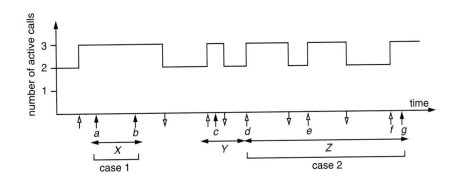

Figure 6.7 Example of the overflow process. Upward arrows are arrivals, and downward arrows are departures. The arrivals at a, $b, \ldots, g$ are all 2-overflows, whereas only those at a, b, c, and g are 3-overflows (indicated by solid arrowheads). The figure does not show link occupancies greater than 3.

and occurs if $t_k < s_{k+1}$. Given the value of t_k, the probability that $t_k < s_{k+1}$ is $e^{-\mu t_k}$. In Figure 6.7 this case is shown as the 2- and 3-overflows at b after the 3-overflow at a. Case 2 implies $t_{k+1} > t_k$, and this happens if $t_k > s_{k+1}$, and, given t_k, its probability is $(1 - e^{-\mu t_k})$. In the figure this is shown as the 2-overflows (but not 3-overflows) at d, e, and f after the 3-overflow at c. Consider the system at c and at d_+, just after the next k-overflow. Because the service times are exponential, the future evolution of the system has the same law at both these instants, and the time until the next $k + 1$-overflow has the same distribution at both of these instants. Thus the density function of the interval dg, shown as Z in the figure, will be $f_{k+1}(y)$ and will be independent of the distribution of the segment cd, shown as Y in the figure. Note that in case 2, the time between two $k + 1$-overflows is the sum of segments Y and Z, as shown in the figure. Let $f_{k+1}(y|t_k)$ be the probability density function of the time between $k + 1$-overflows conditioned on the time from the $k + 1$-overflow until the next k-overflow being t_k, and let $\tilde{f}_{k+1}(s|t_k)$ be its Laplace transform. From the foregoing discussion, we can write

$$\tilde{f}_{k+1}(s|t_k) = (e^{-\mu t_k})(e^{-s t_k}) + (1 - e^{-\mu t_k})(e^{-s t_k} \times \tilde{f}_{k+1}(s))$$

$$= e^{-(s+\mu)t_k} - e^{-(s+\mu)t_k}\tilde{f}_{k+1}(s) + e^{-s t_k}\tilde{f}_{k+1}(s)$$

Unconditioning on t_k and solving for $\tilde{f}_{k+1}(s)$, we get

$$\tilde{f}_{k+1}(s) = \tilde{f}_k(s + \mu) - \tilde{f}_k(s + \mu)\tilde{f}_{k+1}(s) + \tilde{f}_{k+1}(s)\tilde{f}_k(s)$$

$$\tilde{f}_{k+1}(s) = \frac{\tilde{f}_k(s + \mu)}{1 + \tilde{f}_k(s + \mu) - \tilde{f}_k(s)} \qquad (6.10)$$

When the arrivals to the primary path form a Poisson process, the interval between successive arrivals has a *Phase type* distribution. Section D.1 describes this derivation.

Recall our earlier example to show that the overflow process is bursty. One mechanism to characterize the burstiness of a call arrival process is to specify the stationary probability of the number of active calls in an infinite-capacity link with the given arrival process. For example, if the call arrivals are Poisson and the service times are exponential, we know from Equation 6.2 that the stationary distribution of the number of active calls is Poisson with mean ρ and variance ρ. We now obtain the variance of the number in the system in a link with infinite capacity when the call interarrivals are a renewal process with arbitrarily distributed renewal densities of mean $1/\lambda$ and the holding times have an exponential distribution of mean $1/\mu$. Such a system is called a GI/M/∞ queue, with the GI denoting the i.i.d. interarrivals with a general distribution, the M signifying that the service times are Markovian (exponential distribution), and the ∞ indicating that there is no blocking. We will compare the mean and variance that we obtain for this system with ρ, the mean and variance when the arrival process is Poisson.

Let $X(t)$ be the number of active calls at time t, $A(t)$ the number of call arrivals in the interval $(0, t)$, and $N_j(t)$ the number of arrivals in the interval $(0, t)$ that find j active calls on arrival. Define

$$\pi_j := \lim_{t \to \infty} \frac{1}{t} \int_0^t I_{X(u)=j} \, du$$

$$p_j := \lim_{t \to \infty} \frac{N_j(t)}{A(t)}$$

where $I_{X(u)=j}$ is the indicator function of the event $\{X(u) = j\}$ that takes a value 1 when the event occurs and a value zero otherwise; π_j is the time average probability that $X(t) = j$, and p_j is the probability that an arriving call finds j active calls.

Now let $U_j(t)$ be the number of upward $j \to (j+1)$ transitions of $X(t)$, and $D_j(t)$ the number of downward $(j+1) \to j$ transitions in $(0, t)$ made by $X(t)$. Observe that $|U_j(t) - D_j(t)| \leq 1$. Hence, dividing across by t, if the limits exist, we can write

$$\lim_{t \to \infty} \frac{U_j(t)}{t} = \lim_{t \to \infty} \frac{D_j(t)}{t} \tag{6.11}$$

An arriving call that sees j active calls will make $X(t)$ go from j to $(j+1)$, and hence $U_j(t) = N_j(t)$. The left side of Equation 6.11 can therefore be written as

$$\lim_{t \to \infty} \frac{N_j(t)}{t} = \lim_{t \to \infty} \frac{N_j(t)}{A(t)} \frac{A(t)}{t} = \lambda p_j$$

A departure causes a downward transition from state $j+1$ to state j. When $X(t) = j + 1$, because the service times are all i.i.d. exponential, the rate of a departure is $(j + 1)\mu$. Thus the right side of Equation 6.11 can be written as

$$\lim_{t \to \infty} \frac{D_j(t)}{t} = (j + 1)\mu \pi_{j+1}$$

Thus we have

$$j\pi_j = \frac{\lambda}{\mu} p_{j-1} \tag{6.12}$$

Note that this argument is essentially the level crossing analysis that is used to derive Equation 1.1 in Section D.1.9. It requires singleton arrivals and departures.

Multiplying both sides of Equation 6.12 by $j - 1$ and summing from $j = 2$ to ∞, we get

$$\sum_{j=2}^{\infty} j(j - 1)\pi_j = \sum_{j=2}^{\infty} (j - 1)\frac{\lambda}{\mu} p_{j-1}$$

Expanding the left side and noting that from Little's law, the average number of active calls is ρ (i.e., $\sum_{j=0}^{\infty} j\pi_j = \rho$), we get

$$\sum_{j=2}^{\infty} (j - 1)j\pi_j = \sum_{j=2}^{\infty} j^2\pi_j - \sum_{j=2}^{\infty} j\pi_j = (\mathsf{Var}\,(X) + \rho^2 - \pi_1) - (\rho - \pi_1)$$

where $\mathsf{Var}\,(X)$ is the variance of the number of active calls. Now we can write

$$\mathsf{Var}\,(X) + \rho^2 - \rho = \frac{\lambda}{\mu} \sum_{j=0}^{\infty} j p_j \tag{6.13}$$

The sum on the right side represents the average of the number of busy circuits seen by an arriving call. Let $\hat{X}_n$ be the number of busy circuits seen by the nth arrival, and A_n the time between the arrivals of the nth and $(n + 1)$th calls. Given A_n, each of the $\hat{X}_n + 1$ (including the nth arrival) calls will finish service in the interval independently with probability $(1 - e^{-\mu A_n})$. Thus the number that remain in the system at the end of A_n has a binomial distribution, with mean $(\hat{X}_n+1)e^{-\mu t_n}$. We can thus obtain the following.

$$E\left(\hat{X}_{n+1}|\hat{X}_n, A_n\right) = \left(\hat{X}_n + 1\right)e^{-\mu A_n}$$

Taking expectations on both sides of the equation while noting that X_n and A_n are independent, and then taking limits as $n \to \infty$, we get

$$E\left(\hat{X}_{n+1}\right) = \left(E\left(\hat{X}_n\right) + 1\right)\tilde{f}(\mu)$$

$$E\left(\hat{X}\right) = \frac{\tilde{f}(\mu)}{1 - \tilde{f}(\mu)}$$

where $\hat{X}$ is the number of active calls seen by an arbitrary arrival, and $\tilde{f}(s)$ is the Laplace transform of the probability density function of the interarrival times. Using this and Equation 6.13, we get

$$\text{Var}(X) = \rho\frac{\tilde{f}(\mu)}{1 - \tilde{f}(\mu)} - \rho^2 + \rho \tag{6.14}$$

In Exercise 6.4, we apply Equation 6.14 to the special case of Poisson call arrivals.

Exercise 6.4

Consider a primary path with capacity C, and a secondary path with infinite capacity. Call arrivals form a Poisson process of rate λ, and holding times have an exponential distribution of mean $1/\mu$. Using Equation 6.10, which gives the call interarrival time distribution to the overflow path in Equation 6.14, show that the mean and variance of the number of active calls on the overflow path (ρ_C and V_C, respectively) are given by

$$\rho_C = \rho P_B(\rho, C) \tag{6.15}$$

$$V_C = \rho_C \left(1 - \rho_C + \frac{\rho}{\rho_C + C + 1 - \rho}\right) \tag{6.16}$$

For any call arrival process to a link with infinite capacity, the ratio of the variance V of the number of active calls to its mean ρ is often called the *peakedness* Z of the call arrival process (i.e., $Z := \frac{V}{\rho}$). For a Poisson call arrival process, $Z = 1$.

Exercise 6.5

For the overflow from a path of capacity C, show that $V_C > \rho_C$. This implies that the variance of the overflow calls is higher than that of a Poisson process having the same mean.

6.3.2 Approximations for Blocking Probability with Non-Poisson Traffic

In a network, the offered load to a link is the superposition of Poisson first-offered traffic and non-Poisson overflow traffic from many links. Calculating the exact blocking probability on such links is complicated because the call arrival process—a superposition of many arrival processes of different characteristics—is complex. We take recourse to approximations, with the motivation coming from the foregoing analysis. The simplest of these, *Hayward's approximation*, is obtained using the following argument. Consider a link of capacity C with calls requiring k circuits rather than one circuit. If the offered load is ρ/k and if C is infinite, the mean number of active circuits would be ρ, and the variance would be $k\rho$. If C is finite, then the blocking probability is $E_B(\rho/k, C/k)$. If call arrivals require only one circuit but arrive in groups of k, then an offered load of ρ would result in a variance of the call arrival process being $k\rho$ (i.e., a peakedness of k). Hayward's approximation assumes that call arrivals with a peakedness of Z can be approximated as arrivals in which calls arrive in groups of size Z, and the approximate call-blocking probability, $\tilde{P}_B(\rho, Z, C)$, is given by

$$\tilde{P}_B(\rho, Z, C) = E_B\left(\frac{\rho}{Z}, \frac{C}{Z}\right)$$

Observe that if the call arrivals form a Poisson process, $V = \rho$, and substituting this we get back the Erlang-B blocking formula. Because C here can be a noninteger, interpolation can be used. In fact, Equation 6.9 has been extended to allow C to be real rather than an integer.

A second technique is to approximate the actual traffic process by a process for which the blocking probabilities can be obtained. The parameters of the approximating process are chosen to match some moments of the actual process. Typically, the first two moments are matched. The moment-matching technique that we will discuss has a simple idea and is called the *equivalent random method*. We match the first two moments of the number of active calls when the capacity is infinite to the first two moments of the overflow process. Given the mean ρ and peakedness Z of the offered traffic to a link, we construct a virtual primary path with mean offered load ρ' and capacity C' such that the mean and peakedness of the overflow process are ρ and Z, respectively; in other words, we match the moments of the actual process to those of an "overflow process." This involves solving Equation 6.16 for ρ' and C'. Solve

$$\rho = \rho' P_B(\rho', C')$$

$$Z = \left(1 - \rho + \frac{\rho'}{\rho + C' + 1 - \rho'}\right)$$

for ρ' and C'. Numerical solutions are typically used, and tables are available to obtain the ρ' and C' from ρ and Z. Alternatively, we can use Rapp's approximation for ρ' and C':

$$\rho' = \rho Z + 3Z(Z - 1)$$

$$C' = \frac{\rho'(\rho + Z)}{\rho + Z - 1} - \rho - 1$$

If the capacity of the link is C, then the "overflow" path in this approximation has capacity C, and, as can be seen in Figure 6.8, the blocking probability for the peaked traffic can be approximated by $P_B(\rho', C' + C)$.

In our discussions, we have considered single-service links. Describing the overflow process from a multiservice link is a hard problem, and neither exact solutions nor good approximations are known.

6.4 Multiclass Networks

We are now ready to use the link models from the preceding sections to consider networks. The networks that we consider could be multiclass circuit-multiplexed networks like those that were proposed for ISDN networks. They could also be integrated packet networks, where the calls declare their offered load and

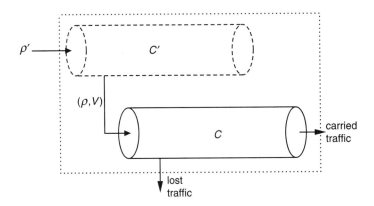

Figure 6.8 Calculating the blocking probability on a link that is offered non-Poisson traffic using the equivalent random method. The offered load ρ' and the capacity C' of the upper link are obtained to generate an overflow traffic of mean of ρ and variance V. The blocking probability for the original traffic is equal to the blocking probability of the system enclosed in the dashed lines; that is, $E_B(\rho', C + C')$.

QoS requirement by specifying the effective bandwidth calculated as described in Section 5.7.

The model of the physical topology of the network is that of a set of nodes interconnected by links of integer capacities. Let J be the number of links in the network, numbered $1, 2, \dots, J$. Link j has integer capacity C_j. As in our discussion of a multirate link, we assume that there is a basic rate and that all capacities and bandwidth requirements are expressed as integral multiples of this basic rate. The traffic offered to the network belongs to many classes, and each class has its own bandwidth requirement, arrival rate, mean holding time, and a fixed route defined by a loop-free sequence of consecutive links over which the call is carried. We assume that there are K classes of traffic indexed by k. A class of traffic is described by a Poisson call arrival process of rate λ_k, an exponentially distributed holding time of mean $1/\mu_k$, the route from the source to the destination, and the bandwidth b_k that will be required on all the links over which the traffic is to be routed. The routes for the different traffic classes are captured by a $J \times K$ matrix $\hat{R} = \{\hat{R}_{j,k}\}$ defined as follows.

$$\hat{R}_{j,k} := \begin{cases} 1 & \text{if link } j \text{ is used by class } k \\ 0 & \text{otherwise} \end{cases} \qquad (6.17)$$

We also define $\mathbf{R} = \{R_{j,k}\}$, which is also a $J \times K$ matrix defined as follows.

$$R_{j,k} := \begin{cases} b_k & \text{if link } j \text{ is used by class } k \\ 0 & \text{otherwise} \end{cases}$$

See Figure 6.9 for an example of the routes in a network. The $\mathbf{R}$ matrix for this network is

$$\mathbf{R} = \begin{bmatrix} b_1 & 0 & 0 & b_4 & 0 & b_6 \\ 0 & b_2 & 0 & 0 & 0 & 0 \\ 0 & b_2 & b_3 & 0 & 0 & 0 \\ b_1 & 0 & 0 & 0 & b_5 & b_6 \\ 0 & 0 & 0 & b_4 & b_5 & 0 \end{bmatrix}$$

When a class k call arrives, it is admitted if and only if at that time, b_k units of bandwidth are available on each link j for which $\hat{R}_{j,k} = 1$. Otherwise, the call is blocked and lost forever.

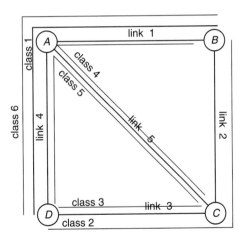

Figure 6.9   Routes in a network for the different classes of traffic. A four-node network with five links and six classes is shown. Classes 1 and 6 have the same route but have different bandwidth requirements and offered loads. Classes 1 and 2 are between the same node pair, as are classes 3 and 5.

As before, define $\rho_k := \lambda_k/\mu_k$. The following column vectors are used to denote the parameters more concisely: $\mathbf{C} := [C_1, C_2, \ldots, C_J]^T$, $\mathbf{b} := [b_1, b_2, \ldots, b_K]^T$, and $\rho := [\rho_1, \rho_2, \ldots, \rho_K]^T$.

Let n_k denote the number of active calls of class k, and define $\mathbf{n} := [n_1, n_2, \ldots, n_K]^T$. Clearly, not all $\mathbf{n}$ are feasible, and let $\mathcal{S}$ denote the set of all such feasible vectors. $\mathcal{S}$ will be given by

$$\mathcal{S} = \{\mathbf{n} : \mathbf{R} \cdot \mathbf{n} \leq \mathbf{C}\}$$

Let $X_k(t)$ be the number of active calls of class k at time t. Define $\mathbf{X}(t) := [X_1(t), X_2(t), \ldots, X_K(t)), \ t \geq 0]$ to be the vector of active calls in the network at time t. Because the arrivals are Poisson and because service times are exponential, it is easy to see that $\{\mathbf{X}(t), \ t \geq 0\}$ evolves as a continuous time Markov chain. The number of states over which $\mathbf{X}(t)$ evolves is finite, and if the arrival and holding rates are finite and nonzero, it is also irreducible. Hence $\{\mathbf{X}(t)\}$ is positive recurrent and has a stationary distribution. Let $\pi(\cdot)$ denote the stationary probabilities of $\{\mathbf{X}(t)\}$ with $\pi(\mathbf{n})$, $\mathbf{n} \in \mathcal{S}$, denoting the stationary probability that $\mathbf{X}(t) = \mathbf{n}$. Mimicking the arguments leading up to Theorem 6.1 in Section 6.2, we can obtain $\pi(\mathbf{n})$ for the multiclass network and state the following theorem.

Theorem 6.2

$$\pi(\mathbf{n}) = \frac{1}{G} \prod_{k=1}^{K} \frac{\rho_k^{n_k}}{n_k!} \tag{6.18}$$

where

$$G = \sum_{\mathbf{n} \in \mathcal{S}} \prod_{k=1}^{K} \frac{\rho_k^{n_k}}{n_k!} \tag{6.19}$$
∎

We illustrate the theorem using the simple two-link, three-class network shown in Figure 6.10. For this network, let $C_1 = C_2 = 2$, $b_1 = b_2 = 2$, and $b_3 = 1$. The state space for the network and the transition rates are as shown.

Exercise 6.6

Consider a two-link network with three classes of traffic like the one shown in Figure 6.10, with $C_1 = C_2 = \infty$. The admissible states can be written

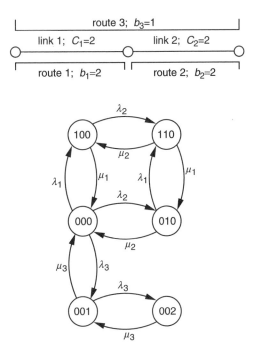

Figure 6.10 Example of a two-link, class-route network. The states of the network and the corresponding CTMC representation are shown in the lower panel.

as a tree with $[0, 0, 0]$ when there are no calls in the network at the top, $[1, 0, 0]$, $[0, 1, 0]$, and $[0, 0, 1]$ when there is one call in the network, at the next level, $[2, 0, 0]$, $[0, 2, 0]$, $[0, 0, 2]$, $[1, 1, 0]$, $[1, 0, 1]$, and $[0, 1, 1]$ when there are two calls in the network, at the level below it, and so on. Write the transition rates for this CTMC, and verify the Kolmogorov criterion for reversibility.

As with Theorem 6.1, Theorem 6.2 is true even if the holding times had a general distribution—say, $F_k(\cdot)$ and mean $1/\mu_k$ for $1 \leq k \leq K$—rather than the exponential distribution assumed earlier.

Remark: Although treating the network as a collection of routes gives us a simple model, it increases the problem size significantly, and even for medium-sized networks, the number of dimensions of the Markov chain can become very large. This is because the the number of possible source–destination pairs increases as the square of the number of nodes. Furthermore, for each node pair and class of traffic, a number of routes may need to be specified.

6.4.1 Analysis of the Network

To analyze the network and dimension the links, we must obtain the blocking probabilities for calls of each class in the multiclass network, just as we did for a single link. Throughout this discussion, we assume that the network is in steady state. Although the stationary distribution has a product form, the blocking states for the various classes of calls do not have a well-defined structure. Obtaining the normalizing constant and the blocking probabilities is difficult except for small networks, where enumeration can be used. Therefore, we need approximation techniques. In the following, L_k denotes the probability that an arriving class k call is blocked, and $\beta_{j,k}$ denotes the probability that link j cannot accept a class k call because it does not have b_k units of free bandwidth.

A simple approximation is to assume that the links are independent and then obtain the blocking probability of a class k call as the probability of not being blocked on any link along the route. Denote the offered load of class k calls to link j by $v_{j,k} = \hat{R}_{j,k}\rho_k$. The approximate blocking probability for class k on link j, $\hat{\beta}_{j,k}$, can be calculated using $v_{j,k}$ and C_k in the multirate link-blocking formula of Equation 6.5. The approximate blocking probability for a class k call, $\hat{L}_k$, will then be

$$\hat{L}_k = 1 - \prod_j (1 - \hat{\beta}_{j,k})^{\hat{R}_{j,k}}$$

When $b_k = 1$ for all k, the offered load before blocking to each link is $v_j = \sum_k \hat{R}_{j,k}\rho_k$. The approximate blocking probability on link j, denoted by $\hat{\beta}_j$, is $\hat{\beta}_j = E_B(v_j, C_j)$, and the approximate blocking probability on route k (i.e., of a class k call) is obtained to be

$$\hat{L}_k = 1 - \prod_j (1 - \hat{\beta}_j)^{\hat{R}_{j,k}} = 1 - \prod_j (1 - E_B(v_j, C_j))^{\hat{R}_{j,k}} \qquad (6.20)$$

where $E_B(\cdot)$ is the Erlang-B blocking formula of Equation 6.9. Note that in a network, a call is not offered to a link if any of the other links on the route cannot accept the call. Hence the actual offered load of class k calls to link j is less than the $\nu_{j,k}$ defined here. It turns out that if $b_k = 1$ for all k, then $\hat{\beta}_{j,k}$ is an upper bound for $\beta_{j,k}$, the actual blocking probability of class k. Hence, $\hat{L}_k$ is an upper bound for L_k. The proof is complicated, and we do not present it here. The basic idea of the proof is to show that the vector of blocking probabilities $\boldsymbol{\beta} = [\beta_1, \ldots, \beta_K]$ is an increasing function of the offered load vector $\nu = [\nu_1, \ldots, \nu_J]$.

Not much is known about the tightness of the bound obtained here. In fact, in many cases the bound can be very loose. See Problem 6.6 for a numerical example. Intuitively, the bound will be good when the bound is small, the routes have a small number of links, and K is large. The argument in favor of this is that when the bound is small, very few calls are lost, and the load offered to the links will be close to ν_j. If the number of links in the routes is small, blocking on one link does not significantly affect the offered load on the other links.

The bound from Equation 6.20 does not extend to the general case of arbitrary b_k. This is because for a multiclass link, the blocking probability is not monotonic with respect to the vector of offered load for that link. Consider, for example, a link with three classes of traffic. Let $b_1 = 1$, $b_2 = 2$, $b_3 = 3$, and $C = 3$. Increasing ρ_1, the offered load from a narrowband call, will increase the blocking probability of the class 3 broadband calls. This in turn means that the link is less occupied and can probably accept more class 2 or class 1 calls and hence decrease their blocking probability.

In the next section we describe a more accurate approximation technique.

6.5 Erlang Fixed-Point Approximation

We continue to assume that the links are independent but "correct" the offered load to a link by accounting for the blocking of a call by the other links on the route. This correction is applied by reducing, or *thinning*, the offered load of each class on each link by an amount corresponding to that blocked by the other links used by that class. We first consider single-rate networks with $b_k = 1$ for all k and then consider multiclass networks with a more general b_k.

6.5.1 Single-Rate Networks

The bandwidth requirement of each class is identical, so from our discussion in Section 6.2, all the classes that use the same set of links can be considered together. Therefore, in the following discussion, we use the terms *route* and *class* synonymously. Let $\mathcal{S}_j \subset \mathcal{S}$ denote the set of **n** in which link j is blocking. Note that because we are considering single-rate networks, when a link is in a blocking state all calls on routes using that link will be blocked. The blocking probability of link j, β_j, is given by

$$\beta_j = \sum_{\mathbf{n} \in \mathcal{S}_j} \pi(\mathbf{n})$$

where $\pi(\mathbf{n})$ is the probability that the number of active calls in the network is represented by the vector **n** and $\mathcal{S}_j := \{\mathbf{n} : n_j = C_j\}$. If $\hat{R}_{j,k} = 1$, consider the k calls that are blocked by links other than link j on route k. We can say that such calls are not offered to link j and that these calls *balk* from link j. If call arrivals to the network are Poisson and if we assume that balking is independent, then the call arrival process to each link is Poisson. Thus $v_{j,k}$, the offered load by calls using route k to link j, would be the fraction of calls not blocked by the other links on route k:

$$v_{j,k} = \hat{R}_{j,k}\, \rho_k \prod_{j' \neq j} (1 - \beta_{j'})^{\hat{R}_{j'k}}$$

and the offered load to link j, v_j is given by

$$v_j = \sum_{k=1}^{K} v_{j,k} \tag{6.21}$$

We continue to assume that the links block independently of each other. Furthermore, we assume that every call is independently offered to each link on its route. Because call arrivals to the network are from a Poisson process and because the links block independently, the call arrivals to each link form a Poisson process. The approximate link-blocking probability, $\hat{\beta}_j$, is obtained using the Erlang-B formula of Equation 6.9 as

$$\hat{\beta}_j = E_B(v_j, C_j) \tag{6.22}$$

Combining Equations 6.21 and 6.22 and writing out the terms, we see that the $\hat{\beta}_j$ are obtained as the solution of the set of equations

$$\hat{\beta}_j = E_B \left(\sum_{k=1}^{K} \hat{R}_{j,k}\, \rho_k \prod_{i \neq j} (1 - \hat{\beta}_i)^{\hat{R}_{i,k}},\ C_j \right) \tag{6.23}$$

for $1 \leq j \leq J$. The approximate route-blocking probability, $\hat{L}_k$, is obtained as follows.

$$\hat{L}_k = 1 - \prod_j (1 - \hat{\beta}_j)^{\hat{R}_{j,k}} \tag{6.24}$$

Writing $\boldsymbol{\beta} = [\hat{\beta}_1, \ldots, \hat{\beta}_J]$, the foregoing equations can be more compactly written as

$$\boldsymbol{\beta} = \mathbf{E_B}(\boldsymbol{\rho}(\boldsymbol{\beta}), \mathbf{C}) \tag{6.25}$$

where $\boldsymbol{\rho}(\boldsymbol{\beta})$ is the vector of reduced loads on the links obtained from the link loss probabilities $\hat{\beta}_j$, and $\mathbf{C}$ is the vector of link capacities. Equations 6.24 and 6.25 together constitute the approximation for blocking probabilities called the *reduced load approximation*. This name reflects the fact that the load offered to each link by a route is appropriately reduced by accounting for the blocking by the other links on the route.

Given $\mathbf{C}$, Equation 6.25 has the form of a *fixed-point equation*. Furthermore, $\boldsymbol{\beta}$ is a vector whose elements are β_j, the link-blocking probabilities. Hence $\boldsymbol{\beta}$ takes values in $[0,1]^J$. Also, by definition, $\mathbf{E_B}(\cdot, \cdot)$ is a probability and has range $[0,1]^J$. Thus, given $\mathbf{C}$, $\mathbf{E_B}(\cdot, \cdot)$ is a function of $\boldsymbol{\beta}$ and hence its support, or domain, is also $[0,1]^J$. Furthermore, E_B is a continuous function of $\hat{\beta}_j$. Hence, $\mathbf{E_B}(\cdot, \cdot)$ maps $[0,1]^J$ onto itself, and we can apply Brouwer's fixed-point theorem and conclude that Equation 6.23 has a fixed point. (See Section B.2 for a discussion of Brouwer's fixed-point theorem.)

An immediate question is the uniqueness of the fixed point. Uniqueness implies that there is only one result from the approximation. Also, it helps in simplifying the iterative solution technique that we present later because irrespective of the initial values for the iteration, the solution will be the same.

Note that if this were an exact calculation and if the fixed point is not unique, then for a given offered load to the network, the network could exist in many states, each having possibly very different blocking behavior. The different states in this case would be identified by different occupancies, and the network could also demonstrate oscillatory behavior between these states. It turns out that the fixed point is unique.

6.5.2 Uniqueness of the Erlang Fixed Point

The proof of the uniqueness of the Erlang fixed point of Equation 6.25 uses a novel method. The solution to the fixed-point equation is shown to be equivalent to the minimization of a strictly convex objective function. Because the minimum of a strictly convex function is unique, the fixed point of Equation 6.25 will also be unique.

Let $\rho(x)$ be such that $E_B(\rho(x), C) = x$; that is, $\rho(x)$ is the offered load to a link with C circuits that will cause a blocking probability of x.

For $y \geq 0$, $0 \leq e^{-y} \leq 1$ and we can treat e^{-y} as a probability. For $y \geq 0$, define an implicit function $U(y, C)$ as follows:

$$U(y, C) := \left(\rho(1 - e^{-y}) \right) e^{-y}$$

Let e^{-y} represent a blocking probability on a link, and $\rho(1 - e^{-y})$ the offered load to it. In this interpretation, $U(y, C)$ as defined here is the carried load on a link with capacity C when the the offered load is $\rho(1 - e^{-y})$ Erlangs and the blocking probability is $1 - e^{-y}$. Although it can be formally shown that for a fixed C, $U(y, C)$ is an increasing function of y, we provide only an intuitive argument here. Increasing the offered load from 0 to ∞ increases the blocking probability and also the carried load. Thus as y increases from 0 to ∞, implying increasing blocking probability, $U(y, C)$ will be an increasing function. It turns out that $U(y, C)$ is a strictly increasing function in y and has the form shown in Figure 6.11. Hence the integral of $U(y, C)$

$$\int_0^y U(z, C) \, dz$$

is a strictly convex function.

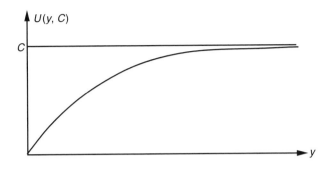

Figure 6.11   The form of $U(y, C)$ as a function of y.

Consider the following minimization problem.

$$\text{minimize } \psi(\mathbf{y}) := \sum_{k} \rho_k \exp\left(-\sum_{j} y_j \hat{R}_{j,k}\right) + \sum_{j} \int_0^{y_j} U(z, C_j)\, dz$$

subject to $\mathbf{y} \geq 0$

The first term on the right side is a sum of strictly convex functions with positive multipliers and hence is a strictly convex function. We have argued that the integral in the second term is a strictly convex function in y. Thus as $\| \mathbf{y} \| \to \infty$, $\psi(\mathbf{y})$ is strictly convex on $[0, \infty)^J$ and hence has a unique minimum. Also, $\psi(\mathbf{y})$ is differentiable, and the partial derivatives of $\psi(\mathbf{y})$ with respect to y_j will be 0 at its minimum value. Taking the partial derivatives of $\psi(\mathbf{y})$ and equating them to 0, we obtain, for $1 \leq j \leq J$,

$$\sum_{k=1}^{K} \hat{R}_{j,k}\, \rho_k \exp\left(-\sum_{i=1}^{J} y_i \hat{R}_{i,k}\right) = U(y_j, C_j) \tag{6.26}$$

at the minimum of $\psi(\mathbf{y})$.

Let us now go back to Equation 6.23 and make the one-to-one transformation $\hat{\beta}_j = 1 - e^{-y_j}$. For $j = 1, \ldots, J$, we can write the following.

$$E_B \left(\sum_{k=1}^{K} \hat{R}_{j,k} \, \rho_k \prod_{\substack{i=1 \\ i \neq j}}^{J} (e^{-y_i})^{\hat{R}_{i,k}}, \ C_j \right) = 1 - e^{-y_j}$$

$$E_B \left(e^{y_j} \sum_{k=1}^{K} \hat{R}_{j,k} \, \rho_k \, e^{-\sum_{i=1}^{J} y_i \hat{R}_{i,k}}, \ C_j \right) = 1 - e^{-y_j}$$

$$\sum_{k=1}^{K} \hat{R}_{j,k} \, \rho_k \, e^{-\sum_{i=1}^{J} y_i \hat{R}_{i,k}} = e^{-y_j} E_B^{-1} \left(1 - e^{-y_j}, C_j \right)$$

Here $E_B^{-1}(p, C)$ denotes the inverse function $E_B(\cdot, C)$—that is, the offered load required to achieve a blocking probability of p on a link of capacity C. The right side of the preceding equation is the carried load when the blocking probability is $1 - e^{-y_j}$ and is therefore $U(y_j, C_j)$. Thus we have

$$\sum_{k=1}^{K} \hat{R}_{j,k} \, \rho_k \, \exp \left(-\sum_{i=1}^{J} y_i \hat{R}_{i,k} \right) = U(y_j, C_j)$$

which is Equation 6.26. Thus the solution to fixed-point equation $\beta = E_B(\rho(\beta), c)$ is the unique minimum of $\psi(y)$ defined here, and the fixed point of Equation 6.25 is unique.

6.5.3 Evaluating the Fixed Point

The next issue is to efficiently obtain the solution to the fixed-point equation. There are many ways to solve a fixed-point equation, but a solution using repeated substitution, also called the *relaxation* method, is appealing in its simplicity. In this method, we start with an initial set of values for β_j, which are used in Equation 6.25 to obtain the next set of β_j and so on iteratively until a convergence criterion is met. For efficient computation, it is necessary that these should converge and converge quickly. We answer the convergence question next.

Let $\beta^{(m)}$ be the value of β after the mth iteration and is obtained as $\beta^{(m)} = E_B(\rho(\beta^{(m-1)}), C)$. Although relaxation does not work in general, it works for the special case of what are called *contraction* functions. A contraction function $f(\cdot)$: $X \rightarrow X$ is defined as follows: $|f(x) - f(y)| < c|x - y|$ for $x, y \in X$ and $0 < c < 1$. This means that applying $f(\cdot)$ to two points in X brings them closer. See Section B.2 for a discussion of the use of relaxation methods to solve fixed-point equations. However, $E_B(\cdot)$ is not a contraction function.

Let $\boldsymbol{\beta}^*$ be the fixed-point solution of Equation 6.25. Since a fixed-point solution exists, $\mathbf{0} < \boldsymbol{\beta}^* < \mathbf{1}$, where $\mathbf{0} = [0, 0, \ldots, 0]$ and $\mathbf{1} = [1, 1, \ldots, 1]$. That the inequalities are strict can be seen from the Erlang-B formula for finite nonzero load.

To evaluate the approximate link-blocking probabilities using Equation 6.25, the offered load to a link is *reduced* by the blocking probabilities (obtained from $\boldsymbol{\beta}$) on the other links, and this reduced load is used in the Erlang-B formula. Now suppose that we have two initial blocking probability vectors: $\boldsymbol{\beta} < \boldsymbol{\beta}'$. Clearly, using $\boldsymbol{\beta}$ to reduce the link loads would yield a higher offered load on the links than using $\boldsymbol{\beta}'$. Hence the blocking probability obtained using $\boldsymbol{\beta}$ would be higher than that obtained using $\boldsymbol{\beta}'$:

$$E_B(\rho(\boldsymbol{\beta}), C) > E_B(\rho(\boldsymbol{\beta}'),\ C)$$

$$E_B{}^2(\rho(\boldsymbol{\beta}), C) < E_B{}^2(\rho(\boldsymbol{\beta}'), C)$$

where $E_B{}^2(\rho(\boldsymbol{\beta}), C) := E_B(\rho(E_B(\rho(\boldsymbol{\beta}), C)), C)$. $E_B{}^2(\rho(\boldsymbol{\beta}'), C)$ is defined similarly. Suppose $\boldsymbol{\beta}^{(2m+1)} < \boldsymbol{\beta}^*$. Then, from above, we get $\boldsymbol{\beta}^{(2m+2)} > E_B(\rho(\boldsymbol{\beta}^*), C) = \boldsymbol{\beta}^*$. The equality follows from the definition of the fixed point. Similarly if $\boldsymbol{\beta}^{(2m)} > \boldsymbol{\beta}^*$, then $\boldsymbol{\beta}^{(2m+1)} < \boldsymbol{\beta}^*$, and if $\boldsymbol{\beta}^{(2m+1)} < \boldsymbol{\beta}^{(2m+3)}$, then $\boldsymbol{\beta}^{(2m+2)} > \boldsymbol{\beta}^{(2m+4)}$.

If we start with $\boldsymbol{\beta}^{(0)} = \mathbf{1}$ and $\boldsymbol{\beta}^{(1)} = \mathbf{0}$, we will get

$$\boldsymbol{\beta}^{(1)} < \boldsymbol{\beta}^* < \boldsymbol{\beta}^{(2)} < \boldsymbol{\beta}^{(0)}$$

and from induction we can write

$$0 = \boldsymbol{\beta}^{(1)} < \boldsymbol{\beta}^{(3)} < \cdots < \boldsymbol{\beta}^{(2n+1)} < \boldsymbol{\beta}^* < \boldsymbol{\beta}^{(2n+2)} < \cdots < \boldsymbol{\beta}^{(2)} < \boldsymbol{\beta}^{(0)} = 1$$

From this discussion, we see that $\boldsymbol{\beta}^{(2m+1)}$ (respectively, $\boldsymbol{\beta}^{(2m)}$), $m = 0, \ldots$ is monotonically increasing (respectively, decreasing) and is bounded above (respectively, below) by $\boldsymbol{\beta}^*$. Therefore, from monotone convergence, $\boldsymbol{\beta}^{(2m+1)}$ and

$\beta^{(2m+1)}$ will converge to a limit. Hence, if $\lim_{n\to\infty} \beta^{(2n+1)} = \underline{\beta}$ and $\lim_{n\to\infty} \beta^{(2n)} = \overline{\beta}$, then

$$\underline{\beta} \le \beta^* \le \overline{\beta}$$

Thus, for a single-service network, the Erlang fixed-point solution to the reduced load approximation is unique, and the relaxation procedure can be used to iteratively obtain bounds on the solution. Note that the iterations here will only give upper and lower bounds to β^* but says nothing about the tightness of the bounds. In some cases it is possible that the fixed-point solution is not a good approximation of the actual blocking probabilities or that the iterations may not converge. However, it has been found that the iterations do converge for most networks of practical interest and that the fixed-point solution is indeed a good approximation of the actual blocking probabilities.

6.5.4 Multiclass Networks

We now extend the results from Section 6.5.3 to a multiclass network with different classes of traffic on different routes. We must relax the assumption that $b_k = 1$ for all k and allow it to be an integer from a prescribed finite set. The link-blocking probabilities must now be specified for each k, for a specified capacity and vector of offered loads. Let $\beta_{j,k}$ denote the blocking probability of class k calls on link j. A possible approximation of the blocking probability of a class k call, $\hat{L}_k$, is

$$\hat{L}_k = 1 - \prod \left(1 - \beta_{j,k}\right)^{\hat{R}_{j,k}}$$

The link-blocking probabilities can be approximated by $\hat{\beta}_{j,k}$, which are obtained from the multiclass link-blocking formula of Equation 6.5, with the offered load of class k to link j approximated by $\hat{R}_{j,k}\rho_k(1 - \hat{L}_k)/(1 - \hat{\beta}_{j,k})$.

Just as in the case of the single-service network, the set of equations relating the $\beta_{j,k}$ as defined earlier is a continuous mapping from the set $[0,1]^{J \times K}$ to itself. Brouwer's fixed-point theorem is applicable, and a fixed point exists. As we have seen before, for a multiclass link, $P_B(\cdot)$ is not monotonic in ρ, and we cannot use the relaxation technique that we used for the single-service case. Thus, although we have closed-form expressions for the stationary distribution of the system and the loss probabilities, it is very difficult to numerically evaluate them and use them in actual design. However, for low ρ, the blocking probability is small enough, and the independence approximation can be used to get very good approximations of the route-blocking probabilities. When ρ becomes large—specifically, when it is close to C—numerical solutions are not available. However, asymptotic analysis shows that when the link capacities are high relative to the bandwidth required for the calls, and when the offered load is close to the capacity, the reduced load approximation is asymptotically correct. Furthermore, the link-blocking probabilities become independent, and we can obtain the call-blocking probability by considering the links on the route independently.

6.6 Admission Control

In all our discussions so far, we admit a class k call if there are b_k or more circuits available on the links—that is, the link bandwidths are available for all the classes to share in any arbitrary manner. This is called *complete sharing* of the resources. Many times this may not be the best thing to do. An obvious reason would be to give differential grades of service, specified by the blocking probability, to different classes of calls. Another reason could be economics. It might be that a call that generates less revenue for the system, if admitted, might increase the probability of losing revenue from another call that, if it arrives and is admitted, will provide higher revenue. For a simplistic example, consider the case of $C = 30$ and $K = 3$, $b_1 = 1$, $b_2 = 2$, and $b_3 = 6$. Assume that the offered loads from the three classes of traffic are of the same magnitude but the charge for class 3 calls is significantly higher than that for class 1 calls. If at some time, 24 channels are busy and a class 1 call arrives, if admitted, it will block a potential class 3 call until another call finishes. It might be a good idea to block this class 1 call in anticipation of a higher gain by admitting a class 3 call that might arrive "soon."

Yet another reason not to admit an admissible call would be that if the call is accommodated, then the system might get into a *bad state* that will eventually decrease revenue, as in the following example. Recall that in most networks a call can be routed on one of a multiple set of paths. Consider a network that has two routes between a node pair: the direct route, and an indirect route that may involve more links than the direct route. Consider the three-node network shown in Figure 6.12. Calls between nodes A, B, and C are typically carried along the one-hop route (shown as dashed lines). Assume that for node A, if the direct route is not available, the longer two-hop route may be used. It is easy to construct a sequence of call arrivals and departures such that all the calls involving node A are along the two-hop routes. This will lead to BC being occupied with AB and AC calls, leaving no room for BC calls. As you can see, it is an inefficient use of the network resources. From a revenue perspective, the AB and AC calls provide

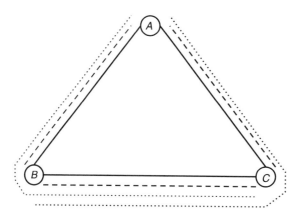

Figure 6.12   An example of a bad state in a circuit-multiplexed network. Dashed lines indicate one-hop routes, and dotted lines indicate two-hop routes.

the same revenue whether they are carried on the direct or the indirect route. The network may lose revenue due to lost BC calls because of the state being inefficient.

The specific problem that we have described is an example of dynamic nonhierarchical routing, which is known to take the network into bad states. The problem is overcome by *trunk reservation*. In trunk reservation, the network does not admit a call on the alternative route if the occupancy on the link exceeds a specified threshold—say, M.

More generally, an admission control policy is described by functions $f_k :$ $S \to \{0,1\}$, for $k = 1,\ldots,K$. For every $\mathbf{n} \in S$, $f_k(\mathbf{n}) = 1$ would mean that a class k call arriving when the system state is $\mathbf{n}$ would be admitted. Otherwise, $f_k(\mathbf{n}) = 0$, and the class k call is rejected.

Typically, admission policies are defined on links. The network admission policy is fairly straightforward: Admit the call if all the links on the route will admit the call. A few link admission policies can be immediately described. The *complete sharing* admission policy that we have implicitly discussed so far is to admit a class k call on the link if there are b_k channels available and otherwise to reject the call. No bandwidth is reserved for any class. The $f_k(\mathbf{n})$ for this policy is described by

$$f_k(\mathbf{n}) = \begin{cases} 1 & \text{if } \mathbf{n} \cdot \mathbf{b}^T + b_k \leq C \\ 0 & \text{otherwise} \end{cases}$$

A *complete partition* policy can also be defined, where capacity c_k is reserved for class k calls, $\sum_{k=1}^{K} c_k = C$. The system model here is that there are K separate single-service knapsacks, each with capacity c_k. If an arriving class k call sees a capacity less than b_k in its part of the knapsack, then the call is lost. The admission policy function is

$$f_k(\mathbf{n}) = \begin{cases} 1 & \text{if } n_k b_k \le c_k - b_k \\ 0 & \text{otherwise} \end{cases}$$

This scheme loses the efficiency advantages of sharing resources among the call classes. However, it offers protection to the classes from possibly misbehaving classes by reserving resources for them.

An intermediate *partial sharing* policy, which offers some protection to the classes while retaining some of the multiplexing gains, can also be defined. A part of the capacity is reserved for exclusive use by each class. The rest of the capacity is pooled.

Exercise 6.7
Write the admission policy function for the partial sharing case defined as follows. The total capacity C is divided into $K + 1$ partitions $c_0, c_1, \ldots, c_K$, with c_k reserved for class k, and the circuits in c_0 are to be allocated on a FCFS basis to arrivals that find their class to be full. Note that it is not necessary to reserve the same set of circuits to the classes.

An important class of admission policies that lends itself to easy analysis and yet includes a large class of useful admission policies is called the class of *coordinate convex* policies. A set $X \subset \mathcal{R}^n$ is convex if, for all $x, y \in X$, $\lambda x + (1 - \lambda)y \in X$ for $0 \le \lambda \le 1$. This essentially means that all the points on the line joining two elements x and y of the set X are also in X. Along similar lines we define a coordinate convex set as follows. Let $\mathbb{Z}^+$ denote the set of positive integers. A set $\Omega \subset \mathbb{Z}^{+^K}$ is coordinate convex if $\mathbf{n} \in \Omega$, and $n_k > 0$ implies $\mathbf{n} - \mathbf{e}_k \in \Omega$. A policy is coordinate convex if it permits admission of a class k call in state $\mathbf{n}$ if $\mathbf{n} + \mathbf{e}_k \in \Omega$; that is, $f_k(\mathbf{n}) = 1$ if $\mathbf{n} + \mathbf{e}_k \in \Omega$ for coordinate convex Ω. An important advantage of coordinate convex policies is that they retain the product form for the stationary distribution of the system state:

$$\pi(\mathbf{n}) = \frac{1}{G} \prod_{k=1}^{K} \frac{\rho^{n_k}}{n_k!} \qquad \mathbf{n} \in \Omega$$

where G is the normalization constant obtained as before. In Problem 6.4 we explore how a non–coordinate convex policy can result in the loss of the product form solution for the stationary distribution. The admission policy affects the performance of cellular networks quite significantly.

In Section 6.8 we compare two schemes using some of the techniques that we develop in the previous sections to show a very counter intuitive result.

6.7 Waiting Room and Retrials

Blocking and admission control imply that calls are refused service. A convenient assumption in the previous analyses is that a customer refused admission is lost to the network. That is obviously not realistic. For example, what does a bank do if there is no teller available to serve an arriving customer? It provides a waiting room to an arriving customer who cannot be served immediately. In an analogous manner, when the network cannot admit a customer it may choose not to return a busy tone immediately. Rather, the arriving call can be made to wait for a short time, with the expectation that one of the many active calls will finish and the circuit will soon become free. In the case of a link in a single-service network, this case is modeled as a queueing system that has a "waiting room" in which calls that could not be assigned a circuit on arrival can wait until a circuit becomes available. In the model, the waiting room capacity may be finite or infinite.

An alternative to modeling calls arriving to a full link is to consider what we do when we get a busy tone when making a telephone call (i.e., when the call is rejected by the network). We redial soon, usually immediately. This means that the rejected call is not lost but decides to *retry*. The rejected call can be assumed to enter into an *orbit* where it waits for a random amount of time—say, exponentially distributed with mean $1/\gamma$—before another attempt, which is independent of the other calls in the orbit. Of course, the orbit has infinite room. An immediate generalization is to say that a rejected call enters the orbit with probability α_1, and a retrying call that is blocked again enters the orbit with probability α_2. We can easily write a two-dimensional Markov chain with state (i, j), where i is the number of active calls and j is the number of calls in the orbit. A closed-form expression for the stationary distribution is not known, but many approximations are available.

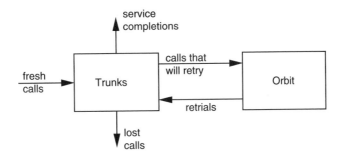

Figure 6.13 The retrial queue system.

Some stability conditions are fairly straightforward, and in the following, we discuss one of them. Figure 6.13 summarizes a retrial queue.

Stability in the retrial queue does not imply the stability of only the queue but of the queue and the orbit together. If $\alpha_2 < 1$, calls in the orbit will leave the system in a finite amount of time, and the number in the orbit will be finite. It can be shown that if $\alpha_2 = 1$ (i.e., all the calls that enter the orbit stay until they get a service), then the system is stable if and only if $\alpha_1 \lambda < C\mu$. Clearly, the condition is true for $\alpha_1 = 1$. An intuitive argument is that when the blocking probability increases, due to an increase in λ, γ, or α_1, the fraction of customers who stay in the system is approximately $\lambda \alpha_1$, and this should be less than the maximum service rate for the stability of the system. Observe that γ, the retrial rate, does not contribute to the stability of the system. It does, however, contribute to the delay and blocking probabilities.

6.8 Channel Allocation in Cellular Networks

So far in this chapter we have considered only wireline networks. It turns out that much of the theory that we have developed can be fairly simply used for analyzing and dimensioning cellular networks. In this section we consider some issues in cellular networks.

6.8.1 Overview

Recall the discussion of cellular networks in Section 2.3.1. The spectrum that is available to a cellular network typically is divided into smaller segments called channels, each of which is sufficient to carry a small number of calls. In this discussion we assume that a channel can carry exactly one call. Because a cellular telephony call needs a full duplex channel, a pair of channels must be allocated for each call: one for the uplink (the link from the mobile station to the base station), and the other for the downlink (the link from the base station to the mobile station). Typically the set of channels for the uplink and the downlink are paired, and we can assume that each call gets one paired channel.

Recall from the discussion in Section 3.5 that the wireless channel is a broadcast channel; that is, in a given area only one transmitter–receiver pair may use any part of the radio spectrum. Also, as we will see in Chapter 8, the received power decays as $d^{-\alpha}$, where d is the distance from the transmitter and α depends on the radio environment and is almost always greater than 2. To increase the utilization of the spectrum, cellular networks divide an operational area into smaller cells, with each cell being serviced by a base station. The *channel reuse constraint* specifies the *channel reuse distance*, the minimum distance at which a channel can be used simultaneously by another mobile station. For the purpose of this discussion we assume that there are a finite number of channels—say, C—and that each call requires exactly one channel while it is active.

Consider a cellular system that is divided into cells as shown in Figure 6.14. The base station is at the center of each cell. Assume that the channel reuse distance is specified by the distance between the base stations. For hexagonal cells, the following algorithm can be used to specify the cells in which the same channel can be used simultaneously, the *channel reuse group*. Proceed along a line of hexagonal cells to the ith cell, turn 60 degrees counterclockwise, and then proceed to the jth cell in the new direction.

Consider a graph G constructed as follows to represent the network. Cell i in the network corresponds to vertex v_i in G. Draw an edge between vertices v_i and v_j in G if the distance between cells v_i and v_j is less than the reuse constraint. When a call arrives in cell i, it is allocated a channel that is not in use in any of the neighbors of v_i in G. When a call moves from cell i to a neighboring cell j, it should be allocated a new channel in j. This is called handoff. If a channel cannot be found, the call must be dropped, and in this case we say that the handover failed. In either case the channel that was in use in cell i is released for use by another call.

Clearly the cellular system is a circuit-multiplexed network, and the performance measure of interest is the call-blocking and the handover failure probabilities. Of course, both of these measures depend on the offered load and the channel allocation algorithm. In this section, our primary aim is to study some simple channel allocation schemes and also to introduce the techniques that can be used in the dimensioning of cellular networks. Hence, we use simple models and do not consider handover explicitly.

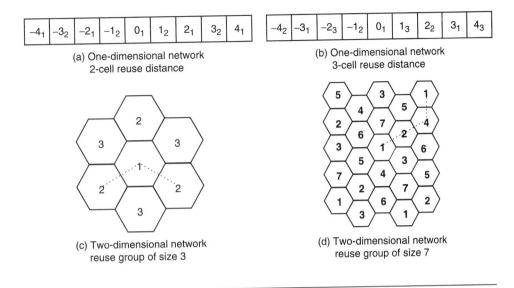

(a) One-dimensional network
2-cell reuse distance

(b) One-dimensional network
3-cell reuse distance

(c) Two-dimensional network
reuse group of size 3

(d) Two-dimensional network
reuse group of size 7

Figure 6.14 One- and two-dimensional cellular networks with different reuse distances. Panels (a) and (b) show one-dimensional networks in which channels can be reused in cells at distances of 2 and 3, respectively. The numbers on the lower right of each cell indicate the channel groups that can be used in the cell, and the numbers in the middle indicate a sequence number for the cell. Panels (c) and (d) show two-dimensional networks with reuse groups of size 3 ($i=1, j=1$) and 7 ($i=2, j=1$), respectively. Numbers in the cells indicate the number of the channel group that is allocated to the cell.

Many channel allocation schemes have been proposed in the literature for use in cellular networks. An obvious strategy is to use a fixed allocation scheme. The total number of available channels is partitioned into subsets, and each cell is statically allocated a subset. The allocation should satisfy the reuse constraint. For example, in the network shown in Figure 6.14, each cell is allocated a group of $C/2$ channels in the network in (a), $C/3$ channels in the networks in (b) and (c), and $C/7$ channels in the network in (d). In this channel assignment scheme, a new call is admitted if the number of active calls in the cell is less than the number of channels allocated to it. This is called *fixed channel assignment* (FCA).

Alternatively, a greedy algorithm called the *maximum packing* (MP) scheme can be used. In this scheme, a new call is accepted if there exists *any* channel allocation for the system state **n** with this call admitted. Channels allocated to the active calls may need to be reallocated to accommodate an incoming call. We do not assume any limit on the number of reallocations. To accommodate a new call, the MP scheme requires knowledge of the state of the entire network—that is, the number of active calls in each of the cells. The number of rearrangements that may be necessary to accommodate a new call grows unbounded in the number of cells. Thus, this scheme is difficult to implement in practice. However, understanding its behavior can help us in comparing the performance of dynamic and static channel allocation schemes.

A third channel allocation alternative is to use *clique packing*. In a graph, a *clique* is a fully connected subgraph, a subset of nodes and edges of the original graph such that every node in the subgraph has an edge to every other node. In the graph of the cellular network we have described, a clique will be a set of cells all of which interfere with each other. In this algorithm, an incoming call is accepted if the sum of the number of active calls in every clique of which the cell is a member, is less than C. This too is a greedy algorithm but requires knowledge of only the local state—that is, the number of active calls in the neighboring cells. Channel allocation according to this algorithm may violate the reuse constraint. Again, this is a model only to help us understand the behavior of greedy algorithms and is an analyzable alternative to the MP algorithm.

6.8.2 Comparing Dynamic and Fixed Channel Allocation Schemes

Let us now obtain the loss probabilities in a cellular network with different channel allocation schemes. Let $\mathbf{n} := [n_1, \cdots, n_M]$ (n_i being the number of active calls in cell i) denote the state of the system. First, consider the FCA with C_i channels allocated to cell i. Calls are accepted independently of the occupancy in the other cells. If calls arrive according to a Poisson process of rate ρ_i to cell i and if they have i.i.d. holding times of unit mean, then the probability of a lost call from cell i is given by the Erlang-B formula

$$L_{\text{FCA}} = E_B(\rho_i, C_i)$$

Now consider the MP channel allocation strategy. Let $\mathcal{S}_{\text{MP}}$ denote the set of states of **n** that this channel allocation strategy accepts. Note that this is a coordinate convex admission policy, and the stationary probabilities have a product form and can be written as

$$\pi_{\text{MP}}(\mathbf{n}) = \frac{1}{G} \prod_{i=1}^{M} \frac{\rho_i^{n_i}}{n_i!} \quad \text{for } \mathbf{n} \in \mathcal{S}_{\text{MP}}$$

where G is the normalizing constant. As mentioned earlier, it is difficult to evaluate the blocking probability for the MP channel allocation scheme in a general network, and we do not pursue that here. However, let us consider the special case of the one-dimensional network shown in Figure 6.14(a). Let us assume that the same channel cannot be used in adjacent cells. For this channel allocation scheme, $\mathcal{S}_{\text{MP}}$ is defined by

$$\mathcal{S}_{\text{MP}} = \{\mathbf{n} : n_i + n_{i+1} \leq C \quad \text{for } i = 1, \ldots, M - 1\}$$

Let the set of $\mathbf{n}$ in which the number of active calls in cell i is n be denoted by $\mathcal{S}_{\text{MP}}(i, n)$. The stationary probability of there being n active calls in cell i can be written as

$$\pi_i(n) = G^{-1} \sum_{\mathbf{n} \in \mathcal{S}_{\text{MP}}(i,n)} \frac{\rho_i^n}{n!} \prod_{\substack{j=1 \\ j \neq i}}^{M} \frac{\rho_j^{n_j}}{n_j!} = G^{-1} \frac{\rho_i^n}{n!} \phi(i, n)$$

where

$$\phi(i, n) := \sum_{\mathbf{n} \in \mathcal{S}_{\text{MP}}(i,n)} \prod_{\substack{j=1 \\ j \neq i}}^{M} \frac{\rho_j^{n_j}}{n_j!}$$

From Problem 6.2, we know that the carried load in cell i is the average number of active calls in cell i and is given by $\sum_{n=1}^{C} n \pi_i(n)$. For the rest of the discussion we assume that the offered load is spatially homogeneous (i.e., $\rho_i = \rho$ for all i). Subtracting the carried load from the offered load and expressing it as a fraction of the offered load, the loss probability, L_{MP}, is

$$L_{\text{MP}} = \frac{\rho - G^{-1} \sum_{n=1}^{C} n \frac{\rho^n}{n!} \phi(i, n)}{\rho} = 1 - G^{-1} \sum_{n=0}^{C-1} \frac{\rho^n}{n!} \phi(i, n+1)$$

Obtaining $\phi(n)$ is complicated, and we omit that here. For $C = 2$ and $n \to \infty$, we can show L_{MP} to be given by

$$L_{MP} = \frac{p^2(14 - 10p - 5p^2 + 3p^3)}{2(2 + p^2 - 2p^3)}$$

where p is the solution in $(0, 1)$ to the cubic equation $\rho(1 - p)(2 - p^2) = 2p$.

For fixed channel allocation in the linear network with a reuse distance of two cells and $c = 2$, $L_{FCA} = \frac{\rho}{1+\rho}$ because each cell gets exactly one channel. Figure 6.15 shows L_{MP} and L_{FCA} as a function of the offered load ρ. Notice at about $\rho = 2.6$ the fixed channel assignment outperforms the MP dynamic channel assignment! More interestingly, as c becomes large the crossover happens at lower values of ρ, indicating that for high-capacity cellular networks, the fixed channel allocation scheme will outperform the dynamic channel scheme when the load is time- and space-homogeneous (i.e., ρ_i is independent of i and does not change with time). The result is definitely counterintuitive; we expect dynamic schemes to be better than static schemes. A heuristic explanation for this effect is that the MP allocation

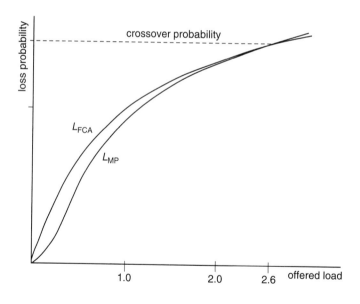

Figure 6.15 L_{MP} and L_{FCA} as a function of ρ for $c = 2$. Adapted from [162].

scheme can upset the tight packing of the channels and calls that it was designed to achieve. In a way, FCA behaves like the trunk reservation schemes discussed earlier.

We can conclude from this analysis that it might be better to reject some calls, especially at high loads, to improve the overall system performance. The MP scheme will accept a call if the channels can be rearranged to accommodate it, whereas the FCA will reject a call if all the channels allocated to the cell are busy; it will not borrow channels from other cells to fulfill a request. However, we cannot conclude that dynamic channels do not have advantages at all. The advantages are realized if the offered load is nonhomogeneous in space and is nonstationary.

6.9 Wavelength Allocation in Optical Networks

Recall the discussion in Chapter 2 on the bit-carrier infrastructure, specifically the discussion in Section 2.3.1 with reference to Figure 2.18. Different layers of the infrastructure offer different levels of aggregation and different granularities of bandwidth allocation. Much of the discussion in the previous sections can be seen to be applicable at the top three layers of the hierarchy shown in Figure 2.18. In this section we briefly discuss the extensions of what we have learned in the previous sections to the analysis of all optical networks. Such networks are expected in the the second layer from below in the infrastructure hierarchy of Figure 2.18.

Consider an all-optical network using wavelength division multiplexing (WDM), in which each link in the network has a capacity to carry a maximum of C wavelengths. A lightpath is created between a node pair by allocating a wavelength on each link of the route for exclusive use by the lightpath. The network is transparent to the content of each lightpath. Thus in such optical networks, capacity is allocated to calls at the granularity of a wavelength. To keep the lightpaths separated on the common links, they should be allocated different wavelengths on the common links. For example, consider the all-optical network shown in Figure 6.16 with lightpaths $LP_1, \ldots, LP_4$. On link BC, lightpaths LP_1 and LP_2 cannot be allocated the same wavelength, whereas on link CD, LP_1–LP_3 should be allocated different wavelengths.

A network may not be able to route all lightpath requests that arrive, and requests may be blocked because C is finite. Consider a network with full wavelength conversion. *Full wavelength conversion* means that at a node, a lightpath on wavelength c_i on the input line can be converted to any wavelength c_j, $j = 1, \ldots, C$ on the corresponding output link. In this case the call can be admitted if on each of the links on the route there is at least one wavelength free.

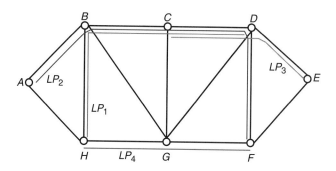

Figure 6.16 An all-optical network showing four lightpaths.

This is exactly like a circuit-multiplexed network except for the terminology, and the analysis techniques from the previous sections are directly applicable.

In the present state of the art, full wavelength conversion is expensive and needs to be minimized. If no wavelength conversion is possible, a lightpath must be allocated the same wavelength on *all* the links on its path. This is called the *wavelength continuity constraint*. In the example of Figure 6.16, satisfying this constraint means that lightpaths LP_1–LP_3 must be allocated different wavelengths, whereas LP_4 can be allocated any wavelength. Assume that LP_1 is allocated wavelength c_1, LP_2 is allocated c_2, LP_3 is allocated c_3, and LP_4 is allocated c_3. If a new lightpath request must be routed along $GHBC$, it must be assigned a wavelength c_4. If the network has only three wavelengths and if wavelength reallocation is not permitted, then this lightpath must be blocked if there were no wavelength conversions, whereas it could be accepted if wavelength conversions were allowed.

In the rest of the section we consider a network having no wavelength conversion and in which the routing of the lightpaths is predefined. There are J links and K predefined routes in the network, and the routes are defined by the route-link incidence matrix $\hat{R}$ as defined in Equation 6.17. Lightpath requests on route k arrive according to a Poisson process of rate λ_k, and the holding times are all assumed to have an exponential distribution of unit mean. If the lightpath request is accepted, then it will be allocated wavelength c_i on *all* the links on the route. We assume that the allocated wavelength is chosen randomly from among those that are available. Our interest is in the probability of a lightpath request on route k being blocked either because there was no wavelength available on at least one of the links on its route or because the same wavelength could not be made available to the lightpath on all the links. We will illustrate one fairly intuitive approximation procedure to obtain the blocking probability. There are many more reported in the literature.

Let the random variable indicating the number of free, continuity-constrained wavelengths on route k be denoted by Y_k, and that indicating the number of free wavelengths on link j be X_j. Let $q_j(m) := \Pr(X_j = m)$ for $m = 0, \ldots, C$. This is shown in Figure 6.17. We assume that the X_j are independent random variables, much as we the assume that the links block a call independently in our discussion in Section 6.5. Clearly, the rate at which a new lightpath is established through link j depends on the number of free wavelengths on it and on the other links, and also on the arrival rate of the lightpath requests on the routes using link j. We capture these dependencies using a very simple model in which we assume that the time until the establishment of the next lightpath through link j is dependent on m, the number of free wavelengths on the link. Furthermore, we assume that this time is exponentially distributed with mean $1/a_j(m)$. From the link independence assumption and from the assumption of state-dependent Poisson lightpath arrival processes and exponential holding times, the state of link j, expressed as the number of free wavelengths, is approximated by a simple birth–death process with state-dependent transition rates as shown in Figure 6.18. Solving this CTMC, we obtain $q_j(m)$ to be

$$q_j(m) = q_j(0) \prod_{i=0}^{m-1} \frac{C - i}{a_j(i + 1)}. \tag{6.27}$$

$q_j(0)$ is obtained from the normalizing condition of $\sum_{m=0}^{C} q_j(m) = 1$.

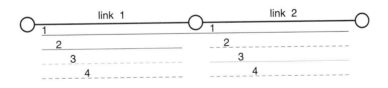

Figure 6.17 Sample two-link route with link capacity $C=4$. Solid lines below the link indicate a wavelength in use, and dashed lines indicate free wavelengths. In this example, $X_1 = 2$, $X_2 = 2$, and $Y = 1$.

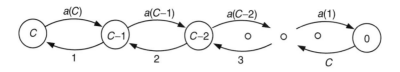

Figure 6.18 **The CTMC for the number of active routes using a link. The state is the number of free wavelengths on the link. $a(m)$ is the state-dependent rate of new call acceptances on the link.**

A lightpath is accepted on route k if $Y_k > 0$; that is, there is at least one continuity-constrained wavelength free on the route, which we have assumed to be a function of the number of free wavelengths on the links of the route. We capture this dependency through the conditional probability of finding a continuity-constrained wavelength on the route, given the number of free wavelengths available on link j, and we write

$$a_j(m) = \begin{cases} 0 & \text{for } m = 0 \\ \sum_{j=1}^{J} (\hat{R}_{j,k}\lambda_k) \, \Pr(Y_k > 0 | X_j = m) & \text{for } m = 1, \ldots, C \end{cases} \qquad (6.28)$$

If there is only one link in route k—say, link j—then $L_k := \Pr(Y_k = 0) = q_j(0)$. If there are two links on route k—say, links j and j_1—then we first obtain the conditional probability of a free wavelength on the route conditioned on m and m_1 wavelengths being free on links j and j_1, respectively, and then uncondition it.

$$\Pr(Y_k > 0 | X_j = m) = \sum_{m_1=0}^{C} \Pr(X_{j_1} = m_1 | X_j = m)$$

$$\times \Pr(Y_k > 0 | X_j = m; X_{j_1} = m_1)$$

$$= \sum_{m_1=1}^{C} \Pr(X_{j_1} = m_1)$$

$$\times (1 - \Pr(Y_k = 0 | X_j = m; X_{j_1} = m_1))$$

The last equality follows from the assumption that the link occupancies are independent. The index of the summation can start at $m_1 = 1$ because $m_1 = 0$ implies that there are no free wavelengths on route k.

We now derive the conditional probability of finding a continuity-constrained wavelength on a route of two independent links, arbitrarily called links 1 and 2. Let m_1 and m_2 be the number of free wavelengths on links 1 and 2, respectively. Define

$$p(m_1, m_2) := \mathsf{Pr}(\text{No common free wavelength on links 1 \& 2} \mid m_1, m_2)$$

Let C boxes represent the C wavelengths available on each link. Let the free wavelengths on link 1 be represented by m_1 black balls, and the m_2 free wavelengths on link 2 be represented by m_2 brown balls. Place each set of colored balls randomly in the C boxes, with at most one of a given color in a box. The event of not finding a common free wavelength on links 1 and 2 is the same as having no box that contains both a brown and a black ball. This is possible only if $m_1 + m_2 < C$. We obtain $p(m_1, m_2)$ as follows. First, place the m_1 black balls in the C boxes. The m_2 balls can be distributed in the C boxes in $\binom{C}{m_2}$ ways. To have no box containing both brown and black balls, the m_2 brown balls should be distributed in the $C - m_1$ boxes that are empty. Thus we get

$$p(m_1, m_2) = \frac{\binom{C - m_1}{m_2}}{\binom{C}{m_2}} \quad \text{for } m_1, m_2 > 0,\ m_1 + m_2 < C$$

We can now write the probability of finding a free wavelength on route j conditioned on there being m and m_1 free wavelengths on the two links of the route:

$$\mathsf{Pr}(Y_k > 0 \mid X_j = m; X_{j_1} = m_1) = \sum_{m_1 = 1}^{C} q_j(m)\,(1 - p(m, m_1)) \qquad (6.29)$$

This is used in Equation 6.28 to obtain the $a_j(m)$. Notice that Equations 6.28 and 6.29 form a set of dependent equations. Notice also that Equation 6.28 calculates the reduced load on link j due to blocking by the other links. An iterative procedure similar to the Erlang fixed-point approximation can be used with the initialization $\Pr(Y_k > 0|X_j = m) = 1$ and hence $a_j(m) = \sum_{k=1}^{J} \hat{R}_{j,k}\lambda_k$. This is clearly an approximation. The convergence properties of the iterative procedure on Equations 6.28 and 6.29 are not known.

The loss probability on route k, L_k, is obtained by unconditioning on the number of free wavelengths the links in the route.

$$L_k = \begin{cases} q_j(m) & \text{only link } j \text{ on route } k \\ \sum_{m=1}^{C} \sum_{m_1=1}^{C} q_j(m)\, q_{j_1}(m_1)\, p(m, m_1) & \text{two links } (j \text{ and } j_1) \text{ on route } k \end{cases}$$

This procedure can be extended for routes having more than two hops.

6.10 Summary

We start the chapter by considering a circuit-multiplexed link that is offered multiclass traffic according to a stationary Poisson process and obtain a product form expression for the stationary probabilities of the occupancy of this link. We then derive the Kaufman–Roberts recursive algorithm for efficient numerical evaluation of performance measures using the product form stationary probabilities. The classical Erlang model for the link-blocking probabilities in single-class networks is presented as a special case of the multiclass link. Blocked calls overflow. The interval between overflow times from a link with finite capacity is first characterized. Then we describe the equivalent random method for approximating the link-blocking probabilities when the offered load to a link is non-Poisson and bursty.

Networks are modeled as a vector of routes, with each route representing a class of traffic. The Erlang fixed-point approximation is used to study the blocking behavior in circuit-multiplexed networks. Using a novel technique of showing that the fixed point is the solution to the unique minimum of a strictly convex function, we show that the Erlang fixed point is unique. Furthermore, we show that a relaxation technique can be used to solve the fixed-point equation. The concept of admission control and the retrial of blocked calls is discussed briefly.

Although much of our discussion uses the framework of circuit-multiplexed wireline networks, the basic principle of circuit multiplexing is more widely used. The cellular telephone system is one example. We briefly discuss models for some channel allocation algorithms in cellular networks that can be derived from the theory of circuit multiplexing and analyzed the relative performance of two channel allocation algorithms. We also discuss the blocking probability for lightpath requests in a WDM optical network with no wavelength conversion. This analysis is also derived from the theory of circuit multiplexing.

6.11 Notes on the Literature

The theory of circuit multiplexing was developed in the context of telephone networks and is nearly a hundred years old. The past couple of decades, though, have seen significant advances in the theory and also a wider variety of applications for it. We mention only those from which we have borrowed significantly. A large number of fairly important resources must be omitted because of lack of space.

Two of the more recent books on circuit multiplexing are those by Girard [119] and Ross [256]. The references in these books are excellent resources. The theory of multiclass traffic on links was first developed independently by Kaufman [159] and Roberts [251]. The discussion on overflows from queues and that on retrials is based on the treatment by Wolff [300]. The equivalent random method and Hayward's and Rapp's approximations are also discussed in Girard [119]. The Erlang fixed-point approximation was used in practice in analyzing circuit-multiplexed networks for quite some time before Kelly [163, 165] showed the uniqueness of the fixed point for single-class networks. Our proof of the convergence of the iterative relaxation method to solve the fixed-point approximation is drawn from Ross [256]. Subsequent to Akinpelu's [8] reporting of simulation evidence of the telephone network existing in multiple states with the same offered load, there have been many proposals for admission control and state protection in a wide variety of circuit-multiplexed networks. Our discussion of the blocking probability analysis of one-dimensional cellular networks is based on the work of Kelly [162]. Rappaport [244] and Stuber [278] are two good references for cellular networks. The clique-packing algorithm was first described by Raymond [245]. After the paper by Everitt and MacFayden [95], it was believed that clique packing and maximum packing were equivalent. It is now known that the constraints for clique packing are necessary conditions for maximum packing but not sufficient. Kulshrestha and Sivarajan [180] analyze the maximum packing strategy in arbitrary networks. The discussion on the optical networks is based on Birman [36]. Subsequently, Tripathi and Sivarajan [286] have obtained the blocking probability for a network with limited-range wavelength conversion.

Problems

6.1 Two classes of calls are offered to a link of capacity C. The arrival processes are Poisson with rates λ_1 and λ_2. Each call of each class requires 1 circuit, and the holding times are exponentially distributed with means μ_1^{-1} and μ_2^{-1}; $\mu_1 \neq \mu_2$. Let $(X_1(t), X_2(t))$ represent the joint process of the number of busy channels.

   a. Obtain the stationary distribution of the CTMC $\{(X_1(t), X_2(t)), t \geq 0\}$.

   b. Obtain an expression for the mean number of channels occupied by each type of call.

6.2 Using Little's law, show that the mean number of active calls on a single-service link is equal to the carried load by the link.

6.3 Consider a single-service link with Poisson call arrivals of rate λ and exponential call holding times with mean μ^{-1}.

   a. Find the conditional probability that an arrival following a blocked call is blocked.

   b. Show that this probability is higher than the unconditional probability given by the Erlang-B formula.

6.4 A link of capacity C is offered two classes of traffic and uses the following admission policy: Only class 1 calls are accepted if $X_1(t) + X_2(t) \geq C - k$ for some fixed k (≥ 1, reservation parameter). This is called trunk reservation where the "last k circuits" are reserved for class 1 calls.

   a. Show that the trunk reservation admission policy is not coordinate convex.

   b. For $C = 3$, $k = 1$ and $b_1 = b_2$, find the stationary probability that n circuits are free, and show that it does not have a product form.

6.5 Two classes of calls share a link with 6 units of bandwidth. Class 1 requires 2 units per call, whereas class 2 requires 1 unit per call. Calls of class i arrive in a Poisson process of rate λ_i, and have exponentially distributed holding times with mean μ_i^{-1}. Three types of sharing policies need to be compared: (i) complete sharing, (ii) maximum of 2 calls of class 1 and 4 calls of class 2, and (iii) reservation of the *last* 2 units of bandwidth for class 2; i.e., a class 1 call is not accepted if it would make bandwidth utilisation greater than 4 units.

a. For each sharing strategy, show the state space and Markov transition diagram of $\{(X_1(t), X_2(t))\}$, the process of active calls.

b. For the policies for which product form is available, show the form of the stationary distribution of $\{X_1(t), X_2(t)\}$.

6.6 Consider the network shown in the figure below. Let $K = 1$, $b_1 = 1$ and $C_j = C$ for $j = 1, \ldots, J$. Find the exact blocking probability and also the bound from the Erlang fixed-point approximation. Compare the bound and the exact value as J becomes larger.

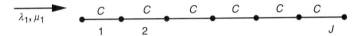

6.7 For simulation or for any other purpose, if we want to start a system in steady state, we pick the initial state according to the steady state distribution. In the case of a multiclass network, if we want to start the system in steady state, if the call arrivals are Poisson and the holding times are exponential, it is sufficient to assign the number of active calls on each route according to the stationary distribution and pick the holding times according to the corresponding exponential distribution. If the holding times, are general, then it is not sufficient to start off the system in state **n**, picked according to the stationary distribution $\pi(\cdot)$. How would you pick the remaining call holding times?

6.8 Consider a circuit-multiplexing network with N links and M routes. The route-link incidence matrix is R. The capacity of link n is C_n units. Calls arrive on route m at rate λ_m in a Poisson process; the Erlang offered load on route m is ρ_m. Each call requires 1 unit capacity on each link on its route. Denote by $B(\rho, C)$ the Erlang blocking formula; you are given that $\frac{\partial}{\partial \rho} B(\rho, C) > 0$.

a. Write down the fixed-point iteration for approximately computing the call-blocking probabilities.

b. If the link busy probability operator in (a) is denoted by $T(\underline{L})$, show that if $\underline{L}_1 < \underline{L}_2$, then $T(\underline{L}_1) > T(\underline{L}_2)$.

6.9 Obtain a closed form expression for $U(y, 1)$ and verify that it indeed has the form shown in Figure 6.11. Also verify that $\int_0^y U(z, 1)\, dz$ is convex.

6.10 Let ρ be the offered load vector, $\mathbf{b}$ the bandwidth requirement vectors and C the capacity of a multiservice link. Let $G(\rho, \mathbf{b}, C)$ be the normalizing constant as given in Theorem 6.1. Find an expression for $\frac{\partial G}{\partial \rho_k}$. Use this expression to obtain a closed form expression for $\frac{\partial P_B}{\partial \rho_k}$.

6.11 In a cellular network with a fixed channel assignment, a cell is allocated C channels. The offered load is ρ. The channel conditions make some of the free channels unusable. Assume that any time a free channel is independently unusable with probability θ and that channels do not go bad while in use. Find the blocking probability if

a. The network has complete knowledge of the channel conditions and allocates a usable channel to an arriving call.

b. An unused channel is randomly allocated to an arriving call. The call is lost if the allocated channel is unusable.

6.12 Repeat Problem 6.11 for a two-cell system with two channels and maximum packing channel allocation algorithm.

CHAPTER 7

Adaptive Bandwidth Sharing for Elastic Traffic

I n Chapter 3, Section 3.4, we explain that elastic traffic essentially comprises the transfers of files between computers, such transfers being generated by applications such as e-mail, Web browsing, and FTP. The network bandwidth must be shared in some way between the ongoing file transfers. Because there is no intrinsic rate at which a file must be transferred, the sending rate of data from the file can adjust to the bandwidth that the network can provide to the transfer. The rate available to transfers from a server to a client can vary from time to time, and may not be constant even during any particular transfer. To achieve network bandwidth sharing and dynamic rate control of transfers, the network requires feedback-based distributed algorithms. In this chapter, we analytically discuss the issues involved in such algorithms. A considerable part of this chapter is devoted to the analysis of the transmission control protocol (TCP), which is an end-to-end protocol that implements the bandwidth-sharing algorithms for elastic traffic in the Internet.

7.1 Elastic Transfers in a Network

In Section 3.4, to motivate the issues, we discuss elastic transfers over a single wide area link interconnecting two high-speed LANs. The general problem is, of course, much more complex. Figure 7.1 shows a more general network topology that interconnects several Web servers and clients at several locations. A user at a client can request a file transfer from any of the servers. Each transfer occurs over whatever routing the network has in place. In Figure 7.1, such routes are shown by dotted or dashed lines between some of the servers and users.

Notice in Figure 7.1 that the various transfers share certain links in the network. The routing governs which links are shared for a given set of elastic transfers. The rate at which each transfer can send depends on the bandwidths of the links along its path, the other transfers that it shares these links with, and the *fairness* objectives in the bandwidth-sharing algorithm. On a single link, fair sharing may imply that all transfers obtain equal rates, but this may not be the

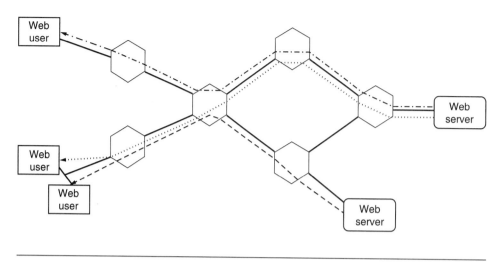

Figure 7.1   Elastic transfers between Web servers and clients over a network.

correct objective to aim for in a network; we examine sharing objectives in a network in Section 7.7.

The problem is further complicated by the fact that the number of transfers along each route is time-varying. This is because transfer requests arrive randomly over time, and each has a finite volume of data to transfer. When a new request arrives on a route, the number of transfers on that route increases, and when a transfer completes on a route the number of transfers on that route decreases. Because the transfer volumes can range from only a few bytes to many hundreds of megabytes, these changes can take place over very short to very long timescales.

Furthermore, in a general network, there is no central entity that knows which transfers are going on at any point in time and which routes they are flowing over. Hence practical solutions of the dynamic bandwidth-sharing problem in a network must necessarily be based on distributed algorithms.

Clearly a mathematical model that captures all the complexities would be intractable. Yet it is important to examine the issues analytically to gain an understanding of how bandwidth-sharing algorithms behave, and how the system parameters determine the performance of elastic transfers. Because this kind of traffic forms the bulk of traffic carried by packet networks, such understanding is essential.

The simplest models assume only a single link shared among a fixed number of elastic transfers, each with an infinite amount of data to send. We too begin our discussion with such a model. Thinking again in terms of timescales, this model can be useful to analyze a situation in which the network carries only large file transfers and the number of transfers varies slowly. Then in between changes in the number of ongoing transfers, the network carries only a fixed number of transfers, and these periods last long enough so that steady state analysis can be performed. If we can also obtain the probability with which each number of transfers is active, then we can obtain the average transfer throughput. Later in the chapter we introduce models with a single link and randomly arriving finite-volume transfers. Then we turn to general networks with fixed routes, a fixed number of transfers per route, and infinite-volume transfers. Finally, we discuss the limited work that exists on the most general problem: an arbitrary topology and randomly arriving finite-volume transfers.

7.2   Network Parameters and Performance Objectives

We begin by discussing certain parameters and related terminology that we will be encountered repeatedly in this chapter. We then provide an understanding of the main performance measures in the context of elastic file transfers.

7.2.1   Propagation Delay and Round-Trip Time

Consider elastic transfers taking place between clients and servers on high-speed LANs that are connected by a relatively low-speed wide area link (see Figure 7.2). In the following discussion we assume that the LANs are infinitely fast and hence do not cause any additional delay. In addition to the finite link speeds that limit the

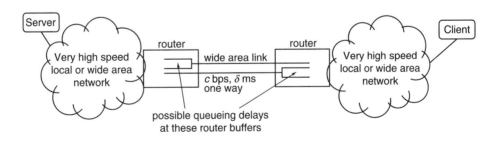

Figure 7.2   An elastic transfer from a server to a client over a wide area path involving a single bottleneck link.

rate at which data transfers can take place, there is a propagation delay between the clients and the servers. Conceptually, a bit inserted into one end of the link will start emerging from the other end after the propagation delay; if the bit rate is c bps, then the bit is fully received after an additional $\frac{1}{c}$ seconds. A simple rule of thumb for cable media is to compute propagation delay as 5 ms per 1000 km link length between the end points.(Of course, this is not the distance "as the crow flies," but the actual cable length needs to be accounted for.)

In the context of elastic transfers the *round-trip propagation delay* (RTPD) is an important parameter that affects the transient and steady state behavior of such transfers. Hence, going by our rule of thumb, a campus link would have an RTPD of a few microseconds, a metro link would have an RTPD of a fraction of a millisecond, a statewide link, a few milliseconds, a nationwide link, a few tens of milliseconds, and an intercontinental link (say, between Europe and the United States or between Asia and the United States), 100 ms to 200 ms. If the link involves a geosynchronous satellite segment, then 250 ms must be added for each direction of a satellite hop. Let us denote the one-way propagation delay by δ (see Figure 7.2). In some of the discussions we tacitly assume that the propagation delay is the same each way. (This need not be the case in practice; for example, it is possible for an intercontinental link to be partly over cable and partly over satellite, and the satellite segment may be only a single hop in *one* direction, leading to an extra 250 ms delay in that direction of the link.)

Now consider a packet of length L bits being sent by a server at one end of the link to a client at the other end. It will take $\frac{L}{c}$ seconds for the transmitter to clock out the packet over the link. The leading edge of the first bit will arrive at the other end after an amount of time δ, and the packet will be fully received by the physical receiver after a further delay of $\frac{L}{c}$. If an acknowledgment (ACK) is sent back for this packet and if the ACK packet is of length a bits, then the ACK would be received back at the server after an additional time $\frac{a}{c} + \delta$. The minimum time between sending a single packet and getting its ACK back is called the *base round-trip time* (BRTT); that is, BRTT $= \frac{L}{c} + \text{RTPD} + \frac{a}{c}$.

The discussion so far assumes that when the server sends a packet it is immediately transmitted onto the link. However, the packet may encounter other packets in the buffer at the link (see Figure 7.2); these could be from its own transfer or from other transfers. Similarly, the ACK packet could also encounter a *queueing delay* in the link buffer in the reverse direction. If we now consider the time between a packet transmission and the receipt of its ACK by the server, we need to add the *round-trip queueing delay* (RTQD). The total *round-trip time* (RTT) for a packet is thus given by RTT = BRTT + RTQD. Although this expression certainly applies to each packet, it is most useful when we think in terms of the mean queueing delay and the mean round-trip time. These means are also denoted by RTT and RTQD, and the same relation holds.

7.2.2 Performance Measures: Transfer Throughput

A user downloading a file or browsing a Web site is basically interested in obtaining the requested information as quickly as possible. If the transfer volume is V and it takes an amount of time S to transfer this file, then the throughput for this transfer is $\frac{V}{S}$. For very large transfers (abstracted as infinitely large in models, in which case the definition will be via the limit as time goes to infinity), clearly *transfer throughput* is the only meaningful measure. Given the throughput that the network provides, for some given large file, the user can determine how much time to allow for the file to be fully received. For short transfers, however, one could consider the average throughput and the average transfer time as valid performance measures. We focus, however, on average throughput, because this is a measure of performance that can be provided to a user and is roughly independent of the actual file sizes that an individual user may wish to transfer. For finite-volume transfers, several measures of throughput can be defined. We defer these details until Section 7.6.7.

Suppose that one user begins to download a file. Ideally this single user should be able to obtain a file download rate of C bps, because the entire link is devoted to this single transfer. In practice, the transfer will start with some signaling between the end points in order to create the transfer association. For example, in the Internet, TCP connection establishment involves a so-called three-way handshake protocol. These signaling exchanges take a few RTTs. From a user's point of view these RTTs will add to the total time taken to transfer the file, and hence for short file transfers the user will perceive a transfer throughput less than C bps; the reduction in throughput is more the larger the RTT and the smaller the file. For much of our discussion we take the file transfer throughput to be computed over the interval starting from the instant that the server application sends the first byte of the file, and ending at the instant when the server receives an acknowledgment that the last byte of the file has been received by the user.

There is a second important issue in the context of defining what we mean by throughput. During its transfer from the server to the user's machine, the file is split into packets, and headers are attached to each packet. It is these packets with headers that are transported over the network. Thus to achieve the file transfer, the network transfers more data than in the original file. The user is not interested in the additional data, and hence again perceives a throughput smaller than that being offered by the network to this file transfer. Because we are basically interested in studying the performance of the network transport mechanisms (multiplexing, scheduling, and switching), we do not "penalize" the network for the additional headers that the network protocols attach to the actual data. We view the file as including the packet headers, and we take throughput as meaning the *network-level throughput* rather than the *user-level throughput*. We can obtain the user throughput by multiplying the network throughput by the fraction of transferred bits that correspond to user data.

7.3 Sharing a Single Link

For a single link and no particular distinction between the ongoing file transfers, it seems reasonable to take fairness to mean equal sharing. In this section we discuss techniques and algorithms for fair sharing of the bandwidth of a single link among elastic transfers.

The assumption of a single link is actually not as restrictive as it may at first seem. If several transfers share a route, and if one of the links along the path has smaller bandwidth than those of all the other links (i.e., is the bottleneck), then, ideally, this link will govern the fair bandwidth to be provided to the transfers; thus with n transfers sharing the path, the fair rate of each transfer should be $\frac{1}{n}$ of the bandwidth of the bottleneck link. Because the packets of each transfer also must flow through the other links, the transmission times at these links will add to the RTPD. This is an ideal view, however, and applies if the packet emissions from the sources can be perfectly paced at $\frac{1}{n}$ of the bottleneck bandwidth. In practice, the sources adapt their rates based on feedback, and this leads to randomness in the arrival processes that results in nontrivial effects at the other links. The present discussion, then, applies to situations in which the nonbottleneck links have speed so much higher (or relative load so much less) than that of the bottleneck link that queueing at the nonbottleneck links can be ignored.

Consider the transfer of a large file from a server to a client over a wide area network link (WAN link) as shown in Figure 7.2. A bottleneck link constrains the flow of traffic between the local area or wide area networks to which the server and client are attached. These networks are assumed to be of a speed much higher than the bottleneck link, so we can assume that data is transported instantaneously between a router of the WAN link and the server or the client on the same local network as the router.

We examine the two common techniques for controlling the transmission of data from the server: *rate-based control* (RBC) and *window-based control* (WBC). A source that uses RBC, adaptively sets the rate at which it emits data; in principle there is no bound on the amount of data that could be outstanding between a source and its receiver. With window-based transmission control, a source adaptively sets the maximum number of packets that could be outstanding between the source and its sink; as will be seen, this indirectly controls the average rate of data transmission from the source.

7.3.1 Control Objectives

Clearly, given the RBC or WBC transmission mechanism, a variety of adaptation algorithms can be designed. These differ in the way the network state is fed back to the transmission controllers and in the way these controllers adapt to the feedback provided. The following are the qualitative objectives for designing such an adaptive system.

- *Efficient link utilization:* Wide area telecom links are expensive resources. A control algorithm that does not efficiently utilize a link, while giving poor throughputs to the transfers, would not be acceptable.

- *Minimum queueing delay:* The sources would not know the parameters of the network (the bottleneck link rate, the number of other transfers, the round-trip propagation delay, the available buffer at the bottleneck link). Hence the sources will adaptively seek an operating point (sending rates, or windows). In this process the queue occupancy, and hence the queueing delay, may become large. Thus although the link may be well utilized, the packets of the file transfers would see a large additional delay at the link. This would have a detrimental effect on the performance of interactive applications (e.g., chat or remote editing) or on the transfer times of short file transfers. The latter point can be seen by noting that the throughput of a one-packet file transfer will be governed mainly by the RTT, which includes the queueing delay.

- *Fairness between the rates given to transfers:* As discussed earlier, on a single link with one class of traffic, fair sharing is just equal sharing. In a network of links, fairness is more complicated, as we discuss in Section 7.7.

- *Stability and transient response:* For example, the objective could be to maintain the average queueing delay at the bottleneck link buffer at some given value q^* and to have equal sharing between the transfers. As for any control system, a requirement is that this operating point be an *equilibrium* point, which should be stable; furthermore, if the network parameters change (e.g., the number of transfers changes), the system should reach the new equilibrium point rapidly.

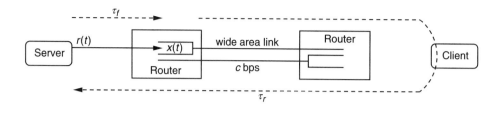

Figure 7.3   A deterministic dynamic system model for rate-based control of a single transfer over a bottleneck link in a wide area network.

7.4   Rate-Based Control (RBC)

Figure 7.3 shows a model for studying the rate-based control of a *single (long-lived) transfer* from the server to the client in the network shown in Figure 7.2. For simplicity we consider a fluid model and view $r(t)$ as the rate at which the source emits data at time t. Such fluid analysis suffices for capturing average behavior, but does not capture fine packet-level behavior (for such a packet-level analysis in the TCP context, see Section 7.6.4). The amount of data in the bottleneck link buffer (in the data transmission direction) is $x(t)$ at time t. We assume that in the reverse direction the bottleneck carries only small control packets, and hence, in the reverse direction, the fluid transmission rate is small, permitting us to ignore the queueing at the link buffer in this direction. The figure shows that τ_f is the propagation delay from the server to the bottleneck link buffer, and τ_r is the propagation delay over the return path from the bottleneck link buffer to the server. Thus the RTT is $\tau_f + \tau_r$ plus the queueing delay at the router queue. Define $\tau = \tau_f + \tau_r$.

Note that the fluid that arrives at the buffer at time t is the one that left the source at time $t - \tau_f$. The following is the dynamical equation for $x(t)$, for any $r(t)$ process.

$$\dot{x}(t) = \begin{cases} r(t - \tau_f) - c & \text{for } x(t) > 0 \\ (r(t - \tau_f) - c)^+ & \text{for } x(t) = 0 \end{cases} \qquad (7.1)$$

The first part of Equation 7.1 is clear: When the queue is positive it increases or decreases at a rate given by the difference between the arrival rate and the link rate. The second part arises because, in a fluid model, when the buffer is empty it stays at 0 if the arrival rate is less than the link rate (in the fluid model the arriving data pass through instantaneously), and it increases if the arrival rate exceeds the link rate.

7.4.1   Explicit-Rate Feedback

If special control packets are employed, then (with reference to Figure 7.3) the simplest control mechanism is to feed back the capacity, c, of the link to the source. This works as follows (see Section 7.4.5 for the ABR service in ATM networks where such an approach can be used). The source emits control packets in the forward direction. The link sets the value of its capacity in the control packets, which then proceed to the receiver. The receiver returns the capacity value to the source in reverse-direction control packets. The source then sends data at the rate c, the link is fully utilized, and a queue never builds up. If the number of connections is $n \geq 1$ and if the link knows this number, then the explicit-rate feedback $\frac{c}{n}$ is provided to each source. This ideal situation is not practical, however, for the following reasons:

- The link capacity available to the elastic transfers would, in practice, be time-varying. This is because the link could be carrying other, higher-priority stream or elastic flows. Hence the link capacity available to the transfers that we are analyzing will be less than c and will be time-varying. To formally develop and analyze control algorithms, we can model the available capacity as a random process $C(t)$.

- The number of transfers (in the class of transfers that we are analyzing) would be time-varying. This again means that the capacity available to each transfer is time-varying. We can analyze such a situation by modeling an arrival process of transfers, with each transfer bringing a certain random and finite amount of data to transfer.

The propagation delays between the source, the link, and the destination further complicate the design and analysis of rate control mechanisms.

Let us start by making a simple observation. Consider the case of a single connection and a stochastic available link capacity process $C(t)$. Assume that this process is stationary and ergodic. Then $\mathsf{E}(C)$ is the average available capacity—that is, with probability 1,

$$\mathsf{E}(C) = \lim_{t \to \infty} \frac{1}{t} \int_0^t C(u) \, du$$

For this stochastic capacity model, if we follow an approach parallel to the one leading to Equation 5.15 in Section 5.6.2, we can obtain the following result.

Exercise 7.1

Consider a buffer fed by a constant-rate fluid arrival process with rate r (i.e., for this process $A(t + s) - A(t) = r \cdot s$). The buffer can be drained at time t at the rate $C(t)$, a stationary and ergodic process. Show that the buffer-level process $X(t)$ converges in distribution to $\sup_{s \geq 0}(r \cdot s - (C(0) - C(-s)))$ and that this limit is finite, with probability 1, if $\mathsf{E}(C) > r$, where $\mathsf{E}(C)$ is as defined earlier.

Suppose that we can design an explicit rate-based algorithm that succeeds in setting the rate of the source to the fixed value $\mathsf{E}(C)$. Then, under fairly general conditions, the link's queue-length process will be unstable in the following sense. Let $X(t)$ denote the queue-length process; then $\lim_{t \to \infty} \Pr(X(t) \leq x) \to 0$ for any finite value x. Recall that this is similar to the situation in an M/M/1 queue when the arrival rate is equal to the service rate (the case $\rho = 1$); in this case, the queue-length process is a null recurrent Markov chain (see Appendix D).

One approach to solve this problem is for the link to use an *available capacity* number in its calculations, where this available capacity is some statistic of the process $C(t)$, such that if the arrival rate were equal to this value then the steady state queue-length distribution would be "acceptable." In this direction, the simplest idea is to scale down the average of $C(t)$ by some factor α ($0 < \alpha < 1$—say, 0.9) and thus to take $\alpha\mathsf{E}(C)$ as the available capacity. Although this would eliminate the instability, any such arbitrary factor will not yield predictable queue behavior. To see this, let us denote the stationary queue length by the random variable X; one QoS objective could be to require that $\Pr(X > x_{max}) < \epsilon$, for some given x_{max} and ϵ, or to require that $\mathsf{E}(X) < x_{max}$. If the available capacity is simply the mean of $C(t)$ scaled by a fixed factor α, then $\Pr(X > x_{max})$ will depend on the correlation structure of $C(t)$ and hence will not be fixed by only fixing α. This can be understood by recalling Example 5.5, where we studied a Markov modulated process feeding a buffered server with a fixed capacity. We learned there that the correlation structure of the Markov modulated arrival process affects the queueing, and it is not sufficient to look at the average arrival rate. In the present problem, if the $C(t)$ process is a Markov modulated process, then what we have is a *dual* of the situation in Example 5.5, with a fixed-rate source feeding a Markov modulated server. One technique for dealing with this issue is to use the notion of *effective service capacity* (ESC) provided in the following subsection.

Effective Service Capacity (⋆)

Instead of using an arbitrary scaling factor, we can follow an approach that is a dual of the one leading to Equation 5.23. Let us assume that, for $\theta > 0$, the following log-moment generating function exists:

$$\Gamma_C(\theta) = \lim_{t \to \infty} \frac{1}{t} \ln E_\nu e^{-\theta \int_0^t C(t)\, dt}$$

where E_ν denotes that the expectation is with respect to the marginal distribution ν of the process $C(t)$. The log-moment generating function would exist, for example, if $C(t)$ is modeled as a finite state Markov process.

Exercise 7.2

Calculate $\Gamma_C(\theta)$ for a finite state Markov $C(t)$.

With $E(C) > r$, let $X(r)$ denote the stationary queue-length random variable when the arrival rate is r (see Exercise 7.1). Then it can be shown that, for $\theta > 0$,

$$r < \frac{-\Gamma_C(\theta)}{\theta} \iff \lim_{x \to \infty} \frac{1}{x} \ln(\Pr(X(r) > x)) \leq -\theta$$

Suppose that the QoS requirement on the buffer at this link is that $\Pr(X(r) > x_{max}) < \epsilon$. Define

$$\theta^* = \frac{-\ln \epsilon}{x_{max}}$$

From this result (see also Figure 7.4) we infer that, for a large x_{max}, we could satisfy the requirement $\Pr(X(r) > x_{max}) < \epsilon$ by requiring that

$$\lim_{x \to \infty} \frac{1}{x} \ln(\Pr(X(r) > x)) \leq -\theta^*$$

for which it suffices that

$$r < \frac{-\Gamma_C(\theta^*)}{\theta^*}$$

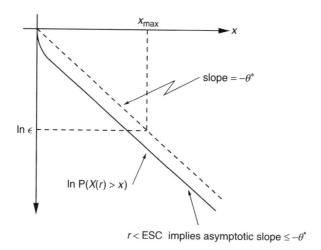

Figure 7.4   Relation between Pr($X(r) > x$) and the QoS requirement Pr($X(r) > x_{\max}$) $< \epsilon$ when r < ESC.

Hence we can call $\frac{-\Gamma_C(\theta^*)}{\theta^*}$ an effective service capacity for the service process $C(t)$ for the QoS requirement $\Pr(X > x_{\max}) < \epsilon$. In other words, the ESC is the r^* such that the asymptotic slope of $\ln(\Pr(X(r^*) > x_{\max}))$ is $-\theta^*$. Observe that, given a QoS requirement, if the available capacity of the link is taken to be r^*, then the link need only estimate r^* and feed this explicit value back to the source. If the source sends at this rate, then the argument shows that the desired buffer behavior will be achieved.

Explicit-Rate Control Algorithms

For the single-bottleneck-link case and n homogeneous sources, we can then consider the design of explicit rate control algorithms. Given a QoS requirement, we can estimate the ESC r^*. Consider periodic instants $t_k, k = 1, 2, \ldots$. Suppose at t_k we have an estimate r_k such that $\lim_{k \to \infty} r_k = r^*$ (with probability 1); then, if the number of connections n is known, a simple strategy is for the sources to be fed back the explicit rate $\frac{r_k}{n}$. Eventually the source rates will converge to $\frac{r^*}{n}$, as desired. In some packet-switching technologies, the network switches are not explicitly involved in connection setup and hence may not easily have the information on the number of connections. In addition, as would be the case in practice, the number of connections would be varying with time; let us assume for now that this variation of the number of connections with time is slow.[1] The problem then is for the sources to learn the explicit rate $\frac{r^*}{n}$ without the network knowing the number of sources. We defer further discussion of explicit-rate feedback algorithms until Section 7.7, where we consider the more general problem of distributed algorithms for fair bandwidth sharing among long-lived elastic transfers over a network of links.

7.4.2 Queue-Length-Based Feedback

In Section 7.4.1 we determine an effective service capacity such that if the source sends at this rate, then the buffer length is controlled to some desired objective. This available capacity is estimated, and each source is explicitly fed back its fair share of this capacity. As can be seen, the available capacity notion depends on a model for the capacity process and involves substantial computation. A simpler and more direct alternative is to provide feedback based on the queue-length error with respect to the target buffer length, and then to attempt to adjust the source rate so that this error is driven to zero. We discuss this approach in this section.

Let us assume that the network feeds back to the source the error between the instantaneous queue length and a target queue length x^*. The control packets that carry this information are assumed to be small and are emitted by the source in some small, fixed proportion of the data (e.g., RM cells in ATM's ABR service; see Section 7.4.5). We also assume that at the bottleneck link buffer these control packets are served with a high priority and hence do not see the queueing delay.

7.4.3 A Simple Rate Adaptation: ODE Model

Let us first consider the following naive adaptation of $r(t)$.

$$\dot{r}(t) = -ar(t - \tau)(x(t - \tau_r) - x^*) \tag{7.2}$$

This equation has the following explanation. At time t, the source receives control packets at a rate proportional to $r(t - \tau)$, because $r(t - \tau)$ is the rate at which

---

[1]How slow? Right now we are discussing algorithms that assume that there is a fixed configuration (number and routes) of connections and that these last forever. For this approach to be useful in predicting the performance of a practical network, obviously the connection configuration should change so slowly that the algorithm should be able to track the desired performance; the changes should be slower than the convergence times of the traffic control algorithm. Later in the book, we discuss the traffic control problem in the situation when the connection configuration is rapidly varying, as is the case in networks that carry predominantly finite-volume file transfers (e.g., Web traffic).

the source emitted data one round-trip time back. It is assumed that the control packets experience only the fixed round-trip delay, τ (because they are given priority handling at the packet switches). A control packet received at t carries the error between the queue length at time $t - \tau_r$ (at which instant the control packet would have observed the link's buffer) and the target queue length, x^*. For each such control packet, the source adjusts the sending rate *proportional* to the error that is fed back in the control packets. The constant a accounts for the proportion of control packets in the data and a factor indicating the change in rate as a proportion of the error.

Equations 7.1 and 7.2 can be written compactly as the following *ordinary differential equation* (ODE).

$$\begin{bmatrix} \dot{r} \\ \dot{x} \end{bmatrix} = \mathbf{f}(r(t - \tau), r(t - \tau_f), x(t - \tau_r)) \tag{7.3}$$

Here $\mathbf{f}(\cdot)$ is a vector function (a pair of functions), each of which is nonlinear. Hence we have a *nonlinear, ordinary, delay differential equation*. We see that if $r(t) = c$ and if $x(t) = x^*$, for all t, then the right side of Equation 7.3 is 0, and hence this is an equilibrium solution (or fixed point) of the ODE. If the system is started with these values of rate and queue length, or reaches these values after a perturbation, then it will continue to stay in this state.

We would like the system to operate at this equilibrium point and return to this equilibrium if a disturbance pushes it away. This is a question of *stability of the ODE*. To study the stability of this equilibrium point, one technique is to linearize the ODE about the equilibrium point and study the stability of perturbations of the resulting linear ODE. Although this does not lead to sufficient conditions for *global* stability (i.e., stability under arbitrarily large perturbations), it does lead to necessary conditions and provides useful insights. We limit ourselves to examining *stability of linearizations about the equilibrium point*.

The linearization is done by writing the first two terms of the Taylor series of the vector function $\mathbf{f}(\cdot)$ at the equilibrium point. The evaluation of the function at the equilibrium point, of course, yields 0. The linear term is the gradient of $\mathbf{f}(\cdot)$, evaluated at the equilibrium point, multiplied by the vector of perturbations of the delayed variables from the equilibrium values. Note that, in the ODE under discussion, even though there are only two variables ($r(t)$ and $x(t)$), several delayed values of these variables appear as arguments of $\mathbf{f}(\cdot)$ (i.e., $r(t)$ appears with two delays, and $x(t)$ appears with only one delay). In the process of linearization, $\mathbf{f}(\cdot)$ is taken to be a function of as many arguments (i.e., three arguments in the current example). Carrying out this linearization at the equilibrium point, we approximate the right side of Equation 7.3 as follows.

$$\mathbf{f}(c, c, x^*) + \nabla \mathbf{f}(c, c, x^*) \begin{bmatrix} (r(t - \tau) - c) \\ (r(t - \tau_f) - c) \\ (x(t - \tau_r) - x^*) \end{bmatrix}$$

Exercise 7.3
Show that

$$\nabla \mathbf{f}(c, c, x^*) = \begin{bmatrix} 0 & 0 & -ac \\ 0 & 1 & 0 \end{bmatrix}$$

and hence (noting that $\mathbf{f}(c, c, x^*) = (0, 0)^T$) the linearization of $\mathbf{f}(\cdot)$ around the equilibrium point yields the following linear delay ODE:

$$\dot{r}(t) = -ac\,(x(t - \tau_r) - x^*)$$

$$\dot{x}(t) = r(t - \tau_f) - c$$

We now consider perturbations of the system around the equilibrium point; that is, we will now examine the variables $(x(t) - x^*)$ and $(r(t) - c)$ because we are interested in studying the stability of the equilibrium point. We *simplify the notation by using the same variables (i.e.,* $r(t)$ *and* $x(t)$*) to denote the deviations of the two variables from the equilibrium point.* From the linearized equations displayed earlier, this yields

$$\dot{r}(t) = -ac\,x(t - \tau_r)$$

$$\dot{x}(t) = r(t - \tau_f)$$

where we emphasize that $r(t)$ and $x(t)$ are perturbations from the equilibrium point (c, x^*), and this linear delay ODE characterizes the behavior of the system around the equilibrium point.

Taking derivatives in the equation for $\dot{r}(t)$ and substituting from the equation for $\dot{x}(t)$, we get (after recalling that $\tau_r + \tau_f = \tau$)

$$\ddot{r}(t) = -ac\,\dot{x}(t - \tau_r)$$

$$= -ac\,r(t - \tau) \qquad\qquad (7.4)$$

This is a *second-order, linear, ordinary delay differential equation*. We immediately observe that there is a problem with the simple negative feedback control. Suppose we take $\tau = 0$ (i.e., zero propagation delay); then we get $\ddot{r}(t) + ac\,r(t) = 0$. It follows that, for $ac > 0$, the solution $r(t)$ will oscillate because there is no "damping" term. Delay in a negative feedback often makes matters worse for stability. For example, observe that the solutions of the first-order system $\dot{y}(t) = -\frac{\pi}{2}y(t)$ decay exponentially to 0, whereas the system $\dot{y}(t) = -\frac{\pi}{2}y(t-1)$ has the oscillatory solution $y(t) = \cos(\frac{\pi}{2}t)$.

7.4.4 Additive Increase, Multiplicative Decrease: ODE Model

We can obtain the damping or "friction" term that is missing in the preceding control by causing the sending rate to suffer a *multiplicative decrease* when the queue length exceeds the target value. Instead of sending back the actual deviation in the queue length from the target, let us also simplify the feedback to a single bit,[2] which is set as follows. In each control packet the network node sets the bit to 1 with probability $\eta \cdot (x(t) - x^*)^+$, where η is a factor that scales the error. We assume that η is chosen so that this expression yields a probability. When the source receives an unmarked control packet (i.e., with a 0 congestion control bit), it additively increases its rate; on the other hand, when the source receives a marked control packet, it multiplicatively decreases its rate (i.e., the rate is decreased by an amount *proportional to the current rate*). The following dynamic equation for the source rate evolution, along with the queue dynamics of Equation 7.1, models such a rate-control algorithm.

$$\dot{r}(t) = a\,r(t - \tau)\,(1 - \eta \cdot (x(t - \tau_r) - x^*)^+)$$

$$- b\,r(t)\,(r(t - \tau)\,\eta \cdot (x(t - \tau_r) - x^*)^+) \qquad\qquad (7.5)$$

---

[2]Note that even single-bit feedback can be used in a much more powerful way than only to indicate that the queue is above or below target. If, for example, the information is in the fraction of 1's (or 0's), then, over sufficient time, essentially any real number in [0, 1] can be fed back!

The first term on the right side corresponds to an additive increase in the rate each time an unmarked control packet is received by the source. The rate of receipt of control packets at the source at time t is proportional to $r(t - \tau)$, and a fraction $(1 - \eta \cdot (x(t - \tau_r) - x^*)^+)$ of these are not marked at the bottleneck link buffer. The constant a accounts for the proportion of control packets in the data emitted by the source and also for the additive increase factor. The second term on the right side corresponds to multiplicative decrease. The constant b accounts for the proportion of control packets sent by the source and also for the multiplicative decrease factor. $(r(t - \tau) \eta \cdot (x(t - \tau_r) - x^*)^+)$ is the rate of receipt of marked control packets at time t, and a marked control packet at t causes a decrease in rate that is proportional to the rate at t (i.e., $r(t)$).

It is easily seen that $r = c$ and $x = x_0$, such that $(x_0 - x^*) = \frac{a/\eta}{a + bc}$ is an equilibrium point for this nonlinear, delayed ODE. Linearizing the ODE about this equilibrium point (as we did in Section 7.4.3), we obtain

$$\dot{r}(t) = (-ac\eta - bc^2\eta)\,(x(t - \tau_r) - x_0)$$

$$+ (r(t - \tau) - c)\,\left(a\left(1 - \frac{a}{a + bc}\right) - bc\,\frac{a}{a + bc}\right)$$

$$- bc\,\left(\frac{a}{a + bc}\right)(r(t) - c)$$

$$\dot{x}(t) = (r(t - \tau_f) - c)$$

A little algebra shows that the coefficient of $(r(t - \tau) - c)$ is 0. Consider perturbations of the system about the equilibrium point, and use the same variables (i.e., $r(t)$ and $x(t)$) to denote the deviations of the two variables from the equilibrium point. This yields

$$\dot{r}(t) = (-ac\eta - bc^2\eta)(x(t - \tau_r)) - bc\left(\frac{a}{a + bc}\right)r(t)$$

$$\dot{x}(t) = r(t - \tau_f)$$

Taking a derivative in the equation for $\dot{r}(t)$ and then substituting from the equation for $\dot{x}(t)$ (i.e., $\dot{x}(t - \tau_r) = r(t - \tau)$), we obtain the following second-order, delayed, linear ODE for the perturbation of the source rate from the equilibrium.

$$\ddot{r}(t) + \gamma\dot{r}(t) + \beta r(t - \tau) = 0 \qquad (7.6)$$

where

$$\beta = c\eta(a + bc)$$

$$\gamma = \frac{abc}{a + bc}$$

We immediately notice that if $b = 0$, then the damping term $\gamma \dot{r}(t)$ disappears, and we are back to the form of Equation 7.4.

If the round-trip propagation delay τ is zero, $\gamma \neq 0$ and $\beta \neq 0$, then it is well known that the second-order linear differential equation is stable. If the system is perturbed from its equilibrium point, the system will eventually return to equilibrium. The time taken to do so can also be studied from the coefficients of the differential equation.

With $\tau > 0$ the analysis is considerably more involved. Solutions of this linear ODE will be of the form e^{st}, for some values s, which can possibly be complex numbers. If the ODE is stable there should not be a solution for which the s has a positive real part, because such a solution will blow up with increasing time t. We can obtain all the possible values of s by examining the roots of the *characteristic equation*, which is obtained by substituting $r(t) = e^{st}$ into Equation 7.6. Substituting $r(t) = e^{st}$ yields the equation

$$(s^2 + \gamma s + \beta e^{-s\tau})e^{st} = 0$$

Hence the possible values of s are roots of the characteristic equation

$$(s^2 + \gamma s + \beta e^{-s\tau}) = 0$$

Applying the change of variable $s\tau = \lambda$, we obtain the following equivalent equation in λ.

$$(\lambda^2 + \gamma \tau \lambda)e^{\lambda} + \tau^2 \beta = 0$$

In general, such an equation will have an infinite number of roots. We would like to obtain conditions on the coefficients of this equation so that all the roots have negative real parts, because this is necessary and sufficient for the stability of the system described by the delayed differential equation. The following is an application of a theorem in the classic book [25, page 449].

Theorem 7.1

The linear delayed differential equation in Equation 7.6 is asymptotically stable if and only if

$$\tau\beta \, \frac{\sin \omega_1}{\omega_1} < \gamma \tag{7.7}$$

where ω_1 is the unique solution in $(0, \pi)$ of

$$\cot \omega = \frac{\omega}{\tau\gamma} \tag{7.8}$$

∎

Because, for $\omega \in (0, \pi)$, $\frac{\sin \omega}{\omega} < 1$, it is sufficient that $\tau\beta < \gamma$. Using the expressions of β and γ displayed here, we see that it is sufficient for stability of the delayed differential equation in Equation 7.6 that the following inequality is satisfied.

$$\tau\eta < \frac{ab}{(a + bc)^2}$$

We see that when the round-trip propagation delay τ increases, the feedback gain parameter η must be proportionately reduced to stabilize the system. Notice also, from the form of the equilibrium solution of the nonlinear dynamical system Equation 7.5, that even though we will be able to stabilize the system by using a small η, the equilibrium queue length will deviate substantially from the target queue length.

Exercise 7.4

Consider n connections, indexed by i, $1 \leq i \leq n$, sharing a bottleneck link of capacity c. Assume that all the connections have the same forward and reverse propagation delays (τ_f and τ_r), with round-trip delay τ. Let $r_i(t)$ denote the transmission rate of source i. The dynamics of the queue length process $x(t)$ at the bottleneck buffer are then given by

$$\dot{x}(t) = \begin{cases} \sum_{i=1}^{n} r_i(t - \tau_f) - c & \text{for } x(t) > 0 \\ \left(\sum_{i=1}^{n} r_i(t - \tau_f) - c\right)^+ & \text{for } x(t) = 0 \end{cases} \tag{7.9}$$

The rate update equations are

$$\dot{r}_i(t) = a_i \, r_i(t - \tau) \, (1 - \eta(x(t - \tau_r) - x^*)^+)$$

$$- b_i \, r_i(t) \, (r_i(t - \tau) \, \eta(x(t - \tau_r) - x^*)^+) \qquad (7.10)$$

Show that the equilibrium rates and the bottleneck buffer level are given by $r_{oi} = \frac{a_i}{b_i}\xi$ and $x_o = x^* + \frac{1/\eta}{1+\xi}$, where $\xi = \frac{c}{\sum_{i=1}^{n} \frac{a_i}{b_i}}$. Take $a_i = a$ for all i.
Linearize the delayed differential Equation 7.10 around this equilibrium point, and show that the perturbation of the total rate process $r(t) = \sum_{i=1}^{n} r_i(t)$ from its equilibrium value, c, satisfies the delayed differential equation

$$\ddot{r}(t) + \gamma \dot{r}(t) + \left(\sum_{i=1}^{n} \beta_i \right) r(t - \tau) = 0 \qquad (7.11)$$

where $\gamma = a\frac{\xi}{1+\xi}$, and $\beta_i = \frac{a^2}{b_i}\eta \, \xi(1 + \xi)$. Finally, show that a sufficient condition for stability of this system is given by

$$\tau \eta < \frac{1}{a(1 + \xi)^2 \sum_{i=1}^{n} \frac{1}{b_i}}$$

The final result of Exercise 7.4 helps in explaining the effect of the number of connections on the stability of the system. Observe that if $b_i = b$ for all i, then the number of connections n appears in the denominator of the right side of the stability condition. It follows that as the number of connections increases, for a given propagation delay, the value of η should decrease in order to ensure stability.

7.4.5 The ABR Service in ATM Networks

In ATM network technology (see Section 2.3.6), the Available Bit Rate (ABR) service is used for transporting elastic transfers. This technology uses rate control of sources. In this section we describe the various kinds of protocol support that are available to implement rate-control algorithms such as those discussed in Section 7.4.

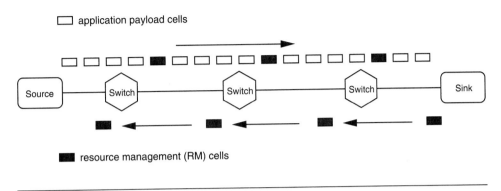

Figure 7.5 Flow of resource management (RM) cells over an ABR ATM connection.

The ATM Traffic Management Specification standardizes the way that the source and destination of a connection should be implemented. The mechanisms for exchange of control information are also standardized. This level of standardization ensures the interoperability of equipment from different vendors. What remain are the switch algorithms that determine when there is congestion, that set the congestion bits in the data cells and the RM cells, and that compute explicit rates. These algorithms are not standardized and are left for the switch developers to design and implement.

Resource Management Cells

Figure 7.5 shows an ABR ATM connection between a source and a sink, traversing four links and three ATM cell switches. The source emits data cells, and interspersed with the data cells are resource management (RM) cells. The data and RM cells are received by every switch in the path of the connection. Whereas the data cells are switched through without any modification of their payload, the RM cells are meant for signaling associated with the rate-control algorithm. Thus the RM cells need not be transmitted in sequence with the data cells; in fact, RM cells would follow a fast path in the switch. It is important, however, that the switch not reorder the sequence of the RM cells.

Figure 7.6 shows the format of an RM cell. Each row of the format corresponds to 1 byte; there are 53 bytes in every ATM cell. Notice that the 4th byte has the fixed Payload Type field with value 110, indicating that this is an RM cell. The payloads of these cells are read by the switches and possibly modified. The RM cells are received by the sink, and backward RM cells are generated to feed control information back to the source.

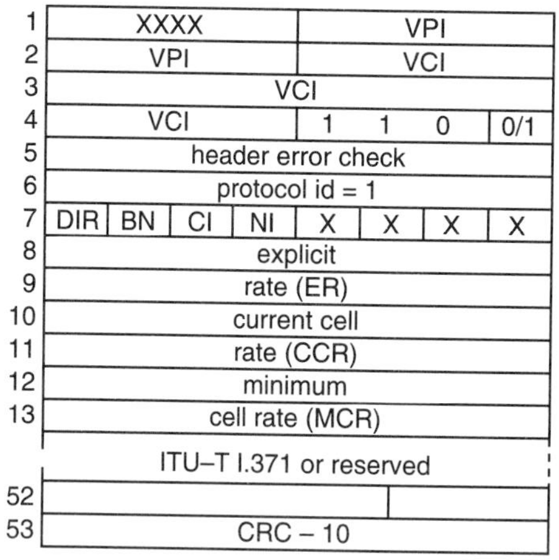

1	XXXX		VPI		
2	VPI		VCI		
3	VCI				
4	VCI		1	1	0	0/1
5	header error check				
6	protocol id = 1				
7	DIR	BN	CI	NI	X	X	X	X
8	explicit				
9	rate (ER)				
10	current cell				
11	rate (CCR)				
12	minimum				
13	cell rate (MCR)				
	ITU–T I.371 or reserved				
52					
53	CRC – 10				

Figure 7.6   The format of an RM cell of an ABR ATM connection. Each X denotes an unused bit.

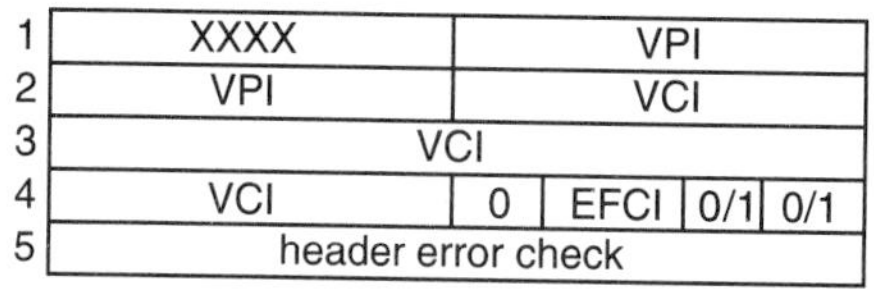

1	XXXX		VPI		
2	VPI		VCI		
3	VCI				
4	VCI		0	EFCI	0/1	0/1
5	header error check				

Figure 7.7   The format of an ATM data cell header, showing the EFCI bit.

Notice in Figure 7.6 that there is an Explicit Rate field in the RM cell. This field can be used by the network switches to feed an explicit-rate value back to the source. The RM cell also contains a Minimum Cell Rate (MCR) field, which can be used by the switches in their computations to ensure that the rate fed back does not fall below the MCR. Admission control of ABR connections should ensure that a connection demanding an MCR can actually get at least this rate after it is admitted. In Section 7.7 we discuss explicit-rate algorithms that can take care of MCR requirements.

Recall that in Section 7.4.2 we discuss a control algorithm that probabilistically marks a bit in control packets, the marking probability depending on the extent of the deviation of the switch queue length from a target queue length.

Such a bit is available in the ATM cell header itself; see the EFCI (Explicit Forward Congestion Indicator) bit in Figure 7.7 (the 0 bit just before the EFCI bit denotes that this is a data cell). When the sink sees data bits with EFCI bits marked, it can send backward RM cells to the source with the Congestion Indication (CI) bit marked (see Figure 7.6). Alternatively, the source can generate forward RM cells; these can be turned around by the sink, and on the way back the Backward Notification (BN) bit can be marked by the switches; this constitutes a *backward explicit congestion notification* (BECN).

The direction bit (DIR) is set to 0 in RM cells generated by the source, and set to 1 in RM cells sent (in the opposite direction) by the destination or the switches. The No-Increase (NI) bit in conjunction with the CI bit permits a more refined congestion indication to the source; for example, $CI = 0$, $NI = 1$ would inform the source that although the network is not congested ($CI = 0$), the source should not increase its rate ($NI = 1$), perhaps because measurements show that the derivative of the congestion is positive and the congestion limit is quite close.

ABR Connection Source and Destination Behaviors

In Figure 7.8, we provide a schematic of an ABR source adapting its rate in response to the network feedback received via RM cells. If the feedback is not explicit, then a computation must be done in the source to determine the rate at which it should send. The application above the ATM layer sends data into the ATM layer. These could be large packets, which are split up by the ATM adaptation layer (typically AAL5 for services over ABR). These AAL cells queue up in an ABR source buffer, which then emits them at the current rate.

The following is an outline of the standardized ABR source behavior:

1. At all times the source has an *allowed cell rate* (ACR) at which it can send cells (i.e., data or RM cells). The initial value of ACR is ICR (the *initial cell rate*). Each connection has a peak cell rate (PCR) and a minimum cell rate (MCR). At all times the ACR is such that the following inequality is satisfied: $MCR \leq ACR \leq PCR$. After a cell is emitted, the next cell can be emitted only after a time $\frac{1}{ACR}$.

2. When the source is actively transmitting cells, every N_{rm}th cell emitted is an RM cell. There is a provision to force the transmission of RM cells periodically if the source does not have data cells to send. The RM cell has the *current cell rate* (CCR) set to the current value of ACR; the MCR value of the connection is inserted in the corresponding field; in octet 7, $DIR = 0$, $BN = 0$, $CI = 0$, and $NI = 0$.

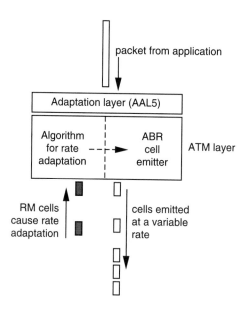

Figure 7.8 ABR rate adaptation at the source of an ATM connection, based on feedback in RM cells.

3. If the network is not being probed often enough and feedback RM cells are not being received, then the ACR is decreased. This is a precautionary measure.

4. There are two control parameters—RDF (*rate decrease factor*) and RIF (*rate increase factor*)—associated with each connection; both are numbers in the interval $(0, 1]$. When a backward RM cell is received, if CI=1, then the ACR is reduced by an amount ACR $\times$ RDF, and if CI = 0, NI = 0, then the ACR is increased by the amount PCR $\times$ RIF. Note that this corresponds to *multiplicative decrease and additive increase* (see Section 7.4.4).

5. If explicit-rate feedback is used in the network, then the ER field in outgoing RM cells is set to PCR. This can be viewed as the source asking the network for a rate equal to PCR. Of course, the network will feed back the rate it can actually provide. This will be in the ER field of backward RM cells. Finally, upon receiving an RM cell and computing the ACR based on the CI and NI fields, the ACR is taken as max {MCR, min {ACR, ER}}, where

in this expression the ACR is the one that was computed by the procedure described.

The following is an outline of the standard behavior of the destination node of an ABR connection.

1. The destination maintains a congestion state for the connection. This is set equal to the EFCI bit in received data cells. A data cell passing through a congested switch has its EFCI bit set to 1. Because a data cell is emitted, by its source, with the EFCI bit set to 0, a data cell arriving at its sink with the EFCI bit still equal to 0 indicates lack of congestion in the forward path of the connection.

2. The destination can *generate* RM cells in the backward direction. These can carry the congestion state of the connection back to the source via the CI bit. The fact that these RM cells are generated by the network is indicated by BN = 1.

3. If the source generates RM cells, these are "returned" by the destination with their direction bits changed to 1 (backward RM cells). The control information values in the forward RM cell are copied into the backward RM cells. If the congestion state is set to 1, then the CI bit in the backward RM cell is set to 1. End-to-end flow control can also be performed by the destination, reducing the ER value (in the backward RM cell) to match what the receiving application can handle.

An ABR Switch Algorithm: Queue-Length-Based Binary Feedback

Consider the single-link scenario in Figure 7.2. The source generates forward RM cells at a rate proportional to the rate of transmission of data cells. The EFCI bits in the data cells are marked by the LAN–WAN switch on the server side (see Figure 7.2). The marking algorithm the switch follows is to probabilistically mark the EFCI bits with a probability proportional to the error between the queue length at the WAN link and a target queue length. The congestion state conveyed by the EFCI bits is recorded in the destination node. The RM cells are turned around by the destination, with their CI bits set to 1 if congestion is indicated. Notice that this simple switch and destination behavior, along with the default source behavior, corresponds to the rate update dynamics shown in Equation 7.5. The analysis following that equation therefore applies.

7.5   Window-Based Control (WBC): General Principles

When data must be transferred reliably from one entity to another over an unreliable communication medium (unreliable because of bit errors or congestion), the receiver must continuously return acknowledgments (ACKs) for successfully received data. A common technique for controlling the transfer of data between the two entities is to fix the amount of unacknowledged data that can be outstanding between them. This limit can be fixed in terms of a number of bytes or packets and is called the transmission *window*. Let us denote this number by W. When W units of data are outstanding, the source stops until an ACK is received.

If we view the data being transferred as being laid out in sequence on a line, the window of outstanding data can be thought of as a contiguous interval of data on this line. The left edge of the window is the oldest unit of unacknowledged data, and the right edge is the latest such unit. Receipt of the ACK of the data unit at the left edge of the window results in shifting the window one unit to the right. This *opens* the window, and one more new unit of data can be sent. Generally, the protocol is that if a data unit covered by the window is acknowledged but is not the leftmost unit, then the window does not shift to the right, and hence new data transmission stays stalled.

This mechanism is widely used and appears in *automatic repeat request* (ARQ) protocols in link layers. As is evident from the description, a window protocol can be used to control the rate of transmission of data units from the source. For example, if the receiver decides to return the ACKs slowly, then the source will automatically slow down. This idea of using the window to control the transmission rate is exploited in window-based control (WBC) mechanisms for elastic traffic.

Let us now consider the window-based transmission of a long file. Figure 7.9 shows a more detailed schematic of the elements involved in such a file transfer. Conceptually, the window-based packet transmission and recovery protocol sits between the file and the network. The protocol transmitter takes chunks of data (bytes or segments) from the file and keeps the file transfer application updated as to the point to which the data has been successfully transmitted (and received by the client). Figure 7.10 depicts a large file being transferred in this way. The data are taken from the file in fixed-length segments. If we think of the file as composed of a concatenation of these segments, then at time t, $A(t)$ is the index of the segment to the left of which all segments have been transmitted and confirmed to have been received by the client; also, $A(t) + M(t)$ is the index of the next segment that will be given to the WBC protocol (when it asks for additional data to transmit). Thus the $M(t)$ segments indexed $A(t), A(t) + 1, A(t) + 2, \ldots, (A(t) + M(t) - 1)$ have been

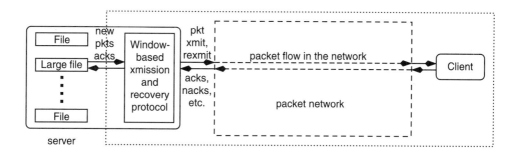

Figure 7.9 A window-controlled elastic transfer from a server to a client over a packet network.

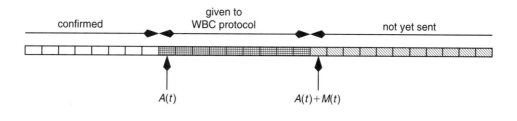

Figure 7.10 A file is transferred in segments; *A*(*t*) points at the next segment for which a confirmation is expected, and *A*(*t*) + *M*(*t*) points at the next segment that will be given to the WBC protocol for transmission.

given to the WBC protocol for transmission, but their status is unknown to the file transfer application.

Note that this discussion assumes that the WBC protocol confirms the successful transmission of segments in sequence, and thus $A(t)$ can be viewed as a nondecreasing function that increases in integer steps. In general, $M(t)$ can be constant, or it can increase or decrease. $M(t)$ is constant for a fixed window protocol. In a protocol such as TCP, the ACK of the packet corresponding to $A(t)$ may lead to a window increase; such an ACK will cause $A(t)$ to increase and $M(t)$ also to increase. It will become clear in Section 7.6 that in the case of TCP, $M(t)$ must be distinguished from the TCP transmitter's *congestion window* $W(t)$; the reason is that after accepting $M(t)$ packets from the file, the TCP transmitter may reduce its window because of network congestion.

Let D_k denote the time elapsed between segment k being given to the WBC protocol and the file transfer application receiving a confirmation of its successful transmission. Define γ to be the file transfer rate (in segments per second); we assume that γ exists. Also let us assume that the following limits exist:

$$E(M) = \lim_{t \to \infty} \frac{1}{t} \int_0^t M(u)\,du$$

$$E(D) = \lim_{n \to \infty} \frac{1}{n} \sum_{k=1}^n D_k$$

With reference to Figure 7.9, $E(M)$ is the time average number of segments in the dashed box, and $E(D)$ is the average delay between a segment entering the box and the file transfer application learning that the segment was successfully transferred to the file receiver at the other end. It then follows from Little's theorem (Theorem 5.1) that

$$\gamma = \frac{E(M)}{E(D)} \tag{7.12}$$

Remarks 7.1

When the transmission window used by the WBC transmitter is fixed (say, W) and there are *no packet losses*, then $E(M) = W$ and $E(D)$ is the average RTT in the network. Thus in this case $\gamma = \frac{W}{\text{mean RTT}}$.

If packet losses do occur in the network (because of buffer overflow or transmission errors), then these lost packets must be resent. Often, because of lack of complete information, in the process of recovering packet losses the WBC transmitter may resend packets that have reached the receiver successfully. Thus, in general, the actual rate of transmission of packets by the WBC transmitter will be larger than γ. Hence the packet transmission rate by the WBC transmitter is often called the *throughput*, and γ would be called the *goodput*. ∎

Exercise 7.5

Consider the classical Go-Back-N window-based control, which uses a fixed window of N. The receiver accepts packets only in sequence; it does not buffer packets received out of order. Consequently, if the transmitter sends

N packets and if the first is lost, then it must resend all the N packets starting from the first one. Hence the name Go-Back-N. Suppose that the probability of packet loss is p, the packet loss events being independent from transmission to transmission. Use the approach of Equation 7.12 to obtain the well-known formula for the goodput $\gamma = \frac{(1-p)N}{2\delta(1+(N-1)p)}$, where δ is the one-way propagation delay. It is assumed that there is no queueing delay. (Hint: Observe that for packet k, D_k is the sum of two terms: the time until that transmission of k such that all packets up to $k-1$ have been successfully transmitted, and, after this, the time until packet k is itself successfully received and ACKed. The means of these two terms are $2\delta\frac{(N-1)p}{1-p}$ and $2\delta\frac{1}{1-p}$, leading to $\mathsf{E}(D) = \frac{2\delta(1+(N-1)p)}{1-p}$. Note that in this example the throughput of the WBC transmitter is $\frac{N}{2\delta} \geq \gamma$.

Again with reference to Figure 7.9, we now turn to the procedures that the WBC transmitter will need to implement. Suppose that the transmitter has a single packet of length L bits to send. The link's transmission rate is C bits per second. In Figure 7.11 we trace the sequence of events involved in sending this single packet and then receiving its acknowledgment at the transmitter. At time $t = 0$ the transmitter begins to clock the bits of the packet out onto the link, and $\frac{L}{C}$ seconds later the packet is entirely in the link. The figure assumes that $\frac{\delta}{L/C} = 5$. At time $t = \delta$ the leading bit of the packet arrives at the bit receiver, which begins to extract the bits from the link. A further $\frac{L}{C}$ seconds later, the entire packet has been received by the router at the other end of the link. As assumed earlier, the packet instantaneously reaches the client, where we assume that instantaneous processing of the packet takes place by the WBC receiver, and at the same instant an ACK packet (of length a bits) gets queued at the router for transmission toward the server. The ACK packet is short compared with the data packets (e.g., $L = 1500$ bytes, and $a = 40$ bytes). At $t = \delta + \frac{L}{C} + \delta$ the leading bit of the ACK packet reaches the receiver of the router at the server end. A further $\frac{a}{C}$ seconds later, the server's WBC transmitter comes to know that the packet was successfully received. Thus, if a fixed window of 1 is used by the WBC transmitter, then at this instant another data packet can be taken from the file and queued for transmission. It thus takes $\delta + \frac{L}{C} + \delta + \frac{a}{C}$ seconds for a packet to be sent and its ACK received. Recall that this is what we call the base RTT. In terms of packet times (i.e., $\frac{L}{C}$) this is $\frac{2C\delta}{L} + 1 + \frac{a}{L}$. Define $\Delta := \frac{2C\delta}{L}$ as the *bandwidth delay product* (BDP) in packets; more precisely, this is the bandwidth RTPD product. For example,

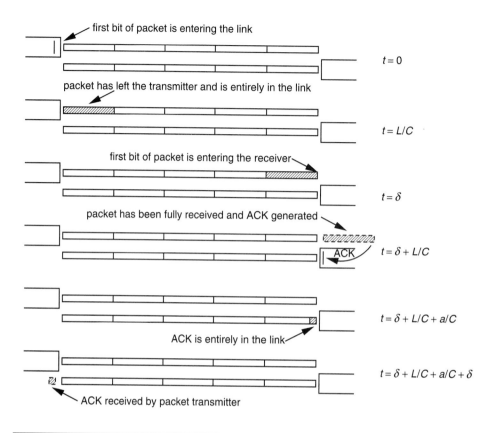

first bit of packet is entering the link

$t = 0$

packet has left the transmitter and is entirely in the link

$t = L/C$

first bit of packet is entering the receiver

$t = \delta$

packet has been fully received and ACK generated

ACK　　$t = \delta + L/C$

$t = \delta + L/C + a/C$

ACK is entirely in the link

$t = \delta + L/C + a/C + \delta$

ACK received by packet transmitter

Figure 7.11　Sequence of events and time instants involved in transmitting a single packet and receiving its acknowledgment.

take $L = 12,000$ bits, and $2\delta = 10$ ms (corresponding to 1000 Km). Then for $C = 10$ Mbps, $\Delta = \frac{100}{12} \approx 8.33$ packets; for $C = 100$ Mbps, $\Delta = \frac{1000}{12} \approx 83.3$ packets; and for $C = 1$ Gbps, $\Delta = \frac{10000}{12} \approx 833$ packets.

Suppose Δ is known to the WBC transmitter; then, ignoring the transmission time of ACKs, it becomes clear from Figure 7.12 that if the transmitter used a window of $\Delta + 1$, the round-trip pipe would be full, and the full capacity of the link would be utilized. In this situation, $\frac{\Delta}{2}$ packets would reside in the forward link, $\frac{\Delta}{2}$ ACKs would reside in the reverse link (one just entering the link, and the rest propagating across), and one packet would be ready to be transmitted (having just been generated from the just received ACK; note that we are ignoring

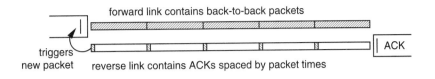

Figure 7.12   A WBC transmitter working ideally with a full round-trip pipe.

the ACK transmission time in this argument). Note that, after ignoring the ACK transmission time, $\Delta + 1$ is the BRTT normalized to the packet transmission time, $\frac{L}{C}$.

Exercise 7.6

The preceding discussion is with respect to Figure 7.2, assuming that the server and the client are connected to the bottleneck wide area link by high-speed local area networks. Suppose instead that the server and the client are connected to the bottleneck link by paths in high-speed wide area networks. The round-trip propagation delay from the server to the bottleneck link is δ_s, and from the bottleneck link to the client the round-trip propagation delay is δ_c. The round-trip propagation delay over the bottleneck link is δ_b. Define $\delta = \delta_s + \delta_b + \delta_c$. As before, define $\Delta := \frac{C\delta}{L}$. Similarly define Δ_s, Δ_b, and Δ_c corresponding to the RTPDs δ_s, δ_b, and δ_c. Assume that Δ, Δ_s, Δ_b, and Δ_c are all integer valued. Develop a picture like the one in Figure 7.11 and show that, ignoring the ACK transmission times on the bottleneck link, a TCP window of $\Delta+1$ is required to fill up the bottleneck link. Show where the packets reside if this window is used.

In practice, however, the WBC transmitter (which resides in the server; see Figure 7.9) would not know the value of Δ. Furthermore, in general, the link would be shared, in the same direction, with WBC transfers from other servers. In the latter case, the capacity of the pipe (i.e., $\Delta + 1$) would need to be shared between the various ongoing transfers. In the simple case in which all transfers travel over the same path (and hence have the same bottleneck link and the same RTPD), for equal sharing of the link's bandwidth the window of each transfer would need to be an equal share of the capacity of the round-trip pipe. Any particular WBC transmitter would not know how many other transfers were sharing the pipe, and furthermore this number would be time-varying. In view of this, the WBC transmitter must *learn* the correct window it should use in order to ensure that the link is efficiently utilized, and is shared fairly among the ongoing transfers. An adaptive window-learning algorithm is at the core of the transmission procedures in the Internet's TCP.

We do not develop a general theory and analysis for window-based elastic transfer protocols in this section. Instead, in the next section we turn our attention to TCP's adaptive window algorithm, which is the major practical example of a WBC protocol. It is for this specific algorithm, and related variations, that we develop the models and analysis.

7.6 TCP: The Internet's Adaptive Window Protocol

The Internet's Transmission Control Protocol (TCP) resides in end systems at the transport layer in the OSI model (see Section 2.3.5). Thus TCP sits between the applications and IP, the Internet's packet-routing and -forwarding protocol. TCP is connection-oriented, which means that an association must be established between the end points before data transfer can start, and this association has to be torn down when the data transfer completes. As discussed in Section 2.3.5, TCP enhances the unreliable, nonsequential packet transport service provided by IP to create a reliable and sequential packet transport service. It uses a window-based ARQ mechanism to achieve this function. In addition, the window-based mechanism is employed for two other major functions that TCP provides: (1) sender–receiver flow control, which prevents a fast source of packets (at the application level) from overwhelming a slow sink, and (2) adaptive bandwidth-sharing in the network. Thus internets employ a WBC for adaptive bandwidth-sharing.

In Sections 7.6.1, 7.6.2, and 7.6.3 we first describe TCP's packet transmission mechanisms in some detail. These mechanisms have evolved considerably since the first implementations, and we provide some insights into the reasons behind the evolutionary steps. We then turn to explaining the performance of TCP-controlled elastic transfers. Because the main application of TCP is for transferring files, the transfer throughput is the main performance measure of interest. We first examine, in Section 7.6.4, a single large file being transferred over a single-bottleneck link. We find that the evolution is essentially deterministic. This analysis leads to useful insights into how the various network parameters (such as the normalized bandwidth delay product and the buffer size) affect the evolution and the transfer throughput. Some of the TCP design decisions also become clear by such an analysis.

In Section 7.6.5 we analyze throughput of a TCP transfer in a wide area network with a random loss model. In Section 7.6.6 we examine explicit mechanisms by which the network can provide congestion feedback. In practice, file transfers are of finite lengths and are initiated at random times. Thus the network bandwidth is shared among a time-varying random number of file transfers. In Section 7.6.7 we discuss the throughput performance of randomly arriving, finite-volume, TCP-controlled file transfers. Some studies have reported that traffic in the Internet has the property of long-range dependence; we introduce the concept of long-range dependence earlier in Section 5.11. In Section 7.6.8, we examine the origins of this phenomenon and the implications of such traffic.

7.6.1   Slow-Start and Congestion Avoidance

In Section 7.5 we saw that for a single WBC transfer on a wide area link, the optimal window is $\Delta + 1$, where Δ is the BDP in packets. It is important to note that even if the WBC transmitter knew the value of this window, it should not *start* with this window, because that would entail the WBC transmitter at the server taking $\Delta + 1$ packets from the file and sending them all at once. With reference to Figure 7.2, because these packets would be transported instantaneously from the server to the router, all these $\Delta + 1$ packets would arrive at the router queue at the same instant. If the buffer space available is less than $\Delta + 1$, some packets would be dropped at the router, thus requiring their recovery. Thus even if the WBC transmitter knows $\Delta + 1$, the ideal situation depicted in Figure 7.12 would be obtained by the transmitter increasing the window until a window of $\Delta + 1$ is reached. This phase is called *slow-start*. In the course of this window increase, the bottleneck link transmitter would automatically space the packets out so that the situation in Figure 7.12 is gradually achieved.

In practice, of course, Δ is not known, and hence a TCP transmitter increases its window starting with the value of 1 packet. The idea is to avoid a potential buffer loss if packets are sent in a burst, and to discover how large a window this connection can use. When the window is only 1 packet, the transfer rate can be 1 packet per RTT. For a 1.5-KB TCP packet and 0.5 seconds RTT, the transfer throughput with a window of 1 is 3 KBps. This would usually be unacceptably small, and hence there is a need to increase the transfer rate by increasing the window.

When acknowledgments are received, the TCP transmitter assumes that there is spare capacity in the network and increases its window in the hope of getting a larger transfer rate. During slow-start, the TCP transmitter's window increases by 1 packet for each acknowledgment received. Thus the window increases rapidly,

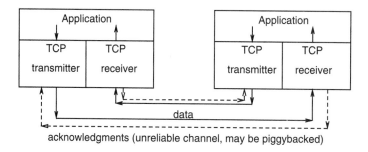

Figure 7.13 A TCP connection is full duplex and involves a transmitter and a receiver in each direction. Although the data is reliably transferred, the acknowledgments can be lost.

the attempt being to quickly ramp up the transmission rate. There is another variable, the *slow-start threshold* (denoted by W_{th}), which is a dynamic boundary between the rapid window increase phase and a more gradual increase phase called *congestion avoidance*. During congestion avoidance, if an acknowledgment is received at time t, then the window increases by $\frac{1}{W(t)}$. Thus, in slow-start, if the window is W then the successful transmission of these W packets results in a window of $2W$. In congestion avoidance, the resulting window would be $\left(W + \frac{1}{W}\right)$ after the first ACK, $\left(W + \frac{1}{W}\right) + \frac{1}{\left(W + \frac{1}{W}\right)}$ after the second ACK, and so on. For a large W, the final window after W ACKs will be approximately $W + W \times \frac{1}{W} = W + 1$; this is the approximation that is commonly used in analysis.

7.6.2 Receiver Buffer and Cumulative ACKs

For each direction of data transfer in a TCP connection, there is a transmitter–receiver pair (see Figure 7.13). During connection setup, several parameters are negotiated between the transmitter and the receiver in each direction; for example, the maximum segment size (MSS; i.e., the maximum packet length excluding the TCP/IP header). Another connection parameter that is negotiated is W_{max}, the maximum transmission window. A typical value of W_{max} is 32 KB, which, for a TCP packet length of 1500 B, is about 21 packets. A larger maximum window can be negotiated when the BDP is large.

One of the constraints on W_{max} is the amount of buffer space in the TCP receiver to store out-of-order packets. Consider the example shown in Figure 7.14. By using a *resequencing* buffer at the receiver, TCP converts the nonsequential

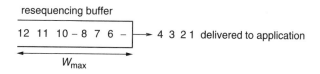

resequencing buffer

12 11 10 – 8 7 6 – ⟶ 4 3 2 1 delivered to application

W_{max}

Figure 7.14 The resequencing buffer at a TCP receiver. Out-of-order packets are accepted but are delivered to the application only in order.

packet delivery service provided by the IP layer to a sequential delivery service. Such a buffer is shown in Figure 7.14. Packets 1, 2, 3, and 4 are received in sequence and are passed to the application. At this point the receiver sends back an ACK(5), which tells the receiver that all packets up to and including packet 4 have been received, and the next packet required has the number 5. Note that, whereas the actual implementation deals with bytes (i.e., byte counts are acknowledged, and the window is also kept in bytes), for simplicity we work with packets. Thus the TCP receiver's acknowledgments are *cumulative ACKs*; each ACK acknowledges everything in the past. This is how TCP deals with an unreliable acknowledgment channel.

Along with ACK(5), the receiver also sends back a *window advertisement* of W_{max}. In the example, we have taken $W_{\mathrm{max}} = 8$. From these two pieces of information the transmitter infers that the receiver is willing to accept packet numbers $\{5, 6, \ldots, 12\}$ (i.e., W_{max} packets starting from packet number 5).

Let us now continue following the example in Figure 7.14. For some reason packet 5 does not show up, but packets 6, 7, and 8 do. These are accepted and stored after space is reserved in the buffer for packet 5. For each of these packets the receiver sends back an ACK(5), which tells the transmitter that (1) a packet was received, but (2) only data up to, and including, packet number 4 have successfully arrived in order. Furthermore, the window advertisement is kept at W_{max}. Note that by this point in time, the transmitter will have received the *first ACK(5)*, and its window's left edge is at packet number 5. Packet 9 also does not show up, but packets 10, 11, and 12 do. With these packets the receiver buffer is now full (having left space for the missing packets). The transmitter cannot send any more packets because the window is closed. Packet 5 then arrives (perhaps it took a circuitous route) and is placed in its position in the buffer. The receiver now returns ACK(9).

The receiver is now ready to pass packets 5, 6, 7, and 8 to the application. However, the application is not ready (e.g., it is a slow line printer with a small internal buffer and has not yet finished printing the previous data). An interlayer flow control will therefore operate and prevent the TCP receiver from sending the packets to the application. Hence the TCP receiver sends back a window advertisement of 4 with the ACK(9), thus telling the transmitter that it can accept only packets 9, 10, 11, and 12. In this situation, at the transmitter, the net effect of receiving this ACK will be to shift the left edge of the transmission window to 9, drop the window to 4, and stay stalled. If packet 9 is really lost, then loss recovery mechanisms (discussed next) will eventually come into play. We thus see that the window advertisement mechanism implements the function of *sender–receiver flow control*; that is, it prevents a fast transmitter from flooding a slow receiver that has limited storage.

7.6.3 Packet Loss Recovery and Congestion Control

Now returning to the window increase phase, suppose that the window increases to $W_{\max}$ without loss; then the window will simply stay at this value. TCP will behave as a fixed-window mechanism. In practice, the initial value of W_{tb} is large (e.g., 64 KB), and if no loss occurs during slow-start, then the window will rapidly increase to $W_{\max}$ (e.g., 32 KB) and stay there.

In practice, however, when there are several competing connections, as each source increases its window the round-trip pipe (of capacity $\Delta + 1$ packets) begins to fill up, and after the pipe fills up, the link buffer also begins to fill up. Ultimately, further increase in the windows causes buffer loss. This buffer loss must be detected, and the lost packet must be retransmitted. At this point, however, TCP's *congestion control function* is invoked. The buffer loss is taken as an indication of congestion, and an indication that the window has grown larger than what the network can handle. Hence TCP takes a corrective step by decreasing its window.

To continue this description, we take recourse to Figures 7.15, 7.16, and 7.17, which show the evolution of data transfer under TCP for various scenarios and TCP versions. In these figures, the variable $A(t)$ denotes the left edge of the transmitter's window. The figures are drawn from the point of view of the transmitters. Time is viewed as increasing in steps equal to the transmission time of a TCP packet. All the figures start at a point at which $A = 11$, $W = 5$, and $W_{tb} = 8$. The RTT (which includes 1 packet transmission time) is assumed to be 4 packet transmission times. This means that the ACK for packet 11 is received just before the transmission of packet 14 has been completed. The slanting arrows indicate this. Note that the tail of the first arrow in Figure 7.15 is at packet 11, and its tip is at packet 14; the number at the tip is the packet number carried in the ACK. If a transmission is lost, it is marked with X.

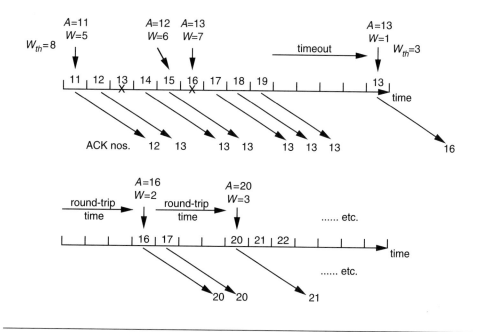

Figure 7.15 Evolution of packet transmission, $A(t)$, $W(t)$, and $W_{th}(t)$ in an example with timeout-based loss recovery. Lost transmissions are marked with X.

Let us now follow the example in Figure 7.15, which is for an early version of TCP in which loss recovery was only by timeout. At the start, $A = 11$ and $W = 5$, so packets 11, 12, 13, 14, and 15 are sent right away. Packets 11 and 12 are successfully sent and elicit ACKs 12 and 13, which arrive during the transmissions of packets 14 and 15, respectively. When ACK(12) is received at the transmitter, the transmission window is 5, which is less than $W_{th} = 8$. This ACK has two effects: (1) $A = 12$, and (2) $W = 6$. This permits the transmitter to send packets 12, 13, 14, 15, 16, and 17. With packets 12, 13, and 14 already having been sent at this time, packets 15, 16, and 17 can also be sent. Receipt of ACK(13) results in $A = 13$ and $W = 7$, permitting packets 18 and 19 to be sent.

The figure shows that packets 13 and 16 are lost. Hence, packets 14, 15, 17, 18, and 19 all elicit ACK(13) from the receiver. We observe that the receiver obtains six ACK(13)s; one is the *first ACK* for packet 12, and the remaining five are all *duplicate ACKs*. After packet 19 is transmitted, the transmitter's window is closed and a *retransmission timeout* (RTO)[3] timer is started. The RTO is assumed to have a value of 4 packet transmission times in the example. All the duplicate ACKs are ignored. When the RTO expires, the window is dropped to 1, and W_{th} is set to half of the value that the window had reached at the time the loss was detected by the transmitter; that is, $W = 1$, $W_{th} = \lfloor \frac{7}{2} \rfloor = 3$, and, of course, $A = 13$, because this is the place where retransmissions should start. Packet 13 is sent and arrives successfully, generating ACK(16) (see Section 7.6.2). Because only one packet could be sent, the transmitter is idle for a whole RTT. When ACK(16) is received, we have $A = 16$ and $W = 2$, and packets 16 and 17 are sent; notice that packet 17 is resent, even though it was received correctly the first time. This is a consequence of the limited feedback in the ACK packets in this protocol. The duplicate packet is discarded by the receiver. Another RTT elapses, the receipt of packet 16 generates ACK(20), which results in $A = 20$ and $W = 3$, and packets 20, 21, and 22 are then sent.

Notice that, in the protocol as just described, after a packet loss there are long time periods during which the transmitter is stalled and the link remains unutilized. The first improvement that we can seek is quicker loss discovery. Notice that, following the successful receipt of packet 12 and the loss of packet 13, five duplicate ACKs were received, all asking for packet 13. These duplicate ACKs can be interpreted by the transmitter as follows: "Some of my packets are getting through (so the network has bandwidth for me), but packet 13 seems to have not reached the destination." This interpretation can lead to an earlier loss discovery. But we must be careful, because duplicate ACKs can be received even if packets are reordered. The approach taken is to set a threshold K (usually $K = 3$) such that if K duplicate ACKs are received then loss recovery is started. The recovery can be very conservative (TCP Tahoe's *fast-retransmit*), or less conservative (TCP Reno's *fast-retransmit and recovery*).

In Figure 7.16 we show the evolution of the same example, but this time the fast-retransmit feature is assumed to be implemented in the TCP transmitter. To simplify the diagram, we assume that only packet 13 is lost. Let us take the duplicate ACK threshold K to be 3. Following the loss of packet 13, the third duplicate ACK is caused by the successful receipt of packet 16. This duplicate ACK is received when packet 19 is being sent. Hence following the transmission of packet 19 a fast-retransmit can take place. The TCP transmitter takes the

---

[3]An RTO is estimated by the TCP transmitter by observing some RTTs of the packets it sends and obtaining a statistical estimate of the mean RTT and a measure of variability in the RTT. These two statistics are used to obtain an RTO. The RTO is calculated only as a multiple of a timeout granularity (e.g., 500 ms). There is also a minimum RTO in most implementations.

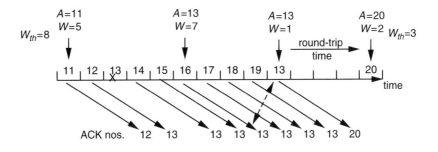

Figure 7.16   Evolution of packet transmission, *A(t)*, *W(t)*, and *W$_{th}$(t)* in an example with triple duplicate ACK (fast retransmit)–based recovery. The dashed double arrow indicates that the third duplicate ACK causes the retransmission of packet 13.

same actions that it would if this were a timeout; $A = 13$, $W = 1$, and $W_{th} = 3$. The receipt of packet 13 generates ACK(20), which results in the left edge of the transmitter window shifting to 20 ($A = 20$), and the window increasing to 2 ($W = 2$). Comparing this evolution with that in Figure 7.15, we see that a whole RTT of dead time is reduced. However, because the window has been reduced, it will take a few RTTs to build it up again so as to fill up the link's bandwidth.

   The fact that duplicate ACKs are being received can reassure the transmitter that congestion is not severe. Therefore, we can take a less drastic action than dropping the window to 1. With this in mind, in addition to implementing fast-retransmit, a TCP transmitter can implement fast-recovery. The evolution with this improvement is shown in Figure 7.17. When three duplicate ACKs are received, loss recovery is triggered. The transmission window is halved. Both W and W_{th} are set to this value (unlike in fast-retransmit, where W is set to 1). However, during the loss recovery phase, W is increased by 1 for each duplicate ACK received. Because $K = 3$ duplicate ACKs have already been received, W is immediately increased by 3, yielding $W = 3 + 3 = 6$ in the example. Packet 13 is now retransmitted. Meanwhile three more duplicate ACKs are received, and each of these causes the window to increase by 1 (i.e., 7, 8, 9). Because $A = 13$ and packets have been sent up to number 19, packets 20 and 21 can also be sent when the window becomes 8 and then 9. By the time the transmission of 21 completes, the acknowledgment for 13 (ACK(20)) returns. This was the expected ACK at the time that 13 was sent. Hence loss recovery is complete and the transmitter can return to normal operation. It does this by setting W equal to the stored value of W_{th}, which

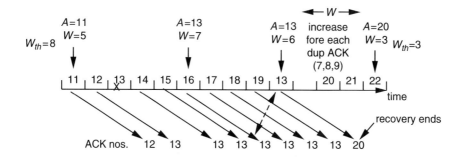

Figure 7.17 Evolution of packet transmission, *A(t)*, *W(t)*, and *W$_{th}$(t)* in an example with fast loss recovery. The dashed double arrow indicates that the third duplicate ACK causes the retransmission of packet 13.

in this example is 3. Notice that at this point packets 20 and 21 are outstanding, and with $A = 20$ and $W = 3$, exactly one packet (i.e., number 22), can be sent. Because $W = W_{th}$, further window increase takes place by congestion avoidance. In this simple illustration we see that more dead time has been eliminated by the combination of fast-retransmit and fast-recovery. We emphasize, however, that it is no accident that the recovery ends with exactly one packet needing to be transmitted; we describe this in detail in relation to Figure 7.19.

As mentioned before, fast-retransmit is implemented in TCP Tahoe and in all versions since then. TCP Reno implements fast-recovery as well, but if there are multiple losses (as in Figure 7.15), then this version goes through a window reduction and recovery process for each lost packet. In fast-recovery, as the loss is recovered transmission can continue because the window increases as duplicate ACKs for the lost packet are received. With multiple losses, however, there may not be sufficient duplicate ACKs for the later losses, and hence a timeout would need to occur for recovery of these losses. A newer version, TCP NewReno, is more aggressive in the fast-recovery phase and therefore reduces the frequency of timeout-based loss recovery.

A further improvement in the TCP protocol is to enhance the information that ACKs provide to the transmitter. As described in Section 7.6.2, cumulative ACKs indicate only the number of the first missing packet. However, an ACK could carry more information, and informing the transmitter of the entire pattern of received and missing packets in the receiver's buffer. Such an ACK is called a *selective ACK* (SACK) and is implemented in the latest versions of TCP.

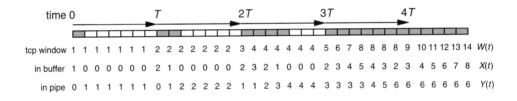

Figure 7.18 Packet transmission schedule during TCP's slow-start; shown are the sequence of transmission windows and buffer occupancies, and the number of packets in the pipe. The diagram assumes that $T = \Delta + 1 = 7$.

7.6.4 Single Connection: Analysis with Buffer Overflow

We denote by $W(t)$ the transmitter's congestion window, and we assume that the TCP transmitter and the file transfer application (see Section 7.5) have the same value of $A(t)$, the index of the first packet in the window. In general, $M(t)$ and $W(t)$ have different values, but you will see that always $M(t) \geq W(t)$. We assume that the TCP connection has been set up, and hence we do not include connection setup time in the following analysis. When a connection starts, $W(0) = 1$. One packet is taken from the file and is transmitted. Because the source has a high-speed connection to the wide area link, this first packet immediately enters the WAN link buffer.

We now follow Figure 7.18, where a fixed packet length of L bits is assumed, and time is divided into slots of length $\frac{L}{C}$. We take time to be measured in these units. The shaded areas in the horizontal bar in the figure indicate times during which the link transmitter (on the server side) is transmitting; the unshaded times correspond to an idle transmitter. Below the horizontal bar are three sequences of integers, showing the values of $W(t)$, the link buffer content (in packets) $X(t)$, and the number of packets in the pipe $Y(t)$. Thus, at $t = 0$, $W(t) = 1$, $X(t) = 1$, and the pipe is empty (there are no other connections in this discussion). At $t = 1$ the buffer is empty and the packet is in the pipe.

Analysis of the Slow-Start Phase

As discussed in Section 7.5, the ACK for the first packet arrives at the sender at time $t = T := \Delta + 1$, which in Figure 7.18 is 7. Upon receiving this ACK, $A(t)$ moves to the right by 1, and the TCP transmitter (at the server) increases the window to 2. This results in TCP taking two more packets from the file (so $M(t)$ is also 2) and sending them. Thus, at $t = 7$, the router queue $X(7) = 2$, and the pipe is empty. After another packet time, the buffer depletes by 1 and one packet has entered the pipe. At time $2T = 14$ the ACK of the packet sent in the interval $[7, 8)$ results in $W(t) = 3$. One packet, however, is still in the pipe, and hence only two packets can be sent by the TCP transmitter. This results in $X(t) = 2$. Notice that the equality $W(t) = X(t) + Y(t)$ has been maintained throughout because the outstanding packets are either in the buffer or in the pipe. When the packet sent in the interval $[8, 9)$ is ACKed, the window grows again by 1 at $t = 15$, and the buffer receives two more packets: one because the ACK causes the left edge of the window to advance by one packet, and the second one because the window has also grown by one packet. However, over the interval $[14, 15)$ one packet has been transmitted from the buffer, and hence $X(15) = X(14) - 1 + 2 = 3$. In two more intervals the buffer empties out, the link transmitter idles, and $W(t) = 4 = Y(t)$.

Notice that after $t = 3T = 21$ the link is no longer idle because before the buffer can empty out, ACKs begin to return, causing a window increase and insertion of more packets. The contents of the pipe saturate at 6, because this is all that the link can hold, and as the window grows the contents of the link buffer increase. There is a limit to this, and eventually the buffer will overflow. Note that this discussion assumes that the file has a large (infinite) amount of data.

Short Transfer Throughput

It is instructive to calculate the throughput that would be obtained by a *finite*-size file that is transmitted entirely in slow-start. Let $t_k = k\frac{L}{C}, k \in \{0, 1, 2, \ldots\}$, denote multiples of the packet transmission time. Assuming $L = 1500$ bytes, consider a 40.5-KB file (i.e., 27 packets). Each time an acknowledgment arrives the window grows by 1, and 2 packets are taken from the file. It can be seen from Figure 7.18 that all the 27 packets would have been handed over to TCP at t_{33}; 19 packets have been transmitted until this time, and 8 packets remain in the buffer. These will take an additional 8 time units to transfer, and hence the file would be fully transmitted at t_{41}. In another 6 time units the TCP transmitter would receive the last ACK, and the file transfer would be complete. Thus if the buffer available is at least 8 packets, then the file comprising 27 packets would be fully transmitted and acknowledged at t_{47}. Hence the normalized throughput for this file transfer would be $\frac{27}{47} \approx 0.57$; that is, the transfer could achieve a throughput of 57% of the link capacity. We see that short file transfers can obtain throughputs that are substantially smaller than the available link rate. With sufficient buffer space available, longer files would be able to achieve higher throughputs.

For the purposes of analysis, two cases can be distinguished.

- *Small Δ:* This is depicted in Figure 7.18. Initially, the transmitter's window doubles in each round-trip. This follows from the fact that during slow-start the window is increased by 1 for each ACK. In the initial cycles the buffer builds up and empties out, the maximum value increasing in each cycle. Define $K = \lceil \log_2(\Delta + 1) \rceil$. In the $(K+1)$th base RTT cycle, the round-trip pipe is full, and the buffer does not empty out in subsequent cycles. In Figure 7.18, $K = 3$.

- *Large* Δ: The window keeps doubling in each cycle, the pipe does not become full, and in each cycle the buffer fills up and empties out. Furthermore, the maximum buffer level reached increases in successive cycles until there is a cycle in which there is a buffer overflow.

Because this evolution is deterministic, it is straightforward to write formulas for the various processes. In the viewpoint presented here, the various processes evolve only at the t_k instants. We can write expressions for $W(t_k), X(t_k)$, and $Y(t_k)$. Because the formulas for the small Δ case are complicated and do not really add much insight, we discuss here only the formulas for the large Δ case.

When Δ is very large, the window and the maximum buffer level continue to grow with each base RTT cycle, and the round-trip pipe does not fill up. Let us index the cycles by $m \in \{0, 1, 2, \ldots\}$, the first cycle being given the index 0. Observe that up to the end of the mth cycle the window grows to 2^m. The window grows by 2^{m-1} in the mth cycle (for $m \geq 1$); in each cycle the buffer starts at two packets and then grows by one packet each time the window grows. Hence the buffer level reaches a maximum of $1 + 2^{m-1}$ in the mth cycle (for $m \geq 1$). Each time the window grows, two packets are taken from the file; hence, up to the mth cycle (for $m \geq 1$) a total amount of data equal to $2^{m+1} - 1$ packets has been transferred from the file. Also notice that, in each cycle, the window and the buffer reach their maximum values at the same instant. We can see that if the amount of buffer available at the link is B packets, with $B \geq 2$, then an overflow will occur in cycle m, where m (≥ 2) is such that $1 + 2^{m-2} \leq B$, and $1 + 2^{m-1} > B$; that is, $m = \lfloor \log_2(B-1) \rfloor + 2$.[4] Define this value to be $m(B)$. Hence, buffer overflow will occur when the window is in the interval $(2^{m(B)-1}, 2^{m(B)}]$.

Exercise 7.7

For the large Δ case, show that the buffer overflow occurs at t_k, where $k = m(B)(\Delta + 1) + (B - 1)$; at this instant, the window is $W(t_k) = 2^{(m(B)-1)} + B$; and the total number of packets from the file transmitted until loss occurs

---

[4]If n is an integer and x is a real number, such that $n \leq x$ and $n + 1 > x$ then $n = \lfloor x \rfloor$.

is $(2^{m(B)} - 1) + 2B$. For example, with reference to Figure 7.18, if $B = 4$, then $m(B) = 3$, loss occurs at t_{24}, the window at this instant is $W(t_{24}) = 8$, and the number of packets given to the TCP transmitter from the file is 15. If the initial congestion avoidance threshold is W_{th}, obtain a condition on B so that buffer overflow does not occur in slow-start.

Slow-Start: Evolution after Buffer Overflow

Because we are considering a single connection in slow-start, packet arrivals into the link buffer occur in bursts of exactly two packets. Hence when buffer overflow occurs, exactly one packet is dropped; this is because at the beginning of the preceding interval the buffer would have been full, one more packet would be transmitted, and two arrive from the file, causing one to be stored and the other to be dropped. At this time, if the transmitter's congestion window is W_1, then there would be B packets in the link buffer and $(W_1 - 1) - B$ packets in the round-trip pipe (because a total of $(W_1 - 1)$ packets are outstanding in the network, the last one sent in the current window of W_1 having been lost). Acknowledgments for these $W_1 - 1$ packets will return, causing the left edge of the window ($A(t)$) to keep moving to the right and causing the window to grow to $W_1 + (W_1 - 1) = 2W_1 - 1$. The acknowledgments will result in the transmission of additional packets by the TCP transmitter and in possible additional losses at the link buffer. If the window is growing when the loss occurs, then additional losses will certainly occur. In fact, in this case, as the window increases, every alternate packet transmitted by the TCP sender will be dropped at the link buffer.

The value of $A(t)$ will become stuck at 1 more than the index of the last correctly received in-order packet (i.e., at the first lost packet). The last of the acknowledgments that cause the window to grow will be followed by duplicate acknowledgments, all acknowledging the last correctly received in-order packet. These duplicate ACKs will not cause the window to grow any further. Because the window does not grow, no further packets will be taken from the file by the TCP sender; $M(t)$ (see Figure 7.10), the number of packets outstanding from the file, will be equal to $2W_1 - 1$. After the TCP transmitter has sent the last packet that it can, a timer is started (this is the retransmission timeout timer). We are assuming here an implementation that recovers from a loss in slow-start only by timeout. When this timer expires, the transmitter sets its window to 1, and the slow-start threshold, W_{th}, to half of the maximum window reached; that is, $\lfloor \frac{2W_1 - 1}{2} \rfloor = W_1 - 1$. Note that at this point $M(t) = 2W_1 - 1$, whereas the TCP window is 1. The TCP transmitter now must ensure that the packets taken from the file are reliably delivered to the receiver.

For large Δ, from the formulas derived earlier, we see that if the connection starts with a window of 1 and if in the initial slow-start the window grows without bound (in practice, the initial slow-start threshold is large), then the first loss occurs at the window $W_1 = 2^{(m(B)-1)} + B$ (see Exercise 7.7), and then the window grows to $2^{m(B)} + 2B - 1$. Hence the second slow-start will have a threshold $W_{th} = 2^{(m(B)-1)} + B - 1$, which is less than $2^{(m(B)-1)} + B$. Hence *loss will not take place in the second slow-start phase.*

The number of packets taken from the file in the first slow-start phase is the number taken until loss occurs—that is, $(2^{m(B)} - 1) + 2B$ (see Exercise 7.7)—plus the number taken after loss, $2(W_1 - 1) = 2^{m(B)} + 2B - 2$. Thus, it is important to note that, if B is large enough, then a short file would have finished transmission in slow-start itself.

> ### Exercise 7.8
> Assuming that loss occurs during the first slow-start and that the file transfer is long enough, obtain the duration of the second slow-start and the number of packets transmitted during it.

Let us denote the total number of packets sent in slow-start by n_{ss}, and the duration of the slow-start phase by t_{ss}.

Analysis of the Congestion Avoidance Phase

If the first slow-start completes without loss, then the congestion avoidance phase is entered with the initial window of W_{th} (the initial slow-start threshold). If there is loss in the first slow start, then there is no loss in the second slow-start, and the congestion avoidance phase is entered with $W_{th} = 2^{(m(B)-1)} + B - 1$, as explained in the preceding section.

Recalling that the congestion window is always less than W_{max}, let us take $W_0 < W_{max}$ to be the initial window for a congestion avoidance phase. Also, let us assume that $W_0 < \Delta + 1$. In congestion avoidance, if the window at time t is $W(t)$ and if the first acknowledgment of a packet is received, then the transmitter increments its window by $\frac{1}{W(t)}$. Thus the window now grows by one packet for each RTT. With this model, the window reaches $\Delta + 1$ (and the round-trip pipe becomes full) after a time $t_{ca\text{-}pipe\text{-}fillup} = (L/C)(\Delta + 1)(((\Delta + 1) - W_0) + 1)$. During this time the number of packets transmitted is $W_0 + (W_0 + 1) + \cdots + (\Delta + 1)$, which simplifies to $n_{ca\text{-}pipe\text{-}fillup} = \frac{((\Delta+1-W_0)+1)(\Delta+1+W_0)}{2}$.

When the pipe becomes full, acknowledgments return at the rate at which the bottleneck link can serve packets (see Figure 7.18); the buffer begins to grow, the RTT increases, and the rate of window increase decreases. In this phase, the window increases by $\frac{1}{W(t)}$ every packet transmission time, which is $\frac{L}{C}$. A discrete time model with discrete window values becomes cumbersome and does not add any insight. Hence, during the congestion avoidance phase, it is convenient to model the window process $W(t)$ as taking continuous values and evolving continuously in time. We can model this process by using the following differential equation:

$$\frac{dW}{dt} = \frac{C/L}{W(t)} \tag{7.13}$$

The pipe fills up when the window reaches $\Delta + 1$, and buffer overflow occurs when the window reaches $\Delta + 1 + B$. Hence the differential equation can be used to obtain the duration of the buffer fill-up phase.

$$t_{ca\text{-}buffer\text{-}fillup} = (L/C)\frac{1}{2}\left((\Delta + 1 + B)^2 - (\Delta + 1)^2\right)$$

Because the pipe is full, the number of packets transmitted in this phase is

$$n_{ca\text{-}buffer\text{-}fillup} = \frac{1}{2}\left((\Delta + 1 + B)^2 - (\Delta + 1)^2\right)$$

Congestion Avoidance: Evolution after Buffer Overflow

Buffer overflow occurs when the window reaches $\Delta + 1 + B$. Let the sequence number of the missing packet be n_l. At such an instant, the TCP transmitter window would have grown so as to allow only one more segment. Two packets are sent by the TCP transmitter, one of which is dropped. Hence, the number of remaining packets in the buffer and in the pipe is $\Delta + B$. Each of these packets, as it is acknowledged, causes the window to increase, but the total increase over all these $\Delta + B$ packets is less than 1, permitting no further window increase. Each acknowledgment received after the buffer overflow results in a single new packet being sent by the TCP transmitter. This keeps the buffer and the pipe full, but there is no further loss. We see that $\Delta + B$ new packets are transmitted after the lost packet. Each of these results in a duplicate ACK (requesting the sole missing packet). At the end of this process, the value of $A(t) = n_l$ (the sequence number of the missing packet), and $M(t) = W(t) = \Delta + 1 + B$. There are two basic alternatives for loss recovery: timeout-based recovery and fast recovery.

Congestion Avoidance: Timeout-Based Recovery

We see from the foregoing discussion that when buffer overflow occurs, the transmitter's window is at $\Delta + 1 + B$ and does not grow any further. Following the transmission of $\Delta + B$ more packets (each taking time L/C), the transmitter cannot proceed any further, the RTO timer is running, and $\Delta + B$ packets are outstanding. When the RTO timer runs out, all these packets would have been acknowledged by duplicate ACKs, and the system would be empty (the RTO would typically be a small multiple of the RTT). The TCP transmitter's window is then set to 1, and the missing packet is retransmitted. After a time $(L/C)(\Delta + 1)$, this packet is acknowledged, $A(t) = n_l + \Delta + 1 + B$, the window grows to 2, and two more new packets are taken from the file. The TCP transmitter executes a slow-start, and the cycle is repeated deterministically, a slow-start phase repeating in each cycle. The average throughput, in packets per second, can be computed as follows (here "ss" stands for "slow start" and "ca" stands for "congestion avoidance"):

$$\text{average throughput} = \frac{\text{no. of packets sent and acked in a cycle}}{\text{duration of a cycle}}$$

where

$$\text{no. of packets sent and acked in a cycle} = n_{ss} + n_{ca\text{-}pipe\text{-}fillup}$$
$$+ \; n_{ca\text{-}buffer\text{-}fillup} + \Delta + B,$$

and

$$\text{duration of a cycle} = t_{ss} + t_{ca\text{-}pipe\text{-}fillup} + t_{ca\text{-}buffer\text{-}fillup}$$
$$+ \; (L/C)(\Delta + B) + RTO$$

Congestion Avoidance: Fast Recovery

Alternatively, the returning duplicate ACKs can be taken as an indication of a missing packet. Let t_l denote the time instant at which there was buffer overflow. The TCP window at this instant is $W(t_l) = \Delta + B + 1$. The window grew to this value at some acknowledgment, and two packets were inserted into the buffer, the second of which was dropped, leaving $\Delta + B$ packets in the system. We illustrate the evolution of fast recovery using Figure 7.19, where it has been assumed that $\Delta = 6$ and $B = 6$, and hence $W(t_l) = 13$.

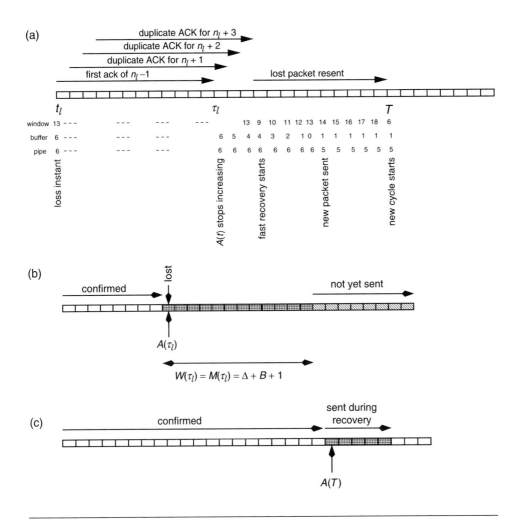

Figure 7.19 Packet transmission schedule during TCP's fast recovery: (a) The sequence of transmission windows, buffer occupancies, and the number of packets in the pipe; the diagram assumes $B = 6$, $\Delta = 6$. (b) A view of the packets transferred from the file at time τ_l. (c) A view of the packets transferred from the file at time T.

Let the index of the dropped packet be n_l. Hence the packets in circulation are $n_l - 1, n_l - 2, \ldots, n_l - (\Delta + B)$, with $n_l - 1$ being the last packet in the buffer, and $n_l - (\Delta + B)$ being the next packet whose ACK will be received. During the interval $(t_l, t_l + (L/C)(\Delta + B)]$, ACKs for these packets are received at the TCP transmitter. At time $\tau_l = t_l + (L/C)(\Delta + B)$, the *first ACK* of packet $n_l - 1$ is received at the TCP transmitter; this is the packet just before the lost packet. As discussed earlier, because the sender is in congestion avoidance, the window does not grow by a full TCP segment until τ_l; hence, we take $W(\tau_l) = W(t_l) = \Delta + B + 1$. When this first ACK is received, $A(\tau_l)$ is set equal to n_l, the sequence number of the lost packet. There are $\Delta + B$ packets outstanding, the window is closed, and no further first ACKs will be received until loss recovery is initiated. This is the situation depicted in Figure 7.19(b).

All the outstanding packets are after the lost packet (i.e., these are packets $n_1 + 1, n_l + 2, n_l + \Delta$), and hence their ACKs will be duplicate ACKs. Unless the missing packet is successfully acknowledged, $A(\cdot)$ will remain stuck at n_l. In this situation, unless $W(\cdot)$ grows beyond $W(\tau_l)$, no new packets will be transmitted from the file.

In fast recovery, when K (common implementations take $K = 3$, and this is assumed in Figure 7.19(a)) duplicate ACKs have been received (at time $\tau_l + K(L/C)$), the packet with sequence number n_l is resent, and the transmission window is modified as follows:

$$W(\tau_l + K(L/C)) = \left\lfloor \frac{1}{2} W(\tau_l) \right\rfloor + K$$

In the example shown in Figure 7.19, the window drops to $\lfloor \frac{13}{2} \rfloor + 3 = 9$. As additional duplicate ACKs return, the window is increased until it exceeds $\Delta + B + 1$, at which time new packets are taken from the file and transmitted. In Figure 7.19(a) this instant is marked as "new packet sent." Notice that the number of packets in circulation is now $\lfloor \frac{1}{2}(\Delta + B + 1) \rfloor$ (6 in the example).

The missing packet is sent at $\tau_l + K \frac{L}{C}$. At this time $(\Delta + B) - K$ packets are in circulation. Hence the ACK for the missing packet is received at $\tau_l + K \frac{L}{C} + ((\Delta + B) - K + 1) \frac{L}{C} = \tau_l + \frac{L}{C}(\Delta + B + 1)$, where the 1 in the left side accounts for the transmission time of the resent packet. Let this time instant be denoted by T; see Figure 7.19. At this time the number of packets in the pipe is $\lfloor \frac{1}{2}(\Delta + B + 1) \rfloor - 1$. This number is obtained as follows. The recovery window becomes equal to $\Delta + B + 1$, after $\Delta + B + 1 - \lfloor \frac{1}{2}(\Delta + B + 1) \rfloor$ duplicate ACKs have been received.

Then $\Delta + B - (\Delta + B + 1 - \lfloor \frac{1}{2}(\Delta + B + 1) \rfloor)$ more duplicate ACKs are received, and as many new packets are sent; this is equal to $\lfloor \frac{1}{2}(\Delta + B + 1) \rfloor - 1$.

When the ACK of the missing packet is received, this ACK packet will also acknowledge the $\Delta + B$ packets that were sent after the lost packet but before fast recovery started (see Figure 7.19(c)). On the other hand, none of the $\lfloor \frac{1}{2}(\Delta + B + 1) \rfloor - 1$ packets sent after fast recovery started has yet been acknowledged; hence, $A(T) = A(\tau_l) + \Delta + B + 1$. At this point, fast recovery is declared to have been completed, the congestion window is reduced to $\lfloor \frac{1}{2} W(\tau_l) \rfloor = \lfloor \frac{1}{2}(\Delta + B + 1) \rfloor$, this is also taken to be W_{th}, and congestion avoidance is entered. Notice that because $\lfloor \frac{1}{2}(\Delta + B + 1) \rfloor - 1$ (5 in the example in Figure 7.19) packets are already in the pipe, exactly one more packet will be sent by the TCP transmitter. This is the purpose of the way the window is allowed to grow for each additional duplicate ACK during the recovery period; if this window growth were not done, then the pipe would empty out during loss recovery, and then there would be a burst of $\lfloor \frac{1}{2}(\Delta + B + 1) \rfloor$ packets into the network at time T. In our example, where $\Delta = B = 6$, this would not cause a problem, but in general with larger Δ it would cause an immediate multiple loss because of buffer overflow.

The congestion avoidance cycle will repeat deterministically for a large file transfer. Note that the window oscillates between the values $\Delta + B + 1$ and $\lfloor \frac{1}{2}(\Delta + B + 1) \rfloor$.

The average file throughput would then be given by:

$$\text{average throughput} = \frac{\text{no. of packets acked in a cycle}}{\text{duration of a cycle}},$$

Notice that we account only for the number of packets ACKed in a cycle. The new packets sent during fast recovery in one cycle are thus counted in the following cycle. It can be seen that

$$\text{no. of packets acked in a cycle} = \left\lfloor \frac{1}{2}(\Delta + B + 1) \right\rfloor + n_{ca\text{-}pipe\text{-}fillup}$$

$$+ n_{ca\text{-}buffer\text{-}fillup} + \Delta + B$$

where the term $\lfloor \frac{1}{2}(\Delta + B + 1) \rfloor$ is the number of new packets that were sent in the fast recovery period of the preceding cycle, and $\Delta + B$ is the number of packets sent after the lost packet but before fast recovery starts in the current cycle). The

$$\text{duration of a cycle} = t_{ca\text{-}pipe\text{-}fillup} + t_{ca\text{-}buffer\text{-}fillup}$$

$$+ \frac{L}{C}(\Delta + B) + \frac{L}{C}(\Delta + B + 1)$$

where with reference to Figure 7.19(a), the penultimate term is $\tau_l - t_l$, and the last term is $T - \tau_l$. Comparing this with the cycle duration for timeout-based recovery, we see that much of the time wasted in RTO (which could be several times $(L/C)(\Delta + B)$) has been saved. Furthermore, slow-start is not repeated in each cycle.

7.6.5 A Stochastic Model for a Wide Area TCP Connection

In practice, of course, hundreds or thousands of TCP connections could share any given route in the network. From the discussions in Section 7.6.4 it should be clear that extending the detailed packet-by-packet analysis to the case of multiple connections will be very difficult, even for the case of a single bottleneck link. One modeling approach is to consider a single connection and to study its performance by approximately incorporating the effect of the network and the other connections.

Consider, therefore, a single TCP connection in a network. The connection is transporting an infinitely large file, and we are interested in obtaining the long-run average throughput that this transfer gets. The TCP protocol builds up its adaptive window over round trip times, and hence we assume that the average round-trip time for this connection is d; this includes the fixed round-trip propagation delay and the average queueing delay in the path of the connection. TCP resets the adaptive window on experiencing a packet loss, and hence it is important to include a model for the way losses occur in the TCP connection we are interested in.

We model the packet loss experienced by this connection by using a packet-level loss model. Packet losses could occur because of *tail drop* at one of the routers in the path of the connection, or because of bit errors in some link over which the connection traverses. The simplest model for packet losses is the Bernoulli process (each packet is lost independently with equal probability). Typically, errors are correlated in time; tail drop losses occur in a burst when some router buffer is full, and bit errors in links are well known to be correlated, particularly in the case of wireless links because of the fading phenomenon (see Section 8.1.4). Such correlations in the packet loss process can be captured by a Markov model for packet losses. In either case (the Bernoulli or the Markov loss model), we denote the average packet loss probability by p.

In a network that is already carrying a large number of connections, it can be expected that the addition of another connection would not have any significant effect on the statistics (such as the average queueing delays and packet loss processes). Thus if d and the loss probability are estimated for a path in the network, then a model such as that just described would help in predicting the performance of a new TCP connection along that path. In some situations, such as a fading wireless channel, the packet loss process can be obtained from some physical characterization of the channel. In the technique of *random early discard* (RED) (see Section 7.6.6), the packet discard probability is related to the average queue length. It is clear, therefore, that the approach of analyzing a single TCP connection, while capturing all other effects by simple stochastic models, would be useful in several situations.

In this setting, one analysis methodology proceeds as follows. Let us assume that there is some random process $Z(t)$ (a vector process in general) that is associated with the TCP connection and that the analysis of this process will lead to the average bulk transfer throughput. For example, the process $Z(t)$ could involve the congestion window and the slow-start threshold. In general, direct analysis of $Z(t)$ is not tractable. So we consider the processes embedded at certain random time points denoted by $(T_0, T_1, \ldots, T_k, \ldots)$. Let the values of $Z(t)$ sampled at these times be denoted by $X_k, k \geq 0$ (i.e., $X_k = Z_{T_k}$). The analysis then proceeds as follows:

1. It is shown that $\{X_k, k \geq 0\}$ is a Markov chain. Let $\pi(x)$ denote the stationary probability distribution of the Markov chain $\{X_k, k \geq 0\}$.

2. It is further shown that, given X_k, the distribution of T_{k+1} can be computed without knowledge of any additional history. Hence, the process $\{(X_k, T_k), k \geq 0\}$ is a *Markov renewal process* (MRP) (see Section D.1.5).

3. Given the Markov renewal process $\{(X_k, T_k), k \geq 0\}$, with the cycle (T_k, T_{k+1}) is associated a "reward" V_k equal to the number of packets successfully transferred in the interval. Define $U_k = T_{k+1} - T_k, k \geq 0$, as the cycle length process. Then, by the renewal reward theorem (see Sections D.1.4 and D.1.5), the throughput γ is given by

$$\gamma = \frac{\mathsf{E}_\pi V}{\mathsf{E}_\pi U} \tag{7.14}$$

where $\mathsf{E}_\pi(\cdot)$ denotes the expectation w.r.t. the stationary distribution $\pi(x)$.

As an illustration, let us perform an analysis of a single TCP connection that always *evolves in congestion avoidance*. We must keep track of the congestion window process, which we denote by $W(t)$ (thus, with reference to the foregoing general discussion, $Z(t)$ is $W(t)$ in this specific case). Let the initial window be W_0 and $T_0 = 0$, and examine the evolution of the congestion window process at multiples of d (the RTT), denoted by $t_1, t_2, \ldots$. We assume d to be large so that an entire window of packets can be sent by the TCP transmitter during any RTT. The TCP sender transmits W_0 packets in the interval $(0, t_1)$, and if there is no packet loss, then the window at time t_1 becomes $W_0 + 1$ (because the evolution is in congestion avoidance). Suppose that N_0 RTTs elapse without any loss, and there is a packet loss in the $(N_0 + 1)$th RTT. Hence after the RTT in which the loss occurs, the congestion window has grown to $W_0^{(l)} := W_0 + N_0$. We assume that there is no window growth in the RTT in which the loss occurs (because there are not enough ACKs to cause the full increase by one packet). The number of packets transmitted in the interval $(0, d(N_0 + 1))$ is

$$W_0 + (W_0 + 1) + \cdots + (W_0 + (N_0 - 1)) + (W_0 + N_0)$$

which simplifies to

$$(N_0 + 1) \left(W_0 + \frac{N_0}{2} \right) \tag{7.15}$$

In the RTT in which loss occurs, if the loss occurs at the M_0th packet, then $M_0 - 1$ ACKs will return to the TCP sender, causing $M_0 - 1$ more packets to be transmitted even after loss occurs (again, there will be no further window increase with these ACKs). All packets transmitted *subsequent* to the lost packet will result in duplicate ACKs. Let us assume that the number of these is sufficient to result in fast recovery (the analysis can be refined to include the case when there is an insufficient number of duplicate ACKs). Thus, fast recovery will commence in the interval $(d(N_0 + 1), d(N_0 + 2))$, with the lost packet being retransmitted. The transmission window at the instant when fast recovery begins is $W_0^{(l)}$, because no window increase takes place after the loss.

During fast recovery, $\frac{W_0^{(l)}}{2} - 1$ additional packets are sent (see Section 7.6.4). We can take the recovery as ending at $d(N_0 + 2)$, with the congestion window set to $\frac{W_0^{(l)}}{2}$. The $\frac{W_0^{(l)}}{2} - 1$ packets sent during the preceding recovery period can be assumed to be sent, starting at $d(N_0 + 2)$. Notice that we are back in a situation similar to the one where we started at $T_0 = 0$, except that the window is now $\frac{W_0^{(l)}}{2}$; let us denote this time instant by T_1, and the congestion window at this instant by W_1. We can view this as the start of a new cycle, albeit with a different initial condition.

Thus, associated with this cyclical evolution, we have a random vector process $\{(W_k, N_k, M_k, T_k), k \geq 0\}$, for which the following expressions can easily be obtained. Suppose $W_k = w$. Then loss takes place in the $(n + 1)$th RTT if the first $\sum_{i=0}^{n-1}(w + i)$ packets are successful and there is loss among the next $w + n$ packets (the ones that are transmitted in the $(n + 1)$th RTT). With the Bernoulli loss model and a loss probability of p, it follows that

$$\Pr(N_k = n \mid W_k = w) = (1 - p)^{\left(\sum_{i=0}^{n-1}(w+i)\right)}(1 - (1 - p)^{w+n}) \qquad (7.16)$$

The initial window in each cycle is half the window when loss occurs in the preceding cycle; hence we can write

$$W_{k+1} = \frac{W_k^{(l)}}{2} = \frac{W_k + N_k}{2} \qquad (7.17)$$

Finally, we also have

$$T_{k+1} = T_k + d(N_k + 2)$$

Because the distributions of N_k, M_k, and T_{k+1} are determined entirely by W_k and because W_{k+1} is a function of W_k and N_k, it follows that W_k is a Markov chain, and that (W_k, T_k) is a Markov renewal process. The stationary probability distribution $\pi(w)$ of W_k can then be obtained by standard numerical techniques.

The number of packets successfully transmitted in (T_k, T_{k+1}) (i.e., V_k) satisfies

$$(N_k + 1)\left(W_k + \frac{N_k}{2}\right) - (W_k + N_k) + (M_k - 1) \leq V_k$$

$$\leq (N_k + 1)\left(W_k + \frac{N_k}{2}\right) - 1 + (M_k - 1)$$

where the lower bound is the number of packets transmitted before the first lost packet in (T_k, T_{k+1}), and the upper bound is 1 less than the total number of transmissions in (T_k, T_{k+1}). With reference to the notation, we also have $U_k = d(N_k + 2)$. Hence upper and lower bounds on the average throughput γ can be obtained, as explained earlier (see Equation 7.14). This approach is perfectly amenable to a numerical computation. The following simple and approximate calculation, however, provides more insight into the effect of the system parameters.

Let Y_k be the number of packets transmitted until (i.e., including) the first lost packet in the interval (T_k, T_{k+1}). Following this, the number of additional transmissions is $W_k^{(l)} - M_k$ (in the same RTT), and $M_k - 1$ in the following RTT. Using Equation 7.17, we find that the total number of packet transmissions in (T_k, T_{k+1}) is

$$Y_k + 2W_{k+1} - 1$$

The same number can be obtained another way, from Equations 7.15 and 7.17, as

$$\frac{(N_k + 1)}{2}(W_k + 2W_{k+1}) + (M_k - 1)$$

Equating the two expressions, we have

$$Y_k + 2W_{k+1} - 1 = \frac{(N_k + 1)}{2}(W_k + 2W_{k+1}) + (M_k - 1) \qquad (7.18)$$

We need expectations. Clearly, $\mathsf{E}(Y_k) = \frac{1}{p}$. Taking an expectation on the right side, however, poses a difficulty because of the dependence between N_k and W_k. Hence, proceeding further requires detailed computations using the probabilities from Equation 7.16. As a simplification, let us assume that the variations in W_k are small, so that we can approximately take it to be a constant—say, w. Then, from Equation 7.17, we have that $N_k = w$. We make the further approximation that when a loss occurs, its position is uniformly distributed over the $W_k^{(l)}$ packets sent in that RTT; this leads to approximating $M_k - 1$ by $\frac{W_k^{(l)}}{2}$, or, in the present context, by w. Then the previous expression gets simplified to

$$\frac{1}{p} + 2w - 1 = \left(\frac{w + 1}{2}\right)3w + w$$

Solving the resulting quadratic we find that

$$w = \sqrt{\frac{1}{36} + \frac{2}{3}\left(\frac{1-p}{p}\right)} - \frac{1}{6} \qquad (7.19)$$

For small p, we find that $w \approx \sqrt{\frac{2}{3}\frac{1}{p}}$. It can be seen that small changes in the assumptions leading to Equation 7.18 will lead to differences in the exact expression for w in Equation 7.19. However, the essential behavior is that in each cycle the window grows linearly from w to $2w$, and hence the average window over a cycle is $\frac{w+2w}{2}$. This growth occurs over w RTTs and hence involves $w\left(\frac{w+2w}{2}\right)$ transmissions. The number of packets sent over this period is roughly $\frac{1}{p}$. Equating these we get the approximate formula $\approx \sqrt{\frac{2}{3}\frac{1}{p}}$, which will be unchanged by small variations in the accounting of transmitted packets.

As stated earlier, $U_k = d(N_k + 2)$, and hence we can approximate $\mathsf{E}(U) = d(w + 2)$. Furthermore, the same approximation used with the bounds on V_k (see earlier) yields $\frac{3w^2}{2} \le \mathsf{E}(V) \le \frac{3w^2}{2} + 2w - 1$. Hence for small p, using $w \approx \sqrt{\frac{2}{3}\frac{1}{p}}$ and applying Equation 7.14, we obtain

$$\gamma \approx \frac{1}{d}\sqrt{\frac{3}{2p}} \qquad (7.20)$$

where the units of γ are packets per second. As an illustration, consider $d = 0.5$ second, and $p = 0.01$, which yields $\gamma = 24.5$ packets per second; for a 1.5 KB packet, this translates to 36.7 KBps, a number that could be considerably less than the link speeds over which the packets are being carried. Observe that, in this example, increasing the maximum window size $W_{\max}$ will not increase throughput because the window is loss-limited.

The approximate formula in Equation 7.20 appears to say that the throughput goes to infinity as $p \to 0$. Of course, the throughput will be limited by the maximum window $W_{\max}$ to $\gamma \le \frac{W_{\max}}{d}$. If we can identify a bottleneck link, then γ will be further limited by the capacity of that link in packets per second.

This analysis assumes that each packet received at the TCP receiver results in an ACK being sent back to the TCP sender. Many TCP receiver implementations try to minimize the utilization of bandwidth by TCP ACKs by attempting to piggyback acknowledgments onto data packets being sent in the reverse direction. A TCP receiver implements this optimization by delaying an ACK on receipt of a packet until the earliest of three events: (1) the receipt of another packet to be ACKed, (2) the generation of a data packet in the reverse direction, and (3) the expiration of an "ACK delay" timer. If data are being transferred in only one direction (as is assumed in our analysis and as is usually the case), alternative (1) will most often be the cause of generation of an ACK. This implies that the sender will receive ACKs for alternate packets that it sends. Because TCP ACKs are cumulative, acknowledging alternate packets also acknowledges the packets received in between. Generalizing this idea to acknowledging every bth packet, this analysis can be extended to obtain the following result.

Exercise 7.9

Show that if the TCP receiver returns a cumulative ACK for every bth packet, then (with all the assumptions made in the foregoing analysis) the packet throughput is approximately given by $\gamma \approx \frac{1}{d}\sqrt{\frac{3}{2bp}}$.

The analysis we have presented makes several assumptions:

- *The RTT for the transfer being analyzed is not affected by the window growth of this transfer:* Recall that this is motivated by the wide area network situation and that this is one among many connections. As a consequence, the window growth of this connection is assumed to have a negligible effect on the RTT along its path. There are situations, however (for example, a single mobile user downloading a file from a Web server collocated with a wireless base station), when we must incorporate into the model the increase in the connection's RTT as the window of the connection increases.

- *The congestion window is unbounded:* In practice, the congestion window is bounded by the receiver's advertised window, W_{max}. A typical value is $W_{max} = 32$ KB (for Linux implementations of TCP receivers).

- *The loss probability is provided as an extraneous parameter:* If the window growth of the transfer being analyzed is the main cause of buffer overflow (as would be the case in a model with only a single transfer), then the model should itself capture the loss mechanism, and the analysis should yield the packet loss rate as one of the outputs.

- *Timeouts are never required for recovery from losses:* The number of packets that are transmitted after a lost packet may not be large enough to generate three duplicate ACKs. In addition, there may be multiple losses in a round-trip time, leading even TCP Reno to require a timeout for loss recovery.

This analysis technique can be used to provide an analysis of TCP with all the preceding assumptions relaxed. For example, the following is the celebrated PFTK formula [230] for the goodput of a TCP Reno connection over a wide area network with RTT $= d$, maximum window $W_{\max}$, RTO $= T$, and loss probability p.

$$\gamma = \begin{cases} \dfrac{\frac{1-p}{p} + \frac{W(p)}{2} + Q(p, W(p))}{d(W(p) + 1) + \frac{Q(p, W(p))G(p)T}{1-p}} & \text{for } W(p) < W_{\max} \\[3ex] \dfrac{\frac{1-p}{p} + \frac{W_{\max}}{2} + Q(p, W_{\max})}{d\left(\frac{W_{\max}}{4} + \frac{1-p}{pW_{\max}} + 2\right) + \frac{Q(p, W_{\max})G(p)T}{1-p}} & \text{for } W(p) \geq W_{\max} \end{cases} \tag{7.21}$$

where

$$W(p) = \frac{2}{3} + \sqrt{\frac{4(1-p)}{3p} + \frac{4}{9}}$$

$$Q(p, w) = \min\left\{1, \frac{\left(1 - (1-p)^3\right)\left(1 + (1-p)^3(1 - (1-p)^{(w-3)})\right)}{1 - (1-p)^w}\right\}$$

$$G(p) = 1 + p + 2p^2 + 4p^3 + 8p^4 + 16p^5 + 32p^6$$

This formula accounts for the situations in which packet loss recovery is initiated after a timeout (i.e., three duplicate ACKs are not received). The term $G(\cdot)$ has the particular form because each repeated timeout results in a doubling of the timeout period.

At the end of this chapter we provide references of more general analyzes. Often it is often not possible to obtain simple closed-form approximate formulas. Numerical computations can, however, be done, and these can be used to explore the effects of parameters. Such analyses, although computationally cumbersome, do lead to useful insights and can also be used to validate simulation programs or experimental results.

7.6.6   TCP with Explicit Feedback (RED and ECN)

The foregoing analysis assumes that TCP obtains congestion feedback as a result of buffer overflow in the network. Notice that this implies that TCP must grow its window to fill up the router buffer, thus causing a loss. This tail drop loss then serves as the congestion indication. There are two problems with this behavior:

- One of the objectives of a bandwidth-sharing algorithm for elastic traffic is to keep the network links occupied while sharing the bandwidth fairly. For this it is sufficient to keep only one packet in the buffer of the bottleneck link. The blind window adaptation algorithm that TCP uses must cause a loss in order to get feedback. Hence the queueing delay in the network becomes much larger than it needs to be just to keep the links occupied. As we have seen, the RTT has a direct impact on the TCP throughput. Furthermore, interactive applications such as chat are adversely affected by large queueing delays. Hence it is important to limit the queueing delays while keeping the network links occupied.

- A second consequence of relying only on tail drop loss as the congestion indication is that if several connections share the same bottleneck link, then, when the buffer fills up, many connections incur loss at the same time. This causes all these connections to back off their windows at the same time, thus very probably resulting in an underutilization of the bottleneck link. This is called *loss synchronization*.

A more controlled behavior can be obtained if the network *explicitly* feeds back congestion information to the TCP sources. Such feedback can be achieved if a network node deliberately drops packets that are arriving at a congested (but not yet overflowing) link buffer. An implementation of this idea was first proposed as the random early discard (RED) algorithm. As shown in Figure 7.20, a measure of the average queue is maintained at a link buffer. A simple averaging procedure is *exponential averaging*. Suppose that the average is updated periodically and is denoted by q_k. Then at an update time the new average relates to the preceding average and the observed queue length x_{k+1} as follows:

$$q_{k+1} = (1 - \alpha)q_k + \alpha x_{k+1} \qquad (7.22)$$

where $0 < \alpha < 1$ is the *forgetting factor*. A value of α close to 1 puts more weight on the new observation and hence rapidly "forgets" the history. As long as this average queue length is less than q_{lo} nothing is done. However, if the average queue

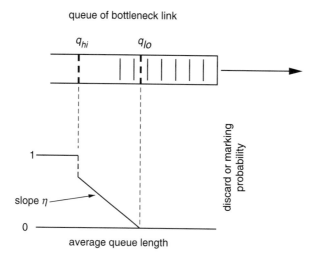

Figure 7.20   Explicit congestion feedback in the Internet: random early discard (RED) and explicit congestion notification (ECN).

length exceeds q_{lo} (but is less than q_{hi}), arriving packets are dropped randomly, with an average probability that is proportional to the difference between the measured queue average and q_{lo}; in Figure 7.20, η is shown as the proportionality constant. If the average queue length exceeds q_{hi} then all arriving packets are dropped. The function that maps the average queue length into a drop probability is called a *drop profile*.

A better approach is not to drop packets, but instead to reserve a bit or two in the packet header for signaling congestion information back to the connection end points. One practical way of doing this is for the congested node to set a congestion bit in a data packet in the forward direction; on receiving such congestion information, the receiver signals it back to its source by using a congestion indication bit in the acknowledgment packets it returns. This is called explicit congestion notification (ECN) and can use a marking profile just like the one in RED (see Figure 7.20).

A Dynamic System Model
Motivated by RED and ECN, let us use a deterministic dynamic system model (as in Section 7.4) to study TCP with explicit congestion feedback. We assume

that the TCP sender is in congestion avoidance and hence that an ACK packet not carrying a congestion indication causes the window to increase by $\frac{a}{w(t)}$ when the window is $w(t)$. The node randomly marks congestion bits with a probability proportional to the amount by which the buffer exceeds the target buffer level; the factor η is used to control this feedback. When the source receives a congestion indication bit, the window is multiplicatively decreased; that is, it is reduced by an amount proportional to the current window. The proportional decrease factor is b.

The forward propagation delay is τ_f, the reverse propagation delay is τ_r, the round-trip propagation delay is τ, and the queuing delay in the link buffer is τ_q. Because there are no special control packets in this system, the queueing delay must be considered in the control loop. We simplify the analysis and assume that there is a mean queueing delay τ_q, and hence, at time t, ACKs arrive to the source at the rate $r(t - \tau - \tau_q)$. We assume that there is one ACK for each packet sent by the source.

Putting this description together, we obtain the following differential equation for the window (see also Equation 7.5 and the material in Section 7.4.3).

$$\dot{w}(t) = \frac{a}{w(t)} \, r(t - \tau - \tau_q) \cdot (1 - \eta \cdot (x(t - \tau_r - \tau_q) - x^*)^+)$$
$$ - bw(t) \, r(t - \tau - \tau_q) \cdot \eta \cdot (x(t - \tau_r - \tau_q) - x^*)^+ \qquad (7.23)$$

To understand this equation, note that the first term on the right side corresponds to the increases in the window when unmarked packets are received at the source, and the second term corresponds to the multiplicative decreases in the window when marked packets are received at the source. At time t, the probability that an ACK arriving at the source is marked is $\eta \cdot (x(t - \tau_r - \tau_q) - x^*)^+$.

Note that in writing Equation 7.23, for simplicity we assume that the marking is based on the deviation of the *instantaneous* queue length from the target x^*. Recall that in the earlier discussion of RED we state that actually an exponentially averaged queue length is used (see Equation 7.22). This can be incorporated in the ODE analysis as follows. First we write the average queue length $q(t)$ as a continuous exponential average of the instantaneous queue length $x(t)$, as follows:

$$q(t) = \alpha \int_{-\infty}^{t} e^{-\alpha(t-u)} x(u) \, du$$

Notice that at time t the queue length at time $u, u \leq t$, is given a weight of $e^{-\alpha(t-u)}$ (i.e., the older the queue length, the smaller the weight). It can then be shown that $q(t)$ satisfies the following differential equation:

$$\dot{q}(t) = \alpha(x(t) - q(t)) \tag{7.24}$$

And an ODE analysis can be developed by writing $q(t)$ in place of $x(t)$ in Equation 7.23. Note the similarity of Equations 7.22 and 7.24. We, however, proceed here with the assumption of instantaneous queue-length-based feedback.

Denote by $d := \tau + \tau_q$ the round-trip delay for the connection. Let us assume that the connection window is related to the source rate by the relation $w(t) = r(t)d$; note that, by Little's theorem (Theorem 5.1), this is true for average values.

With this approximation, Equation 7.23 can be rewritten in the following way.

$$\dot{r}(t) = \frac{a}{r(t)d^2} r(t-d)(1 - \eta(x(t-d+\tau_f) - x^*)^+)$$

$$- br(t)r(t-d)(\eta(x(t-d+\tau_f) - x^*)^+) \tag{7.25}$$

The dynamics of the bottleneck link buffer are the same as in Equation 7.1. Let us view this ODE as follows:

$$\dot{r}(t) = f(r(t), r(t-d), x(t-d+\tau_f))$$

The equilibrium point of this ODE is found by seeking the constant values $r(t) = r_o$ and $x(t) = x_o$ such that $f(r_o, r_o, x_o) = 0$. It can easily be seen that $r_o = c$, and $(x_o - x^*) = \frac{a/\eta}{a + bc^2 d^2}$.

Linearizing around this equilibrium point (as we do in Section 7.4.3), substituting from the dynamics of $x(t)$, and using $r(t)$ again to denote the perturbation of the rate from the equilibrium, we obtain the following delay differential equation.

$$\ddot{r}(t) + \gamma \dot{r}(t) + \beta r(t-d) = 0 \tag{7.26}$$

where

$$\gamma = \frac{2abc}{a + bc^2 d^2}$$

$$\beta = \frac{\eta}{d^2}(a + bc^2 d^2)$$

It follows from the same arguments as in Section 7.4.4 that a sufficient condition for stability of this system is given by $d\beta < \gamma$:

$$d\,\frac{\eta}{d^2}\,(a + bc^2 d^2) < \frac{2abc}{a + bc^2 d^2}$$

or, equivalently,

$$dc < \frac{1}{\eta}\,\frac{2abc^2 d^2}{(a + bc^2 d^2)^2}$$

Observe that cd is the product of the bottleneck bandwidth and the round-trip time. For large cd, it can easily be seen that the condition becomes

$$(dc)^3 \eta < \frac{2a}{b}$$

A typical implementation is to take $a = 1$ and $b = \frac{1}{2}$, which will yield the sufficient condition

$$(dc)^3 \eta < 4$$

One insight from this last result is that if the delay bandwidth product doubles, then η will need to be scaled by $\frac{1}{2^3}$ to maintain stability. It can be easily checked (as in Exercise 7.4) that if there are n connections with the same round-trip times and the same parameters a and b, then we obtain the same stability condition with η replaced by $n\eta$. Thus η needs to be further scaled down for increasing number of connections.

Although the condition $(dc)^3 \eta < 4$ was obtained from the linearized delayed ODE, it is instructive to examine the behavior of the nonlinear delayed ODE, Equation 7.23, when this condition is satisfied and when it is violated. Figure 7.21 shows plots from a simulation of Equation 7.23, in conjunction

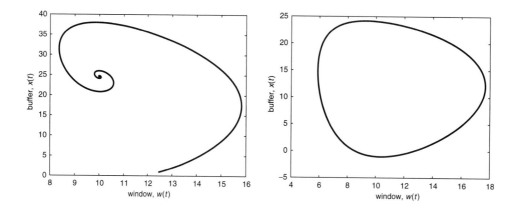

Figure 7.21   Plots of x(t), the buffer occupancy, vs w(t), the window, showing stability and instability of the nonlinear delayed ordinary differential Equation 7.23 for $(dc)^3\eta < 4$ and $(dc)^3\eta > 4$; $a = 1$, $b = 0.5$, $c = 1$, $d = 10$, $x^* = 5$; $\eta = 0.0010$ in the plot on the left, and $\eta = 0.0042$ in the plot on the right.

with Equation 7.1. In this example, $dc = 10$, and hence the simple condition requires $\eta < 0.004$; we find that for $\eta = 0.001$, the window and buffer converge to the equilibrium values $w_o = r_o \times d = 10$, and $x_o = 24.6$, whereas for $\eta = 0.0042$, the window and buffer both oscillate. For $\eta > 0.0042$ we find that the oscillations are of increasing amplitude. In, general, however, we cannot draw conclusions about the global stability of the original nonlinear delayed ODE from the linearized version, and we must resort to more sophisticated analysis techniques.

Random Marking
Let us consider what happens if the TCP packets are randomly marked at the link buffer *without any regard to congestion*. Again, we use a deterministic dynamic model in which the marking rate is taken to be the *average* marking rate, p. We obtain the following dynamics of the sender window process $w(t)$.

$$\dot{w}(t) = \frac{a}{w(t)}\frac{w(t-d)}{d}(1-p) - bw(t)\frac{w(t-d)}{d}p$$

where we note that the factor $\frac{w(t-d)}{d}$ is an approximation of the rate at which ACKs arrive at the source at time t. It is then easily seen that the equilibrium point is

$w_o = \sqrt{a(1-p)/bp}$. For small p, we notice that we have the form of the simplified formula $w \approx \sqrt{\frac{2}{3p}}$ obtained from Equation 7.19.

Linearizing the nonlinear ODE around the equilibrium point and using $w(t)$ to denote the deviation from w_o, we get the following differential equation.

$$\dot{w}(t) = -\left(\frac{a}{d}(1-p) + \frac{b}{d}pw_o^2\right)\frac{1}{w_o}\,w(t)$$

which, upon substituting for w_o, simplifies to

$$\dot{w}(t) = -\frac{2}{d}\sqrt{abp(1-p)}\,w(t)$$

This first-order differential equation is always stable. We observe that, for small p, the *time constant* of convergence for this equation is approximately given by $\frac{d}{2\sqrt{abp}}$. Thus we see that for large d and small p, the convergence to equilibrium will be quite slow.

Throughput Analysis with Multiple Long-Lived Connections

In Section 7.6.5 we present an analysis of a single connection with given RTT, d, and loss probability, p, along its path. Suppose that there are n connections sharing a path with a bottleneck link; each connection has an infinite amount of data to send. The RTPD is δ, and the bottleneck link rate is C. Let L be the data packet length of each TCP transfer. There is a random dropping or marking probability of p at the bottleneck link buffer. We would like to determine the average throughput of the connections and the mean queue length at the bottleneck buffer.

Suppose that a formula such as the one provided in Equation 7.20 is available to us. Note that more detailed formulas than this are available in the literature. These formulas express the TCP throughput in terms of the parameters p (the loss probability) and d (the RTT); denote such a formula by $\gamma(p,d)$ in packets per second. Let us suppose that we can write such a formula in the form $\gamma(p,d) = \frac{f(p)}{d}$ (in packets per second); Equation 7.20 is in this form. Let τ_q be the mean queueing delay at the bottleneck buffer, and hence we write the RTT as $d = \tau_q + \delta$. Suppose that the queueing delay is 0; this will happen when p is large and hence drops or marks are frequent, and thus the TCP windows are not large enough even to fill up the round-trip pipe. Then $L\frac{f(p)}{\delta}$ is the throughput of each transfer, and the total transfer rate is $nL\frac{f(p)}{\delta} \leq C$. If, however, $nL\frac{f(p)}{\delta} > C$ it means that there is queueing delay. To obtain the queueing delay we can use a simple heuristic. We know that when the queueing delay is positive the total throughput should be equal to C, with $\frac{C}{n}$ being the throughput of each transfer. (This is an approximation because it ignores the probability of the queue emptying out; thus the approximation will not work well at the "transition point" between the two regimes.) The mean queueing delay, τ_q, can then be obtained as the value that satisfies $nL\frac{f(p)}{\tau_q+\delta} = C$.

Define $\Delta := \frac{C\delta}{L}$ (i.e., the bandwidth delay product in packets). We can put together the observations earlier to conclude that the mean queueing delay normalized to the RTPD is given by

$$\frac{\tau_q}{\delta} = \left(\frac{nf(p)}{\Delta} - 1\right)^+ \tag{7.27}$$

It is easy to see that the same analysis can also be performed if the TCP sessions sharing a bottleneck link have different propagation delays (see Problem 7.4).

In Figure 7.22 we show simulation results that can be explained with the analysis that we have described. There are $n = 10$ (or 20 or 40) TCP-controlled file transfers sharing a bottleneck link of speed $C = 100$ Mbps, in a path of bandwidth delay product $\Delta = 400$ packets. The packet length is $L = 1$ KB; hence it can be seen that $\delta = 32$ ms. Let us consider the case of 40 file transfers. When $p = 0$ there is a large queue length in the buffer, the link is full, and each transfer obtains a throughput of $\frac{C}{40} = 2.5$ Mbps $= 312.5$ KBps. As the drop probability p increases, the TCP windows decrease and consequently the queue length in the buffer decreases. Beyond $p = 0.015$ the buffer empties out and the transfer throughput begins to decrease because of inadequate window per transfer. Note that the throughput curve for each n starts flat for small p and then meets a common curve; this is the curve $\frac{Lf(p)}{\delta}$.

From the foregoing discussion it is clear that, for each n, we would wish to operate the system at a value of p a little less than the value at which the link buffer empties out, because this minimizes queueing while obtaining maximum utilization of the link. One can imagine the RED drop profile or the ECN marking profile (see Figure 7.20) superimposed on the mean queue-length plots in Figure 7.22. Let us write this profile as

$$p = g(q) \tag{7.28}$$

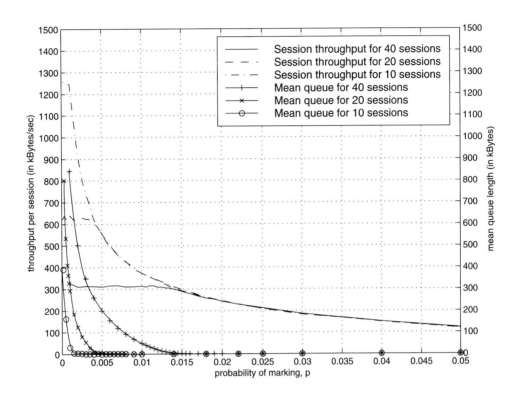

Figure 7.22   Performance of several file transfers at a bottleneck link with random packet drop; simulation results. Average throughput per transfer (left scale) and the mean queue length (right scale) are plotted as a function of the random drop probability p. The link speed is $C = 100$ Mbps, the packet length $L = 1$ KB, $W_{max} = 50$ KB, and the bandwidth delay product in packets is 400. There are 10, 20, or 40 connections. Adapted from Cherian [57, Figure 3.8].

That is, when the average queue is q, then the dropping or marking probability is $g(q)$. Writing $\tau_q = \frac{q}{C}$, we can seek a pair of values (p, q) that simultaneously solves Equations 7.27 and 7.28. Note that the problem of obtaining such a solution can also be expressed as solving for q in the following equation:

$$q = L\Delta \left(\frac{nf(g(q))}{\Delta} - 1 \right)^+$$

which has the form of a *fixed-point* solution. If such a solution exists, then the system can be expected to stabilize at this point. The slope of the profile, η, will need to be chosen to achieve a stable operating point, as discussed earlier in this section in the dynamical system model. We have discussed the analysis for a single link, but it can be performed for a network with arbitrary topology, carrying infinite-volume transfers on arbitrary routes, with each link implementing RED or ECN. In this general situation, the variables are (p_l, q_l) at each link l. As before, a fixed-point problem can be set up and, if its solution exists, the system can be expected to operate at that point.

7.6.7   Randomly Arriving, Finite-Volume Transfers

All the models for elastic transfers discussed thus far are concerned with the transfer of a fixed number of infinitely long files. Such models can be useful when the network carries large file transfers and the number of file transfers changes infrequently, allowing the TCP dynamics to converge in between the changes. In such a situation, for each number n of files, we can obtain the throughput obtained by each transfer. If we can then obtain the distribution of the random number of transfers, then we can obtain the average throughput over time. In practice, however, requests for file transfers (FTP requests, e-mails, and Web transfers) arrive randomly in time, and each such request transfers a finite random volume of data that can range from a few bytes to hundreds of megabytes. The number of transfers sharing a link varies randomly and rapidly.

Let us consider the simplest network scenario, as shown in Figure 7.23. Several clients are transferring files from several servers over a wide area link;

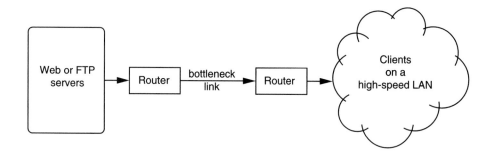

Figure 7.23 Several file transfer clients downloading files from servers across a link, the transfers being controlled by TCP.

the servers and clients are attached to the link by very high-speed local area networks. Hence the performance of the transfers is governed by the link connecting the high-speed LANs. The clients randomly generate requests for transfers of finite volumes of data from randomly chosen servers. In general, the request generation and file size processes are complex. For example, while browsing the Web, users begin *sessions* at random times. During a session, a user may visit several Web sites. At each such site the user might *download* several Web pages, each page comprising a main text part and several associated images. Between downloading each Web page, the user typically spends some time reading the page and thinking about it; this is often called a *think* time.

A considerable number of measurements have been made, and sophisticated statistical models are now available for modeling the manner in which users visit Web pages, the distribution of the number of *objects*, and their sizes, on each Web page, and the distribution of the think times between Web page accesses. There is one general conclusion that these statistical modeling efforts have reached: The instants of arrivals of user sessions are well modeled by a Poisson process. Another observation made in these studies is that the distribution of the file transfer sizes is *heavy-tailed* (i.e., $\Pr(V > v) \sim \frac{1}{x^\alpha}$ for $1 < \alpha < 2$). Such a distribution is called *Pareto* and has a finite mean but infinite variance.

In this context we would like to obtain some measure of performance of the transfers over the link. Because the transfers are of finite amounts of data and because the number of simultaneous transfers over the link is a random process, the definition of a performance measure is itself an interesting question.

Let us begin by defining some notation we use to define some performance measures and carry out some initial analysis on them. For $t \geq 0$, and $k \geq 1$, define the following:

a_k: the arrival instant of the kth transfer request.

$A(t)$: the number of transfer request arrivals in $(0, t]$.

λ: the transfer request arrival rate (i.e., $\lambda := \lim_{t \to \infty} \frac{A(t)}{t}$).

V_k: the volume of the kth requested transfer (in bytes); if some requests arrive in batches, then they are assumed to be indexed in some way.

ρ: the normalized offered byte rate into the link (i.e., $\rho = \frac{\lambda E(V)}{C}$, where C is the link speed in bytes per second).

$N(t)$: the number of ongoing transfers at time t.

L: the packet length used in the window-controlled transfer.

S_k: the transfer time of the kth transfer; that is, the kth transfer exists in the interval $(a_k, a_k + S_k)$.

$W_k(t)$: the window of the kth transfer at time t (in packets); this is 0 if the transfer does not exist at t.

$W(t)$: the total window of all the ongoing transfers at time t (in packets) (i.e., $W(t) = \sum_{k=1}^{\infty} W_k(t) I_{\{t \in (a_k, a_k + S_k)\}}$).

D_j: the delay between the insertion of the jth packet into the network and the receipt of its ACK.

$r_k(t)$: the transfer rate obtained by transfer k at time t; we define $r_k(t) = 0$ for $t \notin (a_k, a_k + S_k)$.

$R(t)$: the total transfer rate applied to all the transfers at time t (i.e., $R(t) = \sum_{k=1}^{\infty} r_k(t) I_{\{t \in (a_k, a_k + S_k)\}}$).

The most natural definition of the average throughput of the transfers is the average throughput obtained by each transfer, averaged over all the transfers. We refer to this measure as γ (bytes per second):

$$\gamma = \lim_{n \to \infty} \frac{1}{n} \sum_{k=1}^{\infty} \frac{V_k}{S_k}$$

If the sequence of random vectors $(V_k, S_k), k \geq 1$, converges in distribution to the random vector (V, S), then we will have

$$\gamma := \mathsf{E}\left(\frac{V}{S}\right) \tag{7.29}$$

In general, it is difficult to characterize the joint distribution of (V, S) and to calculate γ from this distribution. In some cases a numerical calculation can be carried out.

A simpler performance measure is the following one (again, the units are bytes per second):

$$\theta := \frac{\mathsf{E}(V)}{\mathsf{E}(S)} \tag{7.30}$$

For a specific model for the bandwidth-sharing between the ongoing transfers, we later obtain a relationship between γ and θ (see Exercise 7.11). The following result is more general.

We use Brumelle's theorem (Theorem 5.2) to write the following equation:

$$\lim_{t \to \infty} \frac{1}{t} \int_0^t W(u)\,du = \lambda \lim_{n \to \infty} \frac{1}{n} \sum_{k=1}^{n} \int_{a_k}^{a_k+S_k} W_k(u)\,du$$

$$= \lambda \lim_{n \to \infty} \frac{1}{n} \sum_{k=1}^{n} \left(\frac{1}{S_k} \int_{a_k}^{a_k+S_k} W_k(u)\,du \right) \cdot S_k$$

Define

$$W_{avg} := \lim_{t \to \infty} \frac{1}{t} \int_0^t W(u)\,du$$

$$\overline{W}_k := \frac{1}{S_k} \int_{a_k}^{a_k+S_k} W_k(u)\,du$$

Thus, W_{avg} is the time average of the sum of the windows of the ongoing transfers, and $\overline{W}_k$ is the average window of the kth transfer over its sojourn in the system. Assuming that the random vector $(\overline{W}_k, S_k)$ converges in distribution to $(\overline{W}, S)$ and that $\lim_{n \to \infty} \frac{1}{n} \sum_{k=1}^{n} \left(\frac{1}{S_k} \int_{a_k}^{a_k+S_k} W_k(u)\,du \right) \cdot S_k = \mathsf{E}\left(\overline{W} \cdot S \right)$, the preceding expression can be written more compactly as

$$W_{avg} = \lambda \mathsf{E}\left(\overline{W} \cdot S \right) \tag{7.31}$$

Because L is the packet length, and assuming that every file is an integral number of packets, the packet arrival rate is given by $\frac{\lambda \cdot \mathsf{E}(V)}{L}$. (In general, this expression is approximate because the last packet generated by a transfer often is not of length L.) Because W_{avg} is the time average number of packets in the network, writing $\mathsf{E}(D) = \lim_{n \to \infty} \frac{1}{n} \sum_{k=1}^{n} D_k$ as the average RTT over packets, we obtain (using Little's theorem (Theorem 5.1))

$$W_{avg} = \frac{\lambda \cdot \mathsf{E}(V)}{L} \mathsf{E}(D)$$

which, on using Equation 7.31, yields

$$\frac{\mathsf{E}(V)}{L} \cdot \mathsf{E}(D) = \mathsf{E}\left(\overline{W} \cdot S\right) \tag{7.32}$$

Hence, in general, we have the following result for the measure θ for window-controlled file transfers:

$$\theta = \frac{\mathsf{E}\left(\overline{W} \cdot S\right)}{\mathsf{E}(D) \cdot \mathsf{E}(S)} \cdot L \tag{7.33}$$

In particular, if the window protocol uses a fixed window size w, then $\overline{W} = w$, and we obtain

$$\theta = \frac{w}{\mathsf{E}(D)} \cdot L \tag{7.34}$$

Note that even though we might expect the expression to be of this form, it cannot be obtained by applying Little's theorem to a file transfer, because here the file transfers are finite and hence the limits in Little's theorem cannot be taken.

Another performance measure that can be defined is the time average bandwidth share per ongoing transfer, averaged over those times when there is an ongoing transfer. For specific bandwidth-sharing models, this measure has useful properties. We denote this measure by σ and define it as

$$\sigma := \lim_{t \to \infty} \frac{\int_0^t \frac{1}{N(u)} \left(\sum_{k=0}^{\infty} r_k(u)\right) I_{\{N(u) \geq 1\}} \, du}{\int_0^t I_{\{N(u) \geq 1\}} du}. \tag{7.35}$$

In this expression, the integrand in the numerator is the total transfer rate at time u divided by the number of ongoing transfers, if there are any sessions at time u. The denominator is the total duration, in the interval $[0, t]$, during which there are ongoing transfers.

Exercise 7.10

Suppose that time intervals between the transfer request arrival instants are i.i.d. (i.e., the arrival instants form a renewal process; see Appendix D) and that the transfer volumes are i.i.d. Show that $\lim_{\lambda \to 0} \sigma = \lim_{\lambda \to 0} \theta$. Hint: Notice that as $\lambda \to 0$ the transfers take place individually, with no overlap between the transfers.

Ideal Bandwidth Sharing: The Processor-Sharing Model

Let us assume that in each session only one file is downloaded, after which the user goes away. Thus the transfer request model reduces to the following simple form: Requests for file transfer arrive in a Poisson process; the request sizes are independently and identically distributed with some distribution.

Figure 7.24 shows a schematic of how traffic flows in the network shown in Figure 7.23. We can view the requested files as being queued at the servers, waiting as they are gradually transferred to the clients. In Figure 7.24, the portions of the files remaining in the servers are shown broken up into fixed-length packets. Some of the packets are in transit; these would be either in the link buffer or in "flight." The total number of packets in transit is the sum of the windows of the transfers. When the propagation delay between the servers and the clients is zero (and the reverse direction of the link is not carrying any other traffic), we can assume that the sender gets the acknowledgment of a packet as soon as the packet is served at the link, and hence the sender releases the next packet as soon as one of its packets gets served. Furthermore, if we assume that timeouts rarely occur, then the link bandwidth wasted in timeouts can be ignored. These assumptions lead to two conclusions:

- The sum of the windows of all the transfers is in the link buffer.
- The link does not idle if there are unfinished files to be transferred.

Suppose that the window-based packet transfer protocol uses a window of 1 packet. Then at all times there would be exactly as many packets in the link buffer as there are unfinished files, and the files would be served in a *round-robin* fashion. Even when TCP's adaptive window protocol is used, it has been observed that when the file size distribution is not heavy-tailed, then even though the packets are served in the order of their arrival to the link buffer, the packets from the active files are interlaced in such a way that the data from these files in the link buffer is served roughly in a round-robin manner. This motivates the assumption that the

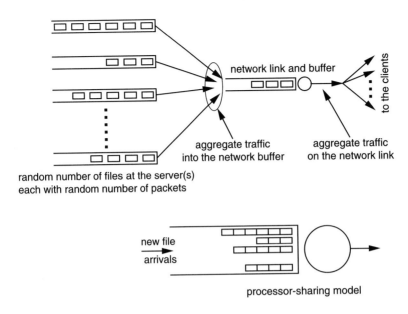

Figure 7.24 A schematic of the actual flow of packets in the network, showing the unfinished files, the flow of packets into the link buffer, the packets in the link buffer, and the departure process of packets into the clients (top). The bottom part shows a processor sharing model for bandwidth-sharing at the file level.

service of the data of the active transfers in the link buffer occurs in a processor-sharing (PS) fashion; that is, when there are $N(t) = n$ ongoing transfers, each file transfers its packets at $\frac{1}{n}$th of the link rate (i.e., $r(t) = \frac{1}{n}C$, where C is the link speed). Thus, considering a *fluid* model, we model the file transfer process, with *ideal* bandwidth-sharing, by a PS queue model. The unsent fragments of the files (which are partly in the link buffer and partly in the server) constitute the "customers" queued in the PS model (see the lower part of Figure 7.24). In standard queueing terminology, we have the M/G/1 PS model (see Appendix D).

Observe that as the propagation delay on the link increases, the time wasted for transmission of a packet over the link increases; this effect depends on the window size of a transfer and the number of other transfers sharing the link. Essentially, the situation with zero propagation delay can be viewed as the *bandwidth-limited* case. The PS model basically asserts that, at all times, the entire link rate is shared equally between the time-varying number of sessions; hence the limit on the transfer rates is imposed by the link's transmission rate.

On the other hand, when propagation delay becomes large we need to consider the *window limitation*. Consider, for example, a very light load, so that transfers do not overlap. In this case, when a new transfer arrives, its window gradually grows, with the link idling in between (recall Figure 7.18). Thus, with a large propagation delay, clearly we will not have a work-conserving server. In fact, data from the transfers will be served as if the server is of some time-varying rate $R(t)$, less than the actual link capacity, and this service is shared equally among the ongoing transfers. The PS model then corresponds to $R(t) = C$ for all t. This idea of a large propagation delay showing up as a time-varying service rate has been used in developing an analysis by studying $R(t)$ as a Markov regenerative process, but we do not present this analysis here.

Note that the restriction to small propagation delays still results in useful situations because, typically, large-capacity, small-length links are used for intracity connections. We proceed to use the PS model to analyze situations in which the propagation delays are small.

At any time instant t, an active transfer will have successfully transferred some data to its client, some of its data is in the link buffer (which is the current window size of TCP controlling the transfer of the file), and the remaining data is in the server waiting to be transferred. At any time instant we use the term *age* for the amount of data of a file successfully transferred to the client, and by *residual file*, or *residual service*, we mean the amount of data of the file yet to be transferred (i.e., the sum of the data in the link buffer and that remaining in the server). As data from a file is served (along with other files, in a PS manner), more data is pulled into the link buffer from the file server to replenish the windows and to account for any window growth. Eventually the amount of data of the file in the server reduces to zero, and the entire residual file is in the link buffer. Note that as long as the file is not fully transferred, a positive amount of it is always in the link buffer. Thus in terms of the PS queue model, the server is the link, and each customer in service is split between the file server and the link buffer.

Throughput vs Load

For an M/G/1 PS queue, we have $\mathsf{E}(S) = \frac{\mathsf{E}(V)}{(1-\rho)}$, yielding $\theta = 1-\rho$. This is insensitive to the distribution of V (see Appendix D).

Exercise 7.11
Show that for the PS model, $\gamma \geq \theta$. Hint: Use Jensen's inequality in the expression for γ, and then use the conditional mean sojourn time for the PS model given the file size (see Appendix D).

With the PS model for bandwidth sharing, $r_k(u) = \frac{C}{N(u)}I_{\{N(u) \geq 1\}}$; hence,

$$\sigma = C \lim_{t \to \infty} \frac{\int_0^t \frac{1}{N(u)} I_{\{N(u) \geq 1\}} \, du}{\int_0^t I_{\{N(u) \geq 1\}} \, du} \qquad (7.36)$$

Assuming that there is a steady state ($\rho < 1$), let $\pi(n)$ denote the probability of n transfers active in steady state. By dividing the numerator and denominator terms inside the limit in Equation 7.36 by t and then letting $t \to \infty$, it follows that $\sigma = \frac{1}{1 - \pi(0)} \sum_{n=1}^{\infty} \frac{\pi(n)}{n}$. It is well known (see [300]) that, for the PS queue, the stationary probability distribution $\pi(n)$ is insensitive to the distribution of service time (in this problem this translates to insensitivity to the distribution of file sizes) and is given by $\pi(n) = (1 - \rho)\rho^n$, $n \geq 0$. A little algebra then shows that (after normalizing by link capacity)

$$\sigma = \frac{1 - \rho}{\rho} \ln \frac{1}{1 - \rho} \qquad (7.37)$$

Thus σ has a simple, closed-form formula for the PS model, whereas γ as defined in Equation 7.29 does not have a closed-form expression even for the PS model. We can, however, numerically compute γ from the conditional distribution of the sojourn time given the file size; this is known in the literature in the form of a Laplace transform.

Figure 7.25 shows σ and γ (normalized to link capacity) for different file size distributions for an ideal PS queue with a unit capacity server. The values for γ are obtained from simulating the PS queue, and σ is obtained from Equation 7.37. The figure also shows θ. Whereas the measure σ has been proved to be insensitive to the file size distribution, γ is not. It is seen from Figure 7.25, however, that γ is very close to σ and also is not very sensitive to the file size distribution (and this holds even for heavy-tailed file size distributions). We thus see that σ, although being computationally simple and insensitive to file size distribution, is also a close approximation of γ for a variety of service distributions, including some approximations of heavy-tailed distributions. Notice also that θ is a rather poor approximation of the more meaningful measure—namely, γ.

We can draw a couple of additional insights from the plots in Figure 7.25.

- *Increasing efficiency with increasing C:* Note that the throughput is normalized to the link capacity, C. If there is a certain target throughput

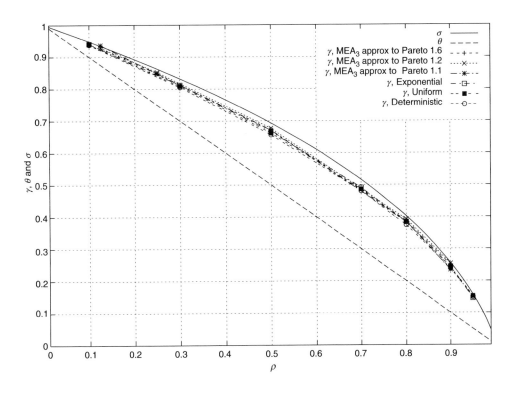

Figure 7.25 σ (from Equation 7.37), θ ($= 1 - \rho$) and γ (from fluid simulations of the PS model), all normalized to the link speed, C, vs ρ for various file size distributions. "MEA$_3$ approx to Pareto $x.x$" means a mixture of three exponentials approximation to a Pareto distribution with parameter $x.x$.

value (unnormalized) that an operator wishes to give its customers (e.g., 20 KBps), then for a given value of the link capacity, the curves yield the maximum normalized load that the link should carry. If the link capacity is increased, however, then the *normalized* target throughput decreases and the curves show that a *larger* normalized load can be carried. Thus higher-speed links yield better statistical multiplexing efficiency.

- *Throughput goes to 0 as $\rho \to 1$:* As the offered load approaches the capacity of the link, the average transfer throughput goes to 0. This can be understood from the PS model, from which we can infer that the number

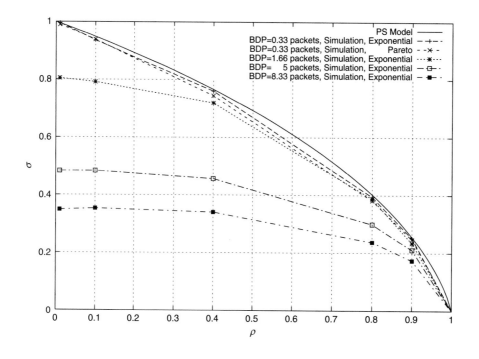

Figure 7.26 σ vs ρ (σ **normalized to** *C,* **the link capacity) obtained from the PS model and from** *ns* **simulations for a 10-Mbps link with various propagation delays. "PS Model" means results from Equation 7.37. The file size distribution and BDP used are given in the legend.**

of ongoing transfers $N(t)$ blows up as $\rho \to 1$, and hence the throughput that each transfer obtains goes to 0.

Figure 7.26 shows plots of σ (normalized to the link speed, C) versus the normalized link load, ρ. The solid line is Equation 7.37, and the other plots are obtained from simulations (using the popular network simulation package *ns*). In the simulation experiments, the packet size of TCP is fixed at 1500 bytes, the mean file transfer size is fixed at 30 Kbytes = 20 packets, and the TCP acknowledgment size is 40 bytes. We use a link of capacity C = 10 Mbps, and the link buffer is B = 125 packets. We vary the load on the link (ρ) by varying the file transfer request arrival rate (i.e., λ). We estimate σ from the simulation by estimating the time averages in the definition of σ.

We find from Figure 7.26 that the formula for σ obtained from the PS model matches very well with simulations when the BDP is small; note that for a 10 Mbps link and a TCP packet size of 1500 bytes, a BDP of 0.33 packets is obtained for a link of length 40 Km ($= \frac{1500 \times 8 \times 0.33}{10 \times 10^6}$). On the other hand the PS model becomes progressively worse if BDP is increased to 1.66 packets (200 Km), 5 packets (600 Km), or 8.33 packets (1000 Km). As discussed earlier, although motivating the processor-sharing model for this problem, this behavior is as expected, because the non-idling assumption, inherent in the PS model, no longer holds for large BDPs. We discuss the large-BDP case further in the last part of this section. Analytical models for the medium- and large-BDP cases are also available. We emphasize that although the discussion that motivates the PS model is presented only in terms of the propagation delay, the parameter that really matters is the BDP. Thus a 100 Mbps link of length 4 Km would also be well approximated with the PS model.

It can also be noted from Figure 7.26 that the performance for exponentially distributed files and for Pareto distributed files is almost the same for a small BDP. This is also the conclusion we would reach from the PS model, because the formula for σ is insensitive to the distribution of the file size. This observation does depend on the buffer size being sufficient to prevent excessive loss and hence timeouts. It has been found, however, that for larger BDPs and for small buffers, the performance *is* sensitive to the distribution of the file sizes.

Very-High-Speed Wide Area Networks

An interesting feature of the plots shown in Figure 7.26 is that as BDP increases, the performance flattens out as a function of ρ. This insensitivity of performance with ρ is even more striking when the BDP is very large; see Figure 7.27, where we plot the measure γ vs the link load ρ. The throughput scale in this plot is logarithmic, and note the sensitivity of throughput to ρ when BDP $= 0$. The throughput does drop significantly as BDP increases, but it remains constant up to ρ beyond 0.9; note, however, that 0.004 of 1 Gbps (see the curve for BDP $= 1000$) is still a substantial file transfer throughput. The insensitivity of throughput with ρ for a large BDP is explained as follows. With randomly arriving, finite-volume transfers, each arriving session goes through slow-start, with the corresponding rapid window increase. As we saw in Figure 7.18, queueing occurs because a window increment causes two packets to be inserted into the bottleneck link buffer. For large BDPs, the successive bursts of packets (caused by TCP's window increments) are spaced far apart, and the window increments of the different TCP transfers are unlikely to overlap. Hence the queueing in the link buffer is small. Thus, for a given ρ, as

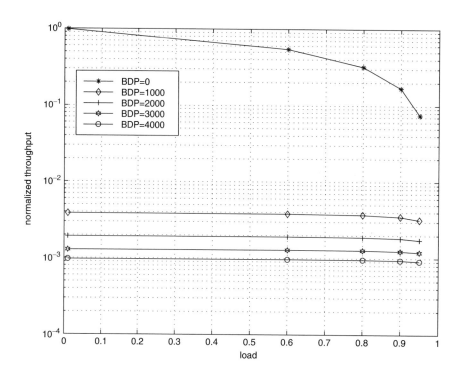

Figure 7.27 Average throughput (γ) normalized by link capacity vs load, ρ, for a file size distribution that is a mixture of three exponential distributions with means 2, 20, and 200 packets, such that the resulting mean is 20 packets. Results obtained from *ns* simulations. The BDP is in packets.

the BDP increases, the queueing delay becomes small compared with the RTPD, and hence the throughput values stay close to the 0-load value, for a wide range of loads. For a given BDP, when the load ρ becomes very close to 1, however, then queueing delay becomes large and comparable to the BDP. From Figure 7.27 we observe that, for large BDPs, this does not happen until the load is well beyond 0.9. Because heavier-tailed distributions (more typical of the Internet) grow their windows for a longer time, such transfers cause larger queue lengths, and we need to have larger BDPs for the queueing delay to become insensitive to BDP. Note that a BDP of 1000 packets will be obtained over a 1-Gbps link spanning a distance of 1200 Km, with a TCP packet size of 1500 bytes. Links of such speeds over these long distances are becoming common with the widespread deployment of optical backbones.

These observations suggest that we can engineer very-high-speed wide area backbones simply by taking the capacity of each link to be, say, 90% of its actual capacity and then packing routes into the backbone topology. Another way of saying this is that we can aim for 10% *overprovisioning*. We can then expect that the performance of the TCP transfers will be governed by the BDP alone. (Of course, to obtain substantial TCP throughputs over large BDP paths, TCP end points must negotiate the extended window option.) This viewpoint motivates the *packing* approach that is often used in the design of routing in packet networks (see Chapter 15).

7.6.8   Long-Range Dependence of TCP-Controlled Traffic

In Section 5.11 we introduce the concept of long-range dependence in the context of stream traffic, such as that emanating from a real-time video source. In practice, the Internet predominantly carries TCP-controlled elastic traffic, and measurements have shown that traffic on Internet links is long-range-dependent. The PS model can be used to show that the origin of this phenomenon can be traced to the heavy-tailed distribution of file transfer sizes. In the following, we assume that conditions for the applicability of the PS model hold (see Section 7.6.7).

There are two aggregate flow processes in our single link (see Figure 7.24): one that flows into the link buffer from the servers, and another one that flows out of the buffer on to the link. The process of packet arrivals entering the link buffer (at the LAN–WAN router interface) comprises transmissions from the servers in response to acknowledgments. This is a complex process, and we refer to papers in the Bibliography for its analysis. On the other hand, the process on the network link is simply the *departure process* of the PS model. The link alternates between carrying traffic at the full link rate and idle periods, and therefore the flow process on it corresponds to the busy–idle process of the PS model. It can be shown that this process is long-range-dependent if the file size distribution is Pareto distributed, and it is short-range-dependent if the file size distribution has a finite second moment. Because the analysis requires advanced techniques, we provide only the outline and the final results.

Long-Range Dependence in the Link Departure Process (⋆)

The busy–idle process of an M/G/1 PS queue is the same as that of any M/G/1 queue having a non-idling (work-conserving) service discipline. Consider the busy–idle process of such an M/G/1 queue. Let $\{X(t), t \geq 0\}$ denote a stochastic process that is 1 when the queue is busy, and 0 when the queue is idle. $\{X(t)\}$ is called an *alternating renewal process*. We want to study the autocovariance function of this process in order to conclude whether the process is long-range-dependent.

Let λ denote the rate of the Poisson arrival process. This corresponds to the arrival rate of file transfer requests (see Figure 7.24). Let B denote the random variable for the work brought by a customer, and $\tilde{b}(s)$ denote the Laplace transform of its density. In the network context, this work is the file size to be transferred divided by the bit rate of the link. Thus $B = \frac{V}{C}$, where V is the random variable for the file sizes, and C is the link capacity. We assume throughout that $\mathsf{E}(B) < \infty$. Let $G(\cdot)$ denote the busy period distribution, and $\tilde{g}(s)$ the Laplace transform of its density function. Finally, as usual, we let $\rho = \lambda \mathsf{E}(B)$.

Except where explicitly noted, we consider the *stationary* version of the process $\{X(t), t \geq 0\}$. This stationary version is characterized as follows (see [67, page 85]): $X(0) = 1$ with probability ρ, and $X(0) = 0$ otherwise. If $X(0) = 1$, then the first 1-period of the process has distribution $G_e(\cdot)$, the excess distribution of the distribution $G(\cdot)$ (see Appendix D):

$$G_e(t) = \frac{1}{\int_0^\infty (1 - G(x))\, dx} \int_0^t (1 - G(u))\, du$$

On the other hand, if $X(0) = 0$, then the first 0-period is exponentially distributed with parameter λ. After the first period is obtained in this way, the process alternates between the states 0 and 1, staying in state 0 for a duration distributed exponentially with mean $\frac{1}{\lambda}$, and staying in state 1 for a duration distributed as $G(\cdot)$.

Define

$$p_e(t) = \mathsf{Pr}(X(t) = 1 \mid X(0) = 1)$$

We wish to study the autocovariance function $r(\tau)$ of $\{X(t)\}$:

$$r(\tau) = \mathsf{E}(X(t)X(t + \tau)) - \rho^2$$

Clearly, because $X(t) \in \{0, 1\}$, then

$$r(\tau) = \rho(p_e(\tau) - \rho)$$

Define $x(s) = 1 - \tilde{g}(s)$. Using some renewal theoretic arguments, it can be shown that

$$\tilde{r}(s) = \frac{\rho(1-\rho)}{s} \left(1 - \frac{x(s)}{\mathsf{E}(B)\,(s + \lambda x(s))}\right)$$

Our aim is now to study the properties of $r(t)$ for large t via its Laplace transform. Such a study can be done via so-called Tauberian theorems, which relate the behavior of the Laplace transform near 0 to that of the time function near ∞. We provide only the results here.

Theorem 7.2
  If $0 < \mathsf{E}(B^2) < \infty$, then

$$\int_0^\infty r(u)\,du = \frac{\rho\mathsf{E}(B^2)}{2\mathsf{E}(B)} \qquad\blacksquare$$

From Theorem 7.2 we conclude that $r(t)$ is summable when the file size distribution has a finite second moment. It follows that in such a situation the aggregate process of (ideally) flow-controlled elastic flows cannot be long-range-dependent (see [30]).

Measurements have shown that the file sizes transferred over the Internet are Pareto distributed. Let us consider the following Pareto distribution:

$$1 - B(x) = \begin{cases} 1 & \text{for } 0 \le x \le 1 \\ x^{-\alpha} & \text{for } x \ge 1 \end{cases}$$

where $1 < \alpha < 2$. For this distribution we have $\mathsf{E}(B) = \frac{\alpha}{\alpha-1}$ and $\mathsf{E}(B^2) = \infty$. It is easily seen that $\tilde{b}(s)$ is given by

$$\tilde{b}(s) = \alpha \cdot s^\alpha\, \Gamma(-\alpha, s)$$

where $\Gamma(\cdot, \cdot)$ is the incomplete Gamma function.[5] Furthermore, by using the series expansion of the incomplete Gamma function (see [127, page 941]), we can get

$$\tilde{b}(s) = \alpha s^{\alpha} \left\{ \Gamma(-\alpha) - \sum_{n=0}^{\infty} \frac{(-1)^n \, s^{-\alpha+n}}{n! \, (-\alpha+n)} \right\}$$

For the M/G/1 busy period, we have the following relationship between the Laplace transform of the density of the busy period and the Laplace transform of the density of the service time (see [173, page 212]):

$$\tilde{g}(s) = \tilde{b}(s + \lambda(1 - \tilde{g}(s)))$$

Combining these two facts, and after some calculations, the following result can be obtained.

Theorem 7.3
  For $1 < \alpha < 2$,

$$\tilde{r}(s) \sim \rho(1-\rho)^{(2-\alpha)}(\alpha-1)\Gamma(-\alpha)\, s^{-(2-\alpha)} \qquad s \to 0_+ \qquad\blacksquare$$

Again, a Tauberian theorem can be used to conclude that

$$\int_0^t r(u)\, du \sim \frac{\rho(1-\rho)^{(2-\alpha)}(\alpha-1)\Gamma(-\alpha)}{\Gamma(3-\alpha)}\, t^{(2-\alpha)} \qquad t \to \infty$$

Thus, as $t \to \infty$, the autocovariance function $r(t)$ behaves as $t^{-(\alpha-1)}$, and hence (with $1 < \alpha < 2$) the process $\{X(t)\}$ is long-range-dependent with $\beta = (\alpha - 1)$ (see Section 5.11).

Buffer Behavior with LRD Traffic
It is pointed out in Section 5.11 that if a buffer has an LRD arrival process, then the stationary distribution will have subexponentially decaying tails. This fact,

---

[5]$\Gamma(a,y) = \int_y^a e^{-u}\, u^{a-1}\, du$

along with the observations and analysis provided in this section, has often been taken to imply that buffers in the internet may need to be very large. However, these conclusions are based on an open-loop queueing analysis of a buffer fed by an LRD source. A buffer in the Internet is in a closed loop (TCP's window control loop), however, and it is the closed-loop traffic into the buffer that is LRD.

The issue is clearly illustrated by the example in Figure 7.28, where we show a queue with a unit rate server. Customers require exponentially distributed service with mean $\frac{1}{\mu}$. In Figure 7.28(a), there is one customer that repeatedly visits the queue; observe that the visit instants of the customer form a Poisson process of rate μ. In Figure 7.28(b), there is a Poisson process of rate μ arriving into the queue. This is an M/M/1 queue with "ρ" $= 1$ and hence is unstable (the number of customers in the queue is a null recurrent Markov chain; see Appendix D). Thus in each case the buffer receives a Poisson process of arrivals, but in the closed-loop case it is very well behaved, whereas in the open-loop case it blows up.

This difference between the open-loop and the closed-loop behavior of the link buffer with Adaptive Window Protocol (AWP)-controlled traffic has been studied. It has been shown that, for a small propagation delay network with a single bottleneck link, even though the input process into the link buffer may be LRD, the queue behavior depends on the way the AWP protocol adjusts its window (see [171] and [172]). An aggressive adjustment (as in slow-start in TCP) can lead to a stationary distribution that is the same as the open-loop case, but a slow window increase (as in congestion avoidance in TCP) will lead to a distribution with tails lighter than what would be expected from an open-loop analysis.

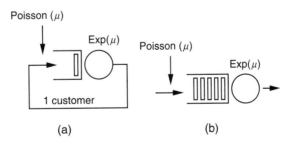

(a)   (b)

Figure 7.28 A simple queueing model illustrating the difference between open-loop and closed-loop buffer behavior for the same input traffic.

7.7 Bandwidth Sharing in a Network

In much of the previous discussion in this chapter, we basically deal with the sharing of a single bottleneck link. We consider the problem of a fixed number of sessions with infinitely large files to transfer, and we discuss the situation of randomly arriving, finite-volume transfers. When a single link is the only limiting resource, it is reasonable to say that equal sharing is fair. *Bandwidth sharing* in a network, however, is a more complex issue. In a network, requiring equal rates leads to unutilized bandwidth in some links. Furthermore, many notions of fairness can be defined as being desirable, or achievable by specific congestion control and bandwidth-sharing mechanisms. We discuss this at length in this section and also motivate the use of distributed algorithms for achieving fair bandwidth sharing.

As a simple example, consider the network and session topology shown in Figure 7.29. A network comprising two links in tandem is shared, over a certain time period, by three sessions. Session 1 and session 2 are transferring data separately in the direction shown on the two links. Session 3 spans both the links and is sharing them in the same direction as the data flow from the other two sessions. The link capacities are 1 Mbps and 2 Mbps, as shown. We notice that if equal sharing is the objective, then all sessions should get a transfer rate of 0.5 Mbps, because with this rate the 1 Mbps link is saturated. This allocation of rates leaves 1 Mbps idle on the 2 Mbps link. No session is any worse off if this 1 Mbps is also allocated to session 2. In fact, for given volumes of data to be transferred by each session, the second allocation will ensure that all transfers are completed no later than with the equal allocation. Hence certainly the second allocation is better than the first. In fact the first allocation does not satisfy a very simple objective of resource sharing, namely, *Pareto efficiency*, which, somewhat informally, is defined as follows.

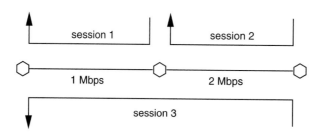

Figure 7.29   Fair sharing is not equal sharing in a network of links and transfers.

Definition 7.1

An allocation of resources in a system is Pareto efficient if there does not exist another allocation in which some individual is better off and no individual is worse off.   ∎

The example of Figure 7.29 illustrates another point. Notice that with the rate allocations of 0.5 Mbps, 1.5 Mbps, and 0.5 Mbps (respectively, for routes 1, 2, and 3), the total amount of user traffic that the network carries is 2.5 Mbps. Suppose we ignored session 3. Then we could allocate 1 Mbps to session 1, and 2 Mbps to session 2, yielding a total carried user traffic of 3 Mbps. We see that there is a trade-off between fairness and the total amount of user traffic that the network carries. In this example, session 3 requires resources from both links, and hence for each unit of session 3 traffic carried we take away a unit of bandwidth from each of the two links. Yet we need to be fair in some sense between the various sessions. This point is explored further via a utility maximization approach in Section 7.7.2.

Figure 7.1 shows a more general network carrying several sessions between Web servers and Web clients. At any time there will be a number of such elastic sessions sharing the network. The routing in the network will determine the paths that the packets of the elastic sessions will take. To begin with, we assume that the network topology, the set of sessions sharing it, and the routes the sessions follow are fixed. *Each session is transferring a single, infinitely large volume of data.* In practice, as we have seen, sessions have finite lifetimes, and new sessions arrive; hence even if the network topology and routing do not change, the session topology is constantly changing. Thus, in a sense, we are considering a situation in which the session topology is *slowly varying*. We make this simplifying assumption to develop an understanding of some of the basic issues in bandwidth sharing and congestion control. To continue the discussion, we first establish some notation.

$\mathcal{L}$: the set of links; each link is assumed to be unidirectional; hence a full-duplex link between two nodes is viewed as two links, one in each direction.

C_l: the capacity of link $l \in \mathcal{L}$ (assumed here to be the same in both directions; C_l is viewed as the capacity of link l *available* to the elastic sessions flowing through link l).

C: the vector of link capacities $(C_1, C_2, \ldots, C_{|\mathcal{L}|})$.

$\mathcal{S}$: the set of sessions; each session is unidirectional and has a route, and hence the following notation can be defined.

$\mathcal{L}_s$: the set of links used by session $s \in \mathcal{S}$.

$\mathcal{S}_l$: the set of sessions through link $l \in \mathcal{L}$.

n_l: the number of sessions through link $l \in \mathcal{L}$.

r_i: the rate of the ith session, $1 \leq i \leq |\mathcal{S}|$; $\mathbf{r} = (r_1, r_2, \ldots, r_{|\mathcal{S}|})^T$ denotes the *rate vector*.

It is assumed that the source of each session has an infinite amount of data to send (called the *infinite backlog* assumption), and hence, for a given rate vector $\mathbf{r}$, source s continuously sends data at the rate r_s. For a rate vector $\mathbf{r}$, and $l \in \mathcal{L}$, we denote the total flow through link l by

$$f_l(\mathbf{r}) = \sum_{s \in \mathcal{S}_l} r_s$$

Note that the 3-tuple $(\mathcal{L}, \mathcal{C}, \mathcal{S})$ characterizes an instance of the bandwidth-sharing problem. Thus we will say, for example, that the rate vector $\mathbf{r}$ is *feasible* for $(\mathcal{L}, \mathcal{C}, \mathcal{S})$ or that $\mathbf{r}$ is the max-min fair rate vector for $(\mathcal{L}, \mathcal{C}, \mathcal{S})$, and so on.

With this notation, it is clear that a rate vector $\mathbf{r}$ is feasible for $(\mathcal{L}, \mathcal{C}, \mathcal{S})$ if, $r \geq 0$ (i.e., element-wise nonnegative), and for all $l \in \mathcal{L}$,

$$f_l(\mathbf{r}) \leq C_l$$

This is a system of $|\mathcal{L}|$ linear inequalities and defines a convex polyhedral set in the $|\mathcal{S}|$ dimensional space of nonnegative real numbers. Clearly none of the rate vectors in the interior of this set is Pareto efficient, whereas a rate vector $\mathbf{r}$ such that for every s there is an $l \in \mathcal{L}_s$ such that $f_l(\mathbf{r}) = C_l$ is Pareto efficient (i.e., every session is bottlenecked at some link, and hence we cannot increase its rate without reducing the rate of some other session). There are several proposals for bandwidth-sharing or fairness objectives that identify one of these rate vectors as being optimal or fair in some sense.

In the foregoing framework, there are several questions that can be addressed.

- What are reasonable bandwidth-sharing, or fairness, objectives?

- For each such objective, can we characterize the vector of fair rates?

- In each case, is there a distributed and asynchronous algorithm that achieves the fair rate?

- Given a congestion control algorithm, such as TCP, can we characterize the rate vector it achieves? Is this the fair rate vector under some bandwidth-sharing objective?

We note that the framework we have set up views the flow of data from the source to the sink of session s as a fluid flowing at constant rate r_s. Given a rate vector $\mathbf{r}$ each session s continuously emits data at the constant rate r_s. In practice, the available capacity at a link will be a stochastic process. (For example, higher-priority stream sessions could be assigned some of a link's bandwidth, and when this bandwidth is not in use it is also dynamically assigned to elastic sessions.) In fact, when the session rates are dynamically adjusted (in an attempt to reach some target fair rates), the rate process itself is stochastic. Hence, queueing of data will be necessary at each link; in practice, all packet switches have buffers that can temporarily store arriving packets that cannot be sent immediately on their exit link. With a stochastic service process, it is clear that we should not aim at making any link a "bottleneck" in the sense that the total input rate into the link is equal to the average service rate of the link, because this leads to instability of the queue (see Section 5.6). Instead, the aim should be to seek session rates so that the rate of arrival of data into a link is *strictly* less than the average available service rate of the link. With this in mind, the capacity C_l, defined earlier for each link, is some statistic of the available capacity process that is less than the mean capacity; for example, this could be some fixed fraction of the mean or an effective service capacity, as defined in Section 7.4.1. The value of C_l can be estimated at each link; such an estimate will be used in an algorithm to be discussed in Section 7.7.1.

7.7.1 Max-Min Fair Bandwidth Sharing and Algorithms

The sharing objective of *max-min fairness* (MMF) was popular in the context of the ABR service in ATM networks (see Section 7.4.5). In this section we provide a characterization of MMF rates. This leads to a class of distributed algorithms for MMF bandwidth sharing in a packet network.

Definition 7.2

A feasible rate vector $\mathbf{r}$ is *max-min fair* (MMF) for $(\mathcal{L}, \mathcal{C}, \mathcal{S})$ if it is not possible to increase the rate of a session s, *while maintaining feasibility*, without reducing the rate of some session p with $r_p \leq r_s$. ∎

It is instructive to compare this definition with that of a Pareto efficient allocation in Definition 7.1. Consider again the example in Figure 7.29.

The allocation of 1 Mbps to session 1, 2 Mbps to session 2, and 0 Mbps to session 3 is Pareto efficient, because increasing the allocation of session 3 will require a decrease in the allocations to the other two sessions. However, this allocation is not MMF because we can increase the rate of session 3 by decreasing the allocations of sessions whose rates are *larger* than that of session 3. It can be checked that the rate allocation of 0.5 Mbps, 1.5 Mbps, and 0.5 Mbps, respectively, to routes 1, 2, and 3 satisfies Definition 7.2 and hence is MMF for that problem.

An important property of the MMF allocation is the following: Consider a feasible rate vector, and look at the smallest rate in this vector. The MMF rate vector has the largest value of this minimum rate. Furthermore, among all feasible rate vectors with this value of the minimum rate, consider the next larger rate. The MMF rate vector has the largest value of the next larger rate as well, and so on.

The MMF rate vector is characterized in terms of the notion of bottleneck links, which we define next.

Definition 7.3

Given a rate vector $\mathbf{r}$, a link l is said to be a *bottleneck link* for a session j if

(i) Link l is saturated (i.e., $f_l(\mathbf{r}) = C_l$) and

(ii) For all the sessions $s \in \mathcal{S}_l$, $r_s \leq r_j$; that is, every session in link l has a rate no more than that of session j.　　■

For example, consider again the network shown in Figure 7.29. The allocation of 0.5 Mbps, 1.5 Mbps, and 0.5 Mbps (respectively, for sessions 1, 2, and 3) is MMF. Now both the links are saturated, but, by definition, the 1 Mbps link is a bottleneck for session 3, whereas the 2 Mbps link is not. To see why the 1 Mbps link is called a bottleneck for session 3, notice that increasing its capacity will permit us to increase the MMF rate for session 3 (while, of course, reducing that for session 2, but because its rate is larger than that of session 3, this is allowed); on the other hand, increasing the capacity of the 2 Mbps link will not result in an increase in the MMF rate for session 3. The following theorem relates the MMF rate vector to the notion of bottleneck links. We leave the proof as an exercise.

Theorem 7.4

A feasible rate vector $\mathbf{r}$ is MMF if and only if every session $s \in \mathcal{S}$ has a bottleneck link.　　■

We note here that with max-min fair flow rates, even though every session has a bottleneck link, *not every link is a bottleneck for some session*. To see this, imagine adding a 10 Mbps link to the network in Figure 7.29, to the right of the 2 Mbps link, through which traffic from both session 2 and session 3 enters the 2 Mbps link. The MMF rates stay the same, and the new 10 Mbps link carries only 2 Mbps, and hence it is not even saturated.

The Link Control Parameters η_l, $l \in \mathcal{L}$

If a central entity knew the network topology, the session topology, and the capacities of all the links, then it could calculate the MMF rate vector using the following centralized algorithm. Initially all transfers are given 0 rates. At step 1, the capacity of each link is divided by the number of sessions through it; define, for $l \in \mathcal{L}$, $\eta_l(1) = \frac{C_l}{n_l}$. For each session s, now compute $r_s(1) = \min_{l \in \mathcal{L}_s} \eta_l(1)$; observe that the rate of session s can be increased to $r_s(1)$ without violating any link capacity constraint. This step will result in some links getting saturated, and hence the sessions using those links will not be able to increase their rates any further.

The capacity allotted to these sessions is now removed from every link that they pass through. The saturated links are also removed, and we have a new problem on the remaining links and the remaining, unbottlenecked sessions. The calculation is repeated, this time yielding $\eta_l(2)$, where l is an unsaturated link. The key observation is that $\eta_l(2) \geq \eta_l(1)$. The reason is that in the first iteration if some sessions and capacity were removed from link l, the amount of capacity removed per session would be less than the average capacity, by virtue of the way $r_s(1)$ is computed. Hence when the residual capacity is divided among the remaining sessions, this average (i.e., $\eta_l(2)$) is larger.

This iterative calculation is repeated, and at each step at least one session gets bottlenecked. The vector of η_ls is a nondecreasing sequence. The algorithm terminates in at most as many steps as there are sessions, yielding a rate vector $\mathbf{r}$ and a vector $\boldsymbol{\eta}$, which we call the vector of *link control parameters* (LCPs). Upon termination of the algorithm, every session passes through a saturated link, but not all links may be saturated. For the example shown in Figure 7.29, it can be seen that the LCP for the 1-Mbps link is 0.5 Mbps, and the LCP for the 2-Mbps link is 1.5 Mbps.

Exercise 7.12
Show that

    a. The rate vector $\mathbf{r}$ obtained by this centralized algorithm is MMF.

    b. The vector $\boldsymbol{\eta}$ is such that, for every s, $r_s = \min_{l \in \mathcal{L}_s} \eta_l$.

Note that the vector of LCPs is not unique if there are unsaturated links at the end of the iteration. Because the LCP of an unbottlenecked link does not constrain the MMF rate vector, any large enough value of LCP for these links will work. Thus we see that every saturated link has a finite, positive LCP, and the LCP of an unsaturated link, l, can be taken to be ∞, or a suitably large number (e.g., the total capacity—say, $C_l^{\max}$—of the link).

The LCPs are crucial to the development of a distributed algorithm. Suppose by some distributed algorithm each link is able to calculate its LCP. Then using a mechanism such as RM cells in the ABR service of ATM (see Section 7.4.5), we can compute the minimum value of the LCPs along each connection's path and feed it back to the source of the connection.

To develop such a distributed algorithm, we provide a useful result about LCPs. We show that a set of MMF rates for a problem $(\mathcal{L}, \mathcal{C}, \mathcal{S})$ can be obtained from a solution of a certain vector equation. This perspective motivates an approach to the design of distributed algorithms for computing the max-min fair share. We now show that a desired (not necessarily unique) vector of link control parameters is a solution of a certain vector equation.

Theorem 7.5

For the max-min fair bandwidth-sharing problem $(\mathcal{L}, \mathcal{C}, \mathcal{S})$, let $\tilde{\mathcal{L}}$ denote the set of links that are not bottlenecks for any session. Consider any vector $(\eta_l, l \in \mathcal{L})$ such that

$$\text{(i)} \quad \min_{j \in \mathcal{L}_s} \eta_j = \min_{j \in \mathcal{L}_s \setminus \tilde{\mathcal{L}}} \eta_j \quad \text{for all } s \in \mathcal{S}$$

$$\text{(ii)} \quad \sum_{s \in \mathcal{S}_l} \min_{j \in \mathcal{L}_s} \eta_j = C_l \quad \text{for all } l \in \mathcal{L} \setminus \tilde{\mathcal{L}}$$

Then the max-min fair rates are obtained as

$$r_s = \min_{j \in \mathcal{L}_s} \eta_j$$

Remark: In this theorem statement, condition (i) states that the minimum of the LCPs along the path of each session s is achieved at one of the bottleneck links that it traverses, and condition (ii) states that for each link l, if for each session that passes through that link the minimum of LCPs along its path is obtained, and if these minima are summed over the sessions passing through link l, then this sum is equal to the capacity of link l.

Proof: By Theorem 7.4 it is sufficient to show that with $r_s, s \in \mathcal{S}$, as defined in the theorem, every session $s \in \mathcal{S}$ has a bottleneck link. Consider any $s \in \mathcal{S}$. Let $l_s \in \mathcal{L}_s \backslash \tilde{\mathcal{L}}$ (this set is not empty because every session has at least one bottleneck link) be such that

$$\eta_{l_s} = \min_{j \in \mathcal{L}_s} \eta_j$$

The link l_s is saturated, by hypothesis (ii) of the theorem. Also,

$$r_s = \eta_{l_s}$$

It follows that, for all $q \in \mathcal{S}_{l_s}$ (i.e., the other sessions that share the link l_s with session s),

$$r_q = \min_{j \in \mathcal{L}_q} \eta_j \leq \eta_{l_s} = r_s$$

Hence by Definition 7.3, l_s is a bottleneck link for $s \in \mathcal{S}$.   ■

Consider the case in which $\tilde{\mathcal{L}}$ (defined in Theorem 7.5) is empty; that is, every link is a bottleneck for at least one session (for example, the network in Figure 7.29). For a vector of LCPs $\boldsymbol{\eta}$, define, for each $l \in \mathcal{L}$, the function $h_l(\boldsymbol{\eta}) = \sum_{s \in \mathcal{S}_l} \min_{j \in \mathcal{L}_s} \eta_j$ (i.e., the flow through link l for the LCP vector $\boldsymbol{\eta}$). Define the vector function $\mathbf{h}(\boldsymbol{\eta}) = (h_l(\boldsymbol{\eta}), \, l \in \mathcal{L})$. Then by Theorem 7.5, the max-min fair allocation can be obtained by solving

$$\mathbf{h}(\boldsymbol{\eta}) = \mathbf{C}$$

We require a distributed algorithm that computes the max-min fair rate and communicates it to the sources. Such an algorithm can be viewed as one that solves the preceding vector equation. The idea is to iteratively adjust the LCP at each link to achieve the equality of the flow rate through the link and the link's available capacity for elastic traffic. The important issue in the design is to choose the amounts of increase or decrease so that convergence to the max-min fair rates is ensured.

Link l updates its LCP η_l at periodic intervals indexed by k, yielding the sequence of LCP iterates $\boldsymbol{\eta}(k)$. Suppose that at each iteration k of the algorithm, link l has an available capacity estimate that is a random variable $C_l(k)$, which we assume can be written as $C_l(k) = C_l + \beta_l(k) + u_l(k)$. Here C_l is as we have defined before (a statistic of the available capacity process at link l), $\beta_l(k)$ is an error in the estimate (which goes to zero as $k \to \infty$), and $u_l(k)$ is zero mean and accounts for short-term variations and measurement noise. The link continuously estimates the available capacity for elastic traffic (i.e., the estimate is not restarted in each update interval), and hence if the available capacity process is stationary and ergodic, then $\beta_l(k)$ will go to 0 as k becomes large. Furthermore, each link l can measure the total flow through it, yielding the measurement $\hat{h}_l(k) = h_l(\eta(k)) + v_l(k)$, where $v_l(k)$ is zero mean measurement noise. In the ABR and ATM context, the link can read the rate of each session passing through it from the CCR (Current Cell Rate) field of the RM cells (see Section 7.4.5).

Motivated by the characterization of the MMF solution in Theorem 7.5, we aim to adjust the vector of LCPs so as to drive $C_l - h_l(\eta(k))$ to zero, given only measurements $C_l(k) - \hat{h}_l(k) = C_l - h_l(\eta(k)) + \beta_l(k) + \omega_l(k)$, where $\omega_l(k) = u_l(k) - v_l(k)$. For such a problem of finding the root of an equation given noisy observations, we can use a *stochastic approximation* algorithm. We write C_l^{max} as the total bandwidth of link l (this is used as a bound on the value of η_l, as explained earlier), and we use the notation $[x]_a^b$ for $\min\{b, \max\{x, a\}\}$. The stochastic approximation iteration for η_l at link l can then be written as

$$\eta_l(k+1) = [\eta_l(k) + a_l(k)(C_l(k) - \hat{h}_l(k))]_0^{C_l^{max}} \tag{7.38}$$

The sequence of gains $a_l(k)$, for each l, is a decreasing sequence and satisfies the following properties for each link l: $\sum_{k=1}^{\infty} a_l(k) = \infty$ and $\sum_{k=1}^{\infty} a_l(k)^2 < \infty$ (e.g., $a_l(k) = \frac{1}{k}$). The reducing gains of the update formula would suppress the effect of the noise term $\omega_l(k)$. It has also been shown that this property of the gains permits *asynchrony* in the updates, and hence the flow through link l at the kth iteration can be of the form $\sum_{s \in \mathcal{S}_l} \min_{j \in \mathcal{L}_s} \eta_j(k - \tau_s^{lj})$, where τ_s^{lj} accounts for the fact that the flow in link l from session s during the measurement interval is based on the value of η_j that is τ_s^{lj} iterations old. The truncation of each iterate is used to keep it within a bounded set. Under certain technical assumptions, the algorithm can be proved to yield session rates that converge to the max-min fair rate corresponding to $C_l, l \in \mathcal{L}$.

The algorithm just outlined is fully distributed and *does not require knowledge of the network or session topology*. Each link l iteratively computes its LCP, η_l. The minimum of the LCPs along a session's path is fed back to the source of the session (e.g., by using RM cells in the ABR and ATM context). The source then sends data at this rate; it is this source behavior that results in $h_l(\eta) = \sum_{s \in \mathcal{S}_l} \min_{j \in \mathcal{L}_s} \eta_j$ for all $l \in \mathcal{L}$. The process continues iteratively, and the source rates converge to the MMF rates. It is possible to track a slowly varying session topology by not using decreasing gains $a_l(k)$ but instead using small constant gains. If these are small, then the rates will converge close to the MMF rates, but there will be a steady state random error.

The algorithm extends to the case in which each session has a minimum required rate, μ_s, subject to the sum of the minimum rates of the sessions traversing each link being less than the link's capacity. It turns out that even in this case, no per-session computation is required at any link; each link computes a single LCP, η_l. The rate computation at each source takes care of the minimum rates. In the ABR and ATM context, such minimum rates were encountered as MCRs (minimum cell rates) in Section 7.4.5.

7.7.2 Network Utility Optimization and Algorithms

Another important and more general approach is to assume that each source of data, s, has a *utility function*, $U_s(\cdot)$, such that when the source receives the rate r_s it obtains a utility $U_s(r_s)$. If the assigned rate vector is $\mathbf{r}$, then the total utility of all the users in the network is $\sum_{s \in \mathcal{S}} U_s(r_s)$. For simplicity, let us assume here that all the sources have the same utility function $U(\cdot)$. With w as a constant *weighting* factor for a session, the following utility functions have been proposed:

$$U(r) := w \log(r) \tag{7.39}$$

and, with $\alpha \neq 1$,

$$U(r) := w \frac{r^{(1-\alpha)}}{1-\alpha} \tag{7.40}$$

Both of these utility functions are nondecreasing and concave functions of r. Concavity is related to the practical fact that any additional rate is of lower incremental value the higher is the rate that a user already has.

In this framework, the optimal bandwidth sharing is provided by the solution of the following utility maximization problem.

$$\max \sum_{s \in \mathcal{S}} U(r_s)$$

subject to

$$\sum_{s \in \mathcal{S}_l} r_s \leq C_l \quad \text{for every } l \in \mathcal{L}$$

$$r_s \geq 0 \quad \text{for every } s \in \mathcal{S} \tag{7.41}$$

The first constraint simply requires that the total data rate passing through each link not exceed the capacity of the link. We have a nonlinear maximization problem, with a concave objective function and linear constraints (see Appendix C).

We note here that if $U(r) = \frac{r^{(1-\alpha)}}{1-\alpha}$ and $\alpha = 0$, then the problem is to maximize the total user traffic carried by the network, subject to the link capacity constraints. We saw earlier, in the example related to Figure 7.29, that this may not yield a "fair" solution. We will see later in this section how α can be used to control the trade-off between carried traffic and fairness.

Let us examine the solution of the total utility optimization problem. A rate vector $\mathbf{r}$ will be optimal for the problem if and only if the Karush–Kuhn–Tucker (KKT) conditions hold (see Theorems C.3 and C.4) at this rate vector. Let us denote the dual variables for each link capacity constraint by $p_l \geq 0, l \in \mathcal{L}$. Then the KKT conditions yield the following relationships between the optimal rates $r_s, s \in \mathcal{S}$, and the dual variables.

$$
\begin{array}{ll}
\text{for every } s \in \mathcal{S} & \dot{U}(r_s) = \sum_{l \in \mathcal{L}_s} p_l \\
\text{if } \sum_{s \in \mathcal{S}_l} r_s < C_l & p_l = 0 \\
\text{for every } l \in \mathcal{L} & \sum_{s \in \mathcal{S}_l} r_s \leq C_l
\end{array}
$$

One can interpret the dual variable associated with link l as a *price* charged by the link per unit flow that it carries for a user. The KKT conditions then state that, at the fair rate, the derivative of a source's utility is equal to the total price along its path; the price of a link is 0 if the link has spare bandwidth at the fair rate; and, of course, the rates must be feasible.

A distributed algorithm for obtaining the fair rate can be obtained from the dual of the preceding optimization problem (see Appendix C). Consider the function

$$L(\mathbf{r}, \mathbf{p}) := \sum_{s \in \mathcal{S}} U(r_s) + \sum_{l \in \mathcal{L}} p_l \left(C_l - \sum_{s \in \mathcal{S}_l} r_s \right)$$

where $p_l, l \in \mathcal{L}$, are nonnegative dual variables or Lagrange multipliers. There is one such dual variable for each link, and, as mentioned earlier, it is interpreted as a price per unit flow passing through the link. This Lagrangian function has been obtained by relaxing the capacity constraint in the maximization problem. Instead of solving the original (primal) problem, we look at the following dual problem (see Appendix C, Section C.5):

$$\min \Theta(\mathbf{p})$$

subject to

$$p_l \geq 0 \quad \text{for every } l \in \mathcal{L} \tag{7.42}$$

where the dual objective function $\Theta(\mathbf{p})$ is defined by

$$\Theta(\mathbf{p}) = \sup \left\{ \mathbf{r} \geq 0 : L(\mathbf{r}, \mathbf{p}) \right\}$$

In the optimization problem given by Equation 7.41, the dual objective function has a particularly simple form. For given $\mathbf{p} \geq 0$, define, for $s \in \mathcal{S}$, $p^{(s)} = \sum_{l \in \mathcal{L}_s} p_l$; that is, $p^{(s)}$ is the sum of the dual variables, or the *total price* per unit flow, along the path of session s. Now we see that

$$\Theta(\mathbf{p}) = \sup \left\{ \mathbf{r} \geq 0 : \sum_{s \in \mathcal{S}} U(r_s) - \sum_{l \in \mathcal{L}} p_l \sum_{s \in \mathcal{S}_l} r_s \right\} + \mathbf{pC}$$

$$= \sup \left\{ \mathbf{r} \geq 0 : \sum_{s \in \mathcal{S}} \left(U(r_s) - r_s p^{(s)} \right) \right\} + \mathbf{pC}$$

where we have used the definition of $p^{(s)}$. Because $\mathbf{p}$ is given, each of the terms in the $\sum_{s \in \mathcal{S}}$ can be individually optimized (this *separability* property of the constraints is what leads to a simplification in this problem), and we get

$$\Theta(\mathbf{p}) = \sum_{s \in \mathcal{S}} \sup \left\{ r_s \geq 0 : \left(U(r_s) - r_s p^{(s)} \right) \right\} + \mathbf{pC}$$

The dual objective function now involves an optimization problem for each source. The net *benefit* for each source is the utility minus the price it pays; naturally, given the prices, each source would want to adjust its flow to maximize its net benefit. To evaluate the sup, we differentiate the utility function, and, assuming that the derivative is invertible, we find that, for given prices $\mathbf{p}$, source s would use the rate $(\dot{U})^{-1}(p^{(s)})$. Substituting this into the expression for the dual objective function, we obtain

$$\Theta(\mathbf{p}) = \sum_{s \in S} (U((\dot{U})^{-1}(p^{(s)})) - (\dot{U})^{-1}(p^{(s)}) \cdot p^{(s)}) + \mathbf{p}\mathbf{C}$$

which is an explicit function of $\mathbf{p}$. Now to solve the dual problem of Equation 7.42, we must take partial derivatives of the objective function with respect to p_l, for each $l \in \mathcal{L}$. A simple computation shows that

$$\frac{\partial \Theta}{\partial p_l}(\mathbf{p}) = - \sum_{s \in S_l} (\dot{U})^{-1}(p^{(s)}) + C_l$$

For each set of prices $\mathbf{p}$ chosen by the network, the users will select rates that maximize their net benefit. As seen earlier, they will choose the rate

$$r_s(\mathbf{p}) = (\dot{U})^{-1}(p^{(s)}) \tag{7.43}$$

Hence the partial derivatives become

$$\frac{\partial \Theta}{\partial p_l}(\mathbf{p}) = - \sum_{s \in S_l} r_s(\mathbf{p}) + C_l \tag{7.44}$$

Thus we can solve the dual problem using a gradient descent algorithm. Such an algorithm drives the vector of link prices in a direction opposite to the gradient of $\Theta(\cdot)$, thus aiming to reduce the dual objective function. This yields the following iteration at each link l:

$$p_l(k+1) = p_l(k) - a \left(C_l - \sum_{s \in S_l} r_s(\mathbf{p}(k)) \right) \tag{7.45}$$

where $\mathbf{p}(k)$ is the vector of prices at the kth iteration. Here a is a small *gain* factor, the value of which determines the speed of convergence and the steady state error.

Notice the similarity between this iteration and the one that we obtain for the LCPs, η_l, of the max-min fair problem in Section 7.7.1.

In fact the relationship between these two problems goes further. Consider the utility function of Equation 7.40 with $w = 1$ (i.e., $U(r) = \frac{r^{(1-\alpha)}}{1-\alpha}$), and observe that $(\dot{U})^{-1}(y) = y^{-\frac{1}{\alpha}}$. For each α corresponding to the price p_l, let us define η_l by $p_l = \eta_l^{-\alpha}$; note that the $p_l, l \in \mathcal{L}$, will depend on α, and hence so will $\eta_l, l \in \mathcal{L}$. Now, if the sources adapt their rates according to Equation 7.43, then the gradient of the dual objective becomes (see Equation 7.44)

$$ - \sum_{s \in \mathcal{S}_l} \left(\sum_{j \in \mathcal{L}_s} \eta_j^{-\alpha} \right)^{-\frac{1}{\alpha}} + C_l $$

For large α, only the dominant term $(\min_{j \in \mathcal{L}_s} \eta_j)^{-\alpha}$ matters. Let us assume that the minimum is achieved at exactly one link along the path of session s. Then for large α, the dual gradient can be approximated by

$$ - \sum_{s \in \mathcal{S}_l} \min_{j \in \mathcal{L}_s} \eta_j + C_l $$

which is the same expression as the one that characterizes the LCPs in the MMF problem. We have observed that the stochastic approximation algorithm in Section 7.7.1 has the same structure as the gradient descent algorithm shown in Equation 7.45 but has a direct motivation from the structure of the MMF solution. We also notice that the LCPs are directly related to the link prices. A link that is saturated in the MMF solution has a positive LCP and a positive price, whereas a link that has spare capacity has an infinite LCP and 0 price.

The result suggested by the approximate argument (i.e., the result that as $\alpha \to \infty$ the fairness objective defined by the utility maximization problem of Equation 7.41 approaches MMF) can be formally proved ([217]). We have also noted earlier that with $\alpha = 0$ the solution maximizes the total amount of user traffic carried in the network. Thus we see that, for the utility function defined in Equation 7.40, α controls the trade-off between carried traffic and fairness; a small α puts emphasis on carried traffic, and a large α emphasizes fairness. This point is explored further in Problem 7.8.

7.7.3 Randomly Arriving Transfers in a Network

In Sections 7.7.1 and 7.7.2 we consider a general bit-carrier network that is carrying several long-lived elastic transfers on fixed routes. In practice, as discussed at the beginning of Section 7.6.7, each transfer request is for a finite volume of data. Such requests arrive at random times, share the network with the other ongoing transfers, and then depart after completing their transfers. In Section 7.6.7 we discuss randomly arriving, finite-volume transfers sharing a single link.

Let us now consider the extension to a general network. We retain the same notation for the network links and their capacities as introduced in Section 7.7. The following additional notation is needed:

$\mathcal{R}$: a set of (unidirectional) routes on which elastic transfers occur

$\mathcal{L}_r$: the set of links used by route $r \in \mathcal{R}$

$\mathcal{R}_l$: the set of routes through link $l \in \mathcal{L}$

Let us further assume that on route r transfer requests arrive at the rate λ_r and that the arrival process is Poisson. All these Poisson processes are mutually independent. The transfer sizes are i.i.d., with their volumes denoted by the random variable V, with mean $\mathsf{E}(V)$. Because transfers arrive and depart randomly, the number of ongoing transfers on each route is random. Let $N_r(t)$ denote the number of ongoing transfers on route r at time t, and let $\mathbf{N}(t) = (N_1(t), N_2(t), \ldots, N_{|\mathcal{R}|}(t))$ denote the vector of ongoing transfers over all the routes. The bandwidth-sharing mechanism in the network aims at achieving some fairness objective. As the number of sessions changes, all the transfers sharing the network must adjust their rates to continue to achieve fair sharing. Let us also assume that as the number of sessions changes, the transfer rates can be *instantly* modified so that fair sharing is achieved at all times. The random process $\mathbf{N}(t)$ is then completely specified by the network topology, the routes, the arrival processes, the distribution of transfer sizes, and the bandwidth-sharing objective. Two important questions can be asked in this context:

- Does the process $\mathbf{N}(t)$ have a steady state distribution?

- In steady state, what is the transfer rate performance (e.g., the average transfer rate) obtained by transfers on each route?

For the fairness defined by the utility functions in Equation 7.39 or 7.40, the first question has a very simple answer in the case when the transfer volumes are exponentially distributed. The total bit rate being transferred on route r is clearly

$\lambda_r\mathsf{E}(V)$, and the total bit rate being carried by link l is $\sum_{r\in\mathcal{R}_l}\lambda_r\mathsf{E}(V)$. Define, for each $l \in \mathcal{L}$,

$$\rho_l := \frac{\sum_{r\in\mathcal{R}_l}\lambda_r EV}{C_l}$$

Obviously, it is necessary that for every $l \in \mathcal{L}$, $\rho_l < 1$. With Poisson arrivals and exponential transfer volumes, $\mathbf{N}(t)$ is a continuous time Markov chain, and it can be shown that this Markov chain is positive recurrent (and hence has a stationary distribution) if and only if, for all $l \in \mathcal{L}$,

$$\rho_l < 1 \tag{7.46}$$

The second question, on performance analysis, is harder even for the model with exponential transfer sizes. Explicit results can be obtained for some simple topologies; approximations have been studied for more general topologies. The answer to the first question is still open for the case when the transfer distributions are not exponentially distributed.

In practice, routes between file servers and clients do not change rapidly. This is true with shortest path adaptive routing and is certainly the case with "nailed-down" virtual paths (in the context of ATM VPs and label switched paths (LSPs) in multiprotocol label switching (MPLS); see Part III). Hence over long periods of time, routes across the network remain fixed, and the problem is exactly one of sharing the network bandwidth between randomly arriving, finite-volume transfers over these routes, as captured in the model discussed earlier. There are, however, two limiting assumptions in the model. First, in practice, the bandwidth sharing does not adjust instantaneously as the number of transfers changes; thus the model applies if the number of transfers is slowly varying. Second, even though the session arrival process may be well modeled as Poisson, the arrival instants of individual transfers in a session would be correlated, and thus the overall arrival process of transfer requests would not be Poisson. One might expect that the stability result continues to hold more generally, but this is a topic of ongoing research.

7.7.4 TCP and Fairness

In an internet, there are no explicit feedback packets; instead, feedback is provided by packet loss or by packet marking. This simple mechanism can be used, however, for feeding back the prices in the general setting of Sections 7.7.1 and 7.7.2. For example, consider the MMF objective. Suppose the links have their LCPs. Now each link drops or marks packets with the probability $e^{-\nu \eta_l}$ for some large positive number ν that is known to the sources. Then the probability of marking along the path of session s is, approximately, $\sum_{l \in \mathcal{L}_s} e^{-\nu \eta_l}$, which for large ν can be approximated as $e^{-\nu \min_{l \in \mathcal{L}_s} \eta_l}$. Hence, by measuring the dropping or marking rate, the source of transfer s can obtain $\min_{l \in \mathcal{L}_s} \eta_l$ and adapt its rate. The algorithm for estimating the LCPs can be just as in Section 7.7.1.

In the case of the general utility maximization objective, the same approach of *random exponential marking* (REM) can be used to convey the price along a path to each source s. Knowing its price p_l, each link marks packets with the probability $1 - \phi^{-p_l}$, where ϕ (> 1) is a positive number known to all the sources. The total marking probability along the path of session s is then $1 - \phi^{-p^{(s)}}$, and the source s can infer $p^{(s)}$ by counting the marked packets. The source can then adapt its rate accordingly. The iteration for calculating the prices can proceed as shown in Section 7.7.2.

But which fairness objective is achieved by loss- or marking-based feedback and the standard adaptive window algorithm implemented in TCP sources? Consider an internet in which the dropping or marking probability on link l is ϵ_l. Consider a large-volume TCP transfer, s, along a path in this network. The overall probability of a packet being dropped or marked in session s is $1 - \Pi_{l \in \mathcal{L}_s}(1 - \epsilon_l)$, which, for small probabilities, can be approximated as $\sum_{l \in \mathcal{L}_s} \epsilon_l$, which we write as $\epsilon^{(s)}$. Let d_s be the RTT along the path of transfer s. Then Equation 7.20 yields the transfer throughput to be $\frac{\sqrt{(3/2)}}{d_s \sqrt{\epsilon^{(s)}}}$.

Exercise 7.13

Consider the utility maximization formulation with utility function $U_s(r) = -\frac{1}{d_s^2} r^{-1}$; this is of the form in Equation 7.40 with $\alpha = 2$. Show that the optimal rates are of the form $r_s = \frac{1}{d_s \sqrt{p^{(s)}}}$, where $p^{(s)}$ is the total price along the path of transfer s.

Suppose now that REM is used to feed the prices back to the sources. For small probabilities it can be checked that $\epsilon^{(s)}$ is proportional to $p^{(s)}$. This observation, along with the result of Exercise 7.13, has been used to make the interpretation that TCP aims for a fairness based on maximization of the utility function $U_s(r) = -\frac{1}{d_s^2} r^{-1}$. Such an interpretation (and, in fact, the foregoing model) is valid if the RTT, d_s, $s \in \mathcal{S}$, does not itself depend on the traffic—for example, if the propagation delays are large and dominate the queueing delays.

In Section 7.9 we point out some literature on a more careful analysis of the bandwidth-sharing objectives that TCP achieves, and also on algorithms for fair bandwidth sharing with adaptive window control. We also discuss the literature on the performance analysis of a network of infinite-volume TCP transfers sharing an internet with a general topology. Work has also been done on the throughput performance of randomly arriving, finite-volume TCP-controlled transfers in a general network (see the model in Section 7.7.3); we provide pointers in the notes on the literature. The results presented in Section 7.7.3 suggest that if the total offered load on each network link is less than 1, then the process of the number of ongoing transfers would be stable. For large BDPs, however, the discussion at the end of Section 7.6.7 further suggests that as long as the load on each link is less than 90%, the throughput performance of transfers is insensitive to load. Such an argument lends support to the practice of engineering a network so as to pack its links up to some fraction (e.g., 90%) of their capacities.

7.8 Summary

This chapter deals with the problem of multiplexing elastic transfers into a network topology and dynamically sharing the link bandwidths among the various ongoing transfers to achieve bandwidth-sharing objectives. Such dynamic bandwidth sharing requires implicit or explicit feedback from the network to the traffic sources. We discuss two ways in which the source is controlled: rate control and window control. The former is proposed for the ABR service in ATM networks, and the latter is the mechanism used for TCP-controlled transfers in the Internet.

For rate control we examine explicit-rate feedback and queue-length-based feedback. For the latter we model the feedback-controlled system with an ordinary differential equation and use this to motivate a multiplicative decrease of the source rate. The stability of the resulting control loop is also examined.

We then provide an extensive study of the adaptive window-based TCP protocol. We examine in some detail the packet transmission and loss recovery mechanisms in various versions of TCP. Several models for TCP are then discussed. A deterministic model for a wide area TCP connection gives insight into the process of fast recovery and explains why it yields a performance gain. We also develop the well-known "square root" formula for the throughput of a wide area TCP connection with random packet loss. The idea of active queue management is then introduced. The RED and ECN mechanisms are discussed. An ordinary differential equation model gives insight into how the dropping or marking profile parameters must be tuned to network conditions in order to ensure stability of the control loop. We then examine TCP-controlled transfers of randomly arriving, finite-volume transfers. For the single-bottleneck situation, and with small propagation delay, the processor-sharing model predicts these performance measures well. We discuss the impact of a large bandwidth delay product (BDP) and show results that suggest that for very large BDP networks, traffic engineering can be based on a little overprovisioning.

The model with randomly arriving transfers also permits us to examine the issue of long-range-dependent traffic. If the distribution of the file transfer sizes has a finite second moment, then the network traffic is not long-range-dependent (LRD), but for Pareto distributed file transfer sizes with infinite second moment, the network traffic is LRD. One cannot directly conclude from the fact that traffic into a buffer is LRD that the stationary buffer distribution will have heavy tails; this depends on closed-loop behavior, for which a closed-loop analysis is necessary.

Much of the discussion until this point is in the context of elastic transfers over a single-bottleneck link. We then turn our attention to the problem of several long-lived connections sharing a network topology. Fair sharing no longer means equal sharing, but instead a fairness notion must be developed. We discuss max-min fair sharing and utility optimization. In each case we provide distributed algorithms for achieving the bandwidth sharing in the network. In each case the algorithms work by each link calculating a link parameter or a link price. An appropriate function of the link parameters or prices is fed back to the sources, and based on this feedback the sources adjust their rates. Finally we briefly discuss the situation of randomly arriving, finite-volume transfers in a network with a general topology.

7.9   Notes on the Literature

There is a vast amount of literature on protocols, performance analysis, optimization, and distributed algorithms for elastic traffic in packet networks.

The idea of using rate-based control (Section 7.4) for elastic traffic was pursued mostly in the context of the ABR service in ATM networks. The ATM standards for ABR traffic management are provided in [17]. The idea of using a single bit to feed congestion information back to the sources was used in the DECbit protocol in the middle 1980s and was analyzed by Chiu and Jain in [61]. This article appears to be one of the first efforts to formally study a congestion control protocol. Congestion-based one-bit feedback evolved through several improvements. Surveys have been provided by Jain [151], Ohsaki et al. [227], and Fendick [100]. Specific algorithms have been proposed and studied by Newman [223] and by Siu and Tzeng [262]. A delayed differential equation model for queue-length-based source control was proposed by Bolot and Shankar [37]. The stability analysis of such delay differential equations has been provided by Elwalid in [90].

To obtain more precise control over the source rates, attention has moved to explicit-rate-based algorithms. The effort here was to develop distributed asynchronous algorithms that could compute the fair rates and feed them back to the sources. The notion of fairness favored in this research was max-min fairness. The basic principles of max-min fairness have been discussed in Bertsekas's and Gallager's textbook [33]. The early efforts in developing algorithms attempted to solve the problem by implementing the centralized algorithm in a distributed manner; see Charny et al. [55] and Charny and Ramakrishnan [54]. An algorithm that virtually became the benchmark in the ATM Forum was developed by Kalyanaraman et al. [155]. A related approach was developed by Fulton et al. [112]; the algorithm in Lee and De Veciana [193] is similar, but the authors also examine delays and asynchrony. Subsequent approaches were based on control theoretic formulations, such as those reported in Benmohamed and Meerkov [27] and in Kolarov and Ramamurthy [178]. The solution-of-equation technique discussed in Section 7.7 was developed by Abraham and Kumar [3], and the utility optimization approach was developed by Lapsley and Low [190]. In the latter two cases, the resulting distributed algorithms are shown to be equivalent. Abraham's and Kumar's approach for max-min fairness also handles minimum session rate guarantees (see [1] and [2]).

Window protocols have long been used to control the flow of traffic in packet networks. The adaptive window algorithm of TCP was developed primarily with the aim of managing congestion in the Internet. The 1990s saw an explosion in the deployment of the Internet, and in the past decade a huge amount of research has been done on the modeling, analysis, and optimization of the TCP protocol. See the book by Stevens ([273]) for a comprehensive and detailed description of TCP. One of the earliest attempts to formally model and analyze TCP was in Mishra et al. [213], who observed the cyclical evolution of the adaptive window-based mechanism in TCP. Lakshman and Madhow [187] developed this idea to obtain a throughput analysis for long file transfers. Subsequently this approach was used by Kumar [181] and by Kumar and Holtzman [182] to analyze the performance of some versions of TCP over lossy and fading wireless links. A very general approach to modeling TCP performance in the presence of a stationary and ergodic loss process is provided by Altman et al. in [10]. The analysis provided in Section 7.6.5 is a simplification of the analysis in [230] by Padhye et al.; this is famous as the PFTK model, after the names of the four authors. An analysis approach that does not rely on this cyclical behavior has been developed by Sharma and Purkayastha [260]; they incorporate an exogenous open-loop source (such as UDP traffic) and different RTPDs but need to assume that the bottleneck link is fully utilized. Another technique for analyzing TCP connections with different RTPDs is provided by Brown [44].

When the transfers are short it is important to incorporate the effects of connection setup and slow-start. Such analyses have been provided by Cardwell et al. [48] and by Barakat and Altman [21].

The use of random discard to control queues in the Internet and hence to manage TCP performance was proposed as RED by Floyd and Jacobson in [103]. The ECN protocol was proposed by Ramakrishnan and Floyd in [241]. The analysis of TCP connections over a wide area bottleneck link with random drop (see Section 7.6.6) was done by Firoiu and Borden in [101].

The modeling of TCP via delay differential equations led to new insights into the effects of the various parameters on stability. In the context of TCP, the differential equation model appears to have been first used by Hollot et al. in [142]. Subsequently these models have been extended to cover more complex scenarios than only one traffic class and a single link; see [214], [215], and [50].

The performance analysis of TCP-like protocols with randomly arriving, finite-volume transfers has been done mainly for single-link scenarios. Based on extensive measurements, Floyd and Paxson have reported in [104] that the Poisson process is a good model for the arrival instants of user sessions. In [138], Heyman et al. use a processor-sharing model and conclude that the performance would be insensitive to file size distributions. In [31] Berger and Kogan use a similar model but with a finite number of users cycling between sharing the bandwidth and thinking. A processor-sharing model is used by Ben Fredj et al. in [108], but a considerable amount of detail in the user behavior is modeled by using the theory of quasi-reversible queues. The extension of the processor-sharing model was developed by Kherani and Kumar [170]. The observation that when propagation delays are large, the performance becomes insensitive to load for a large range of loads was also made in this paper, and was studied by many simulations in [291].

The observation that traffic in the Internet is long-range-dependent was made by Leland et al. in [195]; further analysis of the data was presented by Willinger et al. in [299]. That the LRD nature of the traffic could be traced to heavy-tailed file transfers was observed by Crovella and Bestavros in [69]. A study of the effect of queueing with LRD traffic has been reported by Erramilli et al. in [94]. The result that an alternating renewal process with heavy-tailed *on* times is long-range-dependent was proved by Heath et al. in [137]; the argument mentioned in the text was provided by Kumar et al. in [184]. The importance of modeling buffers in the closed loop has been shown by Kherani and Kumar in [171] and in [172]. In these articles, Kherani and Kumar have also analytically established the long-range dependence of the input process into the router buffer of the bottleneck link.

The approach to thinking about these problems in terms of dynamic sharing of the network links to achieve some global sharing objectives began in the context of the ABR service in ATM networks. Several bandwidth-sharing objectives have been proposed, and their implications have been studied; see Roberts and Massoulie [252], Massoulie and Roberts [206], and Bonald and Roberts [39].

The approach to formulating the bandwidth-sharing problem as a utility maximization problem, and then the use of dual variables (interpreted as "shadow" prices) to develop distributed algorithms, originated in the paper by Kelly et al. [166]. In the discussion in Section 7.7 we cover the work of Abraham and Kumar [3] in the context of the max-min fairness objective, and that of Low and Lapsley [201] for more general fairness objectives. The idea of random exponential marking is discussed by Athuraliya et al. in [16]. In [217], Mo and Walrand have developed the result that max-min fair sharing is a limit of fair sharing with a more general utility function; they have also developed distributed algorithms for adaptive window-based transmission. A detailed study of the bandwidth-sharing fairness achieved by TCP-like adaptive window-based algorithms has been provided by Hurley et al. in [144], and by Vojnovic et al. in [292].

The model of a network with randomly arriving transfer requests of i.i.d. volumes of data has been studied for ideal MMF sharing by De Veciana et al. in [80]; for this model, an approximate performance analysis technique is provided by Chanda et al. in [51]. This model has been generalized by Bonald and Massoulie in [38].

In [45], Bu and Towsley provide a fixed-point approximation for obtaining the throughputs of infinite- and finite-volume TCP transfers in a network; the finite-volume transfers are assumed to traverse a single bottleneck link. Another proposal for the modeling and analysis of TCP-controlled finite-volume file transfers has been given by Casetti and Meo [49]; first a detailed model of a TCP session is analyzed, assuming that the RTT is given, and then a model is proposed for obtaining the RTT in a network carrying randomly arriving file transfers. Other modeling and analysis approaches for a fixed number of TCP transfers in a general network are provided by Firoiu et al. in [102] and by Altman et al. in [11].

Problems

7.1 Consider a window-controlled transfer over a connection with an RTPD of 200 ms. The bottleneck link speed on the path is 2 Mbps. The data packet length is 1000 bytes. Assume that there is only one connection over the bottleneck link.

a. Determine the minimum window (in number of packets) required so that the bottleneck link is fully utilised (ignore the ACK transmission times).

b. If a window of 20 packets is used, determine the maximum possible utilisation of the bottleneck link.

c. What happens if a window of 80 packets is used?

7.2 Consider an adaptive window protocol with a window that can be any nonnegative real number. The protocol is used to control fluid transfers. The transmitter increases its window as a function of the amount of data u that has been transferred to the destination and acknowledged. This is what we call the "age" of the transfer. Let the transmission window when the age is u be denoted by $w(u)$. The initial window is 1.

There is no packet loss; hence the window is limited in size because files are of finite size.

a. If the increase in window is equal to the amount of increase in the age of the transfer, then show that $w(u) = u + 1$ for all $u \geq 0$.

b. If the increase in window is $\frac{1}{w(u)}$ per unit increase in age when the age is $u \geq 0$, then show that $w(u) = \sqrt{(2u + 1)}$ for all $u \geq 0$.

c. Relate these results to the TCP protocol.

7.3 Consider a very large file transfer over a link with bit rate C bps and with RTPD 2δ. In congestion avoidance the TCP window increases by 1 packet every RTPD. Hence the transmission rate increases by $\frac{1}{(2\delta)}$ packets per second every RTPD; this is an increase rate of $\frac{1}{(2\delta)^2}$ packets per second. Let us view the window as being increased continuously over time at this rate.

a. For a file transfer of size v packets, show that the time taken to complete the transfer is given by $t = 2\delta\sqrt{2v}$ seconds. Argue that this calculation requires the condition $2\delta C > \sqrt{2v}$.

b. Consider a file of size 1 GB being transferred over a link of speed 2.4 Gbps, with RTPD 200 ms. The packet size, including headers, is 1000 bytes. Determine the time taken to complete the file transfer if: (i) Somehow the full link speed could be used at all times, and (ii) if TCP with the above assumptions is used.

c. Suppose that the file size is exponentially distributed with mean $E(V)$ packets. Show that the mean transfer time is given by $2\delta\sqrt{\frac{\pi E(V)}{2}}$ seconds. Assume that the condition $2\delta C > \sqrt{2V}$ holds with probability close to 1.

7.4 Extend the analysis developed in Section 7.6.6 to the case in which the n infinite-volume TCP transfers sharing the bottleneck link have different RTPDs, $\delta_i, 1 \leq i \leq n$. Show that for each value of the drop probability p, the mean queueing delay q can be calculated, and hence the throughputs for each of the transfers can be computed.

7.5 File transfer requests with i.i.d. volumes, and mean size v bits, arrive to a link of capacity b bits/sec. The arrival instants form a Poisson process of rate λ. Each session is attached to the link by an access link of rate

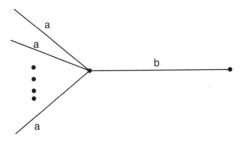

Figure 7.30   The network in Problem 7.5.

$a = b/m$, for an integer $m > 1$ (see Figure 7.30). Assume that the processor-sharing model applies to the shared link.

a. Assuming that the file sizes are exponentially distributed, show that $\{X(t)\}$, the number of active transfers at time t, is a CTMC, and obtain its stationary distribution $\pi(n)$, $n \geq 0$. Use Little's theorem to obtain the mean transfer time of the files.

b. How does the answer in (a) change if the distribution of file size is exponential (mean v_s) with probability p, or exponential (mean v_l) with probability $(1 - p)$ (i.e., it is hyperexponential)? (Hint: Use the concept of *symmetric queues* explained in Appendix D.)

7.6   Consider the throughput model for multiple TCP long-lived connections over a wide area link with a fixed rate of random drops, p, that was developed in Section 7.6.6. The link rate is C bits per second, and the RTPD is δ. Suppose that file transfer requests arrive in a Poisson process of rate λ. The file transfer volumes $(V_1, V_2, \ldots)$ are i.i.d. random variables with an exponential distribution and mean EV. Assume that when there are n file transfers in progress then the throughput per transfer is as obtained in the analysis in the text. Model the number of ongoing transfers $N(t)$ as a Markov chain, and obtain its steady-state distribution. Hence show how to obtain the average file transfer throughput and the time average queue length in the buffer. Discuss the validity of this modeling approach.

7.7   Consider the n hop "line" network shown in Figure 7.31; all the link capacities are equal to C. There are $n + 1$ elastic transfers as shown.

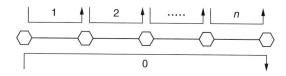

Figure 7.31   Network topology and routes in Problem 7.7.

There is one transfer (indexed $1, 2, \ldots, n$) over each hop, and one transfer (indexed 0) is over all the n hops. Take the utility function of each user to be $U(r) = \log r$, when it is provided a transfer rate of r. Show that the sum-utility maximising rates are $r_0 = \frac{C}{n+1}$ and $r_j = \frac{nC}{n+1}, 1 \leq j \leq n$, and the corresponding link prices are $p_l = \frac{n+1}{nC}, 1 \leq l \leq n$.

7.8   Repeat Problem 7.7 with $U(r) = \frac{r^{(1-\alpha)}}{1-\alpha}, \alpha \neq 1$. Examine the solution for $\alpha \to \infty$, and verify that it converges to the max-min fair solution. Discuss what happens to total carried traffic as α is varied.

CHAPTER 8

Multiple Access: Wireless Networks

In Section 2.2.1, in our discussion of various kinds of multiplexing, we introduce the idea of distributed multiplexing, in which several entities share a multipoint wired link or a wireless medium with no central coordination. In this chapter we discuss issues and problems in distributed multiplexing over a shared radio spectrum; generically this topic is also *called multiple access*. Our discussion is in line with the view of networking over a radio spectrum as presented in Section 3.5. Thus we do not treat wireless links as given bit carriers into which traffic is multiplexed. Instead we begin by developing an understanding of digital communication over radio. This permits us to discuss the area of wireless networks as one of distributed resource allocation over a radio spectrum.

Wireless communication is necessitated when it is difficult or expensive to lay guided media between the communicating nodes, or when the nodes are arbitrarily mobile. Figure 8.1 shows a taxonomy of wireless networks.

Fixed wireless networks include line-of-sight microwave links, which until recently were very popular for long distance transmission. In this category we can also include satellite networks, which are basically line-of-sight microwave links that employ a communications satellite as a "reflector-in-the-sky" to bounce signals off.

As discussed in Section 2.3.1, currently the most important role of wireless communications technology is in mobile access to wired networks. We can further classify such access networks into two categories: cellular networks and random access networks. Cellular wireless networks were introduced in the early 1980s as a technology for providing access to the PSTN by mobile users. Until recently these networks have been driven primarily by the needs of circuit-switched voice telephony; on demand, a mobile phone user is provided a wireless digital communication channel on which is carried compressed telephone-quality (although not *toll*-quality, such as in the wired PSTN) speech. Recently, with the growing need for mobile Internet access, there have been efforts to provide low-bit-rate data access on these networks as well.

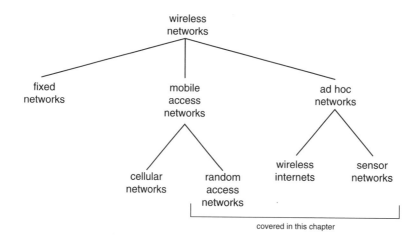

Figure 8.1   A taxonomy of wireless networks.

Cellular networks employ a centrally coordinated mechanism for sharing the radio spectrum; on demand, the system provides capacity to each telephone call, and demands can be blocked. Cellular networks have evolved rapidly since their first deployment in the early 1980s. There are now several cellular standards vying in the marketplace, and new standards are evolving. The emphasis in the technology evolution of cellular networks appears to be primarily on physical layer issues. In this chapter we focus mainly on distributed resource allocation issues of emerging interest. We do not explicitly address cellular networks, for which several textbook treatments are available.

Random access networks utilize some kind of distributed mechanism for sharing the radio spectrum. Such networks originated in the Aloha experiments more than 30 years ago, and transport of bursty packetized data was their primary objective. These classical technologies and some of the associated analysis are reviewed in Section 8.5. An important traditional application of random access is over signaling and control channels in cellular networks and satellite networks. Spurred by advances in physical wireless digital communication, random access networks can now support bit rates close to those of desktop wired Ethernet access. Hence random access wireless networks are now rapidly proliferating as the technology of choice for wireless Internet access with limited mobility. In Section 8.6 we discuss this topic in the context of the IEEE 802.11 standard.

Wireless access networks basically provide the last hop between a mobile node and the wired infrastructure. All communication between a mobile node and another node (mobile or fixed) passes through a *base station* or *access point* that has a radio interface and also an interface into the wired network. At any point in time a mobile node must associate itself with at least one base station or access point. In contrast, *wireless ad hoc networks* are aimed at supporting arbitrary communication between randomly located mobile nodes with no a priori well-defined physical association between them. Each node in such a network is equipped with a digital radio, which can communicate with other similar radios over some range. Nodes must discover their neighbors, establish associations, and thus set up a radio network topology. Packets going from one node to another may need to be forwarded by other nodes. Thus these are *multihop* wireless packet radio networks, and they have been studied as such for several years. Interest in such networks has again been revived in the context of *multihop wireless internets* and *wireless sensor networks*. We discuss these topics in Sections 8.7, 8.8, and 8.9.

8.1 Bits over a Wireless Link: Principles, Issues, and Trade-Offs

Recall from Figure 3.9 that when studying information transport networks over wireless communication, we do not take the links as given bit carriers but are also concerned with the sharing of the wireless spectrum resource. The strictly layered approach (see Section 2.3.2) views the wireless physical layer as providing a bit-carrier service to the link layer. The link layer simply offers packets to the physical layer, which does the best it can. If, on the other hand, there is interaction between the layers then the link layer can be aware of the time-varying quality of the wireless communication, and hence it can prioritize, schedule, defer, or discard packets in an attempt to meet the QoS requirements of the various flows. It is therefore important to gain an understanding of how digital radio communication is performed and of the issues, constraints, and trade-offs that are involved. The material in the remainder of this section is well established and is available in great detail and much more generality in many books on digital communications; readers familiar with this topic can skip this section with no loss of continuity.

The primary resource that is shared in a wireless network is the *radio spectrum*. We limit ourselves to the situation in which the communicating nodes share a radio spectrum of *bandwidth* $2W$, centered at the carrier frequency f_c (see Figure 8.2). It is assumed that $f_c \gg W$; for example, $f_c = 2.4$ GHz and

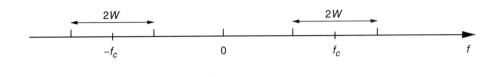

Figure 8.2 The nodes in a wireless network share a portion of the radio spectrum.

$2W = 5$ MHz. All communication between any pair of nodes can utilize this entire spectrum.

8.1.1 Simple Binary Modulation and Detection

As shown in Figure 8.3, we achieve digital communication over the given radio spectrum by *modulating* a sequence of *pulses* by the given bit pattern. The pulse, $p(t)$ (also called the *baseband pulse*), is chosen so that when translated to the carrier f_c its spectrum fits into the given radio spectrum; in this case, the spectrum of the baseband pulse occupies the frequencies $(-W, +W)$. The pulses are repeated every T seconds. It is assumed that the energy of the pulse is 1 (i.e., $\int_0^T p^2(t)\,dt = 1$). In the situation depicted in Figure 8.3, the modulation is very simple: Each pulse in the pulse train is multiplied by $+\sqrt{E_s}$ if the bit to be transmitted is 1, and by $-\sqrt{E_s}$ if the bit to be sent is 0. It is said that the modulator maps bits into channel *symbols*. Thus, in this example, the symbol set is $\{-\sqrt{E_s}, +\sqrt{E_s}\}$. In general, there could be more than two possible symbols; for example, four symbols would permit

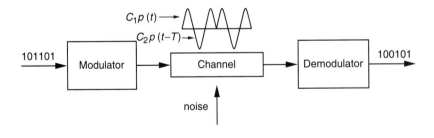

Figure 8.3 A sequence of pulses is modulated with the bits to be transmitted. The basic pulse is $p(t)$. Notice that the bit sequence 101101 is transmitted as $+\sqrt{E_s}\,p(t)$, $-\sqrt{E_s}p(t-T), +\sqrt{E_s}p(t-2T), \ldots, +\sqrt{E_s}p(t-5T)$. There is an error in the third bit, so after detection, the received sequence is 100101.

two incoming bits to be mapped into each channel symbol. Continuing our simple example, let C_k denote the symbol into which the kth bit is mapped. If the pulses are repeated every T seconds, the modulated pulse stream can be written as

$$\sum_{k=-\infty}^{\infty} C_k \, p(t - kT)$$

This baseband signal is then translated to the radio spectrum shown in Figure 8.2 by multiplying it with a sinusoid at the carrier frequency. The resulting signal can be written as

$$X(t) = \sqrt{2}\,\text{Re}\left(\sum_{k=-\infty}^{\infty} C_k \, p(t - kT) \, e^{j2\pi f_c t}\right) \tag{8.1}$$

where Re denotes the real part of the complex number. The multiplication by $\sqrt{2}$ makes the energy in the modulated symbols equal to E_s, as discussed later. In Figure 8.3 we do not show the translation of the signal by the carrier. It is as if the channel has been shifted to the baseband. The generality of the way in which $X(t)$ has been written is useful a little later. In this case we have the simplification

$$X(t) = \sqrt{2} \sum_{k=-\infty}^{\infty} C_k \, p(t - kT) \, \cos(2\pi f_c t)$$

This shows that we have multiplied the real-valued baseband signal by the sinusoid $\cos(2\pi f_c t)$. To see why we have chosen the symbols C_k to be $\pm\sqrt{E_s}$ and why we have used the factor $\sqrt{2}$, notice that the energy in each transmitted pulse is

$$2\int_0^T (C_k)^2 p^2(t) \cos^2(2\pi f_c t) dt = E_s \left(\int_0^T p^2(t)dt + \int_0^T p^2(t) \cos(4\pi f_c t)dt\right)$$

$$\approx E_s$$

where we have used the facts that $2\cos^2\theta = 1 + \cos 2\theta$, $\int_0^T p^2(t)dt = 1$, and $\int_0^T p^2(t) \cos(4\pi f_c t)dt \approx 0$, the latter approximation being accurate for $f_c \gg \frac{1}{T}$. Thus the *symbol energy* in the transmitted signal is E_s joules/symbol, and because the symbol rate is $\frac{1}{T}$ symbols/second, the transmitted signal power is therefore $\frac{E_s}{T}$ watts.

As shown in Figure 8.3, as the modulated signal passes through the channel and is processed in the front end of the receiver, it is corrupted by noise. This is taken to be zero mean *additive white Gaussian noise* (AWGN), which means that noise just adds to the signal and is a Gaussian random process with a power spectrum that is constant over the radio spectrum of the system (hence the term "white," because all frequencies ("colors") have the same power). Because the signal occupies a band of $2W$ Hz around the carrier frequency f_c (W Hz below and W Hz above $\pm f_c$; see Figure 8.2), we need be concerned only with noise that occupies this band. Such *bandpass* (i.e., limited to the band of $2W$ Hz around f_c) *white Gaussian noise*, with a power spectral density of $\frac{N_0}{2}$, is mathematically represented as follows (see [238]):

$$N(t) = U(t)\cos(2\pi f_c t) - V(t)\sin(2\pi f_c t) \qquad (8.2)$$

where the processes $U(t)$ and $V(t)$ are independent zero mean white Gaussian processes with power spectral density N_0, band-limited to $(-W, +W)$. We can view the noise processes $U(t)$ and $V(t)$ as baseband noise processes that are modulated and placed in the radio spectrum of the channel. This intuition becomes clear from the analysis here and from the analysis of modulation with two-dimensional symbols in Section 8.1.2.

At the receiver, the signal is translated back to the baseband by multiplying the received signal by $\sqrt{2}\cos(2\pi f_c t)$, which yields

$$\sum_{k=-\infty}^{\infty} C_k \, p(t-kT) \left(\cos(4\pi f_c t) + 1\right) + U(t)\frac{\left(\cos(4\pi f_c t) + 1\right)}{\sqrt{2}} - V(t)\frac{\left(\sin(4\pi f_c t)\right)}{\sqrt{2}}$$

The receiver filters the signal to the bandwidth W Hz (i.e., to the frequency interval $(-W, +W)$). Hence, the high-frequency terms are filtered out (because $f_c \gg W$), and we are left with

$$\sum_{k=-\infty}^{\infty} C_k \, p(t-kT) + \frac{1}{\sqrt{2}} \, U(t) \qquad (8.3)$$

Notice that the noise term $V(t)$ in Equation 8.2 has disappeared, because we are using only real-valued symbols C_k; this term is needed when we use complex-valued symbols, as we will see in Section 8.1.2. The noise $\frac{1}{\sqrt{2}}U(t)$ is white Gaussian with power spectral density $\frac{N_0}{2}$. Because the signal is now band-limited to the interval $(-W, +W)$, the average noise power is $2W \times \frac{N_0}{2} = WN_0$ watts; this means that

$$\lim_{t \to \infty} \frac{1}{t} \int_0^t \left(\frac{U(x)}{\sqrt{2}}\right)^2 dx = WN_0$$

where the integrand on the left side is the power dissipation if the noise was put across a 1 ohm resistor; the integration yields energy over $(0, t)$, and the division by time yields the average power.

The receiver also needs to *synchronize* to the pulse boundaries. After this is done, the demodulator then looks at each received pulse and determines which symbol it is carrying. This step is called *detection*. Let us now see how the kth symbol is detected—that is, how it is determined whether $C_k = +\sqrt{E_s}$, or $C_k = -\sqrt{E_s}$. For ease of discussion, we conceptually shift the kth symbol to the interval $[0, T]$. The received signal is then multiplied by the pulse $p(t)$ and integrated over $[0, T]$, the pulse $p(t)$ being assumed to be known at the receiver. Because $\int_0^T p^2(t)dt = 1$, this yields

$$C_k + \int_0^T \frac{U(t)}{\sqrt{2}} p(t)dt$$

Because $U(t)$ is a zero mean Gaussian process, using the fact that a linear combination of Gaussian random variables is again Gaussian, we conclude that $\int_0^T \frac{U(t)}{\sqrt{2}} p(t)\, dt$ is a zero mean Gaussian random variable with variance

$$\mathsf{E}\left(\left(\int_0^T \frac{U(t)}{\sqrt{2}} p(t)dt\right)^2\right) = \frac{1}{2}\mathsf{E}\left(\int_0^T \int_0^T U(t)p(t)U(x)p(x)dt\, dx\right)$$

Exercise 8.1

Use the double integral representation, and the fact that $\mathsf{E}(U(t)U(u)) = N_0\delta(t-u)$ (where $\delta(\cdot)$ is the Dirac delta function), to show that $\mathsf{E}\left(\left(\int_0^T U(t)p(t)dt\right)^2 = N_0\right)$.

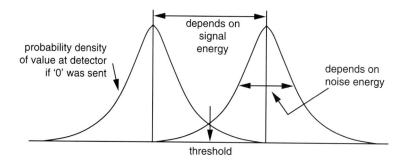

Figure 8.4   The probability densities of the statistic Y_k under the two possible symbols.

Hence we find that the inference is based on the *statistic*

$$Y_k = C_k + Z_k$$

where Z_k is a zero mean Gaussian random variable with variance $\frac{N_0}{2}$. Figure 8.4 depicts the probability density of Y_k under the two possible values of C_k. These are both Gaussian densities with variance $\frac{N_0}{2}$. The detector concludes that the bit sent was 0 if the value of Y_k is smaller than the threshold, and was 1 if the value of Y_k is more than the threshold. An error occurs if 1 is sent and Y_k falls below the threshold, and vice versa. When the source produces 0's and 1's with equal probabilities, then the threshold is midway between the means of the two densities (i.e., the threshold is 0). The probability of error if a 0 was sent is then given by

$$\Pr(Y_k > 0 \mid 0 \text{ was sent}) = Q\left(\sqrt{\frac{2E_s}{N_0}}\right)$$

where $Q(\tau) := \int_\tau^\infty \frac{1}{\sqrt{2\pi}} e^{-\frac{x^2}{2}} \, dx$. This can be seen to be the same as the probability of error if a 1 was sent. Hence the probability of error of the binary modulation scheme that we have described, under AWGN, is given by

$$P_{error\text{-}AWGN} = Q\left(\sqrt{\frac{2E_s}{N_0}}\right) \tag{8.4}$$

Note that in this example, because each symbol is used to send one bit, the error rate obtained is also the bit error rate (BER).

Exercise 8.2

Show that $P_{error\text{-}AWGN}$ decreases exponentially with $\frac{E_s}{N_0}$. Hint: For large x,

$$Q(x) \approx \frac{1}{x\sqrt{2\pi}}e^{-\frac{x^2}{2}}.$$

We see that the probability of correct detection depends on $\frac{E_s}{N_0}$, which is the ratio of the symbol energy to the noise power spectral density. Increasing the symbol energy increases the separation between the two Gaussian probability densities in Figure 8.4 and hence, for given noise variance, reduces the probability of Y_k falsely crossing the threshold. Similarly, decreasing the noise reduces the width of the two Gaussian probability densities, thus also reducing the error probability for a given signal energy.

8.1.2 Getting Higher Bit Rates

In the simple example in Section 8.1.1, because each pulse is modulated by one of two possible symbols and because the symbol rate is $\frac{1}{T}$, the bit rate is $\frac{1}{T}$ bps. One of the goals in designing a digital communication system over a radio spectrum is to use this spectrum to carry as high a bit rate as possible. With the binary modulation example in mind, there are two possibilities for increasing the bit rate.

- Increase the symbol rate (i.e., decrease T).

- Increase the number of possible symbols from 2 to $M > 2$.

Then, in general, the bit rate will be given by $\frac{\log_2 M}{T}$. There are, however, limits on both possibilities.

Note that if the pulse bandwidth is limited to W (the channel bandwidth), then the pulse duration will not be time-limited, and in fact the received signal in a symbol interval will be the sum of the pulse in that interval and parts of pulses in neighboring intervals. The pulses therefore must be designed appropriately to take care of this effect. This leads to the so-called *Nyquist criterion*, which limits the pulse rate to no more than $2W$ (i.e., $T \leq \frac{1}{2W}$).

Before we proceed, it is useful to make an observation. We saw in Section 8.1.1 that the probability of error for that binary signaling system depends on the ratio $\frac{E_s}{N_0}$. If the signaling rate is $\frac{1}{T}$ then the average power in the transmitted signal is $E_s \times \frac{1}{T}$. The noise power in the channel bandwidth is $2WN_0$. Hence the *signal power to noise power ratio* (SNR) is given by $\frac{E_s}{T2WN_0}$. If, in addition, the symbol rate is such that $T \times 2W = 1$ then the SNR is $\frac{E_s}{N_0}$. Thus for this example the probability of error depends on the SNR.

Let us now consider the other alternative for increasing the bit rate: increasing the number of possible symbols that can modulate the pulses. Figure 8.5(a) shows the binary symbol set that we have discussed. This is called binary *pulse amplitude modulation* (PAM), or 2-PAM. An example of the simplest possibility is shown in Figure 8.5(b); this is called 4-PAM. Because each of the 2-bit patterns 00, 01, 10, and 11 can be mapped to one of the symbols, this scheme can transmit 2 bits per symbol. However, to achieve a particular probability of error with a given noise power, the distance between the symbols must be retained as in the binary case; to see this, consider Figure 8.4; add a Gaussian density for each new symbol added, and then consider the probability of error between neighboring symbols. This means that the symbol energy when transmitting the leftmost and rightmost symbols in Figure 8.5(b) will be 3^2 times larger than for the other two symbols. This in turn implies a larger average signal power and hence a larger SNR (assuming the same noise power) for achieving the same probability of error.

Yet another alternative is shown in Figure 8.6(a), where we have *two-dimensional symbols*. Each symbol can be written in the form $ce^{j\theta}$, with $c \in \{-1, +1\}$ and $\theta \in \{0, \frac{\pi}{2}, \pi, \frac{3\pi}{2}\}$. This symbol set is called QPSK (*quadrature phase shift keying*) because all the symbols have the same amplitude but they have

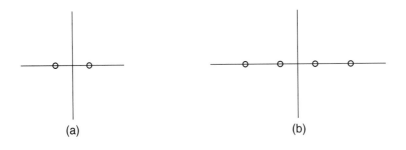

(a)　　　　　　　　　　　　　(b)

Figure 8.5　**Some symbol sets: (a) binary antipodal, (b) four-level amplitude modulation.**

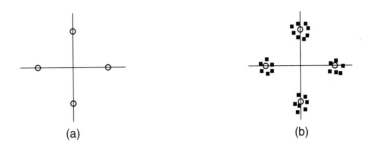

Figure 8.6   (a) A complex symbol set with four symbols; (b) the symbol set with noise added.

different phases. We can now take recourse to the general form shown in Figure 8.1 to write

$$X(t) = \sqrt{2}\,\mathrm{Re}\left(\sum_{k=-\infty}^{\infty} C_k e^{j\Theta_k}\, p(t-kT)\, e^{j2\pi f_c t} \right)$$

$$= \sqrt{2} \sum_{k=-\infty}^{\infty} C_k \cos(\Theta_k) p(t-kT) \cos(2\pi f_c t)$$

$$- \sqrt{2} \sum_{k=-\infty}^{\infty} C_k \sin(\Theta_k) p(t-kT) \sin(2\pi f_c t)$$

Here the sequence (C_k, Θ_k) is a random sequence that depends on the modulating bits. Thus, basically, the x-coordinate of the symbol modulates the carrier $\cos(2\pi f_c t)$, and the y-coordinate of the symbol modulates $-\sin(2\pi f_c t)$, which is also called the *quadrature* carrier (because it is $\frac{\pi}{2}$ out of phase with the *in-phase* carrier). In Figure 8.6(b) we show the received symbols after corruption by noise; the noise now has a two-dimensional Gaussian density, which is circularly symmetric about each symbol. At this point it is useful to recall the bandpass white Gaussian noise representation in Equation 8.2; we can interpret $U(t)$ and $V(t)$ as the in-phase and quadrature noise processes, respectively.

Notice from the geometry in Figure 8.6(a) that, by utilizing both dimensions, for a given probability of error, we can use a *smaller symbol spacing* than for the symbol set in Figure 8.5(b), and hence we can achieve a given BER with less average power. In fact, notice that the QPSK signal shown in Equation 8.5 is the superposition of two orthogonal 2-PAM signals; that is, the in-phase and quadrature signals are both 2-PAM signals. Hence after *down-conversion* (multiplying the signal by $\sqrt{2}\cos(2\pi f_c t)$ and also by $-\sqrt{2}\sin(2\pi f_c t)$ and filtering out the high-frequency terms) and multiplication and integration with the pulse $p(t)$, we obtain the following pair of statistics:

$$Y_k^{(i)} = C_k \cos(\Theta_k) + Z_k^{(i)}$$

$$Y_k^{(q)} = C_k \sin(\Theta_k) + Z_k^{(q)}$$

where the superscripts (i) and (q) denote the in-phase and quadrature components. Thus we have noisy observations of the two coordinates of the transmitted complex symbol, from which the transmitted symbol must be detected. Because in each symbol only one of the phases is used (and the other is 0 because of the simple QPSK symbol set), the average signal power is that of a 2-PAM signal.

As is evident from Figure 8.6, many more symbol sets are possible. If the amplitude as well as phase of the symbols can vary, then it is called QAM (*quadrature amplitude modulation*), whereas if only the phase can vary, it is called a PSK symbol set. Symbol sets are also called *constellations*. The probability of error in all the digital modulation and demodulation schemes based on the basic ideas we have discussed can be expressed as a function of the SNR at the receiver.

8.1.3 Channel Coding and a Fundamental Limit

In a given situation, because of physical limitations it may not be possible to increase the SNR so as to achieve the desired BER. The application being transported on the wireless link may require a lower BER in order to achieve reasonable performance. For example, if the link is used to transport packets and the packet length is L bits, then a BER of ϵ yields a packet error rate of $1 - (1 - \epsilon)^L$. We will see in Section 8.3 that a high packet error rate can seriously affect the performance of TCP transfers. Hence this may place a requirement on the minimum BER required from the link.

For a given digital modulation scheme, the BER as seen by the data source can be reduced by *channel coding*. The simplest viewpoint is shown in Figures 8.7 and 8.8. The channel with the given modulation scheme is viewed as an error-prone binary channel. Blocks of the incoming bits of length K are coded into *code words* of length $N(>K)$, thus introducing redundancy. If the code length and the codes

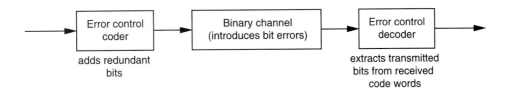

Figure 8.7 Channel coding: adding redundant bits to protect against channel errors.

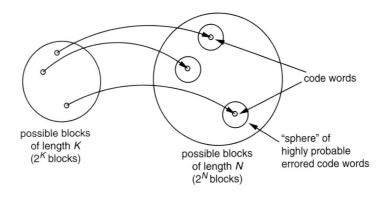

Figure 8.8 A channel code maps source bit strings into longer code bit strings (or code words); decoding involves identifying the code word nearest to the received bit string.

are chosen judiciously, then even after the channel introduces errors, an errored code word can be expected to stay close to the original code word. In Figure 8.8 we show source bit strings of length K being mapped into code blocks of length N. Because the number of possible code strings (2^N) is larger than the number of possible source strings (2^K), the code words can be chosen so that there is sufficient spacing between them. Now even if the channel causes errors, the errored code words will occupy spheres of high probability around the transmitted code words. Hence, by using *nearest code word decoding*, we can infer the transmitted code word, and hence the original source string, with a small residual error probability. The trade-off is that the information bit rate of the communication link becomes $\frac{K}{N}$, which is less than 1 information bit per code bit. This is called the rate of the code, denoted by R.

One trivial way of improving error performance is to increase N, because this results in the code words being spaced farther apart; but this reduces the information rate. It is possible, however, to increase K with N, keeping the information rate R constant while reducing the bit error rate to arbitrarily small values. Shannon's noisy channel coding theorem states that there is a number C, called the *channel capacity*, such that if $R < C$, then, as the block length increases, an arbitrarily small bit error rate can be achieved (albeit at the cost of a large block coding delay). If we attempt to use $R > C$, then the bit error rate cannot be reduced to 0.

A coder is followed by a digital modulation scheme that maps the code bits into channel symbols. As discussed in Section 8.1.1, the modulator maps a certain number of code bits (e.g., 2 in QPSK) into each channel symbol. Thus the capacity of the overall system (coder-modulator-channel-demodulator-decoder) can be expressed in terms of bits per symbol. At this point, it is obvious that to achieve this capacity *the receiver must know the channel coding and the modulation scheme that the transmitter is using*.

Shannon also provided the fundamental relationship between the channel capacity (C) and the SNR for an additive white Gaussian noise (AWGN) channel. We state this important result here so that we can use it later in our discussion. Suppose that, after down-conversion at the receiver, the received signal $Y(t)$ can be written as

$$Y(t) = X(t) + N(t)$$

where $X(t)$ is the received signal component with bandwidth W and average power P_{rcv}, and $N(t)$ is a white Gaussian noise process with power N_0 watts/Hz. An example of such a representation can be seen in the binary modulation scheme discussed in Section 8.1.1 (see Equation 8.3).

The following is the celebrated *Shannon capacity formula* for an AWGN channel.

$$C = \frac{1}{2} \log_2 \left(1 + \frac{P_{rcv}}{2W N_0} \right) \text{ bits/symbol} \tag{8.5}$$

Note that $\frac{P_{rcv}}{2W N_0}$ is the *received* SNR.

If the symbol rate is $\frac{1}{T}$ then Shannon's formula yields the bit rate

$$\frac{1}{T} \frac{1}{2} \log_2 \left(1 + \frac{P_{rcv}}{2W N_0} \right) \text{ bits/second}$$

For a pulse bandwidth W (or radio spectrum width $2W$), the bit rate is therefore limited to

$$W \log_2 \left(1 + \frac{P_{rcv}}{2WN_0} \right) \text{ bits/second} \qquad (8.6)$$

Furthermore, if the average received energy per symbol is E_{rcv}, then $P_{rcv} = E_{rcv} \times \frac{1}{T}$. Hence, if $\frac{1}{T} = 2W$ then the received SNR can also be written as $\frac{E_{rcv}}{N_0}$, yielding the capacity formula

$$W \log_2 \left(1 + \frac{E_{rcv}}{N_0} \right) \text{ bits/second}$$

We have seen the ratio $\frac{E_s}{N_0}$ (symbol energy divided by noise power spectral density) before, in Section 8.1.1, where it appears in the probability of error for the system discussed therein. Note that, in general, $E_{rcv} \neq E_s$ because of attenuation in the channel. We turn to this issue next.

8.1.4 Delay, Path Loss, Shadowing, and Fading

In the foregoing discussion we assume that the transmitted signal is contaminated only by additive white Gaussian noise. However, in practical channels, signals undergo attenuation and delay; in wireless channels, because of multipath, the attenuation can vary with time and with the relative distance between the transmitter and the receiver. We have seen that the BER performance of a digital communication system depends on the *received* SNR. Hence we are interested in the received signal power after the signal has passed through the channel. We proceed by first motivating a simple characterization of the received signal.

Radio waves are scattered by the objects they impinge on. Hence, unless a very narrow antenna beam is used, the receiver's antenna receives the transmitted signal along several paths. There is often a direct, or line-of-sight, path, and there are several paths along which the signal reaches the receiver after one or more reflections from various objects. Energy is lost in reflections and is absorbed by media through which the signal passes (partitions and walls). Hence the received signal is the sum of attenuated and delayed versions of the original signal.

Superposition of the delayed signals from the various paths can cause a symbol from one path to overlap a neighboring symbol from another path. Let us examine this issue first. These are electromagnetic signals, and hence they travel at the speed of light; let us take the propagation time to be roughly 0.33 μsec per 100 meters. Hence this is the kind of *delay spread* that can be expected if the various path lengths differ by no more than 100 m.

If the symbol time is several µseconds (e.g., 100,000 symbols per second), then there will not be significant overlap between the neighboring symbols, and we can assume that the symbols are still separately discernible, except that each is multiplied by a complex attenuation. If this happens then the channel is said to have *flat fading*. We can then write the kth received symbol after down-conversion as

$$Y_k(t) = A_k e^{j\Phi_k} \, C_k e^{j\Theta_k} \, p(t) + N(t)$$

where A_k is the random attenuation of the symbol, and Φ_k is the random phase. The symbol energy is thus multiplied by A_k^2. Let us write $H_k = A_k^2$; note that $\{H_k\}$ is a random process, and we must characterize it in order to understand the effect of the channel on the received signal power and hence on the SNR. We note that the H_k are also called *channel gains*.

When the delay spread is not very small compared with the symbol time, then the superposition of the signals received over the variously delayed paths at the receiver results in *intersymbol interference* (ISI) (see Figure 8.9). In some systems, this can be combated by passing the received signal through a *channel equalizer*, which can compensate for the various channel delays, making the overall system (i.e., the channel followed by the equalizer) appear to be a fixed-delay channel. In a mobile wireless situation, because of mobility, the paths that a signal takes between a transmitter and a receiver may keep changing, and hence a channel equalizer must be adaptive. In some systems the problem of signals arriving over

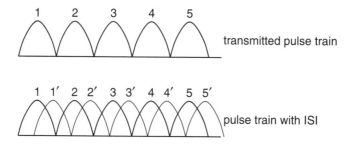

Figure 8.9   A simple illustration of the intersymbol interference that occurs when the same pulse train is received over two paths: one with zero delay, and the other with a large delay. The pulses labeled with ′ are the ones that arrive over the large delay path.

multiple paths is turned into an advantage. If the paths can be *resolved*, their powers can be added to increase the received SNR; such a receiver is called a *Rake receiver*.

Another approach to handling ISI is to partition a wide spectrum into several subchannels and send data over each at a slower rate (i.e., the symbols occupy more time) so that the delay spread is small compared with the symbol time. The arriving bit stream is partitioned over each of these channels. Basically, the incoming serial bit stream is converted to parallel bit streams that are then modulated over the various subchannels. The receiver detects the transmissions over the subchannels and recombines them to generate an output serial bit stream. Such a technique is called *orthogonal frequency division multiplexing* (OFDM). We do not discuss these topics further.

A Characterization of the Signal Attenuation

We characterize the process $\{H_k\}$ by writing it in terms of three multiplicative components:

$$H_k = \left(\left(\frac{d_k}{d_0} \right)^{-\eta} \cdot G_k \cdot R_k^2 \right) \tag{8.7}$$

Let us write the marginal terms of the stationary random processes in this expression by dropping the symbol index k. We now discuss each of these terms.

The term $\left(\frac{d}{d_0} \right)^{-\eta}$ is the path loss factor. Here, d is the distance between the transmitter and the receiver when the kth symbol is being received, d_0 is the *reference distance* at which the transmitter power is measured, and η is the path loss exponent, which is typically in the range 2 to 5. The distance d_0 is taken to be large enough so that the *power law* formula applies for distances more than d_0, but small enough so that a receiver would be unlikely to be at a distance less than d_0.

If the attenuation is measured at various points at a distance d from the transmitter, then the attenuation will be found to be random, because of variations in the terrain and in the media through which the signal may have passed. Empirical studies have shown that this randomness is captured well if the second factor G, in Equation 8.7, has the form $10^{-\frac{\xi}{10}}$, with ξ being a Gaussian random variable with mean 0 and variance σ^2. This is called the *shadowing* component of the attenuation, and, because $\log_{10}$ of this term has a Gaussian (or normal) distribution, it is called *log-normal shadowing*. It is often convenient to express values of power and power ratios in the *decibel* (dB) unit, which is obtained by taking $10 \log_{10}$ of the value. Hence the shadowing attenuation in signal power is $10 \log_{10} G = -\xi$ dB, which is zero mean Gaussian with variance σ^2. Shadow fading is spatially varying, and hence if there is relative movement between the transmitter and the receiver then shadow fading will vary. The correlation in the shadow fading in dB between two points separated by a distance D is given by $\sigma^2 e^{-\frac{D}{D_0}}$, where D_0 is a parameter that depends on the terrain. Some measurements have given $D_0 = 500$ m for suburban terrains, and $D_0 = 50$ m for urban terrains. Hence if the distance varies by a few meters per second (note that 36 Kmph = 10 meters/second), then the shadowing will vary over seconds, and this means that the variations will occur over hundreds of thousands of symbols.

We now turn to the third factor in the expression for attenuation in Equation 8.7. Typical carrier frequencies used in mobile wireless networks are 900 MHz, 1.8 GHz (e.g., these two frequency bands are used in cellular wireless telephony systems), or 2.4 GHz (e.g., used in IEEE 802.11 wireless LAN systems), and hence the carrier wave periods are a few picoseconds. Hence when the transmitted signal arrives over several paths, very small differences in the path lengths (a few centimeters) can cause large differences in the phases of the carriers that are being superimposed. Thus, although these time delays may not result in ISI, the superposition of the delayed carriers results in constructive and destructive carrier interference, leading to variations in signal strength. This phenomenon is called *multipath fading*. This is a random attenuation that has strong autocorrelation over a time duration called the *coherence time*; the attenuations at two time instants separated by more than the coherence time are weakly correlated. The coherence time is related to the *Doppler frequency*, f_d, which is related to the carrier frequency, f_c, the speed of movement, v, and the speed of light, c, by $f_d = f_c \frac{v}{c}$. Roughly speaking, the coherence time is the inverse of the Doppler frequency and, from the point of view of the receiver, the channel's fade level can be assumed to be varying at the Doppler frequency. For example, if the carrier frequency is 900 MHz and $v = 20$ meters/sec, then $f_d = 60$ Hz, leading to a coherence time of tens of milliseconds. In an indoor office or home environment, the Doppler frequency could be only a few Hz (e.g., 3 Hz), with coherence times of hundreds of milliseconds. The marginal distribution of R^2 depends on whether all the signals arriving at the receiver are scattered signals or whether there is also a line-of-sight signal. In the former case, assuming uniformly distributed arrival of the signal from all directions, the distribution of R^2 is exponential with mean $E(R^2)$:

$$f_{R^2}(x) = \frac{1}{E(R^2)} e^{(-x/E(R^2))}$$

The distribution of the amplitude attenuation (i.e., R) is Rayleigh, and hence this is also called *Rayleigh fading*. On the other hand, if there is a line-of-sight component so that $\frac{K}{K+1}$ of the signal arrives directly and if the remaining signal arrives uniformly over all directions, then

$$f_{R^2}(x) = \frac{K+1}{E(R^2)} e^{\left(-K - \frac{(K+1)x}{E(R^2)}\right)} I_0 \left(2\sqrt{\left(\frac{K(K+1)x}{E(R^2)}\right)}\right)$$

where

$$I_0(x) = \frac{1}{2\pi} \int_0^{2\pi} e^{-x \cos(\theta)} d\theta$$

This is called the Ricean distribution.

With this characterization of the attenuation in the received signal power, we can now write the received SNR in terms of the transmitted SNR (i.e., the ratio of the transmitted signal power to the noise power); let us denote these by Ψ_{rcv} and Ψ_{xmt}, respectively. Then

$$\Psi_{rcv} = \Psi_{xmt} \cdot H$$

$$= \Psi_{xmt} \cdot \left(\frac{d}{d_0}\right)^{-\eta} \cdot 10^{\frac{-\xi}{10}} \cdot R^2 \qquad (8.8)$$

In dB we can write the received SNR as

$$(\Psi_{rcv})_{dB} = (\Psi_{xmt})_{dB} + 10 \log_{10} H$$

$$= (\Psi_{xmt})_{dB} - 10\eta \log_{10}(\frac{d}{d_0}) - \xi + 10 \log_{10} R^2 \qquad (8.9)$$

BER and Channel Capacity with Fading

We now turn to the calculation of the performance of the wireless link in the presence of fading. We have seen that even though the transmitter may send at a fixed power, in the presence of fading, the received power (and hence the SNR) is time-varying. The rate of variation of the SNR depends on the mobility of the receiver. A receiver that moves short distances over the duration of a *conversation* (e.g., a voice call or a file transfer) would sample the distribution of the Rayleigh fading but would see roughly constant values of path loss and shadowing. On the other hand, a receiver that makes large movements during a call duration would see variations in all three attenuation factors during the call.

Let us consider the former situation. In this case the *in-call* performance depends on the value of path loss and shadow fading sampled by the call, and on the distribution of Rayleigh fading, but the performance *across calls* also depends on the variation in path loss and shadowing. We would like the performance not to fall below some value. For example, there could be a desired upper bound on BER; to exceed this bound would be termed an *outage*. Let us examine this point in the context of the binary modulation scheme discussed in Section 8.1.1. The BER for this modulation scheme is given by Equation 8.4:

$$P_{error\text{-}AWGN}(\Psi_{rcv}) = Q\left(\sqrt{2\Psi_{rcv}}\right)$$

If the path loss and shadowing factors during a call are fixed, then we can calculate the *in-call BER* as follows.

$$\int_0^\infty P_{error\text{-}AWGN}\left(\left(\frac{d}{d_0}\right)^{-\eta} \cdot 10^{\frac{-\xi}{10}} \cdot \gamma \Psi_{xmt}\right) f_{R^2}(\gamma) d\gamma$$

where, as mentioned earlier, $f_{R^2}(\cdot)$ is the exponential distribution with mean $E(R^2)$. In many cases it can be shown that this expression can be simplified to the following form:

$$P_{error\text{-}fading}\left(\left(\frac{d}{d_0}\right)^{-\eta} \cdot 10^{\frac{-\xi}{10}} \cdot \left(E\left(R^2\right) \Psi_{xmt}\right)\right)$$

Let us write $\left(\frac{d}{d_0}\right)^{-\eta} \cdot 10^{\frac{-\xi}{10}} \cdot \left(E(R^2) \Psi_{xmt}\right)$, the average SNR during the call, as $\overline{\Psi}_{rcv}$. Then, for example, for the binary modulation scheme discussed earlier, it can be shown that $P_{error\text{-}fading}\left(\overline{\Psi}_{rcv}\right) = \frac{1}{2}\left(1 - \sqrt{\frac{\overline{\Psi}_{rcv}}{1+\overline{\Psi}_{rcv}}}\right)$, which for large $\overline{\Psi}_{rcv}$ can be shown to decrease as the reciprocal of the average SNR, rather than exponentially, as for unfaded AWGN (see Exercise 8.2; see also Problem 8.2).

We can write the average SNR during a call (i.e., $\overline{\Psi}_{rcv}$) in dB as

$$\left(\overline{\Psi}_{rcv}\right)_{dB} = \left(\Psi_{xmt}\mathsf{E}\left(R^2\right)\right)_{dB} - 10\eta\log_{10}\left(\frac{d}{d_0}\right) - \xi$$

The term $\left(\Psi_{xmt}\mathsf{E}(R^2)\right)_{dB}$ is the *Rayleigh faded SNR "referred to"* d_0. We see that the received SNR, in dB, at a distance d from the transmitter is Gaussian with mean $\left(\Psi_{xmt}\mathsf{E}(R^2)\right)_{dB} - 10\eta\log_{10}\left(\frac{d}{d_0}\right)$ and variance σ^2. To achieve a certain BER—say, ϵ—the received SNR must be greater than a threshold (say, β):

$$\overline{\Psi}_{rcv} > \beta \Rightarrow P_{error\text{-}fading}\left(\overline{\Psi}_{rcv}\right) < \epsilon$$

Violation of this requirement would be an outage, the probability of which we would like to limit to P_{outage}. The BER and outage requirement can then be expressed in the following form:

$$\mathsf{Pr}\left(\left(\overline{\Psi}_{rcv}\right)_{dB} < (\beta)_{dB}\right) < P_{outage}$$

Equivalently,

$$\mathsf{Pr}\left(\left(\Psi_{xmt}\mathsf{E}\left(R^2\right)\right)_{dB} - 10\eta\log_{10}\left(\frac{d}{d_0}\right) - \xi < (\beta)_{dB}\right) < P_{outage}$$

Example 8.1

Given that $\frac{d}{d_0} = 10$, $\eta = 3$, the shadowing standard deviation $\sigma = 8$ *dB*, the received SNR threshold is $\beta = 10$ *dB*, and $P_{outage} = 0.01$, the earlier displayed requirement is satisfied if

$$\left(\Psi_{xmt}\mathsf{E}\left(R^2\right)\right)_{dB} - 30 - 2.3 \times 8 = 10$$

where the factor 2.3 is obtained from a table of the Gaussian distribution. This yields

$$\left(\Psi_{xmt} \mathsf{E}\left(R^2 \right) \right)_{dB} = 58.4 \; dB \qquad \blacksquare$$

How does a time-varying channel attenuation affect the Shannon capacity formula? If the channel attenuation is h and if the noise is AWGN, then, for transmitted power P_{xmt}, the channel capacity is given by Equation 8.6:

$$W \log_2 \left(1 + \frac{h P_{xmt}}{2 W N_0} \right)$$

Suppose that the transmitter is unaware of the extent of the channel fading and uses a fixed power and a fixed modulation and coding scheme. Suppose also that the fading level varies slowly. Then, for a given level of fading, the receiver must know h in order for the communication to achieve the Shannon capacity. To see this, let us look at Figure 8.5(b). If the channel's power attenuation is h, the received symbols are multiplied by $\sqrt{h}$. This results in the symbols being "squeezed" together or spread apart. Obviously, the detection thresholds must depend on the level of fading.

Suppose that H_k is a stationary and ergodic process. It can then be shown that, if the transmitter cannot adapt its coding and modulation but the receiver can exactly track the fading, then the channel capacity with fading is given by

$$C_{fading\text{-}CSIR} = \int W \log_2 \left(1 + \frac{h P_{xmt}}{2 W N_0} \right) g_H(h) dh \qquad (8.10)$$

where $g_H(\cdot)$ is the marginal density of the channel attenuation process H_k. For example, $g_H(h)$ is exponential for Rayleigh fading. CSIR stands for *Channel state (or side) information at the receiver*. Thus the transmitter can encode at any *fixed-rate* $R < C_{fading\text{-}CSIR}$, and for large enough code blocks the error rate can be made arbitrarily small, provided that the receiver can track the channel. Bear in mind that this is an ideal result; to achieve it the channel fades must be averaged over, and this will result in large coding delays.

> **Exercise 8.3**
> Use Jensen's inequality to show that $C_{fading\text{-}CSIR} \leq W \log_2 \left(1 + \frac{E(H)P_{xmt}}{2W N_0} \right)$
> (i.e., the capacity with fading is less than that with no fading with the same average SNR).

With fading, there will be times when the SNR is higher than the average, and other times when the SNR will be lower than the average. Yet the exercise shows that the resulting channel capacity is less than that without fading, as long as the same average SNR is maintained.

8.2 Bits over a Wireless Network

In Section 8.1 we discuss the issues in transmitting bits over a radio spectrum between a transmitter–receiver pair. A wireless network would comprise several transmitters and receivers sharing an assigned portion of the radio spectrum (see Figure 8.2). We would like to carry as much traffic in this network as possible while keeping all transmissions within this spectrum. An inefficient way would be to permit exactly one transmitter and receiver pair communicate at one time. Obviously the attempt should be to let several transmitter and receiver pairs communicate simultaneously.

Let us assume that all the nodes use identical transmitters and receivers, and that a minimum SNR of β is required by a receiver to decode a transmitted packet. We can then define the *transmission range* (or *decoding range*) of a transmitter as the distance over which its transmissions can be received by another node, in the absence of any other interference. Suppose that we assume that the only signal attenuation is due to path loss (see Section 8.1.4) and that the path loss exponent is η. Then with transmitter power P, we obtain the transmission range r from the following equation:

$$\frac{P \left(\frac{r}{d_0} \right)^{-\eta}}{2W N_0} = \beta \tag{8.11}$$

where we account only for Gaussian noise because no interference is assumed. Thus r is the minimum distance between a transmitter and a receiver for successful communication. It is normally assumed that node i (as a receiver) is in the transmission range of node j, if and only if node j (as a receiver) is also in the transmission range of node i. It is important to emphasize that this is in the absence of interference and assumes the same radio electronics at both the nodes. With this discussion in mind, it is customary to represent the "within transmission range" relation between nodes in a wireless network as an undirected *graph* (see Section 8.7.1). In this graph we would have an edge between nodes i and j if they are within transmission range of each other. Again we emphasize that this means that a transmission from i to j will be decoded by j in the absence of transmissions from any other node. Transmissions from other nodes may or may not cause sufficient interference at j to jeopardize the reception of i's transmission.

Now let us consider multiple nodes sharing the same geographical area and communicating over the same spectrum, using the same kinds of transmitters and receivers. If a node receives signals from different transmitters over the same frequency band, then there is a possibility of cochannel interference, which may lead to the inability at the receiver to detect either of the signals. There are two (complementary) ways in which the spectrum can be simultaneously used by several transmitter–receiver pairs.

The first approach is to design the network and the radio receivers so as to deal with cochannel interference. This leads to the concepts of capture, spatial reuse, and multiuser detection. Even if there are simultaneous transmissions in the same channel, it may be possible to decode one or more of them at a receiver. Suppose a transmitter i is transmitting at power P_i to node j; let us denote the power attenuation from i to j by $H_{i,j}$. Other nodes $k, 1 \leq k \leq M$, are also transmitting at powers P_k (to other receivers), and their attenuations to node j are $H_{k,j}$. The total interference power at node j is then $\sum_{k=1}^{M} H_{k,j} P_k$. Hence, the SNR (or, more appropriately, the SIR, or *signal to interference ratio*) for the signal from i at receiver j is given by

$$\Psi_{i,j} = \frac{H_{i,j} P_i}{2 W N_0 + \sum_{k=1}^{M} H_{k,j} P_k} \tag{8.12}$$

where in the denominator k runs over only the interfering nodes (i.e., not including the node i). As an approximation, the interference is also taken as white Gaussian noise, and hence this expression can be used along with the distributions of the channel attenuations, and the BER formula for the modulation and detection scheme, to determine the bit error rate for the i to j communication. Alternatively, for a given desired BER, the target SIR can be obtained from such formulas (or tables or graphs), and it can be checked whether the SIR given by Equation 8.12 is greater than the target. The phenomenon in which a receiver can decode one transmission in the presence of several others has been called *capture*.

The idea of treating cochannel interference as noise, along with the geographical attenuation of power discussed in Section 8.1.4, can be used to deliberately let several transmitter and receiver pairs simultaneously use the same radio spectrum as long as they are sufficiently separated in space. Called *spatial reuse*, this is an important way of optimizing the utilization of a radio spectrum. At each receiver there would be some target value of SIR, $\Psi_{i,j}$ (see Equation 8.12), and hence there may be a need for power control of the transmitters to achieve the desired SIR targets. For example, if some other (cochannel) transmitter k is transmitting to receiver l, and if $H_{k,l}$ is small, then P_k may need to be increased. At the same time, however, $H_{k,j}$ may be large, thus resulting in a lot of cochannel interference at j, which really wants to decode i's transmission. This may necessitate an increase in the transmission power P_i. The problem would then be to obtain a set of consistent powers so that all SIR targets are met. In a mobile wireless network, the attenuations or channel gains would be time-varying, and hence there would be a need for dynamic power control of the transmitters, based on SIR or BER measurements.

We can expect that with enough sophistication at the receiver it may be possible for it to detect all or many of the interfering signals. Intuitively, if there is a strong signal and a weak signal, the receiver can first detect the strong one, treating the others as noise, and then it can subtract the detected signal from the received signal, thus being able to detect the weak signal in nothing more than the ambient noise. Such a receiver is said to be carrying out *multiuser detection*. A Shannon theory for such an approach has also been worked out; this is known as *multiuser information theory*. We do not pursue this topic further in this book.

The second way to allow use of a spectrum by several transmitter–receiver pairs is by *signal orthogonalization*. The interfering signals that are transmitted and that can potentially interfere can be arranged in such a way that they can be separated and detected by their respective receivers (which may be the same receiver). This can be done by separating the signal transmissions in time (time division multiple access (TDMA)), in frequency (frequency division multiple access (FDMA)), or in signal space (code division multiple access (CDMA)). As the names suggest, in TDMA different transmissions are interleaved in time; in FDMA, the spectrum is divided into sub-bands that are then assigned to different transmitter–receiver pairs. Many systems use a combination of FDMA and TDMA; that is, within an FDMA sub-band, TDMA is used. In CDMA, during a symbol time the signals sent by the various transmitters are chosen to be orthogonal, so that even if the transmissions are in the same frequency and also overlap in time, the receiver(s) can separate them.

Interference and Packet Collision

It is usually very difficult to work with the general approach of calculating the SIRs (see Equation 8.12) for all possible transmitter–receiver pairs, and, hence, in analytical work certain approximations are used. A worst case assumption that is often made in analyzing multiple access protocols is that whenever two transmissions in the same frequency band arrive at a receiver, neither can be detected; this is called a collision. A collision can be taken to occur even after a receiver has successfully begun to decode a transmission and is then hit by one or more interfering signals. Such an assumption is made because analysis of multiple access protocols involves models of other processes, such as arrivals of packets. The simplifying assumption of collisions makes the analysis tractable. The results of the performance analysis can be expected to yield a lower bound on what is achievable. If several nodes share a spectrum and all use the same power, then a less conservative model is to assume a *collision threshold*—say, $K \geq 1$. If the number of interfering transmissions at a receiver is less than K, then one of them can be captured.

In this context, we can define the *interference range* of a receiver. This is the range within which a transmitter can cause interference; transmitters outside this range will not even be "heard" by the receiver electronics. Thus the simplest model assumes that any interfering transmission in the interference range of a receiver causes a collision. A less pessimistic model is the following. Consider a transmitter sending to a receiver at a distance d. Then we can say that any interfering transmitter that is at a distance greater than $(1+\Delta)d$, for some appropriate positive Δ, is too far from the intended receiver to affect the ongoing transmission. Any interfering transmitter closer than this will cause a collision.

8.3   TCP Performance over Wireless Links

Sections 8.1 and 8.2 discuss the physical transport of bits over a shared radio spectrum. We now turn to a study of information transport over such a point-to-point bit carrier.

An important application of wireless links is to connect mobile users to access the Internet. The predominant use of the Internet is for the transport of data files (e.g., e-mail, Web browsing, and file transfer), as discussed in Chapter 7. The TCP protocol is used to control the transport of packets in such applications. Hence we first turn to a study of performance issues that arise when TCP operates over a point-to-point lossy and fading wireless link. Such a link could be between, for example, a laptop computer and a base station that is attached to a wired network.

In Section 7.6, we discuss TCP at length and describe models for evaluating the performance of TCP-controlled file transfers in several situations. In Section 7.6.5 we studied a model that can be used to obtain the performance of TCP-controlled file transfers with random packet loss. We saw that the performance of TCP can be significantly affected by packet loss. In these discussions the concern is with congestion-related loss; either a packet is lost because of buffer overflow, or a packet is deliberately dropped at a router queue because of imminent congestion (see the discussion on the RED mechanism in Section 7.6.6). In those discussions we are not concerned with the possibility of packet loss in the physical bit carriers. In a sense, we are assuming a wired physical infrastructure. Wired links can be properly established so that they have small BERs. On the other hand, as discussed in the review in Section 8.1, mobile wireless links are subject to random variations in their quality. It is therefore of interest to study the performance of TCP transfers over a wireless access link, particularly in light of the growing importance of mobile wireless access to the Internet.

Figure 8.10 shows a simple scenario in which a mobile host is making a TCP-controlled file transfer from a file server on a wired LAN. The propagation delay between the LAN–wireless router and the mobile host is small, thus modeling a typical home or office environment. The BER on the wireless link is such that packets are lost with probability p. The packets are lost independently (i.e., correlated losses due to channel fading are not modeled here). Only ACKs are sent from the mobile host to the LAN, and because these are small (40 bytes) their

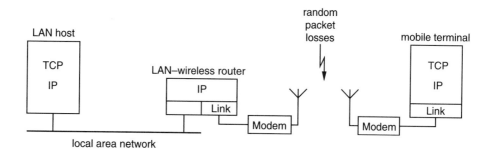

Figure 8.10   A mobile host transferring data over a wireless link from a host on the wired LAN.

loss probability is ignored; recall that TCP uses cumulative ACKs, and this further limits the effect of ACK loss. No link layer protocol is assumed, and hence any packet loss must be recovered by TCP. As the TCP transmitter on the file server grows its window, the wireless link buffer builds up. The buffer can hold as many packets as needed (i.e., there is no buffer loss). Eventually a loss occurs in the wireless link, and one of the loss recovery mechanisms is invoked.

We can analyze the throughput of a large file transfer using the analysis approach described in Section 7.6.5. A sample of results obtained from this analysis is shown in Figure 8.11. The parameters and the results are normalized. We plot the file transfer throughput versus the packet loss probability. The throughput is normalized to the bit rate of the wireless link. One set of parameters that would correspond to the results is LAN speed (10 Mbps), wireless link bit rate (2 Mbps),

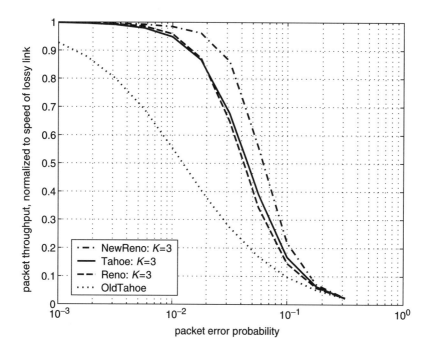

Figure 8.11 File transfer throughput (normalized to the link's bit rate) vs packet loss probability for various versions of TCP; OldTahoe refers to a version that recovers losses only by timeout. K is the duplicate ACK threshold for fast-retransmit in the TCP loss recovery protocol. Adapted from Kumar [181].

TCP packet length (1500 bytes) (hence the packet transmission time is 6 ms), timeout granularity (420 ms), minimum timeout (600 ms), and $W_{max} = 24$ packets (see Section 7.6.1 for a discussion of these parameters). The performance of four versions of TCP is compared: OldTahoe (which is the name we give to the version of TCP that predates Tahoe and always requires timeouts to recover losses; see Figure 7.15), Tahoe, Reno, and NewReno (see Section 7.6). We observe that even with a packet loss probability of 0.001, the throughput with OldTahoe is less than the full link rate and drops to just more than 50% of the link rate for a packet loss probability of 0.01. The other three TCP versions implement fast-retransmit, and they yield 100% throughput at $p = 0.001$, and more than 95% throughput up to $p = 0.01$. Beyond 1% packet loss, the performance of these versions too begins to drop, and it is not much better than OldTahoe for a 10% packet loss rate. Reno is slightly better than Tahoe up to $p = 0.02$, but it becomes worse for large loss rates because multiple losses cause it to waste more time than does Tahoe. The more aggressive fast-recovery of NewReno results in this version yielding almost 90% throughput up to $p = 0.03$. A broad observation we can make is that random packet loss probabilities larger than 1% can significantly affect the performance of TCP, with the foregoing parameters. Note that the coarse timeout and minimum timeout values are large in this example. Smaller values for these parameters will yield better performance, because losses will then result in less waste of link capacity.

If a packet loss probability of p is desired and if the packet length is L bits, then the BER on the wireless link ϵ should satisfy the requirement $p = 1 - (1 - \epsilon)^L$. Hence we see that the performance of the application we wish to carry on the wireless link puts a requirement on the performance of the link. Given the wireless spectrum, the problem is then to design the coding and modulation scheme so as to obtain the desired bit error rate in a cost-effective way. We saw in Section 8.1 that the BER on a wireless link is a function of the SNR. Because mobile devices run on batteries, one of the design considerations is to use as little power as possible to achieve a desired link performance. If the spectrum is being spatially reused in the network, then reduction in transmitter power also helps in reducing cochannel interference (see Section 8.2).

We now turn to the performance of TCP-controlled file transfers over a fading channel. In Section 8.1.4 we discuss models for channel fading, pointing out that the fading is correlated. Thus for a given average BER there would be periods when the BER is greater than the average, and periods during which the BER is less than the average. A similar statement can be made for the packet error rate if fixed-length packets are being used, as is typically the case with large file transfers over TCP. A simple approach is to model the channel as being in one of two

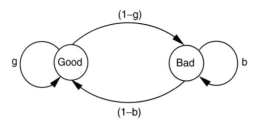

Figure 8.12   Transition structure of the two-state Markov model for a fading channel.

states: a Good state (during which a packet transmission is successful) and a Bad state (during which a packet transmission is unsuccessful). A further simplification is to model the state process as a two-state Markov process on the state space $\{Good, Bad\}$; see Figure 8.12. The durations in each state are taken to be multiples of the packet transmission time. We obtain the transition probabilities of the Markov chain by specifying the amount of fading that leads to a bad transmission (at other times a good transmission is assumed). The marginal distribution of the fading process, as well as results about correlations in the fading process, can be used to obtain the transition probabilities. For a packet length L and channel bit rate C, Doppler frequency f_d, the parameter $f_d \frac{L}{C}$ is a measure of the fade durations relative to the packet transmission time; thus $f_d \frac{L}{C} = 0.01$ means that channel coherence time is roughly 100 packet transmission times.

Using the Markov model for the channel state, we can analyze the throughput of a long file transfer under TCP using the approach outlined in Section 7.6.5. In performing this analysis, in addition to the state of the TCP window adaptation process, we must maintain the state of the channel. Figure 8.13 shows typical numerical results with Rayleigh fading. The normalized throughputs with TCP Tahoe and Reno are plotted versus the average packet error probability, with and without fading. For the results with fading, the parameter $f_d \frac{L}{C} = 0.01$. The other parameters are the same as in Figure 8.11 except that the local area network is taken to be infinitely fast. Notice that the performance without fading is similar to that depicted in Figure 8.11. With fading, the performance is significantly different. For the same probability of error, we find that the performance of TCP Tahoe increases substantially, whereas that of TCP Reno drops for $p < 3 \times 10^{-2}$ and improves for large packet loss probabilities. This can be understood as follows. With independent losses, the repeated reductions in the window lead to a small effective window; hence when a loss occurs there are not enough packets in

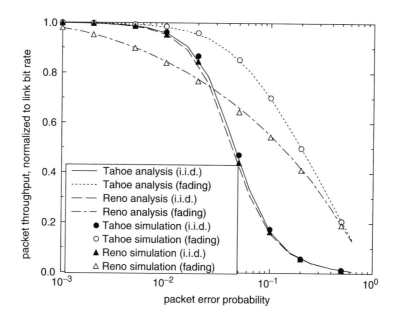

Figure 8.13 **File transfer throughput (normalized to the link's bit rate) vs packet loss probability for TCP Tahoe and Reno, with independent losses (denoted as "i.i.d") and with Rayleigh fading with $f_d \frac{L}{C} = 0.01$. Adapted from Zorzi et al. [311].**

circulation to generate the number of duplicate ACKs required for a fast retransmit. Thus with uncorrelated losses, timeouts are more frequent. When packet errors are clustered (as in the case of fading), the durations between packet loss events are larger. Hence with correlated packet losses, the TCP transmitter can grow its window to larger values than in the independent packet loss case (for the same average packet error probability). When a loss does occur, it is more likely that there are enough successful packets sent subsequently in the window to trigger a fast-retransmit. Even if a timeout does occur, it is long enough to last out the fade, so that when transmission resumes the channel is likely to be in the Good state. For small values of p the performance of Reno is worse because Reno requires additional duplicate ACKs for recovering each lost packet. With correlated losses, multiple losses are more likely, and this results in Reno wasting more time than Tahoe. Reno attempts to perform a fast-retransmit for each lost packet, spends time in this process waiting for duplicate ACKs, and then times out anyway. For large values of p the two protocols have similar behavior because with the high loss rate the window grows to small values, and the number of duplicate ACKs is insufficient to trigger a fast-retransmit; hence it is very likely that both protocols recover with a timeout.

This discussion illustrates the effect of correlated errors on TCP-controlled file transfer performance, but it is important to make a comparison by fixing the average SNR. The same two-state Markov model can be used. The SNR that corresponds to a Bad state is first fixed. Then for each SNR and Doppler frequency, the two-state Markov model can be parameterized. Sample results for TCP Tahoe are shown in Figure 8.14. To compare with the results presented earlier, no channel coding or link-level retransmissions are taken into account. The normalized throughput is plotted against the average SNR in dB. We observe that without fading, an SNR of about 12 dB suffices to obtain a TCP throughput

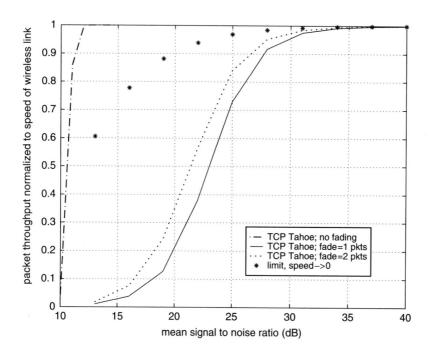

Figure 8.14   File transfer throughput (normalized to the link's bit rate), with TCP Tahoe vs SNR in dB, with no fading (AWGN only) and with Rayleigh fading. The legend "fade = *n* pkts" means that the mean Bad state duration is *n* packets, where a Bad state occurs if the SNR < 10 dB. Adapted from Kumar and Holtzman [182].

of more than 90% of the link rate. This is because the packet error probability itself is very small without fading. With fading, however, much larger Rayleigh-faded SNRs are required—between 25 and 30 dB for a throughput of 90% of link rate. We notice that slower fading (and hence more correlated errors) improves the TCP throughput, but the throughput even with this improvement is much worse than without fading. We also show the speed $\to 0$ case, which corresponds to the fade level being constant during the entire TCP transfer; the channel is either good throughout the transfer or bad throughout. This is a bound on the achievable throughput with fading. As the faded SNR decreases, the probability of the Good state reduces, and hence the bound rapidly decreases for decreasing average faded SNR.

8.4 Adaptive and Cross-Layer Techniques

In Sections 8.1 and 8.2 we discuss the issues involved in carrying bits over a wireless spectrum. Then, given such a wireless link, in Section 8.3 we examine the performance of a TCP-controlled file transfer application over the link. We find that TCP performance can be very poor unless the packet error rate on the wireless link is small. We suggest that this could be a way of putting a design constraint on the wireless link. This is, however, a typical layered viewpoint; each layer performs its own function independently of the state of any other layer, but the need for each layer to perform its functions well may place performance requirements on the other layers.

There could, however, be benefits (e.g., the ability to handle more traffic in a given spectrum for given QoS requirements) if the behavior of procedures in certain layers could adapt to the state of the other layers. As a simple example, consider the operation of TCP over a fading channel. When a packet loss occurs, TCP infers a congestion loss and drops its window. We have seen the detrimental effect of this on TCP throughput in a local network scenario in Section 8.3. In satellite networks, where there are large propagation delays, the effect is even more severe. Now suppose that the lower layers could inform the TCP layer about the poor condition of the channel; then instead of dropping its window, the TCP transmitter could defer transmission (until, perhaps, informed of an improvement in the channel condition by the lower layers). This not only would result in better TCP performance but also would reduce unnecessary transmissions and retransmissions that would deplete the battery of a mobile device.

Let us begin by examining adaptive techniques at the physical layer. We have seen that the capacity of an AWGN channel with stationary and ergodic fading, with fixed power and modulation and coding at the transmitter, and with perfect channel tracking at the receiver is given by Equation 8.10:

$$C_{fading\text{-}CSIR} = \int W \log_2 \left(1 + \frac{hP_{xmt}}{2W N_0}\right) g_H(h) dh$$

To achieve this channel capacity, large block lengths are required so that the ergodic probability law of the fading process is adequately sampled. For short block lengths and for fixed transmit power, poor performance may be obtained for small values of h, and good performance may be obtained for large values of h. Suppose, however, that the transmitter can also determine the channel state. At first thought it may seem reasonable to increase the transmitter power when h is small and reduce the power when h is large. An alternative could be to use more power when h is large (the channel is good) and not send at all when the channel is poor. Which is the better approach? Suppose we have an average power constraint, P_{avg}; this could be a way to model a battery drain rate constraint. We could then seek a power control policy (under the average power constraint) such that the capacity of the link is optimized. This yields the following optimization problem.

$$\max \quad W \frac{1}{\ln 2} \int \ln \left(1 + \frac{hP(h)}{2W N_0}\right) g_H(h) dh \qquad (8.13)$$

subject to

$$\int P(h) g_H(h) dh \leq P_{avg}$$

$$P(h) \geq 0 \quad \text{for every } h$$

Exercise 8.4
Consider a channel that takes a finite number n of fading states with gains $h_i, 1 \leq i \leq n$, the state i occurring with probabiliy $g_H(i)$ (i.e., $\sum_{i=1}^{n} g_H(i) = 1$). Write the foregoing problem for this channel, and then use the Karush–Kuhn–Tucker sufficient conditions (see Appendix C)

to establish the following optimal power control. There is a number $\lambda > 0$ such that

$$P(h_i) = \left(\frac{1}{\lambda} - \frac{2W N_0}{h_i} \right)^+ \tag{8.14}$$

where λ is then obtained from

$$\sum_{i=1}^{n} P(h_i) g_H(i) = P_{avg}$$

Show that the optimal capacity is then given by

$$C_{fading\text{-}optimal}(P_{avg}) = W \sum_{\{i:h_i > 2W N_0 \lambda\}} \log_2 \left(\frac{h_i}{2W N_0 \lambda} \right) g_H(i) \tag{8.15}$$

The general solution with a continuous state-space channel has exactly the same form as in the exercise, except that we obtain the average power and the optimal channel capacity by integrating over the channel states with respect to the density $g_H(\cdot)$. Note that states with very small values of h are not assigned any power (i.e., the policy recommends not transmitting at all if the channel is very poor). Also, we see that the channel capacity is positive.

Observe from this solution that for each channel state, the policy prescribes a power, and to achieve channel capacity the transmitter must use a modulation and coding scheme that achieves the capacity for that channel state; furthermore, the receiver must also know the channel state to know how to demodulate and decode the received signal.

Before we further discuss the solution, let us consider an alternative power control. This power control aims to maintain a constant received power, $P_{constant}$, thus keeping the capacity of the channel constant. It can be called a *constant-rate* or *constant-SNR* power control. Clearly, this policy is

$$P(h) = \frac{P_{constant}}{h} \tag{8.16}$$

where, to satisfy the average power constraint, $P_{constant}$ must satisfy

$$\int \frac{P_{constant}}{h} g_H(h)dh = P_{avg}$$

It follows that $P_{constant} = \frac{P_{avg}}{\mathsf{E}\left(\frac{1}{H}\right)}$. The resulting channel capacity is given by

$$C_{fading\text{-}constant\text{-}rate}(P_{avg}) = W \log_2 \left(1 + \frac{\frac{P_{avg}}{\mathsf{E}\left(\frac{1}{H}\right)}}{2\,W\,N_0} \right) \qquad (8.17)$$

It is easy to see that for Rayleigh fading, for which H is exponentially distributed, $\mathsf{E}\left(\frac{1}{H}\right) = \infty$, and hence the constant-rate power policy yields 0 rate for Rayleigh fading.

Let us now compare the optimal power control and constant-rate power control to see how the policies differ. Figure 8.15 depicts the two policies. The x-axis scale is a scaled reciprocal of the channel gain. Observe that the optimal power control (Equation 8.14) fills power into the triangular region between the flat threshold line (the threshold is $\frac{1}{\lambda}$), the slanting "y = x" line, and the "x = 0" line. Thus more power is allocated to the good channel states. The threshold is adjusted so that the average power constraint is met with an equality. On the other hand, the constant-rate power control (Equation 8.16) allocates power in proportion to $\frac{1}{h}$, thus needing to allocate increasingly more power to poor channel states. Because of the appearance of filling up the triangular space above the slant line, until the average power constraint is met, the optimal power control is said to have a *water-pouring form*, which occurs frequently in such constrained optimization problems. Notice that because we are averaging over the states that the channel takes over time, the power control varies over time, and hence this is also called *water pouring over time*.

In this power control problem the concern is only with optimizing the physical channel capacity subject to an average power constraint. The optimal policy assigns no power to certain channel states and assigns low power (and hence a low rate) to some other channel states. Although this might optimize the channel capacity, it might not be completely satisfactory from the point of view of the application that is sending data on the channel. This application may have a delay constraint for its data. Deferring transmission during a poor channel state

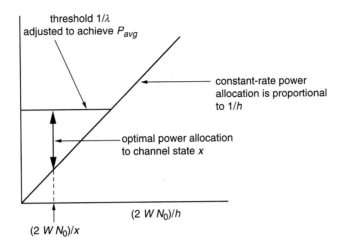

Figure 8.15 Power control policies under an average power constraint. The x-axis is a scaled 1/h. The optimal policy "fills" power up to the level 1/λ, which is chosen to meet the average power constraint; an example for channel gain x is shown. The constant-rate policy allocates power in proportion to 1/h.

would make the data wait longer for transmission. Thus there would be a trade-off between data delay and satisfying the power constraint. The linkage between the amount transmitted, the channel state, and the power required to do so would be through a formula such as Shannon's AWGN channel capacity formula. If we formulate and solve such a problem, we can expect that when the data backlog is large then even for poor channel states the policy would recommend transmission. The following is such a formulation.

Time is slotted into intervals of length τ, and each slot can accommodate N channel symbols. The kth slot is the interval $(k\tau, (k+1)\tau)$, $k \in 0, 1, 2, \ldots$ (see Figure 8.16). At the end of each slot, new data arrives; denote these arrivals by $A_k, k \geq 0$. The data buffer occupancy at the beginnings of slots is denoted by $S_k, k \geq 0$. The channel gain is H_k during the kth slot, $k \geq 0$. The amount of data transmitted from the buffer is R_k in slot k. Thus the evolution equation for the buffer is (for $k = 0, 1, 2, \ldots$):

$$S_{k+1} = S_k - R_k + A_k$$

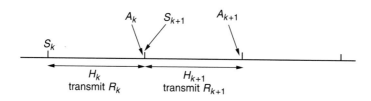

Figure 8.16   The processes in the model of joint buffer and power control.

The number of bits transmitted in the kth slot (i.e., R_k) and the power required (P_k) are assumed to be related by Shannon's capacity formula:

$$R_k = N\frac{1}{2}\log_2\left(1 + \frac{H_k P_k}{2W N_0}\right)$$

Let us denote the power required to transmit r bits when the channel gain is h by $P(h,r)$, which is obtained by inverting the preceding formula, and which, for fixed h, is exponentially increasing in r. The problem is to determine a power control policy so as to minimize the average queue

$$\lim_{n\to\infty}\frac{1}{n}E\left(\sum_{k=0}^{n-1}S_k\right)$$

subject to the average power constraint:

$$\lim_{n\to\infty}\frac{1}{n}E\left(\sum_{k=0}^{n-1}P(H_k, R_k)\right) \le P_{avg}$$

This problem can be solved for the case in which A_k and H_k are independent, i.i.d. processes. This is a constrained average cost Markov decision problem; the solution technique is beyond the scope of this book. The form of the optimum policy is depicted in Figure 8.17. The decision is based on the buffer state s and the channel gain h. We see, as expected, that when the channel is good ($\frac{1}{h}$ is small) and the buffer is not very large, the entire contents of the buffer are served. Even

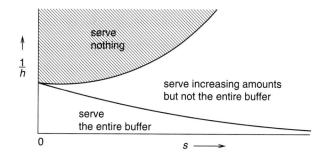

Figure 8.17 The form of the optimum policy for the joint buffer and power control problem.

for a good channel, if the buffer is large not all of it is served. This happens because the power required increases exponentially with the amount to be transmitted. Because of the average power constraint, it is better to serve large amounts of data in two or more transmissions rather than all at once. When the channel gain is small ($\frac{1}{h}$ is large), nothing is served until the buffer crosses a threshold (the upper curve in the figure). Beyond this threshold some of the data is served. It can be shown that for any value of h the amount of data served increases monotonically with the buffer occupancy s.

This problem is an indication of the kinds of results we can expect from formulations of cross-layer design problems, where we jointly optimize actions at multiple layers (in the preceding example, at the link and physical layers). It is important to note that to implement the control policy obtained here the transmitter must observe the channel state and must choose a coding and modulation scheme (which will result in the particular transmission rate given by the optimal policy). The receiver would also need to know the channel state in order to determine which coding and modulation scheme the transmitter has chosen. One practical approach is for the receiver to make SNR measurements from which the channel gain can be derived, and then this information can be provided to the transmitter over a reverse feedback channel.

In this discussion, we have highlighted the beneficial aspects of cross-layer design and optimization, but a few caveats must be borne in mind.

- *Feedback delay and inaccuracy:* In the examples discussed in this section, some state information must be fed back to the controller—for example, the channel state or the buffer level. The channel state is a random process and must be estimated. There will be errors in such estimates, and there could also be a delay in the information fed back to the controller. These errors and delays could render the performance of the policy much poorer than predicted by an idealized analysis that assumes exact, undelayed feedback.

- *Unintended interactions* [160]: Cross-layer designs are motivated by abstracting out a particular aspect of a system, and studying it in isolation. When the resulting technique is implemented in the actual system, there may be unforeseen interactions that could yield poorer performance than without the cross-layer optimization, or, worse, instabilities could be introduced.

8.5 Random Access: Aloha, S-Aloha, and CSMA/CA

In Sections 8.3 and 8.4 our primary focus is on point-to-point wireless links. In Section 8.3 we discuss the performance of TCP-controlled elastic file transfers over a wireless access link. In Section 8.4 we examine how the capacity of, and delay performance at, a fading point-to-point wireless link can be optimized by dynamic power control. We now start considering wireless networks. Wireless networks were first developed as multiple access networks, in which a large number of nodes, each with a relatively low average throughput requirement, access a single, high-bandwidth channel using a distributed packet multiplexing scheme. The distributed multiplexing algorithm for sharing the multiple access channel is called a medium access control (MAC) protocol. If the number of nodes is large or if the set of nodes can vary with time (or both), controlled channel access can cause significant protocol overheads and can even be difficult to implement. In such cases, simple random access MAC protocols are used. We discuss these schemes in Section 2.2.1. In this section we study some random access protocols in the context of their application to wireless networks.

8.5.1 Non-Carrier-Sensing Protocols: Aloha and Slotted Aloha

The earliest random access MAC protocol is called Aloha. The idea is as simple as it can be: If a node has a packet to transmit, it just transmits! To understand why this is not a bad idea, consider an extreme example: a satellite network in which every node transmits to a satellite, which then reflects the transmission back for every node to receive it. Let the data rate on the network be 1 Mbps and the packets be 1000 bits long so that the packet transmission time is 1 ms.

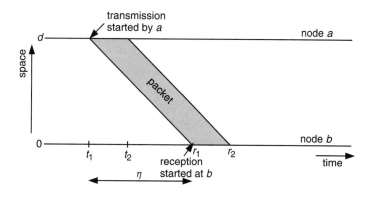

Figure 8.18 Space–time schematic of a transmission and reception in an Aloha network. Node a is transmitting during $[t_1, t_2]$, and node b is receiving after a significant delay (η) during $[r_1, r_2]$. Node b need not defer transmission during $[r_1, r_2]$ and cannot defer during $[t_1, t_2]$ because it does not know of the transmission from a.

Now consider the propagation delay in the network. Every packet must travel approximately 75,000 Km to reach the receivers, resulting in a propagation delay of approximately 250 ms. Thus what a transmitter is hearing on the channel is actually a transmission from 250 ms before the present time. Hence there is no use deferring to a carrier, and it is best to simply transmit the packet and hope that no other node is transmitting to the destination at the same time. Figure 8.18 shows a space–time diagram of a transmission and reception in such a network and the futility of deferring to a carrier. This random access protocol is called *pure Aloha*.

Now let us model the spatial distribution of the nodes in the network. To keep the analysis simple, we assume that an infinite number of nodes are distributed along a straight line of length d. Let η be the end-to-end propagation delay on the line along which the nodes are arranged. At any time instant, each node has at most one packet to transmit. This models a large network with very little traffic rate per node. We assume that all packets have a fixed length and a fixed transmission time of one unit. Thus time is normalized to the packet transmission time.

Each packet transmission attempt is characterized by an ordered pair of numbers (t, y), where $t \geq 0$ is the time at which the transmission started, and $0 \leq y \leq d$ is the location of the node that initiated it. We model this *space–time attempt process* as a two-dimensional Poisson process of rate G/d (attempts per

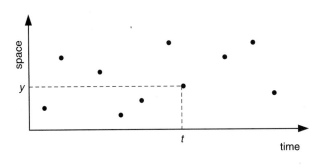

Figure 8.19 A sample realization of packet attempt instants in space and time. The points indicate the times and locations of packet arrivals; for example, a node at location _y_ attempted a packet at time _t_.

meter-second) viewed in the region $[0, \infty) \times [0, d]$ (see Figure 8.19). It follows that the aggregate of transmission attempts from all the nodes is a Poisson process of rate G attempts per second.

Corresponding to the reception of a transmission at its intended destination, a _collision time window_ can be defined at each location in $[0, d]$ such that if a transmission is begun at that location during that window of time, then it will collide with the desired transmission. Extending this collision window to all points in the network gives us a _collision cone_ corresponding to a reception. Any transmission attempt in this collision cone causes a collision at the receiver and a reception failure. See Figure 8.20 for an example. A transmission is begun at space–time point A and is intended to be received at the location corresponding to space–time point a. For this transmission to succeed, no transmission can be initiated at location d during the time interval (b, c), and no transmission can be initiated at location 0 during the time interval (f, e). Note that these intervals are of length 2 time units. The entire collision cone is also shown in the figure. The space–time area covered by the collision cone is $2d$.

Because G/d is the space–time arrival rate, multiplying it by $2d$ gives the parameter of the Poisson distribution of the number of arrivals in the collision cone. Thus, the probability that a reception is successful, P_s, is given by

$$P_s = \text{Pr}\big(\text{No transmission attempt in collision cone}\big) = e^{-(\frac{G}{d} \times 2d)} = e^{-2G}$$

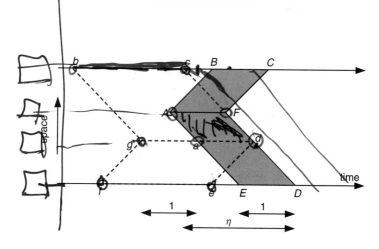

Figure 8.20 A space–time diagram showing a packet transmission and its collision cone in pure Aloha. A packet transmission is shown from space–time point A. The transmission (of duration 1) is shown propagating along ABCFDEA. If a successful reception is desired at space–time point a, then no other transmission should be initiated in the collision cone bcdefgb. The area of this collision cone is 2d. The line segment ga has length 1. The horizontal distance from c to D is η.

Defining the throughput, S, as the number of successful attempts per unit time, we get $S = GP_s = Ge^{-2G}$. The maximum value of S is achieved for $G = 0.5$, and $S_{\max} = 1/(2e) \approx 0.18$.

> ### Exercise 8.5
> To simplify the analysis of random access protocols it is sometimes assumed that all nodes are the same distance from each other. In such a case we do not need the space–time model. An example of a network where this is valid is a satellite network. Show that the throughput is still $\frac{1}{2e}$ under this simplifying assumption.

There is a simple way to make pure Aloha more efficient. Instead of allowing a node to begin transmission at any time, let time be slotted with slot length equal to the sum of the packet transmission time (unity) and the maximum propagation delay in the network (η). As before, assume that an infinite number of nodes are arranged on a line of length d. The nodes begin transmission only at slot boundaries, and the transmission and reception of a packet are completed in one slot. Thus for two packets to collide at the receiver, they must begin transmission

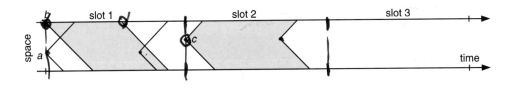

Figure 8.21   Time slotting in slotted Aloha. In slot 1 there are two transmissions from *a* and *b*, and these two transmissions collide at all points in the network. There is one transmission from *c* in slot *1*, which is successfully received everywhere. There is no transmission in slot 3.

in the same slot. This is illustrated in Figure 8.21, which shows a collision, a successful transmission, and an idle slot. If we assume that packets for transmission arrive according to a Poisson process of rate G and that the source node is uniformly distributed in $[0, d]$, then for a packet transmission to be successful, no other packet should be attempted in the same slot (i.e., no other packet should have arrived in the preceding slot). The collision cone is now a rectangle, and the Poisson rate of packet arrivals in a slot is $(G/d) \times (1 + \eta)d = G(1 + \eta)$, where G/d is the rate of the two-dimensional space–time attempt process and $(1 + \eta)d$ is the area of the collision cone. Thus the probability that a packet transmission is successful is $P_s = e^{-G(1+\eta)}$, the probability that the Poisson arrival process of intensity G did not generate an arrival in the interval $(1 + \eta)$. The throughput, S, is the product of the attempt rate, G, and probability of success, P_s, and is given by $S = Ge^{-G(1+\eta)}$, and the maximum achievable throughput, S_{max}, is obtained as $S_{max} = 1/(e(1+\eta))$. This is the *slotted Aloha* (S-Aloha) MAC protocol and has found wide applications in practice, especially when bandwidth cannot be allocated to a node statically.

A sample use of the slotted Aloha protocol is in the GSM (Global System for Mobile communication) cellular networks (see Section 2.3.1 for a description of this type of cellular network). In the GSM network the TDM channels either are traffic channels carrying voice or are control channels carrying control and signaling information to or from the mobile stations. A control channel on the reverse link from the mobile station to the base station, called the random access channel (RACH), is used by the GSM mobile stations to send messages to the network. The types of messages include those to initiate new calls, register locations of the mobile stations, and reply to paging queries. The messages are small and are generated at a very low rate compared with the capacity of the RACH channel. The number of mobile stations in a cell is not fixed and also is quite large, and signaling bandwidth cannot be to them allocated statically the nodes. Hence, slotted Aloha is used on this channel. After transmitting on the RACH using the slotted Aloha protocol, the mobile station waits for a fixed duration to know whether the transmission was successful. If an acknowledgment is not received before this duration expires, a retransmission is attempted.

Another application of slotted Aloha is in VSAT networks, which consist of a hub and a number of remote nodes. Such a network is shown in Figure 2.10 in Chapter 2. The nodes can communicate only with the hub, and all internode communications occur over two hops via the hub. When a remote node wants to transmit, it first requests a reservation on the frame over which nodes transmit to the hub. This reservation request is made using the slotted Aloha protocol on the uplink from the remote station to the hub. This can be very efficient if the bandwidth allocated for the reservation requests is small and if the amount of reserved bandwidth is comparatively large.

The DOCSIS (Data Over Cable Service Interface Specifications) standard for digital data transmission over cable TV networks is also similar to the VSAT network. All communication is controlled by the head end of the cable TV network. Separate parts of the spectrum over the cable are reserved for upstream (node to head end) and downstream (head end to node) traffic, and the control of the channels in both directions is with the head end. The upstream spectrum, in which the nodes transmit data to the head end, is available only by reservation that is obtained as follows. This spectrum is divided into *contention* slots, in which the nodes request bandwidth on the upstream channel, and *transmission* slots, during which the nodes transmit actual data. A node with data to transmit requests that the head end reserve some of the upstream transmission slots for it. The requests are made using a slotted Aloha protocol.

In deriving the throughput of slotted Aloha, we make the slot length equal to $1 + \eta$ rather than 1. This is done to absorb the variations in the propagation delays between the nodes, but it is not always necessary. In fact it is rarely necessary. In a satellite network the propagation delays between any pair of nodes is very nearly the same, approximately 250 ms, and we can use a slot length of 1. In terrestrial networks, such as the cellular and cable networks, the nodes usually transmit to a central node. The nodes use ranging to determine the propagation delay to the central node, and they advance or delay their transmission times to approximate a TDM link, absorbing the differences in the propagation delays.

Instability of Aloha

In the throughput discussion of Aloha and S-Aloha, we overlook an important feature of random access MAC protocols: A packet that has suffered a collision stays in the network and makes retransmission attempts until it is successful.

Such packets are called backlogs. Let B_k denote the backlog at the beginning of slot k. We assume that all fresh arrivals during a slot will attempt a transmission at the beginning of the next slot. We can write the evolution equation for B_k as follows:

$$B_{k+1} = B_k + A_k - D_k$$

where A_k is the number of arrivals, and $D_k \in \{0, 1\}$ is the number of departures in slot k. If new packet arrivals form a Poisson process of rate λ and if the backlogs attempt retransmission independently in each slot with probability r, then $\{B_k\}$ is a discrete time Markov chain. Consider the *drift* from a state n denoted by $d(n)$ and defined as

$$d(n) := \mathsf{E}\big(B_{k+1} - B_k | B_k = n\big) = \mathsf{E}((A_k - D_k)|B_k = n)$$

$d(n)$ is the average change in the backlog in one slot when the backlog is n. The backlog decreases by 1 if no new arrivals occur and if only one of the backlogs attempts transmission in the slot. The backlog increases by 1 if exactly one arrival occurs and if at least one of the backlogs attempts a retransmission. If the number of new arrivals is more than 1, the backlog increases by that amount. For all other combinations the backlog does not change. Thus we can write

$$\mathsf{Pr}(A_k - D_k = +1|B_k = n) = \lambda e^{-\lambda}(1 - (1 - r)^n)$$

$$\mathsf{Pr}(A_k - D_k = +m|B_k = n) = \frac{\lambda^m}{m!}e^{-\lambda} \qquad \text{for } m \geq 2$$

$$\mathsf{Pr}(A_k - D_k = -1|B_k = n) = e^{-\lambda}nr(1 - r)^{n-1}$$

Using this and simplifying, we get

$$d(n) = \lambda e^{-\lambda}(1 - (1 - r)^n) + \sum_{m=2}^{\infty} m\frac{\lambda^m}{m!}e^{-\lambda} - e^{-\lambda}nr(1 - r)^{n-1}$$

$$= \sum_{m=0}^{\infty} m\frac{\lambda^m}{m!}e^{-\lambda} - \lambda e^{-\lambda}(1 - r)^n - e^{-\lambda}nr(1 - r)^{n-1}$$

$$= \lambda - e^{-\lambda}(1 - r)^n \left(\lambda + \frac{nr}{1 - r}\right)$$

Clearly, as n becomes large, irrespective of the value of λ and r, the second term becomes very small and $d(n)$ is positive. Thus the mean drift for all large values of backlog (except finitely many values) is nonnegative. Because at most one packet can depart the system in each slot, we can apply Theorem D.10 to conclude that the backlog Markov chain is not positive recurrent. This means that if the backlog becomes large, the network has a tendency to increase the backlog rather than decrease it. This in turn implies that the Aloha protocol, if it is running for a long time, can develop a large backlog that may never be cleared.

The assumption of an infinite number of nodes is for analytical convenience. It is also a worst case analysis because with finite nodes, packets from the same node do not compete with each other, and in that case the performance can only improve. However, it can be shown that even when the number of nodes in the network is finite, the behavior is qualitatively similar to that we have derived; that is, if the backlog becomes large, the network has a tendency to operate with a large backlog for very long times. An obvious issue now is to design mechanisms to make the network stable. We do this by making the retransmission probabilities adaptive. To see how this can be done in the S-Aloha network, assume that all the nodes know the size of the backlog at the beginning of every slot and also know the stationary packet arrival rate. A packet is successfully transmitted in a slot either if exactly one new packet arrives and none of the backlogs attempt retransmission, or if no new packet arrives and exactly one of the backlogs attempts a retransmission. Thus, P_s is given by

$$P_s = \lambda e^{-\lambda}(1 - r)^n + e^{-\lambda}nr(1 - r)^{n-1}$$

Given $B_k = n$, the r that will maximize the probability of success—say, $r(n)$—is given by

$$r(n) = \frac{1 - \lambda}{n - \lambda}$$

If an adaptive retransmission probability $r(n)$ is used instead of fixed r, the drift will become

$$d(n) = \lambda - e^{-\lambda}\left(\frac{n - 1}{n - \lambda}\right)^{n-1}$$

In this case $d(n) \to \lambda - e^{-1}$. Hence, using Theorem D.9 with $f(i) = i$, we conclude that for $\lambda < \frac{1}{e}$ the Markov chain $\{B_k\}$ is positive recurrent; that is, the S-Aloha network is stable.

It is not practical for the nodes to know B_k, and a node should learn the network state from the events that it can observe. Let Z_k be the event in slot k, with the possible events being idle (denoted by 0) if no transmission was attempted in it; success (denoted by 1) if exactly one transmission was attempted; and an error (denoted by e) if more than one transmission was attempted. Typically two kinds of event observations are used. For example, we could use

$$S_{k+1} = \max\{1, S_k + aI_{\{Z_k=0\}} + bI_{\{Z_k=1\}} + cI_{\{Z_k=e\}}\}$$

and transmit with probability $\frac{1}{S_k}$ in slot k if the node has either a fresh packet or a backlog. Note that a, b, and c are fixed constants. A possible choice for the parameters is $a = -1$, $b = 0$, and $c = 1$. Alternatively,

$$S_{k+1} = \max\{1, a(Z_k) \times S_k\}$$

where $a(Z_k)$ are predetermined constants for the network. It has been shown that $a(Z_k)$ exist to achieve the maximum possible throughput of $\frac{1}{e}$.

Using the feedback from the network in adapting the retransmission times requires that all nodes be active all the time. Also, they all should see the same channel state at all times. This is clearly a very strong requirement, but it is avoidable. To make the protocol more robust, in many random access protocols nodes use their transmission attempt history to adapt the retransmission times. After a collision is detected by a transmitting node, the node stops the transmission and does not attempt another transmission for a backoff period of x units of time. Here x is a uniformly distributed random integer in the interval $[0, B - 1]$. B is updated by the node at every event (collision or success). A typical update equation has the form

$$B = \begin{cases} \min(a \times B, B_{\max}) & \text{If a transmission collides} \\ \max(B - b, B_{\min}) & \text{If a transmission is successful} \end{cases} \tag{8.18}$$

where a, b, $B_{\min}$, and $B_{\max}$ are predefined.

Maximum Throughput in S-Aloha

In the foregoing discussion, S-Aloha is stabilized to provide a maximum throughput of $\frac{1}{e}$. Clearly this maximum throughput is a bit too low, and the natural question is how much more throughput can be achieved. Toward answering that question, observe that the randomized retransmission strategy is a mechanism to resolve collisions and assigning transmission times to the nodes when there are multiple contenders for the multiaccess channel. The key to improving the throughput is to resolve collisions quickly. Many collision resolution algorithms have been proposed, their stability has been proved, and the corresponding maximum throughput has been calculated. An algorithm that leads to the best known stable throughput is described here.

The basic idea of this collision resolution algorithm is to define a time interval for each slot and mandate that all the packets that have arrived during this *enabled interval* transmit in the slot. A successful transmission in slot k or no transmission in it (idle slot) implies that all the arrivals in the enabled interval have been successfully transmitted by the end of the slot. If there is a collision in the slot, we resolve the collision by considering a subset of the enabled interval in the next slot, with the hope that only one packet is enabled. The following is a formal description of the algorithm.

For each time slot k we define the following three parameters: T_k is the left boundary of the interval, α_k is the duration of the interval, and σ_k indicates the part of the starting enabled interval (left or right) that must be resolved. These parameters evolve as follows.

$$
\begin{aligned}
&\text{If } Z_{k-1} = e \\
&\quad \text{then } T_k = T_{k-1} \qquad\qquad \alpha_k = \frac{\alpha_{k-1}}{2} \qquad\qquad\qquad \sigma_k = L \\
&\text{If } Z_{k-1} = 1 \text{ and } \sigma_{k-1} = L \\
&\quad \text{then } T_k = T_{k-1} + \alpha_{k-1} \qquad \alpha_k = \alpha_{k-1} \qquad\qquad\quad \sigma_k = R \\
&\text{If } Z_{k-1} = 0 \text{ and } \sigma_k = L \\
&\quad \text{then } T_k = T_{k-1} + \alpha_{k-1} \qquad \alpha_k = \frac{\alpha_{k-1}}{2} \qquad\qquad\quad \sigma_k = L \\
&\text{If } Z_{k-1} = 0/1 \text{ and } \sigma_k = R \\
&\quad \text{then } T_k = T_{k-1} + \alpha_{k-1} \qquad \alpha_k = \min(\alpha_0, k - T_k) \quad \sigma_k = R
\end{aligned}
$$

In slot k all arrivals in the interval $(T_k, T_k + \alpha_k)$ are enabled. Observe that the enabled intervals in successive slots are such that the packets are made to depart in the order in which they arrived. Hence this is also called the first-come-first-served (FCFS) collision resolution algorithm. By appropriately choosing α_0 and with minor modifications to the basic idea, a maximum throughput of 0.487760 is known to be achievable.

8.5.2 Carrier-Sensing Protocols

In networks having small propagation delays compared with the packet transmission time, rather than use a pure random access protocol like Aloha, it is reasonable to infer the channel state (busy or idle) through carrier sensing and use a more efficient random access protocol. If the maximum propagation delay is small compared with the packet length, then if a node senses the channel as busy and yet transmits, it can cause a collision at the receiver of the destination of the ongoing transmission. Furthermore, it is likely that the ongoing transmission is being heard at the intended destination of the new transmission, and a collision will occur there as well. Thus a node should listen to the channel before beginning to transmit and defer to an ongoing transmission. This is called the Carrier Sense Multiple Access (CSMA) protocol. Once a node begins transmission, it transmits the complete packet.

As with the Aloha protocol, we can define a collision window. This is the time since the beginning of a transmission during which another node (not having heard the ongoing transmission) can begin its own transmission and hence collide with the first transmission. In CSMA, the collision window is equal to the maximum propagation delay in the network. The reason is that after this interval, the carrier would have reached every node in the network, and all nodes will defer a transmission attempt until the end of the packet transmission. Two or more nodes can begin transmission within a short time of each other (less than the collision window) and collide. In this case, all the colliding transmissions will be lost. Note that during a collision in CSMA, the entire packet is transmitted by all the colliding nodes. The duration of a collision is the time from the beginning of the first transmission in the collision until the earliest time at which another transmission can begin. Clearly, the maximum duration of a collision is $t_P + 2t_{\mathrm{pd}}$, where t_P is the packet transmission time and t_{pd} is the maximum propagation delay. Figure 8.22(a) shows a time–space representation of a collision and a successful transmission.

There are many variations of the CSMA protocol based on what the node does on finding a busy channel when it wants to transmit (i.e., on sensing a carrier on the channel). In *nonpersistent* CSMA, the node senses the channel again after a random amount of time. In p-persistent CSMA, with probability p, the node continues to sense the channel and begins transmission immediately after the

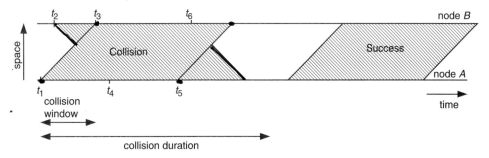

(a) A collision and a successful transmission in a CSMA network

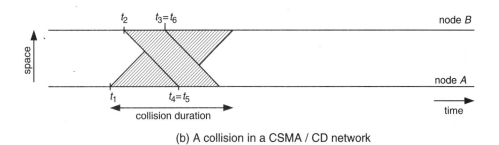

(b) A collision in a CSMA / CD network

Figure 8.22   Collisions in CSMA and CSMA/CD networks. A starts transmission at t_1, and B at t_2. The transmission from A reaches B at t_3, and that from B reaches A at t_4. Node A stops transmitting at t_5, and B stops at t_6. Collision ends when there is no carrier on the network. During a collision in a CSMA network, $t_5 = t_1 + 1$, and $t_6 = t_2 + 1$ whereas in CSMA/CD $t_5 = t_4$ and $t_6 = t_3$.

current transmission. The Ethernet and the IEEE 802.3 LAN and IEEE 802.11 wireless MAC protocols are 1-persistent.

To further improve CSMA, the node an continue to monitor the channel after beginning transmission. If it senses a collision on the channel, then the node can immediately stop transmission and minimize the loss of channel capacity. This is called CSMA with Collision Detection, or CSMA/CD. Figure 8.22(b) shows a collision in a CSMA/CD network. The maximum collision duration in this network is reduced to $(t_2 - t_1) + 3t_{pd}$. The CSMA/CD protocol was invented for the popular Ethernet local area network. Collision detection is typically done by the analog front end of the receiver detecting a higher level of energy on the medium than a single transmission could have generated. Clearly, for successful collision detection the longest signal propagation distance in a CSMA/CD network must be limited. We can construct larger networks by using physical signal repeaters or store-and-forward packet switches.

The Ethernet uses the backoff scheme of Equation 8.18 with $a = 2$, $b = B$, $B_{min} = 2$, and $B_{max} = 1024$. This means that the backoff period is reduced to zero after every successful transmission, and the interval is doubled after every collision. Because of its doubling of the backoff interval after a collision, this is also called a binary exponential backoff algorithm. In a busy network, this can lead to one node getting access to the channel more often than others and causing excessive delays to the other nodes. The unit of the backoff period in Ethernet, called a slot, is equal to twice the maximum round-trip delay in the network. This is seen to be a "natural" unit of time because when a node begins transmission, it will know whether its transmission was successful within the maximum round-trip time. In 10- and 100-Mbps Ethernet, the slot duration is equal to 512-bit transmission times.

As discussed in Section 8.2, in wireless networks, spatial reuse allows the spectrum to be used simultaneously in various parts of the network and significantly increases the traffic-carrying capacity. Spatial reuse requires that the transmission range of a node be much smaller than the geographical spread of the network; in other words, not all nodes should be within receiving range of each other. Using a carrier-sensing protocol in such a network can be inefficient because of hidden and exposed terminals (discussed shortly). An additional source of inefficiency is that, in contrast to the wired Ethernet, in wireless networks the received signal energies are very low, and it is very difficult to design reliable collision detection hardware. We now turn to a discussion of these issues and the design of protocols that help in addressing them.

Let two nodes a and b have transmission ranges A and B, respectively, as shown in Figure 8.23, and let X denote the intersection of A and B. Consider an ongoing transmission from a. Because b is out of the transmission range of a, it cannot sense the carrier from this transmission and can decide to transmit. If node b transmits at the same time as node a, the transmissions from a and b will be received at all nodes in X, and there will be a collision at these receivers. If a was transmitting to c in X, then c will not be able to decode the packet. However, a will not know of the collision at c and will continue to transmit; recall that collision detection is not practical in wireless communication. In the scenario just described, we say that b is *hidden* from a with reference to a transmission to c.

Now consider the situation of node d, whose transmission range is shown as D. This node wishes to send a packet to node e when a is transmitting to c. Node d is within range of a, and hence d can sense the signal while a is transmitting to c.

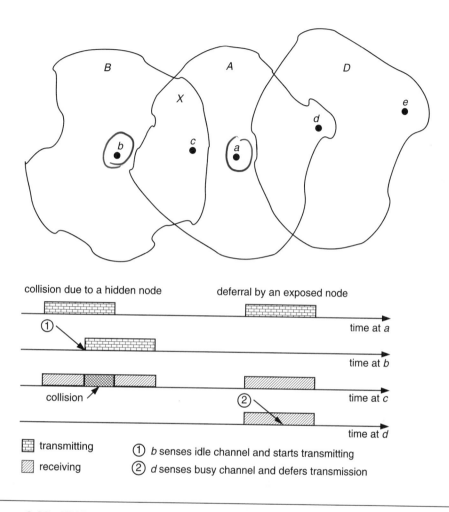

Figure 8.23 Hidden and exposed terminals in a wireless network. The illustration assumes that the propagation delays are zero. The transmission ranges of nodes *a*, *b*, and *d* are shown as regions *A*, *B*, and *D*, respectively.

If node *d* were to transmit to *e* while the *a*-to-*c* transmission was going on, the two transmissions could coexist, because *c* (the receiver of *a*'s transmission) is outside the range of *d*, and *e* (the proposed receiver of *d*'s transmission) is outside the range of *a*. Yet, on hearing the signal from *a*, node *d* will be forced to keep silent because it does not know what the consequence of its intended transmission would be.

In this case we say that node *d* is *exposed* to a transmission from *a*. The lower part of Figure 8.23 illustrates the collision due to a hidden node and an unnecessary deferral by an exposed node.

In summary, in a wireless network, hidden nodes reduce the capacity by causing collisions at receivers without the transmitter knowing about it, and exposed nodes force a node to be more conservative in its transmission attempts, thus reducing spatial reuse.

A well-known solution to the hidden node problem is the use of a narrowband auxiliary signaling channel in addition to the data channel. A node actively *receiving* data on the data channel transmits a *busy tone* on the signaling channel to enable the hidden nodes to defer to receiving nodes in their transmission ranges.

It is cumbersome and inefficient to divide the available spectrum into two parts with a sufficient spectral gap between them so as to enable the busy tone signaling mechanism to work properly. Hence, a different strategy is used in wireless local networks. Actual data transfer is preceded by a *handshake* between the transmitter and the receiver, and this is used to convey an imminent reception to the hidden nodes. Before transmitting a data packet, a source node transmits a (short) *request to send* (RTS) packet to the destination. If the destination receives the RTS correctly, it means that it is not receiving any other packet, and it acknowledges the RTS with a *clear to send* (CTS) packet. The source then begins the packet transmission. If the CTS is not received within a specified timeout period, the source assumes that the RTS had a collision at the receiver (most likely with another RTS packet), and a retransmission is attempted after a random backoff period. The binary exponential backoff of Ethernet may be used.

In the handshake protocol just described, the RTS from the transmitter serves as a way to tell nodes in the transmission range (see the discussion at the beginning of Section 8.2) of the transmitter node about the imminent transmission of a packet, and the CTS from the receiver serves the same purpose for nodes in the transmission range of the receiver. Thus nodes that are not in the transmission range of the transmitter (the one that intends to send a packet and hence sent the RTS) but are within the transmission range of the receiver (the one that responded with the CTS) are informed of the imminent packet transmission. If the packet length information is included in the RTS and CTS packets, then nodes that decode these packets can determine the time until the completion of the reception and can then schedule their own transmissions accordingly. Thus this is a *collision avoidance* scheme. After the RTS/CTS exchange completes, then, as far as the nodes in the union of the transmission ranges of the transmitter and receiver are concerned, the medium is reserved. This channel access scheme can be termed a *Carrier Sense Multiple Access with Collision Avoidance* (CSMA/CA) scheme.

Note that collision avoidance helps to reduce the inefficiency that is introduced by the inability to do collision detection in wireless networks. In principle, only RTSs collide; these are short packets, and hence the time wasted is small. This basic channel access mechanism, is called *Multiple Access with Channel Acquisition* (MACA), is an adaptation of the handshake protocols used in RS-232-C (between terminal equipment such as personal computers and peripheral equipment such as modems and printers) and Appletalk (between communicating terminal equipment). In these protocols the handshake phase precedes actual data transfer.

The RTS/CTS scheme just described helps ameliorate the hidden terminal problem but does not eliminate it. The transmission range of a transmitter is smaller than the range over which the transmitter can cause interference. Hence nodes that are hidden from the transmitter of the RTS and that are not in the transmission range of the transmitter of the CTS (but are within its interference range) will not realize that a packet transmission is imminent and how long it will last. However, they continue to use carrier sensing to defer to ongoing transmissions. Hence when such nodes hear the CTS, they defer for some length of time. When the data packet transmission starts, however, they cannot hear it; having exhausted their deferral period, one or more of them could attempt to transmit its own CTS, thus colliding with the ongoing packet transmission.

Note that we have not yet solved the exposed node problem. In fact, in the CSMA/CA scheme, it seems difficult to allow an exposed node to transmit. Any node in the interference range of the transmitter of the ongoing packet is exposed. Even if such a node were allowed to transmit an RTS to a node (outside the interference range of the transmitter of the ongoing packet), it cannot itself receive the subsequent CTS, and hence it will not know if whether it can transmit.

It should be clear that the CSMA/CA protocol would work well, provided that the nodes do not cover a large geographical area. Hence this protocol has been adopted for wireless LANs in the IEEE 802.11 series of standards, which we discuss at some length in Section 8.6.

Many improvements to this basic protocol have been suggested, with the most popular being MACAW, or MACA for Wireless. The major difference between MACA and MACAW is the use of an acknowledgment from the receiver after a successful reception of a packet. In addition, it is suggested that the mean backoff period be varied much more slowly than that of MACA. A multiplicative increase and linear decrease in the mean backoff period (i.e., $a < 2$ and $b = 1$) is also suggested. Another feature of the MACAW protocol is the transmission of a short *data sending* (DS) packet preceding the actual data transfer. This is because it is possible that an exposed node has heard the RTS and not the CTS.

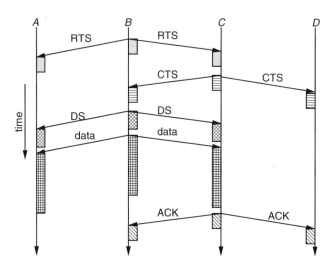

Figure 8.24   The handshake and data exchange sequence in MACAW. *A* **is within range of** *B***,** *B* **is within range of** *A* **and** *C***,** *C* **is within range of** *B* **and** *D***, and** *D* **is within range of** *C***.** *B* **wants to transmit data to** *C***. RTS from** *B* **is heard by** *A***, and it defers transmission until CTS may be received at** *B***. CTS from** *C* **is heard by** *D***, and it defers transmission until data exchange is complete. DS is heard by** *A***, and it knows that the RTS/CTS handshake was successful and that data exchange is in progress. It also knows the duration of the data exchange. The ACK from** *C* **is heard by** *D***, and it can then infer that data exchange between** *C* **and** *B* **is complete, and** *D* **can now enable its transmitter.**

If such a node is not sensing the channel, it will not know whether the RTS–CTS handshake was successful and may attempt to transmit an RTS. This in turn could collide with the ACK at the transmitter of the packet. To avoid this situation and in the presence of nodes that may not have carrier-sensing capability, the DS packet provides information to the exposed nodes about the beginning and end of transmission times. It has since been argued that this is not necessary, and the IEEE 802.11 standard does not include this message in its handshake protocol. The handshake and the data exchange sequence of the MACAW protocol are shown in Figure 8.24.

The basic ideas from MACA and MACAW have been formalized in wireless LAN standards, and we discuss them in the next section.

8.6   Wireless Local Area Networks

In the previous sections in this chapter we discuss and analyze generic issues in digital communication over mobile wireless links, as well as random multiple

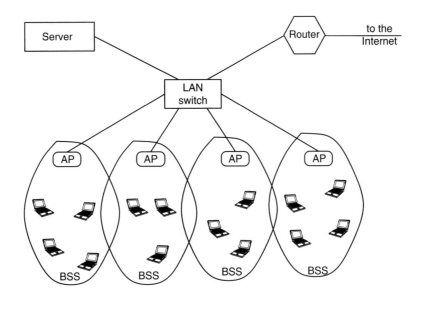

Figure 8.25   A typical IEEE 802.11 ESS architecture.

access over such links. At various points certain technologies become so widely accepted and deployed that it becomes imperative to understand their detailed mechanisms, their performance, and various related open issues. Such is the case with the CSMA/CA-based wireless LAN technologies standardized by IEEE under the IEEE 802.11 series of standards, and by ETSI (European Telecommunications Standards Institute) as the Hiperlan 1 and Hiperlan 2 standards. Here we limit ourselves to the IEEE 802.11 standards. We briefly list the various components of the standard, and then we describe the specifics of the 802.11 MAC. Finally we provide a simple performance analysis and discuss the insights that flow from it.

8.6.1   The IEEE 802.11 Standards

Figure 8.25 shows the typical architecture in which users access a wired network, and services on the network, over an IEEE 802.11 wireless LAN. Each mobile device or mobile station (MS), shown as laptop computers in the figure, associates itself with one of the *access points* (APs). The APs are attached to the wired LAN and could actually be fixed to a wall. An AP, together with its associated MSs,

constitutes a *basic services set* (BSS).[1] The various BSSs are schematically shown
to overlap, indicating that the areas "covered" by the various APs overlap this
permits the movement of MSs over large distances within a building as long as
they stay within the range of some AP. There is an *association* and *disassociation*
protocol that facilitates such movement of MSs between BSSs. A collection of
several interconnected BSSs is called an *extended services set* (ESS); Figure 8.25
shows an ESS with several BSSs interconnected by a LAN switch.

The desktop Ethernet access speed of 100 Mbps is now very common. Hence,
for wireless access to replace Ethernet as a way to access the wired network, there
have been several efforts to provide increasingly higher physical layer bit rates in
the IEEE 802.11 series of standards. Three major physical layer standards have
been defined. The following is a chronological list and brief description of the
physical layers.

- *802.11b* operates in the 2.4 GHz unlicensed band and provides the bit
  rates 1, 2, 5.5, and 11 Mbps. Two forms of *direct sequence spread
  spectrum* (DSSS) modulation are used to achieve these four data rates.
  In the allowed spectrum in the 2.4 GHz band, there are 14 overlapping
  channels, each with a bandwidth of 5 MHz. Each BSS operates, at any
  point in time, on one of these channels. Overlapping BSSs must use
  nonoverlapping channels.

- *802.11a* operates in the 5 GHz unlicensed band and provides the bit
  rates 6, 9, 12, 18, 24, 36, 48, and 54 Mbps. Each BSS operates in a
  20 MHz spectrum in the 5 GHz band. This spectrum is further divided into
  30 carriers, which are modulated with the data using OFDM technology
  (see Section 8.1.4).

- *802.11g* operates in the 2.4 GHz unlicensed band and provides speeds up
  to 54 Mbps.

Notice that each of these physical layer standards provides for several bit rates.
These rates are obtained by various combinations of coding and modulation
schemes, and each of them requires a certain minimum SNR to operate
satisfactorily. Recalling our discussion in Section 8.4, note that this feature of the
physical layers provides scope for the design of channel state adaptive techniques.
If a transmitter–receiver pair determines that the signal quality between them is

---

[1]This is not to be confused with base station subsystem, described in Chapter 2 and also abbreviated BSS.

poor, then they can agree to use a more robust modulation scheme with a lower
bit rate.

In addition to the physical layer standards mentioned here and multiple access
control (described in the next section), the IEEE 802.11 series of standards covers
issues such as security, as well as country-specific concerns regarding utilization
of the unlicensed bands and coordination with other communication technologies
that use the unlicensed radio spectrum. We do not discuss these details further in
this book.

8.6.2   Multiple Access in IEEE 802.11

The 802.11 specifications define a *polling-based* standard, called the *Point
Coordination Function* (PCF), and a distributed random access standard, called the
Distributed Coordination Function (DCF). The DCF random access procedures
are based on the CSMA/CA mechanism discussed in Section 8.5. In addition, the
RTS/CTS (request to send/clear to send) mechanism (also discussed in Section 8.5)
is added to reduce the chances of collisions and to reduce the wasted time if a
collision does occur. In the IEEE 802.11 MAC protocol, the backoffs are measured
in multiples of a basic slot time, which depends on the physical layer standard over
which the MAC is operating. Thus, for example, in IEEE 802.11a the slot time is
9 μsec.

Figure 8.26 shows the events during data transfer between four nodes. It is
assumed that all nodes are in transmission range of each other; thus, each node
can decode the RTS and CTS packets sent by any other node. At the beginning
of the fragment of time shown in the figure, node 1 is transmitting a MAC data
packet, and all the other nodes have deferred to this transmission. At the end
of the data transmission, there is a *short interframe space* (SIFS) that allows the
receiving node to *turn around its radio* and send back a MAC-level ACK packet.
When this ACK transmission ends, the channel is sensed to be idle by all the
nodes, and each one of them starts a *DCF interframe space* (DIFS) timer. The
DIFS duration is more than SIFS (e.g., in 802.11a, SIFS = 16 μsec and DIFS =
25 μsec). Thus even though nodes 2, 3, and 4 did start their DIFS timers when
node 1's data packet completed transmission, because SIFS < DIFS, the channel
became busy again (with the ACK packet) before the DIFS timers could expire.
Thus, we see that the channel is essentially reserved for the receiver and transmitter
between which a data packet is being transferred. In the light of our discussions
in Section 8.5.2, such reservation is valid only if there are no hidden nodes that
are within interference range of the receiver but cannot decode the CTS and also
cannot hear the transmitter. Such nodes will hear the CTS, and hence will defer

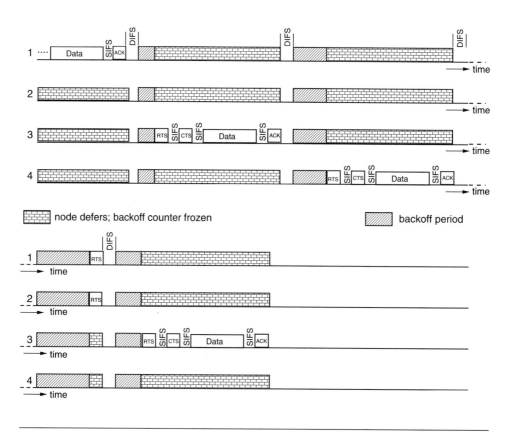

Figure 8.26 Events during data transfer in the IEEE 802.11 DCF MAC protocol with the RTS/CTS mechanism. There are four nodes. There is one timeline for each node; the timelines start in the top left side of the figure, proceed to the right, and then continue from the left in the lower part of the figure.

transmission for a DIFS after the CTS signal ends. After this, they are free to transmit and hence could nullify the supposed reservation.

When the DIFS timers expire, each node enters a backoff phase. Even though the channel is idle, the random backoff is used in an attempt to order the transmissions so that a collision does not occur. The node that just completed its data transmission samples a new random backoff value. If a node was already in a backoff when node 1 started its data transmission, then during node 1's data transmission this node's backoff timer is frozen. Upon completion of node 1's transmission, each deferring node continues the remainder of its backoff.

If a node was idle when node 1 started its data transmission and if a packet arrived during node 1's transmission, then the node with the new arrival defers until the completion of the transmission; then it waits for the DIFS period and starts a new backoff. The backoff durations are multiples of the basic slot time. When a new backoff is sampled, this multiple is sampled uniformly from the integers $\{0, 1, \ldots, CW_{min} - 1\}$.

When a backoff period of some node expires (for example, node 3 in Figure 8.26), then this node transmits an RTS packet to its destination node. Upon hearing activity on the medium, all other nodes freeze their backoff timers. The node to which the RTS was directed then sends back a CTS packet (an SIFS elapses in between the two). Node 3 then waits for an SIFS and sends its data packet, after which an ACK is sent by the receiver node. Figure 8.26 shows another successful transmission from node 4.

A collision occurs if two nodes finish their backoffs within one slot of each other. It is assumed that the extent of the network is such that all nodes can sense a transmission within one slot time. In this case, both nodes send RTS packets, which collide. A CTS timeout then follows, after which the colliding nodes sample a backoff from a doubled collision window—that is, from the window $\{0, 1, 2, \ldots, 2 \cdot CW_{min} - 1\}$. After the collision event, the nodes that were not involved in the collision continue their backoffs with their residual backoff timers. A collision event (between RTS packets of nodes 1 and 2) is shown in the continued timelines the bottom of Figure 8.26, where we have taken the additional time after the RTS transmission to be DIFS, before the backoff durations are started. Repeated collisions lead to a doubling of the collision window until it reaches CW_{max}, after which the collision window remains fixed. In the standard, $CW_{min} = 32$, and $CW_{max} = 1024$.

8.6.3 Performance Modeling and Throughput

Here we provide a simple analysis based on the foregoing description of the IEEE 802.11 MAC protocol. It is important to note that we are considering a single so-called *cell*; that is, all the nodes are in transmission range of each other, and hence, there can be only one successful transmission at a time. This also implies that there are no exposed nodes or hidden terminals (see Section 8.5.2).

Let us assume that all nodes have packets to send at all times; this is the *infinite backlog* assumption. Assume that all packets are of the same length L. The MAC protocol permits packet fragmentation into MAC fragments, but let us assume that the entire packet given by the link layer is transmitted as a whole by the MAC. We further assume that the backoff times are sampled from an exponential distribution with mean $\frac{1}{\beta}$, which is the same for all the nodes.

Note that as the number of nodes increases, collisions will increase in the way the protocol is actually implemented, and the average backoff durations will increase; in a sense, $\frac{1}{\beta}$ captures this average backoff time. We will see a little later how the average backoff time can itself be analyzed. Let us list the relevant notation.

n: the number of nodes (each is infinitely backlogged).

r: the data rate (assumed to be fixed and the same for all the nodes).

L: the fixed packet length used by all the nodes.

δ: the slot time.

β: the parameter of the exponential backoff duration; thus $\frac{1}{\beta}$ has units of time.

T_o: the packet transmission overhead time (i.e., the time for which the medium is reserved for a transmission in addition to $\frac{L}{r}$). In Figure 8.26 this is shown to be the time for the transmission of an RTS, a CTS, and an ACK, plus three SIFSs and one DIFS. In addition, the transmission of each data or control packet is preceded by a physical layer preamble. Thus, in addition to the other listed durations, T_o includes the time for four such preambles: one for the RTS, one for the CTS, one for the data packet, and one for the ACK.

T_c: the time spent in a collision before the next backoff period. In Figure 8.26 this is shown to be the time for the transmission of an RTS plus one DIFS. In addition, T_c contains the time for the one physical layer preamble that precedes the RTS.

From Figure 8.26 we can identify certain renewal instants (see Appendix D)—namely, those at which the DIFS after a successful transmission ends, or a collision period ends. Because of the assumption of exponential backoff times, the residual backoff times and the "fresh" backoff times are all exponentially distributed. Hence the time until the first backoff completes is exponentially distributed with mean $\frac{1}{n\beta}$. A collision takes place if a second backoff completes within δ of the first one.

Exercise 8.6
Show that the probability that a transmission attempt suffers a collision is given by $\gamma = 1 - e^{-(n-1)\beta\delta}$.

Now the mean time between successive renewal times is given by

$$\frac{1}{n\beta} + (1-\gamma)\left(\frac{L}{r} + T_o\right) + \gamma T_c$$

Let us denote the network throughput by Θ. Hence, using the renewal reward theorem (Theorem D.15), the network throughput is given by

$$\Theta(\gamma,\beta) = \frac{(1-\gamma)L}{\frac{1}{n\beta} + (1-\gamma)\left(\frac{L}{r} + T_o\right) + \gamma T_c} \qquad (8.19)$$

The normalized throughput can be written as

$$\frac{\Theta(\gamma,\beta)}{r} = \frac{(1-\gamma)\frac{L}{r}}{\frac{1}{n\beta} + (1-\gamma)\left(\frac{L}{r} + T_o\right) + \gamma T_c} \qquad (8.20)$$

where the right side is the fraction of time that the network is carrying data packets.

From the preceding expression for the normalized throughput we observe an interesting trade-off. Notice that the backoff rate parameter, β, appears in two ways: in the mean time until an attempt (i.e., $\frac{1}{n\beta}$), and in the collision probability, γ. Using a large value of β (i.e., a high attempt rate) reduces the mean time between attempts but increases the probability of collision, and using a small value has the opposite effect. We observe that as $\beta \to 0$, the normalized throughput goes to zero, because in the limit there are no attempts; furthermore, as $\beta \to \infty$ the normalized throughput again goes to 0, because the probability of collision goes to 1. It is therefore interesting to seek the value of β that maximizes the normalized throughput.

Exercise 8.7

Show that the normalized throughput is maximized for

$$\beta = \frac{1}{n}\frac{1}{2T_c}\left(\sqrt{1 + \frac{4nT_c}{(n-1)\delta}} - 1\right)$$

We observe from Exercise 8.7 that the optimal value of β scales inversely with n, the number of transmitting nodes. This exercise clearly motivates the need for adaptive backoff in the IEEE 802.11 MAC protocol. In fact, we find that the mean backoff time should be proportional to the number of nodes. As standardized, the MAC protocol always drops the contention window to CW_{min} after a successful transmission. This is a good approach when a few nodes are intermittently active. If several continuously active nodes are sharing the network, it may be better for the nodes to adapt to the optimal value of the contention window and then retain this value.

We now turn to the problem of obtaining the average backoff duration (i.e., $\frac{1}{\beta}$) or, equivalently, a node's attempt rate, β, in the IEEE 802.11 MAC protocol framework. Let us assume now that we have the collision probability γ. We then obtain an expression for a node's attempt rate in terms of γ. This expression, along with the equation for collision probability in terms of β (obtained in Exercise 8.6), can be solved to obtain both of these quantities and hence to calculate the throughput using Equation 8.19.

We observe from Figure 8.26 that, after removing the periods during which there are successful transmissions or collisions (including the associated overheads), we are left with only the periods during which the nodes are in backoff (the shaded periods in Figure 8.26). Let us consider a node and focus only on those time periods when its backoff timer is counting down. Let t represent this *conditional* time, and let $G(t)$ be the number of attempts by a node up to t. Then formally the backoff rate would be the following limit:

$$\beta = \lim_{t \to \infty} \frac{G(t)}{t}$$

Let us generalize, and introduce the following notation:

K: the MAC protocol at a node discards the packet being attempted on the $(K + 1)$th successive collision for this packet.

b_k: the mean backoff duration of a node after the kth collision, $k = 0, 1, 2, \ldots, K$; note that b_k has units of time (i.e., milliseconds if all other times, such as the slot time δ, are in this unit, or units of slot time, in which case $\delta = 1$ and all time intervals are in units of slot time).

Thus, for example, if $K = 1$, then each packet is attempted at most twice. In the first attempt the mean backoff is b_0; if this attempt collides, then one more attempt is made after a mean backoff b_1. Failure of this second attempt leads to the packet being discarded. In practice, b_k increases with k. In the IEEE 802.11 standard, $K = 12$, $b_k = \left(\frac{2^k CW_{\min} - 1}{2}\right)\delta$, for $0 \leq k \leq 5$, and $b_k = \left(\frac{2^5 CW_{\min} - 1}{2}\right)\delta$, for $6 \leq k \leq 12$, where $CW_{\min} = 2^5$.

Having captured the effect of the other nodes by the collision probability γ, we assume (as an approximation) that the instants at which the node makes a successful attempt or discards a packet are renewal instants. With this is in mind, let us further define the following:

B: the random amount of *cumulative backoff* time between successive successful packet transmissions or packet discards at a node

A: the number of attempts that a node makes between successive successful packet transmissions or packet discards (including the attempt that resulted in a success or a discard)

With the renewal assumption, and viewing the attempts as a *reward* in the renewal cycle, the average attempt rate of a node (we are still working within the conditional backoff time) is given by (using the renewal reward theorem, Theorem D.15)

$$\beta = \frac{\mathsf{E}(A)}{\mathsf{E}(B)}$$

We now use our definitions of K; the mean backoff durations, b_k; and the collision probability, γ, to obtain expressions for $\mathsf{E}(B)$ and $\mathsf{E}(A)$. The mean of the random variable B is at least b_0. Furthermore, if there is a collision (with probability γ) then we must add b_1 plus the additional time for further collisions, and so on. This argument yields

$$\mathsf{E}(B) = b_0 + \gamma(b_1 + \gamma(b_2 + \gamma(\cdots(\cdots(\gamma b_K)))))$$

$$= b_0 + \gamma b_1 + \gamma^2 b_2 + \cdots + \gamma^k b_k + \cdots + \gamma^K b_K$$

In an identical manner we also obtain

$$\mathsf{E}(A) = 1 + \gamma + \gamma^2 + \cdots + \gamma^K$$

As desired, we have thus expressed β in terms of quantities specified by the standard and the assumed collision probability γ. Let us define the two functions

$$G(\gamma) := \frac{1 + \gamma + \gamma^2 \cdots + \gamma^K}{b_0 + \gamma b_1 + \gamma^2 b_2 + \cdots + \gamma^k b_k + \cdots + \gamma^K b_K}$$

$$\Gamma(\beta) := 1 - e^{-(n-1)\beta\delta}$$

Solving these two equations together is equivalent to asking for a solution to the following fixed-point equation:

$$\gamma = \Gamma(G(\gamma))$$

Because $\Gamma(G(\cdot))$ is a continuous function that maps the interval $[0, 1]$ into the interval $[0, 1]$, there must be a solution to this fixed-point equation (see Appendix B, Section B.2).

Exercise 8.8

Assume (as in [35], where the basic idea of the foregoing analysis originated) that $K = \infty$ (i.e., a packet is attempted until it succeeds; in other words, a packet is never discarded) and that there is an $m \geq 1$ such that $b_k = \left(\frac{2^k CW_{\min} - 1}{2}\right)\delta$, for $0 \leq k \leq m - 1$, and $b_k = \left(\frac{2^m CW_{\min} - 1}{2}\right)\delta$, for $k \geq m$. Show that

$$G(\gamma) = \frac{2(1 - 2\gamma)}{(1 - 2\gamma)(CW_{\min} - 1) + \gamma CW_{\min}(1 - (2\gamma)^m)} \frac{1}{\delta}$$

Show also that $G(\gamma)$ is a decreasing function of γ.

Remark: The mean backoff durations shown here are obtained by assuming that when the collision window is CW, the backoff is uniformly distributed over $\{0, 1, 2, \ldots, CW - 1\}$ slot times. If, however, in an implementation the backoff is uniformly distributed over $\{1, 2, \ldots, CW\}$ slot times, then $b_k = \left(\frac{2^k CW_{\min} + 1}{2}\right)\delta$, for $0 \leq k \leq m - 1$, and $b_k = \left(\frac{2^m CW_{\min} + 1}{2}\right)\delta$, for $k \geq m$. In this case, the formula for β has $(CW_{\min} + 1)$ in the denominator rather than $(CW_{\min} - 1)$.

Obviously, $\Gamma(\cdot)$ is increasing in its argument. For case in Exercise 8.8 we have seen that $G(\cdot)$ is decreasing in its argument. It follows that $\Gamma(G(\gamma))$ is decreasing in γ and hence that there is a unique fixed point. Let this fixed point be γ_0. The attempt rate can then be calculated as $G(\gamma_0)$, and the network throughput is calculated as $\Theta(\gamma_0, G(\gamma_0))$.

The results from this model were found, in [35], to be highly accurate in predicting the performance of the IEEE 802.11 MAC protocol. We provide here a simplification and generalization of the analysis in [35]. Thus the simple approach of modeling the "coupling" between the nodes by the collision probability, and using averages rather than the exact distributions in Equation 8.19 has yielded a simple and accurate model. Notice that in the computation of β we do not use the distributions of the backoff times, but only the averages (i.e., the b_k, $k = 0, 1, \ldots, K$); this suggests that the performance may depend on the backoff distributions only through their means.

In fact, the approach to obtaining the expression in Equation 8.19 can be used to derive other expressions that yield valuable insights. For example, because the nodes are mobile, the channel quality between transmitters and receivers will vary, and nodes must adapt their transmission rates to suit the channel conditions. Thus instead of the common rate r in Equation 8.19, a node-dependent rate r_k would have to be used. Such rate adaptation, as nodes move around, is possible within the IEEE 802.11 standards. See the discussion of the physical transmission standards in Section 8.6.1; in addition, the standard provides a mechanism for nodes to autonomously vary their rates as they perceive poor or improved performance. We note that the exponentially distributed backoff model can be easily extended to more complicated situations, such as different transmission rates and different backoff rates for the different nodes. This is explored in Problems 8.4 and 8.5.

8.7 Wireless Ad Hoc Networks

Infrastructure-based networks build information transport services by building most of these services into a set of devices that have reasonably static relationships between them. The sources and sinks of information flows essentially "plug in to" these services. For example, in an internet, packets that need to be carried to a destination are handed to a fixed router, which *routes* them appropriately in cooperation with other routers in the network. Fixed wireless and satellite networks (see Figure 8.1), which are essentially like wireline networks except for using wireless links, can be said to follow this paradigm. In mobile access networks, nodes are mobile and plug in to the information transport service by associating themselves with devices that connect to an infrastructure-based network.

For example, in wireless cellular networks, a mobile node obtains its transport services from a base station. Data from the mobile node is first transmitted to a base station and is then carried over a wired network toward the destination. In 802.11 WLANs, the nodes connect to an access point, which in turn is connected to a wired network. Although the association between the mobile nodes and the network devices is less static than with wireline networks, an underlying infrastructure is assumed to provide the transport services.

An ad hoc network is a complete contrast to infrastructure-based networks. There is no association between the nodes and any set of "static" devices. No assumptions are made about a fixed networking infrastructure nor about the timescale over which the network characteristics are static. An ad hoc network builds information transport services over a set of mobile, arbitrarily located nodes. Every node cooperates in providing the information transport services. In the language of the Internet and the cellular network, we can say that every node in the network carries its own router or base station (or both) with it. An even harsher assumption in these networks is that the network topology can be highly time-varying.

Applications of wireless ad hoc networks are many. They can be used to provide communication services by being deployed quickly in emergency areas where the infrastructure has been damaged, such as areas affected by storms, floods, and earthquakes. They can provide connectivity among a group of vehicles, ships, aircraft, or even people on foot, operating in a geographical area with no networking infrastructure. The military applications are obvious.

Ad hoc networks necessarily use wireless communication and hence are called wireless ad hoc networks, or WANETs. Nodes may be mobile, in which case they may also be called *mobile ad hoc networks*, or MANETs. We prefer to use the more general WANET. Although it is possible to have WANETs using point-to-point wireless links, a multiple access protocol is more convenient in such networks. In WANETs, nodes are assumed to be mobile, and mobility implies that the energy source is a battery. Because the energy supply from a battery is finite, many energy conservation mechanisms are used, and this in turn imparts certain characteristics to WANETs that affect their performance. These characteristics are (1) the multihop nature of connectivity between the nodes and (2) a highly time-varying topology. In the following we briefly discuss how these characteristics come about in WANETs.

The transmission range of a node in a WANET does not, and should not, cover the entire operational area. Small transmission ranges allow spatial reuse (i.e., simultaneous use of the same spectrum in geographically separated locations) and hence increase network capacity. They also conserve battery energy by having a low transmission range and hence low power consumption. If a destination node is not within the transmission range of a source node, intermediate nodes cooperate in the communication process by relaying the packets. Every node in the network is a peer of every other node, and the WANET provides multihop connectivity between cooperating peer nodes. If connectivity to an infrastructure network is needed, at least one node should have access to it.

Node mobility, the wireless channel characteristics, and the battery energy conservation mechanisms contribute to the high time variability in the topology. As we have seen earlier in this chapter, the wireless channel characteristics change quite rapidly with time, space, and relative motion of the communicating nodes. Thus the ability of node pairs to directly communicate with each other changes with time even when the nodes are not mobile. It is known that a battery recovers some energy when it is not in use. Hence, mobile nodes can be switched off fairly often to conserve and also to recover some energy. Even when a mobile node is switched on, depending on its communication needs and battery level, the transmission power may vary with time. Thus, in a WANET, the number of active nodes at any time and their connectivity are random.

In addition to the time-varying topology, the spectrum allocated to a WANET is small. The same spectrum must be used to transfer both data and control information. Thus the information transport service that is to be built for a WANET must be bandwidth-efficient, especially with reference to disseminating network topology information.

An important aspect of the information transport service is the routing protocol. The issues in routing in ad hoc networks give rise to many trade-offs that must be addressed. The time-varying topology requires that the topology information be exchanged frequently so that every node is updated with the latest topology. However, this is a drain on the node's resources, especially if the frequency of using a route is comparable to the frequency of updates. An alternative to frequent exchanges of topology information is to discover a route between a source–destination pair on demand—that is, whenever a source has a packet to transmit to the destination. However, this increases the delay in transferring the packet to the destination. On-demand routing is accomplished by the source flooding a *route request* packet in the network and waiting for a *route reply* packet that contains the routing information. The routing protocol could also be proactive and discover routes in anticipation of future usage. Routing protocols that use geographic location information can also be used. Routes that are discovered on demand or proactively can be cached for future use. Routes can also be *maintained* to ensure that they are still active by listening to ongoing transmissions. As can be seen in this short discussion, there are many issues that need to be considered in a routing protocol to make it efficient in terms of energy and spectrum usage. Many such routing protocols for ad hoc networks have been defined. We do not discuss the routing issues in this chapter.

In the remainder of this section we discuss two important performance issues in WANETs: the connectivity and capacity of a random network.

8.7.1 Topology and Connectivity

Before we begin the study of the connectivity properties of ad hoc networks let us first consider how an ad hoc network is brought about. For such a network, we can define an operational area in which nodes are randomly placed, and the locations may follow a spatial distribution. These nodes now must communicate with neighbors. Each node can transmit with a certain power. Now recall the notion of reception range defined at the beginning of Section 8.2. We consider two nodes i and j to be neighbors if they are within transmission range of each other.

Let G be the graph that represents a WANET with node set $\mathcal{N}$ and edge set $\mathcal{E}$, with an edge between nodes indicating that direct communication between the nodes is possible; that is, the nodes are within transmission range of each other. There is randomness in the node locations, and hence the set $\mathcal{E}$ is also random. Hence to study its topological properties, G is best modeled as a *random graph*. We obtain the statistical behavior of the connectivity property by modeling a realization of a WANET as a realization of a random graph. The importance of the study of the connectivity property comes from the following observation: A low transmission range reduces the energy consumption and increases spatial reuse (see Figure 8.27) but if it becomes too low then the network can become disconnected (i.e., a sequence of hops between at least one node pair may not be available). Thus it is necessary to keep the transmission range high enough to keep the network connected. Clearly, if we keep the operational area of the ad hoc network fixed, the minimum transmission range to keep the network connected should be a function of the number of nodes in the network: A sparse distribution would require a higher transmission range, whereas a dense distribution could use a lower range. In the following discussion we let the transmission range of every node be $r(n)$, where n is the number of nodes in the network.

A random graph with a given number of nodes or edges (or both) could be constructed in many ways. Following the classical random graph model, we can assume that any pair of nodes are within transmission range of each other independently with probability $p(n)$, which can be obtained from $r(n)$. This is clearly not a reasonable model for a WANET: If node pairs (a, b) and (b, c) are within transmission range of each other, then the relative locations of node

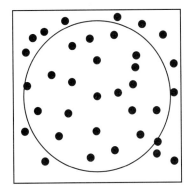 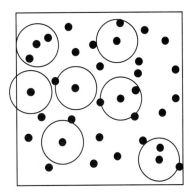

Figure 8.27 Increased spatial reuse with small transmission range. The transmission range in the network on the left is large, and at any time at most one transmission can occur. In the network on the right, with smaller transmission ranges, many transmissions can occur simultaneously.

pair (a, c) are defined, and we cannot say that the existence of link (b, c) is independent of the existence of links (a, b) and (a, c). Thus a random graph model for an ad hoc network must model the randomness in the graph resulting from the randomness of the geographical locations of the nodes. This is done using *geometric random graphs*, which are constructed as follows: Nodes are distributed over an operational area according to a spatial distribution, and two nodes are connected by an edge if the Euclidean distance between them is less than $r(n)$. The spatial distribution of the nodes can capture the randomness in the location that is the result of the mobility of the nodes.

First consider a two-node, one-dimensional network, with the location of each node uniformly distributed in $[0, z]$ and chosen independently of each other. Let the transmission range of both nodes be r. We now obtain the probability that the two nodes are connected. Without loss of generality, let x_1 be the location of the first node, and x_2 that of the second node (i.e., $x_1 \leq x_2$). The two-node network is connected if $x_2 - x_1 \leq r$. This is graphically shown in Figure 8.28. The set of values that (x_1, x_2) can take is denoted by the area OAB. The set of (x_1, x_2) that would give a connected network is given by the shaded area S in the figure. S is the region satisfying $x_1 < x_2$ (by definition of x_1 and x_2) and $x_2 - x_1 < r$ (the connectivity requirement). Because the nodes are distributed uniformly in $[0, z]$, the probability that the network is connected is the ratio of the area of S to the area of OAB, which can be seen to be $(z^2 - (z - r)^2) = 2zr - r^2$. This analysis can be

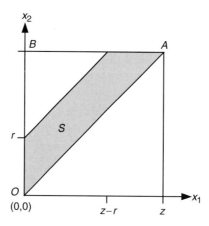

Figure 8.28 The feasible region of a random, connected, two-node, one-dimensional ad hoc network.

generalized to obtain $p_c(n, z, r)$, the probability that n nodes distributed uniformly in $[0, z]$, each with a transmission range of r, form a connected network. It is shown in the appendix to this chapter that

$$p_c(n, z, r) = \sum_{k=0}^{n-1} \binom{n-1}{k} (-1)^k \left(1 - k\frac{r}{z}\right)^n u\left(1 - k\frac{r}{z}\right) \quad \text{for } r < z \qquad (8.21)$$

where $u(x)$ is the unit step function.

Figure 8.29 plots $p_c(n, z, r)$ against r for $z = 1$ and different values of n. Observe that as n becomes large, for small values of r the network is disconnected with high probability, and for large values of r it is connected with high probability. The range of r over which the probability that the network is connected takes intermediate values is very small. This indicates some kind of a *thresholding* behavior for the connectivity probability: For large n, as r is increased from 0, the network is disconnected with probability very nearly 1, and at a threshold transmission range, it becomes connected with probability very nearly 1. This is indeed true. In fact it is true for a network with the operational area extending in any number of dimensions. We now provide arguments to obtain this threshold transmission range as a function of the number of nodes in a two-dimensional network.

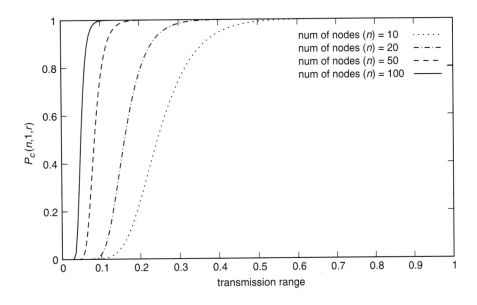

Figure 8.29 The probability of connectivity in a one-dimensional network as a function of r for different n, the number of nodes in the network. The operational area is [0,1].

For a fixed n, as $r(n)$ increases, the probability of the network being connected increases. Thus connectivity is a *monotone* property in $r(n)$. Threshold functions are defined for monotone properties. We can define the transmission range threshold function for connectivity as that $t(n)$ that satisfies the following properties: As n increases, $r(n)/t(n) \to 0$ implies that the network is almost surely disconnected, and $r(n)/t(n) \to \infty$ implies that the network is almost surely connected. This means that if $r(n)$ is a function that decreases faster than $t(n)$, then choosing the transmission range as $r(n)$ will make the network almost surely disconnected, and if $r(n)$ is a function that decreases slower than $t(n)$, then the network will almost surely be connected.

For the network to be connected there should be no isolated nodes. Let n nodes be distributed uniformly in the square operational area $[0, 1]^2$. Let $r(n)$ be the transmission range of the nodes when there are n nodes in the network. As the number of nodes n increases, we would like to decrease $r(n)$ so as to increase spatial reuse. However, $r(n)$ should not become so small that the network becomes disconnected. We wish to study, as n increases, how fast $r(n)$ can decrease so that

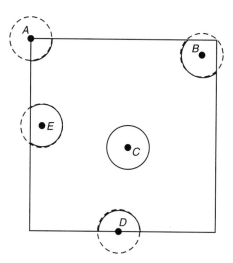

Figure 8.30 **The areas (inside the operational area) covered by the transmission ranges of nodes at various locations. Each circle has a radius of *r(n)*. The area bounded by the solid lines is the area covered by the transmission range that is also within the operational area. Node *A* at the corner has the minimum coverage inside the operational area.**

the network stays connected. Let X_i be the event that node i is isolated (i.e., it is not in the transmission range of any other node), and let $X := \cup_i^n X_i$ be the event that there is at least one isolated node in the network. A node is in isolated if the other $n - 1$ nodes are distributed outside the area covered by its transmission range. The area not covered by a transmission range of $r(n)$ is at most $1 - \pi r^2(n)/4$ (see point A in Figure 8.30). Therefore, we have

$$\Pr(X_i) \leq \left(1 - \frac{1}{4}\pi r^2(n)\right)^{n-1}$$

From the union bound, $\Pr(\cup_{i=1}^n X_i) \leq \sum_{i=1}^n \Pr(X_i)$, and we can write

$$\Pr(X) \leq n \left(1 - \frac{1}{4}\pi r^2(n)\right)^{n-1}$$

$$= e^{\left(\ln n + (n-1)\ln\left(1 - \frac{1}{4}\pi r^2(n)\right)\right)}$$

$$\leq e^{\left(\ln n - (n-1)\left(\frac{1}{4}\pi r^2(n)\right)\right)}$$

$$= e^{\ln n\left(1 - \frac{(n-1)}{\ln n}\left(\frac{1}{4}\pi r^2(n)\right)\right)}$$

$$= e^{\ln n\left(1 - \frac{\pi}{4}\left(\frac{r(n)}{\sqrt{\frac{\ln n}{n-1}}}\right)^2\right)}$$

where, in writing the second inequality, we have used the fact that $\ln(1 + x) \leq x$. For there to be no isolated nodes with probability 1, we want $\Pr(X) \to 0$ as $n \to \infty$. From the final equation, this can be seen to happen if $r(n)/\sqrt{(\ln n)/n} \to \infty$; that is, $r(n)$ is made to decrease strictly slower than $\sqrt{(\ln n)/n}$, with the ratio going to ∞ as $n \to \infty$. Thus $\sqrt{(\ln n)/n}$ is a possible threshold function for $r(n)$. In fact, it can be shown that if $r(n)/(\sqrt{(\ln n)/n}) \to 0$, then $\Pr(X = 1) \to 1$, implying that $\sqrt{(\ln n)/n}$ is actually a threshold function for the disappearance of isolated nodes in the random WANET.

In this discussion we have obtained the threshold function for there to be no isolated nodes in the network. It can be shown that as $r(n)$ is increased from 0, the value at which the isolated node almost surely disappears is also the value of $r(n)$ at which the network becomes connected; that is, the threshold transmission range for there being no isolated nodes is also the threshold transmission range for the network becoming connected. Thus, if $r(n)$ decreases slower than $\sqrt{(\ln n)/n}$, asymptotically, the network is almost surely connected. This can be achieved by having $r(n) = O\left(\sqrt{(\ln n + c(n))/n}\right)$, with $c(n) \to \infty$ as $n \to \infty$.

> **Exercise 8.9**
> Using a similar argument to the one just made, show that for a one-dimensional network, $r(n)$ should decrease slower than $(\ln n)/n$.

Recall that the transmission range of a node is a function of the transmitter power, the channel characteristics, and the receiver sensitivity. Transmission range can be controlled by controlling the transmission power.

8.8 Link Scheduling and Network Capacity

Section 8.7 explains how to think about the topology and connectivity in a multihop wireless ad hoc network. In this section we consider scheduling of transmissions in multihop wireless networks and derive fundamental limits on the capacity of such networks.

In the following we give a brief summary of the issues in link scheduling in wireless networks. Time could be slotted as in slotted Aloha; then the transmission start times are restricted to the beginnings of slot boundaries. The network could also be unslotted. In a slotted WANET, a transmission schedule is used to specify the set of nodes that are allowed to transmit in each slot. In addition to specifying the transmitter, the schedule may specify the destination of the transmissions. The schedule can be static or dynamic. Static scheduling is analogous to a TDM link, where a set of slots forms a frame and the transmission schedule determines the nodes that transmit in each slot of a frame. The same transmission sequence repeats in every frame. In dynamic scheduling a schedule is determined separately for each slot, and the decision can use the current state of the network and possibly the recent history. In the case of dynamic scheduling, the algorithms can be executed in a centralized or in a distributed manner. In distributed dynamic scheduling, each node decides when to transmit, much as in the random access networks that you studied earlier. A centralized scheduler may require complete network topology information and queue length information at the beginning of every slot.

Scheduling deals with the problem of optimizing the performance (e.g., throughput or delay) of a given WANET (with a given topology and given transmission constraints). A poorly chosen schedule may not permit much traffic to be carried. The question of optimal scheduling is typically addressed in the context of specific assumptions about the way the network is operated. On the other hand, if a large number of nodes are distributed in space and are equipped with radio transceivers, it is also of interest to understand what the traffic-carrying capacity of the network is, and how this capacity scales with the number of nodes, irrespective of how the network is operated. We briefly discuss an important recent result on this topic.

8.8.1   Scheduling Constraints

Consider a multihop wireless network whose topology has already been discovered and is described by the directed graph $G(\mathcal{N}, \mathcal{E})$, where $\mathcal{N}$ is the set of nodes and $\mathcal{E}$ is the set of directed edges. Recall that an edge $(i, j) \in \mathcal{E}$ means that a transmission from i, addressed to j, can be decoded by j, provided that the SIR at j is adequately high. In particular, the presence of an edge (i, j) means that if no other transmission is taking place then the signal to noise ratio for a transmission between i and j is

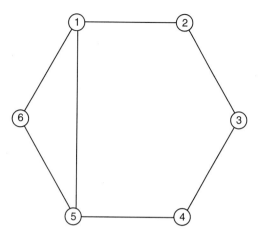

Figure 8.31   The connectivity graph of a wireless ad hoc network.

sufficient for it to be decoded. For a directed edge $e = (i, j)$, we denote its tail (i.e., transmitter) node by $t(e) = i$, and its head (i.e., receiver) node by $h(e) = j$.

Let us assume that time is divided into slots and that there is a centralized scheduler that decides which links should be activated in each slot. The edges can be grouped into subsets such that the edges in a subset can be activated in the same slot and that the receiver in each edge can decode the transmission from the tail node of the edge. Thus when such a set—say, S—is activated, one packet can be sent across each edge in S. Basically these sets respect any radio operation constraints (e.g., a node cannot send and receive at the same time) and interference constraints. We use the term *independent set* to mean a set of edges that can be activated simultaneously. Note that this notion of an independent set is different from the one used in graph theory. Let us denote by $\mathcal{S}$ the set of all independent sets; subsets of independent sets are also independent sets.

The sets $S \in \mathcal{S}$ must be obtained from the imposed *link activation constraints*. These could be imposed as topological constraints. For example, it could be required that a set of links can be activated as long as none of the edges has common vertices. Thus in the wireless network topology shown in Figure 8.31, the following sets of edges are independent sets: $S_1 = \{(1, 2), (5, 6), (3, 4)\}, S_2 = \{(2, 3), (1, 5)\}, S_3 = \{(2, 3), (4, 5), (1, 6)\}$. We observe that these sets are also *maximal* independent sets, because none of the sets can be enlarged without violating the activation constraints. With the simple activation constraint specified, in each independent set each link can be activated in either direction.

Exercise 8.10

Independent sets can be obtained by *graph coloring*. Thus the activation constraint that none of the activated edges can have common vertices can be viewed as coloring the *edges* of the network graph so that adjacent edges do not have the same color. Color the edges of the network shown in Figure 8.31 such that adjacent edges do not have the same color. List all the independent sets.

The foregoing simple topological constraints are not adequate in general. For example, in the network in Figure 8.31, even though edges $(1, 2)$ and $(3, 4)$ do not have any vertices in common, it may not be possible to transmit a packet from node 1 to node 2 and simultaneously from node 3 to node 4, because the signal from node 3 may cause excessive interference to the reception at node 2. More generally, the independent sets can be obtained from the requirement that at every receiver the SIR for the transmission intended for it is greater than some threshold β. Thus if $S = \{e_1, e_2, \ldots, e_n\}$ is an independent set, then it must be the case that, for every i, $1 \leq i \leq n$,

$$\frac{P \left(\frac{d_{t(e_i), b(e_i)}}{d_0} \right)^{-\eta}}{2 W N_0 + \sum_{\{k : 1 \leq k \leq n, k \neq i\}} P \left(\frac{d_{t(e_k), b(e_i)}}{d_0} \right)^{-\eta}} > \beta$$

where we are assuming only a path loss model, P is the transmit power at every node, d_{n_1, n_2} denotes the distance between nodes n_1 and n_2, and d_0 is the reference distance (see Section 8.1.4).

8.8.2   Centralized Scheduling

In the foregoing framework, a schedule basically specifies a sequence of independent sets to be activated. Let us first consider a static link activation schedule. Allocate m_S slots to independent set S. Transmissions are scheduled as follows: Time is divided into frames of $M := \sum_{\{S \in \mathcal{S}\}} m_S$ slots, and m_S slots in each frame are allocated for transmission on the edges in S. Then the bandwidth allocated to edge e will be

$$b_e = \frac{\sum_{\{S \in \mathcal{S}\}} m_S I_{\{e \in S\}}}{M}$$

where, as usual, $I_{\{\cdot\}}$ is an indicator function.

In general the sequence of independent sets activated is a random process. Let us denote it by $\{S_n, n \geq 1\}$, with $S_n \in \mathcal{S}$, for every n. Let us assume that the process is ergodic so that time averages exist. Now for an edge $e \in E$, consider

$$b_e = \lim_{n \to \infty} \frac{1}{n} \sum_{i=1}^{n} I_{\{e \in S_i\}}$$

For the given activation schedule, b_e is the fraction of time that edge e is active, and hence b_e is the packet rate that the edge e can carry. Thus, for a given schedule we can think of the wireless network as a *capacitated network* represented by the directed graph $G(\mathcal{N}, \mathcal{E})$ and edge capacities $b_e, e \in E$. If the network must transport packet flows, then, with the given schedule, the flow allocated to edge e can be no more than b_e.

We can also write b_e in another way. Define, for $S \in \mathcal{S}$,

$$\phi_S = \lim_{n \to \infty} \frac{1}{n} \sum_{i=1}^{n} I_{\{S_i = S\}}$$

That is, ϕ_S is the fraction of slots in which the independent set S is activated. Then clearly $b_e = \sum_{\{S : S \in \mathcal{S}, e \in S\}} \phi_S$. A schedule can now be thought of in terms of the probabilities $\phi_S, S \in \mathcal{S}$; define $\mathbf{\Phi}$ to be the vector of these probabilities. In fact, given a probability vector $\mathbf{\Phi}$ it is easy to see that we can obtain a corresponding schedule by simply using $\mathbf{\Phi}$ to pick an $S \in \mathcal{S}$ independently in each slot. This is a static randomized schedule. If we consider all possible probability vectors $\mathbf{\Phi}$ we obtain all the possible capacitated network topologies. Let $\mathbf{B}(\mathbf{\Phi})$ be the capacity vector for the network with the schedule probability vector $\mathbf{\Phi}$; there is an element in the vector $\mathbf{B}(\mathbf{\Phi})$ for each edge in E.

Consider now several packet flows $j, 1 \leq j \leq J$, that must be carried across the network. Flow j has source node $s(j)$ and destination node $d(j)$. The offered rate of flow j is λ_j; $\mathbf{\lambda}$ is the vector of flow rates. Let us assume that each packet flow can be split arbitrarily across paths in the network, as long as all these subflows

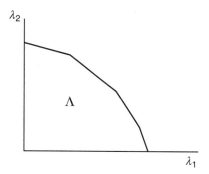

Figure 8.32 A sketch of a possible schedulable region for two flows through a multihop wireless ad hoc network.

converge at the destination node. Clearly a given vector of flow rates λ can be carried by the network only if there exists a schedule, with probability vector Φ, and an allocation of all the flows to paths such that the total flow allocated to edge e is *less* than $b_e(\Phi)$. Let us denote the set of all such flow rates λ by Λ; this is called a *schedulable region*. Figure 8.32 sketches a possible schedulable region for the case $J = 2$. The following general properties are easy to check. Note the requirement that the flow on each link be *less* than the average link capacity under the schedule. Hence Λ does not include the boundary of the set.

Exercise 8.11

Show that Λ is a convex set, and if $\lambda \in \Lambda$ then for any λ' such that $\lambda'_j < \lambda_j$, for all j, $\lambda' \in \Lambda$.

Let us now observe that if we are given a set of flow rates $\lambda \in \Lambda$ then by definition there exists a Φ such that if the links are scheduled with this probability vector then the given flow rates can be carried. In fact such a probability vector can be characterized rather simply, but note that (1) the flow rates λ must be known a priori to obtain an appropriate Φ, and (2) if the flow rates change, then the static randomized schedule must also be recomputed. Does there exist a schedule that does not need to know the offered flow rates λ and yet can carry any $\lambda \in \Lambda$? The answer is yes and is provided by the following very important result, which has applications in several other problems (see the switching example in Chapter 1).

There is a centralized scheduler that knows the network topology. Also, the queue occupancies in the nodes are known to the scheduler at the beginning of every slot, and the scheduler uses this information to determine the set of active links and the flows whose packets should be transmitted on these links in each slot. Each node maintains a separate logical queue for each flow, and $Q_{n,j}(k)$ denotes the number of packets of flow j at node n at the beginning of slot k. It is assumed that the arrivals of the various flows are independent Bernoulli processes with rates $\lambda_j, 1 \leq j \leq J$. With these assumptions, the vector of queue length processes $(Q_{n,j}(k), n \in \mathcal{N}, 1 \leq j \leq J, k \geq 0)$ is a Markov chain. We would like to schedule the links so that for the given offered rates λ the Markov chain eventually, with probability 1, enters a positive recurrent class of states (see Section D.1). We will then say that the schedule *stabilizes* the vector of flow rates λ.

At the beginning of slot k, for each edge e and each flow j, define

$$w_{e,j}(k) := Q_{h(e),j}(k) - Q_{t(e),j}(k)$$

Note that $w_{e,j}(k)$ denotes the difference between the number of packets of flow j at the head of the edge e and at its tail. (Note that if $t(e)$ is the destination node of flow j then no packets of j would be queued there.) Denote $\tilde{w}_e(k) = \max_j w_{e,j}(k)$ to be the weight of edge e. Obtaining the set of active links in a slot means that a $S \in \mathcal{S}$ must be chosen. For each S define $\hat{w}_S(k)$ as

$$\hat{w}_S(k) := \sum_{e \in S} \tilde{w}_e(k)$$

For an illustration see Figure 8.33.

It has been shown that the following transmission strategy stabilizes every $\lambda \in \Lambda$. Let $S^*(k)$ denote the independent set for which the $\hat{w}_S(k)$ is the maximum in slot k. The links in $S^*(k)$ are enabled during slot k. For each $e \in S^*(k)$, choose j^* as that j with the maximum $w_{e,j}$, and transmit a packet from queue $Q_{h(e),j^*}$ to node $t(e)$. Note that this algorithm also performs dynamic routing of the packets.

The complete result (of which we have presented a simplified version) allows for multiple destinations for each flow and also allows for link errors. These lead to simple modifications of the scheduling policy. The results have been generalized to fading channels.

Although there are some strong assumptions about the complete knowledge of buffer occupancies and network topology in determining the schedule in every slot, this is an important result because it tells us what is the best that

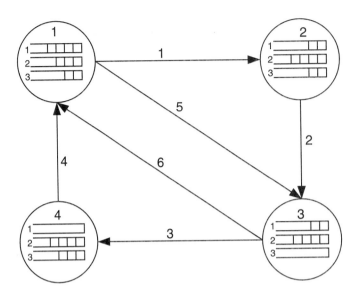

Figure 8.33 Illustrating the variables for dynamic scheduling in a WANET with six links and three flows of packets. The link number is marked against the links. The queue occupancies at the beginning of a slot are shown. Some sample values for the variables are $Q_{1,1} = 4$, $Q_{2,1} = 2$, $Q_{1,2} = 3$, $Q_{4,3} = 3$, $w_{1,1} = 4 - 2 = 2$, $w_{1,2} = 3 - 4 = -1$, $w_{1,3} = 2 - 2 = 0$, $\bar{w}_1 = 2$.

we can do. It can be used to compare the performance of other, possibly more practical, scheduling algorithms with the best possible performance. Two important questions remain, however, as important problems of current research:

- Are there *distributed* scheduling algorithms that can also stabilize every $\lambda \in \Lambda$ with only local information at a node or perhaps its neighborhood?

- If we are interested not only in stabilizing schedules but also ones that achieve some performance objective (e.g., minimizing the mean delay), then what are good scheduling algorithms, both centralized and distributed?

8.8.3   Capacity of a WANET

We have argued that having a low transmission range (greater than the threshold at which the network is connected) increases spatial reuse and can increase capacity when the nodes are uniformly distributed in the operational area and when the packets choose their destinations randomly. However, reducing the transmission radius increases the number of hops between source–destination pairs, and the number of times a packet must be transmitted increases in inverse proportion to the transmission range. This offsets some of the gains made by spatial reuse. Let us investigate this trade-off and describe a scaling law for the per-node capacity of the network.

A Bound from Spatial Reuse

Let $r(n)$ be the transmission range of all the nodes in an n-node network, $h(n)$ the average number of hops that a packet must make before reaching its destination, and $\lambda(n)$ the arrival rate of packets per node. Hence, the total required rate of packet transmissions in the network is $n\lambda(n)h(n)$. We observe that decreasing $r(n)$ increases spatial reuse by a factor proportional to the area covered by a transmission, and hence the spatial reuse is $O(1/r^2(n))$. It follows that we must have

$$n\lambda(n)h(n) \leq O\left(\frac{1}{r^2(n)}\right)$$

Hence

$$n\lambda(n) \leq \frac{1}{h(n)} O\left(\frac{1}{r^2(n)}\right)$$

Now the number of hops is directly proportional to the distance between the source and destination nodes, and a packet can cover a distance proportional to $r(n)$ in one hop. Therefore, $h(n)$ scales inversely with $r(n)$. It then follows that

$$n\lambda(n) \leq O\left(\frac{1}{r(n)}\right)$$

However, we have seen that for the network to remain connected, $r(n)$ should decrease slower than $\sqrt{\frac{\ln n}{n}}$. Thus, we obtain finally that

$$\lambda(n) \leq O\left(\frac{1}{\sqrt{n \ln n}}\right)$$

This means that as the number of nodes increases, the per-node capacity (i.e., the maximum rate at which a node can generate and transmit data) decreases at least as fast as $\frac{1}{\sqrt{n \ln n}}$. Intuitively, this is because as the number of nodes increases, each node spends more time relaying packets from other nodes, and the benefits of reduced transmission range and the consequent increased spatial reuse are not quite realized.

It should be clear that the "derivation" provided here is rough. Because we have randomly distributed nodes and because packets choose random destinations, a precise statement of the result should be probabilistic. Furthermore, in deriving the result we should carefully model the conditions under which packet transmissions are successful. Such an analysis has been performed ([132]), and results of the following form have been obtained. There exist constants $0 < c < c' < \infty$, such that

$$\lim_{n \to \infty} \Pr\left(\lambda(n) = c \left(\frac{1}{\sqrt{n \ln n}}\right) \text{ is achievable}\right) = 1$$

$$\lim_{n \to \infty} \Pr\left(\lambda(n) = c' \left(\frac{1}{\sqrt{n \ln n}}\right) \text{ is achievable}\right) = 0$$

This result has a form similar to the thresholding behavior of connectivity (with respect to the transmission range) observed in Section 8.7.1.

Summarizing, we can say that reducing $r(n)$ increases spatial reuse and hence increases the capacity of the network, but there is a limit as to the spatial reuse we can achieve, and this limit is imposed by the transmission range requirement to keep the network connected. These observations suggest that very dense wireless ad hoc networks should not be used for communication between arbitrary pairs of nodes (as is the case in a general internet); instead such networks may be more useful for nearest neighbor communication, as would be required for distributed instrumentation applications over ad hoc wireless sensor networks. We turn to this topic in Section 8.9.

8.9   Wireless Sensor Networks: An Overview

The networks that we have described as wireless LANs or wireless internets are basically meant for carrying point-to-point Internet applications. The objective of such communication is exactly the same as it is in a wired internet. Entities on two hosts associate themselves and then transfer data to each other. There are individual QoS objectives for each instance of such a transfer, and the network

Figure 8.34 The Berkeley Mote, which includes a light and temperature sensor, a digital radio, and an 8-bit, 4 MHz microprocessor. The U.S. 1-cent coin is shown for size comparison. Photograph taken from http://www-bsac.eecs.berkeley.edu/~shollar/macro_motes/macromotes.html; courtesy Seth Hollar.

objective is to carry all the transfers efficiently while meeting the individual QoS objectives. Such would be the requirements from an ad hoc internet based on IEEE 802.11 nodes that is set up in a disaster area where the wired infrastructure has been damaged—for example, because of a massive earthquake.

In recent years there has been growing interest in a new application of wireless ad hoc networks. Advances in microelectronics have made possible the development of tiny devices that can sense, compute, and communicate. Each such device might have one or more sensors (e.g., acoustic, chemical, light, etc.), a microcomputer, a digital radio, and, of course, a tiny battery. Figure 8.34 shows the Berkeley Mote, which was developed at the University of California, Berkeley, and is a popular platform for sensor network research. It is expected that in the future the size of such smart sensors will decrease significantly. A large number of such devices (hundreds or even thousands) can be deployed randomly in a given geographical area. For example, they could be dropped out of an airplane into a field or forest, or they could be worn by people, or they could be randomly embedded in large buildings or vehicles (e.g., railway trains). Once they are deployed, the idea is for these sensors to organize themselves into an ad hoc network and then carry out some global computational task. Some examples of such tasks are

- Detecting fires in large forests

- Measuring and tracking levels of chemical or radioactive contamination resulting from industrial accidents

- Identifying locations of survivors of building and train disasters

- Serving as environmental instrumentation in large factories and buildings

We notice, first of all, that each of these tasks has a global objective, and there is no individual objective of point-to-point communication between pairs of nodes (as in the case of an ad hoc wireless internet). In fact we can think of a sensor network as a *distributed instrument*. To carry out tasks such as those listed, we can identify several procedures that a random sensor network would have to implement:

- *Neighbor discovery and self-organization:* The need for such a step is common to all ad hoc networks. The particular objectives of a sensor network may render certain discovered topologies inferior to others. For example, if a sensor discovers all its neighbors and attempts to exchange measurements with all of them, it may cause so much interference that communication will rarely succeed. A connected but sparse topology may lead to better performance.

- *Distributed computation algorithms:* The sensor network's task would be to provide some useful inferences based on the measurements at the sensors. Notice that this requirement is different from that in traditional sensor arrays, which simply send all the measurements to a central signal processor for computation. For example, if the maximum value of the measured quantity is needed, then the approach of sending all the individual measurements to a central operator would be extremely inefficient in communication complexity ($O(n^2)$ for n sensors placed in one dimension, with the operator node at one end). A more efficient way would be for the maximum to be recursively computed in the network as the sensors exchange their measured values ($O(n)$ communication complexity for this example).

- *Capacity optimization:* The distributed computation would require communication between neighbors, and a high communication capacity of the network would mean that the distributed computation converges quickly and that time-varying data can be tracked. The sensors would have a limited battery life and hence would need to carry out their communication tasks efficiently. There would be a built-in multiple access mechanism for sharing the radio spectrum that their digital radios are designed to operate in. As we saw in Section 8.6.3, some parameters of the multiple access mechanism may need to be tuned in order to optimize the communication capacity.

- *Localization:* Consider the example of sensors being used to detect a forest fire or to identify the presence of injured people in a building collapse. It is not enough for the network to detect the occurrence of the event; it must also inform the operator as to the location of the event. Because sensors are randomly placed, they cannot have programmed coordinates but instead must learn their coordinates from external references or from a few *beacon* nodes that may have been deployed with the other sensors. These beacons would have the capability, for example, of determining their own locations by using the global positioning system.

8.10 Summary

We begin this chapter with a taxonomy of wireless networks and deal with random access networks and wireless ad hoc networks. We view wireless networking as being concerned with the problems of sharing a radio spectrum among traffic flows. Hence, to study wireless networking, an understanding of digital communication over radio is essential. We provide an overview of this topic in Section 8.1; we introduce digital wireless communication, channel coding, the Shannon capacity formula, fading in wireless channels, and the capacity of fading channels. Then we discuss special issues in and techniques for sharing a radio spectrum among many users. Having set up an understanding of how bits are transported over point-to-point and shared radio links, we turn our attention to analyzing wireless networks. The simplest situation is of Internet access over a point-to-point wireless link. A common application being file transfer, we describe the performance of TCP-controlled bulk transfers over a lossy wireless channel in which the losses can be independent or correlated. Next we discuss the emerging area of cross-layer design and optimization of wireless communication links; in particular, we discuss channel state and buffer adaptive power control. We then turn our attention to random access wireless networks, reviewing the classical Aloha protocol and developing an understanding of carrier-sensing protocols, a version of which is used in the popular IEEE 802.11 standard for wireless LANs. We then discuss the IEEE 802.11 multiple access protocol and provide some simple models to understand how various parameters affect its performance. Moving on from single-hop random access networks, we then examine multihop ad hoc wireless networks. Here we discuss issues such as connectivity, topology, link scheduling, and network capacity. Finally, we provide a brief overview of the emerging area of ad hoc sensor networks.

8.11   Notes on the Literature

In keeping with our view of wireless networks presented in Section 3.5, we begin this chapter with a review of basic concepts in digital communication over wireless channels. We discuss only the basic concepts and provide some insight into commonly used terms such as carrier frequency, spectrum bandwidth, symbol rate, bit rate, signal to noise ratio, coding, channel capacity, and Shannon's theorem. This subject is treated in several excellent textbooks; two texts that we have used are the ones by Proakis [238] and by Lee and Messerschmitt [194]. The area of mobile multiuser communication over wireless channels has made rapid progress in recent years. We have discussed some of the engineering issues and trade-offs, mentioning some of the technologies (such as DSSS and OFDM) only in passing. Yet these are implemented in some of the wireless network technologies that we have discussed. Again, we provide only the basics in order to motivate and discuss the issues; extensive coverage is provided in the texts by Stuber [278] and by Rappaport [244].

The material on analysis of TCP-controlled file transfer throughput over lossy wireless links has been taken from the papers by Kumar ([181], which assumes i.i.d. packet loss) and by Zorzi et al. ([311], which accounts for correlated packet losses). An approach for two-state Markov modeling of a fading channel is provided by Zorzi et al. in [312]. Additional references on TCP throughput analysis with correlated packet losses in the wireless setting are Kumar and Holtzman [182], Zorzi and Rao [310], and Anjum and Tassiulas [14]. Researchers have also found that the performance of TCP can be very poor in multihop IEEE 802.11 networks; two references are Gerla et al. [116] and Xu and Saadawi [303, 304].

Adaptive and optimal control problems in wireless communications and networks have been getting increasing attention in the research literature. The ergodic capacity of a fading channel and the related coding theorem were obtained by Goldsmith and Variaya in [122]. The optimal power control problem for a buffered system was studied by Berry and Gallager in [32]; the form of the optimal control we have shown was obtained by Goyal et al. in [125]. Additional related references are Telatar and Gallager [283], Goldsmith and Chua [121], Jindal and Goldsmith [154], Zhang and Wasserman [309], and Holliday et al. [141].

The Aloha protocol was first described by Abramson in [4]. The CSMA protocol was first analyzed by Kleinrock and Tobagi [175] and by Lam [189]. The instability of the slotted Aloha with nonadaptive feedback has been shown by many authors, including Kleinrock [174], Fayolle et al. [96], and Kelly [162]. Space–time models for local networks were first presented by Molle et al. [219]. Many collision resolution algorithms for use in random access networks have been proposed. Bertsekas and Gallager [33] summarize and analyze many of the well-known ones. Rom and Sidi [253] discuss the performance analysis of a large class of random access protocols. Ethernet was first proposed by Metcalfe and Boggs [212]. It has been standardized as the IEEE 802.3 CSMA/CD local area network protocol by the IEEE [147]. This standard has undergone many revisions, and the original specification for 10 Mbps networks over coaxial cables has now evolved to provide 1 Gbps over fiber optic and copper cables. See Shoch et al. [261] for an early history of Ethernet. An analytical model for the performance of CSMA/CD was first given by Tobagi and Hunt [285]. The MACAW protocol was proposed by Bharghavan et al. [34], and it in turn was an adaptation of the MACA protocol first described by Karn [157].

The authoritative source for the IEEE 802.11 MAC protocol is the standards document published by IEEE and downloadable from the site of this working group. In [203], Mangold et al. have provided a concise description of the basic MAC protocol as well as the 802.11e extension, which provides QoS differentiation. The detailed performance analysis that accounts for uniform sampling of the backoff multiplier was provided by Bianchi in [35]. The analysis presented in this text is a simplification and generalization of the analysis in [35]. The idea of adapting the backoff to optimize the throughput was studied by Cali et al. in [47]. There is now considerable interest in adaptive and cross-layer techniques for IEEE 802.11 networks; some references are Qiao and Choi [239], Yuen et al. [306], and Chevillat et al. [59]. A cautionary note on the pitfalls that can accompany cross-layer optimizations has been provided by Kawadia and Kumar [160].

Perkins [236] discusses applications, addressing and the proposed routing protocols for ad hoc networks. The connectivity analysis of the finite one-dimensional network is by Desai and Manjunath [76].

The threshold transmission range was derived by Gupta and Kumar [131]. Also see [231, 183]. Penrose [235] has some results on geometric random graphs directly relevant to WANETs.

Multihop ad hoc wireless networks were studied as packet radio networks more than two decades ago. An early modeling of transmission constraints via independent sets in graphs was done by Nelson and Kleinrock [221]. Centralized static scheduling in multihop wireless networks has been well studied. See, for example, Ramanathan and Lloyd [242]. The queue-length-based dynamic scheduling algorithm that stabilizes the network is by Tassiulas and Ephremides [282]. An early study of the spatial capacity of multihop packet radio networks was provided by Nelson and Kleinrock [222]. The notion of transport capacity and the capacity bounds was developed by Gupta and Kumar [132]. An analysis of information theoretic limits on the capacity of wireless ad hoc networks has been provided by Xie and Kumar [302].

We have provided only a brief overview of the emerging area of wireless sensor networks. A survey has been provided by Akyildiz et al. in [9].

Appendix: Probability of Connectivity in a One-Dimensional Ad Hoc Network

Let there be n nodes in the network, and let the location of node i be denoted by x_i. Note that x_i are i.i.d. with uniform distribution in $[0, z]$. Thus a random network is represented by a random vector $\mathbf{x} = [x_1, x_2, \ldots, x_n]$. Let $p_c(n, z, r)$ be the probability that $\mathbf{x}$ forms a connected network when each node has a transmission range of r. Let $\hat{\mathbf{x}} = [\hat{x}_1, \hat{x}_2, \ldots, \hat{x}_n]$ be the nodes ordered according to their positions on $[0, z]$ (i.e., $\hat{x}_1 < \hat{x}_2 < \cdots < \hat{x}_n$). Define $\hat{x}_0 = 0$. The condition $\hat{x}_{i+1} - \hat{x}_i < r$ for $i = 1, \ldots, (n-1)$ must be satisfied for $\mathbf{x}$ to be connected.

The set of all realizable networks is contained in the polytope, A_n, defined by $0 \leq \hat{x}_1 \leq \hat{x}_2 \leq \cdots \leq \hat{x}_n \leq z$. The set of connected networks is contained in the polytope, $A_c(n, z, r)$, defined by $\hat{x}_{i+1} - \hat{x}_i < r$ for $i = 1, \ldots, (n-1)$. Let $V_n(z)$ and $V_c(n, z, r)$ be the volumes of the polytopes A_n and $A_c(n, z, r)$, respectively. Because we assume that the node locations are uniformly distributed in $[0, z]$, the probability that the network is connected, $p_c(n, z, r)$, is $V_c(n, z, r)/V_n(z)$.

We obtain $p_c(n, z, r)$ as follows. Define $y_i = \hat{x}_{i+1} - \hat{x}_i$, $i = 0, \ldots, n - 1$. Let $U(n, z)$ be the volume of the set

$$\left\{ y_0, y_1, \ldots, y_{n-1} : y_i \geq 0 \text{ for } i \geq 0, \sum_{0}^{n-1} y_i \leq 1 \right\}$$

and let $U_c(n, z, r)$ be the volume of the set

$$\left\{ y_0, y_1, \ldots, y_{n-1} : y_i \geq 0 \text{ for } i \geq 0, y_i \leq r \text{ for } i > 0, \sum_{0}^{n-1} y_i \leq 1 \right\}$$

Notice that because $\hat{x}$ and y are related by a linear invertible transformation, we have $U_n(z) = K V_n(z)$ and $U_c(n, z, r) = K V_c(n, z, r)$ for some constant $K > 0$. Thus

$$p_c(n, z, r) = \frac{U_c(n, z, r)}{U_n(z)}$$

We calculate $U_n(z)$ using the following recurrence relation.

$$U_n(z) = \int_0^z U_{n-1}(z - t) \, dt = \frac{z^n}{n!}$$

From Figure 8.35, $U_c(n, z, r)$ can be written as follows:

$$U_c(n, z, r) = \int_0^r U_c(n - 1, z - t, r) \, dt \qquad (8.22)$$

Equation 8.22 essentially means that the n node network is connected if two conditions are satisfied: The $n - 1$ node network formed after taking out the first node is connected, and the node that was taken out is within r of the rightmost node in the $n-1$ node network. This essentially means that the $n-1$ node network will span over at most $z - r$.

Defining

$$h(z) := \begin{cases} 1 & 0 \leq z \leq r \\ 0 & \text{otherwise} \end{cases}$$

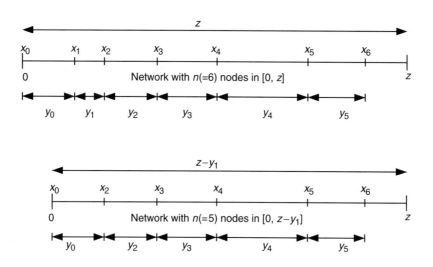

Figure 8.35 Illustrating Equation 8.22. For the *n* node network to be connected, the *n* − 1 node network without the node at *x₁* should also be connected and *y₁* should be at most *r*. Thus the *n* − 1 node network will be spread over *z* − *y₁*, with *y₁* taking values in (0,*r*).

we can write the right side of Equation 8.22 as the convolution of $U_c(n-1,z,r)$ and $h(z)$. Thus

$$U_c(n,z,r) = h * U_c(n-1,z,r) = \cdots = h^{((n-1)*)} * U_c(1,z,r)$$

where $h^{((n-1)*)}(\cdot)$ is the $(n-1)$-fold convolution of $h(\cdot)$ with itself. Note that $U_c(1,z,r) = zu(z)$, where $u(z)$ is the unit step function.

Let $\tilde{h}(s)$ and $\tilde{U}(n,z,s)$ denote the Laplace transform of $h(t)$ and $U_c(n,z,r)$, respectively.

$$\tilde{h}(s) = \frac{1-e^{sr}}{s}$$

$$\tilde{U}_c(n,z,r) = \left(\frac{1-e^{sr}}{s}\right)^{n-1}\frac{1}{s^2}$$

$$= \frac{(1 - e^{sr})^{n-1}}{s^{(n+1)}}$$

$$= \frac{\sum_{k=0}^{n-1} \binom{n-1}{k} (-1)^k e^{srk}}{s^{(n+1)}}$$

Taking the inverse Laplace transform,

$$U_c(n, z, r) = \sum_{k=0}^{n-1} \binom{n-1}{k} \frac{(-1)^k (z - kr)^n u(z - kr)}{n!} \qquad (8.23)$$

We thus get

$$p_c(n, z, r) = \frac{U_c(n, z, r)}{U_n(z)}$$

$$= \sum_{k=0}^{n-1} \binom{n-1}{k} (-1)^k \frac{(z - kr)^n}{z^n} u(z - kr)$$

Problems

8.1 Consider the binary modulation scheme analyzed in Section 8.1. Obtain the bit error rates for various SNR values $\gamma = 12$ dB, 11 dB, 10 dB, and 9 dB. In each case, calculate the probability of packet error for 1500-byte packets. Hence compare the plots in Figure 8.11 with the AWGN plot in Figure 8.14. Hint: Use the approximation $Q(x) \approx \frac{1}{x\sqrt{2\pi}} e^{-\frac{x^2}{2}}$.

8.2 For the same situation as Problem 8.1, consider Rayleigh fading. For average (Rayleigh-faded) SNRs $\gamma = 12$ dB, 24 dB, and 36 dB, obtain the fraction of time that the SNR is less than 9 dB. Hence explain why a very large SNR is required in Figure 8.14 to obtain a high throughput.

8.3   Show that with $P_{xmt} = P_{avg}$ the capacity given by Equation 8.10 is less than the optimal capacity with transmitter power control given by Equation 8.15.

8.4   For the IEEE 802.11 protocol description given in the text, assume that backoff times are real numbers and that the channel propagation delay is zero, and hence that no collisions occur. Ignore the fixed time overheads (such as SIFS and DIFS). Assume that all nodes have the same packet length. Consider an n node network. Node k has an exponentially distributed backoff with mean $\frac{1}{\beta_k}$ and uses the data rate r_k. Under these simplifications show that the normalized throughput for node k is given by $\frac{\Theta_k}{r_k} = \frac{\eta_k L}{1 + \sum_{j=1}^{n} \eta_j L}$, where $\eta_k = \frac{\beta_k}{r_k}$. Hence argue that when the transmission rates are different (for example, because of different distances of the nodes from the AP), to achieve fair normalized throughputs the mean backoff times of nodes should be adapted to be inversely proportional to their transmission rates.

8.5   Repeat the derivation in Problem 8.4 but without assuming that there are no collisions, and also accounting for the overhead durations T_o and T_c.

8.6   As a model of a wireless ad hoc network consisting of a large number of randomly strewn devices, consider a Poisson field of points of intensity λ per m^2 on the plane, with each point denoting the location of a wireless transceiver. All transceivers transmit on a common frequency with unit power using omnidirectional antennas. For the propagation model, assume only path loss with exponent η. Time is divided into fixed-length slots. In each slot, a device decides to transmit with probability α and decides to receive with probability $(1 - \alpha)$ independent of past transmissions and other devices. Find the distribution of interference power received at a randomly selected device in a slot. Show that the mean interference power is finite only if $\eta > 2$. Show also that when $\eta > 2$, the mean is linear in λ and α.

8.7   Consider a pure Aloha network with an infinite number of nodes on a straight line of length a. Each packet transmission is of unit length, and transmissions are attempted according to a Poisson process of rate G. Assume that in this network there are only broadcasts and that a transmission should be received at all the nodes. A transmission is a success only if it is received correctly at all the nodes.

    a. Find the conditional probability that a transmission starting at x, $0 \leq x \leq a$, is successful.

    b. Find the broadcast throughput of the network.

8.8   For a time-slotted network, where the slots are small compared with the packet lengths, consider the following variation of slotted Aloha. When a node has a packet to transmit, it begins transmission at the beginning of a slot. If there was no collision in the first slot, then it has captured the next $X - 1$ slots, where X is the packet transmission time and all other stations will defer. If there was a collision in the first slot, then the node makes a randomized retransmission attempt (as in slotted Aloha) and continues to do so until it succeeds. All nodes will know of the end of transmission of this packet when they sense the channel idle again. If slotted Aloha with an adaptive protocol were to yield a throughput of η when the packet length is equal to the slot length, what would be the throughput of this network? Note the similarities with the CSMA/CD protocol.

8.9   Consider a slotted Aloha network where the attempt process is Poisson with rate G. Under the condition of maximum throughput, what is the fraction of empty, successful and collision slots? If it is observed that the network is not operating under maximum throughput conditions and that the fraction of idle slots is 0.1, what is the throughput of the network? Is this network overloaded or underloaded? Explain.

8.10 Consider the Aloha protocol for multiple access networks. Let X be the number of nodes that are backlogged at the beginning of a slot. Assume X has a Poisson distribution with mean $\hat{x}$. Now assume that each backlogged node transmits in the slot with probability $1/\hat{x}$ independent of the others.

    a. Obtain the joint probability of k nodes being backlogged and r transmitting. Also obtain the unconditional probability that the slot is idle.

    b. If the slot was observed to be idle, what is the *a posteriori* probability that k nodes were backlogged at the beginning of the slot?

c. Similarly, find the *a posteriori* probability that given that there was a successful transmission in the slot, there were k backlogged nodes at the beginning of the slot.

d. Using these results suggest a method to continuously estimate $\hat{x}$ based on the event in a slot–success or idle. Suggest an estimation method for $\hat{x}$ when a collision is observed in a slot.

8.11 Consider a CSMA/CD protocol (not the Ethernet standard). For convenience assume that time is divided into slots. Consider a situation where n nodes are involved in a collision. Each of them transmits in the next slot with probability p. Find the probability that there is a success in the next slot. If p were a design parameter, what would be the optimum value of p? Assuming all nodes transmit in a slot with the optimum value of p, what is the probability of success as $n \to \infty$?

8.12 Consider a multiple access channel in which the following situation arises. Two nodes A and B are ready to send a packet at the same time. This typically happens immediately following a successful transmission. In the kth round after $k - 1$ collisions have occurred, the nodes wait for a random period of $w \in [0, 1, \ldots, 2^{k-1} - 1]$ slots of time, with each of the 2^{k-1} choices being equally likely. Let c_k be the probability of a collision in round k, given that the previous $k - 1$ rounds had a collision.

a. Find c_k as a function of k for all k.

b. What is the probability that round k lasts n slots? What is its mean? Assume that a collision occurred in the round.

c. Find the probability that round k lasts n slots. Do not assume that collision occurred. Also find the mean number of slots in round k.

d. Find p_k, the probability that exactly k rounds are needed to resolve a collision involving only two nodes and no new nodes transmitting until the collision is resolved.

e. Assume that the collision is resolved in favor of A in the third round. In this case A will reset its collision counter. Assume that the packet being transmitted by A is longer than the backoff time chosen by B. Because of 1-persistence, B will transmit soon after A's transmission. Now if A has another packet to transmit, there will be a collision immediately following A's successful transmission and both A and B

will increase their collision counters. What is the probability that this collision is resolved in favor of A in the second round?

8.13 Consider any scheduling policy—say, π—for the slotted WANET described in Section 8.8.2. Let Λ_π denote the set of arrival rate vectors λ that are stabilized by this policy. Let π_0 denote the queue-length-based centralized scheduling policy described in Section 8.8.2. Argue that $\Lambda_{\pi_0} = \cup_\pi \Lambda_\pi$. This means that if there exists any policy that stabilizes the queues under the arrival rate vector λ, then the queues will be stable for this arrival rate vector under policy π_0.

Part II

Switching

CHAPTER 9

Performance and Architectural Issues

Transmission capacity is an expensive resource, and it needs to be utilized efficiently. Switching helps improve the efficiency of link usage by enabling a greater degree of multiplexing. We saw in Chapter 2 that switching enables an information flow to traverse multiple links on its path from a source to a destination by moving (switching) flows among the links that are interconnected through the switch.

As discussed in earlier chapters, a network has either packet-multiplexed links or circuit-multiplexed links. Switch design and performance issues differ accordingly. Packet-multiplexed links terminating at a switch could be asynchronous, with no time slots and arbitrary packet lengths—for example, Ethernet. Or links could be time-slotted, or synchronous, with packet transmissions beginning and ending on slot boundaries. Such time-slotted links are also a convenient abstraction in studying packet switch architectures. Circuit-multiplexed networks use time division multiplexed links—for example, T-1, E-1, or SONET (or SDH) links such as OC-3 (or STM-1).

In Chapter 2 we discuss switching and switches in packet- and circuit-multiplexed networks. Clearly, the performance measures and architectural choices for a packet switch are significantly different from those for a circuit switch. In this chapter we discuss the performance measures and the design choices for both types of switches. In the next three chapters we discuss some of the design choices and their effect on the performance of the switch.

9.1 Performance Measures

Recall from Figure 2.22 that network functions can be classified into data, control, and management planes. The functions in a switch can be classified similarly. Correspondingly there are performance measures from the point of view of each of these types of functions. We focus on the data and control planes only.

9.1.1 Packet Switches

First consider the data plane functions in a packet switch. These functions are performed on every packet and are thus high-volume, fast-timescale functions.

We immediately see that the *packet-processing capacity* of the switch is an important performance measure. The packet-processing capacity should minimize packet-processing delays even when the packet arrival rates are high. The packet arrival rate to the switch depends on the data rates on the links that the switch interconnects. Furthermore, if the links allow variable-length packets, the packet length distribution also influences the packet arrival rate. To appreciate the dependence between the packet length distribution and the packet arrival rate, it is instructive to see the distribution of packet lengths in a real packet-switched network. Figure 9.1 shows the packet length distributions from the packet traces collected at NASA Ames Internet Exchange (AIX) over one week in May 1999

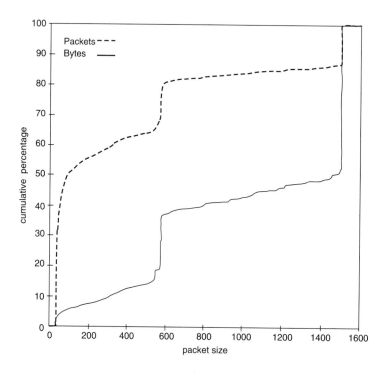

Figure 9.1 Example of a cumulative packet length distribution collected at the NASA Ames Internet Exchange. Adapted from the graphs in http://www.caida.org/ analysis/AIX/plen_hist/ **as seen on 03 Aug 2003. The dashed plot is the fraction of packets with length less than *x* bytes, and the solid plot is the fraction of bytes in packets with size less than *x* bytes.**

and is representative of the packet length distribution seen on the Internet. We see from the figure that whereas nearly 50% of the bytes are from 1500-byte packets (read the dashed curve), nearly 50% of the packets are small, 40-byte packets (read the solid curve). This means that although small packets do not contribute much to the utilization, they consume significant packet processing power. The packet-processing rate of a switch must be dimensioned with this in view.

> ### Exercise 9.1
> Derive the solid plot from the dashed plot in Figure 9.1; that is, given the distribution function of the packet lengths, obtain the fraction of bytes contributed by packets of a given size. Observe the *length biasing* (see Section D.1.4) that occurs: If a random packet is observed on the link, it is more likely that a longer packet is observed than a shorter packet.

Figure 9.2 summarizes the discussion from Chapter 2 on the functions of a packet switch. To keep the discussion general, assume a QoS-enabled switch that can provide delay and loss rate guarantees to the packets. We will discuss the performance and design issues associated with each of the blocks shown in Figure 9.2. The receiving and transmission of bits and the extraction of packets are what can be called "physical layer functions," and we do not discuss these functions in this book.

Packet header processing typically has two important functions: *route lookup*, which determines the output link for the packet, and *packet classification*, which determines the specific service grade that is to be provided to the packet. Packet classification may be based on source and destination addresses of the packet and the application that generated it. In virtual-circuit networks, such as ATM networks, route lookup and packet classification are typically a table lookup and, with very high probability, can be completed with a small, constant delay.

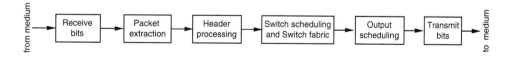

Figure 9.2   Block diagram of a view of the passage of a packet through a switch.

In Internet-like datagram networks, route lookup and packet classification are more complex, and the time taken to perform these tasks could be variable, possibly even exceeding packet interarrival times. We discuss the algorithms and performance of route lookups in Chapter 12. The point to note now is that the variability in processing time can cause a packet queue to build up at the input, and a packet will experience variable delay before its service category is determined. Also, because the buffer space available for the packet queue is finite, some packets may be lost. In providing delay and loss guarantees to the packet, it is imperative that the header processing be completed with a constant latency after the arrival of the packet. Otherwise the switch might excessively delay packets with stricter guarantees and not even know it! Thus the variability of packet header processing time at the input becomes an important performance issue.

After the output link and the service category for the packet have been determined, the packet must be switched to a queue on the output link. There are three issues associated with the switching of packets to the output link: placement of a packet queue, scheduling of packets to be switched to the output queue, and the design of the fabric itself. Recall from Chapters 1 and 2 that output contention occurs when two or more packets arriving on different input links want to leave from the same output link at the same time. If the links are not time-slotted, a packet may arrive for an output while a packet transmission is in progress on that output. With time-slotted links, two packets may arrive in the same slot from different inputs and may want to be output on the same link. All but one of these contending packets must be queued, and a packet queue must be maintained. Thus, output contention necessitates queueing of the packets. If the packets are queued at the inputs, then the switching capacity should be such that the queueing delay at the input is minimized. The queueing delay and the loss probabilities in the input or the output queue are important performance measures for the switch and are a function of the switching capacity, the packet buffer sizes, and the packet arrival process. If packets are queued at the input to the switching fabric, then a scheduler must be used to decide when to offer a packet to the switch fabric.

Now consider the design of the switch fabric. In a slot, the switch fabric may or may not be able to switch all permutations of destination requests. Consider an $N \times N$ switch, with $\mathbf{d} = [d_1, \ldots, d_N]$ representing the destination vector of the packets at the N inputs. In a *nonblocking* switch, if the d_i are unique (i.e., $\mathbf{d}$ is *any* permutation of $[1, \ldots, N]$), then the packet at input i can be switched to output d_i. If the switch is blocking, the blocking probability must be evaluated.

Remark: We have seen in Chapters 4 and 5 that QoS for stream sessions can be guaranteed by appropriate packet scheduling on the output link. Thus, if a switch were to provide QoS to the packets in terms of delay and loss guarantees, then the packets should reach the output queue with constant latency (or with a tightly upper-bounded delay) after its arrival at the switch. Alternatively, we could say that the packet should be available for transmission scheduling on the output link with an approximately constant latency. This is the primary aim of packet switch design.

The control plane functions in a packet switch are slow-timescale functions and include functions such as executing the signaling protocols and evaluating the network state (e.g., network topology) from the signaling messages exchanged. The details of the control plane functions depend on the type of the network. An example of a control plane function is the execution of a routing protocol and the determination of the routing table. In executing a routing protocol, the switch exchanges network topology information with its neighbors, and the routing table is computed after all the information is collected from all the neighbors. How often the routing table can be computed depends on the size of the network and the processing capability of the control processor and is thus a measure of the ability of the switch to adapt to changing network conditions. Typically, the control plane functions are performed in software on general-purpose processors. The MIPS (million instructions per second) rating of the processor is a good measure of control plane performance of the switch.

9.1.2   Circuit Switches

The links in a circuit-multiplexed network are time division multiplexed. Recall from Figure 2.17 that after a call is accepted, the mapping is determined between the input link and slot in the frame and the output link and slot. In an $N \times N$ switch fabric, at any time up to N connections could be active. It can be very expensive to design the switch fabric to support these many simultaneous connections for any combination of source–destination pairs and any sequence in which the call requests arrive. To see this we interpret each of the ways in which the N input and N outputs may be interconnected as a *state* of the switch fabric. Allowing any input–output interconnection pattern would require the switch fabric to have $N!$ states. Note that $N!$ grows exponentially with N. We can economize by taking advantage of the fact that it is highly unlikely that this would be necessary, and the maximum number of connections that the switch fabric can simultaneously support can be made less than N. Also, we can support only specific combinations of source–destination pairs and call arrival sequences. An arriving connection request that cannot be supported is blocked. We can thus define a performance parameter of the switch fabric called the switch-blocking probability, which is a measure of the inability to accept a call request because of a lack of resources inside the switch fabric; the output port of the switch for the call is free, but the switch fabric cannot establish a connection inside the switch between the input and output ports.

It must be emphasized that the switch-fabric-blocking probability is distinct from the network-blocking probability. Network blocking occurs when sufficient capacity cannot be found on the links in the network to complete the call. Network-blocking probability is a function of the load in the network, the routing algorithm, and the topology of the network. These factors affect the offered load to a switch which, along with the switch fabric architecture, determines the switch-fabric-blocking probability. We discuss network blocking in Chapter 6.

The control plane functions provide another set of performance measures for the switch in a circuit-multiplexed network. The time from when the calling phone goes offhook until the time that the call is completed and the ringback is heard is called the *call setup delay* and is one performance measure. This is the time required by the switch to apply the dial tone to the calling phone, accept and process dialed digits, find the route on which the call will be set up, and set up the path. The *call-processing capacity* of the switch, specified as the number of *busy hour call attempts* (BHCA) that the switch can handle without significantly dropping the calls, is another measure of performance. Note that while the call is being set up, the call holds up the signaling resources, and a large call setup delay could exhaust these resources, leading to dropping calls before they are processed. In circuit-multiplexed networks, after a call is set up the delay through the switch is just a fixed delay.

Observe that the control plane performance in a circuit-multiplexed network has a direct impact on the quality of the service seen by the user, whereas in packet-multiplexed networks its impact is indirect.

In virtual-circuit networks, links are packet-multiplexed, but calls must be set up so that the network can determine the path for the flow of packets and also statistically or deterministically reserve resources (bandwidth on the links, and buffer space in the switches). Thus the control plane performance measures of such a switch should include those of the circuit switch (call setup delay and call-processing capacity) in addition to those of the packet switch. The data plane performance measures are the same as those of packet switches because these are essentially packet switches.

9.2 Architectural Issues
9.2.1 Packet Switch Organization

We can construct a simple packet switch by having all the functions performed by a general-purpose processor, with a network interface card (NIC) extracting

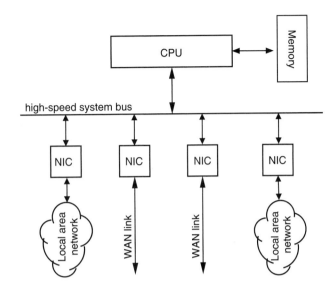

Figure 9.3 A packet switch using a general-purpose CPU and network interface cards (NICs).

the packets from the medium and transmitting them. The system bus is used to exchange packets and control information. Typically a separate NIC is used for every link. Such a packet switch architecture is shown in Figure 9.3.

Note that every packet must cross the bus twice: once from the input NIC to the main memory of the processor, and then from the processor memory to the output NIC. Software in the CPU handles packet processing, switching decision, and output scheduling. Two bottlenecks clearly emerge: the CPU speed and the bandwidth of the bus over which packets are to be switched across networks. In a single-CPU system, the CPU must also perform other control plane and operating system functions. Consequently, the fraction of time available for packet processing is limited, and such switches have a low packet-processing capacity. The following exercise helps illustrate the achievable capacity of such switches. Note that in such a switch, the status of the packet queues for each input and output is easily available to the CPU, and providing QoS services is straightforward.

Exercise 9.2
A packet switch using a general-purpose processor rated at 4000 MIPS is to be constructed. The NICs and the processor communicate over the

system bus, which is 32 bits wide and operates at 33 MHz. Half (50%) of the packets are 64 bytes, 35% are 512 bytes, and 15% are 1024 bytes. Assume that each packet requires the equivalent of 10,000 instructions for processing. If 25% of the CPU cycles are available for packet switching, what is the packet-processing capacity of this switch?

Obvious improvements to this basic architecture are to minimize the amount of processing that the CPU performs on each packet and to reduce the number of bus traversals made by a packet. One way to do this is to include a processor on each NIC and locally store enough control information to process a large fraction of the incoming packets in the NIC itself. If the NIC does not contain the information needed to process the packet, the packet is be transferred to the central processor. For example, the NICs could maintain a routing table cache and transfer the packet to the central CPU only if a route lookup for the packet cannot be performed from the cache. Such an architecture achieves a *fast path* for most of the packets, and a *slow path* for a small fraction of the packets, much like the memory cache in a computer system. If the packet is processed on the fast path, it can be transferred directly to the output NIC, thus eliminating one transfer over the bus. This significantly increases the packet-processing capacity of the switch. There are, however, a few trade-offs. Because centralized information is not available, complex services must be implemented in a distributed manner. This requires significant communication among the processors. Also, the central CPU must communicate control information, such as routing tables, to the NICs. The system bus must be used for these communications, thereby using some of its capacity.

For both of the architectures, bus arbitration mechanisms play a crucial role in the performance of the system. These issues are well understood in computer architecture literature, and we pursue them only minimally.

A further improvement in performance compared with the previous two architectures is to replace the shared bus with a *space-switching fabric*, possibly a *nonblocking* fabric. To switch a packet, a space-switching fabric creates a physical path between the input and output ports, and it can create such paths between many input–output pairs simultaneously. Such a packet switch architecture is shown in Figure 9.4. (Contrast this with the bus-based architectures described earlier.) The central CPU and the processors in the NICs may also communicate through the switching fabric.

An immediate design issue is the rate at which packets can be switched by the fabric relative to the line rates. Assume a time-slotted switch, and in one slot,

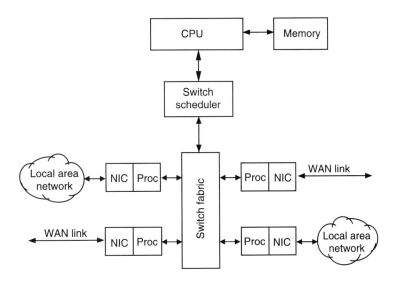

Figure 9.4 A packet switch with a space-switching fabric for communication among the NICs and the central processor. The NICs are shown to have their own processor. A switch scheduler is used to select the packets to be offered to the switch fabric.

let p denote the maximum number of packets that the fabric can switch from an input, and let q denote the maximum number of packets that it can switch to an output. Recall from our discussion in Section 9.1, that a packet queue must be maintained in the switch. The obvious choices for the location of this queue are at the input to the switching fabric or at its output. If $p = 1$ and $q = 1$, we have an input-queued (IQ) switch; because at most one packet can be switched to the output in a slot, and if more than one packet arrive for an output in a slot, all but one must be queued at the input. Because at most one packet is being transmitted to the output in a slot, there need be no queue at the output. Clearly, $p = 1$ and $q = N$ corresponds to an *output-queued* (OQ) switch, because any packet that arrives in a slot is switched to the output in the same slot, and no queue need be maintained at the input. Other combinations of p and q can also be used. Note that if $p < N$, a packet queue should be maintained at the input, and if $q > 1$ a packet queue should be maintained at the output. Thus if $p < N$ and $q > 1$, we get a *combined-input-output-queued* (CIOQ) switch. The placement of the queues (at the input, output, or both) and the buffer spaces available to these queues are important design issues. Delay distributions and packet loss probabilities as a function of the packet buffer size and of p and q are important performance measures. We can achieve $p, q > 1$ by operating the switching fabric at a rate higher than the line rate (i.e., by a *speedup* of the fabric). We can also achieve it by *parallelism*, in which the switch fabric is allowed to simultaneously select more than one packet from an input or to an output (or both). A combination of speedup and parallelism can also be used.

For most of the rest of our discussion on space switches, we consider an $N \times N$ time-slotted cell switch having identical link rates on all the inputs and outputs. The trade-off between input queueing and output queueing is discussed briefly in Chapter 1, and we discuss it in more detail in Chapter 10. Briefly, the output-queued switch can provide the maximum throughput but requires high switching capability and may be technologically infeasible, whereas an input-queued switch requires lower switching speeds but has significantly reduced throughput. We can have p or q greater than 1, maintain queues at both the input and the output, and achieve throughputs comparable to the OQ switch. To achieve high throughput and also provide QoS capabilities when $q < N$, the packets at the input to the switch fabric should be scheduled using a switch scheduler. As you will see in Chapter 10, the computational complexity of the algorithms can be significant. Thus there is the additional implementation trade-off between the computational complexity of the switch scheduler of the IQ switch and the fabric speed of the OQ switch. In Chapter 10, we discuss the performance implications of queue placement and the choice of p and q, along with the switch-scheduling algorithms.

Another design issue is the handling of variable-length packets. Time-slotted switches can switch variable-length packets by breaking a packet into small, fixed-length packets called cells and then switching the cells. This simplifies the design of the switch but introduces additional overheads, such as disassembly and reassembly of each packet into cells, padding of some of the cells to fit into a slot, and adding a header for each cell that contains information about the packet to which it belongs. This trade-off is examined in Chapter 10.

As mentioned earlier, the best way to provide QoS is to have an OQ switch and a scheduler for the output queue. From our argument, an OQ switch may be technologically infeasible. An interesting issue is the ability of the IQ switch to emulate the QoS capabilities of the combination of an OQ switch and the QoS scheduler. In Chapter 10 we discuss some results of the emulation of an OQ switch by an IQ switch.

There are many architectural choices for large space switches. To operate at close to link rates and yet achieve sufficient switching rates, these switches should efficiently *guide* the packet from the input NIC to the output NIC in hardware. The basic hardware element that is used is a small $m \times m$ switch, and a number of them are appropriately interconnected. Typically $m = 2$ is used. These elements use a distributed control mechanism to perform the switching function. In addition to performing *unicast* switching, the ability of the switch to *multicast* the packets (i.e., switch copies of the packet to multiple outputs) is a design issue. We discuss the architectures for these switches in Chapter 11.

As we have seen before, the data plane functions in a packet switch must be performed at high speed. Increasingly, special-purpose hardware structures are being explored to perform these functions at wire speed (i.e., at the same rate at which packets can arrive at the switch inputs). An important example is route lookup in IP networks. For quite some time, this was considered to be a bottleneck function in a packet switch of an IP network. Many schemes to perform a route lookup at wire speed have been proposed, and it is believed that this problem has been solved. Packet classification is another function for which a hardware implementation has been explored in detail. This has become important because of services such as *application-aware* switching, in which the packet switch processes the packet header to identify the application that generated the packet. Many hardware structures to perform classification at wire speed are now available.

Another development in high-speed packet processing is the availability of special-purpose *network processors* (NPs) along the lines of digital signal processors. Network processors are designed to allow a high degree of parallelism in packet processing by providing a large number of processing elements. An NP may also provide additional parallelism by allowing each processing element to run multiple program threads. We discuss packet processing structures in Chapter 12.

9.2.2 Circuit Switches

In a circuit switch, two important elements stand out: the switch fabric and the control processor. In fact the theory of switch fabric architectures was first developed in the context of circuit switches, and much of our discussion on packet-switching fabrics is prefaced by a discussion on circuit-switched fabric architectures. An $N \times N$ space switch can be built using an $N \times N$ crossbar with N^2 crosspoints, as shown in Figure 9.5. The number of crosspoints is a measure of the complexity of the switch. An important design issue is to reduce the number of crosspoints in the switch fabric while keeping the blocking probability within specified limits.

The design of the control plane processing architecture in a circuit switch like that in a telephone network has been quite challenging. There are many subtasks associated with call setup, and each of them can have a variable delay. During times of congestion, some of the subtasks may experience large delays, and the user may

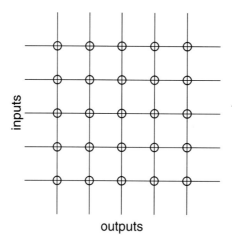

Figure 9.5   Crossbar switch.

abandon the call midway through the call setup procedure. This leads to a loss of the work that was performed toward that call setup before its abandonment. If not controlled properly, this can lead to a significant decrease in the number of successful call attempts in spite of a high processor utilization. The overload control of these control plane processors has been studied extensively. We do not pursue this further in this book.

CHAPTER 10

Queueing in Packet Switches

C hapter 9 introduces the performance and architectural issues in the design of packet switches. We are now familiar with the concept of a cell switch and the placement of the queues relative to the switch fabric. Specifically, we have discussed the input-queueing (IQ), output-queueing (OQ), and combined input-output-queueing (CIOQ) options for packet queue placement in cell switches. This chapter first analyses the switching capacity and delay performance of IQ, OQ and CIOQ switches assuming that the queues are all FIFO. We then consider alternatives to FIFO scheduling and discuss the emulation of an OQ switch by an IQ switch and an appropriate switch scheduler.

Much of this chapter assumes cell switches. Recall that a cell switching fabric operates in a time-slotted manner, where the slot duration is equal to the cell transmission time. In our analysis, we assume that cell arrivals occur at the beginning of a slot and that cell departures are completed at the end of the slot. Thus, an arrival in a slot will be available for transmission in the same slot. Also, in this chapter, we consider only nonblocking switches and do not worry about how such a switch may be constructed.

10.1 FIFO Queueing at Output and Input

Let us first recollect what we discuss in Chapters 1 and 9: In a slot, if k active inputs have cells for the same output—say, output port j—and if there are to be no queue at the inputs and no cells are to be dropped at the inputs, the switch should be capable of switching all the k cells to output j. Of the k cells that reach the output, only one may be transmitted on the output link, and the other $k - 1$ must be put in a queue at the output port. It could be that at the output there are packets that are waiting from the previous slots, in which case all the k must be buffered. To handle the worst case situation, the switch should be capable of switching up to N cells to their respective outputs in one slot time if cells are not to be dropped at the input (i.e., the switch should operate at N times the line rate). Observe that the cells destined for different outputs do not interfere with each other at the inputs and are not delayed at the input. Thus the switch is work-conserving in the sense that no output link is idle while there is a cell to be transmitted on that link in the switch. This is true because every arriving cell is sent to the queue on the output link in the same slot that it arrived. Thus the OQ switch can achieve 100% throughput on the output links.

The queueing abstraction for an output port—say, j—of this switch is a single-server, discrete time queue with fixed-length service times and an input process that is a superposition of the arrival processes from each of the inputs to output j.

Now consider the IQ switch with one FIFO queue at each input. The switch can transfer at most one packet from an input, and at most one packet to an output in a slot. If, at the beginning of a slot, more than one head-of-line (HOL) cells from the input queues have the same destination, then only one of them is switched and transmitted on the output link in the slot. The other HOL cells continue to be queued at their inputs. If any of these inputs contains a non-HOL packet whose destination is free, it is not switched, because the queue is FIFO and the packet at the head of the queue is blocked. Thus, packets in an IQ switch can experience *head-of-line blocking*, in which a blocked HOL cell blocks the cells behind it in the input FIFO queue even though the destination ports of these other cells are free and are idling. Thus the IQ switch with FIFO discipline is *non-work-conserving*, in the sense that there may be cells queued in the switch that are to be transmitted on an output port but cannot be, and the output port idles. Because the IQ switch is non-work-conserving, its capacity is less than one cell per port per slot.

Given that the OQ switch has greater capacity than the IQ switch, why should we be interested in the IQ switch? To answer this, consider the construction complexity of both architectures. Because the queues are maintained at the inputs and because only one cell need be transmitted in the event of a destination conflict, the switch can operate at the same rate as the input and output links. This means that in an IQ switch each input should be capable of sending at most one cell in a slot, and each output should be capable of receiving at most one cell in a slot. Furthermore, the maximum transfer rates from memory (the rate in bits per second at which data can be read from or written to) used for the input queue should be twice that of the link rate. This is because, in a slot, at most one write operation (corresponding to an arriving cell) and one read operation (corresponding to reading the packet from the input queue and switching it to the output) are performed. However, in the case of the OQ switch, to handle the worst case situation the switch should operate at N times the line rate; that is, it should be capable of transferring up to N cells from the inputs to an output. Furthermore, it should allow a cell to be transmitted on the output link. This means that the memory used for the output queue should be capable of a memory transfer rate of $N + 1$ times the line rate for the OQ switch.

Exercise 10.1

a. What is the memory transfer rate required for an $N \times N$ IQ and OQ switch for $N = 16$, 32, and 64? Assume 64-bit cells and 10-Gbps line rates. Also assume that the internal organization of the switch uses a header of 32 bits per cell.

b. Find out about available memory technologies and their access times. Obtain information about the cost of, say, 1 MB of SRAM memory, and plot the access time versus cost function for this. Extrapolate and guess the cost of memory.

c. Repeat for 16- and 32-bit cells.

Although the switch speedup of N times the line rate may not be technologically infeasible, clearly the memory transfer rates required make it infeasible at high line rates. Thus, in terms of construction complexity, the IQ switch is probably the only technologically feasible option in the core of the Internet, where the number of ports required on the switches and the line rates are both very high. However, as we remarked earlier, the non-work-conserving property of the IQ switch means that its capacity is less than 100%. The question then is, how much less than 100% is the maximum achievable throughput of an IQ switch? We answer this question next.

We first consider an IQ switch with saturated inputs. Input saturation means that the input is always active and has a cell to transmit to an output in every slot; in other words, there is always a cell behind the HOL cell to take its place when the HOL cell departs from the input queue. If all the inputs are saturated, the rate at which cells depart from the switch is called the *saturation throughput* of the switch. There are two reasons for considering saturation throughput. First, the analysis is comparatively easy. Second, the results of this analysis give us insight into the capacity of the IQ switch. In fact, for a special case, we show that the saturation throughput is a lower bound for the capacity by showing that if the arrival rate is less than the saturation throughput, the input queues are stable. We then present an approximation argument to derive the saturation throughput and hence the capacity of the IQ switch.

10.1.1   Saturation Throughput and Capacity (⋆)

Consider a saturated $N \times N$ cell switch with uniform routing (i.e., the destination of each cell is independently and randomly chosen from among the N outputs). Assume that the input queues are FIFO and that only the packets at the head of the queues at the beginning of a slot can be switched to the output in the slot. Now consider the HOL cells destined for a tagged output (defined soon)—say, output O_j. Conceptually, we can view these cells as being in a queue to get to O_j. Of course there is no such physical queue. Call this the HOL queue for output j, denoted by HOL_j for $j = 1, \ldots, N$. Let $Q_j^{(H)}(t)$ be the number of cells in HOL_j at the beginning of slot t. If the inputs are saturated, $0 \le Q_j^{(H)}(t) \le N$ and $\sum_{j=1}^{N} Q_j^{(H)}(t) = N$.

An example of the evolution of the HOL queues for a 2×2 switch is shown in Figure 10.1. In this example, if the inputs are saturated, we can say that cell a "came back" to HOL_1 as cell b at the beginning of slot 2. Similarly, cell e "went" as cell f to HOL_2 at the beginning of slot 3. We can say that in the HOL queue after a cell finishes service, it goes into any of the N HOL queues, with probability equal to that of a cell having that queue as its destination. This is exactly like a *closed queueing network*: a network of queues in which the total number of

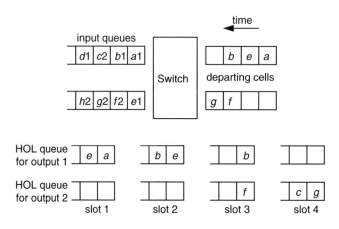

Figure 10.1   Evolution of the HOL queues in a 2 x 2 switch over four consecutive slots. The bottom panel shows occupancy of the HOL queues at the beginning of the slots. Letters are used to name a cell, and numbers indicate their destinations.

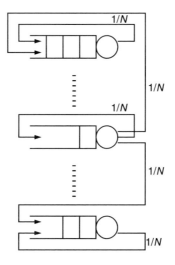

Figure 10.2   The closed queueing network representation of the HOL queues.

customers remains constant. In such a network, customers finishing service at one queue join another queue. There are no departures from the network nor external arrivals to it. Thus, for saturated inputs we can represent the set of HOL queues as a closed queueing network, shown in Figure 10.2. This means that the problem of finding the saturation throughput of the switch is the same as that of finding the throughput from any of the queues of such a closed queueing network. The closed queueing network is synchronous, and an exact analysis is intractable except for small N. We use this closed queueing network model in our analysis of the saturated IQ switch.

First, let us see what happens when an increasing number of inputs are saturated in an $N \times N$ switch. Consider input I_i, with n of the inputs (including I_i) being saturated. Define $\gamma^{(i)}(n)$ as the throughput from input I_i when n inputs are saturated and when the other $(N - n)$ inputs do not have any cell arrivals. In the closed queueing network, this corresponds to having n cells and N queues. We show that as n increases toward N, the throughput from input I_i decreases monotonically.

Lemma 10.1

For $1 \leq n \leq N$, $\gamma^{(i)}(n) \geq \gamma^{(i)}(n+1)$.

Proof: Consider a closed queueing network with n cells and N queues (like that shown in Figure 10.2) and cell C_i from input I_i. The throughput from I_i is the number of times C_i is switched to its desired output per unit time. (This in turn is the reciprocal of the average time for C_i to be switched once to its output.) The total throughput is $n\gamma^{(i)}(n)$. Now add one more cell, but give it the lowest priority. Clearly, the addition of this new cell does not affect the throughput of the n cells that are already in the system, and their throughput is still $\gamma^{(i)}(n)$. Because the cells are indistinguishable from each other in their routing behavior, the total throughput with $(n+1)$ cells is $(n+1)\gamma^{(i)}(n+1)$, and this is higher than that with n cells. The increase in the throughput is caused by the contribution of the new cell, and because its priority is the lowest, its throughput is less than or equal to that of the others. That is,

$$(n+1)\gamma^{(i)}(n+1) - n\gamma^{(i)}(n) \leq \gamma^{(i)}(n)$$

$$(n+1)\gamma^{(i)}(n+1) \leq (n+1)\gamma^{(i)}(n)$$

∎

We remark here that the preceding argument is based on random selection of the cells from the HOL queues. If the HOL queue contains only the low-priority cell, that cell is selected. If there are others cells, any one of the others is randomly selected. Because the cells are indistinguishable, the total throughput is the sum of the throughputs of all the cells.

Lemma 10.1 suggests that an arrival rate less than $\gamma^{(i)}(N)$ should be a sufficient condition for the stability of the input queue at I_i and defines the per-port capacity of the IQ switch when all the inputs have the same packet arrival rate and when each packet chooses its destination independently and randomly. Before we show that this is indeed the case, we present a few definitions.

Definition 10.1

In a slotted service system of N queues, let $Q_i(t)$ be the number of cells in queue i at the beginning of slot t. Let $\mathbf{Q}(t) := [Q_1(t), \ldots, Q_N(t)]$ denote the queue-length vector at the beginning of slot t. The queueing system is considered stable if the distribution of $\mathbf{Q}(t)$ as $t \rightarrow \infty$ exists and is proper:

$$\lim_{t \rightarrow \infty} \Pr(\mathbf{Q}(t) \leq \mathbf{m}) = Q_{\mathbf{m}} \text{ and } \lim_{\mathbf{m} \rightarrow \infty} Q_{\mathbf{m}} = 1$$

Here **m** is an N-dimensional vector and $\mathbf{Q_m}$ is the limiting distribution. Furthermore, $\mathbf{Q}(t)$ is said to be *substable* if the following is true:

$$\lim_{\mathbf{m}\to\infty} \liminf_{t\to\infty} \Pr(\mathbf{Q}(t) \le \mathbf{m}) = 1$$

∎

If $\{\mathbf{Q}(t)\}$ is substable, it means that $\mathbf{Q}(t)$ is finite with probability 1; that is, there is no "escape of probability mass to infinity as $t \to \infty$." However, the probabilities need not converge to a single probability distribution. A substable, aperiodic, irreducible, discrete time Markov chain is stable.

Because all inputs have identical statistical behavior, we choose to concentrate on one input, called the *tagged input*, and characterize its behavior. Let input I_i be the tagged input of an $N \times N$ switch with the other $(N-1)$ inputs being saturated. Let the cell arrivals to this tagged input be from a Bernoulli process of rate λ. Assume that whenever the queue at I_i becomes empty, a dummy cell with a uniformly assigned destination is placed in the queue. If a new cell arrives before the dummy cell departs, it takes the place of the dummy cell and also adopts its destination. Thus all the inputs are saturated, and the closed queueing network model of Figure 10.2 with N queues and N cells can be used. Now consider the instants, $\{t_m, m \ge 1\}$, at which a cell departs from input queue I_i. Let $Q(t_m)$ be the number of cells in input queue I_i at instant t_m. Define $S_m = t_m - t_{m-1}$, and let A_m be the number of cell arrivals to input queue I_i in S_m. These are shown in Figure 10.3. The evolution of input queue I_i, embedded at these instants, can be written as

$$Q(t_{m+1}) = \max\{(Q(t_m) + A_{m+1} - 1), 0\} \qquad (10.1)$$

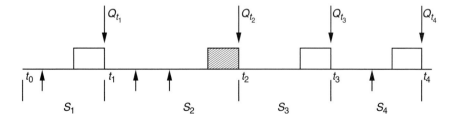

Figure 10.3   Queue evolution in the tagged input illustrating $\{t_m\}$, $\{S_m\}$, and $\{Q(t_m)\}$. The shaded cell is a dummy cell and the others are real cells. The upward arrows indicate packet arrivals.

In Section 5.6.2 we discuss how, for a stationary process $\{A_m\}$, $Q(t_m)$ as defined here almost surely converges in distribution to a proper random variable if $\sum_{r=1}^{m}(A_r - 1) \to -\infty$. Thus, intuitively, the input queue is stable if the sequence $\{A_m\}$ is stationary and if $\mathsf{E}(A_1) < 1$. The latter condition implies that the average number of cell arrivals between the times that a cell is being serviced by the output is less than 1.

In fact, it is sufficient to have $\{A_m\}$ to be asymptotically stationary; that is, the finite dimensional random vectors $(A_{J+r+m_1}, A_{J+r+m_2}, \ldots, A_{J+r+m_k})$ should not depend on r for arbitrary k, and $m_1, m_2, \ldots, m_k$ as $J \to \infty$. If the service from the HOL queues is in random order, then it can be shown that the sequence $\{A_m\}$ is indeed asymptotically stationary. We skip that proof here. We are now ready to derive the stability condition for the tagged input queue and the system of N input queues of the switch to be stable.

Theorem 10.1

(i) The tagged input queue is stable if $\lambda < \gamma(N)$.

(ii) The system of N queues is stable if $\lambda_i < \gamma(N)$ for $i = 1, 2, \ldots, N$.

Proof:

(i) For the tagged input queue evolving according to Equation 10.1, the following are almost surely true:

$$\lim_{m \to \infty} \frac{\sum_{r=1}^{m} A_r}{\sum_{r=1}^{m} S_r} = \lambda \qquad \lim_{m \to \infty} \frac{\sum_{r=1}^{m} 1}{\sum_{r=1}^{m} S_r} = \gamma(N)$$

The left expression is the arrival rate of packets to input I_i. The right expression is the definition of the saturation throughput of input I_i. This implies that if $\lambda < \gamma(N)$, then almost surely

$$\lim_{m \to \infty} \frac{\sum_{r=1}^{m}(A_r - 1)}{\sum_{r=1}^{m} S_r} < 0$$

and hence, almost surely $\sum_{r=1}^{\infty}(A_r - 1) = -\infty$. This shows that the process $Q(t)$ is stable in the sense of Definition 10.1. Because $Q(t_m)$ converges in distribution to a proper random variable and because $\{A_m\}$ is a stationary random process with a proper marginal distribution, it follows that $Q(t)$ is substable if $\lambda < \gamma(N)$. Note that we cannot conclude that $Q(t)$ is stable because we are only upper-bounding it by a stable process. Let $X^{(j)}(t)$ be the output port of the HOL cell in input I_j. Note that $\{(Q(t), X^{(1)}(t), \ldots, X^{(N)}(t)),\ t > 0\}$ is a multidimensional, irreducible, and aperiodic discrete time Markov chain. Hence substability implies stability of $Q(t)$, and a sufficient condition for the stability of the tagged input queue is $\lambda < \gamma(N)$.

(ii) Now consider the system of N input queues described by the N-dimensional vector process $\{\mathbf{Q}(t) = (Q^{(1)}(t), \ldots, Q^{(N)}(t)),\ t > 0\}$. Let λ_i be the Bernoulli cell arrival rate to input I_i. From part (i), $\lambda_i < \gamma(N)$ is sufficient for queue I_i to be stable. When $\lambda_i < \gamma(N)$,

$$\lim_{m \to \infty} \lim_{t \to \infty} \Pr\left(Q^{(i)}(t) \le m_i\right) = 1, \quad 1 \le i \le N \qquad (10.2)$$

Define $\mathbf{m} = (m_1, m_2, \ldots, m_N)$ and we have

$$1 \ge \lim_{\mathbf{m} \to \infty} \lim_{t \to \infty} \Pr\left(Q^{(i)}(t) \le m_i, i = 1, 2, \ldots, N\right)$$

$$\ge 1 - \sum_{i=1}^{N} \lim_{m_i \to \infty} \lim_{t \to \infty} \Pr\left(Q^{(i)}(t) > m_i, i = 1, 2, \ldots, N\right)$$

$$= 1$$

The second inequality follows from De Morgan's theorem, and the union bound and the last equality follow from Equation 10.2. This means that

$$\lim_{\mathbf{m} \to \infty} \lim_{t \to \infty} \Pr(\mathbf{Q}(t) \le \mathbf{m}) = 1$$

Thus $\mathbf{Q}(t)$ is substable when $\lambda_i < \gamma(N)$ for $i = 1, 2, \ldots, \gamma(N)$. If $X^{(i)}(t)$ is as defined earlier, then $\{(\mathbf{Q}(t), X^{(1)}(t), X^{(2)}(t), \ldots, X^{(N)(t)}, t > 0)\}$ is an aperiodic, irreducible Markov chain, and part (ii) of the theorem is proved.   ∎

We reiterate that packet arrival rate for which the input queues will remain stable, as derived earlier, is for Bernoulli arrivals at the inputs with uniform routing and random order of service from the HOL queues. It is generally believed that in the cases not covered here, the input queues will be stable if the arrival rate is less than the saturation throughput of the switch, although this has not been formally proved for any other case. Specifically, the case of FIFO service of the HOL queues has not been considered, nor has the case when more than one cell can be switched from each input or to an output. This leads us to make the following definition.

Definition 10.2

For a specified distribution of the cell destinations, we say that the arrival process satisfies the *saturation stability property* if the input queues are stable whenever the expectation of the number of arrivals in each slot to each input is less than the saturation throughput from that input.   ∎

We use this property in discussing the delays in the switches.

10.1.2 Saturation Throughput of an IQ Switch

In Chapter 1 we consider a Markov chain model of an input saturated $N \times N$ IQ switch for arbitrary finite N. Recall that the saturation throughput per port decreases as N increases (see Table 1.1). The numerical values suggest that there is probably a limiting value of the saturation throughput as $N \to \infty$.

Consider what happens in the saturated IQ switch in each slot. From each nonempty HOL queue, one cell is transmitted to the output, and a total of $D(t)$ cells are transmitted in slot t. We assume that the cells depart at the end of a slot, that fresh packets arrive at the beginning of a slot, and that the arrivals in a slot are available for departure in that slot. Let $Q(t)$ denote the total number of HOL packets in the input queues at the end of the slot, after the switching of the packets. This is shown in Figure 10.4. Because the input queues are saturated, these $D(t)$ cells are replaced by fresh cells in slot $(t + 1)$. Thus into the HOL queues of the outputs, $D(t)$ cells arrive in slot $(t+1)$, each independently choosing any of outputs $1, \ldots, N$ with probability $1/N$ (i.e., they are uniformly routed). Let $A_j(t + 1)$ be the number of new cells arriving into HOL_j in slot $t + 1$. $A_j(t + 1)$ has a binomial distribution with mean $D(t)/N$.

$$\Pr\bigl(A_j(t + 1) = k\bigr) = \binom{D(t)}{k} \left(\frac{1}{N}\right)^k \left(1 - \frac{1}{N}\right)^{D(t) - k}$$

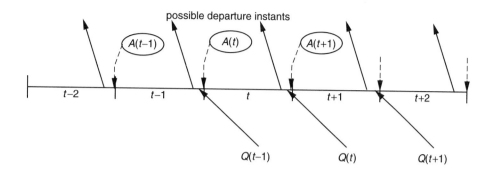

Figure 10.4 The arrival instants, the departure instants, and the instants at which the queue state is observed in each slot. Solid lines show possible departure instants, and dashed lines show arrival instants. $A(t)$ is the number of new arrivals in slot t. Slots $t-2, \ldots, t+2$ are shown.

$\sum_{j=1}^{N} Q_j(t)$ is the number of cells remaining in the inputs at the end of slot t, and we can write

$$D(t) = N - \sum_{1}^{N} Q_j(t) = \sum_{j=1}^{N} A_j(t+1) \qquad (10.3)$$

As before, let $\gamma(N)$ denote the saturation throughput from each output port. Note that $\gamma(N)$ can also be interpreted as the probability of a cell departing from an output—say, output j. Under stationary conditions, $\gamma(N) = E(D)/N$, where $E(D)$ is the stationary average of the number of cells departing from the switch in a slot. Taking expectations in the first equality of Equation 10.3, dividing by N, and using the fact that all outputs are statistically identical, we get

$$\gamma(N) = 1 - E(Q_j) \quad \text{or} \quad E(Q_j) = 1 - \gamma(N) \qquad (10.4)$$

Under input saturation, the average number of fresh cell arrivals to an HOL queue in a slot is equal to the saturation throughput from an input port and thus is $\gamma(N)$. To obtain $\gamma(N)$ as $N \to \infty$, we use the following lemma.

Lemma 10.2

Let $\gamma(\infty) = \lim_{N\to\infty} \gamma(N)$. In the steady state, if exactly one cell is switched from each nonempty HOL queue, then as $N \to \infty$, $A_j(t)$ has a Poisson distribution with mean $\gamma(\infty)$ and is independent of the number in the queue. ∎

Let us now consider the evolution equation for $Q_j(t)$. From Figure 10.4, we can write the following.

$$Q_j(t+1) = \max\left(0, Q_j(t) - 1 + A_j(t+1)\right)$$

$$= Q_j(t) + A_j(t+1) - \Delta_{Q_j(t)+A_j(t+1)} \tag{10.5}$$

where

$$\Delta_Q \triangleq \begin{cases} 1 & \text{if } Q > 0 \\ 0 & \text{if } Q = 0 \end{cases}$$

Taking expectations on both sides of Equation 10.5 we get

$$\mathsf{E}\big(Q_j(t+1)\big) = \mathsf{E}\big(Q_j(t)\big) - \mathsf{E}\Big(\Delta_{Q_j(t)+A_j(t+1)}\Big) + \mathsf{E}(A(t+1))$$

In steady state, the statistics of $Q_j(t + 1)$ and $Q_j(t)$ will be identical, and $\mathsf{E}\big(Q_j(t+1)\big) = \mathsf{E}\big(Q_j(t)\big)$, leaving us $\mathsf{E}\Big(\Delta_{Q_j(t)+A_j(t+1)}\Big) = \mathsf{E}(A(t+1))$. Define p_0 to be the steady state probability that $(Q_j(t) + A_j(t + 1)) = 0$.

Exercise 10.2

Show that in steady state, $\mathsf{E}\Big(\Delta_{Q_j(t)+A_j(t+1)}\Big) = 1 - p_0$ and $p_0 = 1 - \gamma(\infty)$.

Denote the moment-generating function of $Q_j(t)$ and $A_j(t)$ by $Q_j(t, z)$ and $\mathcal{A}_j(z, t)$, respectively. They are defined by

$$Q_j(t, z) := \mathsf{E}\Big(z^{Q_j(t)}\Big) \qquad \mathcal{A}_j(t, z) := \mathsf{E}\Big(z^{A_j(t)}\Big)$$

To obtain $Q_j(t, z)$, we proceed as follows.

$$Q_j(t+1, z) := \mathsf{E}\left(z^{Q_j(t+1)}\right)$$

$$= \mathsf{E}\left(z^{Q_j(t)-\Delta Q_j(t)+A_j(t+1)+A_j(t+1)}\right)$$

$$= p_0 + \sum_{k=1}^{\infty} \Pr\left(Q_j(t) + A_j(t+1) = k\right) z^{k-1}$$

$$= p_0 + \frac{1}{z}\left(Q_j(t, z)A_j(t+1, z) - p_0\right)$$

We obtain the third equality by separating the term for the case of $Q_j(t) = A_j(t+1) = 0$. The last equality follows from the assumption that the number of new packets into an HOL queue is independent of the current occupancy. In steady state, $Q_j(t, z) = Q_j(t+1, z)$, and we can drop the dependence on t to write the moment-generating function as $Q_j(z)$. Similarly for $A_j(t+1, z)$. The last equation then simplifies to

$$Q_j(z) = \frac{p_0(1-z)}{A_j(z) - z} \tag{10.6}$$

Substituting $p_0 = 1 - \gamma(\infty)$ and noting that because the number of new arrivals to an HOL queue in a slot is from a Poisson distribution with mean $\gamma(\infty)$, $A_j(z) = e^{\gamma(\infty)(z-1)}$, we get

$$Q_j(z) = \frac{(1-\gamma(\infty))(1-z)}{e^{\gamma(\infty)(z-1)} - z}$$

Differentiating with respect to z, putting $z = 1$, and using Equation 10.4 as $N \to \infty$, we get

$$\mathsf{E}(Q_j) = \frac{\gamma(\infty)^2}{2(1 - \gamma(\infty))} \tag{10.7}$$

Solving for $\gamma(\infty)$ from Equations 10.4 and 10.7, $\gamma(\infty) = 2 - \sqrt{2} \approx 0.586$.

For an $N \times N$ IQ switch under independent, uniform routing and FIFO buffering at each input, as $N \to \infty$, the saturation throughput is $2 - \sqrt{2} \approx 0.586$.

10.1.3   Discussion

From the foregoing results, we see that queue placement in a cell switch must trade off the simple implementation and lower throughput of the IQ switch and the complex, possibly infeasible implementation and higher throughput of the OQ switch. We can say that an $N \times N$ OQ switch needs $O(N)$ times more resources than an $N \times N$ IQ switch and provides approximately twice the capacity. Many proposals have been made to design a switch whose complexity is a constant times that of the IQ switch. These designs essentially take one of two approaches. The maximum number of cells that can be transmitted from an input or to an output port in a slot is increased. This method necessarily requires that queues be maintained at both the input and the output (CIOQ switches). The second approach is to change the FIFO scheduling at the input buffers. In the next section we consider these design choices for CIOQ switches.

10.2   Combined Input–Output Queueing

10.2.1   The Knockout Principle

A simple way of achieving the throughput characteristics of an OQ switch but with an implementation complexity that does not grow with N as compared with that of an IQ switch is to have the capacity to resolve a smaller number of output conflicts rather than N. Consider a switch that can handle up to L conflicts to an output port; that is, if, in a slot, n cells arrive for an output port, then if $n \leq L$, all n cells are switched to the respective outputs, whereas if $n > L$, then an arbitrary L of these are switched to the output. The *knockout* switch takes this approach and drops the cells that are not switched.

Consider an $N \times N$ knockout switch. Assume that the cell arrival to each input is a Bernoulli process with mean λ. Also assume uniform routing. In a slot, the number of arrivals for a tagged output will be a binomially distributed random variable with mean $\frac{\lambda}{N} \times N = \lambda$:

$$\text{Pr}\big(k \text{ cells with destination } j\big) = \binom{N}{k} \left(\frac{\lambda}{N}\right)^k \left(1 - \frac{\lambda}{N}\right)^{N-k}$$

For large N we can approximate a binomial random variable by a Poisson random variable. With this approximation, the average number of dropped packets per slot per output port is

$$\sum_{k=L+1}^{\infty} (k-L)\frac{\lambda^k e^{-\lambda}}{k!}$$

The average number of packet arrivals per slot per output port is λ. The approximate probability of dropping a cell at the input is obtained in the following exercise.

Exercise 10.3

Using the Poisson approximation to the binomial distribution, show that the approximate probability of dropping a cell at the input in a knockout switch is given by

$$\mathrm{Pr}(\text{cell drop}) = \left(1 - \frac{L}{\lambda}\right)\left(1 - \sum_{k=0}^{L}\frac{\lambda^k e^{-\lambda}}{k!}\right) + \frac{\lambda^L e^{-\lambda}}{L!} \qquad (10.8)$$

Observe that this is different from the probability that in a slot, a packet for a tagged output port is dropped. This will be simply the second term in brackets in Equation 10.8.

10.2.2 Parallel Fabrics

Choosing $L = 7$ in Equation 10.8 yields a loss probability of less than 10^{-7} for an arrival rate (λ) of 0.90. This means that even at very high loads and for large switches, the ability to resolve a small number (L) of conflicts to a destination, which is independent of N, allows us to switch most cells to the output in the same slot as they arrive. This suggests that with L smaller than 7 and by queueing cells at the input instead of dropping them, we could achieve switch capacities close to 1.0. Of course, we should also maintain a queue at each output port because more than one cell may be switched to the output in a slot.

There are two ways to send more than one cell to an output in a slot: parallelism and speedup. Whereas parallelism is a space-based solution, speedup is a time-based solution. For both of these, we describe schemes that are easy to analyze and visualize. There are, of course, a number of proposals in the literature that cannot be strictly classified into these categories.

Parallelism can take many forms. In the following, we assume that in a switch with parallelism of degree P, we have P parallel switching fabrics and that up to P cells from the HOL of the input queues are switched to the output in a slot. This allows each output to receive up to P cells from the HOL positions of the input queues. To find the saturation throughput of such a switch with parallelism P, we first generalize Equation 10.5 as follows:

$$Q_j(t) = \max\left(0, Q_j(t-1) + A_j(t) - P\right) \qquad (10.9)$$

From this we can obtain an expression for $\mathsf{E}(Q_j)$ under steady state as follows:

$$\mathsf{E}(Q_j) = \frac{\gamma(\infty)^2 - P(P-1)}{2(P - \gamma(\infty))} + \sum_{i=1}^{P-1} \frac{1}{1 - z_i} \qquad (10.10)$$

In deriving this, we make the approximation that as $N \to \infty$, the packet arrivals to the HOL queue is a Poisson random variable with mean $\gamma(\infty)$. (Note that Lemma 2 does not hold because, in a slot, more than one cell may be switched from a HOL queue.) Equation 10.10 is a generalization of Equation 10.7 and is the mean queue length in a $D^X/D/P$ queue—that is, a P-server, discrete time queue with Poisson batch arrivals (denoted by X) in each slot and deterministic service time (one slot time). Note that, z_i is the root of the equation $\exp(-\gamma(\infty)(1-z)) - z^P = 0$. Using Equations 10.10 and 10.4, we can obtain the capacity of an $N \times N$ switch as $N \to \infty$. The results are shown in Table 10.1. Observe that a parallelism of 4 gives more than 99% throughput.

10.2.3   Fabrics with Speedup

In a switch with speedup of S, there will be S phases in each slot. During each of these phases, the switch behaves exactly as in a slot in the IQ switch; of all the packets addressed to the same destination at the HOL of the input queues, one is randomly selected and switched to the output, where it may be queued before being transmitted on the output link. Figure 10.5 shows the case of a speedup of 2. The queue state at the end of slot $t-1$ and at end of the two phases of slot t is shown.

In the following, we present an informal argument to derive the saturation throughput of the IQ switch with speedup. Consider an $N \times N$ switch, with the

P	Saturation throughput
1	0.5858
2	0.8845
3	0.9755
4	0.9956
5	0.9993
6	0.9999

Table 10.1 Saturation throughput of an $N \times N$ switch as $N \to \infty$ for different degrees of parallelism (P).

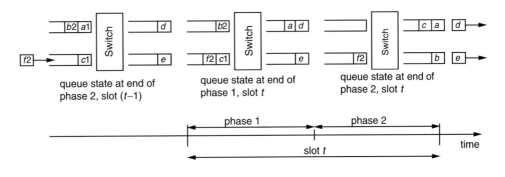

Figure 10.5 The two phases in a slot in a switch. The input and output queue status at the end of slot $t-1$ and the two phases of slot t are shown. The packet f has arrived at the beginning of slot t. The transmissions on the link are completed at the end of slot t.

cell arrival process satisfying the saturation stability property and having a mean cell arrival rate of λ cells per input port per slot. Also assume uniform routing. As usual, let $\gamma(N)$ be the saturation throughput per port of the switch with no speedup (i.e., for $S = 1$). The switch can be considered to be a time-slotted system, with slot boundaries at the phase boundaries. On each input, during each phase of the switch operation, an average of $q = \frac{\lambda}{S}$ cells arrive to the input. If $S > \frac{1}{\gamma(N)}$, then $q < \lambda\gamma(N) < \gamma(N)$ for all $\lambda < 1$ (i.e., the arrival rate to an input in each phase is less than the saturation throughput). Because we have assumed that the arrival process satisfies the saturation stability property, the input queues are stable. This means that the switch input port utilization can be made arbitrarily close to 1. Thus a speedup of $\lceil\frac{1}{\gamma(N)}\rceil$ will suffice to make the switch capacity close to 1 for arrival processes satisfying the saturation stability property.

From the foregoing discussion, speedup does better than parallelism in achieving high throughputs. One reason for this is the modeling assumption. In the parallel switch analysis we assume that at most one packet can be switched from an input in a slot, and also did not allow non-HOL packets to be switched in a slot. This clearly limits the capacity of the switch with parallelism. Thus the difference in the capacities is an artifact of the analysis, and in practice it might be possible to obtain higher throughputs.

10.3 Delay Analyses

The capacity analysis in Section 10.2.3 gives us the maximum load that can be offered to the input and output links of a switch. Recall from Chapter 2 that our divide-and-conquer approach to studying multiplexing and switching is to assume OQ switches while studying multiplexing of packet traffic on links. This assumption allows us to use general models for the packet arrival processes to a link, or equivalently the output queue of a switch. As we have seen, packet queues in most large switches are necessary at the inputs, and there are delays in these queues. Because of the complex interaction among the input and output queues in the switch, delay analysis can be performed only for simple packet arrival models.

In our analysis in this section, we assume an $N \times N$ switch, Bernoulli arrivals to the inputs with mean λ, and independent, uniform routing of the cells. We also assume that when the inputs and outputs have buffers, the buffers are infinite.

10.3.1 Output-Queued Switch

First, consider an OQ switch. The queueing process in an OQ switch is shown in Figure 10.6. Let $A_j(t)$ be the number of cell arrivals to the tagged output queue O_j in slot t.

$$\Pr(A_j(t) = m) := \alpha_m = \binom{N}{m} (\lambda/N)^m (1 - \lambda/N)^{N-m}$$

The moment-generating function of $A_j(t)$, $\mathcal{A}_j(z)$, is given by

$$\mathcal{A}_j(z) = \left(1 - \frac{\lambda}{N}(1 - z)\right)^N \tag{10.11}$$

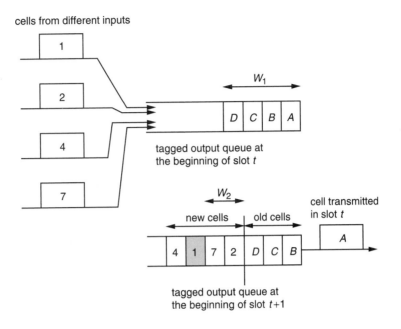

Figure 10.6   Example for the delay process in an output-queued switch. The top panel shows the queue status at the beginning of slot t and the cells arriving from different inputs (1, 2, 4, and 7). The bottom panel shows the queue status at the end of slot $t+1$ when the HOL packet from the preceding slot has been transmitted and the new cells are randomly placed in the output queue. The tagged cell is shown as shaded, and two of the new cells are placed ahead of it. W_1 and W_2 are also shown.

The queue evolution equation describing the relation between the number of cells in the tagged output queue in slot t, $Q_j^{(o)}(t)$, with that in slot $t+1$ is given by

$$Q_j^{(o)}(t+1) = \max(Q_j^{(o)}(t) + A_j(t+1) - 1, 0) \qquad (10.12)$$

This is similar to Equation 10.5, and the steady state moment-generating function of $Q_j^{(o)}(t)$ is the same as Equation 10.6, with $A_j(z)$ from Equation 10.11 and $p_0 = 1 - \lambda$. We therefore have

$$Q_j^{(o)}(z) = \frac{(1 - \lambda)(1 - z)}{\left(1 - \frac{\lambda}{N}(1 - z)\right)^N - z}$$

Now consider a tagged cell in the tagged queue. The output queue is FIFO, and cells arriving in the same slot are enqueued in random order. Thus the waiting time of a cell depends on the number of cells present in the output queue when it arrives and also on the size of the *batch* in which it arrives. We divide the total sojourn time of the tagged packet into two parts: W_1 corresponds to the waiting time until the cells that have arrived earlier depart, and W_2 corresponds to the waiting time until it is randomly selected from its batch. Denote the moment-generating function of W_1, W_2, and $W := W_1 + W_2$ by $\mathcal{W}_1(z)$, $\mathcal{W}_2(z)$, and $\mathcal{W}(z)$, respectively. The arrival of nonempty batches of cells to an output queue in each slot forms a Bernoulli process, with the probability that a nonempty batch arrived in a slot given by $1 - \alpha_0$. We can therefore invoke the *geometric arrival see time averages* (GASTA) property (see Section D.18) to claim that these nonempty batches see the stationary distribution in the queue. Hence, $\mathcal{W}_1(z) = \mathcal{Q}^{(o)}(z)$.

The probability that the tagged cell belongs to a batch of size m is given by $m\alpha_m/\lambda$ (see Problem 10.5). Given that it is in a batch of size m, the tagged cell will be in position k in the batch with probability $\frac{1}{m}$ for $k = 1, \ldots, m$. Thus we have

$$\Pr(W_2 = \tau) = \sum_{m=\tau}^{\infty} \frac{1}{m} \Pr(\text{tagged cell is in a batch of size } m)$$

$$= \frac{1}{\lambda} \sum_{m=\tau+1}^{\infty} \alpha_m = \frac{1}{\lambda}\left(1 - \sum_{m=0}^{\tau} \alpha_m\right) \qquad (10.13)$$

From the last equality, $\mathcal{W}_2(z)$ and $\mathcal{W}(z)$ can be written as follows.

$$\mathcal{W}_2(z) = \frac{1}{\lambda}\left(\frac{1}{1-z} - \frac{A_j(z)}{1-z}\right) = \frac{1}{\lambda}\left(\frac{1 - \left(1 - \frac{\lambda}{N}(1-z)\right)^N}{1-z}\right)$$

$$\mathcal{W}(z) = \frac{1-\lambda}{\lambda}\frac{1 - \left(1 - \frac{\lambda}{N}(1-z)\right)^N}{\left(1 - \frac{\lambda}{N}(1-z)\right)^N - z} \qquad (10.14)$$

We look at the numerical results later in the section.

10.3.2   Input-Queued Switch

Now consider the input-queued switch. As before, we assume FIFO input queues and random order of service from the HOL queues. First, observe that the total cell delay in the switch has two parts: the wait in the input queue until it reaches the HOL of the input queue (and hence the HOL queue of its destination), W_1, and the wait in the HOL queue, W_2. Figure 10.7 shows these components of the delay.

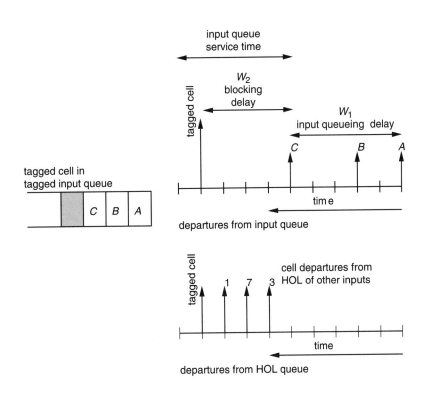

Figure 10.7   Example of the delay process in an IQ switch. Vertical arrows denote cell departure instants. The tagged cell (shaded) must wait until three cells ahead of it in the input queue A, B, and C) depart. This is the input queueing delay. After the cell moves to the HOL of the input, it must wait until the cells from other HOLs are transmitted.

Let us now find the distribution of the time spent in the HOL queue by a tagged cell. Let $p(\tau, k)$ be the probability that the tagged cell will be switched out of the HOL queue in τ slots given that there are k cells in the HOL queue (including the tagged cell). Observe that in every slot new cells may arrive into the HOL queue and when it is nonempty, only one departs. The tagged cell, if it is not switched in the slot, must compete with those left over from the preceding slot and also with those that arrived in the slot. This suggests a recursive relation in τ for $p(\tau, k)$. If there is only one cell in the HOL queue, then it will be transmitted in the slot. If there is more than one, then it will take one slot, with probability $\frac{1}{k}$; with probability $\frac{k-1}{k}$, the tagged cell is not transmitted in this slot and must compete with the $(k-1)$ remaining cells *and* with the new arrivals in the next slot. If we consider the case of a large switch, with $N \rightarrow \infty$, approximating the new arrivals to the HOL queue by a Poisson distribution, we can write the following recursive form.

$$p(\tau, k) = \begin{cases} 0 & \tau > 1, \quad k = 1 \\ \frac{1}{k} & \tau = 1, \quad k \geq 1 \\ \frac{k-1}{k} \sum_{j=0}^{\infty} p(\tau - 1, k - 1 + j) \frac{e^{-\lambda} \lambda^j}{j!} & \tau > 1, \quad k > 1 \end{cases} \qquad (10.15)$$

The last case is obtained by noting that the total number of cells that the tagged cell must compete with in the next slot will be $k - 1 + j$ with probability $\frac{e^{-\lambda} \lambda^j}{j!}$, and it should take $\tau - 1$ slots from then on. Immediately after the cell arrival instant, if there are k cells in the queue (including the tagged cell and those that arrived in its batch), then the probability that it experiences a delay of τ slots is $p(\tau, k)$. Thus we need to find the probability that the cell starts with k (including itself) in the queue. This can happen if there are m, $m < k$, cells in the queue just before the arrival of the tagged cell and if there are $k - m$ cells in its batch.

$$\Pr(W_2 = \tau) = \sum_{k=1}^{\infty} p(\tau, k) \sum_{m=0}^{k} q_m \frac{e^{-\lambda}}{1 - e^{-\lambda}} \frac{\lambda^{k-m}}{(k-m)!}$$

Here q_m is the probability of there being m in the queue when the batch of the tagged cell arrives at the HOL queue. Also, the batch size is conditioned on the batch being nonempty. Now observe that the HOL queue is identical to the output queue and that the queue evolution has the same form as Equation 10.1. Note that the output queue has FIFO service, whereas the HOL queue for an output in the IQ switch is assumed to have random order of service. Because both queues are work-conserving, the evolution equations are the same, and the stationary distribution of the number of packets in the queue is the same for both cases. For large N we use the approximation that in each slot, the number of new arrivals to the HOL queue is a Poisson random variable, and we use Equation 10.6 to obtain the moment-generating function of q_m. The first and second moments can be obtained numerically.

The time spent at the head of the input queue can be treated as the service time of the input queue. The input queue is a discrete time single-server queue, and the Bernoulli arrivals correspond to geometrically distributed interarrival times. Thus the input queue waiting time can be approximated as that in a Geom/G/1 queue, where the G is the service distribution, which in turn is the total delay from the HOL queue that we have modeled as $D^X/D/1$ with Poisson batch arrivals. Thus the mean total delay in the input queue, $E(D)$, is given by

$$E(D) = \frac{\lambda E(W_2)\left(E\left(W_2^2\right) - 1\right)}{2(1 - \lambda E(W_2))} + E(W_2)$$

This is the total delay in a Geom/G/1 queue where the first and second moments of the service time are $E(W_2)$ and $E(W^2)$, respectively.

10.3.3 CIOQ Switch with Parallelism

Now let us look at the FIFO CIOQ switch. We assume that in each slot, up to P cells can be removed from the HOL queue and placed in the output queue. The total delay experienced by a tagged cell has three components: W_1 (the delay in the input queue until it reaches the HOL of the input queue), W_2 (the delay in the HOL "queue" of the destination), and W_3 (the delay in the output queue before it is transmitted from the switch). Note that W_1 and W_2 are similar to the delays in the IQ switch, and W_3 is similar to the delay in the OQ switch. Together, the HOL queue and output queue constitute a single FIFO queue in which the cells in the HOL position get randomly shuffled in every switch slot until they move into the output queue. For large N, using the approximation of Poisson arrivals to the HOL queues, we can obtain the mean of $(W_2 + W_3)$ as the mean delay in the $D^X/D/P$ queue with Poisson batch arrivals. To obtain the mean of W_1, we generalize Equation 10.15 and do exactly as we did with the delay analysis of the IQ switch.

The analysis of the speedup CIOQ is more complex, and we do not pursue it here.

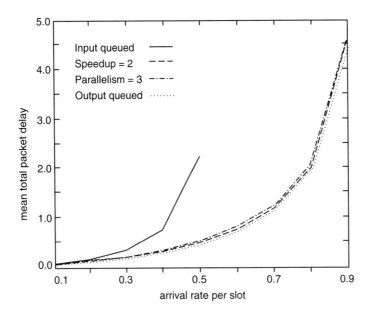

Figure 10.8   Mean cell delay for OQ, IQ, and CIOQ switches with speedup and parallelism.

In Figure 10.8 we show the mean delay as a function of the offered load for OQ, IQ, and CIOQ switches that use speedup and parallelism. The packet arrivals are from a Bernoulli process. Observe that with a parallelism of 3 and a speedup of 2, the mean delay characteristics are almost identical to those of the OQ switch. Thus, with reference to the mean delay and throughput characteristics, a CIOQ switch with either a parallelism of 3 or a speedup of 2 closely approximates the OQ switch. Later in the chapter we consider other, more general ways to describe the emulation of the OQ switch and show how it can be done.

10.4   Variable-Length Packet Switches

With IP becoming the dominant internetworking protocol, we need to consider packet switches in which the packet lengths and packet arrival times are random and are drawn from a general distribution. In this section we look at the results derived earlier and see how they apply to what we can call *continuous time switches*, referring to the continuous distributions of the interarrival times and packet lengths.

The simplest way to build a continuous time switch is to break an arriving packet into fixed-size cells and then use cell switches like those discussed in the previous sections. The first question then is what the cell size should be. The packet must be divided into cells at the input and reassembled at the output of the fabric before being transmitted on the output link. Typically the cells will be attached to a switch header that is used to reassemble them at the output and also to encode other information, such as destination priority and service (or QoS) code. The switch header is used only inside the switch and is removed from the cell before being transmitted on the output link. Also, when the packet is not equal to an integral number of cells, the last cell of the packet should be padded (extra bits added to make the length equal to the cell size). Figure 10.9 illustrates such fragmentation and padding.

In a variable-length packet switch using cell switches, the cell size trade-offs are fairly intuitive. A small cell size will generate a large number of cells per packet and will incur a large switch header overhead, and with a large cell size the padding bits will contribute to the overhead. Let the cells in the switch contain c bits of the packet and h switch overhead bits. For a packet of size p bits, we must transmit $\lceil \frac{p}{c} \rceil \times (c + h)$ bits instead of p bits. Thus the overhead of the cell switch is $\lceil \frac{p}{c} \rceil \times (c + h) - p$. For example, for a packet of length 9600 bits, a cell size of 512 bits, and a switch header of 32 bits, we must transmit 10,336 bits instead of 9600 bits, and the 736 bits is the overhead.

To understand this a bit better we performed the following experiment. The Web site `ita.lbl.gov` contains a large number of packet traces that contain information about the arrival time of packets, their lengths, and other information

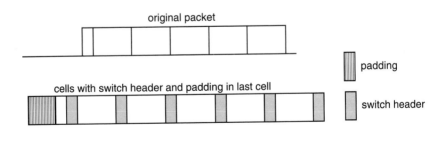

Figure 10.9 A variable-length packet is divided into cells, switch header, and padding bits. There are five full cells and a sixth one that is padded.

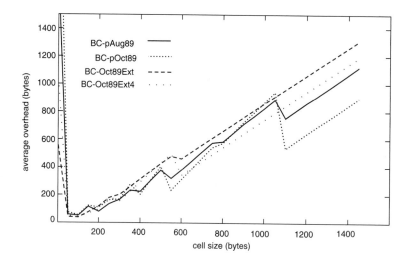

Figure 10.10   The average number of overhead bytes transmitted per packet through the switch for the packets recorded in the Bellcore traces as a function of the cell size. In these graphs we use a switch header of 4 bytes. The four different graphs are for the four different Bellcore traces, available at the Internet traffic archive site ita.lbl.gov.

about source and destination. Figure 10.10 plots the average overhead bits against the cell size for a few of these traces. Observe that the "optimum" cell size seems to be about 50 bytes.

Assume that as the packet arrives on the link, it is divided into cells and these cells are offered to the cell switch at the slot boundaries of the switch. This results in the cells arriving in bursts. An X-byte packet results in a burst of $\lceil \frac{X}{c} \rceil$ c-byte cells. Thus we can obtain the probability mass function of the burst size from that of the packet size using the following relation.

$$\theta_k := \mathsf{Pr}(k \text{ cells in a burst}) = \mathsf{Pr}((k-1)c < X \leq kc) \qquad \text{for } k = 1, 2, \ldots$$

It turns out that if θ_k has a geometric distribution with parameter b (i.e., $\theta_k = (1-b)b^{k-1}$) and if the cells from the same burst are switched contiguously, then as $N \to \infty$, the saturation throughput of the IQ switch is given by $\frac{2-\sqrt{4-2(1-b)}}{1-b}$. We can see that as $b \to 0$, a burst has exactly one cell, and the saturation throughput is $2 - \sqrt{2} = 0.586$, as it should be. If $b \to 1$, then the burst size becomes very large, and the throughput is 0.5.

Rather than use a cell switch for variable-length packets, we might use a continuous time switch, where the switch operations are not synchronized into slots as in a cell switch. Although most commercial switches use some form of internal cell switch to switch variable packets, it helps explain the performance issues in this design choice. To make the analysis simple, we assume that packet lengths have an exponential distribution. The closed-queueing network model that we discussed earlier with reference to cell switches is also valid here, except that it is not time-synchronous. Furthermore, we can actually solve for the saturation throughput of this switch because this closed-queueing network has a product form solution of the same form as the multiclass link that we studied in Chapter 6.

Exercise 10.4

Consider an $N \times N$ variable-length packet switch and its closed-queueing network model. Assume i.i.d. exponential packet lengths with unit mean and independent uniform routing. Using the product form expression for this model, show that saturation throughput is given by $\frac{N-1}{2N-1}$. Observe again that as $N \to \infty$, the throughput is 0.5.

To obtain a delay model for the IQ switch, we proceed exactly as with the time-slotted cell switch. Let packet arrivals to input port I_i be from a Poisson process of rate λ_i. Let $p_{i,j}$ be the probability that output O_j is the destination for a packet from input I_i. Let μ_j be the link rate on output port O_j. We assume FIFO input queues and FIFO HOL queues. The arrival rate to HOL queue of output O_j, Λ_j, is $\Lambda_j = \sum_{i=1}^{N} \lambda_i p_{i,j}$. Now consider the $N \times N$ switch with $N \to \infty$. Again, we approximate the arrival process to the HOL queues by a Poisson process of rate Λ_j, $1 \leq j \leq N$. Because the packet lengths are exponential, the HOL queue is an M/M/1 queue, and the input queue is an M/G/1 queue with service time distribution equal to that of the sojourn time of a FIFO M/M/1 queue.

Exercise 10.5

For the special case of $\lambda_i = \lambda$ for $1 \leq i \leq N$, and $p_{i,j} = \frac{1}{N}$, show that the input queues are stable for $\lambda < 0.5$.

An analysis similar to that of the time-slotted cell switch can be used to obtain the delay characteristics of the switch with parallelism, simultaneously transferring more than one packet from the HOL queues into an output queue. This is explored in Problem 10.7.

10.5 Non-FIFO Input-Queued Switches

Rather than use a switch fabric with speedup, we can increase the capacity of an IQ switch by relaxing the FIFO service discipline at the head of the queues. An example of such a relaxation is to let the first w cells of each input queue compete to be sent through the switch in each slot. Even if not more than one cell is transmitted in a slot from each input queue, using simulation it has been found that the throughput can be enhanced considerably. For example, with $w = 8$, the throughput of a 2×2 switch would increase from 0.75 for FIFO to about 0.96, and for a 128×128 switch the throughput would be 0.88, up from about 0.586. This mechanism of selecting a suitable cell from the first w cells is called windowing, and it requires a scheduler to select one of the first w cells from each input queue. The constraint on the scheduler is that the set of cells selected in each slot not have an output conflict (i.e., no two cells should have the same output port).

10.5.1 Virtual Output Queueing and Maximum-Weight Matching

Windowing increases the capacity by reducing the effect of HOL blocking by allowing another cell from an HOL-blocked queue to attempt to access the switch. In the extreme, it would be good to know the destinations of all the cells in each input queue so that each input queue gets a chance to send a cell to *any* output for which it has a cell. It is easy to check for the existence of cells to each destination at each input if each input maintains a separate queue for each output; in an $N \times N$ switch, each input will have N queues: one for each output. Thus there will be N^2 queues at the input in the switch. This is called virtual output queueing (VOQ), and the corresponding switch is called a VOQ switch. Figure 10.11 shows the components of such a switch. It is easy to see that head-of-line blocking is eliminated in a VOQ switch; a cell is not blocked by a cell that is ahead of it because all cells in a queue go to the same destination.

In each slot, the *matching scheduler* shown in Figure 10.11 finds the best possible set of cells from the input queues to feed to the switch such that a large number of cells are switched. The constraints imposed by the switching fabric in selecting the cells are (1) at most one cell is to be selected from each input, and (2) at most one cell should be selected for each output. This scheduling problem is identical to the well-known *bipartite graph matching* problem described next.

A graph has a vertex set V and an edge set E. An edge $e \in E$ is represented by (i, j), where $i, j \in V$. In a bipartite graph, the vertex set is the union of two

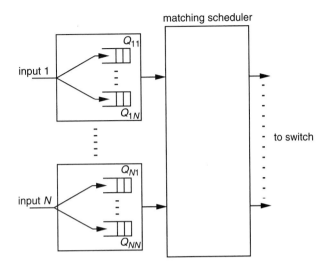

matching scheduler

input 1

Q_{11}

Q_{1N}

Q_{N1}

input N

Q_{NN}

to switch

Figure 10.11 The virtual output queued switch.

disjoint sets—say, V_1 and V_2—and the edges $e = (i, j)$ are such that $i \in V_1$ and $j \in V_2$. A *matching* of the graph is a subset of the edges of the graph such that no two edges are adjacent (have a common vertex). If every vertex of the graph is matched, then we get a perfect matching. For the bipartite graph representation of the VOQ switch, the vertex set V_1 corresponds to the inputs, and the set V_2 corresponds to the outputs. Denoting the number of cells in input i for output j at the beginning of slot t by $Q_{i,j}(t)$, the bipartite graph representation of the VOQ switch at the beginning of slot t has an edge (i, j) if $Q_{i,j}(t) > 0$. See Figure 10.12 for an example. The matching scheduler executes the bipartite graph matching algorithm in every slot. The matching so produced determines the cells from the nonempty queues at the inputs that will be offered to the switch in the slot.

We assume that in a slot each input can get zero or one cell. Let $A_{i,j}(k)$ be the interarrival time between cell k and cell $(k + 1)$ from input i to output j. If $A_{i,j}(k)$ are i.i.d., we say that the cell arrival process is i.i.d. Furthermore, for a fixed i and j, if $A_{i,j}$ satisfy the strong law of large numbers, then we can define $\lambda_{i,j}$ as the cell arrival rate at input i for output j to be the reciprocal of $E(A_{i,j}(1))$. Define $\lambda := [\lambda_{1,1}, \ldots \lambda_{1,N}, \lambda_{2,1}, \ldots, \lambda_{N,N}]^T$ to be the vector of arrival rates. If $\sum_{j=1}^{N} \lambda_{i,j} \geq 1$, then we say that the input i is *oversubscribed*. Similarly, if

inputs   outputs

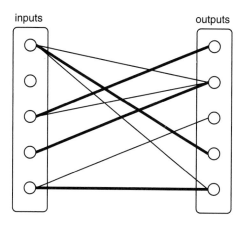

Figure 10.12 Bipartite matching in a 5 × 5 switch. A line from input *i* to output *j* indicates that $Q_{i,j}(t) > 0$. The matched input–output pairs are shown as joined by thick lines.

$\sum_{i=1}^{N} \lambda_{i,j} \geq 1$, then we say that output j is *oversubscribed*. If none of the inputs or the outputs is oversubscribed, we say that the corresponding cell arrival rate vector λ is *feasible*, or *admissible*. Let $Q(t) := [Q_{1,1}(t), \ldots, Q_{1,N}(t), Q_{2,1}(t), \ldots, Q_{N,N}(t)]^T$ be the discrete time queue-length process. An ideal matching scheduler should be able to make the queue-length process stable for all admissible λ. Recall that the queue-length process is stable if the limiting distribution of $Q(t)$ as $t \to \infty$ exists. We next investigate the matching algorithms that can achieve this ideal.

Matching algorithms that produce the maximum number of matching edges are called *maximum-size matching* algorithms. Assume that ties are broken randomly. Maximum-size matching seems intuitively appealing and could lead to high switch throughput. However, it cannot achieve the ideal we have described: a stable queue-length process for all admissible λ. Consider a 3 × 3 switch with Bernoulli arrivals according to the following rates: $\lambda_{1,1} = \lambda_{1,2} = \lambda_{2,1} = \lambda_{3,2} = 0.5 - \delta$ with $0 < \delta < 0.5$. Clearly, λ is feasible. In this case, maximum-size matching may result in an unstable queue at input 1. To see this observe that the arrival rate to input 1 is $1 - 2\delta$, and as δ becomes small, it must be served in almost every slot. When all the queues are nonempty, three maximum-size matchings are possible: $\{(1, 1), (3, 2)\}$, $\{(1, 2), (2, 1)\}$, and $\{(2, 1), (3, 2)\}$. And because ties are broken randomly, input 1 will be served with probability $\frac{2}{3}$. Let us now find the maximum rate at which input 1 will be served. Let $Q_{1,1}(t), Q_{1,2}(t) > 0$, and

$Q_{2,1}(t) = Q_{3,2}(t) = 0$. In this case, input 1 will be served with probability 1 if there is a cell arrival to at most one of the other two inputs (this occurs with probability $(1 - (0.5 - \delta)^2)$), and it will be served with probability $\frac{2}{3}$ if there is a cell arrival to both the other inputs (this event occurs with probability $(0.5 - \delta)^2$). Thus the maximum rate at which input 1 will be served is

$$\frac{2}{3}(0.5 - \delta)^2 + (1 - (0.5 - \delta)^2) = 1 - \frac{1}{3}(0.5 - \delta)^2$$

This implies that input 1 can become unstable for $1 - 2\delta > 1 - \frac{1}{3}(0.5 - \delta)^2$, or $\delta < 0.0358$.

Thus the maximum-size matching algorithm does not yield a 100% throughput for all admissible λ. It can also become unfair if the arrival rates are such that either the inputs or the outputs (or both) are oversubscribed, in which case some of the queues may not be served at all. Although such arrival rates are clearly not desirable because they lead to unstable queues, such arrival rates may occur over extended periods and there can be unfairness during these periods, as shown by the following example. Consider a 2×2 switch with $\lambda_{1,1} = \lambda_{1,2} = \lambda_{2,1} = 1.0$. It is easy to see that the queue $(1, 1)$ from input 1 to output 1 will starve.

We now consider an alternative to maximum-size matching. Recall that if for slot t there is an edge (i, j) in the bipartite graph, it implies that $Q_{i,j}(t) > 0$, and (i, j) is a possible match for slot t. If we could assign weights to the possible matches in each slot such that the weights indicate the preference of the inputs to the outputs to which they want to be matched in the slot, then the edges of the bipartite graph can be assigned these weights. The matching algorithm can then select the matching that maximizes the sum of the weights in the matching. This is called the *maximum-weight matching* algorithm.

There are two simple and obvious ways to assign weights to the edges. The first is called *longest queue first* (LQF). In slot t, the weight $w_{i,j}(t)$ to an edge from input i to an output j is equal to the number of cells in queue $Q_{i,j}$. The use of the queue-length-based maximum-weight matching algorithm in a VOQ switch can lead to extended starvation of some queues. Consider a 2×2 switch, with each queue having one cell at the beginning of slot $t = 0$. Now if the arrival rate to $Q_{1,1}$ is very high, it will get a cell in almost every slot, and while it gets these cells, its queue length will be higher than that of $Q_{1,2}$, which will starve for extended periods of time.

The second weight assignment method is called the *oldest cell first* (OCF) algorithm. In slot n, the weight $w_{i,j}(t)$ to an edge from input i to an output j

is equal to the number of slots that the cell at the head of queue $Q_{i,j}$ has waited. This will not lead to starvation of the queues, because as the cell waits longer, its weight increases, eventually increasing to a value at which it will be served.

The following theorem for maximum-weight matching algorithms says that a VOQ switch with an LQF (or OCF) scheduling algorithm can match the throughput of the OQ switch, albeit with the added computational complexity of performing maximum-weight matching in real time.

Theorem 10.2

(i) In a VOQ switch operating with the LQF maximum-weight matching algorithm, if the cell interarrival times are i.i.d. and the arrival rate vector λ is admissible, the queue-length process $\{Q(t)\}$ is stable. (ii) In a VOQ switch operating with the OCF maximum-weight matching algorithm, if the cell interarrival times are i.i.d. and the arrival rate vector λ is admissible, the queue-length process $\{Q(t)\}$ is stable. ∎

Note that the first part of the theorem is a special case of the link scheduling in ad hoc wireless networks that we discuss in Chapter 8, except that we now have a bipartite graph rather than the general graph that we considered for the wireless network. We do not pursue the proof of this theorem.

Maximum-weight matching algorithms execute in polynomial time, and the best-known algorithm requires $O(N^{2.5})$ time. However, these algorithms are not considered practical for implementation in high-speed switches, and practical algorithms need to be considered.

10.5.2   Maximal Matching for VOQ Switches

We now consider some practical matching algorithms to match the inputs to the outputs. *Maximal matching* algorithms are more practical, and many implementable approximations have been proposed. A *maximal matching* is a matching to which no more matches can be added. Figure 10.13 illustrates the difference between a maximal and a maximum matching. Obviously a maximum matching is also a maximal matching, but the converse is not true.

Theorem 10.3

In a VOQ switch if the cell arrival process $\{A_{i,j}(k)\}$ defined earlier satisfies the strong law of large numbers for all i and j and the rate vector λ is admissible,

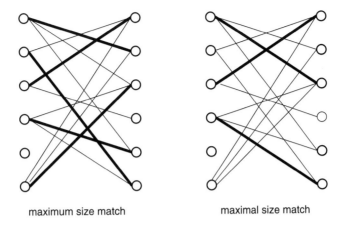

maximum size match   maximal size match

Figure 10.13   A bipartite graph and a maximum size and a maximal size matching. The matches are shown by thick lines. Neither of these matchings is unique.

and if the switch is operating with a speedup of 2, then with any maximal matching algorithm, the long-term departure rate of packets from the queues will be almost surely equal to the arrival rates.   ∎

Theorem 10.3 essentially says that 100% throughput can be achieved in a VOQ switch with a maximal matching algorithm if the fabric speedup is 2. Thus a maximal matching algorithm suffices to achieve OQ-like throughput behavior, with the added complexity of scheduling in real time and operating the switch at twice the line rate. Thus the maximal matching algorithms now assume practical significance. In practice, even maximal matching cannot be computed, and only approximate maximal matching is performed. To see how effective these maximal matching algorithms are, we need to study their properties. In the following we describe two maximal matching algorithms that are popular in the literature.

Parallel Iterative Matching
The *parallel iterative matching* (PIM) algorithm has been used in the GigaSwitch from DEC. (The research prototype was called AN2.) The algorithm converges to a maximal matching rather than to a maximum matching. The proof technique to

describe the rate at which requests are resolved is instructive. Also, by considering an equivalent system, we can theoretically calculate the saturation throughput of the switch.

Each iteration of the PIM algorithm consists of three phases:

1. The **request phase**, in which each input i sends a request to outputs j for which $Q_{i,j}$ is nonempty

2. The **grant phase**, in which each unmatched output randomly selects one request from the inputs and grants the request to the output

3. The **accept phase**, in which the input selects one from the many grants that it has received and notifies the output

Figure 10.14 shows a sample sequence for one iteration of the PIM algorithm. To obtain a maximal matching, only the unmatched inputs and outputs at the end of an iteration are considered in the next iteration. The algorithm runs until a maximal matching is found. In an actual switch, the algorithm is stopped after a finite number of iterations—say, k. Let us now determine a good value for k by obtaining the mean number of unresolved requests after k iterations. The following theorem shows that the algorithm converges very rapidly and that the average number of iterations to obtain the maximal matching is at most $logN + \frac{4}{3}$.

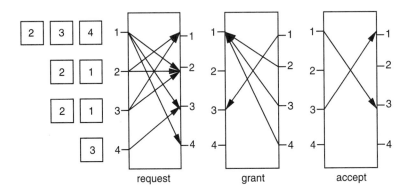

request   grant   accept

Figure 10.14   Phases in the parallel iterative matching (PIM) algorithm.

Theorem 10.4

   (i) The average number of requests resolved in each iteration of the PIM algorithm is at least 75%.

  (ii) The average number of iterations in which PIM converges is less than $\log_2 N + 4/3$.

Proof: The idea of the proof is to consider an output and evaluate the average reduction in the number of requests between successive iterations. We use Figure 10.14 for illustration. Consider an output O_j during iteration k. For O_j to be matched in the iteration, one of the following two things must happen:

- It must grant to an input that does not receive any other grant.

- The grant from O_j must be accepted by the input by selecting it from among competing outputs.

Let m be the number of requests that output O_j received during the slot. (For output 2, $m = 3$.) Of these m, let r inputs not receive grants from any other output. For output 2 in the example, $r = 1$ (i.e., input 2). With probability r/m, O_j will pick an input that does not receive any other grant, and it will be matched in the iteration. This is the first possibility described earlier. However, with probability $1 - r/m$, it will pick an input that will receive more than one grant. In Figure 10.14, output 2 picked the input that received three grants. In this case, O_j will either receive an accept and it will be matched, or it will not receive an accept. O_j not receiving the accept implies that in the next iteration none of the $m - r$ inputs that receives more than one grant from the outputs will be sending requests to output j.

From the foregoing, in iteration $(k + 1)$, an upper bound on the average number of requests received by O_j is derived as follows: With probability r/m, all the requests to O_j are resolved because O_j received an accept and is matched and no other input should match it. With probability $1 - r/m$, at most r requests— those inputs that did not receive a grant from any other output—will come back to O_j in the iteration $k + 1$. Because the $m - r$ inputs have received more than one grant, they will be matched in this iteration. Thus, on the average at most $(1 - r/m)r$ requests will come back to O_j. Because $r \leq m$, the maximum value of this expression is $m/4$. This means that in each iteration, the average number of matched requests is at least 75%. We start with at most N^2 requests, and after k iterations, the average number of unresolved requests is at most $N^2/4^k$.

Let C be the iteration in which the last request is resolved (or no requests are left).

$$E(C) = \sum_k \Pr(C > k)$$

$$= \sum_k \sum_{r>0} \Pr(r \text{ requests unresolved after } k \text{ iterations})$$

$$\leq \sum_k \left(\sum_{r>0} r \Pr(r \text{ requests unresolved after } k \text{ iterations}) \right)$$

The term in the parentheses is the expected number of unresolved requests after k iterations, which we have upper-bounded earlier by $\frac{N^2}{4^k}$. Note that the term in the summation is a probability and cannot be greater than 1. Hence, when $\frac{N^2}{4} > 1$, we can replace the term by 1. Thus we have

$$E(C) \leq \sum_k N^2/4^k \leq \sum_k \min(1, N^2/4^k)$$

$$\leq \sum_{k=1}^{\log_2 N} 1 + \sum_{k \geq \log_2 N} \frac{N^2}{4^k} = \log_2 N + \sum_{k \geq 0} \frac{N^2}{4^{\log_2 N + k}}$$

$$= \log_2 N + \sum_{k \geq 0} \frac{N^2}{N^2 \times 4^k} = \log_2 N + \frac{4}{3}$$

The last inequality is true because at most $\log_2 N$ terms in the summation term will be 1. ∎

It can be seen that on the average more than 99% of the matches will be found in four iterations.

We now obtain the saturation throughput of a VOQ switch using the PIM scheduling algorithm as a function of the number of iterations of the PIM algorithm. To help with the analytical model, we consider the following algorithm, which is equivalent to PIM. Each iteration consists of the following steps:

1. The request phase, in which each unmatched input requests one unmatched output for a match from those for which it has a cell

2. The grant phase, which is as in the original system

Because only one request was sent by each input, the accept phase is not necessary in the equivalent system, and an accept can be assumed by the output.

Exercise 10.6

Argue that the original description of the PIM algorithm is equivalent to the system just described.

Saturated inputs implies that all the N^2 input queues are nonempty in every slot. Let $\gamma(k)$ be the probability that a tagged output, output O_j, is matched in or before iteration k. For $k = 1$, $\gamma(1)$ is equal to the probability that at least one input selects the output:

$$\gamma(1) = 1 - \left(1 - \frac{1}{N}\right)^N$$

For $k > 1$, we must work a little harder. Let m_k be the number of inputs (outputs) matched in or before iteration k, and $n_k := N - \sum_{i=1}^{k} m_i$. Note that n_k is the number of unmatched inputs after iteration k. Let $\theta_k(n_k)$ be the probability that after iteration k, n_k outputs (including O_j) are unmatched. If there are n_{k-1} unmatched inputs, then at the beginning of iteration k, the probability that the tagged output O_j is matched in iteration k is the probability that at least one of the n_{k-1} unmatched inputs selects it. Thus the probability that O_j is matched after iteration k is

$$\gamma(k) = \gamma(k-1) + (1 - \gamma(k-1)) \sum_{n_{k-1}=1}^{N-(k-1)} \left(1 - \left(1 - \frac{1}{n_{k-1}}\right)^{n_{k-1}}\right) \theta_{k-1}(n_{k-1})$$

$$(10.16)$$

The summation goes only up to $N - (k - 1)$ because in each iteration at least one output will be matched.

We now obtain $\theta_k(n)$. We begin by obtaining $\phi(m,n)$, the probability that out of the n unmatched inputs, m are matched in an iteration and O_j is unmatched. Note that m outputs can be selected from the n unmatched inputs such that O_j is not matched in $\binom{n-1}{m}$ ways. Given the set of outputs that are matched, let us find out in how many ways the unmatched inputs can be assigned to them. According to the algorithm, each input sends a request to an arbitrary output. The requests received at the outputs clearly partition the n unmatched inputs into subsets, with subset r containing all the inputs that sent a request to output O_r. The number of nonempty subsets is equal to the number of matchings—that is, the number of outputs that received at least one request that is m. The number of ways in which n unmatched inputs can be partitioned into m nonempty subsets is given by the *Stirling number of the second kind*, S_n^m, defined as

$$S_n^m = \frac{1}{m!} \sum_{k=1}^{m} (-1)^{m-k} \binom{m}{k} k^n$$

Noting that there are $m!$ ways of assigning the partitions to the outputs and that there are n^n ways in which the requests could have been sent from the unmatched inputs, we get

$$\phi(m,n) = \frac{\binom{n-1}{m} m! S_n^m}{n^n}$$

From this, $\theta_k(n_k)$, the probability that n_k outputs are unmatched after k iterations, is given by

$$\theta_k(n_k) = \sum_{n_k=n_{k-1}+1}^{N-(k-1)} \text{Pr}\big(n_{k-1} \text{ unmatched in } k-1 \text{ iterations}\big)$$

$$\times \text{Pr}\big((n_k - n_{k-1}) \text{ are matched in iteration } k\big)$$

$$= \sum_{n_k=n_{k-1}+1}^{N-(k-1)} \theta_{k-1}(n_{k-1}) \times \phi(n_k - n_{k-1}, n_{k-1}) \tag{10.17}$$

From Equations 10.16 and 10.17 we can obtain the saturation throughput of the PIM algorithm in k iterations. It can be easily seen that as N becomes large, the throughput with one iteration is approximately $\frac{1}{e}$. Numerical calculations show that PIM requires three to four iterations to obtain a saturation throughput of nearly 1.0. An important feature of the PIM algorithm is that it uses a significant amount of randomization in each iteration. Random number generators are considered expensive to implement, and deterministic scheduling algorithms are preferred. Next we consider two deterministic alternatives to PIM.

Round-Robin Schedulers and iSLIP

An alternative to the PIM scheduler is a round-robin scheduler that is also a three-phase arbiter and uses a request-grant-accept sequence in each iteration of the scheduler. Rather than use a randomized algorithm, the outputs grant and inputs accept according to a deterministic rule based on priority lists. The inputs and the outputs maintain circular priority lists and a pointer to the highest-priority element of that list (see Figure 10.15). Outputs maintain a grant priority list, which is used to choose the input whose request is to be granted. Inputs maintain an accept priority list, which is used to choose the grant that should be accepted. In the request phase every input sends a request to all the nodes whose VOQ at that input is nonempty. In the grant phase, the output sends an accept to the input that has the highest priority in its list and moves the pointer to point to the input after the one that was given a grant. Similarly, if the input receives one or more grants, it accepts the one with the highest priority from its circular list and moves the pointer to the first input beyond the accepted output.

This round-robin scheduling scheme can cause synchronization of the grant pointers at the outputs. When such synchronization occurs, different outputs will have the same value of the grant pointer and will advance their pointers in lockstep during every slot. This continues as long as they receive packets from the inputs to which they point. If only one iteration is used, then because all the grant pointers point to the same input, only one of these inputs is served in the slot. An example is shown in Figure 10.15. After the first iteration, the grant pointers of outputs 1, 2, and 3 all point to 2. If only one iteration is used and if all the inputs have packets to all the outputs, we can see that in every slot, only one packet will be served. In fact, if the grant pointers are randomly placed and the inputs are saturated, we can see that the probability that an input is served is $(1 - \frac{1}{N})^N$. As N becomes large, the saturation throughput of the IQ switch with the round-robin scheduler using one iteration will be approximately $\frac{1}{e}$.

The iSLIP scheduling algorithm desynchronizes the output grant pointers by changing the grant algorithm of the round-robin scheduler as follows. The output

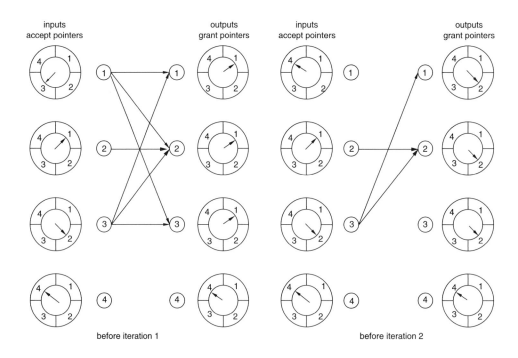

inputs
accept pointers

outputs
grant pointers

inputs
accept pointers

outputs
grant pointers

before iteration 1   before iteration 2

Figure 10.15   The round-robin scheduler. In the first iteration, requests are sent as follows: input 1 to outputs 1, 2, and 3; input 2 to output 2; and input 3 to outputs 1, 2, and 3. All the outputs grant to 1 and advance their grant pointers to 2. Input 1 accepts the grant from output 3 and advances its accept pointer to 4. In the next iteration requests are sent as follows: input 2 to output 2, and input 3 to outputs 1 and 3. Output 1 grants to input 3 and moves its grant pointer to 4. Output 2 grants to input 2 and moves its grant pointer to 3. Both these inputs accept the grants and move their accept pointers appropriately.

advances the grant pointer in the accept phase if and only if it receives the accept from the input. Under low load, the grant pointers at the outputs and the accept pointers at the inputs are in arbitrary positions, and the throughput is similar to that of PIM with one iteration. Under heavy load, each output serves each of the N input FIFO queues in a round-robin manner, and the virtual output queue of an output behaves like a discrete time GI/D/1 queue with a deterministic service time of N slot times. It can be shown that with the iSLIP scheduling algorithm, no VOQ is starved. This is easy to see. Because the outputs do not advance the grant pointer unless they get an accept, an output continues to grant an input until it receives an accept and services the inputs in a round-robin manner. Thus if the VOQs are full, all the queues of a given output will have the same throughput. In other words, the scheduler is fair.

10.6 Emulating Output Queueing with Input Queueing

To transport real-time stream traffic such as voice and video over a packet network, the network should be able to provide QoS support to the packets of a session. QoS support to a real-time stream session requires that the packets of the session be guaranteed an upper bound on the delay that they will experience on each link of the route to the destination. Providing delay guarantees on a link means that the variable component of the packet delay in the switch should be controlled. This is easiest to achieve in an OQ switch: A packet is switched to its output queue as soon as it arrives on the input link, and a link scheduler schedules its transmission time on the output link based on the QoS assurances to the session to which the packet belongs. Thus we can say that the OQ switch, along with the output link scheduler, is an ideal switch. In this section we explain how to emulate an OQ switch using switches that are not pure OQ.

From earlier discussions, we know that a FIFO–CIOQ switch with a speedup of 2 or a parallelism of 3 has throughput and mean delay characteristics almost identical to those of the OQ switch. However, it cannot provide deterministic delay guarantees for the packets because the sojourn time of the packet in the input queue is a random variable, however small its mean.

We will use a general notion of emulation of an OQ switch by requiring the target switch to have the following property: Under identical input cell sequences, the emulating switch should produce the same cell sequence on the output link as that produced by an OQ switch and its output link scheduler. In the FIFO–CIOQ switch, cells with smaller delay guarantees can be blocked by cells with larger delay guarantees ahead of it in the input queue. Hence it cannot emulate the OQ switch, and we must look for alternative approaches. We construct the emulating switch by using a CIOQ switch with speedup S (i.e., we operate S phases) for every input (or output) time slot, and in each phase we find the best set of cells to offer to the switching fabric. The scheduling constraint is that at most one cell be offered from an input and at most one cell be offered to an output in each phase. We need to find a suitable value for S and the algorithm to select the cells in each phase.

The OQ switch that we are emulating is called a *shadow* OQ switch. To illustrate the design, assume that the shadow OQ switch and the emulating switch are both offered the same sequence of cells (i.e., each cell in the emulating switch has a copy in the shadow switch). A cell must depart from the emulating switch in the same slot that it leaves the shadow switch. Assume that the inputs and the outputs in the emulating switch know the departure instants of all the cells in the switch from the shadow OQ switch. With this information, each input can draw up a ranked list of the outputs it wants to be matched in a phase. Similarly, the outputs can also draw up a ranked list of the inputs. These ranked lists are used by the switch scheduler to obtain the matching for each phase. We show that if the scheduler uses a *stable matching* algorithm, then a CIOQ switch with $S = 2$ can transmit packets in the same sequence as an OQ switch with a scheduler.

We make the following assumptions.

1. In an input slot, at most one cell arrives to an input and at most one cell is transmitted from an output.

2. A cell will not be switched in the same slot in which it arrived.

3. The scheduler of the OQ switch that we are emulating is a work-conserving scheduler (i.e., the output link is not idle while there is a packet in that output queue).

4. The scheduling discipline is *monotonic*; that is, an arriving cell does not change the relative ordering of the packets that are already in the queue. Most practical service disciplines are monotonic and work-conserving, FIFO and WFQ (see Chapter 4).

Consider cell c at input I_i with destination O_j. Let $TL(c)$ denote the scheduled departure time of cell c from the output in the shadow switch. We assume that this is known. Let $OQ(c, t)$ be the number of cells at the beginning of slot t that are in the switch and are scheduled to depart earlier than c from O_j. Let $IQ(c, t)$ be the number of cells at the beginning of slot t that are at input I_i that must be transferred to the outputs earlier than cell c. Figure 10.16 illustrates these definitions. Thus, we need to define a mechanism to queue the cells at the input so that the cells are ordered according to their urgency. This is achieved as follows. When cell c arrives at input I_i in slot t, place the cell in position $OQ(c, t)$, or, if there are fewer cells than $OQ(c, t)$, place it at the back of the queue at I_i. This achieves the ordering of the cells at the input according to their urgency.

Recall that our goal is not only to match a large number of the inputs to the outputs but also to deliver a cell to its destination output before its scheduled transmission time. To solve this problem, we use the bipartite graph matching algorithm called the stable matching algorithm. To use the algorithm, the inputs (and outputs) construct a ranked list of outputs (respectively, inputs) to which they would like to be matched. For each phase, the preference list of an input is

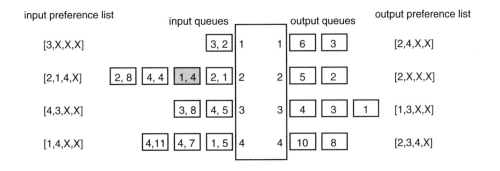

Figure 10.16   Figure illustrating $TL(c)$, $OQ(c, t)$, $IQ(c, t)$, **and** $SL(c, t)$. **At the inputs the cells are marked** $(D, TL(c))$, **where** D **is the destination of the cell. At the outputs the cells are marked** $TL(c)$. **For the shaded cell,** $TL(c) = 4$, $OQ(c, t) = 1$, **and** $IQ(c, t) = 1$. **Also shown are the input and output preference lists, with the** X **indicating a "don't care."**

the ordering of the outputs in the same order as the first cell to that output in the input queue. Similarly, the outputs construct a preference list of inputs as follows. Of the departure times of the cells at input I_i for output O_j, let $D_{i,j}$ be value of the earliest scheduled departure time. Output O_j constructs its preference list of inputs by ordering according to $D_{i,j}$, $i = 1, \ldots N$. In other words, O_j ranks the inputs in the order in which the earliest cells from them are scheduled to depart from O_j. Figure 10.16 illustrates these preference lists.

Let (i, j) and (i', j') be two matched input–output pairs from a matching $\mathcal{M}$. If i prefers j' over j and if j' prefers i over i', then (i, j) and (i', j') form an *unstable match*. If all the matches in a matching $\mathcal{M}$ are stable, then $\mathcal{M}$ is called a *stable matching*. See Figure 10.17 for an example of an unstable matching.

Given the preference lists from each of the input and output ports, Algorithm 10.1 generates a stable match. In the second step of the algorithm, an input can reject a previously matched output if it gets a request from an output for which it has a higher preference.

Exercise 10.7

Construct the stable match for the example in Figure 10.16.

In each switch, we assume that a stable matching is obtained and that the matched cells are transferred from the inputs to the outputs. At the beginning of

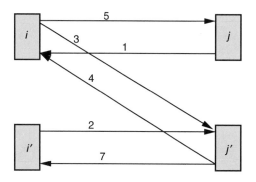

Figure 10.17 Example of an unstable matching. The numbers next to the arrows indicate preferences. The matches (i, j) and (i', j') are unstable.

---

while unmatched outputs not rejected by all inputs exist **do**
{
Every unmatched output requests its most preferred input that has not rejected it;
Every input grants to the output with the most preferred cell;
}

Algorithm 10.1   Stable matching.

---

slot t, consider cell c at input queue I_i for output O_j. In a phase, if input I_i is not matched to output O_j and if cell c stays at the input, then either or both of the following are true:

- A cell ahead of c in the input queue at I_i was transferred, leading to a reduction in $IQ(c, t)$.

- A cell from another input that is scheduled to depart earlier than c from the shadow OQ was transferred, leading to an increase in $OQ(c, t)$.

Defining the *slack* of cell c to be $SL(c, t) := OQ(c, t) - IQ(c, t)$, we see that in both cases, $SL(c, t)$ will increase by 1. Thus for S phases, with no arrivals or departures, the slack decreases by S in one input slot. We now account for the arrivals and departures. An arriving cell to input I_i in the slot may be placed ahead of c in the queue at I_i, and that will increase $IQ(c, t)$ by 1. A departing cell from output O_j will reduce $OQ(c, t)$ by 1. Because in each slot at most one cell may arrive to an input and at most one may depart from the output, an arrival and a departure can together decrease $SL(c, t)$ by at most 2 in a slot. Thus we can state the following lemma.

Lemma 10.3

In a CIOQ switch with a speedup of S and with the input–output matching scheduled using the stable matching algorithm, $SL(c, t + 1) - SL(c, t) \geq S - 2$.   ■

Intuitively, to send the cell in time (i.e., to adhere to the output sequence determined by the shadow switch), we would want the slack of a cell not to decrease in every slot. We now show that $S = 2$ is sufficient to exactly emulate an OQ switch.

Theorem 10.5

In CIOQ switch operating with the input–output matches scheduled using the stable matching algorithm, a speedup of $S = 2$ is sufficient to exactly emulate an OQ switch.

Proof: Consider a cell c arriving at input I_i for output O_j in slot t. If, on arrival, $SL(c, t) > 0$, then $SL(c, t)$ is nondecreasing and the cell will reach its output before it runs out of time. However, if $SL(c, t) = 0$ on arrival and continues to be 0 until its departure slot (say, t'), then in slot t', $IQ(c, t') = OQ(c, t') = 1$. Then input I_i and output O_j will have the highest preference for each other in slot t' and will be matched in the slot. This means that cell c will be transferred to the output in the slot and can be transmitted on the output according to schedule.   ■

We have shown that with sufficient information at the inputs and outputs, the stable matching scheduler, and a speedup of 2 it is possible to emulate an OQ switch that uses any work-conserving, monotonic scheduler. The problem of course is the availability of "sufficient information" at the inputs and the outputs.

Clearly, the stable matching algorithm has a complexity of $O(N^2)$ and is hard to implement in high-speed switches. A distributed implementation of the stable matching algorithm would be better suited. Furthermore, the algorithm requires that the schedule and destination of all the cells be known at the inputs. This is clearly an ideal situation and is not easy to achieve in practice. Thus exact emulation of an OQ switch, although possible in theory, is hard in practice.

10.7 Summary

In this chapter we formally show that asymptotically, as the number of ports becomes large, an IQ switch with FIFO cell scheduling has a capacity of only about 58.6% under uniform routing. Because this is significantly lower than the 100% capacity of an OQ switch, we explore two variations of CIOQ switches using parallelism and speedup. We find that a parallelism of 3 and speedup of 2 can deliver nearly 100% throughput and can also match the mean delay characteristics of an OQ switch when the cell arrivals are from a Bernoulli process and routing is uniform. We also discuss the construction of continuous time switches for variable-length packets using cell switches.

To minimize the effect of HOL blocking in an IQ switch and trading off switching speed for scheduling complexity, we propose the VOQ switch. In a VOQ switch, each input maintains a separate queue for cells of each output. In each slot, if the cells that will be switched from each input are selected according to the maximum-weight matching algorithm, then the queue-length processes at the inputs will be stable for all admissible packet arrival rates. The weights can be assigned according to either queue length (longest queue first) or the waiting time of the HOL cell (oldest cell first). Because maximum-weight matching has a high complexity, we discuss PIM maximal matching. We examine the convergence properties and the saturation throughput as a function of the number of iterations. We also discuss round-robin schedulers and a variant called iSLIP. Finally, we discuss the emulation of an OQ switch by an IQ switch with a speedup of 2 and a stable matching algorithm to select the cells in each of the phases.

10.8 Notes on the Literature

Before its applications in high-speed packet switching, the theory of space switching was developed in the context of telephone switches. It was further developed for use in interconnection networks in multiprocessor systems to interconnect processors and memory units. Patel [234] was among the first to consider the effect of output contention on the throughput of a nonblocking switch. Input queueing was not considered in [234]. Although the derivation of the asymptotic throughput in a FIFO–IQ cell switch is usually attributed to Karol, Hluchyj, and Morgan [158], this result was first shown by Baskett and Smith [22]. Note that the proof of Lemma 10.2 as given in [158] has an error. The correct proof is available in [277]. The proof of the saturation throughput being equal to the IQ capacity of the switch in the special case of uniform loading was proved by Jacob and Kumar [150]. The saturation throughput of IQ switches under bursty traffic has been obtained by Jacob and Kumar [149].

Many CIOQ switch architectures have been proposed. Yeh, Hluchyj, and Acampora [305] proposed the Knockout switch. Oie et al. [228] analyzed the FIFO–IQ switch with parallelism, and Diwan, Guerin, and Sivarajan [84] analyzed the IQ switch with speedup. Delay analysis of the various switching architectures has been reported by many authors under a variety of assumptions. Hluchyj and Karol [139] reported the first delay analysis of IQ and OQ cell switches. Li has analyzed switch performance under nonuniform traffic [198] and correlated traffic [199]. Nong, Muppala, and Hamdi [225] have analyzed switches with multiple input queues. Switches with variable-length queues have been studied by Fuhrman [110] and by Manjunath and Sikdar [204, 205]. Ganjali, Keshavarzian, and Shah [113] discuss the performance throughput trade-offs of cell and packet switches.

When it was understood that HOL blocking caused throughput degradation in IQ switches, a large number of non-FIFO IQ switches were proposed for improving the throughput. Many of these schemes were analyzed using simulation models, and it was known that with appropriate scheduling, 100% throughput could be achieved in an IQ switch. McKeown et al. [210, 211] formally showed that a maximum-weight matching schedule can yield 100% throughput in an IQ switch. It is now clear that the LQF-based maximum-weight matching algorithm for a VOQ switch is a special case of the constrained server system studied in the context of wireless networks by Tassiulas and Ephremides [282]. Dai and Prabhakar [72] proved that with maximal matching a speedup of 2 could yield 100% throughput. Anderson et al. [12] proposed and analyzed the PIM algorithm, and McKeown [209] proposed the iSLIP algorithm and analyzed it using simulations. In practice, only finite iterations of PIM are used, and it provides an approximate maximal match. iSLIP too is a maximal match. Switch performance with realizable approximations of maximum and maximal matching algorithms has been studied by many authors.

Providing QoS guarantees using IQ switches has also received significant attention. Stoica and Zhang [276] and Chuang et al. [65] first gave the proof in the form that we have presented. Other authors have also considered this problem in different settings—for example, Charny et al. [56].

Problems

10.1   Obtain the exact saturation throughput of a 3×2 input-queued switch.

10.2   In the Markov chain model for the $N \times N$ IQ switch under saturation, we mention that the detailed model has N^N states. If the aim is to obtain the saturation throughput under uniform routing, then we can reduce the problem size by realising that the inputs are indistinguishable and that at the beginning of a slot it is necessary to know only the number in the HOL queue. How many states will this Markov chain have? A further reduction is possible by realizing that it is not necessary to associate specific HOL queue lengths with specific outputs. For example, in a 3×3 switch, states $(1, 2, 0)$ and $(2, 1, 0)$ are equivalent. How many states will this representation have? Verify your answers for a 3×3 IQ switch.

10.3   Consider saturation analysis of an IQ switch without buffers at the input. In a slot, if there is more than one cell at the inputs for a destination, only one of them is transmitted to the output, and the others are dropped. In the next slot all inputs get a new cell. As usual, assume independent and uniform routing. What is the throughput per port of such a switch? Why is this system identical to PIM with one iteration?

10.4   Consider an $M \times N$ cell switch with no input or output buffers. Cells arrive at each input according to a Bernoulli process of rate λ. Let $q(i)$ denote the probability that i inputs are active in a slot. Clearly, $q(i)$ has a binomial distribution. Assume that the cells are routed uniformly (all output destinations are equally likely). Let $E(i)$ denote the number of cells switched in a slot when i inputs are active. Show that $E(i) = n \left[1 - \left[\frac{n-1}{n} \right]^i \right]$. Using $q(i)$ and $E(i)$, find S, the average number of cells switched in a slot. Also find P_{succ}, the probability that an arriving cell is switched. Obtain the limiting values of S and P_{succ} by taking limits as $M, N \to \infty$ while M/N is a constant.

10.5   Consider a batch arrival process as the packet arrival process to an input queue. Let a_i be the probability that the batch size is i, $i = 0, 1, 2, \ldots$. What is the probability, p_k, that a randomly chosen packet is the kth packet in a batch? Let $\mathcal{A}(z)$ be the generating function of a_i, and let $\mathcal{P}(z)$ be that of p_k. Show that $\mathcal{P}(z) = \frac{1 - \mathcal{A}(z)}{\lambda(1-z)}$, where λ is the average batch size. This problem is similar to finding the probability that a randomly chosen child is the kth child in a family.

10.6   In an $N \times N$ continuous time IQ switch, as $N \to \infty$ approximate the arrivals to the HOL queues by a Poisson process of rate λ. The packet lengths have an exponential distribution of mean μ^{-1}. Assume that the HOL queues are served according to FIFO. Obtain the average time spent by the packet before reaching the HOL, and the mean sojourn time of the packet.

10.7   Consider a switch as in Problem 10.6. Now assume that up to two packets can be switched simultaneously to the same output, but only one from each input at any time. If the packet arrivals to the HOL queue can be approximated by a Poisson process, what is the capacity of this packet switch? Also obtain the mean sojourn time of a packet before it is switched to the output queue.

10.8   If the packet lengths have a geometric distribution of mean 100 bytes, find the distribution of the number of 50-byte cells in a packet.

10.9   Consider a 2×2 switch. Assume that VOQs $Q_{1,1}$, $Q_{1,2}$, and $Q_{2,1} = 1$ are saturated and that $Q_{2,2}$ has no load. Under PIM, what is the throughput from each of the VOQs? This is the saturation throughput under nonuniform routing. What will be the saturation throughput of the round-robin scheduler if all four VOQs are saturated?

10.10 In Algorithm 10.1 argue that the sequence of inputs selected by an output in each step is nonincreasing in its preference and the sequence of outputs to which an input sends a grant is nondecreasing in its preference. This implies that the matching obtained from this algorithm favors the outputs. Reversing the role of the input and output in Algorithm 10.1 will also result in a stable matching. Execute Algorithm 10.1 on the example in Figure 10.16 with the roles of input and output reversed.

CHAPTER 11

Switching Fabrics

In Chapter 10 we discuss the performance trade-offs in the positioning of the packet queue in a packet switch vis-à-vis the switching fabric. In the process we assume a nonblocking switching fabric. In this chapter we discuss the design issues in the construction of this and other switching fabrics.

Monolithic switching fabrics become expensive and difficult to implement when the aggregate switching capacity (the sum of the transmission rates of the links connected to the switch) increases. We construct high-capacity switching fabrics by interconnecting a number of smaller elementary structures. In this chapter we first examine the elementary structures that are used in packet and circuit switches. We then look at the interconnection of these structures and their performance trade-offs. An important trade-off in large switches is that some amount of blocking is tolerated to reduce the implementation complexity.

In the next section we discuss some of the commonly used elementary switching structures that can be used to build small- and medium-capacity switches having a small number of ports. We then examine the interconnection of these elementary structures to form large switching networks in Section 11.2. Many of these networks are not feasible for use in high-speed packet switches because determining the interconnection pattern in these networks is complex and can become infeasible at high speeds. In Section 11.3 we study a class of networks called self-routing networks, where the packets are guided from the input port to the destination port in a distributed manner. In Section 11.4 we discuss architectures in the design of multicast switches.

11.1 Elementary Switch Structures

11.1.1 Shared-Medium Switches

The first elementary switch structure that we consider is called a *shared-medium* switch. The shared medium is usually a broadcast bus much like the bus in a local area network (e.g., Ethernet) except that the bus spans a very small area—usually a single chip or at most the backplane of the switching system. The input and the output link interfaces write and read, respectively, from this bus. At any time only one port or device can write to the bus. Hence, we need bus control logic to arbitrate access to the bus. Access to the shared medium is typically controlled through one of the following methods.

- *Centralized control using polling*: A central controller polls each input line interface to check for packets that may have been received by it and not yet switched to the corresponding output. If such a packet is present at the input line interface, it is transferred over the shared bus to the appropriate output interface. After the necessary transfer is completed, the controller continues the polling cycle.

- *Asynchronous distributed control using handshake signals between the input and output interfaces*: The input interface, on receiving a packet from the input line, processes it to find out the destination output port for the packet and then generates a request to transfer the packet to the destination output. When the output is ready to accept the packet and the bus becomes free, the condition is indicated through another handshake signal, and transfer of the packet begins over the shared-medium bus. The successful completion of the transfer can also be indicated by appropriate handshaking signals. This is much like the exchange of data among the peripheral devices and the main memory in a computer system over the system bus by using interrupts. In fact, a large number of low-capacity packet switches in the Internet are based on a shared-medium switch over the backplane bus of a computer. Such switches are well suited for networks where the packets arrive at arbitrary times and can have variable lengths.

- *Synchronous distributed control using a time division multiplexed bus*: This is well suited for time-slotted cell switches. For an $N \times N$ switch, time on the bus is divided into N phases in each slot. Each phase is allocated to an input line, which transmits its cell on the bus along with appropriate control information appended to the cell to indicate the destination port for the packet. An address filter on each output determines whether the packet is meant for it and reads it from the bus if it is the intended destination.

A block diagram of a typical shared-medium switch implementation is shown in Figure 11.1. The input interface extracts the packet from the input link, performs a route lookup (either through the forwarding table stored in its cache or by consulting a central processor), inserts a switch header on the packet to identify

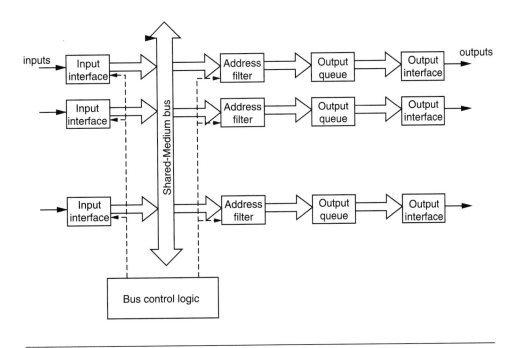

Figure 11.1 A shared-medium switch. Solid arrows indicate the data, or packet, path, and the dashed arrows indicate the path for control information. The switch processor is not shown.

its output port and service class, and then transmits the packet on the shared medium. Only the target output(s) read the packet from the bus and place it in the output queue.

A shared-medium switch is an output-queued switch, with all the attendant advantages and limitations. Observe that multicasting and broadcasting are fairly straightforward in this switch, requiring only that the packet, or cell, contain a *switch header* (the part of the packet header for use only inside the switch) that indicates the output ports that need to read the packet. It is easy to see the limitations of this architecture in terms of scaling properties. The transfer rates on the bus must be greater than the sum of the input link rates. When the link rates are high, such transfer rates are achieved by *wide buses* (i.e., the bus is a parallel bus of width P bits). However, very wide buses are also difficult to implement in practice. In addition to a high transfer rate on the bus, as with any OQ switch, the shared-medium switch requires that the maximum memory transfer rate be at least equal to the sum of the transmission rates on the input links and the transmission rate of the corresponding output.

11.1.2   Shared-Memory Switches

The second commonly used type of switching structure is the *shared-memory* switch. A typical shared-memory switch is shown in Figure 11.2. In its most basic form it consists of a dual-ported memory: a *write port* for writing by the input interfaces, and a *read port* for reading by the output interfaces. It is much like a shared-medium switch, with a memory unit taking the place of the shared bus. The input interface extracts the packet from the input link and determines the output port for the packet by consulting a forwarding table. This information is used by the memory controller to control the location where the packet is enqueued in the shared memory. The memory controller also determines the location from where the output interfaces read their packets. Internally, the shared memory is organized into N separate queues, one for each output. It is not necessary that the buffer for an output queue be from contiguous locations. It is easy to see that the maximum possible transfer rate of the memory should be at least twice the sum of

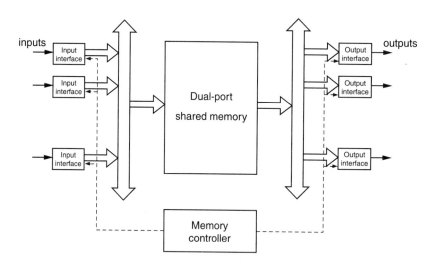

Figure 11.2   A shared-memory switch. Dashed lines indicate the path for flow of control information. Solid arrows indicate the data path.

the input line rates. Also, the memory controller should be able to process N input packets in one packet arrival time to determine their destinations and hence their storage location in the memory.

Notice that in a shared-medium switch, all the output queues are usually separate, whereas in a shared-memory switch, this need not be the case (i.e., the total memory in the switch need not be strictly partitioned among the N outputs). There are many ways in which the total memory space of M bytes can be shared among the different outputs. Assume that the input packets belong to N classes corresponding to the N output queues. Let the M bytes of memory be partitioned into $(N + 1)$ partitions, with M_j bytes in partition j. Partition j, $j = 1, \ldots, N$, is dedicated to packets of output j, and partition 0 forms a common pool. Let $P_{i,j}$ be the number of bytes in the ith packet in partition j, $j = 0, \ldots, N$. For an arriving packet of length $\tilde{P}_j$ bytes for output j, the following procedure is used to store it.

> if $\left(\sum_i P_{i,j} \leq M_j - \tilde{P}_j\right)$ then
>   store in partition j
> else
>   if $\left(\sum_i P_{i0} \leq M_0 - \tilde{P}_j\right)$ then
>   store in partition 0
>   else
>   drop packet

Depending on the allocations M_j, the following three distinct memory-sharing mechanisms are possible.

- *Full sharing* among the N inputs: Here $M_0 = M$ and $M_j = 0$ for $j = 1, \ldots, N$. The memory is not partitioned at all, and the packets from any output can reside in any memory location. The memory controller allocates a memory block to each arriving packet and maintains a linked list of packets for each output queue. This linked list can be arranged according to FIFO or any other scheduling discipline that the output queue chooses to implement.

- *Complete partitioning* among the N outputs in which $M_0 = 0$. This is the easiest to manage: A portion of the memory can be allocated to each output, and packets can be read from the head of the queue in this partition by the output interface. If an arriving packet finds that the memory allocated to its output is fully utilized, then the packet is dropped.

- *Partial sharing*, which is a combination of the preceding two schemes with $0 < M_0 < M$: This scheme reserves a minimum amount of memory for every output and thereby prevents heavily loaded outputs from monopolizing memory.

These buffer-sharing policies are very similar to the link admission control policies that we discuss in Section 6.6, and the analysis methods are also similar. More sophisticated sharing schemes are possible. For example, the actual values of M_j, rather than being static, may be dynamically allocated as a function of the total buffer size and the buffer occupancy. Packets can also be prioritized depending on their QoS requirements, and, in the event of congestion, the low-priority packet can be dropped. Priorities can also be assigned to dictate the order in which the packets are to be transmitted on the output link. Thus two types of priority assignments can be made on the packets: *temporal* priority, where deadlines for the transmission of packets must be adhered to, and *spatial*, where the packets must be given priority in not being dropped by the switch. If buffers are not available to an incoming packet for any of these reasons, it is not necessary to drop a packet. Instead, it can be stored in the buffer but *marked* to be overwritten if another packet arrives and does not find buffer space but is eligible to be stored according to the algorithm. This latter scheme is called the *push out* scheme.

11.1.3 Crossbar Switches

A third type of elementary switching structure that we consider is the *crossbar*. A 4×4 crossbar is shown in Figure 11.3. An $N \times N$ crossbar has N^2 crosspoints at the junctions of the input and output lines, and each junction contains a crosspoint switch. Figure 11.3 also shows a sample design of a crosspoint switch. In this design, the crosspoint switch is a four-terminal device with H_{in} and V_{in} as the two input lines, and H_{out} and V_{out} as the two output lines. The switch can exist in one of two states—the bar state and the cross state—making connections as shown. Assume that the inputs and the outputs are numbered as in Figure 11.3. To set up the path from input I_i to output O_j, the switches $(i, 1)$ to $(i, j - 1)$ and $(i + 1, j)$ to (N, j) are set to cross state, and switch (i, j) is set to bar state. This sets up a physical path for the signal to flow from I_i to O_j, and hence this switch is also called a *space-division switch*. Thus the address of the destination of the packet and its input port can be used to set the crosspoint switches in the crossbar. We look at this capability of a switch in more detail later in the chapter.

> **Exercise 11.1**
> Consider the crosspoint design described here for a 3×3 switch. Route all the 3! permutations in this crossbar switch.

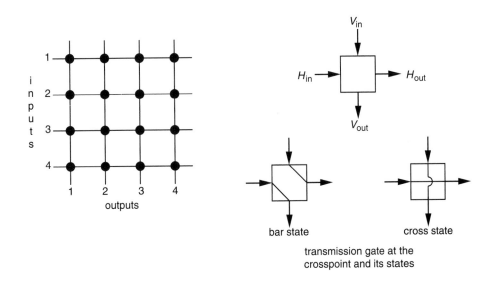

Figure 11.3 A crossbar switching fabric. The right panel shows one possible construction of the transmission gate at a crosspoint. A simple alternative is to have an open or closed switch at the crosspoint.

If there is an output conflict in a packet switch, the crossbar switch as described here transfers only one of the contending packets to its respective destination. Thus, typically, the basic crossbar is an input-queued switch, with queues maintained at the inputs and the crosspoints activated such that at any time an output is receiving packets from only one input. Note that it is not necessary that the input be connected to only one output at any time. Depending on the electrical characteristics of the input interface, up to N outputs can be connected to an input at the same time. This means that performing a multicast and broadcast is straightforward in a crossbar.

The construction complexity of a crossbar switch is a function of the number of crosspoints in the switch and grows as N^2. Thus the crosspoint complexity can become significant even for moderate values of N. Let us now see how this compares with the best possible number of crosspoints. Note that each crosspoint has two states. Thus if each combination of crosspoint settings can be called a switch state, a switch with K crosspoints has 2^K states. An $N \times N$ nonblocking switch must set up the switch crosspoints for $N!$ possible ways to connect the inputs to the outputs. This can be done using $\log_2(N!)$ two-state devices.

Using Stirling's approximation for $N!$, we have

$$N! \approx \sqrt{2\pi} N^{N+\frac{1}{2}} e^{-N}$$

$$\log_2(N!) \approx \frac{1}{2}\log_2(2\pi) + \left(N + \frac{1}{2}\right)\log_2(N) - N\log_2(e)$$

$$= O(N \log N)$$

That is, the order of the growth of the optimum number of two-state crosspoints in an $N \times N$ nonblocking switch is $O(N \log N)$. Thus the crossbar scales poorly compared with the optimum.

   Smaller crossbar switches can be combined to form a larger crossbar such that the larger crossbar has properties similar to the smaller one. A sample interconnection is shown in Figure 11.4. Observe that the number of crosspoints in the larger switch is still N^2.

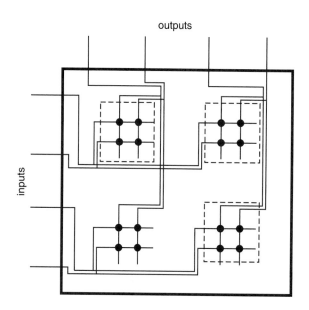

Figure 11.4   Interconnecting smaller crossbars to obtain a larger crossbar. Here, four 2 x 2 crossbars are interconnected to obtain a 4 x 4 crossbar.

We have seen that the three elementary switching structures—shared medium, shared memory, and crossbar—have severe scaling problems and cannot handle a large number of ports over high link transmission rates. In most packet-switched networks, such as the public Internet or enterprise networks, a few tens or even hundreds of links terminate at medium- to large-sized switching *exchanges*. The transmission rates on these links are in the range of a few hundred Mbps to as high as 10 Gbps. This means that the switches should be capable of aggregate throughputs in the range of 10^9 to 10^{12} bits per second.

With the increasing use of packet-switched networks and increasing demand for network connectivity, the number of links terminating at such exchanges or switching centers will increase, and so will the link transmission rates. Clearly, the switching structures considered so far will not suffice, and we need to construct *switching networks* using smaller switches. These switching networks should support a large number of ports per switch and should be capable of aggregate throughputs in the range we have mentioned. As mentioned in Chapter 1, a number of commercially available switches do just that: interconnect smaller switches to obtain a large switch. Examples include the Cisco 12000 series of Internet routers. In the following sections we develop the theory of switching networks that will help us build larger switches from smaller ones.

Before concluding this section, we discuss the *time-slot interchanger* (TSI), an elementary structure for circuit-multiplexed networks using time division multiplexing. A TSI performs switching in time rather than in space, as is done by the three types of switches that we have considered. Recall that in time division multiplexing, time is divided into frames and each frame has a fixed number of slots. A TSI takes the contents of a frame on the input link and produces a frame on the output link whose slots are a permutation of the slots of the input frame. Figure 11.5 shows a TSI for a six-slot frame. This permutation is achieved by writing the slots in the input frame into a local buffer in the order of their arrival and then reading them and transmitting them on the output link in the order in which they are to be permuted. This is repeated for every frame. Clearly, a TSI is nonblocking. Also, note that a TSI introduces a fixed delay of at least one frame duration.

11.2 Switching Networks

In this section we examine the systematic interconnection of a number of elementary switching modules into a switching network to obtain switches having a large number of ports. In the development of this theory, we assume that the switching network is *two-sided*: The inputs and outputs are distinct, and the

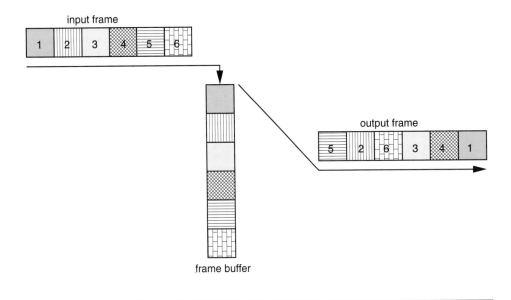

Figure 11.5   A time-slot interchanger for a six-slot frame. Slots are stored in order, are read in the order of the permuted output, and transmitted on the output port.

paths must be set up from the input ports to the output ports according to some connection requirement C. Such an assumption is natural in the context of packet networks, where the packet flows are not symmetric at any time. Even in the case of telephone networks, where the sessions are typically bidirectional, with an equal amount of traffic flowing in both directions, we can treat the "return path" as being from a parallel network with the connections reversed.

The switching network that we construct consists of S *stages* of elementary switching modules, as shown in Figure 11.6. The modules in each stage can be viewed as being arranged in a column. All modules of stage s, $s = 1,\ldots, S$, are identical $m_s \times n_s$ nonblocking switches, with no buffers except at the input of stage 1 and the output of stage S. For ease of exposition we assume that all the modules are crossbars. We could use the other elementary structures in forming the switching network, but the crossbar assumption helps in the exposition of the combinatorial properties of the switching network. The switch can be assumed to operate in a time-slotted manner, with a set of connection requests (or packets) all arriving at the beginning of a slot and departing at the end of the slot (as, for example, in a cell switch). Alternatively the connection requests or packets could

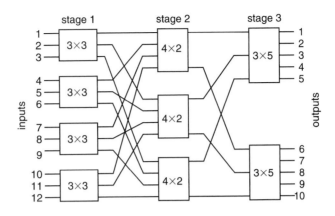

Figure 11.6   An example of a three-stage 12 x 10 switching network with four 3 x 3 crossbars in stage 1, two 4 x 2 crossbars in stage 2, and three 2 x 5 crossbars in stage 3.

arrive and depart at arbitrary times, as in telephone switches or in variable-length packet switches. The connection requests that arrive to the switch can be unicast or multicast. In this chapter we initially discuss only unicast switches, considering multicast switches separately in Section 11.4.

11.2.1 Blocking in Switching Networks

Consider an $M \times N$ switch and let $c = (i, j)$ be a connection request from input i to output j. Let $\mathcal{I}_c$ and $\mathcal{O}_c$ denote the input and the output of a connection request c, and let $C := \{c\}$ be a set of connection requests. C is called a *feasible connection set* if $\mathcal{O}_c$ for all $c \in C$ are distinct and if $\mathcal{I}_c$ for all $c \in C$ are also distinct.

Setting up the paths involves setting up the connections in the smaller switches. Depending on the switching network architecture, there may be many choices in setting up the paths to satisfy a given C, and the chosen set of paths is called a *routing of C* and denoted by $R(C)$. In this chapter, we use the term *routing* to refer to the routing of connections inside the switching network. The set $\{C, R(C)\}$ is called a *state* of the switching network. Figure 11.7 shows some sample states in a three-stage network.

Consider a switch that is set up to satisfy a connection requirement C with a routing $R(C)$. The state $\{C, R(C)\}$ is a *nonblocking state* if for any $c \notin C$ and feasible $\{C \cup c\}$, c can be routed through the switch without disturbing $R(C)$ (i.e., without changing the routing of the existing connections). In a blocking state,

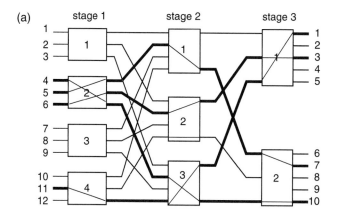

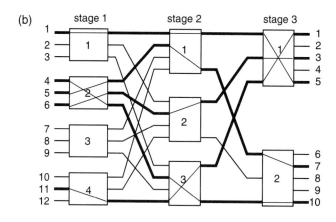

Figure 11.7 Sample states in a three-stage network. The bold lines between the stages show the active connections, and the dashed lines inside the switches correspond to the crosspoint setting of the crossbar. (a) shows the routing for the connection requirement C = {(4,10), (5,7), (6,3), (11,1)}. This is a nonblocking state, and any new connection request can be accommodated. In (b), a new connection (1,5) has arrived and is routed as shown. This is a blocking state because connection requests to outputs 2 and 4 cannot be accommodated in this state.

the connections in C are routed such that there exists a $c \notin C$ and feasible $\{C \cup c\}$ that cannot be routed without disturbing $R(C)$.

If a switch is able to route any feasible C, it is said to be a *nonblocking switch*. As can be imagined, there are many degrees of "nonblockingness." Given a set of active connections C, we say that a new connection request c is valid if $c \notin C$ and $\{C \cup c\}$ is feasible.

Definition 11.1

a. A switching network is *strictly nonblocking* (SNB) if there are no blocking states in the network.

b. A switch is *wide sense nonblocking* (WSNB) under a routing algorithm $\mathcal{R}$ if a new valid connection request c can be routed according to the routing algorithm $\mathcal{R}$ and the new state $\{C \cup c, R(C \cup c)\}$ is also a nonblocking state.

c. A switching network is *rearrangeably nonblocking* (RNB) if a new valid connection request c can be routed after possibly rearranging $R(C)$; that is, $R(C)$ may be changed to $R'(C)$ so as to be able to route c. Equivalently, a switch is RNB if starting with an empty network, any feasible C can be routed. ∎

In an SNB network a new valid connection request can be routed along any free path through the switch, and the new state of the network will be a nonblocking state. In a WSNB network, a route will be available for all new connections, provided that they are all routed according to a predefined algorithm. Thus in a WSNB network, the routing of new connections cannot be arbitrary. In both SNB and WSNB, the existing connections are not rerouted to accommodate the new connection. In an RNB network, when a new connection request arrives, a free path for it, if it does not already exist, is created by rerouting the existing connections. The rearrangement of the existing routes depends on the input and output ports of the new connection.

In a telephone switch, connection requests can arrive at arbitrary times, and active connections can be released at arbitrary times. The same is true in continuous time packet switches, which may be handling variable-length packets, although on a much faster timescale than in telephone switches. SNB or WSNB networks are useful in these environments because in these networks, irrespective of the routing of the existing connections, a new valid connection request can be routed without rearranging $R(C)$. If an SNB switching network were to be used, any free path between the input and output can be chosen, whereas if a WSNB network were to be used, the routing algorithm must be applied. On the other hand, if an RNB network were to be used, although the new valid connection request can be routed, existing connections may need to be rerouted and hence this type of network is not favored in this case. However, for a cell switch, where there is a new set of cells at the inputs in each slot, an RNB may be used.

11.2.2 Clos Networks

Let us now consider the construction of switching networks with the properties we have described. Although there are many ways to interconnect the switching modules to form a switching network, the one that is well understood and widely studied is called a *Clos network*, named after its inventor. An $M \times N$ Clos network is a three-stage network with p_s ($m_s \times n_s$) switching modules in stage s, $s = 1, 2, 3$. From each module of stage s there is exactly one link to *every* module of stage $s + 1$, for $s = 1, 2$. This means that in a Clos network, $p_1 = M/m_1$, $p_3 = N/n_3$, $n_1 = p_2 = m_3$, $m_2 = p_1$, $n_2 = p_3$. Thus a Clos network is completely defined by the 5-tuple $(m_1, p_1, p_2, p_3, n_3)$. Figures 11.6 and 11.7 are examples of Clos networks. Observe that in a Clos network, different connections from (to) the same input (output) module should be routed through different stage 2 modules.

The properties of the SNB, WSNB, and RNB Clos networks help us define the optimal construction of these switching networks. We now look at the defining properties of each of these three types of nonblocking Clos networks.

Strictly Nonblocking Networks

The 5-tuple defining a strictly nonblocking Clos network must satisfy the following theorem, called Clos's theorem.

Theorem 11.1

An $M \times N$ Clos network defined by the 5-tuple $(m_1, p_1, p_2, p_3, n_3)$ is SNB if

$$
p_2 \geq \begin{cases} N & \text{if } m_1 \geq N \\ M & \text{if } n_3 \geq M \\ m_1 + n_3 - 1 & \text{otherwise} \end{cases}
$$

Proof: Clearly, $N \leq m_1$ implies $M \geq N$, and $n_3 \geq M$ implies $M \leq N$. Because at most only $\min(M, N)$ connections will be generated, $p_2 \geq \min(M, N)$ is sufficient. We have proved sufficiency for the first two cases. Consider a valid connection request with input in module I of stage 1 and output in module O of stage 3. In the worst case all the other $m_1 - 1$ inputs of module I and $n_3 - 1$ outputs of module O are busy. Furthermore, in the worst case, each of these connections may be using a different stage 2 module. This means that there are $(m_1-1)+(n_3-1)=m_1+n_3-2$ busy modules in stage 2 with respect to establishing this new connection. One more module in stage 2 can route the new connection. ∎

It is generally believed that for a Clos network to be strictly nonblocking, it is also necessary that p_2 satisfy the condition given in Theorem 11.1. The following is a counterexample to this necessity condition. Consider a 4×3 Clos network with $m_1 = p_1 = 2$, $m_3 = 3$ and $p_3 = 1$. Clearly for this network $p_2 = 3 < m_1 + n_3 - 1 = 4$ is sufficient for the network to be strictly nonblocking.

Wide Sense Nonblocking Networks

The properties of wide sense nonblocking Clos networks have not been well understood, and very few useful results are available. Recall that a WSNB network is defined by a routing rule for arriving connection requests. Many "obvious" routing algorithms are possible, and the most intuitive one is called the *packing* algorithm. In this routing algorithm, a connection request is routed through the *busiest* middle-stage module: the one with the highest number of active connections routed through it. Consider a symmetric network with $M = N$, $m_1 = n_3$, and $p_1 = p_3$. We discuss some properties that define "upper bounds" and that tell us when we cannot construct a WSNB network that is less complex than a SNB network. We also discuss one construction of a WSNB network under the packing rule for routing.

Theorem 11.2

a. A symmetric Clos network with $m_1 = n_3$ and $p_1 = p_3$ is WSNB under the packing rule, if and only if $p_2 \geq 2m_1 - 1$ for $p_1 \geq 3$.

b. A symmetric Clos network with $m_1 = n_3$ and $p_1 = p_3$ and $p_1 \geq (m_1 - 1)\binom{2m_1-2}{m_1-1} + 1$ is WSNB if and only if $p_2 \geq 2m_1 - 1$. ∎

The first part of the theorem says that if we were to use the packing rule to route arriving connections in a symmetric Clos network with more than two modules in stage 1, then the number of stage 2 modules required is the same as that of an SNB switching network. Thus, although it might seem intuitive that a packing rule in routing connections will reduce the number of modules in stage 2, this theorem dispels that intuition, at least in the case of symmetric networks with many stage 1 modules. The second part of the theorem says that no routing rule can reduce the number of stage 2 modules from that required for SNB switches if the number of stage 1 switches is greater than $(m_1 - 1)\binom{2m_1-2}{m_1-1} + 1$. For a given number of inputs in the modules used for stage 1, this upper-bounds the total number of inputs in the switching network. For example, if $m_1 = 3$ and if we want the number of inputs in the switching network to be greater than 36, then we might as well build an SNB switch.

Finally, we describe an interesting network that is WSNB.

Theorem 11.3

For a symmetric Clos network with $p_1 = 2$, the switching network is WSNB under the packing rule if $p_2 \geq \lfloor \frac{3m_1}{2} \rfloor$.   ∎

This network is shown in Figure 11.8 for an 8×8 switch.

From our results and discussion, we see that WSNB networks may not provide significant cost savings as compared with SNB networks. Also, they do not seem to have simple and easily verifiable properties. We do not consider WSNB networks further.

Rearrangeably Nonblocking Networks

To derive the condition on p_2 to make a Clos network rearrangeable, it is sufficient to show that in an RNB network, any feasible C can be routed. This is because

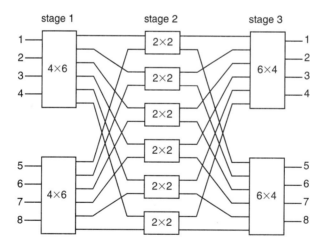

Figure 11.8   An 8 × 8 Clos network that is WSNB under the packing rule for routing arriving connection requests.

when a new connection arrives to add to an existing set of connections, because we allow rerouting of existing connections, we might as well find new routes to all the connections to satisfy the new connection requirement.

Let C be the set of connection requests. Before we state and prove the condition on p_2 for an RNB Clos network, we cast the routing of the connections in C as an edge coloring of a bipartite graph. Recall from Chapter 10 that in a bipartite graph $G = (V, E)$, where V is the vertex set E is the edge set, V is partitioned into V_1 and V_2 such that $V_1 \cup V_2 = V$ and $V_1 \cap V_2 = \phi$. All edges $e \in E$ are defined by 2-tuples (v_1, v_2), where $v_1 \in V_1$ and $v_2 \in V_2$. Now consider a bipartite graph constructed as follows. We represent each module of stage 1 by a vertex in V_1, and each module in stage 3 by a vertex in V_2. For every $c \in C$ we place an edge $e = (v_1, v_2)$, where v_1 represents the stage 1 module of $\mathcal{I}_c$, and v_2 represents the stage 3 module of $\mathcal{O}_c$. Note that in this construction we allow multiple edges between the same set of vertices. Figure 11.9 shows a sample connection requirement on a given Clos network and the bipartite graph representation of that requirement.

To route each connection $c = (\mathcal{I}_c, \mathcal{O}_c) \in C$, we do the following: Find a stage 2 module (say, k) whose input to the stage 1 module of $\mathcal{I}_c$ and output to the stage 3 module of $\mathcal{O}_c$ are free, and route the connection c through k. This routing

$C = \{(1,2), (2,4), (4,3), (5,6), (6,10), (8,3), (9,11), (10,8), (11,7), (12,5)\}$

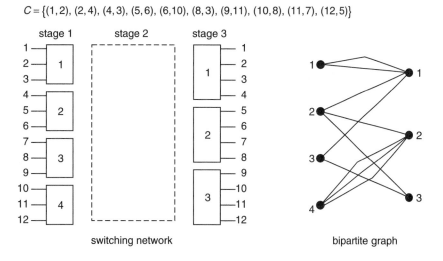

Figure 11.9 An example for converting a connection requirement into a bipartite graph.

problem can be mapped to the following bipartite graph coloring problem. In a graph two edges are said to be adjacent if there is a common vertex between them; edges $e_1 = (v_1^{(1)}, v_2^{(1)})$ and $e_2 = (v_1^{(2)}, v_2^{(2)})$ are adjacent if and only if either $v_1^{(1)} = v_1^{(2)}$ or $v_2^{(1)} = v_2^{(2)}$ or both. Color the edges of the bipartite graph G such that no two adjacent edges have the same color. If the colors are numbered $1, 2, \ldots$, then the color of each edge will be the stage 2 module through which the connection request will be routed. We are now ready to state and prove the following theorem about the p_2 that is required to make a three-stage Clos network RNB.

Theorem 11.4

An $M \times N$ three-stage Clos network defined by the 5-tuple $(m_1, p_1, p_2, p_3, n_3)$ is RNB if and only if $p_2 \geq \max(m_1, n_3)$. This theorem is called the Slepian–Duguid theorem.

Proof: To enable all the inputs of a stage 1 module to be simultaneously active, $n_1 \geq m_1$ is necessary. For all outputs of a stage 3 module to be simultaneously active, $m_3 \geq n_3$ is necessary. Because $n_1 = p_2 = m_3$, the condition is clearly necessary.

From the construction of the bipartite graph, for any connection requirement C, G will have a maximum degree $d = \max(m_1, n_3)$. From graph theory, we know that d colors are sufficient to color the edges of a bipartite graph of maximum degree d. Thus the $p_2 = \max(m_1, n_3)$ modules are sufficient to route any permutation. ■

11.2.3 Construction of Large Switches

So far we have considered only three-stage switches. To keep our discussion simple we consider symmetric $N \times N$ switches with $m_1 = n_3$ and $p_1 = p_3$. Assume p_2 satisfies the condition in Theorem 11.1. Observe that the SNB Clos network has $2p_1(2m_1 - 1)(2m_1 + p_1)$ crosspoints. Because $N = p_1 \times m_1$ can be factorized in many ways, the first question to ask is, what is the best p_1 and m_1? It is easy to see that the smaller the value of m_1, the better it is. However, small m_1 leads to large stage 2 switches. For example, for a 1024×1024 switch, we can choose $p_1 = 512$ and $m_1 = 2$, and we would need three 512×512 switches. Obviously, these three switches could also be built from smaller switches, and this could be done recursively until we use only $m \times n$ switches, where m or n (or both) are prime numbers, and we cannot factorize any further. It can be shown that with recursive construction, the number of crosspoints in a $2^{2^n} \times 2^{2^n}$ SNB switch is $O(6^n N)$.

A number of SNB networks with fewer than $O(6^n N)$ crosspoints have been proposed in the literature. Many of these cannot be easily adapted for use in packet-switched networks, and we do not discuss them further.

Now consider the recursive construction of an $N \times N$ RNB switch. As before, we assume that $p_1 = p_2$ and $m_1 = n_3$. The number of crosspoints in a three-stage RNB network is given by $p_1 m_1 n_1 + p_2 m_2 n_2 + p_3 m_3 n_3$. From Theorem 11.4 we have $p_2 = m_1$, and the number of crosspoints simplifies to $2N(2m_1 + p_1)$. Thus for a given $N = p_1 \times m_1$, with many p_1 and m_1 possible, the least value of m_1 is the best value. If we let $N = 2^n$, we should use $m_1 = n_3 = 2$ and hence $p_1 = p_3 = N/2$. The two modules in stage 2 are $\frac{N}{2} \times \frac{N}{2}$ switches, each of which can be constructed as an RNB switch. Continuing the recursion, we can see that this network will consist of $(2n - 1)$ stages of $\frac{N}{2}$ (2×2) switches and that the number of crosspoints in the switch will be $2N(2 \log_2 N - 1)$, which is $O(N \log N)$. Note that this has the same order as the optimum switch that can be constructed using two-state devices like the 2×2 switches. An RNB switch constructed in this way is called a Benes network. A sample 16×16 Benes network is shown Figure 11.10.

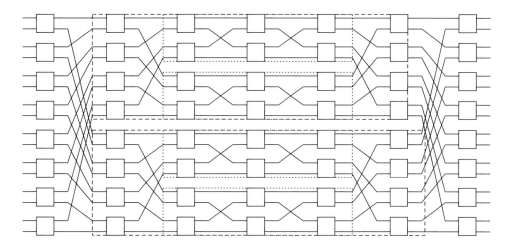

Figure 11.10   A 16 × 16 Benes network with 56 2 × 2 switches. The two dashed boxes are from the first recursion, which requires two 8 × 8 switches. The four dashed boxes are from the second recursion, in which the 8 × 8 switches are built using two 4 × 4 switches.

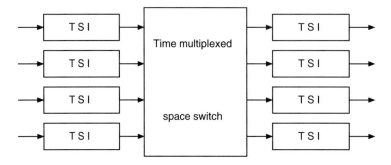

Figure 11.11   The time-space-time switch. The input and output links are time division multiplexed.

In most telephone switches, because the input and output links are time multiplexed, we can replace the input and output stage crossbars by time-slot interchangers and use a space switch only in the middle-stage. If there are M slots in each frame of the input and output links, then the middle-stage space switch will have separate settings for each of the M slots; this space switch is called a time multiplexed space (TMS) switch. Such a combination of TMS and TSI is called a *time-space-time* (TST) switch. Figure 11.11 shows the block diagram of such a switch.

Consider a TST switch with N input and output lines, M slots per frame in the input and output TDM links, and an $N \times N$ TMS switch whose crosspoint settings are changed in every slot. It can be shown that this is equivalent to a Clos network with $m_1 = n_3 = M$, $p_1 = p_3 = N$, and $p_2 = N$.

11.2.4 Setting Up the Paths

Connecting small switches into a larger switching network reduces only the number of crosspoints in the switch. Hence it reduces its hardware complexity but complicates the problem of setting up the paths for the connections. Recall the simple algorithm to set up a path between input I_i and output O_j in a crossbar: Crosspoints $(i, 1)$ to $(i, j-1)$ on row i and $(i+1, j)$ to (N, j) on column j are set to cross state, and crosspoint (i, j) is set to bar state. In a three-stage Clos network, observe that there are p_2 possible paths between I_i and O_j: one through each of the p_2 stage 2 switches. The choice increases with an increasing number of stages in the switching network. In the following we consider only three-stage Clos networks.

Setting up the paths in an SNB is easy because, irrespective of the routing of a set of connections, a free stage 2 switch will be available for a new valid connection. The paths are set up by maintaining a list of free stage 2 switches and arbitrarily choosing a free switch from the list to route a connection. The list is updated by adding to it when a connection is set up and deleting from it when a connection is terminated. Routing a new valid connection in WSNB is similar except that the stage 2 switch must be chosen according to the WSNB routing rule. For example, for the packing rule, the list is ordered according to the number of unused lines in the stage 2 switch. In setting up the path for a connection, the stage 2 switch with the lowest number of unused lines is chosen. Ties are broken arbitrarily. In an RNB switch, setting up the path is more complex, and we look at it in more detail next.

There are two ways to look at the problem of setting up paths in RNB networks.

- As a *rearrangement problem* in which the existing connections are rearranged to accommodate the next connection. The rearrangement algorithm should disturb the lowest number of connections. This is applicable in switches where the connection requests arrive at arbitrary times, such as circuit switches or continuous time packet switches.

- As a *routing problem* in which we assume that the paths for all the connections are to be routed simultaneously. This is applicable to cell switches where in every slot, a new set of cells is available at the inputs.

In the following, we pursue only the routing problem.

To solve the routing problem recall that a set of connection requests C can be represented by a bipartite graph G, with p_1 and p_3 vertices in the two partitions of its vertex set, respectively, representing the input and output stage modules. Each $c \in C$ is represented as an edge. A matching of the bipartite graph will match one input to one output. Thus the connections of a match of the bipartite graph can be routed through the same stage 2 switch. Algorithm 11.1 can be used to route a set of connections C in an RNB network. The bipartite graph representation of C, denoted by G, is the input to the algorithm. The stage 2 switches are numbered $1, \ldots, p_2$. After each matching, an unrouted connection from each of the matched inputs is routed through a new stage 2 switch.

Exercise 11.2

What is the maximum number of iterations of the **while** loop that will be performed?

Input C and G, the bipartite graph representation of C
$color = 1$;
while $(G \neq \phi)$ **do**
   find maximum size matching, denoted by *MSM*, in G
   for (each matched input vertex $v \in MSM$) **do**
     route connection $c \in C$ for which $\mathcal{I}_c = v$ through stage 2 switch numbered *color*
   $C = C \backslash c$
   $G = G - MSM$
   $++ color$

Algorithm 11.1   A routing algorithm for the RNB network.

Another algorithm is available for routing the connections of C. It is called the *looping algorithm*. We describe this algorithm for a three-stage symmetric RNB switch assuming that N is even and $p_2 = 2$, and hence $m_1 = n_3 = 2$. We represent the connections a bit differently, with (i, j) denoting a connection request from an input in module i in stage 1 to an output in module j of stage 3. The algorithm is as follows. Let x and y denote the two $N/2 \times N/2$ switches in stage 2. Route connection $(1, j_1)$ through x. Route the other connection to output module j_1, (i_2, j_1), through y. Route the other connection to input module i_2, (i_2, j_2), through x and continue this procedure. If the procedure stops because the "other" input or output has been routed, start with another free input and continue. Stop when all the connections are routed. Figure 11.12 illustrates the looping algorithm.

The complexity of this routing algorithm is $O(N)$ for an $N \times N$ three-stage switch, and for a Benes network it is $O(N \log N)$.

In high-speed packet switches, the time available for the routing decision is on the order of the packet transmission time—about 500 ns for a 1000-bit cell in a switch that interconnects 2 Gbps links. Note also that the complexity of the looping algorithm to set up the paths is $O(N \log N)$, leading to poor scaling properties. Thus we need to look for other alternatives to guide cells from the inputs to the outputs in high-speed switches.

We look toward parallel computing systems for some solutions. Parallel computing systems interconnect processor banks with memory banks. This is typically done using an *interconnection network*. It turns out that the functional description and performance requirements of cell switches are very similar to those of these interconnection networks. Because there is a well-developed theory

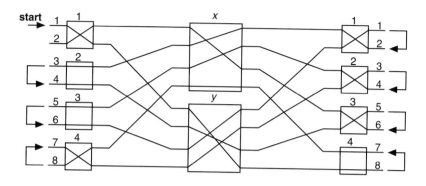

Figure 11.12 Illustration of the looping algorithm in an 8 × 8 RNB Clos network to route the connection requirement {(1,8), (2,6), (3,2), (4,5), (5,4), (6,1), (7,3), (8,7)}, where (I_i, O_j) means that input I_i needs to be connected to output O_j. In terms of the stage 1 switches and stage 3 switches, this translates to {(1,4), (1,3), (2,1), (2,3), (3,2), (3,1), (4,2), (4,1)}.

associated with interconnection networks, we can use it to build large, high-performance packet-switching systems. Specifically, we examine the class of switching networks called self-routing networks and their structural properties.

11.3 Self-Routing Networks

A multistage switching network is a *self-routing network* if the switching modules in the intermediate stages can use source and destination port addresses of the cell to set up the path from the input to the output in a distributed manner. If the paths between all input–output pairs are unique, then the source and destination of a packet uniquely determine the path that the packet will take through the network even in the intermediate stages. In the following we discuss a large class of networks, called *delta networks*, that can be made self-routing.

An $a^n \times b^n$ delta network is built using n stages of $a \times b$ crossbars. The b outputs from each $a \times b$ crossbar of stage k are connected to a separate crossbars of stage $k + 1$. The interconnection pattern between the stages is such that there is a unique path from each input port to each output port. Specifically, a b-ary tree with n levels is constructed from each input stage crossbar, and the leaves of this b-ary tree are the outputs of the switching network. Figure 11.13 shows a sample $3^2 \times 3^2$ delta network and the 3-ary tree rooted at input 1 with leaves corresponding to the nine outputs.

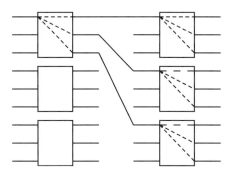

Figure 11.13   A $3^2 \times 3^2$ delta network. The 3-ary tree from input 1 connecting to all the outputs is also shown.

The number of crossbars in the delta network just defined is calculated as follows. There are a^{n-1} crossbars in stage 1, leading to ba^{n-1} output ports from stage 1 and hence ba^{n-2} crossbars in the second stage and so on. Thus the total number of crossbars is

$$\sum_{i=1}^{n} a^{n-i}b^{i-1} = \begin{cases} \dfrac{a^n - b^n}{a - b} & a \neq b \\ na^{n-1} & a = b \end{cases}$$

Before we consider a class of self-routing switching networks, we define the concept of a *shuffle*. An a-shuffle of aR elements is obtained as follows: Divide the aR elements into a groups of R elements each. Starting from the top, pick one element of each group and cycle through the groups. A sample shuffle is shown in Figure 11.14. After an a-shuffle of a list of aR items, an item in position i in the original list will be in position $S(i)$ defined by

$$S(i) = \begin{cases} ai \bmod (aR - 1) & 0 \leq i < aR - 1 \\ aR - 1 & i = aR - 1 \end{cases}$$

$$= (ai + \lfloor i/R \rfloor) \bmod aR$$

This is also illustrated in Figure 11.14.

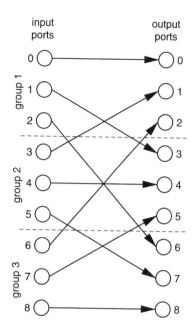

Figure 11.14   Example of a 3-shuffle of nine ports. The three groups are also indicated. The numbers on the arrows from the inputs indicate the sequence in which the port is picked up in the shuffle.

Multistage switching networks in which the outputs from each stage are shuffled and connected to the inputs of the next stage are called *shuffle networks*. We are interested in $a^n \times b^n$ delta networks constructed using a-shuffles as the interstage links. Examples of such shuffle networks are shown in Figure 11.15.

Theorem 11.5

An $a^n \times b^n$ delta network constructed using a-shuffles is self-routing in the following way: Let a cell at an input have $D = (d_{n-1}, \ldots, d_0)_b$ as the destination address of the cell expressed in base b. If the module in stage i switches the cell to output d_{n-i} then the cell will eventually reach the destination.

Proof: We prove the theorem for the special case of $a = b = 2$. Let $S = (s_{n-1}s_{n-2} \ldots s_1 s_0)$ be the source, and let $D = (d_{n-1}d_{n-2} \ldots d_1 d_0)$ be the destination of a cell. We shuffle the inputs before connecting them to the switch. Note that this only relabels the inputs. See Figure 11.16 for an illustration of this theorem.

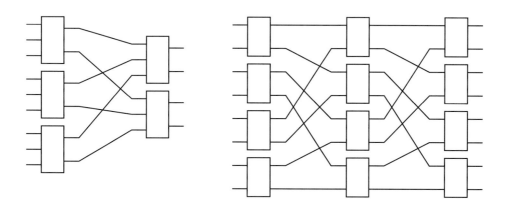

Figure 11.15 Examples of shuffle networks. Shown are a $3^2 \times 2^2$ shuffle switch and a $2^3 \times 2^3$ shuffle switch.

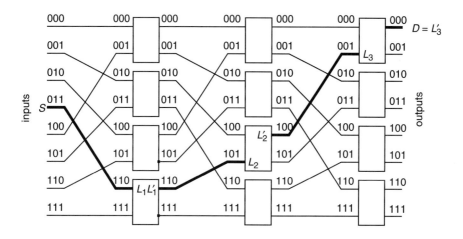

Figure 11.16 Illustration of the proof of Theorem 11.5. The route from $S = 011$ to $D = 000$ is shown.

Input S is connected to link L_1 given by

$$L_1 = \left(S \cdot 2 + \lfloor S/2^{n-1}\rfloor\right) \bmod (2^n)$$

$$= (s_{n-2}s_{n-3}\ldots s_1 s_0 s_{n-1})$$

Link L_1 is connected to module $\lfloor L_1/2 \rfloor$, and according to the self-routing rule, it is switched to line number $L_1' = \lfloor L_1/2 \rfloor 2 + d_{n-1} = (s_{n-2}s_{n-3}, \ldots, s_0 d_{n-1})$ in the output of stage 1. From the definition of the 2-shuffle, the line number of the input to stage 2 that connects to L_1', L_2, is given by

$$L_2 = \left(L_1' \cdot 2 + \lfloor L_1'/2^{n-1} \rfloor\right) \bmod (2^n)$$

$$= \left(L_1' \cdot 2 + \lfloor (s_{n-2}s_{n-3}, \ldots, s_1 s_0 d_{n-1})/2^{n-1} \rfloor\right) \bmod (2^n)$$

$$= (L_1 \cdot 2 + s_{n-2}) \bmod (2^n)$$

$$= (s_{n-2}s_{n-3} \ldots s_1 s_0 d_{n-1} s_{n-2}) \bmod (2^n)$$

$$= (s_{n-3} \ldots s_1 s_0 d_{n-1} s_{n-2})$$

Proceeding along similar lines, we obtain

$$L_k = (s_{n-k-1} \ldots s_1 s_0 d_{n-1} \ldots d_{n-(k-1)} s_{n-k})$$

$$L_k' = (s_{n-k-1} \ldots s_1 s_0 d_{n-1} \ldots d_{n-k})$$

and continuing up to n, we get, L_n', the output from stage n, to be

$$L_n' = (d_{n-1}d_{n-2} \ldots d_1 d_0)$$

as we would like.   ∎

The self-routing shuffle network is not nonblocking. We can see this by looking at the trees from the source ports to the destination ports and constructing a counter example in which two trees overlap. However, there is a condition on the destination set that can make the shuffle network nonblocking. We now explore this condition.

Let D_i be the destination for the cell at input i. The destination addresses are said to be monotonic if $D_1 < D_2 < \cdots D_k$. The active inputs are said to be compact if all the inputs between two active inputs are also active (i.e., all the active inputs are grouped together).

Theorem 11.6

If the active inputs are compact and the destinations of the cells are monotonic, then the self-routing shuffle network is nonblocking.

Proof: Proof is by contradiction. Consider two active inputs S and $\tilde{S}$ with cells having destinations D and $\tilde{D}$, respectively. Without loss of generality, let $S < \tilde{S}$.

From the condition that the inputs are compact and that the destinations of the cells at these inputs are monotonic, we have

$$\tilde{D} - D \geq \tilde{S} - S \tag{11.1}$$

This is true because the inputs increase by 1 (by the compactness condition), whereas the destinations of the cells at these inputs increase by *at least* 1 (from the monotonicity condition).

Assume that the switch is blocking and that the two cells at inputs $S = (s_{n-1} \ldots s_1 s_0)$ and $\tilde{S} = (\tilde{s}_{n-1} \ldots \tilde{s}_1 \tilde{s}_0)$ with destinations $D = (d_{n-1} \ldots d_1 d_0)$ and $\tilde{D} = (\tilde{d}_{n-1} \ldots \tilde{d}_1 \tilde{d}_0)$, respectively, cannot be routed simultaneously to their respective destinations. This means that the output link in some stage—say, stage k—should be the same for both of these cells. From the proof of Theorem 11.5, this means that

$$(s_{n-k-1} \ldots s_1 s_0 d_{n-1} \ldots d_{n-k}) = (\tilde{s}_{n-k-1} \ldots \tilde{s}_1 \tilde{s}_0 \tilde{d}_{n-1} \ldots \tilde{d}_{n-k})$$

We can write

$$\tilde{D} - D = (\tilde{d}_{n-1} \ldots \tilde{d}_{n-k} \ldots \tilde{d}_1 \tilde{d}_0) - (d_{n-1} \ldots d_{n-k} \ldots d_1 d_0) < 2^{n-k}$$

$$\tilde{S} - S = (\tilde{s}_{n-1} \ldots \tilde{s}_{n-k} s_{n-k-1} \ldots \tilde{s}_1 \tilde{s}_0) - (s_{n-1} \ldots s_{n-k} s_{n-k-1} \ldots s_1 s_0) \geq 2^{n-k} \tag{11.2}$$

We can write the first inequality by realizing that D and $\tilde{D}$ are different only in the least significant $n - k$ bits. We can write the second inequality by observing that S and $\tilde{S}$ differ in the most significant $n - k$ bits. Equations 11.2 imply that $\tilde{D} - D > \tilde{S} - S$. This contradicts inequality (11.1), and hence the assumption that the two paths have a common link in stage k does not hold.   ∎

From Theorem 11.6, the self-routing shuffle network, if preceded by a sorter and a compactor (which sort the cells according to their destination addresses and compact the active cells), can become a high-speed cell switch. This means that we need to look for mechanisms to sort and compact at high speeds. Assume that the destination addresses can take integer values in $[0, N - 1]$. Then, if the inactive inputs are assigned dummy cells with dummy destinations greater than $N - 1$, the output of the sorter will also be compact. Thus, from Theorem 11.6, the shuffle network can be made useful in high-speed packet switches if a high-speed sorter, preferably a hardware structure, can be constructed to be used along with the shuffle network. We discuss high-speed sorting next. Specifically, we look at a class of hardware sorters called sorting networks.

Sorting Networks

A *sorting network* is an interconnected network of two input comparators that sort the two inputs. Most well-known sorting algorithms can be represented as sorting networks. Consider the elementary *bubble sort* algorithm. For an N element array A, the following bubble sort algorithm sorts A in descending order and stores it in A.

> **for** $i = 1$ to $N - 1$ **do**
>   **for** $j = 1$ to $N - 1$ **do**
>   **if** $A[j] < A[j + 1]$ **then**
>   Swap $A[j]$ and $A[j + 1]$

The sorting network and its symbolic version for $N = 4$ are shown in Figure 11.17. The number of comparators that are required in this network is

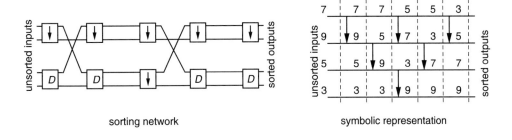

Figure 11.17   Bubble sort network using two input comparators and delay elements. The direction of the arrow indicates the direction of the higher of the two inputs. The D in the box indicates that it is a delay element. The symbolic representation is shown on the right, along with an example.

$N(N - 1)/2$, the same as the number of comparisons in the algorithm. The time delay, corresponding to the number of stages, is $2N - 3$. Although the effect of the $O(N)$ time delay can be absorbed by pipelining (explained in Section 12.5), the number of comparators is $O(N^2)$, and this sorting network can become unviable for large N. Also, it can be shown that an optimal sorting algorithm uses $O(N \log N)$ comparators.

Instead of the bubble sort algorithm, a class of algorithms called *merge-sort* algorithms can be used to sort the cells. The idea of these algorithms is fairly simple: Divide the original list into sublists, sort each sublist separately, and merge the sorted sublists to obtain the sorted version of the original list. This procedure also can be applied recursively to the sublists. Most of the comparisons can be made in parallel, reducing the number of stages significantly. The number of comparators will also be of a lower order. We investigate one such sorting network.

A sorting network is best illustrated using the so-called 0–1 principle, which is stated next without proof.

Theorem 11.7

A sorting network for N elements that can sort all possible 2^N sequences of 0's and 1's can sort all sequences of arbitrary values.   ∎

The 0–1 principle allows us to describe the sorting algorithm using 1-bit inputs to the sorters. Next, we introduce more terminology.

Definition 11.2

A sequence $d_{N-1}, \ldots, d_1, d_0$ with $d_i \in \{0, 1\}$ is called a *bitonic sequence* if there are at most two changes in the values of the sequence; that is, the sequence is such that $d_i = 0$ (or 1) for $i = 0, \ldots K - 1$ and $N - M, \ldots N - 1$, while $d_i = 1$ (or 0) for $i = K, \ldots N - M - 1$.   ∎

Consider a bitonic sequence with N elements (N being an even number) that is input to a sorting network of $N/2$ comparators. The input to comparator p are elements $p - 1$ and $N/2 + (p - 1)$ of the bitonic sequence. This network, called a *half cleaner*, is shown in Figure 11.18.

Lemma 11.1

At the output of the half cleaner, the top half and the bottom half are both bitonic, and every element of the bottom half is greater than every element of the top half.

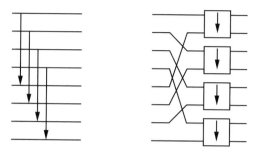

Figure 11.18 The half cleaner in a bitonic sorter.

Input to Half Cleaner	Output of Half Cleaner	
Top Half 0 . . . . .0 0 Bottom Half 0 . . .1 . . . 1 . . .0 . . . 0	0 . . . . .0 0 0 . . .1 . . .1 . . .0 . . . 0	Case 1
Top Half 0 . . . . . 0 1. . . . . . . 1 1 Bottom Half 1 . . .1 0 0	0 . . . . . 0 0. . . . . . . 0 0 1 . . .1 0 . 1 1	Case 2
Top Half 0 . . . . . 0 1. . . . . . . 1 1 Bottom Half 1 . . . . 1 0. . . .0	0 . . . . . 0 1. . .1 0 . . . 0 1 . . . . .1 1. . . .1	Case 3
Top Half 0 . . .1 . . . 1 . . .0 . . . 0 Bottom Half 0 . . . . .0 0	0 . . . . .0 0 0 . . .1 . . . 1 . . .0 . . . 0	Case 4

Figure 11.19 Proof of Lemma 11.1.

Proof: To prove this, consider a bitonic sequence starting with 0's, such as the sequence $d_i = 0$ for $i = 0, \ldots K - 1$ and $N - M, \ldots N - 1$, while $d_i = 1$ for $i = K, \ldots N - M - 1$. We need to consider the following four cases: (1) $K \geq \frac{N}{2}$, (2) $M < \frac{N}{2}$, $K < \frac{N}{2}$, and $M + K \geq \frac{N}{2}$, (3) $K < \frac{N}{2}$ and $M + K \leq \frac{N}{2}$, and (4) $M \geq \frac{N}{2}$. The output of the half cleaner is shown for each of these cases in Figure 11.19. ∎

This lemma means that we can use an $\frac{N}{2}$ bitonic sorter for each half of the output. Because all elements of the top half are lower than those of the bottom half, the two sorted lists can be concatenated. So how do we build the $\frac{N}{2}$ bitonic sorter? We start by having a half cleaner for each half! Continuing this recursively, we get the network of Figure 11.20(a). So far we have been able to sort only a bitonic sequence and not an arbitrary sequence, and thus the problem has been reduced to creating a bitonic sequence from an arbitrary sequence. That is easy. We divide the input sequence into two halves, sort the upper half in ascending order and the lower half in descending order, and concatenate the two sorted sequences. Of course, this can also be done recursively. The resulting network is shown in Figure 11.20(b). Such a sorting network is also called a Batcher sorting network after the person who invented it.

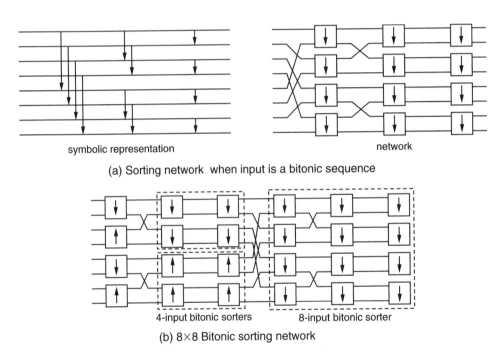

symbolic representation   network

(a) Sorting network when input is a bitonic sequence

4-input bitonic sorters   8-input bitonic sorter

(b) 8×8 Bitonic sorting network

Figure 11.20   (a) The sorting network when the input is a bitonic sequence; (b) an 8 × 8 bitonic sorter. The outputs of the two input comparators are concatenated to form two four-element bitonic sequences and are input to four-input bitonic sorters. The outputs from these are concatenated to form an eight-element bitonic sequence and input to the eight-input bitonic sorter.

Exercise 11.3

How many comparators will a $2^n \times 2^n$ bitonic sorter have?

We have just constructed a self-routing switch using a bitonic sorter (to sort and compact the inputs) and a self-routing shuffle network (to guide the cells from the inputs to the outputs). Figure 11.21 shows a block diagram of the final switch. We are, of course, assuming that there is no output conflict. If output conflict were to be allowed, we would need to trap all but one from every set of conflicting cells and submit them again to the switch in a subsequent slot. A large number of algorithms have been proposed for this and we do not pursue them here.

It turns out that instead of the shuffle for the interconnection pattern between the stages, many other interconnection patterns make the delta switching network self-routing. Examples are omega networks and banyan networks. The interconnection pattern in banyan networks is identical to the interconnection pattern in the right half of a Benes network. In other words, stages n to $(2n - 1)$ of a Benes network can be treated as a self-routing network. If this is the case, it is natural to ask whether we could have made the Benes network self-routing. The following theorem says that we could not have.

Theorem 11.8

An $N \times N$ Benes network with even N and $N \geq 8$ cannot be made self-routing. ∎

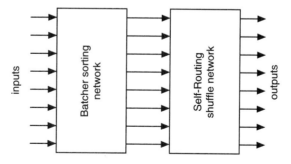

Figure 11.21 The Batcher shuffle switch.

11.4 Multicast Packet Switches

So far we have focused on building unicast switches. We now explore self-routing architectures for multicast switches. We consider cell switches.

In a multicast switch, the set of connection requests C is such that the $\mathcal{I}_c$ for $c \in C$ need not be unique. Let $d_i = \{\mathcal{O}_c; c \in C \text{ and } \mathcal{I}_c = I_i\}$; that is, d_i is the set of destinations to which a cell from input I_i is to be switched. Let δ_i be the cardinality of the set d_i. One way to construct a multicast switch is to make δ_i copies of the cell at input i, assign one destination from the set d_i to each of the cells, and then offer the copies to a unicast switch. This arrangement is shown in Figure 11.22. We thus reduce the problem of designing a multicast switch to that of designing a copy network.

We construct a *copy network* by generalizing the self-routing delta network of the preceding section. Consider an N output copy network with $N = 2^n$. This means that in one slot the sum of the copies of the cells at the inputs can be at most N. In a copy network, the only requirement is that δ_i copies of the cell at input i emerge from the output, and it does not matter which output ports they

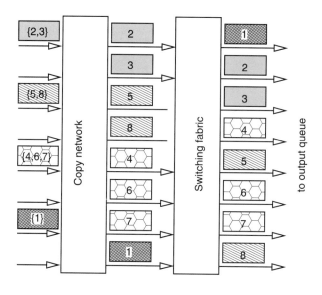

Figure 11.22 Multicast switch using a copy network and a unicast switching network. Numbers on the cells at the input indicate the destination set, and those at the output of the copy network indicate the destination of each of the copies.

emerge from. The copy network architecture is called a *broadcast* network and has a structure identical to those of the self-routing networks in the preceding section. The only difference is that the 2×2 switching elements can copy a cell on an input onto both its output ports and also modify some header values of the cell.

Each cell at the input of the broadcast network is encoded with the interval $[L, U]$, indicating that the copies of the cell should appear at outputs $L, \ldots, U$. Let $L(k)$ and $U(k)$ be the intervals of the cell at the input of the element in stage k. Let $L(k) = [l_{n-1}, \ldots, l_0]$ and $U(k) = [u_{n-1}, \ldots, u_0]$, where l_r and u_r are the $(n - r)$th bits of $L(k)$ and $U(k)$, respectively. The switching element of stage k copies the cell onto one or both of the output ports and modifies the interval of the output cells according to the following *Boolean interval splitting algorithm* on the cell.

> if $l_{n-k} = u_{n-k} = 0$ then
>> the cell is switched to output link 0
>
> if $l_{n-k} = u_{n-k} = 1$ then
>> the cell is switched to output link 1
>
> if $l_{n-k} \neq u_{n-k}$ then
>> the cell is switched to **both** links
>> **for** (the cell switched to link 0) **do**
>>> $L(k + 1) = L(k)$
>>> $U(k + 1) = u_{n-1}, \ldots, u_{n-(k-1)} 01 \ldots 1$
>> **for** (the cell switched to link 0) **do**
>>> $L(k + 1) = l_1, \ldots l_{n-(k-1)} 10 \ldots 0$
>>> $U(k + 1) = U(k)$

Figure 11.23 shows a sample application of this algorithm.

The broadcast network has the following nonblocking property.

Theorem 11.9

Let $x_1, \ldots, x_k$ be the active inputs and Y_i be the set of destinations for the cell at input x_i. Let $Y_1 < Y_2 < \cdots < Y_k$; that is, every output address for the cell at input i is greater than every output address for the cell at input $i+1$, for $i = 1, \ldots, k-1$. If x_i are compact, then the broadcast network is nonblocking.

Proof: The argument is exactly as in the proof of the self-routing property of the shuffle network in Theorem 11.5. For any combination of $(x_1, y_1), \ldots (x_k, y_k)$, where $y_i \in Y_i$, the links from x_i to y_i are nonoverlapping, and all of them can be routed. The copies of the cell at x_i are being guided to the outputs in Y_i along a tree rooted at x_i. Hence the tree from x_i to Y_i does not overlap with any other tree from the set above, and cells can be self-routed. ∎

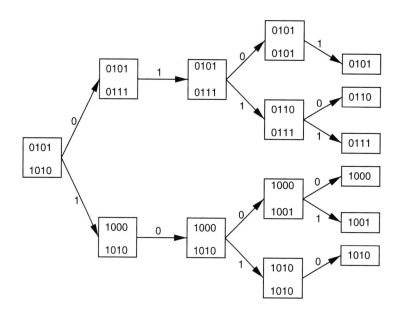

Figure 11.23 Illustration of the Boolean interval splitting algorithm. The cell at the input is to be copied onto intervals 0101–1010 in four stages. The passage of this cell and the copies in the intermediate stages are illustrated.

We can use this theorem to construct a copy network as follows. Without loss of generality, we assume that inputs $1, \ldots, k$ are active and that input i requires δ_i copies. Define

$$
L_i = \begin{cases} 0 & \text{for } i = 1 \\ \sum_{j=1}^{i-1} \delta_i & \text{for } i = 2, \ldots, k \end{cases}
\qquad
U_i = \begin{cases} \delta_i - 1 & \text{for } i = 1 \\ \sum_{j=1}^{i} \delta_i - 1 & \text{for } i = 2, \ldots, k \end{cases}
$$

If the packet at input i is assigned the range (L_i, U_i), then the appropriate number of copies will emerge at the output of the copy network. Note that L_i and U_i can be obtained using a *running adder* network. An N-input running adder network takes δ_i at input i and produces L_i at output i. The copies can then be input to a unicast switch as shown in Figure 11.22, and the multicast is complete. The details of all the components are shown in Figure 11.24.

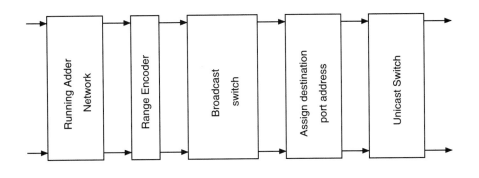

Figure 11.24 The detailed block diagram of the multicast switch.

11.5 Summary

In this chapter we give an overview of shared-medium, shared-memory, and crossbar switches. Shared-medium and shared-memory switches have scaling problems in terms of the speed of data transfer, whereas the number of crosspoints in a crossbar scales as N^2 compared with the optimum of $O(N \log N)$. We then interconnect a number of smaller switches to construct large switches, specifically those that are constructed as Clos networks. Although all the elementary switches are nonblocking, the switching networks can be blocking. We examine switching networks with different degrees of nonblocking: strictly nonblocking, wide sense nonblocking, and rearrangeably nonblocking networks. We derive the properties of Clos networks that have these nonblocking properties. We then use these properties to construct large switching networks, specifically a Benes network. A Benes network reduces the number of crosspoints but requires a complex routing algorithm to set up the paths for a set of connection requests. Thus a Benes network can become infeasible at high speeds.

We also explore self-routing delta networks, in which the smaller switches use the output port address of a cell to set the switch crosspoint to route the packet. Although a self-routing network is a blocking network, it can be made nonblocking by arranging the cells in the inputs in increasing (or decreasing) order of their destination addresses and making the active inputs compact. The input cells can be so arranged by using a sorting network. We examine the class of bitonic sorters and the Batcher sorting network. A sorting network and a self-routing delta network can be combined to build a high-speed nonblocking switch.

Finally we present an extension of the delta network to construct a copy network that is used along with a unicast switch to construct a multicast switch.

11.6 Notes on the Literature

The theory of interconnecting networks has been studied for quite some time. Benes [26] is an excellent exposition on some of the early work. Much of this work was carried out in the context of telephone switches. Advances in parallel computing created the need for high-speed interconnects between memory and processor systems and rekindled interest in these networks and their properties. High-speed packet switches have caused the latest surge of interest in these networks. Hwang [145] is an excellent source for the properties of nonblocking networks. Hui [143] provides a more recent treatment of the material, with applications to packet switches. Utpal Mukherjee provided the counterexample to the necessary condition of Clos theorem. Ahmadi and Denzel [6] and Tobagi [284] provide surveys of the major cell switching architectures proposed in the literature. Turner and Yamanaka [288] provide a survey of some of the more recent developments in switching architectures.

See [26, 145, 143] for a description of the early literature on switching networks. Theorem 11.2 is from [145], and Theorem 11.3 is from [26]. Self-routing delta networks were first proposed by Patel [234], where a general version of Theorem 11.5 is proved. Theorem 11.6 is from Lee [192], who credits it to W. H. Bechmann. Theorem 11.8 is by Douglass and Oruc [87]. Knuth [177] is the best source for sorting algorithms, sorting networks, and their history. The copy broadcast and copy networks were described by Lee [192].

We have not discussed switch-blocking probabilities. Hui [143] is a good reference for the simpler blocking models. Zegura [307] provides a more sophisticated blocking analysis. We also have not discussed the effect of buffering in the stages. Early analysis of such a switching network has been done by Dias and Jump [81] and by Jenq [153]. Many of the concepts described in this chapter are derived from graph theory. There are many excellent textbooks on the subject, such as West [298].

We also have not explored the many architectures that were built as prototypes for fast packet switches. Most of them are covered in the surveys mentioned here.

Problems

11.1 Consider a Clos network with $p_3 = 1$. Show that $p_2 = N$ is a necessary and sufficient condition for the network to be SNB.

11.2 An N-input, N-output strictly nonblocking switch is to be realized using recursive three-stage factorization with parameters $m_1 = n_3 = 3$ and $p_2 = 5$, as a network of 3×5, 3×3, and 5×3 crossbar switches. Assume that $N = 3^n$, where $n \geq 2$ is an integer. Obtain a recursive expression for the total number $f(n)$ of crosspoints in the network and solve for $f(n)$ as a function of n.

11.3 Consider an $N \times N$ switch with $N = 2^{2^l}$. Let $m_1 = p_1 = \sqrt{N} = n_3 = p_3$. For the switch to be SNB $p_2 \geq 2\sqrt{N} - 1 \approx 2\sqrt{N}$ and there are $\sqrt{N}$ $\sqrt{N} \times 2\sqrt{N}$ switches at the input and $2\sqrt{N} \times \sqrt{N}$ at the output. Each of these switches can be treated as equivalent to two $\sqrt{N} \times \sqrt{N}$ switches. Each of the $\sqrt{N} \times \sqrt{N}$ switches can be further constructed as an SNB switch using smaller switches. Continue this recursion, and show that the number of crosspoints in the recursively constructed switch approximately $4N \times 6^l$.

11.4 How many paths are there between any input–output pair in an $N \times N$ three-stage Clos network constructed from r_1 $p \times q$ crossbars in the first stage, r_2 $q \times q$ crossbars in the second stage, and r_3 $q \times p$ crossbars in the third stage? How many paths would there be between an input–output pair in an $N \times N$ Benes network? Assume $N = 2^n$.

11.5 Argue that a TST switch with N input links and M slots in the input and output frame and a TMS switch whose crosspoints are changed P times in a frame is equivalent to a Clos network with $m_1 = n_3 = M$, $p_1 = p_3 = N$, and $p_2 = P$. Map the slots to the input links and the TSI to the input and output stages.

11.6 Using the looping algorithm, argue that in an RNB Clos network with $m_1 = 2$, the first switch in the input stage is not necessary, (i.e., inputs 1 and 2 can be connected directly to the corresponding inputs of the stage 2 switches). Using this result, find the minimum number of 2×2 switches in the recursively constructed $2^n \times 2^n$ RNB switching network.

11.7 A 16×16 self-routing cell switch must be made from 4×4 self-routing crossbars. An appropriate shuffle exchange interconnection must be used between the stages.

a. Show the architecture of the switch; show clearly the indexing of the switch inputs and outputs.

b. Show that your switch is self-routing.

11.8   Consider the connection a three-stage 8×8 switch obtained by connecting 2×2 switches as in stages 1 to 3 of a Benes network. Verify that this interconnection pattern results in a self-routing network. Repeat for a network connected as in stages 3 to 5 of a Benes network.

11.9   Packets arrive according to a Bernoulli process of rate λ to an $M \times N$ copy network. K copies of each active cell are to be made. There are no buffers at the input, and if more than N copies are to be made in a slot, the input cells causing the excess copies are lost. Assuming N to be a multiple of K, find the probability that a cell is lost.

11.10   Let p_k be the probability that a packet at an input requests k copies in a slot, with p_0 being the probability that the input is inactive. In our design of the multicast switch, input i has priority over input j if $i < j$. Assume that in a slot if all the copies requested by an input cell cannot be made, the cell is dropped. Write the expression for L_i, the probability that a cell at input i is dropped. Obtain the Chernoff bound on this probability.

CHAPTER 12

Packet Processing

I n Chapter 11 we consider design choices in the architecture of high-speed switching fabrics, with an emphasis on packet switches. As discussed in Chapters 2 and 9, in packet-multiplexed networks, a considerable amount of processing must be performed on each packet on each link that the packet traverses. As the transmission rates on the links increase, these functions must be performed at increasingly higher speeds. The most important packet-processing function is address lookup, or determining the next hop for the packet. Developments in the addressing mechanisms in IP networks have made the address lookup problem quite complex, and efficient solutions are required for use in high-speed packet switches.

After a preliminary discussion on addressing in networks in general, we discuss addressing in IP networks and efficient address lookup schemes. Increasingly, packets are being classified according to a small set of rules for differential treatment that can include packet filtering, routing, and so on. We discuss some high-speed packet classification algorithms. Other issues in packet switches include high-speed packet scheduling, performing the slower (than packet) timescale functions (such as routing updates), signaling and control protocol execution, management of the network and the switch, and so on.

12.1 Addressing and Address Lookup

12.1.1 Addressing

In most networks, node addresses are hierarchical. Most postal addresses are clearly hierarchical. Telephone numbers are hierarchical, with different parts of the number indicating the calling area, the local exchange to which the phone is connected, and the phone number in the exchange. Similarly, node addresses in the IP network are hierarchical, with the 32-bit address divided into the *network address* part, often called the *network prefix*, and the *node address* part. In an ATM network, the nodes have 20-byte *network service access point* (NSAP) addresses. Although there are different formats for the addresses, all of them have a hierarchical structure.

The node address in a network can be of fixed length or of variable length. Furthermore, inside the address the boundaries separating the hierarchies can be fixed or variable. For example, consider addressing in the telephone network, where the address format is called the *numbering plan*. Although many countries, including those in North America, use fixed-length telephone numbers with a fixed number of digits for each hierarchy, the ITU standard allows for variable-length city codes. For example, in India, the city code can be two to four digits, and the phone number can be five to eight digits. The boundaries are determined by the starting digits in the number. The IP network uses 32-bit fixed-length addresses, but the boundaries separating the network address and the host address are variable. We discuss the IP addressing scheme in detail later in this section. In the ATM network, the first 13 bytes of the 20-byte NSAP address are used identify the location of a specific switch in the network; the next 6 bytes identify the physical end point, and 1 byte is used to identify the logical connection end point on the physical end point. Different NSAP formats have different structures for the first 13 bytes.

In a circuit-multiplexed network, when a connection request arrives, the addresses of the source and the destination determine the path that will be set up for the call. This is a reasonably slow-timescale function. Also, the addresses in most circuit-multiplexed networks are well structured, and finding the path that is to be set up for the connection request is not a very compute-intensive function. In the rest of this section we consider the address lookup problem in ATM and IP networks. The problem is more complex in the IP network, and we discuss some of the design solutions that have been proposed to speed up address lookup in the IP network.

We have seen in Chapter 2 that in an ATM network, the end points set up a connection before actual data transfer. The setting up of the path for the connection is a routing function, which takes into account the resource requirements of the connection and also the current usage in the network. After the path is set up, the switches on the path and the input and output ports on them are determined. All cells belonging to this connection travel along the same physical path and are handled by the same ATM switches for the duration of that connection. Hence the cells of the connection can be identified by a connection identifier rather than the source and destination addresses. In addition, because the node addresses are 20 bytes, it would cause significant overheads if the cells had to carry node addresses. Also recall from Chapter 2 that on each link the connection identifier has two parts: the virtual path identifier (VPI) and the virtual circuit identifier (VCI). The VCI and VPI of a connection may be different on different links on the path. A different set of VCIs and VPIs may be used for traffic in the reverse direction. Thus for each active connection, the switch must maintain information about the VCI and VPI of the incoming cell, the port on which the cell must be output, the VCI of the outgoing cell, and control information indicating the QoS that the cell will receive on the output link. In an ATM network, cells can be switched using the combination of their VCI and VPI or only their VPI. There is no difference in the cell-switching function in the two cases, and we use the term VCI to refer to either the VPI–VCI combination or VPI.

Recall from Chapter 2 that at each switch the forwarding table essentially maps the VCI of an incoming cell on input i to the output port and outgoing VCI for the cell (see Figure 2.33 for a sample map). Thus the route lookup function is essentially a search function through the forwarding table, with the input port and the VCI of the cell as the search keys. Recall that ATM cells are 53 bytes long. At a transmission rate of 2 Gbps, the minimum cell interarrival time is 212 ns, and the access times for the memory that stores the table should be at most this value. When the transmission rate increases to 10 Gbps, the access time requirement decreases to 42.4 ns. Clearly, we cannot do a software-based search through the table because that would require multiple memory accesses.

A simple solution is to use a separate forwarding table for each input port and implement it as a *flat table*, with the VCI of the input cell as the index into the table. The entry in row InVCI in the table contains OutPort (the output port for the cell), OutVCI (the VCI for the cell on the outgoing link), and other control information that the switch can use internally to provide service grades to the cell. This is illustrated in Figure 12.1. There are 24 bits in the VCI, and 16 million words of high-speed memory is expensive. One way to avoid this large size is to limit the number of active connections from each port. For example, we could say that an input will support at most 4096 connections at any time. During connection setup, the switch allocates VCIs with only 12 bits that are unique, and these bits can be used as the address into the table.

An alternative solution is to use *content-addressable memory* (CAM). In random access memory, the user supplies an address and gets back the data that is stored in that address. In content-addressable memory, the user supplies the data, and the address where the data is stored is returned by the CAM in one access time. To do this, each location of the CAM is divided into two parts: a *search field* and a *return field*. It may also contain a *tag* field that contains control information about the location, such as the validity of the entry. Three functions can be performed on a CAM: (1) write into a free location when both the search and the return fields can be modified, (2) search for a matching pattern, where the entire CAM is searched in one step for the location whose search field matches the input pattern, and (3) read the contents of the return field. Figure 12.2 shows the organization and use of a CAM. The input port and the VCI can be stored in the search field, and the

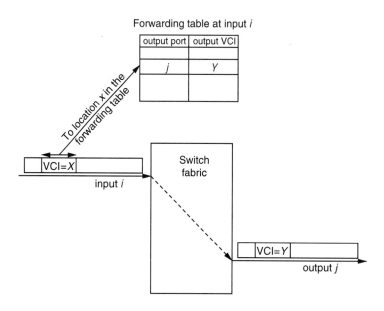

Figure 12.1   The route lookup and cell forwarding in an ATM switch. A cell on input _i_ with VCI _X_ uses this information to index into the forwarding table. From the forwarding table the cell is to be output on port _j_ with VCI _Y_.

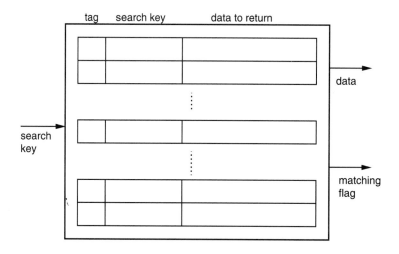

Figure 12.2   Structure and operation of CAM.

outgoing VCI, the output port, and the control can be stored in the return field of the CAM. Because wide CAMs with a large number of bits in the return field are expensive, the return field can supply an index into a lookup table in RAM that contains the outgoing VCI, the output port, and the control information.

In IP networks, datagram routing is used, and each packet should contain the full address of the destination. Also, because hop-by-hop routing is used, the forwarding table of the router contains the address of the next hop node on the path to each destination in the network. Clearly, maintaining 2^{32} entries is infeasible, and searching through them for the next hop of each arriving packet is even more impractical. A simplification is achieved as follows. If the destination host is in the same network as the router, the next hop to the destination host is stored. If the destination host is on a different network, then the next hop to that network is stored in the forwarding table. When a packet arrives, its destination address is extracted, and from its destination address the network address is extracted. Determining the network address of the packet is a complex task.

In the original specification of addresses in IP networks, the following three classes of addresses were identified to enable the router to determine the network address easily.

- *Class A* addresses have 0 in the most significant bit. The next 7 most significant bits form the network address, and the least significant 24 bits form the node address.

- *Class B* addresses have 10 in the 2 most significant bits. The next 14 most significant bits form the network address, and the least significant 16 bits form the node address.

- *Class C* addresses have 110 in the 3 most significant bits. The next 21 most significant bits form the network address, and the 8 least significant bits form the node address.

With this address format, a simple route lookup scheme such as that in an ATM switch described earlier can be used. The forwarding table can be stored as a flat table, with each row containing the next hop to the destination network or the destination host. We determine the class of the destination address from the first bits and then extract the destination network address. If the destination network is not directly connected to the router, we use the destination network address prefix to index the forwarding table entry to get the next-hop information; otherwise, we use the destination host address. Different tables can be maintained for different classes of destination networks and also for each of the networks to which the router is directly connected. This will reduce the size of the table.

Exercise 12.1

a. In the foregoing classification, how many networks are possible?

b. Consider a router in a class B network. What is maximum number of forwarding table entries that this router can have?

12.1.2 Addressing in IP Networks: Subnets and Classless Interdomain Routing

In the foregoing "classful" address specification, IP networks can have either 2^8 nodes, 2^{16} nodes, or 2^{24} nodes. This is too coarse a granularity to specify the network size. Also, for class A and class B networks, the networks are too big, and the forwarding tables contain a large number of entries for destination hosts. This has led to the creation of smaller *subnetworks* out of the larger networks. Thus each router in the large network must know the next hop to the subnetwork to which the destination belongs rather than the next hop to the destination itself. For this reason, routes are aggregated. A *subnet mask* specifies the bits that denote the network address prefix. The destination subnetwork address prefix is obtained using a logical AND of the subnet mask and the 32-bit destination IP address. These subnets can be further subdivided into smaller networks, and subnet masks defined for them too. This division of a class A or a class B network into smaller networks is called *subnetting*.

In a reverse problem, for many organizations a class A or a class B network would be too big, whereas a class C network would be too small. Furthermore, the number of possible class A and class B networks is very small and cannot be freely assigned. Thus it is necessary to group multiple class C networks and assign them to the same organization. This combining of multiple class C networks into one network is called *supernetting*. For example, an organization might be given the addresses of eight contiguous class C networks, allowing its network to have up to 2048 hosts. In routers outside this network, the routing information for this organization might be maintained as eight separate entries corresponding to the eight different class C networks assigned to the organization. Usually, all eight will point to the same next hop, and eight separate entries is clearly a waste of memory resources. This waste becomes especially critical at high speeds, where large, high-speed memories are expensive. Furthermore, as we discussed earlier, searching a table with a large number of entries is time-consuming and is best avoided. Using the same idea as that with subnetting, for each network, a *netmask* is used to indicate the number of bits in the network address prefix. In most real networks we can assume that the 1's in the mask are contiguous (i.e., the forwarding table contains the prefixes of the destination addresses). Because network addresses or network prefixes can have any number of bits, this addressing scheme is called *classless interdomain routing* (CIDR).

Subnetting and CIDR can be illustrated as follows. We use the dotted decimal notation to refer to the 32-bit IP addresses. (In this notation, every group of eight bits is denoted by a decimal number). Consider the class B network 128.16.0.0. It can be divided into four subnets with addresses 128.16.0.0, 128.16.64.0, 128.16.128.0, and 128.16.192.0. The corresponding subnet masks for all of these will be 255.255.192.00. To illustrate CIDR, consider the group of eight class C networks 203.16.64.0–203.16.71.0. They can be grouped into one network with address prefix 203.16.64.0 with 21 bits in the network address prefix. The netmask will be 255.255.248.0.

From the earlier discussion, in an IP network each forwarding table entry consists of a network address prefix, the netmask that should be used to obtain this network address prefix, and the next hop for packets that match the address. Thus the algorithm for each packet is fairly straightforward: Perform a logical AND of the netmask and the 32-bit destination IP address in the packet. If the result of the AND operation matches the network prefix in the forwarding table entry, then the next hop for the packet is the corresponding entry in the table. Thus, route lookup is essentially a search problem, where the search key is the destination address and where the search logic performs the logical AND of the destination address and the subnet mask.

There are many algorithms available for efficient searching. Linear search can become expensive for large forwarding tables. Binary search has a worst case performance of $O(n)$, where n is the number of bits in the address.

Another well-known search technique is called *hashing*. In hashing, the forwarding table entry for destination address D is stored at location $h(D)$, where $h(D)$ is called the *hash function*. The hash function $h(D)$ is chosen such that although more than one value of D maps to $h(D)$, the probability of such collisions in actual practice is low. The entry at $h(D)$ contains additional information to indicate a collision (i.e., there are more than one forwarding table entries that map to $h(D)$). There are many ways of handling hashing collisions, and we do not pursue this here except to say that collisions are resolved using additional memory accesses. Hashing can be efficient because the total number of table entries is significantly smaller than the total number of values that the destination IP address can take. With a well-designed hash function, the average lookup time with hashing can be $O(1)$, but the worst case performance is the same as that of a binary search. Recall from Chapter 10 that it is important to minimize the variation in the input delay. Hence it is necessary to minimize the worst case lookup time rather than the average case.

Another source of complexity in IP route lookup is in the way in which multiple matches of forwarding table entries to a destination IP address are handled. If there are multiple matches to an IP address, then the one that matches the longest network prefix is to be returned by the lookup function. This is because a larger number of matched bits means that the network prefix of the destination is known with "increasing precision." If a network is organized hierarchically, then networks at a lower level in the hierarchy will have more bits in their network prefix than the higher-level networks. Thus the *longest prefix match* (LPM) is the best match, and the packet should be forwarded to the next hop stored against this network prefix. Note that in classful addressing, only one entry will match the destination address. In the next section we consider solutions for efficient route lookups that involve finding the LPM.

12.2 Efficient Longest Prefix Matching

We start with a basic solution and consider the various enhancements to this scheme. The forwarding table is organized as a *binary trie* (pronounced "try"). It is essentially a binary tree, with each vertex at level k corresponding to a specific prefix of k bits. Each vertex has at most two children, corresponding to the two ways in which the k bit prefix can be expanded to a $(k + 1)$ bit prefix. If the forwarding table contains the next-hop information to a network with the k-bit prefix corresponding to a vertex, then a flag and the corresponding next-hop information are stored at that vertex. Route lookup essentially involves tracing the 32-bit destination address in the trie to find the vertex (the entry in the forwarding table) that matches the longest prefix. Algorithm 12.1 summarizes the algorithm to traverse the trie. Each vertex encountered in the traversal involves a memory access to the contents of the vertex.

Figure 12.3 shows a sample forwarding table and its trie representation.

Exercise 12.2

For the forwarding table trie in Figure 12.3, verify the following.

Input Pattern	Memory Accesses	Best Match
1000 0100	5	10000*/F
1010 0000	2	1*/B
0010 0000	5	0*/B

---

input destination address $D = (d_1 d_2 \ldots d_k \ldots d_{32})$
$P(0) = \text{root}$
best match = root
$k = 1$
while (branch d_k exists at vertex $P(k-1)$) **do**
   take branch d_k to matching prefix $P(k)$
   if $P(k)$ is a valid entry **then**
      best match = $P(k)$
      next hop = next hop of $P(k)$
   $++k$
return(best match, next hop)

Algorithm 12.1 Trie traversal to find the longest matching prefix. Note that the bits in the destination address are numbered starting from the most significant bit. *P(k)* is the matching prefix at level *k* and is also used to denote the vertex of *P(k)* in the trie.

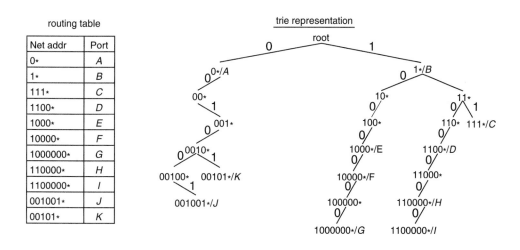

Net addr	Port
0*	A
1*	B
111*	C
1100*	D
1000*	E
10000*	F
1000000*	G
110000*	H
1100000*	I
001001*	J
00101*	K

Figure 12.3 A sample forwarding table and its trie representation. The ∗s indicate that the subsequent bits are "don't-care" bits for the input pattern to match the value in the vertex. The bits next to the branches indicate the value of the next bit for that branch to be taken.

This data structure is quite inefficient in terms of both memory usage and search times. Observe that many vertices contain only pointers and not useful data. More importantly, these must be accessed to go down the binary tree. The maximum number of memory accesses is equal to the number of levels in the trie, which in turn is equal to the maximum number of bits in the network address.

The simplest improvement is obvious: Eliminate one-way branches to vertices that do not contain a forwarding table entry. This is called *path compression*. With path compression, the branching condition from a vertex is a variable string of bits rather than one bit. Thus path compression requires that each vertex that is retained also store the branching sequences to its children. The path-compressed trie of the forwarding table of Figure 12.3 is shown in Figure 12.4. A path-compressed trie is also called a PATRICIA (practical algorithm to retrieve information coded in alphanumeric). Although a path-compressed trie improves storage requirements by eliminating the vertices that do not correspond to a forwarding table entry, the maximum search time is still equal to the number of levels in the compressed trie.

12.2.1 Level-Compressed Tries

A number of improvement algorithms work by *level compression* of the trie. Rather than define a level for each bit of the address, we define a level for groups of contiguous bits. A simple case of level compression is to have a level for every K bits. In this case, if there are N bits in the address string, then the number of

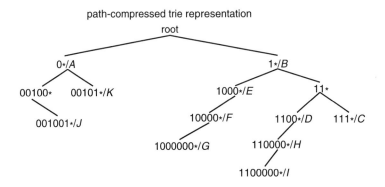

Figure 12.4 The path-compressed trie representation of the sample forwarding table of Figure 12.3.

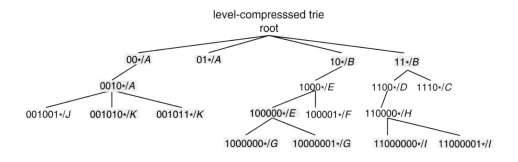

Figure 12.5 **The level-compressed trie for the forwarding table of Figure 12.3. The shaded vertices are the expansion vertices.**

levels is $\lceil N/K \rceil$, and instead of a two-way branch from each vertex of the trie, we have a 2^K-way branch.

Another view of level compression is to say that a *subtree* of height k is compressed into one level. The level-compressed trie representation of the forwarding table of Figure 12.3 is shown in Figure 12.5. Observe that in compressing the level of the trie, the prefixes have been expanded. For example, $0*$ has been changed to $01*$ and $01*$, and both are assigned the same next hop. Thus level compression can also be called *prefix expansion*. The trie traversal algorithm is identical to Algorithm 12.1 except that instead of branching on d_{k+1} at level k of the compressed trie, we branch using the string $(d_{k \cdot K+1}, \ldots, d_{(k+1)K})$.

A Simple Solution

The first method that we consider is motivated by the distribution of the prefixes in the Internet. Although prefixes of all sizes have been observed in the Internet, it has been found that the prefix for a large fraction of packets is much smaller than the maximum. Let as assume that most prefixes are of length K or less. We expand all prefixes of length K or less to K bits (i.e., levels 1 to K in the original trie are compressed into one level). A table, TBL_K, is formed with 2^K words of memory, and the K-bit address of each word acts as an index into the table. All entries in the forwarding table for prefixes of length K or less are captured in TBL_K with appropriate prefix expansion. For prefixes that are longer than K bits, a flag is placed in the location corresponding to the K-bit ancestor in the original trie, indicating that there are other possible matches. The entries for prefixes longer than K bits are stored in a second table, TBL_{long}. To perform a route lookup of an incoming packet, we use the most significant K bits of the destination address to index into TBL_K. If it is the best match, we return the next-hop address. If it is not, we search for the best match in the second table in TBL_{long}. Additional information placed in the location along with the remaining bits in the destination address can be combined to generate an index into TBL_{long}. A block diagram representation of this scheme is shown in Figure 12.6.

Figure 12.7 shows a sample distribution of the prefix lengths from a public Internet service provider in 1998. As shown in the figure, most prefixes are 24 bits or less, and $K = 24$ is a good choice. This scheme requires one memory access for most destination addresses, and two memory accesses and some additional logic to obtain the next-hop information.

The two-table approach just described requires 2^K words of memory and additional memory for the longer prefixes. Assuming $K = 24$, this translates to

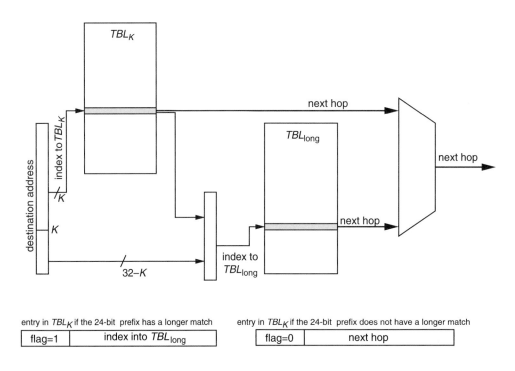

Figure 12.6   The two-table approach to route lookup.

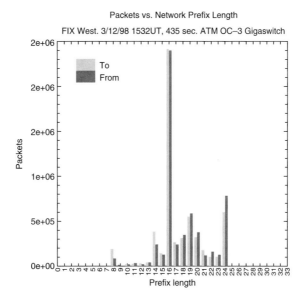

**Figure 12.7 Prefix length distributions. Taken from `http://traffic.caida.org/`
`TrafficAnalysis/Learn/PrefixLength/index.html`. (Reprinted with permission from
Cooperative Association for Internet Data Analysis (CAIDA), www.caida.org;
copyright (c) 1998 The Regents of the University of California.)**

16 million words of high-speed memory. Even in large routers, the maximum
number of forwarding table entries is on the order of a few tens of thousands. Thus
the total memory requirement is two orders of magnitude higher than that required
by the forwarding table. A more systematic optimization is clearly possible. We
discuss two such optimizations next.

Optimal Level Compression to a Fixed Number of Levels

The first optimization that we consider assumes that the number of levels into
which the original binary trie is to be compressed is fixed to, say, M. Clearly,
M determines the worst case route lookup time. The compression is optimized
to minimize the total memory required. In the example of Figure 12.5, $M = 4$.
The optimization problem is to choose the M levels such that the total number of
vertices is minimized because each vertex corresponds to one word of memory.
In the example, we chose the levels to represent 2-, 4-, 6-, and 8-bit prefixes.

The levels in the input binary trie are denoted by i, and level i contains the entries for i-bit prefixes.

Before we define the optimization, let us study the cost of compressing levels $i, i + 1, \ldots, i + j$ of the original binary trie, called the *input trie*, into one level. Observe that if level i in the input trie, with vertices for i-bit prefixes, is compressed into level $i + j$ of the input trie, then each vertex in the input trie at level i must be expanded into 2^j nodes. This is because all prefixes of length i are now expressed as prefixes of length $i+j$. In the example of Figure 12.5 the entry 0* at level 1 of the input trie is compressed to level 2, and 0* is expanded to 00* and 01*. Similarly, 10000* at input level 5 is expanded to 100000* and 100001* at level 6. Thus if we have a level in the level-compressed trie for i-bit prefixes, and if the next level were to have $i + j$-bit prefixes, then the number of vertices in the expanded prefixes at this level would be $(2^j \times V(i + 1))$, where $V(i + 1)$ is the number of vertices at level $(i + 1)$ in the original binary trie. All vertices at level i in the input trie are retained in the level corresponding to the i-bit prefix in the expansion trie. Vertices at levels $i + 1, \ldots, i + j - 1$ are *captured* by the expansion of the vertices at level i; that is, the information in them is included in the prefix expansion of the vertices at level $i + 1$.

We are now ready to define the optimization. First we consider compressing levels 0 to j of the input trie into prefix-expanded levels 1 to r of the level-compressed trie. Here level 0 is the root of the input trie. For the given input trie, let $T(j, r)$ be the minimum number of memory locations required after optimum level compression of levels 0 to j of the input trie into r levels of the level-compressed trie. From our discussion, we can write the following recursive equation:

$$T(j, 1) = 2^j$$

$$T(j, r) = \min_{m \in [r-1, j-1]} \{T(m, r - 1) + V(m + 1) \times 2^{j-m}\} \qquad (12.1)$$

The first equation is clearly true: If the root must be expanded into j bits (level j), then all the 2^j prefixes must be enumerated. The second equation is obtained as follows. For expanding levels 0 to j into levels 1 to r, level j of the original trie should be at level r of the level-compressed trie. If we choose level m of the original binary trie for the $(r - 1)$th level of the expanded prefix, then levels 0 to m of the input trie must be fit into $r - 1$ levels of the compressed trie. This will cost $T(m, r - 1)$ memory locations. The second term is the cost of compressing levels $m + 1, \ldots, j$ into level r of the expanded trie and is derived in the earlier discussion. See Figure 12.8 for an illustration. We can determine the range of m by realizing

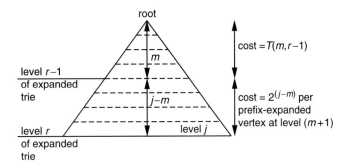

Figure 12.8 Figure illustrating Equations 12.1. Dotted lines indicate the levels of the input trie.

that because there are r levels in the expanded trie, m should be at least $r - 1$. Clearly, it can be at most $j - 1$.

Equations 12.1 form a *dynamic program*. In dynamic programming, the solution to the main problem is obtained by combining the solution to subproblems. These subproblems can in turn be decomposed into even smaller sub-subproblems, and so on. If the solution to the smallest problem is known or given, then the solution to the larger problem can be built from the base solution. The simplest example of a dynamic program is the factorial function, where $N! = N \times (N - 1)!$. Given the base solution that $0! = 1$ we can obtain the solution to $N!$ in N steps. Another well-known dynamic program is the Bellman–Ford algorithm to find the length of the shortest path between node pairs in a weighted graph. This shortest path algorithm is discussed in detail in Chapter 14. The solution of Equations 12.1 is along similar lines. Observe that Equations 12.1 have two variables: j ranging from 0 to N, and r ranging from 1 to M. We form an $N \times M$ table, with entry (n, k) denoting the optimum cost of compressing n levels of the input trie into k levels of the prefix-expanded trie. The first line of Equation 12.1 defines the base equations to fit into row 1. The other elements of the table can be obtained recursively using Equations 12.1. Problem 12.1 considers an example.

To evaluate the complexity of this process, note that there are $O(NM)$ entries in the table and that each entry is obtained by finding the minimum of $O(N)$ quantities. Thus the complexity of the algorithm is $O(N^2 M)$.

Optimal Level Compression into a Fixed-Size Memory

A second type of optimization is to perform the level compression to minimize the average route lookup time while fitting the compressed trie into a given amount of hierarchical memory. The hierarchical memory is assumed to be organized like the different levels of cache in a computer system, with different access times for the different levels. The problem here is to distribute the different *subtries* of the route lookup table among the different memory units.

We consider the case of a two-level memory hierarchy, with the two memory units labeled type 1 and type 2 of sizes C_1 and C_2 and access times t_1 and t_2, respectively. If we let $t_1 < t_2$, it is reasonable to assume that $C_1 < C_2$. Rather than compress all the vertices at a level of the input trie into the same level (as in Equations 12.1), we can generalize and consider the compression of the subtries at each vertex separately. If we compress a vertex by h levels and hence expand the prefixes of the entry at the vertex by h bits, then arguing as before, the prefix expansion will require 2^h bits. Thus the first decision is the level of compression h for each vertex. A second decision that must be made is which of the two memories to place the expanded prefixes into. Putting the compressed part of the trie into type 1 will reduce the average access time but will leave less memory for compressing the other vertices. Figure 12.9 shows a sample trie, memory type in which the entries for the vertices are stored, the number of accesses of each memory type for each possible best matching prefix, and the average memory delay per lookup.

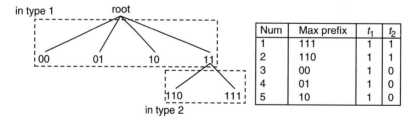

Figure 12.9 Example to calculate the number of memory accesses and average route lookup time for a forwarding table trie. The left panel shows the sample trie and the part of the hierarchical memory where the entries are stored. The table on the right shows the number of memory accesses of each level for each of the best matching prefixes from the trie on the left. The average access time is $(p_1 + p_2 + p_3 + p_4 + p_5)t_1 + (p_1 + p_2)t_2$, where p_i is the fraction of packets for which entry number i in the table is the best matching prefix.

Let $TP_1(i, c_1)$ be the average access time of a prefix lookup when the subtrie rooted at vertex i of the input trie is packed into type 1 memory of capacity c_1 and infinite capacity of type 2 memory. Let p_i be the probability with which vertex i is accessed during a route lookup; that is, p_i is the fraction of packets for which the maximum matching prefix will be the prefix of vertex i or of the vertices of the subtrie rooted at vertex i. If the expanded prefixes are placed in type 1 memory, the access time contribution to the mean access time will be $p_i t_1$ but will leave $c_1 - 2^h$ memory locations for the use of the other nodes. Alternatively, if the expanded prefixes are placed in type 2 memory, the contribution to the mean access time will be $p_i t_2$, and we will retain all the c_1 locations of type 1 memory. This trade-off can be expressed by Equations 12.2. Let H_i be the total number of levels in the subtrie rooted at vertex i, let $L_{h,i}$ be the set of internal (nonleaf) vertices in the subtrie of depth h rooted at vertex i, and let $TP_2(L, c_1)$ be the average access time when the vertices of a tree labeled L are fit into a type 1 memory of capacity c_1 and type 2 memory of infinite capacity. We can write

$$TP_1(i, c_1) = \min_{1 \leq h \leq H_i} \left\{ \min \left[p_i t_1 + TP_2(L_{h,i}, c_1 - 2^h), p_i t_2 + TP_2(L_{h,i}, c_1) \right] \right\}$$

$$(12.2)$$

The first term in the square brackets corresponds to the cost of compressing h levels into storing the expanded prefixes in type 1 memory, and the second term is the cost of putting it in type 2. Note that $p_i t_1$ (respectively, $p_i t_2$) is the contribution to the average access time, and $TP_2(L_{h,i}, c_1 - 2^h)$ (respectively, $TP_2(L_{h,i}, c_1)$) is the cost of packing the remaining part of the subtrie rooted at vertex i into the remaining capacity in type 1 (respectively, type 2) memory.

The initial conditions are

$$TP_1(\cdot, -) = TP_2(\cdot, -) = \infty \qquad (12.3)$$

$$TP_1(\{\}, \cdot) = 0 \qquad (12.4)$$

Equation 12.3 says that if the amount of space in type 1 is less than 0, then its contribution to the average access time is infinity. Equation 12.4 says that if nothing has to be packed (first argument is the empty set), then the cost is 0. Let us now see how to solve TP_2. The first argument for TP_2 is $L_{h,i}$, the set of vertices at level h in the subtrie rooted at node i. If $L_{h,i}$ contains exactly one node, then TP_1 is the same as TP_2. If it contains more than one node, then these nodes are partitioned into two subsets L_1 and L_2. This leads to two questions: (1) How is c_1 divided among the two subsets? (2) Given the division of memory space, how best can we compress the levels in the two subsets to fit into the respective memory spaces? Exactly as in the case of TP_1, a recursive expression can be written for TP_2 as well.

$$TP_2(L, c_1) = \begin{cases} llTP_1(i, c_1) & \text{if } L = \{i\} \\ \min_{0 \le c \le c_1}\{TP_2(L_1, c) + TP_2(L_2, c_1 - c)\} & \text{where } L_1 \cap L_2 = \phi \\ & \text{and } L_1 \cup L_2 = L \end{cases}$$

$$(12.5)$$

There are many ways to split $L_{h,i}$ into disjoint subsets. The solution is simplified if the splitting is done such that each element of L_1 has a common ancestor that is preferably distinct from the common ancestor of the elements of L_2. Equations 12.2 and 12.5 also form a dynamic program and can be solved recursively by solving smaller subproblems. The base solutions to this dynamic program are Equations 12.4. It can be shown that the complexity of the problem is $O(NVC_1^2)$, where N is the number of levels in the input trie and V is the number of vertices.

12.2.2   Hardware-Based Solutions

So far we have looked at mechanisms to optimize the number of memory accesses per route lookup. We now discuss two solutions that are proposed for implementation in hardware.

The first solution is based on *ternary CAMs*, which allow the search field of an entry in the CAM to contain *don't-care* (or wildcard) elements (i.e., a search key of a location could be 1X001, where X is a don't-care bit). This key matches input patterns 10001 and 11001. Allowing don't-cares means that search fields from multiple locations can match the input pattern. To enable the CAM to return exactly one entry for every search request, the entries can be prioritized. If multiple entries match an input, the one with the highest priority is returned. The details of the CAM-based solution must now be obvious: Use 32-bit search fields for the CAMs, and store the forwarding table entries with the prefix in the search field. For a prefix of b bits, the least significant $32 - b$ bits should be made don't-care. The next-hop information is stored in the return field of the CAM, with the longest prefixes getting the highest priority. When a packet is received by the router, the 32-bit destination address of the input packet is extracted and given as the search pattern to the CAM. The contents of the return field are read to obtain the next-hop address and other control information for the packet.

Another, very different solution to the IP lookup problem is to use *reconfigurable hardware* to perform the route lookup. Reconfigurable hardware essentially is made from elements called *field-programmable gate arrays*. The idea is to design a *finite state machine* (FSM) corresponding to the search for the best-matching prefix along the route lookup trie. The basic design principles are very similar to those of standard string matching using Flip Flops. A sample state machine for a small forwarding table is shown in Figure 12.10. For large

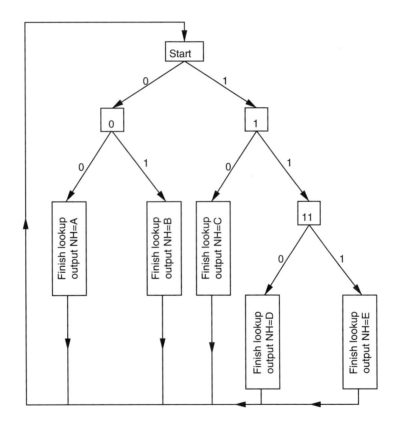

Figure 12.10   A state-machine-based lookup algorithm for the forwarding table in Figure 12.9.

forwarding tables, the number of states in a naive FSM could become quite large. For example, a forwarding table with about 38,000 entries needs more than 170,000 states. Implementing this number of states requires a large number of logic gates and can also slow down the execution of the state machine because of a large number of transmission delays. However, a naive FSM can be efficiently decomposed into smaller state machines and the output from the smaller machines combined to obtain the next-hop information. This method can perform a route lookup in less than one memory access time. The fact that FSMs are reconfigurable allows the changes in the forwarding table to be incorporated into the state machine online.

12.3   Packet Classification

To provide differential grades of service to different information flows through it, a packet switch should classify packets at high speeds. Many emerging services also require that packets be classified at high speeds. For example, a policy-based routing algorithm that routes stream and elastic traffic over different paths must classify the packets into stream and elastic traffic. Within each of these categories, different grades of service may be offered and further classification may be required. Security architectures use firewalls and security gateways, which allow access to only certain classes of traffic on some parts of the network. These devices are placed at the edge of the secure part of the network, and to filter out the ineligible traffic, they require high-speed packet classification. Accounting and traffic monitoring is another obvious application for packet classification. In this section we discuss some of the issues in packet classification as well as some well-known classification algorithms.

A packet may contain headers from the MAC protocol (e.g., Ethernet), the routing protocol (e.g., Internet Protocol), the transport protocol (e.g., TCP), and the application (e.g., streaming multimedia, packet voice, File Transfer Protocol, Hypertext Transfer Protocol for Web downloads, etc.). In IP networks, packets belonging to well-known services are identified by the TCP or UDP port numbers on the packets. A packet can be classified based on any combination of these attributes.

The class of a packet is specified by the rules that match the bit pattern in the packet. An elementary rule, called a cell, is specified by all of the following:

- The *offset* from the beginning of the packet
- The number of bits (*length*) after the offset that should be used to match a pattern

- The *mask*, which specifies the significant bits to be used for matching

- The *value* that the significant bits of the packet obtained earlier should match

A classification *rule* is formed by logically combining the cells.

Most practical applications require use of only a small number of distinct offsets and lengths. In fact almost all packet classifiers assume that the number of distinct pairs of offset and length are five or fewer, and each pair of offset and length is referred to as a *dimension*. A cell as defined earlier is the specification in one dimension. We denote a logical AND of a cell in each dimension as a *basic rule*. Also, the combination of the mask and the value is typically specified as a *prefix* (as in IP route lookup), or as a *range* of values that the set of bits specified by the offset and length can take. Note that a prefix is a special case of a range.

As in the case of route lookup, a packet can match multiple rules, and the rules are prioritized so that the classifier returns the *best matching rule*. A rule is a Boolean expression, with cells as the variables. Because a Boolean expression, can be expressed as a "sum of products" of the variables, a rule—say, rule R—can be replaced by a set of M basic rules—say, $\{r_i : i = 1, \ldots, M\}$. Satisfying any basic rule in $\{r_i\}$ implies satisfying rule R. Extending this concept further, a set of prioritized rules can be replaced by a prioritized sequence of basic rules. Thus the problem of packet classification is the problem of identifying the best basic rule that a packet satisfies.

Packet Classification as a Point Location Problem

For a K-dimensional classifier with N basic rules, let rule n be specified by $r_n = \{e_{1,n}, \ldots, e_{K,n}\}$, where $e_{k,n}$ is the range in the kth dimension. Consider a packet classifier in two dimensions, $K = 2$. Assume that each dimension specifies a range for 4-bit numbers, 0000–1111. Consider three rules $r_1 = \{[6, 9], [3, 12]\}$, $r_2 = \{[4, 10], [4, 10]\}$, and $r_3 = \{[11, 14], [4, 12]\}$. Each of these basic rules can be represented as a rectangle on a two-dimensional plane, with each axis representing one dimension of the classifier. This is shown in Figure 12.11. An input packet has a specific value in the two fields that represents the dimensions of the classifier, and this value can be represented as a point on this two-dimensional plane. Thus the classification problem can be stated as the problem of determining the rectangle to which the point corresponding to the given packet belongs. If the point belongs to multiple rectangles, then the classifier should pick the "best rectangle."

To help speed up the classification in hardware, we do the following. For dimension i, we divide the axis into nonoverlapping intervals $I_{i,j}$ such that each

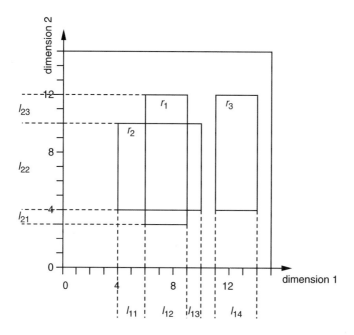

Figure 12.11   Classification as a point location problem. The $I_{i,j}$ are also shown for each dimension. The following can also be verified: $\rho_{1,1} = \{2\}$, $\rho_{1,2} = \{1,2\}$, $\rho_{1,3} = \{2\}$, $\rho_{1,4} = \{3\}$; $\rho_{2,1} = \{1\}$, $\rho_{2,2} = \{1,2,3\}$, $\rho_{2,3} = \{1,3\}$.

interval is a subset of a set of $e_{i,n}$. Because there are N classification rules, there will be at most $2N + 1$ such nonoverlapping intervals (i.e., the range of j is at most $2N + 1$). Let $P_i = \{I_{i,j}\}$ denote the set of these intervals for dimension i. For each $I_{i,j}$, we make a set of rules $\rho_{i,j}$ such that $r_m \in \rho_{i,j}$ if and only if the interval $I_{i,j}$ is a subinterval of $e_{i,m}$. Each $\rho_{i,j}$ can be represented as an N-bit *rule-matching vector*, in which bit m is set to 1 if $r_m \in \rho_{i,j}$.

When a packet arrives, the fields corresponding to the k dimensions are extracted, and for each dimension of the classifier, the $I_{i,j}$ to which the packet belongs is determined. This can be done in parallel. The $I_{i,j}$ determine the $\rho_{i,j}$. A bit-by-bit logical AND of the $\rho_{i,j}$ for $i = 1, \ldots, N$ determines the basic rules that the packet satisfies. If the rules are prioritized and arranged in decreasing order of priorities, the best matching rule is the one specified by the most significant 1 in the result of the logical AND. Consider the following examples. A packet with attributes $(7, 8)$ belongs to $I_{1,2}$ and $I_{2,2}$. The corresponding rule-matching vectors

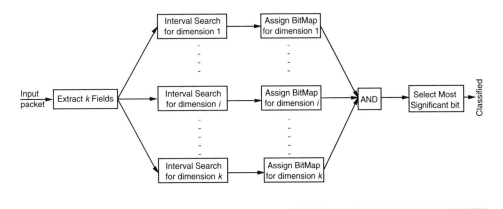

Figure 12.12   A block diagram for the hardware based classification.

are 110 and 111. A logical AND between them yields 110. Applying the higher priority rule we conclude that the packet belongs to class 1. If the packet instead has attributes $(4, 5)$, it belongs to intervals $I_{1,1}$ and $I_{2,2}$ with rule-matching vectors 010 and 111, and the packet is classified as being from class 2. A block diagram of the classification process is shown in Figure 12.12.

To understand the implementation complexity, observe that each dimension requires at most $2N + 1$ intervals and hence $2N + 1$ N-bit vectors. The search operation to determine the interval can be done in parallel for all intervals $I_{i,j}$ using any standard search method that can be implemented in hardware. Each logical AND of d words of N bits must to be performed, and the position of the most significant 1 determines the class of the packet. Thus the complexity is determined by the search for the interval in each dimension.

Heuristic-Based Algorithms

Many heuristic-based algorithms are available for packet classification. We briefly introduce two of them. The first one is called the *tuple space search*. This method is based on the assumption that each rule is specified as a prefix in each of the d dimensions. In practice it has been found that for each dimension, the number of distinct prefix lengths is small, and the number of distinct combinations of prefix lengths is also small. A *tuple space* is defined for each combination of prefix lengths. In a classifier of N rules in d dimensions, the number of tuples—say, m— has been found to be of the same order as N. Each of the tuples is associated with a distinct set of rules. The idea is to exhaustively search each tuple for the matching rules and then pick the best matching rule. The position of the bits in the packet used by each tuple is fixed, and hence hashing can be used to map the relevant bits in the packet to the rules defined by that tuple. The results from all the tuples can be combined to obtain the best matching rule. Heuristics can be used to prune the number of tuples.

A second scheme is called *recursive flow classification*. The idea here is to concatenate all the bits from the packet that are used in the classification and divide them into groups of a smaller number of bits called *chunks*. The value of each chunk is used in a hash function. The hash values can then be used in another hash function and so on until the classification is complete. All the hash functions are implemented as a table lookup. The design of the hash functions makes extensive use of heuristics. Because the number of rules is much smaller than the maximum number of rules possible, the hash function can be designed to eliminate collision. Also, if the hash function is implemented as a table lookup, complex functions can be designed.

12.4   Other Design Issues

Address lookup and packet classification are only two of the many issues in the design of high-speed packet switches for the Internet. These two have been widely studied in recent literature, explaining our extended treatment of these functions in the previous sections. In this section we highlight some of the other design issues for high-speed packet switches.

In QoS-capable packet switches, packet scheduling on the output links must be accomplished at high-speeds. Scheduling in a pure OQ switch seems straightforward because at any time, all the packets in the switch that must be transmitted on an output link are available at the corresponding output buffer. For example, implementing WFQ would mean simply implementing Equations 4.7 and 4.9 (described in Chapter 4), albeit at high speeds. Recall from our discussion in Section 10.6 that in IQ switches, link scheduling becomes more complex, and we must use distributed algorithms with partial information at the inputs. The information is partial because an input does not know the number of backlogged packets of a given class at the other inputs. One way to achieve distributed scheduling in an IQ switch is to use virtual output queueing and maintain a separate queue for each class (i.e., input i maintains queues $Q_{i,j,c}$ for class C packets with destination j). A reasonable approximation of the OQ-based scheduler is to have separate schedulers at each input, each output, and the switch fabric and construct a hierarchical scheduler for each output.

A packet switch in a virtual-circuit network must perform the signaling functions associated with the setting up of a call, maintaining it and tearing it down. The computing resources required for this activity can become significant in large switches. In addition to call setup, the CPU must perform the network-related functions (such as the exchange of routing information), compute the routes to other nodes in the network, and fill up the forwarding tables. Most switches perform the signaling and networking functions in software, although parallel architectures for signaling have been explored in the literature. The call-processing capability of the switch must be analyzed, as in the case of telephone switches, except that the problem is more complex here because the call needs a more complex set of resources.

Each packet switch also has its own operating system to manage the resources of the switch, such as CPU cycles and memory. This operating system typically is executed on a master processor. An important aspect of the operating system design is the interface to the many packet-processing functions. Our discussion in the previous sections concentrates on the algorithms to make packet processing efficient, implicitly assuming that the information required for these functions (e.g., the forwarding table) is available. The operating system must make it available. The routing algorithm must be executed at regular intervals, and the forwarding tables must be stored in the manner required by the lookup algorithm. The operating system must manage this activity such that packet processing is not disrupted during the update.

An important requirement of the operating system, especially in a single-processor packet switch, is to respond to processor overload situations in a manner that minimally affects the packet-processing throughput. In circuit-multiplexed networks (virtual or real), processor overload can occur if the call arrival rate exceeds the call-processing capacity of the switch. This can lead to increased call setup delays. This in turn can cause many calls to time out or be abandoned by the user while call setup is in progress. There may possibly be reattempts. In either case, the processing capacity is used for call-processing but does not result in a call completion. This situation can lead to a condition where the processor is busy but call-processing throughput has decreased because of unproductive CPU cycles. Other overload conditions occur during network transients when excessive control messages (e.g., routing and topology updates) are generated and must be processed by the packet switches, leaving little CPU capacity for other necessary activities. These problems are well known in telephone networks, and a variety of mechanisms have been devised for overload control in telephone switches. Many of the same principles are applicable in packet switches.

12.5   Network Processors

As communication networks become ubiquitous, many new services are being created. Many of these new services need more information about the service that the packet is providing. We make this information available by processing the packet contents beyond the routing headers. We have seen the example of packet classification in Section 12.3. Examining the contents of the packet beyond the routing header is called *deep packet processing*. Although it is possible to design and build specialized hardware for some of these functions, it is clearly impossible to perform all such special tasks in hardware. Performing these packet-processing tasks on general-purpose processors can be inefficient, and it may not be possible to achieve high throughputs even with the more powerful of these processors. One reason for this is that the high processing rate of a general-purpose CPU is due to the efficient exploitation of temporal and spatial locality of reference in programs and data through the memory cache. It has been found that packet-processing functions do not have a high degree of locality of references, and general-purpose processors cannot be made to operate at their full capacity. This is even more so when the link speeds are very high, because at these data rates, a large number of unrelated streams are being multiplexed. Thus, high-throughput, deep packet processing requires special-purpose processors. Many such processors, called *network processors* (NPs), are now available.

From our discussion of the processing in a packet switch and the algorithms that are used for some of the more common complex functions, it must now be clear that packet processing essentially involves processing of event streams, where the events are the arrivals of packets. In this sense they are very similar to digital signal processors that process media streams, and network processors essentially evolved out of this realization. There are, however, significant differences between the two. In this section we highlight the main design features of network processors.

The most important feature of a network processor is the large number of *processing elements* (PEs), sometimes called picoprocessors. Each picoprocessor may have its own local memory for instruction and data and also have access to a common memory. The PEs execute a special-purpose instruction set tailored for packet processing. A master processor, which is more like a general-purpose processor, runs the operating system and the other slow-timescale functions such as participating in network control protocols) by sending, receiving, and processing the related information. Much like a traditional processor, the master processor may have its own data and instruction cache. The instruction set of the master processor is typically similar to that of a general-purpose processor. The input–output interface is specialized to read and write packets.

Many network processors also provide special-purpose hardware to perform the more common tasks, such as CRC (cyclic redundancy check), route lookup, and packet classification. A block diagram of a generic network processor is shown in Figure 12.13.

Because there are a large number of processing elements in a network processor, it is essentially a parallel computer, and, as with parallel computers, different interaction models among NPs will produce different architectures. The two basic architectures of network processors are shown in Figure 12.14. The simplest assumption is to treat each PE individually and to assign the processing of an arriving packet completely to one PE. Arriving packets are stored in a queue at a dispatcher and are assigned in a FIFO manner to the first PE that becomes free. The PEs are identical to each other and can execute the same program but on different data. In parallel computing parlance, this is called a *symmetric multiprocessor* (SMP). If the PEs are powerful but the number is small, they may support *multithreading*, which allows many processes to run simultaneously on the same PE. When a process *stalls*, the PE switches to another active process. A process can stall when performing a function that takes a longer time to execute than a normal instruction (e.g., disk access in conventional computers and possibly main memory access or access to special hardware in an NP).

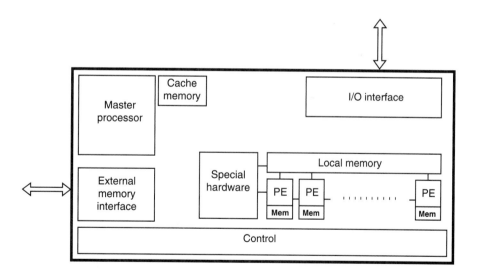

Figure 12.13 Block diagram of a generic network processor.

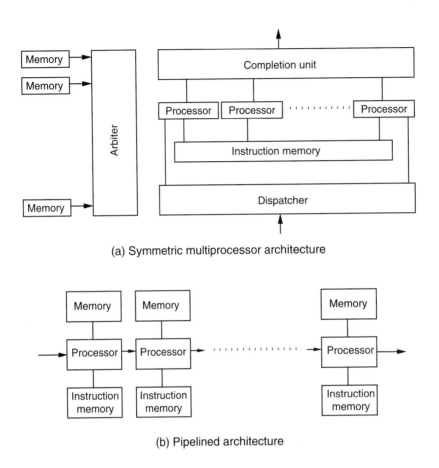

(a) Symmetric multiprocessor architecture

(b) Pipelined architecture

Figure 12.14   High-level view of the two popular network processor architectures.

A second architectural option is to partition packet-processing code into different blocks and to assign each block to a different PE. The PEs are arranged in a *pipeline* manner, with the output of one PE feeding the next one in the pipeline. The pipelining architecture requires that the packet-processing code be partitioned. For the pipeline to be efficient, it is necessary that the partition be such that the execution times are symmetric. This is because the delay in each stage of the pipeline is equal to the maximum delay of the stages. This code partitioning is not easy. SMP is conceptually simpler, and many NPs use this architecture. It can, however, experience degraded throughput if the stalls are not managed properly.

12.6 Summary

In this chapter we provide an overview of the issues in packet processing and examine some algorithms for two specific functions. We begin with an overview of the addressing mechanisms in various networks and a detailed look at CIDR in IP networks. CIDR requires that the destination address of an input packet be matched against the network prefixes stored in the forwarding table and that the longest prefix match be used to forward the packet. We first discuss the trie data structure for storing the forwarding table so that LPM becomes efficient. We then look at ways of optimally compressing the forwarding table trie. One optimization criterion is to minimize the memory usage for a given worst case lookup time, and the other is to minimize the average lookup time while fitting the trie into a given memory. We also discuss some hardware-based lookup schemes.

The second packet-processing function that we discuss here is packet classification. We explain how packet classification can be mapped to the problem of locating a point in a K-dimensional space and discuss an algorithm based on this view. We also briefly discuss other classification algorithms. A brief overview of other design issues in high-speed packet switches is also provided. Finally, we discuss network processors and provide a brief overview of two network processor design paradigms.

12.7 Notes on the Literature

The aggregation of class C addresses into the classless interdomain routing specification is proposed in RFC 1519 [111]. The PATRICIA implementation was first implemented in Berkeley Unix by Sklower [263]. The CAM-based solution is described by McAuley and Francis [207]. Chisvin and Duckworth [60] provide a good overview of CAM technology. The two-memory scheme is from Gupta et al. [134]. The dynamic program to compress the level of the trie to a fixed number of levels to minimize the use of memory is described by Srinivasan and Varghese [265]. The dynamic program to optimize access time is by Cheung and McCanne [58]. The reconfigurable hardware approach to high-speed IP lookup is by Desai et al. [77]. Other IP route lookup schemes described in the literature include those by Doeringer, Karjoth, and Nassehi [85], Degermark et al. [82], Nilsson and Karlsson [224], and Waldvogel et al. [293]. Ruiz-Sanchez, Biersack, and Dabbous [257] provide a good summary of these lookup algorithms.

Mogul, Rashid, and Acetta [218] seem to be the first to have implemented packet classification and filtering. The Berkeley Packet Filter (BPF) by McCanne and Jacobson [208] is a widely used software classifier that is now part of most Unix-like distributions, including Linux. BPF is integrated into the operating system kernel and has been extensively used for network monitoring and debugging. BPF is designed as a virtual machine executing instructions from a specially designed instruction set. The PathFinder, by Bailey et al. [20], is an early hardware packet classifier for a small set of rules. The range-matching algorithm described in this chapter is by Lakshman and Stiliadis [188]. Srinivasan, Suri, and Varghese [266] describe the tuple search method. Gupta and McKeown [133] perform an extensive study of commercial filter rules and have designed the *recursive flow classification* heuristic. Feldmann and Muthukrishnan [98] study the trade-offs in the design of packet classifiers. There are many other algorithms for high-speed packet classification—for example, [267, 301, 19]. Iyer, Kompella, and Shelat [148] describe the commercial ClassiPI classifier.

Varma and Stiliadis [290] describe issues in the hardware implementation of a packet scheduler in ATM networks. Chiussi and Francini [62] and Stephens and Zhang [272] describe a hierarchical architecture to do distributed scheuduling of packets in input-queued switches. Druschel et al. [88] and Ramakrishnan [240] are two good references for network subsystem design. Ghosal, Lakshman, and Huang [117] describe architectures to execute signaling protocols at high speeds using parallel architectures. Hwang, Kurose, and Towsley [146] have studied the call-processing delay in virtual-circuit-based networks. The control architecture of a commercial packet switch is described by Walsh and Ozveren [296].

Manufacturers' data sheets are the best source material for network processors. Some of the popular network processors are PowerNP by IBM, FPX from Cisco, and IXP from Intel. Shah [259] provides a good survey of the available network processors.

Problems

12.1 Consider the routing table of Figure 12.3 and the trie representation shown. If this trie were to be compressed into two levels, how many memory locations would it require? Repeat for compression to three levels.

12.2 Obtain the state machine to extract the network address from the 32-bit IP address if it is known that the addresses are either class A, class B, or class C.

12.3 In packet classification as a point location problem, the interval to which the point belongs must be found for each dimension. Consider an "interval comparator" that has two control inputs H and L. It has three outputs: l, r, and 0. If the input is greater than H, then is l is activated; if it is less than L, then r is activated; and if it is in $[L, h]$, then 0 is activated. Using two-input comparators, construct such an interval comparator. Use interval comparators to construct the block diagram for classification in hardware according to the rule set shown in Figure 12.11. Comment on the classification delay expressed as the number of gate delays as a function of the number of rules.

Part III

Routing

CHAPTER 13

Routing: Engineering Issues

R outing can be viewed as a mechanism for sharing a distributed set of resources among traffic streams generated by sources. The issue of resource sharing has appeared in many places in this book. For example, when we multiplex several traffic streams onto a single link, the basic problem is one of sharing the resource (the link) among the packets in a way that ensures fair sharing among the individual streams, as well as efficient utilization of the shared resource. If we widen the meaning of "resource" to include a complete network, with its specific topology and link capacities, then the same issue appears again: Traffic streams generated by sources must be carried on the network so that the traffic streams see acceptable performance and also that the network resources are efficiently utilized.

Consider Figure 13.1, which shows a network and some sources and sinks of traffic. Inside the dashed oval we have the *backbone* network, consisting of routers (represented by circles) and links (represented by lines). The circles are often called *backbone routers*. The rectangles outside the oval, to which sources and sinks of traffic are attached, are the so-called *aggregation routers*, which collect traffic from a large number of sources before injecting it into the backbone. An aggregation router can be connected to more than one backbone router. This allows the traffic load to be distributed across the links and provides a backup link if the primary one should fail. The sources and sinks of traffic are, as mentioned in Chapter 2, computers at homes and offices, or even human users speaking into a microphone; these are not shown in Figure 13.1. For our purposes in this chapter, the "source" and "destination" of a traffic stream will be a backbone router, to which one or more aggregation routers are attached. These backbone routers are also referred to as *ingress–egress* routers. Evidently, the routing problem that we are concerned with is routing *across the backbone*.

What is the nature of the traffic carried across the backbone network? At this point, recall from Chapter 3 that essentially two kinds of traffic can

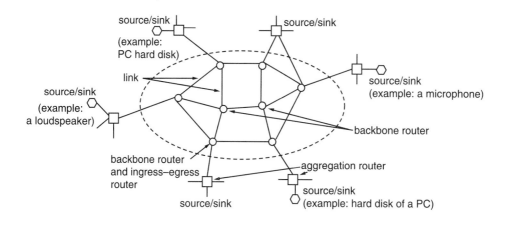

Figure 13.1   A network showing backbone routers and links, aggregation routers, and the "true" sources and sinks of traffic.

be carried: elastic traffic and stream traffic. In Chapters 14 and 15 we consider elastic traffic, and in Chapter 16 we consider stream traffic.

Because the source of the traffic to be routed, in our view, is an ingress router (which is also a backbone router), and not a "true" source such as a computer on a desktop, what is the meaning of the *traffic stream* that the router sends into the network? Let us first consider elastic traffic. As mentioned in Chapter 7, elastic traffic is typically generated by a user who is, for example, browsing the Web by using the services of TCP. Each individual transfer, occurring over TCP, is referred to as a session. Consider a pair of routers on the backbone, with a Web server attached to one router and a number of users attached to the other (through an aggregation router). The backbone router serving the Web server is the source of traffic in our view, whereas the router providing Internet access to the large number of users is the sink. Now, at any instant, there are a great many sessions that are active and are transferring data from the source to the destination. The collection of all these individual sessions constitutes an *aggregate*, with the property that the source and destination routers are common to all the sessions. This is what we refer to as an *aggregate of elastic traffic*. Routing of such elastic aggregates is the topic of Chapter 14.

Alternatively, the true source of traffic could be a corporate entity that is using the backbone network as a *virtual private network* to communicate with its business partners. Here, too, the traffic generated is elastic in nature, but it is not in response to myriad users browsing the Web. Rather, it might be generated by company employees engaged in possibly more specific information exchange, such as transferring sales data from one location of an enterprise to another. As discussed in Chapter 15, this kind of aggregate traffic can impose specific constraints on the routing that is possible through the backbone.

Alternatively, the true source of traffic might be a Voice-over-IP application that takes human speech and then samples, digitizes, and packetizes the voice and sends audio packets into the network. This, of course, is stream traffic, and, as discussed in Chapters 4 and 5, it has specific performance requirements such as bounded end-to-end transfer delay. Chapter 16 shows that this kind of traffic imposes yet another kind of constraint on the routes that can be chosen.

We use the word *demand* to refer to the data transfer requirements of aggregates of elastic traffic. We also use the same word to refer to the data transfer needs of stream traffic sessions. In either case, it is a demand placed on the network's resources that must, if possible, be satisfied.

When the network is presented with a set of such demands to be routed, what is the information available that it can utilize to compute routes? One type of information pertains to the demand itself—for example, whether it is an elastic traffic demand or whether stream-type traffic is to be carried, what the size of the demand is, what the source and destination of the demand are, and so on. The other type of information pertains to the elements of the network—for example, the capacities on its links, the propagation delays across its links, and so on. Routes for demands must be chosen taking this information into account so that the twin objectives of satisfying demand requirements and utilizing network resources efficiently can be met.

In practice, two distinct scenarios call for the use of routing algorithms. The first of these is the *design* scenario, where a network is being designed. Typically, the designer has a predicted traffic matrix giving the expected aggregate traffic between node pairs. At the end of the design, the designer has the capacities of the links interconnecting the nodes and the routes to be taken by traffic between node pairs. The point to note is that the complete set of traffic flows is available before the design begins. A related problem is one in which the network (with links of specified capacities) and a predicted traffic matrix are given, and we must find routes for the demands between node pairs. This is a routing problem only, in contrast to a network design and routing problem, because the link sizes are given. In other words, the algorithm used does not provide link capacities—because they are already specified—but only the routes that should be used by the demands. Routing algorithms used for such problems are offline in nature. They work with information about the complete set of demands to be routed.

In contrast, there is the *operational* scenario, where a network is already deployed and operational and traffic demands arrive to it sequentially. In such a case, a routing algorithm does not have information about *all* the demands that may arrive over time. This can be contrasted with the design scenario, where information about all demands to be routed is available. In many practical situations, there may be no information at all about the arrival process of demands nor about the sizes of the demands. An online routing algorithm must consider the current network state and find an appropriate route for a demand that has just arrived.

Note that the words *offline* and *online* are being used to indicate whether or not complete information about the set of demands to be routed is available. When knowledge about all demands that can arrive is available, it can be taken into account by the routing algorithm. On the other hand, when we do not know the demands that may arrive in the future, the information available to the routing algorithm is necessarily restricted. It must route demands as they arrive, adapting to the present state of the network and the characteristics of the arrived demand.

In almost all cases, the routing problem is modeled as an optimization problem, with a specific optimization objective to be achieved, or as a problem where a point in the feasible set must be identified. When multiple routes for a demand are feasible, the optimization objective typically is used to narrow the choice to one. On the other hand, for some problems, any solution from the feasible set may suffice. An example of an optimization objective is to minimize the aggregate utilized bandwidth in the network. The aggregate utilized bandwidth is obtained by adding, over all links, the aggregate flow passing over each link. Clearly, this is a measure of the total bandwidth resources consumed in supporting the given traffic matrix. Another objective might be to minimize the largest utilization factor over all links, where the utilization factor of a link is the ratio of the aggregate flow on it to its capacity.

Exercise 13.1

Are the two objectives mentioned here—minimization of the aggregate utilized bandwidth and minimization of the maximum utilization factor—equivalent?

In Chapters 14 and 15, we are concerned with a specific performance objective: maximization of the smallest spare link capacity over all links in the network. The significance of this performance metric is discussed in later chapters.

In Chapter 15, where online routing of demands is discussed, an additional performance metric is used to compare different algorithms: the fraction of demands that cannot be routed. Evidently, a routing algorithm that achieves a low *rejection ratio* is desirable. In Chapter 16, the objective is mostly to find feasible paths that satisfy specified bandwidth and delay requirements. We also consider the total weighted carried traffic as a performance metric to compare different routing algorithms.

CHAPTER 14

Shortest Path Routing of Elastic Aggregates

In this chapter, we are given a capacitated network of routers and links and a set of demands for carrying elastic traffic between pairs of routers. For example, the network might be an Internet service provider's (ISP's) operational network, which is used to provide Internet access to customers. The elastic traffic (introduced in Section 3.1 and treated extensively in Chapter 7) that the network carries is generated by end users using applications utilizing the services of the transport layer protocol (see Chapter 2) TCP. Typically, users browsing the World Wide Web generate such elastic traffic. The word *elastic* refers to the property that such traffic does not have any intrinsic transfer rate or transfer delay requirements. The goal is to find routes for these demands so that some performance measure is optimized.

We begin by discussing a scenario that motivates the routing problem formulation taken up in this chapter. Analysis of the problem provides insights into how link weights, used by routing protocols, can be set so as to optimize a specified performance objective. The important conclusion is that optimal routing can be realized by using well-known shortest path algorithms, when link weights are defined appropriately. Next, we discuss the Dijkstra and the Bellman–Ford algorithms for shortest path routing. This is followed by a discussion of a generalized Dijkstra algorithm that is applicable to a wide range of problems. Last, we briefly discuss common routing protocols.

14.1 Elastic Aggregates and Traffic Engineering

Consider Figure 14.1, which shows a network with routers and links. Suppose that a number of Web servers are attached to the network via router s. Similarly, a number of users who are browsing are attached to the network through router t. Users submit file download requests to the Web servers. This causes traffic to flow between s and t.

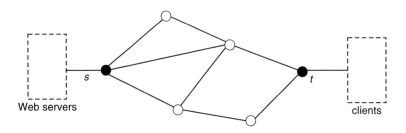

Figure 14.1   A network with routers and links. Several Web servers are attached to the network via the router _s_. Similarly, a number of Web clients are attached to the network through router _t_. Clients send requests for downloading files from the servers. We are interested in characterizing the demand between _s_ and _t_.

Suppose that a request arrives for downloading a file of size V bytes. Then these V bytes must be transferred from s to t. Consider a time interval of length T. Suppose that the number of download requests arriving over this interval is $N(T)$ and that the amounts to be downloaded (i.e., the file sizes) are $V_1, V_2, \ldots, V_{N(T)}$. For simplicity, let us also assume that the file sizes are independent and identically distributed. Then over the interval T, the server receives requests for an aggregate amount $V(T) := \sum_{i=1}^{N(T)} V_i$. This holds for every sample path. So, over the interval of length T, _on the average_, the server receives requests for an aggregate amount given by

$$\overline{V(T)} = (EV)\overline{N(T)}$$

where EV is the average file size and $\overline{N(T)}$ is the average number of requests arriving over the interval T. If we divide both sides of the equation by T, then the quantity on the left ($\overline{V(T)}/T$) is the average rate at which $V(T)$ grows with time. $\overline{V(T)}/T$, called the offered load, is the average rate at which the server emits traffic. On the right side, $\overline{N(T)}/T$ is simply the average rate at which download requests arrive. Thus, denoting the offered load by ρ and the average arrival rate of download requests by λ, we have

$$\rho = \lambda\, EV$$

Note that ρ is expressed in bytes/sec. When we say that there is a demand of size d between a source–destination pair, we mean that the offered load between that pair is d. As mentioned in Section 7.6.7, this is simply the offered byte rate into a link. For a formal treatment of load and stability, see Section 7.7.3.

Suppose that there are a number of source–destination pairs attached to the network and that each pair presents a demand of some size. As mentioned earlier, each demand is an offered load of aggregated elastic traffic. It is important to note that there are many individual sessions, or microflows, that constitute a demand. In our example, each individual download request generates a corresponding session; but *demand* refers to the average offered load generated by all sessions that share the routers s and t. The network of routers and links is a resource that is to be shared among all the demands such that each requirement is satisfied. The choice of routes for the demands determines how effective the sharing will be. For the same set of aggregate demands, poorly engineered routes can lead to poor session transfer throughputs. For a discussion of the relation between load and throughput, see Section 7.6.7. From the end user's perspective, poor throughput means that Web pages take a long time to appear on the screen.

How should the network operator devise a scheme for sharing the network resources among the competing demands? The first point to note is that the operator is at liberty to *split* an aggregate demand among multiple paths between a source and a destination, in case multiple paths exist. Such sharing of the load leads to efficient utilization of network resources and also enables the operator to satisfy larger aggregate demands. Essentially, the network operator's problem is to consider the demands that have been presented and decide how to split these demands, and route each "piece" on the network, so that link capacity constraints are respected.

In this context, an analogy may be useful. The set of presented demands is like a set of items that a person embarking on a journey wishes to carry, and the network is like a bag into which the items must be packed. Obviously, the bag has a finite volume, just as the network has a finite capacity. The items are "flexible" in the sense that they can be split into smaller sizes and rearranged so as to fit into the bag, if possible. Correspondingly, the demands can be split and mapped on to the network resources so that the aggregate demands fit within the network capacity, if possible. For this reason, the problem of routing aggregates of elastic traffic that we consider in this chapter is like a *packing* problem.

If the routing through the network and the link capacities are such that the source router of every source–destination pair can transmit packets at a rate equal to the corresponding offered load *without any link queue becoming unbounded*, then the network is stable. Let us consider a link in the network and assume that it is equipped with an infinite link buffer. Given a routing, there will be several demands passing through that link. If the aggregate offered load of all demands passing through the link is less than the link capacity, then the packet queue at this link is finite with probability 1. If the same is true for all links, then the network is stable.

However, in practice, link buffers are finite. Finite link buffers ensure that no link queue can become unbounded, and so the problem of network stability does not arise. But what will happen when the aggregate offered load on a link approaches its capacity? The link queue will start building up, leading eventually to dropped packets when the buffer is full. Even if packets are not dropped, the average queue length will be comparatively large, and arriving packets will see longer average delays. Noting that we are considering elastic traffic in this chapter (i.e., traffic carried on TCP sessions), we recall from Section 7.6.5 that the performance of individual TCP sessions is dependent on average round-trip times. This shows that larger average queues in a network mean poorer TCP session throughputs. Of course, TCP session performance becomes even worse when queues overflow and packets are dropped.

Hence, a routing that merely ensures that the total offered load on a link is less than its capacity may not be adequate to ensure satisfactory TCP session performance. For this, we need to ensure that average queue lengths are low. We can achieve this by making sure that link capacities are *sufficiently higher* than aggregate offered loads. It has been observed (see Section 7.6.7) that for networks with large bandwidth-delay products, as long as the aggregate offered load is not more than 90% of the link capacity, TCP session throughputs are good. This observation leads to the idea of *de-rating capacity*. This means that when formulating the routing problem, we pretend, for example, that a link's capacity is only $0.9C$ when it is really C. See the remarks at the end of Section 7.7.4. Then a feasible solution to the routing problem (in the sense that the carried traffic on each link is less than its capacity) also automatically ensures good TCP session performance. Thus, in the following discussion, link capacities can be viewed as de-rated link capacities.

We begin by considering a problem in which the demands are given, arbitrary splitting of demands is allowed, and the objective is to maximize the smallest spare link capacity. The analysis leads to the conclusion that a routing that achieves the maximization can be viewed as shortest path routing with appropriate link weights. It has been shown that this general principle—optimal routing can be viewed as shortest path routing with appropriate link weights—holds for a variety of optimization objectives. Thus, when the demands are given and they can be split arbitrarily, a variety of traffic-engineering objectives can be met by simply using shortest path routing with appropriate link weights.

Note that the link weights depend on the demands that are to be routed. When the set of demands changes, the weights must be recomputed. It is well known that average traffic demands can vary significantly from one interval of time to another. For example, during the busy hours from, say, 9:00 AM to 4:00 PM, the average traffic demand is distinctly higher than it is from 8:00 PM to 8:00 AM. Thus, we can possibly recompute the weights to be used during the busy and lean hours. Fortunately, this type of change in the traffic pattern does not occur frequently. Therefore, recomputing weights for changed traffic scenarios is feasible in practice.

Widely deployed routing protocols such as OSPF (Open Shortest Path First, discussed in Section 14.4.2) allow link weights to be assigned by a network administrator. Thus, a mechanism for implementing optimal routing by setting the correct weights already exists in current protocols. However, the traditional approach to link weight assignment has been based on simple heuristics. One such heuristic is simply to set all weights to 1. For this weight setting, shortest path routing reduces to minimum-hop-count routing. Another heuristic strategy recommended by Cisco is to assign a link weight that is inversely proportional to the link capacity. This encourages traffic to use links that have higher capacity.

However, experience has shown that such weight assignments usually lead to congestion on some links of a network, whereas other links have unused capacity. The point is that a link with a relatively low weight may be part of several routes, and the traffic on all these routes causes the link to get congested. Heuristic-based link weight assignment can thus be regarded as a somewhat coarse control mechanism that does not provide sufficient flexibility to design routes for traffic demands in desirable ways. This observation has led to the development of a completely different approach to routing that allows a network operator more control in specifying routes. Here, the idea of setting weights and then routing based on shortest paths according to these weights is abandoned. Rather, the source of a traffic demand specifies the complete sequence of intermediate routers that the demand must traverse. This is referred to as *explicit routing of virtual paths*, where *virtual path* refers to an aggregate demand whose route through the network is explicitly specified. We use *virtual paths* in the generic sense, to mean the path taken by some aggregate demand; thus, we do not refer to ATM virtual paths alone. This approach has led to the creation of new technologies such as multiprotocol label switching (MPLS). The MPLS *label* is an identifier at layer 2 in the OSI model. In an MPLS network, data packets are switched by consulting forwarding tables containing labels instead of IP addresses. The ingress *label-switching router* (LSR) assigns the labels by classifying incoming packets as belonging to one of a set of *forwarding equivalence classes* (FECs). The packets move along *label-switched paths* (LSPs) that have been set up previously. The idea is very similar to that used in an ATM network, where the VPI/VCI field plays the role of the label, and ATM VPs play the role of the LSPs. MPLS is discussed in Section 15.4.

Traffic engineering is the act of causing traffic to follow routes that are desirable, in that they lead to efficient network utilization as well as satisfactory performance perceived by users of the network. The main motivation for developing a technology like MPLS was the improved traffic engineering capabilities it provided. In this context, the insight that many traffic engineering objectives can be achieved almost completely by using familiar shortest path routing *with appropriate link weights* is particularly interesting. The implication is that there may be no need, after all, to devise new technologies with explicit mechanisms for better traffic engineering. Traditional shortest path routing may perform almost as well if the appropriate weights are used.

14.2   Optimal Routing

Let a network be represented as a directed graph $\mathcal{G}(\mathcal{N}, \mathcal{L})$ where $\mathcal{N}$ is the set of routers (or nodes) and $\mathcal{L}$ is the set of directed links. This means that if a and b are two nodes in $\mathcal{N}$, then the link $a \rightarrow b$ and the link $b \rightarrow a$ are distinct. Thus, any link $l \in \mathcal{L}$ has a *head node* $h(l)$ and a *tail node* $w(l)$, and, as the names indicate, the link is directed from the head to the tail. There are K demands that are to be routed on this network. Each demand is associated with an ordered pair of nodes (n_1, n_2), where $n_1, n_2 \in \mathcal{N}$. Note that n_1 is the source of the demand, and n_2 is the destination. We number the quantities as follows: Demands are numbered $1, 2, \ldots, k, \ldots, K$, nodes are numbered $1, 2, \ldots, i, \ldots, N$, so that $|\mathcal{N}| = N$, and links are numbered $1, 2, \ldots, l, \ldots, L$, so that $|\mathcal{L}| = L$.

We assume that traffic corresponding to a demand can be split arbitrarily across multiple paths between the source and the destination. Current routing protocols like OSPF allow only *even* flow splitting among multiple equal-cost paths, and not arbitrary splitting. This means that if there are n shortest paths (of the same length) between a source and a destination, then a fraction $1/n$ of the demand can be routed on each path. The justification for assuming arbitrary splitting is that it allows a formulation from which useful insights can be obtained. Also, future implementations may be able to support arbitrary splitting.

With arbitrary splitting allowed, it is possible that *every* link in $\mathcal{L}$ carries some part of a demand $d(k)$, $1 \leq k \leq K$. This motivates us to define a *flow vector* $\mathbf{x}(k)$ corresponding to the kth demand, with $x(k)_l$ being the amount of the kth demand carried on link l. It is a vector of dimension L. Unless otherwise stated, we assume $\mathbf{x}$ to be a column vector, and $\mathbf{x}^T$ denotes the transpose of $\mathbf{x}$.

The topology of a network can be summarized using its *node–link incidence matrix* **A**. Note that **A** is an $N \times L$ incidence matrix, with a row for each node and a column for each link. Let $\mathbf{A}_{i,l}$ represent the (i,l)th element of **A**. Let $\mathbf{A}_{i,\cdot}$ represent the ith row of **A**, and let $\mathbf{A}_{\cdot,l}$ represent the lth column of **A**. Then the column corresponding to link l has the following entries:

$$\mathbf{A}_{i,l} = \begin{cases} +1 & \text{if } i \text{ is the head node of link } l \\ -1 & \text{if } i \text{ is the tail node of link } l \\ 0 & \text{otherwise} \end{cases}$$

Thus, the column corresponding to link l has a single $+1$ and a single -1, and the rest of the entries are zeros.

With the node–link incidence matrix **A** and the flow vector $\mathbf{x}(k)$ corresponding to demand k thus defined, consider the product $\mathbf{A}\mathbf{x}(k)$. The product is a vector with N elements. The ith element of this vector is the product of the row $\mathbf{A}_{i,\cdot}$ and the vector $\mathbf{x}(k)$. It can be seen that $\mathbf{A}_{i,\cdot}\mathbf{x}(k)$ is nothing but the *net outgoing traffic* corresponding to demand k at node i. All the $+1$'s in $\mathbf{A}_{i,\cdot}$ identify links incident on node i carrying outgoing traffic, and all the -1's in $\mathbf{A}_{i,\cdot}$ identify links carrying incoming traffic into node i.

Let $s(k)$ and $t(k)$ be the source and destination nodes of demand k. Then it is clear that at any node i other than $s(k)$ and $t(k)$, the net outgoing traffic corresponding to demand k is zero, because the node i neither sources nor sinks demand k traffic. The same argument shows that when $i = s(k)$, the product $\mathbf{A}_{i,\cdot}\mathbf{x}(k)$ should be $d(k)$, the size of demand k, and similarly, when $i = t(k)$, the product $\mathbf{A}_{i,\cdot}\mathbf{x}(k)$ should be $-d(k)$. These *flow conservation equations* are as follows:

$$\mathbf{A}_{i,\cdot}\mathbf{x}(k) = \begin{cases} d(k) & \text{if} & i = s(k) \\ -d(k) & \text{if} & i = t(k) \\ 0 & \text{otherwise} \end{cases}$$

If we now consider all the rows of **A** together, then we have the following compact equation:

$$\mathbf{A}\mathbf{x}(k) = \mathbf{v}(k) \tag{14.1}$$

where $\mathbf{v}(k)$ is an $N \times 1$ vector with the following entries:

$$
v(k)_i = \begin{cases} d(k) & \text{if} & i = s(k) \\ -d(k) & \text{if} & i = t(k) \\ 0 & \text{otherwise} \end{cases}
$$

From what we have seen, $\mathbf{v}(k)$ is a vector specifying the amount of net outgoing demand k traffic from each node in the network.

Equation 14.1 holds for all k, $1 \leq k \leq K$. Thus, there are K equations of the form of Equation 14.1—one for each $k \in \{1, 2, \ldots, K\}$. Let us now obtain a single compact equation that expresses the K equalities together. To this end, consider the vector

$$
\begin{bmatrix} \mathbf{x}(1) \\ \mathbf{x}(2) \\ \vdots \\ \mathbf{x}(K) \end{bmatrix}
$$

This is simply a column vector whose first L elements specify the flow vector for demand 1, the next L elements specify the flow vector for demand 2, $\ldots$, and the last L elements specify the flow vector for demand K. Thus, it is a vector of dimension $KL \times 1$. Now consider the matrix

$$
\mathbb{A} = \begin{bmatrix} \mathbf{A} & 0 & 0 & \cdots & 0 \\ 0 & \mathbf{A} & 0 & \cdots & 0 \\ \vdots & \vdots & \vdots & \ddots & \vdots \\ 0 & 0 & \cdots & 0 & \mathbf{A} \end{bmatrix}
$$

There are K block elements in each row and K block elements in each column. $\mathbf{A}$ is the familiar node–link incidence matrix, of dimension $N \times L$. 0 is also a matrix of dimension $N \times L$. Hence $\mathbb{A}$ is a matrix of dimension $KN \times KL$.

With these definitions, consider the equation

$$\begin{bmatrix} A & 0 & 0 & \cdots & 0 \\ 0 & A & 0 & \cdots & 0 \\ \vdots & \vdots & \vdots & \ddots & \vdots \\ 0 & 0 & \cdots & 0 & A \end{bmatrix} \begin{bmatrix} \mathbf{x}(1) \\ \mathbf{x}(2) \\ \vdots \\ \mathbf{x}(K) \end{bmatrix} = \begin{bmatrix} \mathbf{v}(1) \\ \mathbf{v}(2) \\ \vdots \\ \mathbf{v}(K) \end{bmatrix} \tag{14.2}$$

Because $\mathbf{v}(k)$ is a vector of dimension $N \times 1$ for each $k \in \{1, 2, \ldots, k\}$, the vector on the right side is of dimension $KN \times 1$. This is what we expect when a $KN \times KL$ matrix is multiplied with a $KL \times 1$ vector. Equation 14.2 is the compact flow conservation equation we were looking for. Clearly, it is nothing but K equations of the form $\mathbf{A}\mathbf{x}(k) = \mathbf{v}(k)$, with $1 \le k \le K$.

Now consider any link in the network. For a feasible routing, the sum of all flows on a link should stay below the link capacity. Suppose that $\mathbf{C}$ denotes the column vector of de-rated link capacities, with C_l being the de-rated capacity of link l. Then, for a feasible routing, we have

$$\mathbf{x}(1) + \mathbf{x}(2) + \cdots + \mathbf{x}(K) \le \mathbf{C}$$

This is a vector equation. Both sides are vectors with L elements. The vector of *spare capacities*, denoted by $\mathbf{z}$, is given by

$$\mathbf{z} = \mathbf{C} - (\mathbf{x}(1) + \mathbf{x}(2) + \cdots + \mathbf{x}(K))$$

Let $z := \min_{l \in \mathcal{L}} z_l$ be the *smallest* spare capacity corresponding to a given feasible routing. Then the following inequality holds:

$$\mathbf{x}(1) + \mathbf{x}(2) + \cdots + \mathbf{x}(K) \le \mathbf{C} - z\mathbf{1}$$

where $\mathbf{1}$ is a column vector of L elements, all of which are 1.

Again, the foregoing inequality can be written in compact form as follows. Let $\mathbf{I}$ be the $L \times L$ identity matrix. Let $\mathbb{I}$ be defined as

$$\mathbb{I} := \begin{bmatrix} \mathbf{I} & \mathbf{I} & \cdots & \mathbf{I} \end{bmatrix}$$

There are K block elements in the matrix $\mathbb{I}$. Thus, the dimension of $\mathbb{I}$ is $L \times KL$. Then we have

$$
\begin{bmatrix} \mathbf{I} & \mathbf{I} & \cdots & \mathbf{I} \end{bmatrix}
\begin{bmatrix} \mathbf{x}(1) \\ \mathbf{x}(2) \\ \vdots \\ \mathbf{x}(K) \end{bmatrix}
+ z\mathbf{1} \leq
\begin{bmatrix} C_1 \\ C_2 \\ \vdots \\ C_L \end{bmatrix}
\tag{14.3}
$$

As expected, the product of the $L \times KL$ matrix $\mathbb{I}$ and the $KL \times 1$ vector representing the flow vectors of all demands gives the $L \times 1$ vector of link capacities.

Given a network and a set of demands, there may be many feasible routings. To choose one routing from this set, the standard approach is to define an *objective function* and then choose the routing that optimizes the objective function. Let us define the objective function as the foregoing quantity z (i.e., the objective function is the smallest spare capacity resulting from a routing). Then an optimal routing would be the one that maximizes the smallest spare capacity. Defining an optimal routing in this way is reasonable, because *any* link in the network has a spare capacity of at least z. This increases the chance that a future demand between any pair of nodes in the network will find sufficient free capacity. In other words, we avoid routings that lead to a bottleneck link having very little spare capacity. Furthermore, this objective promotes a balanced utilization of capacity and does not create hot spots.

Putting together all the elements, we have the following optimization problem:

$$
\max z
$$

subject to

$$
\begin{bmatrix}
\mathbf{A} & 0 & 0 & \cdots & 0 \\
0 & \mathbf{A} & 0 & \cdots & 0 \\
\vdots & \vdots & \vdots & \ddots & \vdots \\
0 & 0 & \cdots & 0 & \mathbf{A}
\end{bmatrix}
\begin{bmatrix} \mathbf{x}(1) \\ \mathbf{x}(2) \\ \vdots \\ \mathbf{x}(K) \end{bmatrix}
=
\begin{bmatrix} \mathbf{v}(1) \\ \mathbf{v}(2) \\ \vdots \\ \mathbf{v}(K) \end{bmatrix}
\tag{14.4}
$$

$$[\begin{array}{cccc} \mathbb{I} & \mathbb{I} & \cdots & \mathbb{I} \end{array}] \begin{bmatrix} \mathbf{x}(1) \\ \mathbf{x}(2) \\ \vdots \\ \mathbf{x}(K) \end{bmatrix} + z \cdot \mathbf{1} \leq \mathbf{C} \qquad (14.5)$$

$$\mathbf{x}(k) \geq 0, \quad 1 \leq k \leq K, \qquad z \geq 0 \qquad (14.6)$$

We can see that this is a *linear program* (LP), with the variables being $\mathbf{x}(k)$, $1 \leq k \leq K$, and z. The objective is a linear function of the variables, with z being the sole variable determining its value. See Appendix C for an overview of linear programming.

Even though the foregoing representation is compact, it is still not in the usual form of an LP. To represent the LP in a standard form, *all* the variables $\mathbf{x}(k)$, $1 \leq k \leq K$, and z must appear in a single vector, and the constraint Equations 14.4 and 14.5 must be written in terms of this single vector. Recalling the definitions of $\mathbb{A}$ and $\mathbb{I}$, and letting

$$\mathbf{x} := \begin{bmatrix} \mathbf{x}(1) \\ \mathbf{x}(2) \\ \vdots \\ \mathbf{x}(K) \end{bmatrix} \qquad \mathbf{v} := \begin{bmatrix} \mathbf{v}(1) \\ \mathbf{v}(2) \\ \vdots \\ \mathbf{v}(K) \end{bmatrix}$$

consider the following problem:

$$\max z$$

subject to

$$\begin{bmatrix} \mathbb{A} & \mathbf{0} \\ \mathbb{I} & \mathbf{1} \end{bmatrix} \begin{bmatrix} \mathbf{x} \\ z \end{bmatrix} \begin{matrix} = \\ \leq \end{matrix} \begin{bmatrix} \mathbf{v} \\ \mathbf{C} \end{bmatrix} \qquad (14.7)$$

$$\mathbf{x} \geq 0, \qquad z \geq 0$$

On the left side, $\mathbf{0}$ is an all-0's vector of size $KN \times 1$, whereas $\mathbf{1}$ is an all-1's vector of size $L \times 1$.

> **Exercise 14.1**
> Verify that both sides of Equation 14.7 are vectors of dimension $(KN + L) \times 1$.

This is the final form of the optimization problem that defines the optimal routing. This will be called the *primal problem*. Because it is a linear program, efficient algorithms for computing its solution are available, and we can actually obtain the optimal routing.

Interestingly, it turns out that useful insights into the optimal solution can be obtained by considering the *dual* of this problem. See Appendix C for an overview of duality. The primal problem has $(KL + 1)$ variables and $(KN + L)$ constraints. So the dual problem will have $(KN + L)$ variables, one corresponding to each primal constraint, and $(KL + 1)$ constraints, one corresponding to each primal variable.

Let us group the $(KN + L)$ dual variables into two groups: the first corresponding to the first KN equality constraints in Equation 14.7, and the second corresponding to the last L constraints in Equation 14.7. We think of the dual variables as row vectors. Let $\mathbf{u}(1)^T$ be a $1 \times N$ vector of dual variables corresponding to the first N equality constraints, let $\mathbf{u}(2)^T$ be another $1 \times N$ vector of dual variables corresponding to the next N equality constraints, ..., and let $\mathbf{u}(K)^T$ be another $1 \times N$ vector of dual variables corresponding to the last N equality constraints in Equation 14.7. Let

$$\mathbf{u}^T := \left[\mathbf{u}(1)^T, \mathbf{u}(2)^T, \dots, \mathbf{u}(K)^T \right]$$

denote the $1 \times KN$ vector of dual variables in the first group. Clearly, each element of $\mathbf{u}$ corresponds to a (node, demand) pair. Similarly, let

$$\mathbf{y}^T := [y_1 \ y_2 \ \cdots \ y_L]$$

denote the $1 \times L$ vector of dual variables corresponding to the group of L inequality constraints. Each element of $\mathbf{y}$ corresponds to a link. Then the dual problem can be stated as follows:

$$\min \quad (\mathbf{u}^T \mathbf{v} + \mathbf{y}^T \mathbf{C})$$

subject to

$$[\; \mathbf{u}^T \quad \mathbf{y}^T \;] \begin{bmatrix} \mathbb{A} & 0 \\ \mathbb{I} & 1 \end{bmatrix} \geq [\; \mathbf{0}^T \quad 1 \;] \tag{14.8}$$

$$\mathbf{y}^T \geq 0$$

Note that $\mathbf{0}^T$ on the right side of Equation 14.8 is a $1 \times KL$ vector of all zeros. Note that in Equation 14.7, the first KN constraints are actually *equalities*. Hence, the corresponding dual variables in $\mathbf{u}$ can be positive or negative.

The arguments that follow make use of the complementary slackness theorem, which can be roughly stated as follows. (See Appendix C for a formal statement of the theorem). Consider the optimal primal solution. We know from the primal formulation that the primal variables $\mathbf{x}$ and z are nonnegative. Then, if in the *optimal solution* of the primal problem, a primal variable is strictly positive, then the corresponding constraint in the dual formulation, evaluated at the *optimal dual values*, should be tight (i.e., should not be slack).

For example, suppose that the optimal solution of the primal problem $(\mathbf{x}^*, z^*)$ is such that $z^* > 0$. That is, suppose that the maximum value of the least spare capacity is z^*. Let the corresponding optimal dual solution be denoted by $\mathbf{u}^{T,*}$ and $\mathbf{y}^{T,*}$. Then the last of the inequalities in the constraint set in Equation 14.8 should be tight:

$$\sum_{l=1}^{L} y_l^* = 1$$

Optimal Routes Are Shortest Paths

Now suppose that in the optimal solution to the primal problem, there is a positive flow corresponding to demand k, from its source node $s(k)$ to its destination node $t(k)$, along the links $j_1, j_2, \ldots, j_k$. Thus, the links $j_1, j_2, \ldots, j_k$ determine the optimal path of the flow from $s(k)$ to $t(k)$. Clearly, the head of link j_1 $(h(j_1)) = s(k)$, and the tail of link j_k $(w(j_k)) = t(k)$. This means that all the primal variables $x(k)^*_{j_1}$, $x(k)^*_{j_2}, \ldots, x(k)^*_{j_k}$ are strictly positive.

Then, by the complementary slackness theorem, the corresponding constraints in the dual formulation, evaluated at the optimal dual values, should be tight. This means that in Equation 14.8, the corresponding constraints should be equalities. We have already seen that 0^T on the right side of Equation 14.8 is a $1 \times KL$ row vector, with the first L elements corresponding to demand 1, the next L elements corresponding to demand 2, and so on. So we now consider the kth group of L elements and select elements $j_1, j_2, \ldots, j_k$ in this group. These identify the constraints that must be equalities.

Therefore, recalling the structure of $\mathbb{A}$, $\mathbb{I}$, and $\mathbf{u}$, we have

$$\mathbf{u}(k)^{T,*} \mathbf{A}_{\cdot, j_l} + y_{j_l}^* = 0, \qquad l = 1, 2, \ldots, k$$

Using the structure of the j_lth column of $\mathbf{A}$, which has a $+1$ entry corresponding to the head node of link j_l, $h(j_l)$; a -1 entry corresponding to the tail node of link j_l, $w(j_l)$; and zeros elsewhere, we can write

$$u(k)_{h(j_l)}^* - u(k)_{w(j_l)}^* + y_{j_l}^* = 0, \qquad l = 1, 2, \ldots, k \qquad (14.9)$$

Now recall that the links $j_1, j_2, \ldots, j_k$ form a path from $s(k)$ to $t(k)$. Therefore, the tail node of j_l coincides with the head node of j_{l+1}, $1 \le l \le (k-1)$. This allows us to form a telescopic sum of Equation 14.9 for $l = 1, 2, \ldots, k$, and the result is

$$u(k)_{t(k)}^* - u(k)_{s(k)}^* = y_{j_1}^* + y_{j_2}^* + \cdots + y_{j_k}^* \qquad (14.10)$$

This is the relation connecting the optimal dual variables along an optimal path for the flow of demand k in the primal problem.

Now if $\tilde{j}_1, \tilde{j}_2, \ldots, \tilde{j}_k$ are the links on *any* path between $s(k)$ and $d(k)$, then we do not know whether the optimal routing will result in any demand k flow on this path and therefore cannot assert that the optimal primal variables $\mathbf{x}(k)$ are strictly positive on these links. This means that all we can say is that the corresponding inequalities in Equation 14.8 hold with the $\ge$ sign. Then a similar development leads to

$$y_{\tilde{j}_1}^* + y_{\tilde{j}_2}^* + \cdots + y_{\tilde{j}_k}^* \ge u(k)_{t(k)}^* - u(k)_{s(k)}^*$$

By Equation 14.10, we have

$$y_{\tilde{j}_1}^* + y_{\tilde{j}_2}^* + \cdots + y_{\tilde{j}_k}^* \ge y_{j_1}^* + y_{j_2}^* + \cdots + y_{j_k}^* \qquad (14.11)$$

where $j_1, j_2, \ldots, j_k$ define the optimal path, and $\tilde{j}_1, \tilde{j}_2, \ldots, \tilde{j}_k$ define any other path for flow k. Equation 14.11 says that if we treat the dual variables y_l^*, $1 \leq l \leq L$, as the weights of links, then an optimal path for flow k is nothing but the shortest path by these weights. Also, $u(k)_{t(k)}^* - u(k)_{s(k)}^*$ gives the smallest possible path length for a path connecting $s(k)$ and $t(k)$.

It is clear that the argument holds for any k, $1 \leq k \leq K$. Hence the optimal paths for all the demands can be viewed as shortest paths when the optimal dual variables are considered as weights.

Sensitivity

Other interesting insights can be obtained from the dual programming formulation. Most of these are consequences of the *shadow price* interpretation of the dual variables. An example can be given as follows. The primal objective function is the smallest spare link capacity, and the primal problem is concerned with maximizing this quantity. How would this change if the capacity of a link l is increased? In other words, what is the *sensitivity* of the value of the primal LP to changes in link capacity? We can obtain the answer by recalling from linear programming duality theory (see Appendix C) that when both the primal and the dual programs are feasible, their optimal values are equal. Then the optimal value of the primal program is given by

$$z^* = \mathbf{u}^{T,*}\mathbf{v} + \mathbf{y}^{T,*}\mathbf{C}$$

From this equation, we see that, for any link $l \in \mathcal{L}$,

$$\frac{\partial z^*}{\partial C_l} = y_l^*$$

showing that the optimal value of the dual variable, y_l^*, determines the sensitivity of the primal objective function. Because $\mathbf{y} \geq 0$, an increase in link capacity can only cause the largest value of least spare capacity to increase. More importantly, it can be seen that increasing the capacity of only those links with a current spare capacity of exactly z^* can lead to this increase. If the spare capacity on a link is already more than z^*, then increasing its capacity alone causes no change in the optimal value of the primal objective function. Thus, the sensitivity of some links, or the optimal dual variable values corresponding to some links, can be seen to be 0.

Similarly, how would the optimal value of the primal objective function change if a demand size increased? Intuitively, one expects a *nonpositive* sensitivity to demand sizes, because increasing a demand size can lead only to a reduction in the maximum value of least spare capacity. This leads to the following conclusion. The objective function in the dual problem can be written as

$$
\mathbf{u}^{T,*}\mathbf{v} + \mathbf{y}^{T,*}\mathbf{C} = \sum_{k=1}^{K} \left(u(k)^{*}_{s(k)} - u(k)^{*}_{t(k)} \right) d(k) + \sum_{l=1}^{L} y^{*}_{l} C_{l}
$$

Thus, we have

$$
\frac{\partial z^{*}}{\partial d(k)} = u(k)^{*}_{s(k)} - u(k)^{*}_{t(k)}
$$

Therefore, the values of the optimal dual variables must be such that for all $1 \leq k \leq K$,

$$
u(k)^{*}_{s(k)} - u(k)^{*}_{t(k)} \leq 0 \tag{14.12}
$$

Let us now consider for which demands the sensitivity to demand sizes, given by $\left(u(k)^{*}_{s(k)} - u(k)^{*}_{t(k)} \right)$, is nonzero. To begin, we note that with optimal routing, the least spare capacity over the whole network is given by z^{*}, the optimal value of the primal objective function. Consider a demand k and suppose that the optimal route for this demand is over some m paths where $m \geq 1$. Consider the least spare link capacity along each of these m paths. If the least spare capacity *on these paths* is more than z^{*}, then increasing the size of demand k slightly will not change the least spare link capacity *over the whole network*! In other words, the sensitivity of the optimal value of the objective function to the demand size of demand k should be 0. By the same argument, if we find a demand $\hat{k}$ for which the sensitivity given by Equation 14.12 is strictly negative, it follows that this demand $\hat{k}$ sees a spare capacity of exactly z^{*} on all the m paths. Hence, if the size of this demand $\hat{k}$ is increased by amount ϵ, then we can increase the amount of demand $\hat{k}$ traffic on each of the m paths by amount ϵ/m. This would decrease the least spare capacity over the whole network by the same amount ϵ/m. Letting ϵ go to 0, it follows that the sensitivity of the optimal value to demand $\hat{k}$ is $1/m$. Hence, the optimal value of the dual variable $\left(u(k)^*_{s(k)} - u(k)^*_{t(k)} \right)$ is either 0 or $1/m$, where m is the number of paths that the optimal routing of demand k uses.

Our arguments show that the solution of an optimal routing problem yields routes that can be viewed as shortest paths according to appropriately defined weights. We considered a particular objective: maximization of the least spare link capacity. It has been shown, however, that optimal routing problems with other objectives also yield optimal routes that can be viewed as shortest paths. This means that after the weights have been obtained, a routing protocol can simply compute shortest paths using well-known algorithms such as the Dijkstra algorithm, and packets will move along optimal routes. There is no need for explicit routing schemes (for example, MPLS) to ensure that packets flow along optimal routes.

OSPF Routing

If there are multiple shortest paths between a source and a destination, then a demand would have to be split across these paths. When there are m shortest paths between a source–destination pair, optimal routing may require that fractions f_1, f_2, ..., f_m be put on the m paths, where $\sum_{i=1}^{m} f_i = 1$. In general, the optimal split can be arbitrary. That is, there is no reason to expect that the split should be even ($f_i = 1/m$, $1 \leq i \leq m$). However, in practice, forwarding implementations may not allow arbitrary splitting. For the implication of this, consider the simple example in Figure 14.2. A demand of size 20 units from a to c needs to be routed. There are two shortest paths between a and c, each of weight 2. If *even* flow

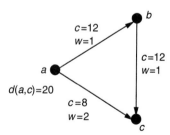

Figure 14.2 A demand of size 20 units between *a* and *c* is to be routed across the network. The link capacities and weights assigned to the links are as shown. Thus, there are two shortest paths between *a* and *c*. However, if the demand needs to be split *evenly* across the two paths, then it cannot be supported.

splitting is required, then each path must carry 10 units, and the demand cannot be supported on the network. But if arbitrary splitting were allowed, then 12 units could be routed on the path $a \to b \to c$ and the remaining 8 units on the path $a \to c$. Thus, the requirement of even splitting can become restrictive.

The OSPF routing protocol works by computing shortest paths between node pairs based on link weights that are assigned by the network operator. Therefore, it seems that OSPF is perfectly equipped for implementing optimal routing. However, typical implementations of OSPF also require that a demand be split evenly across multiple shortest paths between a source and a destination. This characteristic of OSPF implementations makes it difficult to use it for optimal routing.

Can we formulate an optimal *OSPF routing* problem that incorporates the characteristics of OSPF routing in the problem statement itself? Such a formulation would need to consider the link weights and the flow vectors corresponding to each demand as the variables of the problem. The link weights would determine the shortest paths for each demand. For a link l that does not belong to the set of shortest paths for demand k, we require that the flow element $x(k)_l = 0$, because OSPF routes traffic on shortest paths only. When there are multiple shortest paths for a node pair, additional constraints on the flow vectors would appear. For example, if links l_1 and l_2 both belong to the set of shortest paths for demand k and also *share a common head node*, then we would require

$$x(k)_{l_1} = x(k)_{l_2}$$

because demand k must be split evenly. The variables, the constraints, and a chosen objective function would together constitute an optimization problem whose solution would provide the *optimal* OSPF weights that optimize the objective function.

It can be shown that such an optimization formulation cannot result in a linear program. The implication is that the optimal OSPF routing problem promises to be computationally hard. We must resort to heuristic algorithms that search through the space of OSPF weights to come up with a "good" OSPF weight setting. This means that given an OSPF weight vector, we must define a *neighborhood* of it in the weight space and must also devise rules for moving from the given point to another in its neighborhood. Such an approach has been followed in the literature, and good heuristic algorithms have been obtained.

To see the importance of good OSPF weight setting, consider the example in Figure 14.3. All links are bidirectional, and the capacity of each link is assumed

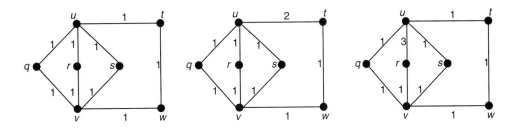

Figure 14.3   A seven-node, nine-link network is shown. All the links are assumed to be bidirectional. Also, the capacity of each link is assumed to be large. Nodes q, r, s, and w generate 1 unit of traffic each. All traffic is destined for node t. The weights assigned to each link are also shown beside the links. The three pictures show three different weight assignments. This example is taken from [107].

to be large. Nodes q, r, s, and w generate one unit of traffic each, for node t. The weights assigned to each link are shown beside the links. Three different weight assignments are considered. The picture on the left shows a weight assignment where all links have unit weight. This causes the traffic of q, r, and s to pass through u, because the path through u is shorter. Of course, the traffic from w takes the single-hop path to its destination t. With this assignment, the link $u \rightarrow t$ carries 3 units of traffic, whereas the link $w \rightarrow t$ carries 1 unit. This significant disparity in carried traffic may not be desirable.

A naive approach to reducing the load on link $u \rightarrow t$ would be to increase the weight of the link to 2. This is shown in the picture in the middle. In this case, each of the nodes q, r, and s finds two equal-length shortest paths to t. This leads to a load of 1.5 units on link $u \rightarrow t$, which is good, but it also means a load of 2.5 units on link $w \rightarrow t$.

The weight setting shown in the picture on the right is the best, because it leads to a load of 2 units each on links $u \rightarrow t$ and $w \rightarrow t$. Because the 4 units of traffic must reach node t, the most equitable division of the load is one in which each of the two incoming links at node t carries an equal amount of traffic. It can be seen that the best weight assignment for this example is not unique.

14.3   Algorithms for Shortest Path Routing

In this section, we consider algorithms that can be used to compute shortest paths, given a weighted network. It is assumed that complete information about the network topology and about all link weights is available to the algorithm. As discussed in a later section, some routing protocols include mechanisms to distribute topology information to all nodes in the network so that each node can have a full view of the network topology. When the full view is available, the algorithms discussed in this section can be utilized.

Prominent algorithms that are used to compute shortest paths are Dijkstra's algorithm and the Bellman–Ford algorithm. We also consider some enhancements of Dijkstra's algorithm in a later section.

14.3.1   Dijkstra's Algorithm

Let there be N nodes and L links in a given network. As usual, the network is represented as a directed graph $\mathcal{G}(\mathcal{N}, \mathcal{L})$, with $\mathcal{N}$ being the set of nodes and $\mathcal{L}$ being the set of links. The weights of the links in $\mathcal{L}$ are stored in an $N \times N$ matrix $\mathbf{W}$. If there is no link between a pair of nodes, the corresponding entry in $\mathbf{W}$ is set to ∞.

Suppose that we want the shortest paths from node 0 to all other nodes $1, 2, \ldots, (N-1)$ in the network. Let $\mathcal{A}$ represent the set of nodes to which shortest paths have been found. Note that $\mathcal{A}$ is initialized to $\{0\}$, because 0 is the only node to which a shortest path is known initially. Let $\delta_j, j \in \{1, 2, \ldots, (N-1)\}$, represent the distance of a shortest path from 0 to j, using only nodes in $\mathcal{A}$. Note that δ_j and $\mathcal{A}$ are updated as the algorithm iterates.

The pseudo-code for the Dijkstra algorithm is shown in Algorithm 14.1. In line 4, the "nearest" node, denoted k, is extracted from the set $\mathcal{A}$. In the next line, $\mathcal{A}$ is augmented to include k. Line 6 reevaluates distances to all nodes that are outside the set $\mathcal{A}$.

1: Input: $\mathcal{G}(\mathcal{N}, \mathcal{L})$, $\mathbf{W}$
2: Initialization: $\delta_0 = 0$, $\delta_j = \mathbf{W}_{0,j}$, $\mathcal{A} = \{0\}$
3: while $\mathcal{A} \neq \mathcal{N}$ do
4:     Define $k = \arg\min_{j \in \mathcal{N} \setminus \mathcal{A}} \delta_j$
5:     $\mathcal{A} = \mathcal{A} \cup \{k\}$
6:     $\forall j \in \{\mathcal{N} \setminus \mathcal{A}\},$ $\delta_j = \min[\delta_j,\ \delta_k + \mathbf{W}_{k,j}]$

Algorithm 14.1   Pseudo-code for the Dijkstra algorithm. $\mathcal{N} \setminus \mathcal{A}$ **refers to the set whose elements are members of** $\mathcal{N}$ **but not of** $\mathcal{A}$.

When the algorithm terminates, the variable δ_j contain? shortest path from node 0 to j, $j \in \{1, 2, \ldots, (N-1)\}$. The a is not recorded in the algorithm, but it is simple to obtain. Th initialized to the direct link between 0 and j. If there is no di initialized to the *null* path, indicated by ϕ. As the algorithm is updated whenever δ_j changes in line 6.

The worst case complexity of Dijkstra's algorithm is $O(N^2)$. This is because the extraction of the minimum in line 4 is of complexity $O(N)$, and there are at most N iterations (line 3).

14.3.2 The Bellman–Ford Algorithm

Let $\delta_j^{(H)}$ represent the shortest distance from node 0 to j when paths with at most H hops are considered. Clearly, as H increases, the set of paths grows larger so that eventually all paths from 0 to j are considered.

The pseudo-code for the Bellman–Ford algorithm is shown in Algorithm 14.2. In line 4, the shortest distance from 0 to j using paths having at most $(H+1)$ hops is found, given the shortest distance using paths having at most H hops. The extra hop considered uses link $i \rightarrow j$, as line 4 shows, and all possible $i \in \mathcal{N}$ are tried. The iterations are repeated until there is no improvement in the shortest distance from 0 to all the other nodes in $\mathcal{N}$ (line 5).

The worst case complexity of the Bellman–Ford algorithm is $O(N^3)$. Each node in $\mathcal{N}$ must be considered when taking the minimum over $i \in \mathcal{N}$ in line 4. This is an $O(N)$ operation. Line 4 also shows that this must be repeated for each $j \in \mathcal{N}$, while obtaining $\delta_j^{(H+1)}$. Combining this with Exercise 14.2, we get a complexity of $O(N^3)$.

1: Input: $\mathcal{G}(\mathcal{N}, \mathcal{L})$, $\mathbf{W}$
2: Initialization: $\delta_0^{(H)} = 0 \ \forall H \qquad \delta_j^{(0)} = \infty, \ j \neq 0 \qquad H = 0$
3: **repeat**
4: $\quad \forall j \in \mathcal{N}, \ \delta_j^{(H+1)} = \min_{i \in \mathcal{N}} [\delta_j^{(H)}, \delta_i^{(H)} + \mathbf{W}_{i,j}]$
5: **until** $\delta^{(H+1)} = \delta^{(H)}$, where $\delta^{(H)}$ is the vector $[\delta_0^{(H)}, \delta_1^{(H)}, \ldots, \delta_{N-1}^{(H)}]$

Algorithm 14.2 Pseudo-code for the Bellman–Ford algorithm.

Exercise 14.2

At most how many iterations of the Bellman–Ford algorithm are needed before the shortest distances to all nodes in $\mathcal{N}$ are found?

14.3.3   A Generalized Dijkstra Algorithm

As discussed in Section 14.3.2, we obtain path lengths by adding predefined link weights of the links on a path. Then polynomial time algorithms, such as the Dijkstra algorithm and the Bellman–Ford algorithm, can be used to find shortest paths from one node in a network to others.

In many situations of practical interest, however, the requirement is to find a path whose *length* (or metric, in general) is not obtainable by adding the lengths of its constituent links. For example, if the need is to find a path between s and t that has the maximum capacity, then the path length obviously cannot be found by adding the link capacities; we must take the minimum of the link capacities to obtain the capacity of the path. Notice that the operation required to obtain a path metric from the link metrics has changed from the usual addition to "taking the minimum of link metrics." Similarly, in some situations, the requirement is to find not the minimum hop-count path but instead the *widest min hop path*; among all paths with the minimum hop count, the need is to identify the one that has the largest capacity. If there are several minimum hop-count paths, choosing the one with the largest capacity leads to lower link loads. In this case, too, the operation required to obtain a path metric given the link metrics is not a straightforward addition.

Can we generalize the Dijkstra algorithm to apply to such cases? The motivation is that if the generalization is possible, a polynomial time algorithm (the "generalized" Dijkstra algorithm) can be used to obtain these paths of interest. We study this briefly in this section.

Optimality of Subpaths of Optimal Paths

Let P be an optimal path from source node s to destination node t. Let $\delta(P)$ be the cost of this path. We obtain $\delta(P)$ by adding link weights along P. Then, for any path P' between s and t, $\delta(P) \leq \delta(P')$. Let P_1 be a *subpath* of P, between nodes a and b, where a and b are two nodes on the path P. Then it is easily seen that P_1 is an *optimal path* between a and b. Suppose that this were not the case. Then there exists a path $\tilde{P}_1$ between a and b such that $\delta(\tilde{P}_1) < \delta(P_1)$. It is then clear that we can strictly improve the cost of the original path P, $\delta(P)$, by including the segment $\tilde{P}_1$ between a and b instead of the segment P_1. This contradicts the initial assumption that P is an optimal path.

The foregoing argument relies on the fact that given real numbers a, b, and c and the usual ordering on the real line $\mathbb{R}$, we have the following: If $a < b$, then $a + c < b + c$. The binary operator $+$ preserves strict inequalities. However, not all binary operators preserve strict inequalities. For example, consider the min operator, which chooses the smaller of two arguments given to it. It is easily seen that all we can say here is that if $a < (>)b$, then $\min(a, c) \leq (\geq) \min(b, c)$.

The bandwidth available on a path through a network is determined by the bottleneck link(s) on the path. We obtain the path bandwidth from the link bandwidths by using the min operator. Figure 14.4 shows a network with four nodes and four links. The link bandwidths are given. Note that $a \to b \to c$ is a widest path between a and c, with the width of the path being 5 units. But the subpath $a \to b$ is not a widest path between a and b, because 10 units of bandwidth are available on the path $a \to d \to b$. If we define "optimal" to mean "widest," then we can state this by saying that even though $a \to b \to c$ is an optimal path between a and c, subpath $a \to b$ is *not* optimal.

Recalling that the intention is to finally obtain a generalized Dijkstra algorithm, let us consider a set $\mathcal{S}$, a binary operator $\oplus : \mathcal{S} \times \mathcal{S} \to \mathcal{S}$, and a *total order* $\preceq$ defined on $\mathcal{S}$. A total order on $\mathcal{S}$ is a relation on $\mathcal{S}$ such that *any two* elements of $\mathcal{S}$ can be related using the order $\preceq$. For a familar example, the usual

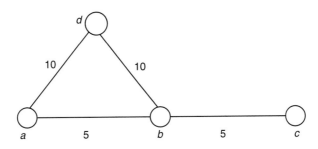

Figure 14.4   **Four nodes and the interconnecting links are shown. The numbers indicate the link bandwidths in some units. The widest path between *a* and *c* is determined by the bottleneck link(s) on paths from *a* to *c*. However, we cannot say that every subpath of a widest path between *a* and *c* is itself a widest path. For example, the path *a* → *b* → *c* is a widest path from *a* to *c*, but the subpath *a* → *b* is not a widest path between *a* and *b*.**

ordering on the real line $\mathbb{R}$, $\leq$, is an example of a total order, because any two real numbers a and b can be related using $\leq$: Either $a \leq b$ or $b \leq a$.

Getting back to the general set S and the total order $\preceq$, let a, b, $c \in S$. Then $\oplus$ is said to be *isotone* over $\preceq$ when $a \preceq b$ implies both $a \oplus c \preceq b \oplus c$ and $c \oplus a \preceq c \oplus b$. Similarly, $\oplus$ is said to be *strictly isotone* over $\preceq$ when $a \prec b$ implies both $a \oplus c \prec b \oplus c$ and $c \oplus a \prec c \oplus b$.

In this terminology, when $S = \mathbb{R}$, $\oplus = \min$, and $\preceq$ $=$ $\leq$, we find that min is isotone over $\leq$, but not strictly so. But when $S = \mathbb{R}$, $\oplus = +$, and $\preceq$ $=$ $\leq$, it is easy to see that $+$ is strictly isotone over $\leq$. Thus, we see that the underlying property that allows us to assert that "for the regular Dijkstra algorithm, every subpath of an optimal path is optimal" is the strict isotonicity of $+$ over $\leq$.

In the general context where S, $\oplus$, and $\preceq$ are defined, the path length of a path P is given by

$$\delta_{l_1} \oplus \delta_{l_2} \oplus \cdots \delta_{l_k}$$

where $l_1, l_2, \ldots, l_k$ are the links on P, and $\delta_{l_1}, \delta_{l_2}, \ldots, \delta_{l_k}$ are elements of S. Suppose that we want shortest paths between s and t, where path lengths are defined as before. We noted that the key property underlying the principle "for the regular Dijkstra algorithm, every subpath of an optimal path is optimal" is the isotonicity of $+$ over $\leq$ when the set is $\mathbb{R}$. This suggests the following: As long as we have an operator $\oplus$ and a total order $\preceq$ over some set S such that $\oplus$ is strictly isotone over $\preceq$, Dijkstra's algorithm, generalized to use the operator $\oplus$ and the order $\preceq$, will work.

When $\oplus$ is merely isotone over $\preceq$ and not strictly so, we know that not every subpath of an optimal path is necessarily optimal. But can the generalized Dijkstra algorithm still be used to find optimal paths? The answer is yes. It can be shown that under isotonicity, there exists at least one optimal path such that every subpath starting at the source node s is optimal, and this is sufficient for the generalized Dijkstra algorithm to work correctly. Consider Figure 14.4. In this example, the binary operator is min, the underlying set is $\mathbb{R}^+$, and two numbers a and b belonging to $\mathbb{R}^+$ are *defined* to satisfy the relation $a \preceq b$ iff $a \geq b$. The reason the relation is defined in this "reversed" way is that we want to think of a higher-capacity link as having a "smaller" length. Clearly, min is isotone over $\preceq$. We see that the path $a \to d \to b \to c$ is a widest path between a and c, such that every subpath starting at a is also a widest path. It is also easy to see that if the generalized Dijkstra algorithm is run with a as the source, the result will be the path $a \to d \to b \to c$.

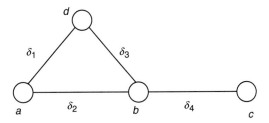

Figure 14.5   Four nodes and four interconnecting links are shown. $\delta_1, \delta_2, \delta_3, \delta_4$ **belong to a given set** S **and represent the weights of the corresponding links. Suppose a binary operation** $\oplus$ **and a total order** $\preceq$ **are defined on** S. **The link weights are such that** $(\delta_1 \oplus \delta_3) \prec \delta_2$, **but** $(\delta_1 \oplus \delta_3 \oplus \delta_4) \succ (\delta_2 \oplus \delta_4)$. **Thus, isotonicity does not hold for this example. (See Problem 14.5 at the end of this chapter.)**

Exercise 14.3

Let $S = [0, 1]$ and the binary operation be $\times$, and suppose that for $a, b \in S$, $a \preceq b$ iff $a \geq b$. Suppose a given network's links are assigned numbers from S and the generalized Dijkstra's algorithm is run. If we interpret the link weight as the reliability of the link, what kind of path does the algorithm yield as output?

On the other hand, consider the example shown in Figure 14.5, where isotonicity of $\oplus$ over $\preceq$ does not hold. We have $(\delta_1 \oplus \delta_3) \prec \delta_2$, but $(\delta_1 \oplus \delta_3 \oplus \delta_4) \succ (\delta_2 \oplus \delta_4)$. The shortest path from a to c is $a \rightarrow b \rightarrow c$, but if the generalized Dijkstra algorithm is run at a, we get the path $a \rightarrow d \rightarrow b \rightarrow c$. Thus, unless isotonicity is verified, the generalized Dijkstra algorithm can give wrong results. This shows that isotonicity is *necessary* for the generalized Dijkstra algorithm to run correctly. Hence, isotonicity is both *necessary and sufficient* for the correct operation of the generalized Dijkstra algorithm.

14.4   Routing Protocols

In this section, we take a brief look at commonly used routing protocols. Our objective is to introduce some of the basic notions that appear in the discussions of these protocols. We make no attempt to be complete.

14.4.1   Distance Vector Protocols

The key feature of distance vector protocols is that each router relies on the reachability information advertised by its neighbors to deduce its best route to a given destination. Advertisements are exchanged frequently so that the most recent information is used in the decision-making process. In contrast to what happens in a link state protocol (discussed later), each router does not build up its own database containing topological information about the whole network.

The *distance* to a given network from a router R usually refers to the number of hops required to reach the network from R. Considering all the networks reachable from R, we get a vector of distances. Hence we have the name "distance vector."

A router R disseminates information by advertising to all its neighbors the currently known shortest distances to different networks. Correspondingly, when router R receives advertisements from its neighbors, it examines the distances to a network N and, if necessary, updates its routing table so that it contains the shortest distance to N. Along with this, the next hop to network N is also updated.

In Figure 14.6, R1 receives an advertisement from R2, announcing that network N2 is H2 hops away. Assuming that H2 is the shortest distance to N2

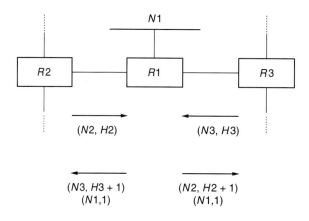

Figure 14.6   *R*1, *R*2, and *R*3 are routers running a distance vector protocol. *R*1 receives an advertisement from *R*2, advertising reachability to a network *N*2 at a distance of *H*2 hops. Suppose that, of all the advertisements received by *R*1, the shortest distance to *N*2 is *H*2. This causes *R*1 to advertise reachability to *N*2 on its other interfaces. The hop count to reach *N*2 becomes *H*2 + 1. Similar actions are taken by *R*1 for the advertisement received from *R*3. *R*1 also advertises reachability to network *N*1 with a hop count of 1.

seen by $R1$ when all received advertisements are considered, $R1$ now updates its routing table. The shortest distance to $N2$ is recorded as $(H2 + 1)$, where 1 is added to $H2$ to take care of the additional hop between $R1$ and $R2$. The next hop to $N2$ is recorded as $R2$. Now when $R1$ sends advertisements to its neighbors, the shortest distance to $N2$ is announced to be $(H2 + 1)$.

Figure 14.7 shows a timeline summarizing the tasks repeatedly executed by a router running a distance vector protocol. For simplicity, we have assumed synchronized operation.

The most well-known example of a distance vector routing protocol is RIP, which stands for Routing Information Protocol. Upon initialization, a RIP router sends a special request message on all its interfaces, asking for the complete routing table of the recipient. Routes to specific networks can also be requested. If the recipient has a route to the network, the hop count is returned. If no route is known, then the hop count is set to 16. In RIP, a hop count of 16 is regarded as "infinity" and indicates that no route to the destination is known to the router. The default interval before advertisements are sent is 30 seconds.

The simplest implementations of RIP can lead to long convergence times and transient routing loops. Consider Figure 14.8, which shows routers A, B, and C, which run RIP. Let us assume synchronized operation as before. Suppose that at T, C's interface to network N goes down. Previously, C was advertising reachability to N with a hop count of 1. Therefore, A and B deduced a hop count of 2 to N. At T, all the routers exchange advertisements. The advertised hop counts to N are shown in Figure 14.8. C sets the hop count to 16 because N is unreachable.

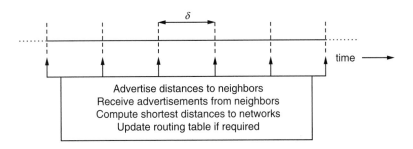

Figure 14.7 Schematic showing the activities of a router running a distance vector protocol. The picture assumes that operation is synchronized. At regular intervals, a router sends advertisements out its own interfaces, receives advertisements, computes shortest paths, and updates the routing table if required.

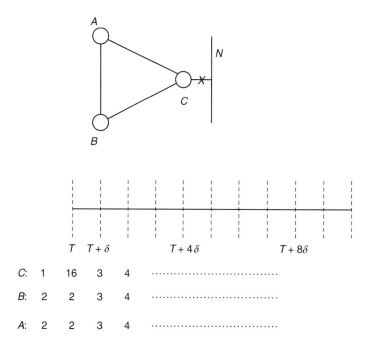

Figure 14.8   A, B, and C are routers running RIP. We assume synchronized operation. At T, C's interface to network N goes down. We are interested in examining the routing tables of A, B, and C at the epochs T, T+δ, ... The lower part shows the distances to network N as calculated by A, B, and C at various epochs.

A and B have not yet received C's advertisement, so they continue announcing that N can be reached through them with a hop count of 2.

Having received one another's advertisements sent at T, the routers execute the algorithm to come to the following conclusions. C finds that A and B are announcing a path to N with a hop count of 2. So even though C's interface to N is down, C concludes that N can be reached via A or B with a hop count of 3. In the same way, A and B also conclude that N is reachable with a hop count of 3. This is what each advertises at $(T + \delta)$. It is clear that at successive epochs, the advertised hop count increases by 1, until it reaches 16, when all routers deduce that N is unreachable. This behavior is called *counting to infinity*. It is the reason the maximum path cost is set to a small value in distance vector protocols.

Note also that during the count to infinity, a routing loop is formed between *A* and *B*, because each stores the other as the next hop toward *N*.

14.4.2 Link State Protocols

Link state protocols are based on the philosophy that each node in a given network should have a detailed view of the current topology of the network. Each router contributes information about the part of the topology that is local to it. Accordingly, information about the links and networks attached to a router is advertised to all other routers within a defined area. Because each router receives the advertisements from all other routers, it is possible to construct a graph representing the current network topology. If there are no changes in the network, then all routers eventually deduce the same topology. Thus, the link state approach leads to a *replicated distributed database*. Each router can then run a shortest path algorithm to find the shortest distances to different networks.

A common example of a link state protocol is the OSPF (Open Shortest Path First) routing protocol. Local topological information is advertised by each router in *link state advertisements* (LSAs). Each LSA has a common LSA header containing several fields, including ones that are used to identify the originating router, to determine whether the LSA is more recent than other instances and to indicate that an old LSA can be deleted from the database because it is advertising out-of-date information. The collection of all LSAs seen by a router constitutes the link state database. An LSA provides data about each link attached to the router. The data includes the *link metric*, which behaves as a link weight that is assigned administratively. Thus, path metrics in OSPF need not be hop counts.

OSPF runs directly over the IP layer without using the services of a transport layer such as TCP or UDP. Communication between OSPF routers occurs using OSPF packets. OSPF Hello packets are part of the Hello protocol used by OSPF to discover neighbor routers and maintain neighbor relationships. OSPF Database Description packets actually carry the LSAs. These packets are used by neighbor routers to exchange databases at initialization. Link State Request packets are used to request specific LSAs. Responses to such requests as well as intimations of changes in local topology are carried in Link State Update packets.

LSAs are distributed throughout a defined area by the process of *reliable flooding*. When a router receives a new LSA in a Link State Update packet, it installs the LSA in its database and then forwards the packet on all its interfaces except the one on which the Update packet was received. Reliability is achieved by using Link State Acknowledgment packets, which are sent by a router receiving a new LSA. If an Acknowledgment packet is not seen, a router periodically retransmits the new LSA until it is acknowledged.

When the link state database is available, the network can be represented as a directed graph. A router then uses the Dijkstra algorithm to compute shortest paths to other routers. We note that because the complete network graph is available, it is possible for a router to find the shortest path not only between itself and other routers but also between *any two* routers in the area.

14.4.3   Exterior Gateway Protocols

An *autonomous system* (AS) is defined as a collection of routers that are under the administrative control of the same entity. Routers within an AS use an *interior gateway protocol* (IGP) to determine how to reach one another. RIP and OSPF, discussed in Sections 14.4.1 and 14.4.2, are examples of IGPs. The administering authority of an AS is at liberty to make operational decisions within the AS without informing any other agency; for example, the entity may decide to change the IGP it uses inside the AS, or even decide to deploy more than one IGP. As an example, the routers in an ISP's backbone constitute an AS. ASs are identified by 16-bit numbers. The Border Gateway Protocol (BGP) is an inter-AS routing protocol.

The overall structure of the Internet can be represented as an arbitrary connected graph, where the nodes are the ASs and the edges are the links connecting the *border routers* or *border gateways* of different ASs. For simplicity, we assume that the border routers are *BGP speakers*; that is, border routers exchange BGP messages for disseminating network reachability information across ASs. More general configurations, where BGP speakers are not routers, are also possible; we do not discuss these. Moreover, there can be multiple BGP speakers in an AS. Therefore, a BGP speaker can have internal as well as external *peers*, with which it corresponds.

Traffic that is generated by or destined to a host belonging to an AS is called *local* traffic. Traffic that is not local is called *transit* traffic. Depending on the type of traffic—local or transit—that an AS carries, three types of ASs are possible.

- *Stub AS:* A stub AS is attached to only one other AS and carries only local traffic.

- *Multihomed AS:* A multihomed AS has connections to more than one other AS but usually carries only local traffic. In other words, a multihomed AS usually refuses to carry transit traffic. Alternatively, a multihomed AS can carry transit traffic for a restricted set of other ASs.

- *Transit AS:* A transit AS is connected to multiple other ASs and carries both local and transit traffic.

BGP uses TCP as the underlying reliable transport protocol. Two BGP speakers open a TCP connection between them and exchange BGP protocol messages over this connection. Initially, the BGP Open message is sent and the complete BGP routing table is exchanged. Thereafter, incremental changes are exchanged in BGP Update messages. To make sure that the peer relationship is alive, BGP KeepAlive messages are exchanged at regular intervals. In case of error events or special conditions, the BGP Notification message is sent.

A *route* is a unit of BGP information that pairs a destination (which can be a set of addresses specified by a *prefix*) with the path attributes of the path used to reach that destination. The destination is specified in the Network Layer Reachability Information (NLRI) field of the BGP Update message, and the Path Attributes field of the same message contains the path attributes. One of the attributes, AS_PATH, is the sequence of ASs to be traversed to reach that destination. Note that in BGP, a route is specified by explicitly mentioning all the ASs that must be crossed to reach the destination. Therefore, routing loops can be avoided in a straightforward manner. The NEXT_HOP attribute specifies the IP address of the border router that must be accessed to reach the destinations advertised in the Update message.

Routing information received, used, and advertised by a BGP speaker is stored in *routing information bases* (RIBs), as shown in Figure 14.9. The Adj-RIB-in database stores routing information received from all other peers. A *decision process* is then employed by the BGP speaker to select routes to these destinations. It is possible that multiple routes to the same destination are available. In that case, preference values are assigned to the routes, and the decision process chooses the route with the highest preference value. In assigning preference values, *routing policies* can be used. For example, suppose that AS i has a policy of not using AS j as a transit AS when sending traffic to AS k. Then, even if a route to AS k through AS j is present in Adj-RIB-in, it will not be selected.

After the decision process has selected routes for various destinations using routing policies as well as other considerations, the routes to be used are stored in the Loc-RIB ("Loc" for "local"). This database contains the routes that are used to forward traffic. The Adj-RIB-out, on the other hand, stores the routes that a particular BGP speaker advertises to its peers using the Update message. It is important to note that a border router in a transit AS can advertise *only* those routes to remote destinations that it uses itself; if a certain route is not used by it,

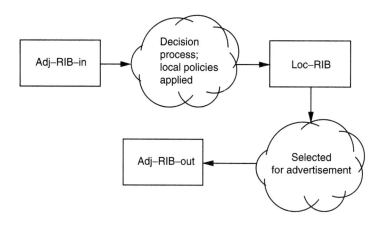

Figure 14.9 BGP uses several routing information bases and uses a decision process to select routes to remote destinations. A subset of the routes used is advertised to BGP peers.

the route will not be advertised. Moreover, not all routes in the Loc-RIB need be advertised, for policy reasons.

For transit ASs, a degree of coordination is required between BGP and the IGP used in the AS so that network reachability information learned by the border routers is also available to the internal (i.e., nonborder) routers. Care needs to be taken to ensure that border and internal routers have synchronized views about how remote destinations can be reached; otherwise, routing loops may result.

14.5 Summary

This chapter considers the problem of optimal routing when the set of demands to be routed is given and demands can be split arbitrarily across multiple routes between a source–destination pair. The optimal routing problem is formulated as a linear program (LP), and the dual of this LP provides valuable insights into how optimal routing can be realized using appropriate link weights. The conclusion is that optimal routing can be realized using well-known shortest path routing, where the shortest path algorithms use the defined link weights. Then we discuss the Dijkstra and Bellman–Ford algorithms, which have seen extensive use in computing shortest paths ever since they were invented. The recent development of a generalized Dijkstra algorithm, which seeks to extend its applicability to many more problems of practical interest, is discussed next. Finally, we take a brief look at distance vector and link state routing protocols, as well as at BGP, the common exterior gateway protocol used in the Internet.

14.6 Notes on the Literature

Optimization formulations of routing problems are natural and traditional. Bertsekas and Gallager [33] provide the formulation and analysis of a routing problem using the optimization framework. Duality theory is a classical topic in optimization. Appendix C provides a discussion of duality theory and provides references. Our analysis based on duality has been developed by us; it was inspired by the analysis presented in [297].

Several works have investigated the problem of OSPF weight setting. In particular, [106] considers the problem of optimal weight setting and gives a heuristic algorithm based on searching the weight space. Several interesting examples and insightful discussion are available in [107]. The example in Section 14.2 is taken from this paper.

The Dijkstra and Bellman–Ford algorithms are classical results that are about 50 years old. An excellent discussion of these algorithms is found in [33]. The work on the generalized Dijkstra algorithm is much more recent, and it was presented in [264]. Our brief treatment in Section 14.3.3 follows this paper. The examples in that section are either taken from that paper or are minor adaptations of examples there.

The routing protocol standards have been developed by the Internet Engineering Task Force (IETF). There are Requests for Comments (RFCs) on the OSPF protocol ([220]) and RIP ([202]). MPLS is a more recent development, discussed in [255]. Our brief discussion of BGP follows [249] and [248].

Problems

14.1 Consider the simple network shown in Figure 14.10. All links are bidirectional and have a capacity of 1 unit in each direction.

a. How many source–destination pairs are there in this network?

b. Give an upper bound to the maximum traffic that this network can carry. Is this upper bound achieved for some set of demands?

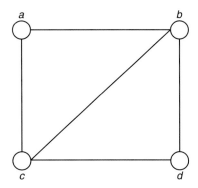

Figure 14.10   A simple network with bidirectional links. The capacity of each link is 1 unit.

---

    c. If all links are to be fully utilized and the demand sizes for all pairs are to be equal, what is the demand size for each pair?

    d. Is the set of demands in part (c) supportable on the network? Why?

14.2 Consider Dijkstra's shortest path algorithm. Let D_n be the distance of the node closest to the source at the beginning of the nth iteration, $n \geq 1$. Show that D_n is monotonically nondecreasing.

14.3 Suppose that Dijkstra's algorithm is used to determine the shortest paths from a node s to all other nodes in a network. Assume that all link lengths are strictly positive. After the algorithm terminates, we note the *predecessor* of every node i on the shortest path from s to i. Let $\mathcal{E}'$ be the set of edges joining the given nodes to their respective predecessors. Show that the subgraph $(V, \mathcal{E}')$ is a *tree*, that is, a graph with no cycles in it.

14.4 How would you use the generalized Dijkstra algorithm of Section 14.3.3 to obtain *widest-shortest* paths? Let the set S be defined as $\{(d, b): d, b \in \mathbb{R}_0^+\}$, where $\mathbb{R}_0^+$ is the set of nonnegative real numbers. Let d give the delay on a link or path, and b give the bandwidth on the link or path.

    a. Obtain an operator $\oplus : S \times S \rightarrow S$ such that if $x, y \in S$, then $x \oplus y$ gives the metrics corresponding to the path obtained by concatenating the paths whose metrics are x and y.

b. Next obtain a total order $\preceq$ on the set S such that given two paths, it is possible to identify the wider-shorter of the two.

c. Now check whether the operator $\oplus$ is isotone over the order $\preceq$.

14.5 We now wish to use the generalized Dijkstra algorithm to find the *shortest-widest* path. Do you see any difficulty if we proceed in the same way as in Problem 14.4?

CHAPTER 15

Virtual Path Routing of Elastic Aggregates

In Chapter 14 we consider the problem of optimal routing when the set of demands to be routed is known a priori. Recall that a *demand* for a source–destination pair of routers refers to the average offered load of an aggregate of elastic traffic sessions between the two. If *knowledge of the set of demands to be routed is available*, and *if demands can be split arbitrarily among multiple routes between a source–destination pair*, then we have seen that the optimal routing algorithm is a shortest path algorithm, with the weights of the links being defined appropriately.

In this chapter, we consider the situation when the complete set of demands to be routed is not known. Furthermore, each demand must be routed as a single unit and cannot be split across multiple routes. We begin by describing a scenario where these characteristics arise naturally. Then we mention a simple routing strategy and show, through an example, how it can cause connection blocking unnecessarily. Next, we introduce the concept of interference experienced by a demand because of the choice of a particular route for another demand. This leads to the idea of designing routes for minimum interference. The problem of obtaining a minimum interference route is then formulated as an integer linear program. We study this formulation and the corresponding heuristic algorithm. This is followed by a second formulation and heuristic that utilizes and builds on the insights gained from the first. Finally, we look at multiprotocol label switching (MPLS), a technique that can be used to implement the routes obtained from the heuristics discussed.

15.1 On-Demand Routing

In practice, prior knowledge of the set of demands may not be available. We consider the example of an Internet service provider (ISP) that has a network deployed and is already providing service to some customers. When the network was being designed (that is, before operations were started), the ISP would possibly have used some predicted traffic values between different source–destination pairs to design the network. Therefore, in this design phase, we see that the complete set of demands that need to be routed across the network can be assumed to be available. But after the network is in operation, additional traffic demands arrive in an unpredictable manner. For example, a large enterprise may, at very short notice, require bandwidth between two locations. Such a requirement is often expressed by saying that the enterprise needs a *virtual path* across the ISP's network. The virtual path could correspond to an ATM VP, a permanent virtual circuit (PVC) in a frame relay network, or tunnel in an MPLS network (MPLS is discussed in Section 15.4).

We consider the problem of routing such virtual paths across a backbone network. We note that the demands that arrive in this scenario are not microflows corresponding to individual elastic traffic sessions (for example, TCP sessions). Rather, a demand is the average offered load of an aggregate of elastic traffic between two locations of an enterprise (as in the earlier example), or between two enterprises that have a business relationship. In either case, the ultimate source and sink of traffic can be identified as corporate entities. In contrast, in Chapter 14, we had hundreds or thousands of traffic end points, not belonging to any single organization, that were grouped into a single aggregate because they accessed the ISP backbone through the same router.

The ISP has no prior knowledge of when demands for virtual paths will arrive nor what the demand sizes will be. Thus, demands now arrive according to an arbitrary arrival process and must be routed such that the fraction of blocked demands is small and network resources are efficiently utilized. The demand arrival process is fundamentally different from that considered in Chapter 14. In particular, we discuss an offline problem in Chapter 14, but here we consider an *online* problem, where routing is done *on demand*.

An additional constraint that often accompanies such demands is that the whole demand must be routed as a single unit across the ISP's network. In other words, the demand is to be considered *unsplittable*. The requirement may arise because the enterprise asking for a virtual path across the network is often interested in a virtual private network for its users. This means that the users should have the illusion that the enterprise has its *own* network resources, meant for their exclusive use. This demands some uniformity in the quality of service experienced by individual sessions within the aggregate; for example, average TCP session throughput performance should be the same for all sessions constituting the aggregate demand. To ensure this, an ISP would like to treat the whole aggregate as a single unit and route it across its backbone without splitting it. The reason is that splitting might cause different sessions to encounter different round-trip times

(RTTs; see Chapter 7) because they can travel on different routes, and that might result in disparities in session performance as perceived by end users.

15.2 Limitations of Min Hop Routing

Minimum hop (min hop) routing has been a popular choice for routing virtual paths. Here, the path from the source router to the destination router is the one that has the smallest number of hops. Evidently, when the weights of links in a network are set to 1, shortest path routing reduces to min hop routing. The obvious advantage of min hop routing is that a demand from a source to the corresponding destination can be routed through the network while consuming the least amount of total bandwidth resources. If a demand requires an amount of bandwidth d, then the total bandwidth consumed on a route is $d \times H$, where H is the number of hops on the chosen route. If H_{min} is the number of hops on the shortest path, then $H_{min} \leq H$, and it follows that min hop routing consumes the least resources.

However, the amount of resources consumed in routing a demand is often not the only important criterion to be considered. An example where min hop routing performs poorly even though it consumes the least resources is shown in Figure 15.1. Suppose that the residual bandwidth on each link is 1 unit and that a demand of 1 unit arrives at x for y. Min hop routing would use the three-hop path $x \to a \to b \to y$. Then the link $a \to b$ is completely full, and, as far as servicing future demands is concerned, the network becomes partitioned. It would have

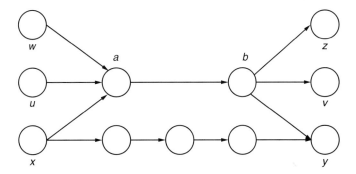

Figure 15.1 Example showing possible problems with min hop routing. Suppose each link has a residual bandwidth of 1 unit. We now have a path setup request between *x* and *y*, asking for 1 unit of bandwidth. If min hop routing is used, the link *a→ b* is fully occupied and the network becomes partitioned. (Example taken from [156].)

been better to route the demand on the four-hop path even though it consumes more resources. The point is that resources on the links that constitute the four-hop path between x and y cannot be used, in any case, by other source–destination pairs (u, v), (w, z). So routing the (x, y) demand along this path does not reduce the capacity available for the other source–destination pairs.

Evidently, min hop routing is a greedy strategy, and, like other greedy strategies, it can be shortsighted. In the example shown in Figure 15.1, the min hop algorithm is oblivious of the importance of avoiding the link $a \rightarrow b$ because of the algorithm's shortsighted nature, which seeks exclusively to minimize resource consumption. Algorithms that have a more global view and objectives can possibly be aware of better alternatives. For example, if the knowledge of the location of source–destination pairs can be incorporated into the algorithm, it may be possible for the algorithm to deduce that resources on the links that make up the four-hop path cannot be used by other source–destination pairs (u, v), (w, z), and so consuming more bandwidth resources does not matter.

15.3   Formulations of the Routing Problem

As discussed in the example, min hop routing can lead to a situation where one link in a network gets congested while unused bandwidth is available on other links. Several attempts have been made over the years to address this shortcoming, and a number of algorithms, including shortest-widest path, widest-shortest path, least-loaded routing, maximally loaded routing, and profile-based routing, have appeared in the literature. In what follows, we discuss two approaches to the problem of routing virtual paths, with the second approach building on the insights gained from the first.

15.3.1   Minimum Interference Routing: Formulation and Heuristic

For convenience, we often refer to a router as a "node." Any chosen route from a node a to another node b on the ISP backbone can possibly reduce the capacity available for demands between other node pairs. This can be regarded as *interference* caused to a node pair because of the chosen route between a and b. Let us consider the *maxflow* from a source to a destination. Maxflow (s, t) is a scalar that indicates the maximum amount of traffic that can be sent from s to t, exploiting *all* possible paths through the network. It is an upper bound on the total bits/sec that can be sent from s to t. As noted earlier, when a demand between nodes a and b is routed through the network, the maxflow from s to t *after routing*

(a, b) can only decrease or remain unchanged. If maxflow (s, t) remains unchanged, then it is clear that the path used for the (a, b) demand does not share any link with the set of paths available for (s, t). The amount of the decrease depends on the path used for the (a, b) demand. This observation suggests the following idea to assess the impact of choosing a certain route for a given demand (a, b) on other node pairs. Suppose (s, t) is another node pair that is not currently injecting traffic into the network but may do so in the future. That is, (s, t) is a potential future demand to be supported on the network. Consider maxflow (s, t). Then the *reduction in maxflow* between s and t as a result of choosing a particular route for the (a, b) demand is a measure of the interference caused to the potential future (s, t) demand by routing (a, b).

Because, in general, there are several (s, t) pairs distinct from (a, b) using the network, we can think of a *vector of interferences* corresponding to a particular routing of the (a, b) demand. The objective would then be to choose a route that causes the least overall interference, in some well-defined sense, to other (s, t) pairs.

There are several possibilities now. We can think of a minimum interference route for a demand (a, b) as a route such that the following holds: After the (a, b) demand has been routed, the *smallest* maxflow value among all other (s, t) pairs is maximized.

For example, consider a network with four source–destination pairs, and suppose that a demand arrives to a for b. Suppose that there are three possible routes for the (a, b) demand and that the vectors of maxflows for the other source–destination pairs are as follows: $(30, 15, 6)$ corresponds to path P_1 for (a, b); $(12, 19, 8)$ corresponds to path P_2 for (a, b); and $(3, 12, 16)$ corresponds to path P_3 for (a, b). Then route P_2 is the minimum interference route for the (a, b) demand.

This criterion considers only the smallest maxflow value in the vector of maxflows. We can think of other criteria that utilize all the elements in the vector rather than only the smallest one. A weighted sum of maxflows is such a criterion.

Yet another criterion that considers all the elements of the vector of maxflows is the *lexicographically maximum* criterion. A *lex-max* vector is one such that the smallest element is the largest possible, among such vectors, the second smallest element is the largest possible, and so on (see Chapter 7). This criterion is also called *max-min fair*.

For simplicity, we consider the first criterion for measuring interferences. That is, we focus on the smallest element of the interference vector. A natural objective now would be to formulate a problem that yields the optimal route for an (a, b) demand. We need some notation first.

Let $\mathcal{G}(\mathcal{N}, \mathcal{L})$ be a directed graph representing the backbone network, where $\mathcal{N}$ is the set of network nodes and $\mathcal{L}$ is the set of links. Let $|\mathcal{N}| = N$ and $|\mathcal{L}| = L$.

Let the node–link incidence matrix be denoted by $\mathbf{A}$. As usual, $\mathbf{A}$ has a row for each node $\in \mathcal{N}$ and a column for each link $\in \mathcal{L}$. Each link $l \in \mathcal{L}$ has a head node $h(l)$ and a tail node $w(l)$. Thus, the link is from $h(l)$ to $w(l)$. This is indicated in the matrix $\mathbf{A}$ by setting $\mathbf{A}_{h(l),l} = +1$, $\mathbf{A}_{w(l),l} = -1$, and $\mathbf{A}_{n,l} = 0$ for $n \neq h(l)$, $n \neq w(l)$, $n \in \mathcal{N}$.

Let $\mathcal{K} \subseteq \mathcal{N} \times \mathcal{N}$ be a subset of the set of ordered router pairs. The set $\mathcal{K}$ denotes the set of potential ingress–egress pairs. All demands that are to be routed across the backbone network belong to the set $\mathcal{K}$. A representative element of $\mathcal{K}$ is denoted by (s, t). Let $|\mathcal{K}| = K$.

Suppose that the L links in $\mathcal{L}$ are numbered in some arbitrary order. Consider an L-vector (i.e., a vector with L elements) $\mathbf{x}(k)$, where $k \in \mathcal{K}$. Note that $\mathbf{x}(k)$ can represent, for example, the amount of flow on each link $l \in \mathcal{L}$, corresponding to the pair k. This is called the *flow vector*, or simply the *flow* corresponding to demand k.

Suppose that a certain number of demands have been routed. Then, each link in the network has a certain *leftover* capacity, after taking into account the capacity used by the demands that have been routed. The original network, with each link capacity being equal to the leftover capacity, is referred to as the *residual network*. Let $\theta(k)$ be a scalar that represents the maxflow that can be sent between nodes $s(k)$ and $t(k)$ in the residual network, where $s(k)$ and $t(k)$ represent, respectively, the source and destination nodes corresponding to demand $k \in \mathcal{K}$. Also, let $\alpha(k)$ be a scalar weight associated with demand k. Furthermore, let $\mathbf{C}$ be an L-vector of residual capacities, and $\mathbf{e}(k)$ be an N-vector corresponding to demand k, with $e_{s(k)}(k) = +1$, $e_{t(k)}(k) = -1$, and $e_n(k) = 0$, $n \neq s(k)$, $n \neq t(k)$. Suppose that the next demand to arrive is for the router pair $\kappa \in \mathcal{K}$, and its size is $d(\kappa)$. Our focus is on finding a minimum interference route for this demand.

At this point, we state a further assumption about the operation of the system. It is assumed that arriving demands never leave (i.e., demands are persistent). This is motivated by the scenario discussed at the beginning, where requests for virtual paths arrive; typically, a virtual path lasts for a very long period.

With the notation defined, we discuss the optimal routing problem formulation, called MIRA (for minimum interference routing algorithm) by its authors, for a demand $\kappa \in \mathcal{K}$, which has just arrived.

Problem **ILP1**:

$$\max z$$

subject to

$$\mathbf{A}\mathbf{x}(k) = \theta(k)\mathbf{e}(k) \quad \forall k \in \mathcal{K} \setminus \kappa \tag{15.1}$$

$$\mathbf{A}\mathbf{x}(\kappa) = d(\kappa)\mathbf{e}(\kappa) \tag{15.2}$$

$$\mathbf{x}(k) + \mathbf{x}(\kappa) \leq \mathbf{C} \quad \forall k \in \mathcal{K} \setminus \kappa \tag{15.3}$$

$$z \leq \alpha(k)\theta(k) \quad \forall k \in \mathcal{K} \setminus \kappa \tag{15.4}$$

$$\mathbf{x}(k) \geq 0 \quad \forall k \in \mathcal{K} \setminus \kappa \tag{15.5}$$

$$\mathbf{x}(\kappa) \in \{0, d(\kappa)\}^L \tag{15.6}$$

Equation 15.1 gives, for each $k \in \mathcal{K} \setminus \kappa$, the flow conservation constraints. For a given $k \in \mathcal{K} \setminus \kappa$, it states that (a) the net maxflow leaving node $s(k)$, along all possible links for which $s(k)$ is the head or tail node, is $\theta(k)$; (b) correspondingly, the net maxflow leaving node $t(k)$, along all possible links for which $t(k)$ is the head or tail node, is $-\theta(k)$; and (c) for all other nodes $n \in \mathcal{N}$, the net maxflow leaving node n, along all links for which n is the head node or tail node, is 0. Equation 15.2 asserts that the flow conservation constraints for the demand κ must be satisfied as well, with the node $s(\kappa)$ sourcing a net maxflow of $d(\kappa)$ and the node $t(\kappa)$ correspondingly sinking a net maxflow of $d(\kappa)$. Equation 15.3 states that the flows for the given $k \in \mathcal{K} \setminus \kappa$ and κ must *together* satisfy the residual capacity constraints. Equation 15.3 also indicates that *one* additional demand from the set $\mathcal{K} \setminus \kappa$, along with the demand κ, is being considered at a time. This is then repeated for all demands $k \in \mathcal{K} \setminus \kappa$. This is the only equation that captures the interaction between different demands. The interaction happens because of limited capacity. As Equation 15.4 shows, the scalar z is obtained as the smallest value of $\alpha(k)\theta(k)$ when this process (of considering a single $k \in \mathcal{K} \setminus \kappa$ along with κ, and then repeating for all pairs in $\mathcal{K} \setminus \kappa$) is complete. Equation 15.5 expresses the requirement that the flow $\mathbf{x}(k)$ must be nonnegative, and Equation 15.6 ensures that the route for the demand κ is a nonsplit route, because a link either does not carry the demand at all or carries the full amount $d(\kappa)$ of the demand.

The optimal routing of the demand κ is such that when (1) each $k \in \mathcal{K} \setminus \kappa$ is considered individually along with κ, and (2) this is repeated for all $k \in \mathcal{K} \setminus \kappa$, then the smallest value of the scalar $\alpha(k)\theta(k)$ is maximized. We emphasize that *all* the $(K-1)$ pairs in $k \in \mathcal{K} \setminus \kappa$ are *not* considered simultaneously along with κ.

In this formulation, all the unknown quantities are real-valued, except for the vector $\mathbf{x}(\kappa)$, which is discrete-valued. Each element of $\mathbf{x}(\kappa)$ takes values in $\{0, d(\kappa)\}$. If all the unknown quantities were reals, the problem formulation shows that we would obtain a linear program, and computationally efficient methods could have been employed to obtain solutions. However, the presence of $\mathbf{x}(\kappa)$ indicates that we have an *integer* linear program instead. This suggests that it may be hard to find optimal solutions with reasonable computational effort. Actually, the problem can be shown to be NP-hard (see Appendix E for a discussion on NP-hardness). A special case of the problem is shown to be nothing but the *directed two-commodity integral flow problem*, which is known to be NP-hard.

Because obtaining an optimal solution involves unacceptably large computational effort, the focus now shifts to the design of heuristics that are fast and provide solutions that are close to optimal.

One option is to solve a linear programming relaxation (referred to as **LP1**) of this problem. This means that in $\mathbf{x}(\kappa)$, each element $x(\kappa)_l$, $1 \leq l \leq L$, is allowed to vary continuously in $[0, d(\kappa)]$. In other words, **LP1** is obtained from **ILP1** by replacing Equation 15.6 with $\mathbf{x}(\kappa) \in [0, d(\kappa)]^L$. If the LP solution is such that each element of $\mathbf{x}(\kappa)$ lies in the set $\{0, d(\kappa)\}$, then this solution is, of course, also the optimal solution to the original problem. However, if each element of $\mathbf{x}(\kappa)$ does not belong to the set $\{0, d(\kappa)\}$, then the LP relaxation gives a split routing for demand κ, because a link can carry an amount of the demand κ that is less than $d(\kappa)$. In that case, we would need heuristics to somehow obtain a single nonsplit route for the demand κ.

Let the optimal objective function values of the problems **ILP1** and **LP1** be denoted by z^*_{ILP1} and z^*_{LP1}, respectively. Clearly, $z^*_{\text{ILP1}} \geq z^*_{\text{LP1}}$, because the relaxed problem allows split routing of demand κ. If z^*_{LP1} is a *tight* upper bound on z^*_{ILP1}, then the linear relaxation provides useful information because we can measure the quality of a proposed heuristic solution by noting the difference $(z^*_{\text{LP1}} - z^*_{\text{heuristic}})$, which closely approximates $(z^*_{\text{ILP1}} - z^*_{\text{heuristic}})$. Lagrangian relaxation methods are often used in the literature to obtain tight upper bounds to the optimal objective function values of integer linear programs.

An alternative heuristic approach seeks to utilize shortest path algorithms, which have long been used in routing problems. Some efficient algorithms for shortest path problems are discussed in Chapter 14. Moreover, as we saw in Chapter 14, some optimal routing problems *where flow splitting is allowed* can be viewed as shortest path routing problems, with the link weights being given by the dual variables when a dual of the original LP formulation is solved. Because flow splitting is allowed, the original problem formulation results in a linear program. However, this line of argument is not directly applicable here, because the original problem formulation is in terms of an *integer* linear program (**ILP1**). Nevertheless, the idea of finding appropriate link weights and using shortest path routing by these link weights is appealing.

How does one go about finding appropriate link weights? We describe the MIRA heuristic approach. Recall the objective: Link weights must be assigned so that the shortest path between nodes $s(\kappa)$ and $t(\kappa)$ by these link weights is a path that causes minimum interference for other pairs $k \in \mathcal{K} \setminus \kappa$. Also, following the basic approach in MIRA, we consider the reduction in maxflow between $s(k)$ and $t(k)$, $k \in \mathcal{K} \setminus \kappa$, as a measure of the interference caused to demand k by the routing of demand κ. This requires that, to begin with, the maxflow corresponding to demand $k \in \mathcal{K} \setminus \kappa$ be computed, when the demand κ is not present. Setting $d(\kappa) = 0$ in (**ILP1**) is equivalent to the demand κ being absent. Under this condition, **ILP1** reduces to $(K-1)$ independent maxflow computations. Recall that in formulation **ILP1**, only one additional demand k from the set $\mathcal{K} \setminus \kappa$ is considered along with the demand κ. So when demand κ is absent, the only demand left in the network is demand k, and the problem is one of obtaining the maxflow corresponding to it. Because this is now repeated for all demands $k \in \mathcal{K} \setminus \kappa$, we have $(K-1)$ independent maxflow computations. Let $\hat{\theta}(k)$ represent the maxflow computed for demand k in isolation.

The maxflow computations are done assuming the demand κ to be absent. If demand κ were present and were routed along some path, then the maxflow value $\hat{\theta}(k)$ would possibly be reduced, because a part of the residual link capacity would possibly be consumed by the demand κ. Thus, $\frac{\partial \hat{\theta}(k)}{\partial C_l}$, representing the rate of change of $\hat{\theta}(k)$ with respect to the residual link capacity C_l of link l, is a natural measure of the interference caused to demand k by a route for demand κ that uses link l. Incorporating the scalars $\alpha(k)$, the weight of a link l in the MIRA heuristic is now defined as

$$y_l = \sum_{k \in \mathcal{K} \setminus \kappa} \alpha(k) \frac{\partial \hat{\theta}(k)}{\partial C_l} \tag{15.7}$$

If $\alpha(k)$ is large for a certain k, then all links that are important in determining the maxflow of k are assigned large weights, and the route selected for the demand κ tends to avoid these links, because ultimately, the weights y_l would be used to find shortest (more correctly, "lightest") paths. Thus, the scalars $\alpha(k)$ determine how important it is to avoid interference to the demand k when routing the demand κ.

We can now use the celebrated maxflow mincut theorem (see the Appendix to this chapter) to simplify Equation 15.7 further. From this theorem, we can conclude that the maxflow value of a particular pair k decreases whenever the capacity of any link in any mincut for that pair decreases. Furthermore, the rate of decrease of the maxflow value with respect to the capacity of a link belonging to a mincut is clearly unity. The MIRA heuristic now defines links that are critical for a pair k as follows.

Definition 15.1

A link l is defined as *critical* for a given pair k if that link l belongs to any mincut for the pair k. ■

If $\mathcal{L}_k$ represents the set of critical links for the pair k, then we can summarize the discussion by noting that

$$\frac{\partial \hat{\theta}(k)}{\partial C_l} = \begin{cases} 1 & \text{if } l \in \mathcal{L}_k \\ 0 & \text{otherwise} \end{cases}$$

Therefore,

$$y_l = \sum_{\substack{k \in \mathcal{K} \setminus \kappa: \\ l \in \mathcal{L}_k}} \alpha(k) \tag{15.8}$$

Thus, the problem of determining link weights has been transformed into one of determining the critical link set for each demand k. The pseudo-code for the MIRA heuristic is shown in Algorithm 15.1.

15.3.2 Another Formulation and Heuristic

Let us consider the operation of MIRA in the network shown in Figure 15.2. Suppose a demand requiring n units of bandwidth has arrived for the pair (s_0, t). For simplicity, we take the weight $\alpha(k)$ to be 1 for each pair k shown in Figure 15.2, even though the following argument is valid irrespective of the weights. The maxflow corresponding to (s_i, t), $1 \le i \le n$, is 1 because the bottleneck link for each (s_i, t), $1 \le i \le n$, is $(s_i \to c)$, and this has a capacity of 1. The maxflow mincut theorem shows that the mincut for (s_i, t) is the link $s_i \to c$, $1 \le i \le n$. Therefore, the link $c \to t$ does not belong to the mincut corresponding to any one of (s_i, t), $1 \le i \le n$. Hence, when a demand arrives at s_0 for t, the MIRA heuristic uses Equation 15.8 and assigns a weight of 0 to the link $c \to t$. By a similar argument, it is seen that each of the links that can possibly be used to go from s_0 to t gets a weight of 0. Thus, the weight of each of the two paths that can be used is 0, and the demand (s_0, t) would be routed along the path with the least number of hops, which is $s_0 \to c \to t$. Clearly, this leaves no residual bandwidth on link $c \to t$, and therefore future calls for pairs (s_i, t), $1 \le i \le n$, would have to be blocked. If the demand (s_0, t) had been routed on the three-hop path, future calls for pairs (s_i, t), $1 \le i \le n$, could have been carried.

1: Input: A graph $\mathcal{G}(\mathcal{N}, \mathcal{L})$ and the residual capacities C_l on all links $l \in \mathcal{L}$, a demand $\kappa \in \mathcal{K}$, with source node $s(\kappa)$ and destination node $t(\kappa)$, with the demand size being $d(\kappa)$
2: Output: A route between $s(\kappa)$ and $t(\kappa)$, having a capacity of at least $d(\kappa)$
3: Compute the maxflow values $\forall k \in \mathcal{K} \setminus \kappa$.
4: Compute the critical link sets $\mathcal{L}_k$, $\forall k \in \mathcal{K} \setminus \kappa$.
5: Compute the link weights

$$y_l = \sum_{\substack{k \in \mathcal{K} \setminus \kappa: \\ l \in \mathcal{L}_k}} \alpha(k)$$

6: Eliminate all links that have residual capacity C_l less than $d(\kappa)$ and form a reduced network.
7: Using Dijkstra's algorithm, compute the shortest path in the reduced network using y_l as the weight of link l.
8: Route the demand κ from $s(\kappa)$ to $t(\kappa)$ along this shortest path and update the residual capacities.

Algorithm 15.1   Pseudo-code for the MIRA heuristic, following [156].

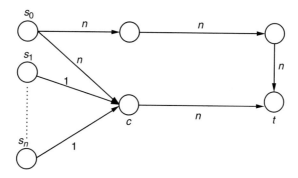

Figure 15.2   Operation of MIRA on a simple network. The link capacities are *n* and 1, as shown. The router pairs between which demands need to be set up are (s_0,t), (s_1,t), ..., (s_n,t). (This sample network is taken from [280]).

This example shows that it is not enough to identify the critical links of the other source–destination pairs. Here, the critical links of (s_i, t), $1 \leq i \leq n$, as defined in the MIRA heuristic, are distinct from the links constituting the possible paths for (s_0, t). Therefore, the MIRA heuristic does not try to avoid the link $c \rightarrow t$.

To address this problem, we follow a different approach. Recall that in the routing problem being considered, demands arrive to the network in an arbitrary arrival process. There is no prior knowledge about when the next demand will arrive. Similarly, the size of a demand is also not known a priori. In such a scenario, we would like to obtain routes that set aside as much resources as possible for future demands.

Let the arrived demand be for source–destination pair $\kappa \in \mathcal{K}$ and suppose that its size is $d(\kappa)$. Consider the vector of demands $(0, \ldots, 0, d(\kappa), 0, \ldots, 0)$; in this vector, there is a single nonzero element $(d(\kappa))$ corresponding to the arrived demand, and the other elements are zeros. Now let us consider a modified demand vector obtained by replacing all the zeros in this demand vector with $\alpha(k)\zeta$, where the index $k \in \mathcal{K} \setminus \kappa$. The modified demand vector assumes that a demand of size $\alpha(k)\zeta$ will be requested in the future for all $k \in \mathcal{K} \setminus \kappa$ that have not yet requested virtual paths. Our objective is to find a route for the modified demand vector such that ζ is maximized. This will not only satisfy the current demand (κ) but will also choose a route such that proportionate amounts of demands can be accommodated between source–destination pairs that can place demands in the future. The numbers $\alpha(k), k \in \mathcal{K}$ are assigned by the network operator. A higher value of $\alpha(k)$ for some k indicates that the operator wishes to set aside more resources to serve future demands from source–destination pair k. This may be required if the customer sending traffic over pair k is a premium customer whose traffic must be carried at a relatively high priority. Thus, in the modified demand vector, we have $d(k) = \alpha(k)\zeta$ for $k \in \mathcal{K} \setminus \kappa$, and the arrived demand is $d(\kappa)$, as usual.

Let us define, for all $k \in \mathcal{K}$ and all links $i \rightarrow j \in \mathcal{L}$, the routing variables

$$q_{i \rightarrow j}(k) = \begin{cases} 1 & \text{if link } i \rightarrow j \text{ is used by pair } k \\ 0 & \text{otherwise} \end{cases}$$

Then, recalling our earlier notation for the node–link incidence matrix $\mathbf{A}$ and the vector $e(k)$, the optimization problem can be stated as follows.

Problem **ILP2**:

$$\max \zeta$$

subject to

$$\mathbf{A}\mathbf{q}(k) = \mathbf{e}(k), \forall k \in \mathcal{K} \tag{15.9}$$

$$\sum_{k \in \mathcal{K}} q_{i \rightarrow j}(k) d(k) \leq C_{i \rightarrow j}, \ \forall (i \rightarrow j) \in \mathcal{L} \tag{15.10}$$

$$\sum_{j \in \mathcal{N}} q_{i \rightarrow j}(k) \leq 1, \ \forall i \in \mathcal{N}, \ \forall k \in \mathcal{K} \tag{15.11}$$

$$\sum_{i \in \mathcal{N}} q_{i \rightarrow j}(k) \leq 1, \ \forall j \in \mathcal{N}, \ \forall k \in \mathcal{K} \tag{15.12}$$

$$\zeta \geq 0 \tag{15.13}$$

$$\mathbf{q} \in \{0, 1\}^{L} \tag{15.14}$$

Equation 15.9 is the flow conservation equation. It asserts that net outflow corresponding to pair k can occur only at the source node of k. Similarly, net inflow for pair k can occur only at the destination node of k. At intermediate nodes, the net outflow corresponding to flow k must be 0. Note that the flow conservation equation (Equation 15.1) in the MIRA formulation **ILP1** is written for the case where split routing for demands $k \in \mathcal{K} \setminus \kappa$ (that is, the future demands in the current problem) is allowed and maxflows are considered. Here, routes are required to be nonsplit. Equations 15.11 and 15.12 ensure that the outflow from any node $i \in \mathcal{N}$ and inflow into any node $j \in \mathcal{N}$ occurs on a single link only. This is because the routing variables $q_{i \rightarrow j}(k)$ take values in the set $\{0, 1\}$; so, for example, if the left side of Equation 15.11 is nonzero, then there can be *only one* nonzero routing variable. This ensures that demand k leaves node i on only one among all the outgoing links from node i. Equation 15.10 requires that the total flow on a link not exceed the link capacity. Note that *all* flows $\in \mathcal{K}$ are being considered simultaneously. This is in contrast to the capacity constraint in the MIRA formulation (Equation 15.3), which considers only *one* flow from the set $\mathcal{K}$, in addition to the flow requesting bandwidth (i.e., κ).

The basic idea of **ILP2** is to maximize the demand sizes of future demands $d(k) = \alpha(k)\zeta$, where $k \in \mathcal{K} \setminus \kappa$. This is achieved in the optimization problem by maximizing ζ. Thus, the route for the arrived demand κ must be found such that there is enough spare capacity to accommodate future demands of size $\alpha(k)\zeta$. Moreover, the formulation explicitly takes care of the requirement that these future demands also be routed without splitting them.

Problem **ILP2** is actually a nonlinear program because, as the capacity constraint given by Equation 15.10 shows, the unknowns ζ and $q_{i \to j}(k)$ occur as factors in a product. Recall that $d(k) = \alpha(k)\zeta$, for $k \in \mathcal{K} \setminus \kappa$; this is how ζ appears on the left side of Equation 15.10. If we assume that all link capacities are integral and that requested bandwidths, including future requests of sizes $\alpha(k)\zeta$, are also integral, then **ILP2** can be converted into an integer *linear* program. We can do this by expressing the integers $d(k)$, $k \in \mathcal{K}$, in binary form, so that each $d(k)$ is expressed as a weighted sum of powers of 2, with the coefficients being 0 or 1. Then the product of each coefficient and the routing variable $q_{i \to j}(k)$ is again a binary variable. This is the key observation that allows the transformation to an integer linear program, in terms of these new variables. Of course, in this process, the number of variables in the integer linear program increases beyond the original number of routing variables. Also, additional linear constraints connecting the routing variables, the coefficients, and the new binary variables must be included.

Like many other integer linear programs, the problem **ILP2** turns out to be computationally hard.

Theorem 15.1

Problem **ILP2** is NP-hard.

Proof: We merely outline the proof. (See Appendix E for a brief discussion of NP-hardness.) The technique of *restriction* is used to claim that **ILP2** is NP-hard. In this method, we show that a special case of the problem at hand is a known NP-hard problem. In other words, it is shown that even a *restricted* version of the given problem is NP-hard.

If split routing is allowed, the problem **ILP2** can also be stated as follows. Let us consider a flow function $\mathbf{x}(k)$ corresponding to demand $k \in \mathcal{K}$ (i.e., $\mathbf{x}(k)$ is a map $\mathbf{x}(k) : \mathcal{L} \to \mathbb{Z}^+$). Note that $x(k)_l$ gives the bit rate of demand k carried on link l. Note also that $x(k)_l$ is a positive integer, because the range space is $\mathbb{Z}^+$. With this notation, problem **ILP2** *with split routing allowed* is now stated as follows: Do there exist flow functions $\mathbf{x}(k)$, $k \in \mathcal{K}$ such that

a. Each demand $k \in \mathcal{K}$ satisfies the flow conservation equations

b. $\sum_{k \in \mathcal{K}} x(k)_l \leq C_l$

c. For demand k, the outflow from $s(k)$ and the inflow into $t(k)$ is at least $d(k)$, the quantity in the modified demand vector?

It is clear that a solution to this problem will provide the routing that we need for **ILP2** with split routes. (**ILP2** with nonsplit routes is even harder, because the routes must satisfy the additional condition of no splitting.) Now if we allow split routes and also restrict ourselves to the case with only two source–destination pairs, the problem that results is the *directed two-commodity integral flow problem*, and this is known to be NP-hard. Hence, **ILP2** is also an NP-hard problem. ∎

Having seen that **ILP2** is an NP-hard problem, we consider a heuristic algorithm to obtain a solution. As in the case of the MIRA heuristic, the objective is to define appropriate link weights and then use shortest path routing, where shortest (lightest) paths are computed using the link weights defined.

The link weight assignment algorithm that we present has two phases: the offline, or preprocessing, phase and the online phase. Offline processing is done only at the start and when there is a change in network topology (such as node or link failure).

The Offline Phase

In this phase, we first compute all paths between all source–destination pairs. Let $\mathcal{P}_k$ represent the set of all paths for $(s(k), t(k))$, and let $\mathcal{P}_k = \{P_{k1}, P_{k2}, \ldots, P_{kk_n}\}$, where the paths for pair k are denoted as P_{k1}, $P_{k2}, \ldots$. We compute $\mathcal{P}_k$ for all $k \in \mathcal{K}$. Each path in $\mathcal{P}_k$ uses certain links. Let $\mathcal{L}_{P_{ki}}$ denote the set of links on path P_{ki}.

The computation required in the offline phase can be substantial. For a fully connected network, the number of paths between a given pair of nodes grows very rapidly with the number of nodes. A practical strategy might be to find M paths in the given network, for some M. A larger value of M means more computation.

Next we compute *link sets* for all pairs $k \in \mathcal{K}$. We define a link set for a source–destination pair k as the set of all links that can possibly be used to route a demand for that pair. That is, if $\mathcal{L}_{(k)}$ represents the link set for pair k, then

$$\mathcal{L}_{(k)} = \bigcup_{i=1}^{k_n} \mathcal{L}_{P_{ki}}$$

The offline weight y_l^{off} of a link l is defined as follows:

$$y_l^{\text{off}} := \sum_{k \in \mathcal{K}: l \in \mathcal{L}_{(k)}} \alpha(k)$$

where $\alpha(k)$ is the weight assigned to pair k at the beginning, and, as discussed before, it reflects the importance of k traffic for the network operator. The intuition for the weight assignment y_l^{off} is as follows. If a link belongs to the link sets of several pairs k, then we want to avoid routing traffic on a path using that link, because such an action is likely to cause a reduction in the objective function ζ of the formulation **ILP2**. In general, the less the extent to which a link is shared by several link sets, the safer it is to route on the link because other demands are affected to a lesser degree. Therefore, links that are shared less should get lower weights. This step is motivated by the objective of keeping ζ in **ILP2** low. Thus, there is a connection between the formulation **ILP2** and the heuristic being developed. Note also that y_l^{off} is determined solely by the topology of the network. Given a network, the link sets corresponding to different source–destination pairs are fixed, and that, by definition, determines y_l^{off}.

The Online Phase

Suppose a demand for $(s(\kappa), t(\kappa))$ has arrived and that the requested size is $d(\kappa)$. All links with residual capacity less than $d(\kappa)$ are eliminated first. If this leaves $s(\kappa)$ and $t(\kappa)$ disconnected, then clearly the demand cannot be routed because of lack of resources.

In the next step, we update link weights as follows: $\forall l \in \mathcal{L}$,

$$y_l \leftarrow y_l^{\text{off}} - \mathcal{I}_{\{l \in \mathcal{L}_{(\kappa)}\}} \alpha(\kappa)$$

where $\mathcal{I}_{\{l \in \mathcal{L}_{(\kappa)}\}}$ is an indicator variable that says whether or not l belongs to link set $\mathcal{L}_{(\kappa)}$. If $l \notin \mathcal{L}_{(\kappa)}$, then the weight is equal to the offline weight. If $l \in \mathcal{L}_{(\kappa)}$, then the offline weight of l is reduced by amount $\alpha(\kappa)$. The motivation for doing this is that the weight of a link should quantify how important it is for pairs k *other* than κ. By subtracting $\alpha(\kappa)$ from y_l^{off} when $l \in \mathcal{L}_{(\kappa)}$, we ensure uniformity among all $l \in \mathcal{L}$ by making the weight depend on all pairs k other than κ. Note that this update of link weights cannot be done in the offline phase because in that phase, we do not know the identity of the demand κ that will arrive.

As a final step, the weight of a link is related to the capacity available on the link, C_l, as

$$y_l \leftarrow \frac{y_l}{C_l}$$

Routing a demand on a path passing through links with low available capacities means that less capacity would be available for future demands, implying a decrease in the value of ζ in formulation **ILP2**. Thus, links with low available capacities get high weights, encouraging shortest path routing to avoid such links.

With the weights of links being defined as before, shortest paths can be computed by the Dijkstra algorithm (see Section 14.3.1). It is also easy to obtain hop-constrained shortest paths by using the Bellman–Ford algorithm (see Section 14.3.2).

Let us now apply the **ILP2** heuristic to the example in Figure 15.2. As noted before, the weight $\alpha(k)$, for each source–destination pair k, is 1, so that there is no preference given to any particular pair. The first step is to obtain the offline weights y_l^{off}. The link sets for the pairs (s_0, t), (s_1, t), ..., (s_n, t), are evident upon inspection. Hence it can be seen that for each link in the three-hop path between s_0 and t, $y_l^{\text{off}} = 1$. Also, $y_{s_0 \to c}^{\text{off}} = 1$, but $y_{c \to t}^{\text{off}} = (n+1)$. Upon arrival of the (s_0, t) demand, we move to the online phase and update weights of links belonging to $\mathcal{L}_{((s_0, t))}$ by subtracting 1 from the offline weights. Hence each link in the three-hop path between s_0 and t now has a weight of 0, and link $s_0 \to c$ has a weight $y_{s_0 \to c} = 0$, but link $c \to t$ has a weight $y_{c \to t} = n$. It is then clear that the weight along the three-hop path is 0, whereas the weight along the path $s_0 \to c \to t$ is strictly positive. Therefore, the **ILP2** heuristic chooses the longer path, preventing the link $c \to t$ from getting choked.

15.3.3 Comparison

In this section we compare the performances of the MIRA and **ILP2** heuristic algorithms. We also include the simulated performance of the min hop routing algorithm because it is widely used. In most cases, substantial improvements in performance are possible by following routing algorithms other than the simple min hop algorithm.

We show a sample network in Figure 15.3. This network is referred to as MIRAnet because it was used by the authors of the MIRA algorithm to evaluate their proposal. All light links are of capacity 1200 units, and all dark links of capacity 4800 units. This is meant to model OC-12 and OC-48 links. OC-12 and

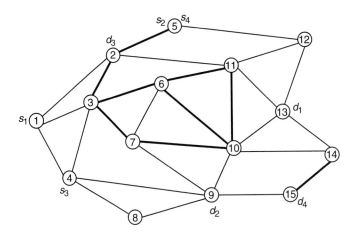

Figure 15.3 A sample network with source–destination pairs shown. Each light link is of capacity of 1200 units, and each dark link of capacity 4800 units. This is meant to model OC-12 and OC-48 links respectively. Also, each link is bidirectional. This topology is referred to as MIRAnet because it was used in evaluating the MIRA proposal.

OC-48 are terms from the SONET standard, which deals with data transmission on optical fibres. OC-12 refers to the optical carrier level 12 signal, which supports a bit rate of 622.08 Mbps. OC-48 supports a bit rate that is 4 times higher—2488.32 Mbps. The source–destination pairs are marked on the network. In the experiments, demands arrive randomly according to a Poisson process and at the same rate for all pairs. Also, for simplicity, we take $\alpha(k) = 1$, for all pairs k. Each arriving demand asks for a guaranteed bandwidth of 1 or 2 or 3 units (corresponding to several Mbps), with equal probability. Thus, the network has the capacity to accommodate thousands of demands.

In an experiment, the network is subjected to a given number of demand arrivals—say, 5000. The number of demands that cannot be routed by the routing algorithm is noted. The experiment is repeated several times. For each run of the experiment, the three routing algorithms—MIRA, **ILP2** heuristic, and min hop—are tried one by one.

Figure 15.4 shows the results of 22 experiments on MIRAnet. For each experiment, the number of demands that cannot be accepted by the routing algorithm is plotted. We see that when 5000 demands arrive to the network, both MIRA and **ILP2** heuristic are able to route all the demands. However, the min hop algorithm rejects roughly 400 of these.

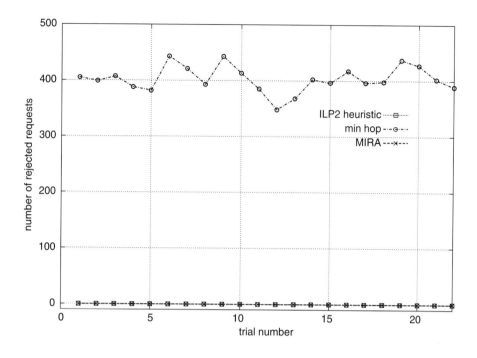

Figure 15.4 Number of demands rejected by various routing algorithms when 5000 demands arrive over a length of time. Each point on the *x*-axis represents a different experiment, but with the number of offered demands being fixed at 5000.

In Figure 15.5, the number of demands arriving to the network is increased to 5500. It can be seen that neither MIRA nor **ILP2** heuristic can accommodate all demands. MIRA appears to reject a slightly higher fraction of demands than **ILP2** heuristic.

It is clear that as the number of offered demands is increased, the number of demands that cannot be routed also goes up. The *fraction* of demands that cannot be routed is a natural performance metric in this context. In Figure 15.6 we show how the average rejection ratio varies as the number of offered demands is increased, for each of the three routing algorithms. Clearly, min hop performs the worst. Also, **ILP2** heuristic provides a slightly lower rejection ratio than the MIRA heuristic. Qualitatively similar results have been observed for other networks as well.

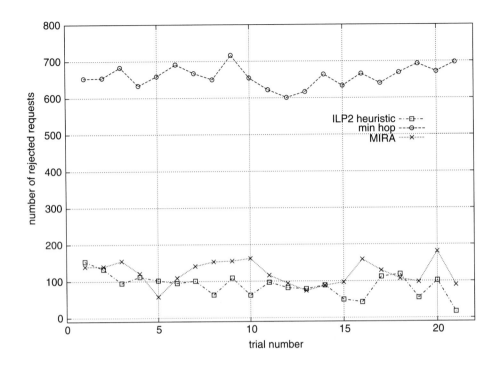

Figure 15.5   Number of demands rejected by various routing algorithms when 5500 demands arrive over a period of time. Each point on the *x*-axis represents a different experiment, but with the number of offered demands being fixed at 5500.

15.4   Multiprotocol Label Switching (MPLS)

IP is a connectionless network protocol. As an IP packet travels in the network, each router on the path decides how to forward the packet *independently* of other routers and other packets. That is, each packet is treated as an independent entity. The forwarding decision depends on the contents of the packet header. In particular, for IP routing, the contents of the Destination Address field determine the next hop. Thus, if the contents of the Destination Address fields of two IP packets are such that the network prefixes are the same, then they are forwarded similarly, even though the contents of the other fields may be quite different.

Therefore, on the data plane, choosing the next hop for a packet can be thought of as a composition of two functions. The first function partitions the set

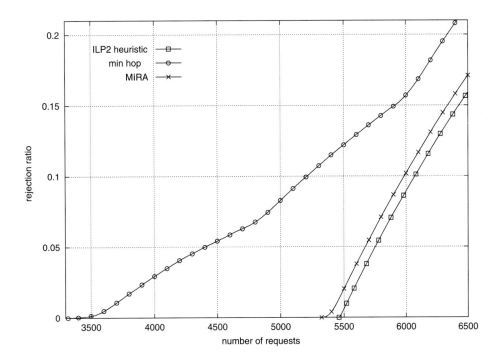

Figure 15.6 The ordinate shows the average fraction of demands that cannot be routed. The number of offered demands is plotted on the x-axis. Results are shown for the min hop, the MIRA, and the ILP2 heuristic routing algorithms. The average is computed over more than 50 experiments, with each value shown on the abscissa.

of possible packets into a set of forwarding equivalence classes (FECs). The second function chooses a next hop for each FEC.

In conventional IP forwarding, a router considers two packets to be in the same FEC when the network prefixes of their destination addresses are the same. As the packet travels in the network, each router reexamines the packet and assigns it to an FEC. In MPLS, however, the assignment of a packet to an FEC is done only once, when the packet enters the MPLS network. The assignment can be based on a rule that considers not only the destination address field in the packet header but also other fields, as well as information that is not present in the network layer header (for example, the port on which the packet arrived). The FEC assigned to the packet is encoded as a short, fixed-length *label*. When a packet is forwarded, the label is sent along with it (i.e., packets are labeled before they are forwarded). At subsequent hops, the packet's network layer header is no longer analyzed. Rather, the label carried by the packet is used as an index into a table that specifies the next hop for the packet as well as a new label. When the packet is forwarded, the new label replaces the old label. This is the familiar label-swapping technique used by ATM networks as well (see Section 2.3.6).

After a label has been assigned to a packet, forwarding decisions are driven by that label. This paradigm of forwarding can be advantageous in several respects. First, MPLS forwarding can be done by switches that are capable of doing label lookup and replacement but not capable of network layer header analysis at high speeds. Second, the MPLS paradigm provides a simple way to force a packet to follow a specified route through the network. A label can be used to represent the route; as the packet travels and labels are swapped, the route gets selected. If this is desired in conventional IP networks, a packet must carry an encoding of the complete route in the packet header (the *source routing* option).

The word "multiprotocol" in MPLS signifies that the procedures specified in MPLS are applicable to *any* network layer protocol. However, MPLS literature often assumes IP to be the network layer protocol. A router that supports MPLS is called a label-switching router (LSR).

As mentioned before, a label is a short, fixed-length, locally significant identifier used to identify an FEC. If r_u and r_d are two LSRs, they can agree that when r_u transmits a packet to r_d, r_u will label a packet with a label value ω if and only if the packet belongs to the FEC f. That is, r_u and r_d agree to a *binding* between the label ω and the FEC f for packets moving from r_u to r_d. Therefore, ω becomes r_u's outgoing label representing FEC f, and r_d's incoming label representing FEC f. With respect to this binding, r_u becomes the *upstream* LSR, and r_d becomes the *downstream* LSR.

In the MPLS architecture, the decision to bind a particular label to a particular FEC is made by a router that is downstream with respect to that binding. The downstream LSR then informs the upstream LSR of the binding. Thus, labels are *downstream-assigned* and distributed in the downstream-to-upstream direction. The set of procedures by which one LSR informs another of its label–FEC bindings is called a *label distribution protocol* (LDP). Some existing protocols are being extended to include the label distribution feature; an example is the Resource Reservation Protocol with Traffic Engineering Extensions (RSVP-TE). New protocols are also being designed with the explicit objective of distributing labels—for example, the Constraint-Routed Label Distribution Protocol (CR-LDP).

In the discussion so far, we have tacitly assumed that a packet carries only a single label. This model can be generalized to one in which a packet can carry a number of labels, organized as a last-in-first-out stack. This is referred to as a *label stack*. If the label stack has depth n, then the label at the bottom of the stack is referred to as the level 1 label, the next one as the level 2 label, and so on, with the label at the top of the stack being the level n label. An unlabeled packet has a label stack of depth 0. Label stacks indicate the presence of a hierarchy in MPLS. However, the processing of a labeled packet is always based only on the top label.

LSRs at the boundary of an MPLS domain are referred to as *ingress* LSRs and *egress* LSRs. An ingress LSR pushes a label onto a previously unlabeled packet and thus begins a label-switched path (LSP). Intermediate LSRs forward the labeled packet based on the label. At the end of the LSP, the egress LSR forwards the packet using conventional IP forwarding. Similarly, we can define a level m LSP as being initiated by an LSR that pushes a level m label onto the packet. Intermediate LSRs make their forwarding decisions based on the level m label. Finally, the egress LSR forwards the packet based on a level $(m - k)$ label, where $0 < k < m$, or forwards the packet using non-MPLS procedures.

As we have seen, labels are used in forwarding data packets across an MPLS domain along LSPs. But before forwarding can commence, the LSPs must be set up, or routed, across the domain. LSPs can be hop-by-hop-routed or explicit-routed. *Hop-by-hop routing* allows each node to independently choose the next hop for each FEC. This is how current IP networks route packets. For *explicit-routed* LSPs, each LSR does not independently choose the next hop; instead, a single LSR, generally the ingress or egress LSR, specifies some or all of the LSRs in the LSP. The explicit route must be specified at the time the labels are assigned.

Sometimes, IP networks use *tunneling* to transfer packets from one router r_u to another router r_d, even though r_u and r_d are not neighbors and r_d is not the packet's ultimate destination. This can be done, for example, by *IP-in-IP encapsulation*, in which the original packet is encapsulated in a new IP packet, with the source and destination IP addresses of the new packet being r_u and r_d, respectively. This creates a *tunnel* from r_u to r_d.

MPLS technology makes it possible to use an LSP as a tunnel and to use label switching rather than IP-in-IP encapsulation to cause packets to travel through the tunnel. The tunnel would be an LSP $\langle r_1, r_2, \ldots, r_i \rangle$, where r_1 is the *transmit end point* of the tunnel and r_i is the *receive end point*. The set of packets to be sent through the tunnel defines an FEC.

Suppose we have an LSP $\langle r_1, r_2, r_3, r_4 \rangle$. Suppose that r_1 receives an unlabeled packet and pushes a label onto it, to cause the packet to follow this path. Let us also assume that r_2 and r_3 are not "true" neighbors but rather are neighbors by virtue of being the end points of an LSP tunnel. The actual sequence of LSRs traveled is, say, $\langle r_1, r_2, r_{21}, r_{22}, r_{23}, r_3, r_4 \rangle$.

When the packet travels from r_1 to r_2, it will have a label stack of depth 1. LSR r_2 first replaces the incoming label with a label that is meaningful to r_3. But then, because the packet must enter an LSP tunnel, r_2 pushes a second label onto the stack. This level 2 label is meaningful to r_{21}. LSRs r_{21}, r_{22}, and r_{23} switch packets using the level 2 label. The penultimate LSR in the tunnel (r_{23}) pops the level 2 label and forwards the packet to r_3, which again sees a level 1 label only. The label stack mechanism allows LSPs to be nested to depths greater than 1.

15.5 Summary

In this chapter, we consider the problem of on-demand routing of bandwidth demands across a backbone network. The demands arrive sequentially, and there is no prior knowledge of when a demand will arrive or what its size will be. Furthermore, each demand must be routed as a unit, with no splitting allowed. We examine the idea of interference imposed on other demands because of the choice of a particular route for a given demand. The problem of minimum interference routing is formulated, resulting in an integer linear program that is NP-hard. Therefore, we consider a heuristic algorithm. The performance of the heuristic in an example leads us to extend the minimum interference routing formulation, which also turns out to be NP-hard, and motivates a second heuristic algorithm. Finally, we take a brief look at MPLS technology, which provides a way to force a particular traffic demand to follow a specified route through the network.

15.6 Notes on the Literature

For routing of virtual paths, the min hop approach has been followed in several works. For example, in [18], min hop routing is assumed while discussing extensions to RSVP (see Chapter 4) for LSP tunnels. Other approaches to routing virtual paths, such as the widest-shortest path algorithm, are discussed in [15]. Similarly, profile-based routing is discussed in [280].

Our discussion of the minimum interference routing algorithm (MIRA) in Section 15.3 follows the important work in [156]. The concept of maxflow was introduced in the classical book [105], and the maxflow mincut theorem was also given there. The insights gained from the MIRA formulation led to the work in the following section, which appeared in [185]. The discussion of the MPLS architecture follows [255].

Appendix: The Maxflow Mincut Theorem

We mention the maxflow mincut theorem and associated concepts very briefly in the appendix. Our discussion follows [7].

Directed Graphs

A *directed graph* $\mathcal{G}(\mathcal{N}, \mathcal{L})$ is defined by a set of nodes or vertices $\mathcal{N}$ and a set of links or edges or arcs $\mathcal{L}$, where the elements of $\mathcal{L}$ are *ordered* pairs of nodes in $\mathcal{N}$. Thus, the link $i \to j$ is distinct from the link $j \to i$, and this accounts for the name "directed."

Cut

A *cut* is a partition of the node set $\mathcal{N}$ into two parts—$\mathcal{S}$ and $\overline{\mathcal{S}} := \mathcal{N} \backslash \mathcal{S}$—and is written as $[\mathcal{S}, \overline{\mathcal{S}}]$. Any cut defines a set of links that have one end point is $\mathcal{S}$ and the other end point is $\overline{\mathcal{S}}$. An alternative definition of a cut is the set of links whose end points belongs to the different subsets $\mathcal{S}$ and $\overline{\mathcal{S}}$. Figure 15.7 shows an example. Here, $\mathcal{S} = \{1, 2, 3\}$ and $\overline{\mathcal{S}} = \{4, 5, 6, 7\}$. The links defined by this cut are $\{(2, 4), (3, 6), (5, 3)\}$.

s-t Cut

An *s-t* cut is defined with respect to two specified nodes s and t. It is a cut $[\mathcal{S}, \overline{\mathcal{S}}]$ such that $s \in \mathcal{S}$ and $t \in \overline{\mathcal{S}}$. In Figure 15.7, if $s = 1$ and $t = 6$, then the cut shown is an *s-t* cut. But if $s = 1$ and $t = 3$, then the cut is not an *s-t* cut.

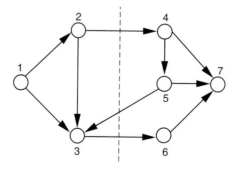

Figure 15.7 An example of a cut in a directed graph.

Capacity of an *s-t* Cut

The capacity $C[\mathcal{S}, \overline{\mathcal{S}}]$ of an *s-t* cut $[\mathcal{S}, \overline{\mathcal{S}}]$ is defined as the sum of the capacities of the forward links in the cut. That is,

$$C[\mathcal{S}, \overline{\mathcal{S}}] = \sum_{(i \to j \in [\mathcal{S}, \overline{\mathcal{S}}])} C_{i \to j}$$

where $C_{i \to j}$ is the capacity of link $i \to j$.

Minimum Cut or Mincut

An *s-t* cut whose capacity is the smallest among all *s-t* cuts is referred to as a *minimum cut*, or *mincut*.

Flow across an *s-t* Cut

Any flow from *s* to *t* must pass across *every* *s-t* cut in the network. This is because *any* *s-t* cut results is *s* and *t* being in disjoint subsets of $\mathcal{N}$. This indicates that the value of the net flow from *s* to *t* can never exceed the capacity of any *s-t* cut.

Maxflow Mincut Theorem

The maximum value of the flow from a source node *s* to a sink node *t* in a capacitated network equals the minimum capacity among all *s-t* cuts.

Problems

15.1 Figure 15.8 shows a network. The residual capacities on the links are marked. A demand of size 3 arrives for the source–destination pair (x, y).

    a. How many routes are possible for the demand (x, y)?

    b. For each possible route for (x, y), find the maxflow value corresponding to the pair (a, y). Repeat for the pairs (a, b) and (c, y), taken one at a time.

    c. Based on part (b) above, identify the best route for (x, y), using the MAX-MIN-MAX criterion of MIRA.

15.2 Consider the network in Figure 15.8 again. As before, a demand of size 3 arrives for the source–destination pair (x, y). The other pairs for which traffic *can* arrive are (a, b), (a, y), and (c, y). (x, y) is called the "current" pair while the others are called "future" pairs.

    a. Write down the possible routes for the demand (x, y), as well as the future demands (a, b), (a, y), and (c, y). A *routing* refers to a collection

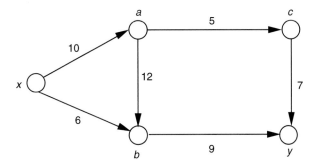

Figure 15.8   A network with several source–destination pairs: (x, y), (a, b), (a, y), and (c, y). The numbers beside the links indicate the capacities available on the links.

of routes for all source-destination pairs, with exactly one route for each pair. How many possible routings are there?

b. For each possible routing, obtain the smallest demand size that can be supported for the future pairs (a, b), (a, y), and (c, y).

c. Hence obtain the best routing from the set of routings.

15.3 Suppose that we wish to apply the **ILP2** heuristic to the network shown in Figure 15.8.

a. Find the offline weights of the links.

b. When the demand (x, y) arrives, how do the weights change?

c. Obtain the length of each path that may be used to route the (x, y) demand, and choose the shortest path. Compare the chosen route with the answer of Problem 15.2.

15.4 Again consider the network in Figure 15.8, and now suppose that the capacity of link $b \rightarrow y$ is only 3 units. As before, a demand of size 3 arrives for the source–destination pair (x, y), and the other pairs are (a, b), (a, y), and (c, y). Suppose that our objective is to maximize the *total amount* of demands (called *sum of demands*) that can be supported in the future.

a. Corresponding to each possible routing, obtain the sum of demands. Hence obtain the best routing.

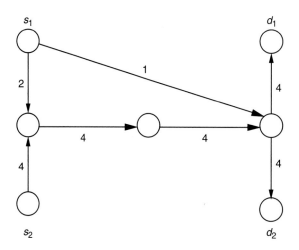

Figure 15.9   A network showing 2 source–destination pairs. The link capacities are marked.

   b. Is the best routing for this problem different from that in Problem 15.2? Why?

15.5 In practice, a network operator may not allow all links in his network to be used by all flows. Let us define a *demand–link incidence matrix* **M** as an $L \times D$ matrix, where L is the number of links and D is the number of demands, follows: $\mathbf{M}_{l,d} = 1$ if demand d is allowed to pass through link l, and 0 otherwise. Consider the "demand space" of dimension D: a point in the demand space represents D demand sizes. Assume that a demand can be split arbitrarily across routes. Corresponding to a given **M** and given link capacities, consider the largest subset of the demand space that is feasible; let us call this the *total routable region*. Show that the total routable region is a convex polytope in $\mathbb{R}^D$.

15.6 Consider the network shown in Figure 15.9. There are two demands: from s_1 to d_1 and from s_2 to d_2.

   a. Plot the routable region assuming that demands can be split arbitrarily across routes.

   b. If demands cannot be split across multiple routes, show how the routable region changes.

CHAPTER 16

Routing of Stream-Type Sessions

In Chapters 14 and 15 we consider demands generated by the need to transfer elastic traffic. A demand represents the average offered load for a pair of routers, which we refer to as the source–destination pair. As mentioned in Chapter 7, elastic traffic has no intrinsic transfer rate or end-to-end delay requirements. Therefore, it is not necessary to consider these explicitly when finding routes for a particular demand.

In this chapter we consider stream traffic that specifies end-to-end bandwidth and delay requirements. Another significant difference in viewpoint is that, in this chapter, a *demand* of size *d* refers to the transfer rate requested by a *single* session or microflow—for example, a packet voice connection set up using RSVP across the Internet (see Section 4.5) or a switched virtual circuit (SVC) across an ATM network (see Section 2.3.6)—rather than an average load generated by an aggregate of elastic traffic transfer sessions (see Section 14.1). We first discuss the problem scenario and some general issues. Then we look at the problem when the end-to-end requirement specified is nonadditive in nature. The next section addresses the additive metrics case and considers separately the problems that arise when rate-based multiplexers and non-rate-based multiplexers are used.

16.1 QoS Routing

A single stream session, with a given bandwidth requirement as well as a specified end-to-end delay requirement, arrives to the network, and the objective of QoS routing is to find a "good" route for the session. Obviously, the route must satisfy the end-to-end bandwidth and delay requirements of the session. Thus, the specified requirements impose a constraint on the choice of routes. If multiple routes satisfying the constraints are available, then the network operator is free to choose one among them. The choice of a particular route is usually guided by the network operator's own objectives. These objectives can be, for example, minimization of total bandwidth consumed or maximization of the smallest spare capacity on the links of the network.

As in Chapter 15, we have an online problem where demands arrive in an unpredictable manner. We do not have the predicted load of stream traffic between different router pairs. In Chapter 14, we assume that the predicted elastic traffic load between different pairs of routers is available, and we proceed to formulate an optimal routing problem. This can be thought of as an offline packing problem. The scenario is different here, because predicted load information is missing and routes must be found as the sessions arrive.

Note that for the scenario being discussed, routing decisions must be made at timescales corresponding to session interarrival times. In contrast, for the packing problem in Chapter 14, the routes found are unchanged for as long as the average offered load values remain at their predicted levels. Thus, if the busy hour lasts from 9:00 AM to 6:00 PM and if the predicted offered load during this time does not change, then we need to set up routes only once in the morning. The routes might be set up, for example, by assigning the appropriate weights to the links of the network. But when routes are to be computed afresh for every arriving session, the frequency of routing decisions is clearly much higher.

We can avoid computing routes at each session arrival by resorting to the following *route precomputation* approach. Here, for example, routes for n packet voice sessions from a source to a destination are computed offline and stored. The number n is chosen by considering the arrival process within each session, the number of hops on the route, and also whether deterministic or statistical QoS guarantees are demanded by the sessions. Models that can be used to determine n in such a case are discussed briefly in Section 5.10.2. After the routes and the number n are precomputed, only admission control is needed. If n sessions are already routed on the chosen path, a fresh session is refused admission; otherwise, the session is routed on the precomputed path. However, the simplicity of network operation comes at a price. For example, it is possible that there are already n ongoing calls on the chosen path, but the newly arrived session *could* be supported on the network by examining the resources available on alternative paths. Because routes are not found *at* session arrivals, the possibility of finding an adequate route outside the reserved set is lost, and we may have an unnecessarily blocked call. This motivates the approach of finding routes at session arrivals. Of course, it is possible to have a hybrid approach, where precomputed routes are used, and, subsequently, when resources on the precomputed routes are exhausted, alternative routes are sought as sessions arrive.

Accordingly, at session arrival, the entity concerned with routing, which we refer to as the *routing module*, examines the network for available resources and chooses an appropriate path. Routing every session as it arrives is also an adaptive strategy, because current network conditions are reflected in the choice of routes. For example, suppose that at time t_1, a stream session S_1 arrives and specifies a bandwidth requirement of B_1 and an end-to-end delay of D_1. A route

R_1 is found that is adequate for this session. If another session specifying the same requirements arrives at a later time t_2, it can possibly be routed on a different path, because R_1 happens to be congested at t_2. Thus, routing adapts to current network conditions.

However, this also means that the routing module must somehow obtain the available bandwidth and delay characteristics of the paths that can potentially be used by the new session. For this to be possible, different routers in the network must advertise available bandwidth and link delay information to others. That is, we would need a routing protocol that disseminates bandwidth and delay information within an area. Efforts in this direction have been initiated by the Internet Engineering Task Force (IETF), and an Internet RFC on "QoS Routing Mechanisms and OSPF Extensions" (RFC 2676) is available. The additional QoS metrics added to the regular link state advertisements (LSAs) (see Chapter 14) are available bandwidth and delay, although the path selection algorithm is limited to considering only bandwidth requirements.

The multiplexing disciplines employed at the routers of a network play a major role in determining the end-to-end delays seen by packets. Multiplexing disciplines fall into two broad groups: rate-based and non-rate-based. A rate-based multiplexer at a link controls the *rates* allocated to connections passing through the link. Larger allocated rates imply lower delays and vice versa. WFQ (see Chapter 4) is an example of a rate-based multiplexing discipline. As discussed in Section 4.5, for a network of WFQ multiplexers, models are available for obtaining the rate to be used at each hop so as to meet the end-to-end delay requirement of a connection. With the appropriate rate known, the routing module finds a path through the network with sufficient unused capacity. This can be done by utilizing the available link capacity information being disseminated by a routing protocol. Rate-based multiplexers are comparatively easier to implement (as discussed in Section 4.6) but can lead to inefficient network operation. On the other hand, multiplexers that can ensure efficient network operation are often not rate-based—for example, the *earliest deadline first* (EDF) multiplexer. When non-rate-based multiplexers are employed, rates and delays are decoupled, and the routing module must consider the bandwidth *as well as* delay information advertised by each node on a candidate path and must select a path that is appropriate. In later sections we consider the problems of setting up a connection in a rate-based and a non-rate-based network.

The advertised resource information (available bandwidth and delay) pertains to the state of the network *before* the newly arrived session has been routed. After the new session has been routed on some path, previously established sessions using links on that path can be affected. For example, link delays on that path can possibly change because of the additional traffic brought in by the new session. Hence, the routing module must *predict* the link delays or link delay distributions and must examine whether adequate performance will be seen by the new connection as well as the existing connections that are impacted by the chosen route.

Each link in the network is associated with multiple metrics that can be classified into two groups: additive and nonadditive. Additive link metrics can be summed over the links of a path to obtain the path metric. Delay is a simple example of an additive link metric. Error-free packet delivery probability on a path is usually computed as a product of error-free packet delivery probabilities on the constituent links of the path.

Exercise 16.1

What is the assumption underlying the computation of packet delivery probability on a path as the product of link delivery probabilities?

In this case, the logarithm of loss probability is an additive parameter. The composition rule to obtain the path metric for nonadditive link metrics is different. For example, the bandwidth available on a path is obtained as the minimum of the link bandwidths.

16.2   Nonadditive Metrics

In this section, we consider a session whose QoS requirements are specified in terms of a nonadditive metric. Specifically, we assume that an arriving session requires d units of bandwidth and therefore that the session can be admitted if the network can find a path such that the *least* available link bandwidth along the path is d. If the *exact* available link bandwidths are known, then this problem has a simple solution. We simply prune the graph representing the network, eliminating links having available bandwidth smaller than d, and observe whether the source and destination remain connected in the resulting subgraph. If the source and destination remain connected, then at least one feasible path exists. When several feasible paths are available, we can choose one of these by considering some other criterion to distinguish among the feasible paths. If the source and destination do not remain connected, then there is no feasible path through the network.

A crucial assumption in the foregoing is that exact available link bandwidths are known. In practice, the link bandwidths are disseminated over the network by a routing protocol, and there may be appreciable delays before all routers have a consistent view. Moreover, by the time the available bandwidth on link l is known at all routers, the bandwidth may well have changed. We cannot exchange state information too frequently, because the overhead involved is too great. The result is that any node's view of the bandwidths available on the links of the network becomes inaccurate. Hence, in practice, a routing algorithm must work with imprecise information on available link bandwidths.

To cope with this situation, the natural approach is to choose the path that has the highest probability of having d units of bandwidth available. Accordingly, let $p_l(d)$ be the probability that the available bandwidth on link l is at least d. The available bandwidth on a path P is given by the smallest available bandwidth among links belonging to this path. Hence, the probability that at least d units of bandwidth are available on path P is given by $\Pi_{l \in P}\, p_l(d)$, where we have assumed that the random variables representing available bandwidths on links $l \in P$ are mutually independent. Then the optimal path P^* can be characterized as

$$\Pi_{l \in P^*}\, p_l(d) \;\geq\; \Pi_{l \in P}\, p_l(d)$$

for all paths P.

Under the *mutual link independence* assumption, the problem of obtaining the path most likely to have at least d units of available bandwidth has a straightforward solution. Taking negative logarithms on both sides of the preceding inequality, it can be seen that path P^* is a shortest path when the length of a link is defined as $(-\log p_l(d))$. Hence, the Dijkstra algorithm, with appropriately defined weights, is sufficient to solve the problem.

16.3 Additive Metrics: Rate-Based Multiplexers

In this section, we consider problems where the performance criterion is *additive* in nature. This means that we obtain the performance metric corresponding to a path by adding link metrics along the path. The typical example is end-to-end delay on a path, which is obtained by adding link delays. The need to consider such additive metrics arises naturally in transporting real-time traffic over packet networks. Applications seeking to transport voice or video over IP are typical examples. In the following, we consider rate-based multiplexers first.

16.3.1 Finding Feasible Routes

Let $\mathcal{G}(\mathcal{N}, \mathcal{L})$ represent the network, with $\mathcal{N}$ being the set of nodes and $\mathcal{L}$ being the set of links. We assume that there are N nodes and L links. Let ξ_l be the sum of propagation delay and maximum transmission time on link l. The maximum transmission time on a link is, of course, determined by the maximum packet size that is seen by the link and the link capacity. As discussed in Section 4.5, ξ_l is simply the sum of the propagation delay and the worst case nonpreemption delay at link l. Also, let C_l be the available capacity on link l, $1 \leq l \leq L$. Let the set of K source–destination pairs in the network be denoted by $\mathcal{K}$. For convenience, we refer to the source–destination pair as "SD pair" in the following.

Consider a path P through the network between a source router and a destination router. The maximum available capacity on path P is clearly $C(P) := \min_{l \in P} C_l$. Also, let $H(P)$ denote the number of hops (i.e., links) on path P, and let $H_{\max}$ be the maximum hop count in the network $\mathcal{G}$.

Connection requests arrive to the network at arbitrary points in time. Suppose that the traffic generated by a connection request for SD pair k, $1 \leq k \leq K$, is characterized by leaky bucket parameters (see Chapter 4) (σ_k, ρ_k), where, as usual, σ_k denotes the maximum burst size and ρ_k denotes the average rate of the source. Notice that we have implicitly taken the peak rate of the source to be ∞. Moreover, let c_k denote the maximum packet length corresponding to session k, and let D_k^{reqd} be the required end-to-end delay. The routing and rate allocation problem is to find, on connection arrival, a route connecting the SD pair and a rate to be allocated on that route, such that the connection's delay and rate requirements are satisfied and also that the capacity constraints of the network links are not violated.

We begin to address this problem by recalling from Chapter 4 the maximum end-to-end delay seen by packets of the connection when a rate r is allocated to a connection at each hop on path P. The upper bound on end-to-end delay can be expressed as

$$D_k(P, r) = \frac{\sigma_k + H(P)\, c_k}{r} + \sum_{l \in P} \xi_l$$

We obtain the upper bound by considering Figure 4.25, which shows a connection passing over a tandem of links, with a rate-based multiplexer (WFQ) at each hop. It can be seen to be the same expression as before, with a simplification resulting from our assumption of infinite source peak rate.

The basic problem can now be stated: Given connections k, $1 \leq k \leq K$, with connection parameters σ_k, ρ_k, c_k and required delay D_k^{reqd}, find *feasible* paths P_k—that is, paths P_k such that all the following are true:

a. $\frac{\sigma_k + H(P_k)\, c_k}{r_k} + \sum_{l \in P_k} \xi_l \leq D_k^{\text{reqd}}, \qquad 1 \leq k \leq K$

b. $\sum_{k:\, l \in P_k} r_k \leq C_l,$ $1 \leq l \leq L$

c. $r_k \geq \rho_k,$ $1 \leq k \leq K$

This is known as a *multicommodity feasible path* problem, where the multiple *commodities* are the $K > 1$ connections to be routed. It is also well known that the problem is NP-hard (see Appendix E for a brief discussion on complexity).

We now consider the simpler case when $K = 1$, that is, the single-commodity feasibility problem. In principle, it is possible to enumerate the paths between the source and the destination and to find out whether any of the paths is feasible. However, enumeration of paths in a network can be a very time-consuming process, because the number of paths can be very large. We seek an algorithm that can provide an answer in polynomial time.

The algorithm tackles the routing and rate allocation problem in two steps. First, a sequence of subgraphs is extracted from the original network graph, and the shortest paths on these subgraphs (according to link metric ξ_l) are found. This gives a set of candidate paths. Next, we check whether any of these paths can satisfy the required end-to-end delay when the maximum possible rate is allocated on the path. Finally, the claim is that if a suitable path cannot be found in this set, then there exists no suitable path in the network.

Let $C^{(1)} < C^{(2)} < \ldots < C^{(M)}$ be the *distinct* values of available link capacities. Clearly, $M \leq L$. Consider Algorithm 16.1. The Bellman–Ford algorithm is discussed in Section 14.3.2. For each m, line 3 *possibly* identifies a shortest path from the source to the destination having at most 1, 2, $\ldots$, H_{max} hops. Of course, it may happen that in the network, there is no path with fewer than $H \leq H_{max}$ hops, and so the first $(H - 1)$ iterations of the Bellman–Ford algorithm yield no paths. Moreover, not all the paths found need be distinct. Nevertheless, we can say that at most H_{max} paths are identified for each value of m, and at most MH_{max} paths overall. We can

---

1: **for all** $m \in \{1, 2, \ldots, M\}$ **do**
2:     Delete all links l for which $C_l < C^{(m)}$.
3:     On the remaining subgraph, run a Bellman–Ford shortest path algorithm with the link metric being ξ_l.

Algorithm 16.1   Algorithm for solving the single commodity problem.

now find a path feasible path P from among these paths by computing the delay $D(P, C(P))$ and checking whether that satisfies the specified requirement D_1^{reqd}. It is also clear that the complexity of the algorithm is polynomial, because the Bellman–Ford algorithm itself belongs to the class of polynomial time algorithms.

How can we show that Algorithm 16.1 works correctly? When the iterations are completed, we have at most $MH_{\max}$ paths. If any of these paths is feasible, then the single-commodity feasibility problem is solved and we are finished. So what we need to show is that if *none* of the (at most) $MH_{\max}$ paths is feasible, then there exists no feasible path.

Consider the first iteration, with $m = 1$. Before this iteration starts, all links with capacity $C_l < C^{(1)}$ have been eliminated. But $C^{(1)}$ is the smallest-capacity link in the network. So, for $m = 1$, none of the links is removed, and the Bellman–Ford algorithm runs on the complete network. For successive iterations $m > 1$, some links are eliminated and the Bellman–Ford algorithm runs on a subgraph of the original graph. Clearly, as the iterations proceed, the set of available paths connecting the source and the destination can only grow smaller, because some available paths "break down" when links are removed.

We can also conclude that the shortest paths returned in iteration $(m + 1)$ have propagation delays that are no smaller than those of the corresponding ones returned in iteration m, $m \geq 1$. This is simply because the set of available paths has grown possibly smaller, because some paths may have been eliminated in going from iteration m to $(m + 1)$. Also, the capacity available on a path returned in iteration m is *at least* $C^{(m)}$, because links of lower capacity have been eliminated. Hence, the *lower bound* on the capacity of a path returned in iteration $(m + 1)$ is no smaller than that of a path returned in iteration m.

Now consider some iteration m, $1 \leq m \leq M$, and some H, $1 \leq H \leq H_{\max}$. So in the mth iteration, we are looking at paths that have at most H hops. Suppose $P_1^{(H)}$ is returned as the shortest path by the Bellman–Ford algorithm, and $P_2^{(H)}$ is some other path. Then the propagation delay on $P_2^{(H)}$ is greater than that on $P_1^{(H)}$. Because none of the at most $MH_{\max}$ paths returned was feasible, we know that $P_1^{(H)}$ is infeasible. Now there are several cases.

- *Case A:* $r(P_1^{(H)}) = C^{(m)}$ and $r(P_2^{(H)}) = C^{(m)}$. In this case, both $P_1^{(H)}$ and $P_2^{(H)}$ are eliminated when links of capacity $C^{(m)}$ are removed. But it is clear that $P_2^{(H)}$ is also infeasible, because its propagation delay is larger than that of $P_1^{(H)}$, whereas its capacity is the same. Consequently, if $P_1^{(H)}$ fails to meet the required delay, then so will $P_2^{(H)}$.

- *Case B:* $r(P_1^{(H)}) = C^{(m)}$ and $r(P_2^{(H)}) > C^{(m)}$. Here, $P_1^{(H)}$ is eliminated but $P_2^{(H)}$ stays, and therefore it may turn out to be the shortest path in some future iteration.

- *Case C:* $r(P_1^{(H)}) > C^{(m)}$ and $r(P_2^{(H)}) = C^{(m)}$. In this case, $P_2^{(H)}$ is eliminated at the beginning of the next iteration. Again, we know that $P_2^{(H)}$ is not feasible, so its elimination does not matter.

- *Case D:* $r(P_1^{(H)}) > C^{(m)}$ and $r(P_2^{(H)}) > C^{(m)}$. Here, both $P_1^{(H)}$ and $P_2^{(H)}$ survive the next round of eliminations.

We note that in cases C and D, the path $P_1^{(H)}$ is eventually eliminated, because it has been found to be infeasible. As the iterations proceed, neither its propagation delay nor its capacity changes, and so it continues to be infeasible. Therefore, at some stage it will be eliminated when its bottleneck link is removed.

From this discussion, it becomes clear that whenever a path is eliminated we already know that it could not have been a feasible path. Thus, a potentially feasible path is *never* eliminated.

When the final iteration ($m = M$) begins, all links of capacity less than $C^{(M)}$ have been eliminated. Thus, the capacity of all paths is $C^{(M)}$. Here, if the shortest path is infeasible, then evidently any other path will also be infeasible because its propagation delay can only be larger.

Summarizing the argument: We begin with the given graph and obtain at most $MH_{\max}$ paths after pruning the graph in M stages. None of the $MH_{\max}$ paths is feasible. Our arguments show that whenever a path is eliminated as a result of pruning, it could not have been a feasible path. In the final subgraph that remains, again, no path is feasible. These observations enable us to conclude that, to begin with, there was no feasible path in the original graph. Hence if none of the at most $MH_{\max}$ paths is feasible, there exists no feasible path in the original graph.

16.3.2   An Upper Bound on Performance

Recall the notation introduced earlier. Let $\mathcal{G}(\mathcal{N}, \mathcal{L})$ represent the network, with $\mathcal{N}$ being the set of nodes and $\mathcal{L}$ being the set of links. We assume that there are N nodes and L links. Let C_l be the available capacity on link l, $1 \leq l \leq L$. Let the set of K source–destination pairs in the network be denoted by $\mathcal{K}$. For convenience, we refer to source–destination pair as "SD pair" in the following. Also, let $\mathcal{P}$ be the set of paths on which connections can be routed, and suppose that J is the number of paths.

Let $\mathbf{G}$ be the $J \times K$ path–SD pair incidence matrix. That is, for $1 \leq j \leq J$ and $1 \leq k \leq K$,

$$\mathbf{G}_{j,k} = \begin{cases} 1 & \text{if path } j \text{ is between SD pair } k \\ 0 & \text{otherwise} \end{cases}$$

Similarly, let $\mathbf{B}$ be the $J \times L$ path–link incidence matrix, where, for $1 \leq j \leq J$ and $1 \leq l \leq L$,

$$\mathbf{B}_{j,l} = \begin{cases} 1 & \text{if link } l \text{ is on path } j \\ 0 & \text{otherwise} \end{cases}$$

Furthermore, let us assume that the arriving connections belong to a set $\mathcal{C}$ containing I classes of traffic. Each class i, $1 \leq i \leq I$, is associated with a triplet of parameters $(\sigma_i, \rho_i, \hat{p}_i)$, where σ_i is the bucket size of the associated leaky bucket (see Chapter 4), ρ_i is the average rate, and $\hat{p}_i$ is the peak rate of connection class i. As seen before, $(\sigma_i, \rho_i, \hat{p}_i)$ characterizes the traffic stream generated by a connection of class i. The end-to-end delay requirement of class i is denoted as D_i^{reqd}.

The delay requirement D_i^{reqd} can be converted into a minimum rate requirement that depends on the path on which the connection is routed. (Again, see Figure 4.25 and the following discussion.) Therefore, we associate a rate r_{ij} with class i and path j; that is, a class i connection that is routed on path j requires r_{ij} amount of bandwidth to be allocated on each link of path j. The dependence of r_{ij} on path j arises as follows. Suppose that there are two paths between an SD pair, one being a two-hop path and the other a four-hop path. The end-to-end delay requirement of a connection can possibly be satisfied on both paths by allocating a higher rate on the four-hop path. Nonpreemption delays are present on four hops on the second path, and we can reduce the end-to-end delay by allocating a higher rate.

In passing, we also note that this model captures, as a special case, the scenario where a connection does not specify an end-to-end delay requirement but only a bandwidth requirement. For this, we have $r_{ij} = r_i$. With the system model as described here, we are now ready to define the problem.

For simplicity, we consider only the offline problem, in which a total of λ connections distributed over the I classes are given to the *routing and rate allocation* (RRA) algorithm at the beginning. We refer to λ as the offered traffic or offered load. Let $p_{ik}\lambda$ denote the number of class i connections for SD pair k. Clearly, $\sum_{i=1}^{I} \sum_{k=1}^{K} p_{ik} = 1$. If not all the connections can be admitted because of capacity constraints, the algorithm's task is to select a subset for admission. The selection is made according to a criterion discussed later. The corresponding online problem can also be addressed in a manner similar to that used for the offline case.

Problem Formulation

Accordingly, let s_{ik} denote the number of class i connections for SD pair k admitted by the RRA algorithm. Note that s_{ik} is simply the carried traffic of class i corresponding to SD pair k. Also, let n_{ij} be the number of class i connections carried on path j. Then we have

$$s_{ik} = \sum_{j=1}^{J} n_{ij} G_{j,k} \quad \forall i \in \mathcal{C}, \ \forall k \in \mathcal{K} \tag{16.1}$$

$$\sum_{i=1}^{I} \sum_{j=1}^{J} r_{ij} n_{ij} B_{j,l} \leq C_l \quad \forall l \in \mathcal{L} \tag{16.2}$$

where $n_{ij} \geq 0$ and integral. This simply says that the link capacity constraints should not be violated.

We define a weight vector as a set of I nonnegative numbers, where each number is associated with a traffic class. The weight associated with a class can be thought of as the revenue generated by a connection of that class. Let $\alpha^f = \left\{ \alpha_1^f, \alpha_2^f, \dots, \alpha_I^f \right\}$, $1 \leq f \leq F$, be F such weight vectors, constituting the set $\mathcal{F}$. With F weight vectors, we have a general framework for analysis that reduces to commonly used performance metrics as special cases. This generality is the motivation for using F weight vectors.

Given s_{ik} for $1 \leq i \leq I$, $1 \leq k \leq K$, the revenue due to a weight vector α^f is $W^f := \sum_{i=1}^{I} \alpha_i^f \sum_{k=1}^{K} s_{ik}$. Thus, we have F revenue values. Let $W_{\min} := \min_{f \in \mathcal{F}} (W^f)$ be the minimum weighted carried traffic.

What is the maximum value of the minimum weighted carried traffic (i.e., $W_{\min}$) that any RRA algorithm can extract from the network? We formulate this as the following optimization problem, referred to as **ILP-R1**.

$$I(\lambda, \mathbf{p}) \;=\; \max_{f \in \mathcal{F}} \min \left(\sum_{i=1}^{I} \alpha_i^f \sum_{k=1}^{K} s_{ik} \right)$$

subject to

$$s_{ik} = \sum_{j=1}^{J} n_{ij} G_{j,k} \quad \forall i \in \mathcal{C}, \; \forall k \in \mathcal{K} \tag{16.3}$$

$$\sum_{i=1}^{C} \sum_{j=1}^{J} r_{ij} n_{ij} B_{j,l} \leq C_l \quad \forall l \in \mathcal{L} \tag{16.4}$$

$$s_{ik} \leq p_{ik}\lambda \quad \forall i \in \mathcal{C}, \; k \in \mathcal{K} \tag{16.5}$$

Equation 16.5 ensures that, for each class and SD pair, the carried traffic is less than the offered traffic. Also, $\mathbf{p}$ represents the vector of p_{ik} values for $1 \leq i \leq I$ and $1 \leq k \leq K$.

Two performance criteria that are often used turn out to be special cases of the general performance criterion having F weight vectors. For the first criterion, there is only one weight vector, and we set $F = 1$. The problem then reduces to maximizing the weighted carried traffic. For the second criterion, $F = I$ and the weight vectors are as follows: For the fth weight vector, α^f,

$$\alpha_i^f = \begin{cases} 1 & \text{if } i = f \\ 0 & \text{otherwise} \end{cases}$$

The weighted carried traffic due to weight vector α^i then corresponds to the carried traffic of class i, and the problem reduces to maximizing the minimum carried traffic. Maximization of the minimum carried traffic is a reasonable fairness criterion. Starting with F weight vectors enables us to study both of these special cases within the same framework.

Optimization problem **ILP-R1** can be equivalently written as follows.

$$I(\lambda, \mathbf{p}) = \max \quad \Gamma$$

subject to

$$s_{ik} = \sum_{j=1}^{J} n_{ij} \mathbf{G}_{j,k} \quad \forall i \in \mathcal{C}, \forall k \in \mathcal{K} \tag{16.6}$$

$$\sum_{i=1}^{I} \sum_{j=1}^{J} r_{ij} n_{ij} \mathbf{B}_{j,l} \leq C_l \quad \forall l \in \mathcal{L} \tag{16.7}$$

$$s_{ik} \leq p_{ik} \lambda \quad \forall i \in \mathcal{C}, k \in \mathcal{K} \tag{16.8}$$

$$\sum_{i=1}^{I} \alpha_i^f \sum_{k=1}^{K} s_{ik} \geq \Gamma \quad \forall f \in \mathcal{F} \tag{16.9}$$

$$n_{ij} \geq 0 \text{ and integral, } \forall i \in \mathcal{C}, \forall j \in \mathcal{P} \tag{16.10}$$

$$s_{ik} \geq 0 \text{ and integral, } \forall i \in \mathcal{C}, \forall k \in \mathcal{K} \tag{16.11}$$

The variable Γ is simply the minimum weighted carried traffic (see Equation 16.9), and the problem seeks to maximize Γ. This problem is an integer linear program that turns out to be hard to solve. Let us consider how to obtain an upper bound on the optimum solution of the foregoing optimization problem. We would also like to design an RRA algorithm whose performance comes close to this upper bound. We will devise an RRA algorithm that achieves the upper bound only asymptotically, as offered load and link capacities increase. The problem of designing an optimal RRA policy, given finite offered load and link capacities, is an open problem.

Our approach is to relax the integer constraints to obtain a linear program instead of an integer linear program. This means that the variables in the problem are no longer constrained to take integer values. That is, the variables become real-valued. Hence, the variables can take values in a *larger* set ($\mathbb{R}^+$ instead of $\mathbb{Z}^+$). The implication is that the optimal objective value of the relaxed problem, the linear program, is *no smaller than* the optimal objective value of the integer linear program.

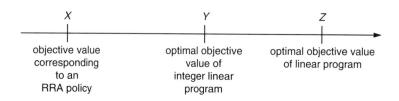

Figure 16.1 The ordering among the objective values corresponding to the linear program, the integer linear program, and any RRA algorithm. Values increase as we move to the right.

This argument is valid for any given set of capacities and load. Solving the linear program for these capacity and load values, we will get an upper bound on the objective value corresponding to *any* RRA algorithm (see Figure 16.1). Ideally, we would like to design an RRA algorithm such that points X and Z in Figure 16.1 are close together. This would imply that the RRA is nearly optimal. However, it is difficult to devise such an almost optimal RRA policy for given finite capacities and load. Therefore, we consider an asymptotic approach and ask this question: Can one get an RRA policy that is close to the optimal *when the capacities and offered load are large?*

To examine this question in a simple setting, we need some additional assumptions. Suppose that C_{min} denotes the smallest link capacity in the network (i.e., $C_{min} = \min_{l \in \mathcal{L}} C_l$). Let $C_1/C_{min} := \eta_1$, $C_2/C_{min} := \eta_2, \ldots$, and $C_L/C_{min} := \eta_L$. Also, let $\lambda/C_{min} = \nu$. Then we have the following.

Assumption 16.1

As link capacities and offered load grow large, the ratios $\eta_1, \eta_2, \ldots, \eta_L$ and ν remain constant.   ∎

Thus, all the link capacities and offered load grow large *in proportion.*

What happens to the required rate r_{ij} when capacities increase? Recall from Chapter 4 that the delay seen by a packet at a node is determined by several factors, one of which is the nonpreemption delay. Clearly, the nonpreemption delay decreases as link capacities increase. This implies that to meet a specified delay requirement, the rate to be allocated to a connection *can decrease* as link capacities increase. This leads us to the following.

Assumption 16.2

As link capacities increase, the required rate for a connection of class i on path j, (r_{ij}) decreases to r_{ij}^{∞}. In other words, $\lim_{C_{\min} \to \infty} r_{ij} = r_{ij}^{\infty}$. ∎

Assumption 16.2 allows us to replace r_{ij} with its asymptotic limit r_{ij}^{∞} in Equation 16.7. Next, we divide the objective function of the ILP and both sides of Equations 16.6 through 16.9 by $C_{\min}$ and relax the integral constraints to obtain the following linear program.

$$L_0(v, \mathbf{p}) = \max \ \gamma$$

subject to

$$w_{ik} = \sum_{j=1}^{J} \beta_{ij} \mathbf{G}_{j,k} \quad \forall i \in \mathcal{C}, \ \forall k \in \mathcal{K} \tag{16.12}$$

$$\sum_{i=1}^{I} \sum_{j=1}^{J} r_{ij}^{\infty} \beta_{ij} \mathbf{B}_{j,l} \leq \eta_l \quad \forall l \in \mathcal{L} \tag{16.13}$$

$$w_{ik} \leq p_{ik} v \quad \forall i \in \mathcal{C}, \ \forall k \in \mathcal{K} \tag{16.14}$$

$$\sum_{i=1}^{I} \alpha_i^f \sum_{k=1}^{K} w_{ik} \geq \gamma, \ f \in \mathcal{F} \tag{16.15}$$

with new real variables $\dfrac{\Gamma}{C_{\min}} := \gamma \geq 0$, $\dfrac{s_{ik}}{C_{\min}} := w_{ik} \geq 0$, $\dfrac{n_{ij}}{C_{\min}} := \beta_{ij} \geq 0$. The optimum value of the foregoing LP is denoted by $L_0(v, \mathbf{p})$. By Assumption 16.1, the link capacities and offered load grow large in proportion, with the ratios remaining constant. So the form of the LP does not change as $C_{\min}$ changes. It is easy to see that Lemma 16.1 holds.

Lemma 16.1

$$\frac{1}{C_{\min}} I(\lambda, \mathbf{p}) \leq L_0(v, \mathbf{p})$$

∎

We now proceed as follows.

1. Consider the optimal solution to the foregoing LP. In general, the solution is given by three real numbers for the three variables. From the three real is numbers, we obtain three integers as a possible solution to the integer linear program **ILP-R1**.

2. We then show that the three integers constitute a feasible solution to the problem **ILP-R1**.

3. Utilizing the feasible solution to **ILP-R1**, we give an algorithm that says, for each SD pair, how many connections of each class should be accepted.

4. We show that the proposed algorithm is asymptotically optimal.

Let β'_{ij}, w'_{ik}, and γ' denote the optimal solution to the $L_0(v, \mathbf{p})$ LP, and let n'_{ij}, s'_{ik}, and Γ' be obtained as follows:

$$n'_{ij} = \lfloor C_{\min} \frac{r^{\infty}_{ij}}{r_{ij}} \beta'_{ij} \rfloor \quad \forall i \in \mathcal{C}, \ \forall j \in \mathcal{P}$$

$$s'_{ik} = \sum_{j=1}^{J} n'_{ij} G_{jk} \quad \forall i \in \mathcal{C}, \ \forall k \in \mathcal{K}$$

$$\Gamma' = \min_{f \in \mathcal{F}} \left(\sum_{i=1}^{I} \alpha^f_i \sum_{k=1}^{K} s'_{ik} \right)$$

This completes step 1. The following Lemma establishes step 2.

Lemma 16.2

$(n'_{ij}, s'_{ik}, \Gamma')$ is a feasible solution to problem **ILP-R1**.

Proof: To show that Equation 16.7 is satisfied, recall that w'_{ik} and β'_{ij} constitute an optimal solution to the LP $L_0(v, \mathbf{p})$. Hence

$$\sum_{i=1}^{I} \sum_{j=1}^{J} r^{\infty}_{ij} \beta'_{ij} \mathbf{B}_{j,l} \leq \eta_l \quad \forall l \in \mathcal{L}$$

$$\Rightarrow \sum_{i=1}^{I}\sum_{j=1}^{J} r_{ij}\frac{r_{ij}^{\infty}}{r_{ij}}\beta'_{ij}\mathbf{B}_{j,l} \leq \eta_l \quad \forall l \in \mathcal{L}$$

$$\Rightarrow \sum_{i=1}^{I}\sum_{j=1}^{J} r_{ij}\lfloor C_{\min}\frac{r_{ij}^{\infty}}{r_{ij}}\beta'_{ij}\rfloor\mathbf{B}_{j,l} \leq \eta_l C_{\min} = C_l \quad \forall l \in \mathcal{L}$$

$$\Rightarrow \sum_{i=1}^{I}\sum_{j=1}^{J} r_{ij}n'_{ij}\mathbf{B}_{j,l} \leq C_l \quad \forall l \in \mathcal{L}$$

To prove Lemma 16.2, we need to show that, in addition, Equation 16.8 is satisfied. To this end, consider

$$w'_{ik}C_{\min} = \sum_{j=1}^{J} C_{\min}\beta'_{ij}\mathbf{G}_{j,k}$$

$$\geq \sum_{j=1}^{J} C_{\min}\frac{r_{ij}^{\infty}}{r_{ij}}\beta'_{ij}\mathbf{G}_{j,k} \quad \text{(using } \frac{r_{ij}^{\infty}}{r_{ij}} \leq 1\text{)}$$

$$\geq \sum_{j=1}^{J} \lfloor C_{\min}\frac{r_{ij}^{\infty}}{r_{ij}}\beta'_{ij}\rfloor\mathbf{G}_{j,k}$$

$$= \sum_{j=1}^{J} n'_{ij}\mathbf{G}_{j,k} = s'_{ik}$$

Hence

$$w'_{ik} \leq p_{ik}v \Rightarrow w'_{ik}C_{\min} \leq p_{ik}\lambda \Rightarrow s'_{ik} \leq p_{ik}\lambda$$

Finally, Equation 16.9 is satisfied by the definition of Γ'. ∎

This completes step 2 on page 756. Consider the following algorithm for accepting connections. Out of the offered λp_{ik} connections of class i on SD pair k, accept s'_{ik} connections (as defined in step 1), and block the rest of the connections.

This is called the *partitioning RRA* algorithm. This is the algorithm referred to in step 3. Note that Γ' represents the minimum weighted carried traffic corresponding to the partitioning RRA algorithm.

For step 4, we consider the following Lemma.

Lemma 16.3

$$\lim_{C_{\min}\to\infty} \frac{1}{C_{\min}}\Gamma' = L_0(v,\mathbf{p})$$

Proof: We first show that $\lim_{C_{\min}\to\infty} \frac{1}{C_{\min}} n'_{ij} = \beta'_{ij}$, which implies

$\lim_{C_{\min}\to\infty} \frac{1}{C_{\min}} s'_{ik} = w'_{ik}.$

To show $\lim_{C_{\min}\to\infty} \frac{1}{C_{\min}} n'_{ij} = \beta'_{ij}$, we have

$$C_{\min} \frac{r_{ij}^{\infty}}{r_{ij}} \beta'_{ij} - 1 \le \lfloor C_{\min} \frac{r_{ij}^{\infty}}{r_{ij}} \beta'_{ij} \rfloor \le C_{\min} \frac{r_{ij}^{\infty}}{r_{ij}} \beta'_{ij}$$

and the proof follows by dividing by $C_{\min}$ and taking the limit as $C_{\min} \to \infty$ and using the fact that $\lim_{C_{\min}\to\infty} \frac{r_{ij}^{\infty}}{r_{ij}} = 1$. Now

$$\Gamma' = \min_{f\in\mathcal{F}} \left(\sum_{i=1}^{I} \alpha_i^f \sum_{k=1}^{K} s'_{ik} \right)$$

and

$$L_0(v,\mathbf{p}) = \gamma' = \min_{f\in\mathcal{F}} \left(\sum_{i=1}^{I} \alpha_i^f \sum_{k=1}^{K} w'_{ik} \right)$$

To prove the lemma, we need to show the following:

$$\lim_{C_{\min}\to\infty} \frac{1}{C_{\min}} \min_{f\in\mathcal{F}} \left(\sum_{i=1}^{I} w_i^f \sum_{k=1}^{K} s'_{ik} \right) = \gamma'$$

This is true if we can interchange lim and min on the left side of the equation. That we can interchange lim and min follows from the following lemma, the proof of which we omit. ∎

Lemma 16.4

Let $h_1(x), \ldots, h_K(x)$ be K functions with the following property: $\lim_{x \to \infty} h_1(x) = h_1, \ldots, \lim_{x \to \infty} h_K(x) = h_K$. Let $h_{min} = \min_{1 \le k \le K} h_k$. Then

$$\lim_{x \to \infty} \min_{1 \le k \le K} h_k(x) = h_{min}$$

∎

Thus, we have come to the end of step 4 and have proved that the partitioning RRA algorithm achieves the upper bound $C_{min} L_0(v, \mathbf{p})$ asymptotically. For given finite link capacity values (that is, a finite C_{min}), we cannot show that the partitioning RRA algorithm is optimal. However, the larger the link capacities and load, the closer the partitioning RRA algorithm is to being optimal.

Our conclusion about the partitioning RRA also has the following implication, which follows directly from Lemmas 16.1 and 16.3. Because we have shown that the scaled objective value (that is, the objective value divided by C_{min}) of a *particular* RRA policy (the partitioning RRA policy) approaches $L_0(v, \mathbf{p})$ as C_{min} approaches infinity, it follows that the scaled *optimal* objective value (that is, $I(\lambda, \mathbf{p})$) also behaves similarly. This is stated next as a corollary.

Corollary 16.1

$$\lim_{C_{min} \to \infty} \frac{1}{C_{min}} I(\lambda, \mathbf{p}) = L_0(v, \mathbf{p})$$

∎

Hence we have shown that $C_{min} L_0(v, \mathbf{p})$ is an asymptotically tight upper bound on the minimum weighted carried traffic of any RRA algorithm in the offline case.

16.4 Additive Metrics: Non-Rate-Based Multiplexers

In Section 16.3, we consider multiplexing disciplines where a connection is allocated a rate, and we examine the single-commodity feasibility problem. We also look at how upper bounds for the performance of any RRA policy can be found. Now we move on to consider non-rate-based multiplexers. Recall that for such multiplexers, the end-to-end delay seen by a connection is obtained by adding the delays at each hop on the path.

We consider the *multiconstrained feasibility* problem, where, in general, m additive constraints are given and the objective is to find a path that satisfies all the m constraints. For example, consider a connection that requires bounds on end-to-end delay as well as end-to-end loss probability. As mentioned in Section 16.1, under certain conditions, we can view end-to-end loss probability as an additive metric by taking logarithms of link loss probabilities. So here, $m = 2$. In case several paths satisfying all m constraints are available, there is no criterion specified for choosing one from this set of paths. In terms of an optimization problem, we see that there is no objective function specified, and it is enough to find a path in the feasible set described by the constraints.

16.4.1 The Multiconstrained Feasibility Problem

Consider a network represented as a directed graph $\mathcal{G}(\mathcal{N}, \mathcal{L})$, with $\mathcal{N}$ being the set of nodes and $\mathcal{L}$ being the set of links. Each link connecting nodes u, $v \in \mathcal{N}$, has an associated link vector $\lambda(u \to v)$, with the m elements being the additive metrics $\lambda_i(u \to v) > 0$, $1 \leq i \leq m$. Given a path $P(s \to k \to \cdots \to l \to t)$, passing from the source node s to the destination node t via $k, \ldots, l$, the corresponding path vector $\lambda(P) = (\lambda_1(P), \ldots, \lambda_i(P), \ldots, \lambda_m(P))$ is given by

$$\lambda_i(P) := \sum_{(u \to v) \in P} \lambda_i(u \to v) \qquad 1 \leq i \leq m$$

Given m positive constraints Λ_i, $1 \leq i \leq m$, the problem is to find a path P from the source node s to the destination node t such that $\lambda_i(P) \leq \Lambda_i$, $1 \leq i \leq m$.

It is known that the feasibility problem with $m \geq 2$ constraints is $\mathcal{NP}$-complete (see Appendix E). A heuristic approach that has been followed is to define a real-valued function that takes the m path metric values $\lambda_1(P), \ldots, \lambda_i(P), \ldots, \lambda_m(P)$, and maps them to a single real value that represents the *effective* path length. For example, letting $\lambda(P)$ denote the mapping function and $m = 2$, we can have

$$\lambda(P) = \alpha_1 \lambda_1(P) + \alpha_2 \lambda_2(P) \qquad\qquad (16.16)$$

where α_1 and α_2 are positive real numbers. Here, a linear function of $\lambda_1(P)$ and $\lambda_2(P)$ is defined. The idea behind the heuristic is to choose a path for which $\lambda(P)$ is minimized. It is hoped that the minimum value of $\lambda(P)$ will be achieved when

$\lambda_1(P)$ and $\lambda_2(P)$ are small, and it is also hoped that these small values will be less than Λ_1 and Λ_2, so that P is a feasible path.

This heuristic can be understood by considering Figure 16.2. The slant lines represent contours of the linear function $\lambda(P) = \alpha_1\lambda_1(P) + \alpha_2\lambda_2(P)$. That is, each line plots the linear equation

$$\alpha_1\lambda_1(P) + \alpha_2\lambda_2(P) = c \qquad (16.17)$$

where c is a constant. Increasing values of c correspond to lines that are farther away from the origin. The dashed box marks the constraint region in $\lambda_1(P)$-$\lambda_2(P)$ space. The coordinates of a dot represent the costs of a path in the two dimensions.

It can be seen that the heuristic essentially *scans* the area within the constraint region by using straight lines as shown in the figure. The slope of the parallel straight lines is given by $-\alpha_1/\alpha_2$. Because straight lines are being used, a part of the $\lambda_1(P)$-$\lambda_2(P)$ space falling outside the region also gets scanned. In fact, in Figure 16.2, the path returned by the heuristic (P) happens to fall outside the constraint region, showing that infeasible solutions can be returned. But this is to be expected, because a simple computationally efficient heuristic for an $\mathcal{NP}$-complete problem cannot guarantee that correct solutions will always be provided. If it could, then the problem would not be NP-complete.

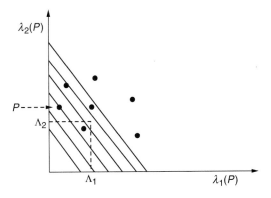

Figure 16.2   We consider the heuristic that combines the path costs $\lambda_1(P)$ and $\lambda_2(P)$ in a linear function. The slant lines represent the contours of the linear function (that is, the lines joining points where the function value is constant). The dots represent the vector costs associated with different paths.

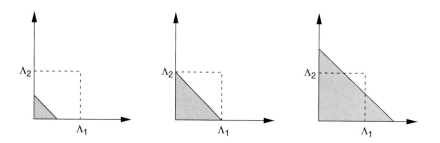

Figure 16.3 The parameters of the straight lines used in scanning the constraint region are chosen to be in the ratio Λ_2/Λ_1. This ensures that the scanned area outside the constraint region is reduced.

If the scanned area *outside* the constraint region is decreased, then the chances of finding an infeasible solution decrease. Suppose that the parameters α_1 and α_2 are chosen such that $\frac{\alpha_2}{\alpha_1} = \frac{\Lambda_2}{\Lambda_1}$. Figure 16.3 shows that as the constant c in Equation 16.17 increases and the straight line moves farther away from the origin, at least *half* the area within the constraint region is scanned before points outside the area are included. So if a feasible solution lies within the shaded part of the middle diagram of Figure 16.3, then a shortest path search with the length being defined as in Equation 16.16 will indeed provide the feasible path.

What can be done to decrease the scanned area outside the constraint region? One possibility is to scan the region using *curved* lines instead of straight lines. In Figure 16.4, we consider again the example shown in Figure 16.2, in which two constraints specify the constraint region. Instead of the straight lines, the constraint area is scanned by curved lines that approach the boundary of the area more closely.

This requires that the metrics $\lambda_1(P)$ and $\lambda_2(P)$ be combined in an expression different from that on the right side of Equation 16.16. Moreover, the expression should be such that the corresponding contours are curved and are close to the boundaries of the rectangular constraint area. Consider the following:

$$\lambda(P) = \left(\sum_{i=1}^{m} \left(\frac{\lambda_i(P)}{\Lambda_i} \right)^q \right)^{\frac{1}{q}} \qquad q > 1 \qquad (16.18)$$

Equation 16.18 is known as the Hölder q-vector norm or L_q norm on $\mathbb{R}^m$.

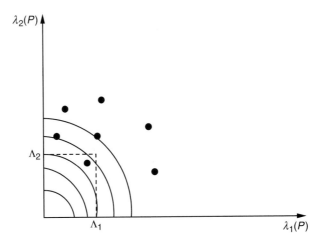

Figure 16.4 The area within the constraint region, which is inside the dashed lines, is now scanned using curved lines instead of straight lines.

Consider a cost function $\lambda^q(P)$, where $\lambda(P)$ is defined in Equation 16.18 and $q \geq 1$. That is, we have

$$\lambda^q(P) = \sum_{i=1}^{m} \left(\frac{\lambda_i(P)}{\Lambda_i} \right)^q \tag{16.19}$$

Suppose that there is an algorithm that returns a path P by minimizing the function in Equation 16.19 for a specific value of $q \geq 1$. Then the following result can be established.

Theorem 16.1

Assume that there is at least one feasible path P^* in the network. Let P be a path that minimizes the cost function $\lambda^q(P)$ for a given $q \geq 1$. Then

a. $\lambda_i(P) \leq \Lambda_i$ for at least one $i \in \{1, 2, \ldots, m\}$.

b. $\lambda_i(P) \leq \sqrt[q]{m}\, \Lambda_i$, for all other $i \in \{1, 2, \ldots, m\}$.

Proof: If the returned path P is feasible, then, for all $i \in \{1, 2, \ldots, m\}$, we have

$$\frac{\lambda_i(P)}{\Lambda_i} \leq 1$$

In this case, the claimed bounds are obviously true.

Now suppose that the returned path P is not feasible. Because the algorithm returns the path that minimizes $\lambda^q(P)$, we have

$$\lambda^q(P) \leq \lambda^q(P^*)$$

Also, because $\lambda_i(P^*) \leq \Lambda_i$ for all $i \in \{1, 2, \ldots, m\}$, we have

$$\lambda^q(P^*) \leq m$$

Therefore,

$$\lambda^q(P) \leq m \tag{16.20}$$

Now if $\lambda_i(P) > \Lambda_i$ for all $i \in \{1, 2, \ldots, m\}$, then, by Equation 16.19, we must have $\lambda^q(P) > m$. Because this contradicts Equation 16.20, it follows that we must have $\lambda_i(P) \leq \Lambda_i$ for at least one $i \in \{1, 2, \ldots, m\}$. We also note that if $\lambda^q(P) > m$, then there cannot be any feasible path P^*, because for every path $\tilde{P}$, there will be at least one $i \in \{1, 2, \ldots, m\}$ such that $\lambda_i(\tilde{P}) > \Lambda_i$.

To prove the second claim, we assume, to the contrary, that for at least one constraint Λ_i, we have $\lambda_i(P) > \sqrt[q]{m}\, \Lambda_i$. This implies that

$$\left(\frac{\lambda_i(P)}{\Lambda_i}\right)^q > m$$

By the definition of $\lambda^q(P)$, it follows that

$$\lambda^q(P) \geq \left(\frac{\lambda_i(P)}{\Lambda_i}\right)^q > m$$

This again contradicts Equation 16.20.   ∎

Corollary 16.2

As q increases, it becomes more likely that a feasible path will be found.

Proof: Let P denote the path returned by the algorithm that minimizes the cost function $\lambda^q(P)$. We have seen that for all $i \in \{1, 2, \ldots, m\}$,

$$\lambda_i(P) \leq \sqrt[q]{m}\, \Lambda_i$$

As q increases, $\sqrt[q]{m}$ decreases toward 1, because $m > 1$. Thus, the upper bound on $\lambda_i(P)/\Lambda_i$ approaches 1 from the right as q increases. Therefore, the chances of obtaining an infeasible path decrease. ∎

It can be shown that the contours of the form $\lambda(P) = c$, where $\lambda(P)$ is defined as in Equation 16.18, are curved lines of the type shown in Figure 16.4. Ideally, the curved lines should match the rectangular boundary exactly. As $q \to \infty$, this is what happens. It can be seen that as q increases to ∞, the sum in Equation 16.18 is dominated by the largest term among the summands. In the limit as $q \to \infty$, we have

$$\lambda(P) = \max \left\{ \frac{\lambda_1(P)}{\Lambda_1}, \ldots, \frac{\lambda_i(P)}{\Lambda_i}, \ldots, \frac{\lambda_m(P)}{\Lambda_m} \right\} \tag{16.21}$$

This is nothing but the usual sup norm in finite dimensions.

Equation 16.21 shows that if a path is feasible, then $\lambda(P) \leq 1$. Conversely, if $\lambda(P) \leq 1$, then the path is feasible. Thus, the metric $\lambda(P)$, where $\lambda(.)$ is as defined in Equation 16.21, characterizes whether or not path P is feasible. So if an algorithm can *efficiently* search the set of paths using the metric $\lambda(.)$, then we can know "quickly" whether a feasible path exists. This is the motivation for considering the sup norm.

As shown in Figure 16.5, when plotted in the $\lambda_1(P)/\Lambda_1$-$\lambda_2(P)/\Lambda_2$ space, a feasible path lies within the unit square, and the effective length is determined by the component closest to the constraints boundary.

Exercise 16.2

Suppose that in Equation 16.16, we redefine $l(P)$ by using $l_1(P)/L_1$ and $l_2(P)/L_2$ on the right side. Also, $\alpha_1 + \alpha_2 = 1$. Can we use the criterion $l(P) \leq 1$ to identify a feasible path?

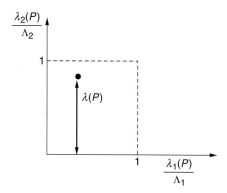

Figure 16.5   The location of a path P in the space $\lambda_1(P)/\Lambda_1(P)$-$\lambda_2(P)/\Lambda_2(P)$ is plotted. When the effective length is defined as in Equation 16.21, the length of a feasible path is determined by that metric which is closest to 1.

A Dijkstra-Type Algorithm in $\mathbb{R}^m$?

Although $\lambda(P) \leq 1$, where $\lambda(P)$ is defined as in Equation 16.21, provides a characterization of feasible paths, it does not help in actually obtaining paths that are feasible. If we have a single metric, then we can find the feasible path by simply obtaining the shortest path by the Dijkstra algorithm. If the path metric of the shortest path happens to be more than the constraint, then we conclude that a feasible path does not exist. The problem now is to devise an algorithm that works when $m \geq 2$ metrics are given.

Can we think of a Dijkstra-type algorithm in $\mathbb{R}^m$? The difficulty is that for a Dijkstra algorithm to work, *any* two paths must be comparable. In other words, if we consider the set of all paths, then there must be a *total order* on that set. With a single additive metric, it is clear that a total order exists among all paths. This is just the usual order on the real line. But when we have $m \geq 2$ additive metrics, there exist many pairs of paths that are not comparable. For example, if the path vectors corresponding to two paths are $(7, 11)$ and $(5, 13)$, respectively, then we cannot say one path is shorter than another.

Recall at this point that in Section 14.3.3, where we discuss the generalized Dijkstra algorithm, one of the assumptions we make is the existence of a total order on the set of paths. Because this assumption is not valid when the path vectors lie in $\mathbb{R}^m$, we have difficulty at the first stage itself.

But now suppose that we define an effective path length as a function from $\mathbb{R}^m$ to $\mathbb{R}$, along the lines of Equations 16.18 and 16.21. Clearly, the problem of the nonexistence of a total order is solved because the range of the function is $\mathbb{R}$. Is it possible now to think of a generalized Dijkstra algorithm?

The difficulty now is of a different kind. Recall that Equation 16.18 is the Hölder *norm*, and similarly, Equation 16.21 is nothing but the sup norm in finite dimensions. By definition, norms always satisfy the *triangle inequality*, which is defined next.

Definition 16.1

If $a, b \in \mathbb{R}^m$ and $\lambda(.)$ is a norm defined on $\mathbb{R}^m$, then

$$\lambda(a + b) \leq \lambda(a) + \lambda(b)$$

Now suppose that we have two path segments P_1 and P_2 such that $\lambda(P_1) < \lambda(P_2)$. Each of P_1 and P_2 is augmented by concatenating path segment P_3 with it. Then we *cannot* say that

$$\lambda(P_1 + P_3) < \lambda(P_2 + P_3)$$

where the notation $(P_i + P_j)$ means that the path vectors corresponding to P_i and P_j are added component-wise. By the triangle inequality, we do have

$$\lambda(P_1 + P_3) \leq \lambda(P_1) + \lambda(P_3)$$

$$\lambda(P_2 + P_3) \leq \lambda(P_2) + \lambda(P_3)$$

But this does not allow any conclusion about the ordering between $\lambda(P_1 + P_3)$ and $\lambda(P_2 + P_3)$.

Exercise 16.3

Suppose $m = 2$. There are two path segments $P_1 = (0.35, 0.55)$ and $P_2 = (0.25, 0.65)$. Let a third path segment be $P_3 = (0.15, 0.20)$. Consider the max norm defined in Equation 16.21. What is the ordering between $(P_1 + P_3)$ and $(P_2 + P_3)$? If the third path segment changes to $P_3 = (0.45, 0.05)$, how does the ordering change?

If it were possible to conclude that $\lambda(P_1 + P_3) < \lambda(P_2 + P_3)$, then the following exercise shows that a crucial condition assumed in the Dijkstra algorithm is satisfied.

Exercise 16.4
Suppose that $\lambda(P_1) < \lambda(P_2) \Rightarrow \lambda(P_1 + P_3) < \lambda(P_2 + P_3)$. Show that we can now conclude that subpaths of optimal paths are optimal.

The principle "subpaths of optimal paths are optimal" is used by Dijkstra's algorithm to construct an optimal path from the source to the destination, utilizing optimal paths from the source to intermediate nodes (see Section 14.3.1). If $\lambda(P_1) < \lambda(P_2) \not\Rightarrow \lambda(P_1 + P_3) < \lambda(P_2 + P_3)$, then subpaths of optimal paths *may not* be optimal.

An example is shown in Figure 16.6. Here, $m = 2$. The link vectors are shown beside the links. Suppose that the max norm defined in Equation 16.21 is used as the path metric. It can be seen that the shortest path from a to d is $a \rightarrow c \rightarrow d$, because $\lambda(a \rightarrow c \rightarrow d) = 0.85$, whereas $\lambda(a \rightarrow b \rightarrow c \rightarrow d) = 0.95$. However, the subpath $a \rightarrow c$ is itself not the shortest path between a and c, because the path $a \rightarrow b \rightarrow c$ is shorter; we have $\lambda(a \rightarrow c) = 0.45$, whereas $\lambda(a \rightarrow b \rightarrow c) = 0.40$.

Dijkstra's algorithm proceeds by storing the optimal path from the source node to intermediate nodes and then augmenting the path to obtain optimal paths to other nodes. In the example of Figure 16.6, the optimal path from a to c ($a \rightarrow b \rightarrow c$) would be stored at c. However, as we have seen, path $a \rightarrow b \rightarrow c$ cannot be extended to obtain the optimal path from a to d. Instead, if *more than one* path from a to c were stored at c, it would be possible to extend one of them, ($a \rightarrow c$), to obtain the optimal path from a to d.

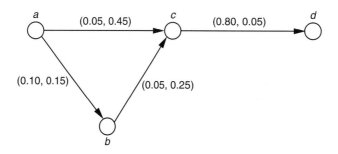

Figure 16.6 This is a network with four nodes and directed links. Each link has associated with it two additive metrics that are shown. The max norm is used to compare paths. The optimal path from a to d is $a \rightarrow c \rightarrow d$. But the subpath $a \rightarrow c$ is not optimal.

This observation suggests that an algorithm that stores *a set* of paths from the source node to intermediate nodes, instead of only a single optimal path, can lead to the desired optimal path. For such an algorithm, we would need to provide the number of paths to be stored—say, J. It is immediate that if J is large, the chances of obtaining an optimal path are higher. The disadvantage is that storage requirements increase, as does computational effort, because it is necessary to check each of the stored paths to determine which one can yield the optimal path. It is also clear that success is not guaranteed. It may happen that none of the J stored paths at an intermediate node leads to the desired optimal path.

The *tunable accuracy multiple constraints routing algorithm* (TAMCRA) is an algorithm based on this idea. Here, at each intermediate node, J paths from the source to that node are stored. "Tunable" refers to the possibility of tuning the accuracy by varying J. TAMCRA also excludes paths that are longer in each component of the path length vector. Let $m = 2$. In Figure 16.7, path P_2 is longer than P_1 on each metric individually. P_2 is said to be *dominated* by P_1.

It is clear that P_2 cannot be extended to yield a path shorter than one obtained by extending P_1. This follows because if $\lambda_1(P_3)$ and $\lambda_2(P_3)$ are the metrics corresponding to a path segment P_3, then

$$\frac{\lambda_i(P_1)}{\Lambda_i} + \frac{\lambda_i(P_3)}{\Lambda_i} < \frac{\lambda_i(P_2)}{\Lambda_i} + \frac{\lambda_i(P_3)}{\Lambda_i}, \qquad i = 1, 2$$

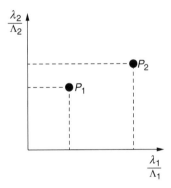

Figure 16.7 Path P_2 is longer than path P_1 on both metrics. P_2 is said to be *dominated* by P_1. For such cases, there is no need to store P_2 among the J shortest paths.

which shows that

$$\max\left\{\frac{\lambda_1(P_1+P_3)}{\Lambda_1}, \frac{\lambda_2(P_1+P_3)}{\Lambda_2}\right\} < \max\left\{\frac{\lambda_1(P_2+P_3)}{\Lambda_1}, \frac{\lambda_2(P_2+P_3)}{\Lambda_2}\right\}$$

Therefore, (P_1+P_3) is shorter than (P_2+P_3) for any P_3.

Thus, if a path P_2 from the source to some intermediate node V is dominated by another path P_1 from the source to the same node, then it is not necessary to store P_2 at node V. This reduces the amount of storage required.

Finally, we study TAMCRA, the proposed algorithm for finding multi-constrained feasible paths. The pseudo-code is given in Algorithm 16.2. Before considering the pseudo-code, we take a brief look at the internal data structures and variables used by the algorithm.

- Associated with each node $n \in \mathcal{N}$ is a counter $ctr(n)$. In Dijkstra's algorithm, in each iteration, the currently known best path from s to node n is stored at n. This path may later be extended to give the optimal path from s to t. As discussed earlier, this approach is not enough for the multiple constraints problem, because a nonoptimal subpath may be extended to finally provide the optimal path from s to t. Therefore, several paths from s to n must be stored at n. Note that $ctr(n)$ counts the number of paths currently stored at n.

- For $i \in \{1, 2, \ldots, ctr(n)\}$, $n[i]$ indicates the ith path from s to n that is stored at n.

- $\mathcal{Q}$ is a set of paths from s to the nodes in $\mathcal{N}$. The complete path metric vector corresponding to each path is stored. $\mathcal{Q}$ is initialized to the empty set ϕ. As the algorithm iterates, paths are inserted into and removed from $\mathcal{Q}$.

- $length(n[i])$ gives the length of the ith path stored at node n, according to the path-length definition in Equation 16.21.

The algorithm begins by initializing $ctr(n)$ to 0 for all $n \in \mathcal{N}$ and initializes $\mathcal{Q}$ to ϕ. Also, $length(s[1])$ is initialized to 0 because, by definition, $s[1]$, which is the first path from s to s, has a length of 0. After this first path is added to the set $\mathcal{Q}$, the algorithm begins extracting paths from $\mathcal{Q}$ and processing them.

At any iteration, there are several paths stored in $\mathcal{Q}$. The subroutine EXTRACT_PATH_HAVING_MIN_LENGTH in line 5 is used to extract the

---

1: Input: $\mathcal{G}(\mathcal{N}, \mathcal{L})$, s, t, Λ_i, $i \in \{1, 2, \ldots, m\}$, J
2: Initialization: $ctr(n) = 0 \; \forall \; n \in \mathcal{N}$, $\quad Q = \phi$, $\quad length(s[1]) = 0$
3: ADD $s[1]$ to Q
4: while $Q \neq \phi$ bf do
5:    $u[i] = $ EXTRACT_PATH_HAVING_MIN_LENGTH from Q
6:    if $u = $ destination node t then
7:       if $length(u[i]) \leq 1$ then
8:          OUTPUT "A feasible path is $u[i]$"
9:          STOP
10:   else
11:       for $n \in$ adjacent_list (u) do
12:          if $n \neq$ predecessor of u on path $u[i]$ then
13:             $PATH = u[i] + (u \rightarrow n)$
14:             $length = $ length of $PATH$
15:             check if $PATH$ is nondominated
16:             if $length \leq 1$ and $PATH$ is nondominated then
17:                if $ctr(n) < J$ then
18:                   $ctr(n) = ctr(n) + 1$
19:                   $j = ctr(n)$
20:                   $n[j] = PATH$
21:                   length of $(n[j]) = length$
22:                   ADD $n[j]$ to Q
23:                else
24:                   n_max $= PATH$ to n in Q with maximum length
25:                   if $length <$ length of n_max then
26:                      REPLACE n_max in Q with $n[j]$
       {Control passes to the following lines if the **while** loop ends without finding a feasible path}
27: OUTPUT "No feasible path"
28: STOP

Algorithm 16.2 Pseudo-code for TAMCRA. Adapted from "TAMCRA: A Tunable Accuracy Multiple Constraints Routing Algorithm" in [75].

shortest path from Q. Recall that a similar action is carried out in Dijkstra's algorithm. There are several paths stored at a node u. The index i, $1 \leq i \leq J$, indicates *which* of the stored paths at u was extracted. If u happens to be the destination node, then the iterations stop. However, we must check whether or not the path provided ($u[i]$) is feasible. If $length(u[i]) > 1$, then the algorithm concludes that a feasible path does not exist in the network. However, a feasible path *can*, in fact, exist. The difficulty is that not enough paths have been stored at the nodes. In other words, J is too small.

When u is not the destination node, the algorithm begins to scan the neighbors of u in line 11. After ensuring that u's neighbor n is not a predecessor of u on the path $u[i]$, we obtain a path to n by simply concatenating the link $(u \rightarrow n)$ to $u[i]$. Again, we notice the similarity to Dijkstra's algorithm. In line 15, the algorithm checks whether the path formed by $u[i] \cup (u \rightarrow n)$ is dominated by any of the paths already stored in n. As discussed before, if $u[i] \cup (u \rightarrow n)$ is dominated by one of the paths already stored in n, then $u[i] \cup (u \rightarrow n)$ should not be stored, because it will not be part of an optimal path to the destination. If the path $u[i] \cup (u \rightarrow n)$ is nondominated and if its length is ≤ 1, which means it is feasible, then it is a candidate path to be stored.

In line 17, the algorithm checks whether the number of paths stored in the node n is already equal to J. When the number of paths already stored is less than J, the path $u[i] \cup (u \rightarrow n)$ is stored, and the counter $ctr(n)$ is incremented. The path is also added to the set Q so that it is available and can be extended to yield an optimal path in future iterations. When the number of paths stored in n is already equal to J, then the longest path already in Q is possibly replaced by $u[i] \cup (u \rightarrow n)$ if the length of $u[i] \cup (u \rightarrow n)$ is less.

16.5   Summary

This chapter considers the problem of routing microflows or sessions carrying stream traffic across a network. Each stream session declares some end-to-end resource requirements on arrival to the network, and *feasible* routes, which support these end-to-end requirements, are needed. We first examine the nonadditive metrics case, and it turns out to have a straightforward solution, even in the presence of uncertainty about resource availability. Then we move to the case where the end-to-end requirement is additive. With rate-based multiplexers in the network, the problem of finding feasible routes turns out to be NP-hard when more than one session is considered. When only a single session is present, we can use a polynomial time algorithm for finding a feasible route. We also examine an approach to obtain an upper bound on the maximum weighted traffic that can be carried in a network, and we discuss a routing and rate allocation strategy that is asymptotically optimal. Finally, we move on to consider a network where the multiplexers are non-rate-based, and the arriving sessions declare end-to-end requirements on more than one additive metric. This problem of finding a multiconstrained feasible route is examined, as is the TAMCRA algorithm.

16.6   Notes on the Literature

The area of QoS routing for stream-type sessions became an active area of research with the advent of technologies for carrying stream traffic over traditional "data-type" networks. The brief treatment of nonadditive metrics in Section 16.2 follows the important work in [129]. This work was the first to systematically evaluate the impact of inaccurate information on routing algorithms. The problem of finding feasible routes in a network using rate-based multiplexers was treated in [229]. The algorithm discussed in Section 16.3.1 was given in this paper. The work in Section 16.3.2, which gives an upper bound on the minimum carried traffic in a network of rate-based schedulers, appeared in [83]. Moving to non-rate-based schedulers, the TAMCRA algorithm was given in [75].

Problems

16.1 A connection arrives to a network and specifies its traffic parameters and end-to-end delay requirement. The source node is s, and the destination node is t. Suppose that all links in the network have equal capacities, and all paths between s and t have the same propagation delay. Without running Algorithm 16.1, obtain a simple test that indicates whether a feasible path exists.

16.2 Consider Algorithm 16.1 for solving the single-commodity problem. The traffic generated by the connection is characterized by a leaky bucket with parameters (σ_1, ρ_1), the maximum packet size is c_1, and the required end-to-end delay is D_1^{reqd}. After the algorithm has terminated, suppose that a path P is found to be feasible. What will be the rate *actually* allocated to the connection on path P?

16.3 Figure 16.8 shows a network with four nodes a, b, c, d and eight directed links $a \rightarrow c$, $a \rightarrow d$, and so on. The capacity of each link is 34 Mbps. There are two SD pairs—(a, b) and (c, d)—and two classes of traffic. Connections belonging to class 1 require 1 Mbps of bandwidth each, and connections of class 2 require 4 Mbps of bandwidth each. There are two paths that can be used to carry traffic for SD pair (a, b): P_1 (composed of links $a \rightarrow c$ and $c \rightarrow b$) and P_2 (composed of links $a \rightarrow d$ and $d \rightarrow b$). Similarly, the paths for carrying (c, d) traffic are P_3 (consisting of links $c \rightarrow b$ and $b \rightarrow d$) and P_4 (consisting of links $c \rightarrow a$ and $a \rightarrow d$).

   a. Obtain the **G** and **B** matrices for this network.

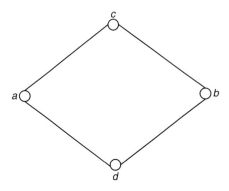

Figure 16.8   A simple network, with four nodes and eight directed links, and two SD pairs. There are four paths through the network.

   b. We consider an offline problem and assume that many connections of each class are available. Suppose that we are interested in maximizing the weighted carried traffic, and therefore must decide how many connections of each class to select from the available pool and carry on the network. There is one weight vector (i.e., $F = 1$), and the weight associated with each class is set to the required bandwidth for that class. Write the optimization problem (integer linear program).

   c. Relax the ILP to obtain an LP, and obtain the maximum possible weighted carried traffic in the network by solving the LP.

   d. From the solution of the LP, obtain the number of connections to be carried according to the partitioning algorithm. What is the corresponding weighted carried traffic? Does this agree with the maximum possible weighted carried traffic obtained earlier?

   e. What can be said about the "fairness" of the solution obtained?

16.4 Consider Equation 16.18. Suppose that each path has two additive metrics associated with it. For a path P, these are denoted as $\lambda_1(P)$ and $\lambda_2(P)$, respectively. Let Λ_1 and Λ_2 be the bounds on the two metrics that must be satisfied. For a given $q > 1$, consider the contour

$$\left\{\left(\frac{\lambda_1(P)}{\Lambda_1}\right)^q + \left(\frac{\lambda_2(P)}{\Lambda_2}\right)^q\right\}^{\frac{1}{q}} = 1$$

a. Suppose that path P is strictly inside the contour for some $q = q_0$, where q_0 is a positive integer. Show that P will remain inside the contour for all integral $q > q_0$.

b. Plot the feasible region in the $\lambda_1(P)/\Lambda_1$-$\lambda_2(P)/\Lambda_2$ space.

c. Show that as $q \to \infty$, the region *below* the contour line approaches the full feasible region.

16.5 In Problem 16.4, show that for each $q \in \mathbb{N}$, the region below the contour line is convex.

Part IV

Appendices

APPENDIX A

Glossary of Terms and Notation

This appendix lists expansions of acronyms along with various technical and mathematical terminology. No attempt is made to explain or discuss the terms or concepts. Such discussion can be found either in the text or in the subsequent appendices (see the index).

A.1 Technical Terms and Expansions of Acronyms

Term/Acronym	Explanation/Expansion
2B + D	two bearer plus one data
AAL	ATM adaptation layer
ABR	available bit rate
ACK	acknowledgment
ACM	address complete message
ACR	(ATM/ABR standard) allowed cell rate
AIX	Ames Internet Exchange
ANM	Answer Message
AP	(IEEE 802.11 standard) access point
AQM	active queue management
ARQ	automatic repeat request
AS	autonomous system
ATM	Asynchronous Transfer Mode
AWGN	additive white Gaussian noise
AWP	Adaptive Window Protocol
B Channel	(ISDN) bearer channel
B-ISDN	Broadband ISDN
B-ISUP	Broadband Integrated Services User Part
BDP	bandwidth delay product
BECN	backward explicit congestion notification
BER	bit error rate
BGP	Border Gateway Protocol
BHCA	busy hour call attempts

BN	(ATM/ABR standard) Backward Notification
BPSK	binary phase shift keying
BRI	Basic Rate Interface
BRTT	base round-trip time
BSC	base station controller
BSS	(IEEE 802.11 standard) basic services set
BSS	base station subsystem
BTS	base transceiver system
CAC	connection admission control
CAM	content-addressable memory
CBR	constant bit rate
CCR	(ATM/ABR standard) current cell rate
CCS	common channel signaling
CDMA	code division multiple access
CEQ	customer equipment
CI	(ATM/ABR standard) Congestion Indication
CIDR	classless interdomain routing
CIOQ	combined input–output queueing
CPE	customer premises equipment
CPU	central processing unit
CR-LDP	Constraint-Routed Label Distribution Protocol
CRC	cyclic redundancy check
CSIR	channel side information rate
CSMA	Carrier Sense Multiple Access
CSMA/CA	Carrier Sense Multiple Access with Collision Avoidance
CSMA/CD	CSMA with Collision Detection
CTMC	continuous time Markov chain
CTS	clear to send
D Channel	(ISDN) data channel
DCF	(IEEE 802.11 standard) Distributed Coordination Function
DEC	Digital Equipment Corporation (acquired by Compaq and then merged with HP)
DIFS	(IEEE 802.11 Standard) DCF interframe spacing
DOCSIS	Data over Cable Service Interface Specification
DRAM	dynamic random access memory
DS	data sending (in MACAW)
DSL	digital subscriber loop
DSS2	Digital Subscriber Signaling System Number 2
DSSS	direct sequence spread spectrum

DTMC	discrete time Markov chain
EBW	effective bandwidth
ECN	explicit congestion notification
EDF	earliest deadline first
EFCI	(ATM/ABR standard) Explicit Forward Congestion Indicator
EGP	Exterior Gateway Protocol
ER	(ATM/ABR standard) explicit rate
ESC	effective service capacity
ESS	(IEEE 802.11) extended services set
ETSI	European Telecommunication Standards Institute
FCA	fixed channel assignment
FCFS	first-come-first-served
FDM	frequency division multiplexing
FDMA	frequency division multiple access
FEC	forwarding equivalence class
FIFO	first in first out
FSM	finite state machine
FTP	File Transfer Protocol
GASTA	geometric arrivals see time averages
GI	general, independent (as in GI/M/1 queues)
GMSC	(GSM standard) gateway MSC
GPS	generalized processor sharing
GSM	(GSM standard) Global System for Mobile communications
HDLC	High-Level Data Link Control
HLR	(GSM standard) home location register
HOL	head-of-the-line
HOLPS	head-of-the-line processor sharing
HTTP	Hypertext Transfer Protocol
HU	high usage
IAM	initial address message
ICR	(ATM/ABR standard) initial cell rate
IETF	Internet Engineering Task Force
IGP	interior gateway protocol
ILP	integer linear program
IN	intelligent network
INI	internetwork interface
IP	Internet Protocol

IQ	input queueing
ISDN	Integrated Services Digital Network
ISI	intersymbol interference
ISP	Internet service provider
ITU	International Telecommunications Union
KKT	Karush-Kuhn-Tucker
LAN	local area network
LAPB	link access procedure, balanced
LAPD	Link Access Procedure for the D-channel
LB	leaky bucket
LCP	link control parameter
LDP	label distribution protocol
LE	local exchange
LP	linear program
LPM	longest prefix match
LQF	longest queue first
LRD	long-range-dependent
LSA	link state advertisement
LSP	label-switched path
LSR	label-switching router
MAC	medium access control
MACA	Multiple Access with Channel Acquisition
MACAW	Multiple Access with Channel Acquisition for Wireless
MANET	mobile ad hoc network
MCR	(ATM/ABR standard) minimum cell rate
MIPS	million instructions per second
MIRA	minimum interference routing algorithm
MMF	max-min fairness or max-min fair
MMPP	Markov modulated Poisson process
MP	maximum packing
MPLS	multiprotocol label switching
MRP	Markov renewal process
MS	mobile station
MSS	maximum segment size
MSC	(GSM standard) mobile switching center
MtoE	mouth to ear
MTP	Message Transfer Protocol
NASA	National Aeronautics and Space Administration
NI	(ATM/ABR standard) No Increase

NIC	network interface card
NLRI	Network Layer Reachability Information
NNI	network node interface
NP	network processor
NSS	network and switching system
NT	network termination
OC	optical carrier
OCF	oldest cell first
ODE	ordinary differential equation
OFDM	orthogonal frequency division multiplexing
OQ	output queueing
OSI	Open Systems Interconnection
OSPF	Open Shortest Path First
PAGCH	paging and access grant channel
PAM	pulse amplitude modulation
PASTA	Poisson arrivals see time averages
PBX	private branch exchange
PCF	(IEEE 802.11 standard) Point Coordination Function
PCI	peripheral component interface (standard I/O bus on PCs)
PCM	pulse code modulation
PCR	(ATM/ABR standard) peak cell rate
PE	processing element
PER	packet error rate
PGPS	packet GPS
PIFS	(IEEE 802.11 standard) PCF interframe sequence
PIM	parallel iterative matching
PNNI	Private Network-to-Network Interface
POTS	plain old telephone service
PPP	Point to Point Protocol
PRI	Primary Rate Interface
PRN	packet radio network
PS	processor sharing
PSPDN	packet-switched public data network
PSTN	public switched telephone network
PVC	permanent virtual circuit
PoP	point of presence
QAM	quadrature amplitude modulation
QPSK	quadrature phase shift keying

QoS	quality of service
RACH	random access channel
RBC	rate-based control
RDF	(ATM/ABR standard) rate decrease factor
RED	random early discard
RFC	Request for Comments
RIB	routing information base
RIF	(ATM/ABR standard) rate increase factor
RIP	Routing Information Protocol
RM	(ATM/ABR standard) resource management
RMON	Remote Monitoring protocol
RNB	rearrangeably nonblocking
RP	repackable (Clos switching network)
RRA	routing and rate allocation
RSVP	Resource reSerVation Protocol
RSVP-TE	Resource Reservation Protocol with Traffic Engineering Extensions
RTO	retransmission timeout
RTP	Real-time Transport Protocol
RTPD	round-trip propagation delay
RTQD	round-trip queueing delay
RTS	request to send
RTT	round-trip time
S-Aloha	slotted Aloha
SACK	selective ACK
SCFQ	self-clocked fair queueing
SCP	signal control point
SD pair	source–destination pair
SDH	Synchronous Digital Hierarchy
SIFS	(IEEE 802.11 standard) short interframe spacing
SIR	signal to interference ratio
SMP	symmetric multiprocessor
SNB	strictly nonblocking
SNIR	signal to noise plus interference ratio
SNMP	Simple Network Management Protocol
SNR	signal to noise ratio
SONET	Synchronous Optical Network
SRAM	static random access memory
SS7	Signaling System 7

SSP	signal switching point
STM	Synchronous Transfer Mode
STP	signal transfer point
SVC	switched virtual circuit
TA	terminal adapter
TAMCRA	tunable accuracy multiple constraints routing algorithm
TCP	Transmission Control Protocol
TDM	time division multiplexing
TDMA	time division multiple access
TE	(ISDN) terminal equipment (types 1 and 2)
TMN	Telecommunication Management Network
TMS	time multiplexed switch
TSI	time-slot interchanger
TST	time-space-time
UBR	Unspecified Bit Rate
UDP	User Datagram Protocol
UNI	user network interface
VAD	voice activity detection
VBR	variable bit rate
VC	virtual channel (or circuit or connection)
VCC	virtual channel connection
VCI	virtual circuit identifier
VLR	visitor location register
VOQ	virtual output queueing
VP	virtual path
VPC	virtual path connection
VPI	virtual path identifier
VSAT	very small aperture terminal
VoIP	Voice over IP
WAN	wide area network
WANET	wireless ad hoc network
WBC	window-based control
WDM	wavelength division multiplexing
WFQ	weighted fair queueing
WFWFQ	worst case fair weighted fair queueing
WLAN	wireless LAN
WRR	weighted round robin
WSNB	wide sense nonblocking
WWW	World Wide Web

A.2   Units

b	bits; hence Mb is megabits, and Kb is kilobits
B	bytes; hence MB is megabytes, and KB is kilobytes
dB	decibel; x in dB is $10\log_{10} x$ dB
Hz	(cycles per second); hence GHz and MHz

A.3   Miscellaneous Operators and Mathematical Notation

$\mathbb{R}$	The set of real numbers.
$\mathbb{Z}^+$	The set of nonnegative integers.
x^+	For a real number x, $x^+ = \max\{0, x\}$.
t_-, t_+	For t, a point in time, t_- is interpreted as "just before t" and t_+ is interpreted as "just after t"; more formally, for $f(t)$ some function of time $f(t_-) = \lim_{\epsilon \downarrow 0} f(t - \epsilon)$, and $f(t_+) = \lim_{\epsilon \downarrow 0} f(t + \epsilon)$.
t_{k-}, t_{k+}	The same interpretation as t_-, t_+ but for the indexed time instant t_k.
$a \wedge b$	For real numbers a and b, $a \wedge b = \min\{a, b\}$.
$x \approx y$	is read as "x is approximately equal to y."
$A \backslash B$	For sets A and B, $A \backslash B$ is the set difference (i.e., $A \cap B^c$).
$|A|$	For a set A, $|A|$ is the *cardinality* of, or the number of elements in, A.
I_A	is the *indicator function* of the set A; if A is a subset of elements of Ω, whose elements are generically labeled by ω, then I_A is a function from Ω to $\{0, 1\}$, with $I_A(\omega) = 1$ if $\omega \in A$, and $I_A(\omega) = 0$ otherwise.

A.4   Vectors and Matrices

a. An element of (or a *vector* in) $\mathbb{R}^n$ is denoted by a boldface lowercase symbol (e.g., $\mathbf{x}$).

b. Vectors are viewed as *column vectors* unless stated otherwise.

c. The unit vectors in $\mathbb{R}^n$ are denoted by e_j, $1 \leq j \leq n$; that is,

$$e_1 = \begin{pmatrix} 1 \\ 0 \\ \cdot \\ \cdot \\ \cdot \\ 0 \end{pmatrix}, e_2 = \begin{pmatrix} 0 \\ 1 \\ \cdot \\ \cdot \\ \cdot \\ 0 \end{pmatrix}, \ldots, e_n = \begin{pmatrix} 0 \\ 0 \\ \cdot \\ \cdot \\ \cdot \\ 1 \end{pmatrix}$$

d. Matrices are denoted by bold uppercase symbols (e.g., $\mathbf{A}$). The element in row i and column j of the matrix $\mathbf{A}$ is denoted by $a_{i,j}$; the ith row of the matrix

A is denoted by $A_{i,\cdot}$ and is viewed as a row vector, by default; the *j*th column of the matrix A is denoted by $A_{\cdot,j}$ and is viewed as a column vector.

e. The transpose of a vector x is denoted by x^T, and the transpose of the matrix A is denoted by A^T.

f. *Gradient vector:* For a function $f : \mathbb{R}^n \to \mathbb{R}$, if f is differentiable at x, its gradient at x is viewed as a column vector $\nabla f(x) = \begin{pmatrix} \frac{df(x)}{dx_1} \\ \cdot \\ \cdot \\ \cdot \\ \frac{df(x)}{dx_n} \end{pmatrix}$.

A.5 Asymptotics: The *O*, *o*, and ~ Notation

We are often interested in expressing compactly the behavior of a function $f(x)$, $x \in \mathbb{R}$, or $f(n), n \geq 1$, for large values of the argument, or for the argument going to zero. Some standard notation has been developed for this. A comprehensive reference for this material is the book on asymptotics by De Bruijn [73]; an extensive discussion on the subtleties of this notation is provided in [128].

Big O Notation

We write

$$f(x) = O(g(x)) \text{ as } x \to \infty$$

if there is a positive number a, such that $|f(x)| \leq a|g(x)|$ for all large enough x. The same statement holds for $f(n) = O(g(n))$. Thus, for example, if $f(x) = x^9 + e^x$, then $f(x) = e^x(x^9 e^{-x} + 1)$, and hence there is an x_0 such that for all $x > x_0, f(x) \leq 2e^x$. Thus we can write $f(x) = O(e^x)$ as $x \to \infty$, which is to say that for large x, $f(x)$ grows exponentially. As another example, $f(x) = ax^2 + bx^9 = O(x^9)$ as $x \to \infty$. Similarly, we write

$$f(x) = O(g(x)) \text{ as } x \to 0$$

to mean that $|f(x)| \leq a|g(x)|$ for all x close enough to zero. Thus, for example, if $f(x) = \ln(x + 1)$, then we can write $f(x) = O(x)$ as $x \to 0$. This follows from the Taylor's series expansion of $\ln(x+1)$ and says that close to the origin $f(x)$ is linear.

Little o Notation
We write

$$f(x) = o(g(x)) \text{ as } x \to \infty$$

to mean $\lim_{x \to \infty} \frac{f(x)}{g(x)} = 0$, and, similarly, we write

$$f(x) = o(g(x)) \text{ as } x \to 0$$

to mean $\lim_{x \to 0} \frac{f(x)}{g(x)} = 0$. For example, $f(n) = an^3 + bn^9 = o(n^{10})$ as $n \to \infty$, and $f(x) = \ln(x+1) - x = o(x)$ as $x \to 0$. That is, the error between $\ln(x+1)$ and its approximation x, near 0, decreases strictly faster than x does near 0.

Asymptotic Equivalence; ~ Notation
We write

$$f(x) \sim g(x) \text{ as } x \to \infty$$

to mean $\lim_{x \to \infty} \frac{f(x)}{g(x)} = 1$, and, similarly, we write

$$f(x) \sim g(x) \text{ as } x \to 0$$

to mean that $\lim_{x \to 0} \frac{f(x)}{g(x)} = 1$. For example, $f(n) = an^3 + bn^9 \sim n^9$ as $n \to \infty$, and $f(x) = \ln(x+1) \sim x$ as $x \to 0$. Note that the latter can be seen by writing $\ln(x+1) = x + o(x)$ as $x \to 0$. Divide the right side by x, and use the definition of $o(x)$ to obtain the result.

A.6   Probability

i.i.d. A sequence of random variables, $X_n, n \in \{0, 1, 2, \ldots\}$, that are mutually independent and in which all the random variables are identically distributed, is said to be independent and identically distributed, abbreviated as "i.i.d."

⊔ A binary operator denoting statistical independence. So if A and B are random variables, then $A \amalg B$ is to be read as A is independent of B.

$\overset{dist}{=}$ If two random vectors $\mathbf{X}$ and $\mathbf{Y}$ have the same joint distributions, we write $\mathbf{X} \overset{dist}{=} \mathbf{Y}$.

w.p. 1 or a.s. With probability 1 or almost surely; thus, for example, saying that a random variable is nonnegative almost surely, or with probability 1, means $\Pr(X \geq 0) = 1$.

marginal If $(X_1, X_2, \ldots, X_n)$ is a random vector with some joint distribution, then the distribution of any of the component random variables X_i, $1 \leq i \leq n$, is called a marginal distribution. The term also applies to random processes, and if, for example, the random process is stationary, then we can refer to *the* marginal distribution of the process.

APPENDIX B

A Review of Some Mathematical Concepts

B.1 Limits of Real Number Sequences

Supremum and Infimum

Let S be a subset of the real numbers, $\mathbb{R}$, such that S is bounded above; that is, there exists $b \in \mathbb{R}$, such that for all $s \in S$, $s \leq b$. Then it can be shown that there is a number $u \in \mathbb{R}$ such that (1) for all $s \in S$, $s \leq u$, and (2) for every b such that b upper bounds S, $u \leq b$; that is, u is the least upper bound (l.u.b.) of S and is called the *supremum* or simply *sup* of the set S. We also write

$$u = \sup\{s : s \in S\} \tag{B.1}$$

Similarly, if S is a subset of the real numbers, $\mathbb{R}$, such that S is bounded below (i.e., there exists $a \in \mathbb{R}$, such that for all $s \in S$, $a \leq s$), then there is a greatest lower bound (g.u.b.), l, which is called the *infimum* or *inf* of S and is written as

$$l = \inf\{s : s \in S\} \tag{B.2}$$

For example, if $S = (-1, +1)$, then $\sup\{s : s \in S\} = 1$, and $\inf\{s : s \in S\} = -1$. Note that in this example the inf and the sup of S are not in S. On the other hand, $S = [-1, +1]$ has the same inf and sup, but in this case these are members of S. When $\inf\{s : s \in S\} \in S$, we also call it the *minimum* of S; in such cases "inf = min." Corresponding comments hold for sup and max of the set S.

Exercise B.1

If $S_1 \subset S_2$ are both bounded below then $\inf\{s : s \in S_1\} \geq \inf\{s : s \in S_2\}$. State and prove the corresponding result for sup.

Limit and Limit Points

A sequence of real numbers $x_n, n \geq 1$, is said to converge to the *limit* $x \in \mathbb{R}$ if for every $\epsilon > 0$, there exists an n_ϵ such that for all $n > n_\epsilon$, $|x_n - x| < \epsilon$; that is, no matter how small an $\epsilon > 0$ we take, there is a point in the sequence (denoted by n_ϵ) such that all elements of the sequence after this point are within ϵ of x (the proposed limit). A limit, if it exists, is clearly unique (why?). This is written as

$$\lim_{n \to \infty} x_n = x$$

If $x_k, k \geq 1$, viewed as a set, is bounded above, and is such that $x_k \leq x_{k+1}$ (i.e., x_k is a nondecreasing sequence), then $\lim_{n \to \infty} x_n$ exists and is, in fact, the sup of the set of numbers $x_k, k \geq 1$. The corresponding result holds if the sequence is nonincreasing and bounded below.

If the sequence $x_k, k \geq 1$, is bounded above and below, then we define the sequence $a_k, k \geq 1$, as follows:

$$a_k = \inf\{x_n : n \geq k\}$$

Then by the earlier discussion on inf, each a_k exists and is a bounded nondecreasing sequence (each bound of $x_k, k \geq 1$, is also a bound for $a_k, k \geq 1$), and hence $\lim_{k \to \infty} a_k$ exists. This limit—say, a—is called the lim inf of the sequence $x_k, k \geq 1$, and we write

$$\liminf_{n \to \infty} x_n = a$$

A corresponding discussion holds for the sequence

$$b_k = \sup\{x_n : n \geq k\}$$

Here $b_k, k \geq 1$, is a nonincreasing sequence, bounded below, and hence it converges. Then $b = \lim_{k \to \infty} a_k$ is called the lim sup of the sequence $x_k, k \geq 1$, and we write

$$\limsup_{n \to \infty} x_n = b$$

We say that $x \in \mathbb{R}$ is a *limit point* of the sequence $x_k, k \geq 1$, if for all $\epsilon > 0$ and for every n, there exists an $m_{n,\epsilon} > n$, such that $|x_{m_{n,\epsilon}} - x| < \epsilon$. Thus for every $\epsilon > 0$ (no matter how small), the sequence comes within ϵ of x *infinitely often*; for if it did not, there would be some n at which x_n comes within ϵ of x for the last time, and there would be no $m_{n,\epsilon} > n$, with $|x_{m_{n,\epsilon}} - x| < \epsilon$.

In general, a sequence can have none (e.g., $x_n = n$), one (any convergent sequence, e.g., $x_n = \frac{1}{n}$), or several limit points (e.g., $x_n = (-1)^n$). If x_n is bounded above and below, there is at least one limit point. The $\limsup_{n\to\infty} x_n$ and the $\liminf_{n\to\infty} x_n$ are both limit points. There could be other limit points as well. Let $\mathcal{X}$ be the set of limit points of $x_n, n \geq 1$. Then it can be shown that

$$\limsup_{n\to\infty} x_n = \sup\{x : x \in \mathcal{X}\}$$

and

$$\liminf_{n\to\infty} x_n = \inf\{x : x \in \mathcal{X}\}$$

In particular, if the lim sup and the lim inf of a sequence are equal then there is only one limit point, which we call the limit of $x_n, n \geq 1$.

B.2 A Fixed-Point Theorem

A function mapping a set C ($\subset \mathbb{R}^n$) into C (i.e., $f : C \to C$) is said to have a *fixed point* at $x \in C$ if $f(x) = x$. Given a function $f : C \to C$, the problem of determining whether there exists an $x \in C$ such that $f(x) = x$ is called a *fixed-point problem*. When faced with such a problem we need to ask questions such as: Does a fixed point exist? If one exists, is it unique?

Before we can state an important theorem we need to understand some elementary concepts from real analysis. We say that a set $C \in \mathbb{R}^n$ is *closed* if every convergent sequence $x_n, n \geq 1$, of points in C has its limit point also in C. Thus, for example, the set $C = (0, 1]$ is not closed, because the limit of the sequence $\frac{1}{n}, n \geq 1$ (i.e., 0), is not in C. We say that a set $C \in \mathbb{R}^n$ is bounded if there is an n-dimensional ball centered at the origin and of radius large enough, but finite, such that C is entirely inside that ball.

We say that a set $C \in \mathbb{R}^n$ is *convex* if whenever $x_1 \in C$ and $x_2 \in C$, then for every $\lambda, 0 < \lambda < 1$, $\lambda x_1 + (1 - \lambda)x_2 \in C$; that is, the entire line segment joining x_1 and x_2 is in C. Thus $C = [0, 1] \cup [2, 3]$ is a closed but nonconvex set.

Theorem B.1   Brouwer's Fixed-Point Theorem

Let $C \subset \mathbb{R}^n$ be a closed, bounded, and convex set. Then a continuous function $f: C \rightarrow C$ has a fixed point in C.   ■

Figure B.1 illustrates the theorem for $n = 1$ and $C = [0, 1]$. Here is the proof for this case. Suppose $f(0) \neq 0$ (i.e., $f(0) > 0$) and that $f(1) \neq 1$ (i.e., $f(1) < 1$). Defining $g(x) = f(x) - x$, we see that $g(0) > 0$ and $g(1) < 0$. Because $g(\cdot)$ is continuous (by a hypothesis in the theorem), there must be a point $x \in (0, 1)$ at which $g(x) = 0$.

The shape of C is not important, but the topological hypotheses in the theorem are necessary. For example, $f(x) = \frac{x}{2}$ has the single fixed point at $x = 0$; if C is defined without this point (and hence is not closed), then $f(\cdot)$ has no fixed points in C.

Exercise B.3

Show that if $f : [0, 1] \rightarrow [0, 1]$ is continuous and is such that $f'(x) \leq 0$ for all $x \in [0, 1]$, then $f(\cdot)$ has a *unique* fixed point in $[0, 1]$. Hint: write $g(x) = f(x) - x$ and use the mean value theorem to obtain a contradiction.

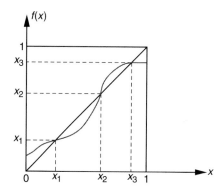

Figure B.1   Illustration of Brouwer's fixed-point theorem in one dimension. Note that x_1, x_2, **and** x_3 **are the fixed points.**

B.3 Probability and Random Processes

The following concepts from elementary probability and random processes are assumed as prerequisite knowledge for this book: random variables, cumulative distributions (discrete, continuous, and mixed), density, moments, characteristic functions, moment-generating functions, elementary inequalities (Markov and Chebychev), and elementary random processes (Bernoulli, Poisson, and Gaussian).

B.3.1 Useful Inequalities

Jensen's Inequality

For a convex function $f(x)$ and a random variable X with finite expectation,

$$\mathsf{E}\big(f(X)\big) \geq f\,(\mathsf{E}(X))$$

with equality if $f(\cdot)$ is a linear function. Equality also occurs for a general convex function if X is constant with probability 1. Note that the left side of the inequality can be possibly infinite. Thus, for example, for a random variable with finite mean $\mathsf{E}(X^2) \geq (\mathsf{E}(X))^2$ with equality only if X is constant with probability 1; also, $\mathsf{E}(e^{\theta X}) \geq e^{\theta \mathsf{E}(X)}$ with equality when $\theta = 0$; if $X \geq 0$, with probability 1, then $\mathsf{E}\big(\frac{1}{X}\big) \geq \frac{1}{\mathsf{E}(X)}$.

Hölder's Inequality

For $p > 1$ and $\frac{1}{p} + \frac{1}{q} = 1$,

$$\mathsf{E}(\mid XY \mid) \leq \big(\mathsf{E}(\mid X \mid)^p\big)^{\frac{1}{p}} \; \big(\mathsf{E}(\mid Y \mid)^q\big)^{\frac{1}{q}}$$

For example, take $p = q = 2$. Hölder's inequality yields

$$\big(\mathsf{E}(\mid XY \mid)\big)^2 \leq \mathsf{E}\big(\mid X \mid^2\big) \; \mathsf{E}\big(\mid Y \mid^2\big)$$

which is called Schwarz's inequality.

B.3.2 Convergence Concepts

Consider a sequence of real-valued random variables $X_n, n \geq 1$. For each sample path ω, we obtain the real number sequence $X_n(\omega), n \geq 1$, which may or may not converge. We would like to talk about the convergence behavior of the sequence of random variables without necessarily requiring convergence for each sample path. Several useful notions of convergence of a sequence of random variables have been defined.

Convergence in Probability

The sequence of random variables $X_n, n \geq 1$, is said to converge *in probability* to a random variable X, if for each $\epsilon > 0$,

$$\lim_{n \to \infty} \Pr(|X_n - X| > \epsilon) = 0$$

That is, the probability that X_n differs from X by more than ϵ converges to 0 as $n \to \infty$. This is written as

$$X_n \xrightarrow{p} X$$

For example, consider $X_n, n \geq 1$, i.i.d., with finite mean μ and finite variance σ^2. Then, by Chebychev's inequality,

$$\Pr\left(\left| \frac{1}{n} \sum_{k=1}^{n} X_k - \mu \right| > \epsilon \right) \leq \frac{\sigma^2}{n \epsilon^2}$$

Hence we can conclude that

$$\frac{1}{n} \sum_{k=1}^{n} X_k \xrightarrow{p} \mu$$

That is, the constant random variable $X = \mu$.

Convergence with Probability 1

The sequence of random variables $X_n, n \geq 1$, is said to converge *almost surely* or *with probability* 1 to a random variable X, if

$$\Pr\left(\lim_{n \to \infty} X_n = X \right) = 1$$

or, more explicitly,

$$\Pr\left(\{\omega: \lim_{n\to\infty} X_n(\omega) = X(\omega)\}\right) = 1$$

That is, the set of sample paths along which $X_n(\omega), n \geq 1$, converges to $X(\omega)$ has probability 1. This is denoted as

$$X_n \overset{a.s.}{\to} X$$

or as

$$X_n \overset{w.p.1}{\to} X$$

We provide an example of convergence almost surely as the strong law of large numbers in Section B.3.3.

Convergence in Distribution

A sequence of random variables $X_n, n \geq 1$, with distributions $F_n(\cdot), n \geq 1$, is said to converge *in distribution* to the random variable X, with distribution $F(\cdot)$, if

$$\lim_{n\to\infty} F_n(x) = F(x)$$

whenever x is not a point of discontinuity (i.e., a "jump" point) of $F(\cdot)$. This is denoted as

$$X_n \overset{dist}{\to} X$$

For example, suppose $X_n(\omega) = \frac{1}{n}$ (i.e., X_n is a constant random variable). Then $F_n(x) = 0, x < \frac{1}{n}$, and $F_n(x) = 1, x \geq \frac{1}{n}$. Let the random variable X be defined by $X(\omega) = 0$, and hence $F(x) = 0, x < 0$, and $F(x) = 1, x \geq 0$. It can then be seen that for all $x \neq 0$,

$$\lim_{n\to\infty} F_n(x) = F(x)$$

However, $F_n(0) = 0$ for all n, and $F(0) = 1$, and hence $\lim_{n \to \infty} F_n(0) \neq F(0)$; but 0 is a point of discontinuity of $F(\cdot)$. Hence, by definition,

$$X_n \overset{dist}{\to} X$$

B.3.3   Laws of Large Numbers and the Central Limit Theorem
Theorem B.2   Weak Law of Large Numbers
$X_n, n \geq 1$, is a sequence of identically distributed uncorrelated random variables with finite mean μ and finite variance. Then

$$\lim_{n \to \infty} \frac{1}{n} \sum_{k=1}^{n} X_k \overset{p}{\to} \mu$$

∎

Theorem B.3   Kolmogorov's Strong Law of Large Numbers
$X_n, n \geq 1$, is a sequence of i.i.d. random variables with finite mean μ. Then

$$\lim_{n \to \infty} \frac{1}{n} \sum_{k=1}^{n} X_k \overset{a.s.}{\to} \mu$$

∎

Theorem B.4   Central Limit Theorem
$X_n, n \geq 1$, is a sequence of i.i.d. random variables with finite mean μ and finite variance σ^2. Then

$$\frac{1}{\sigma \sqrt{n}} \left(\sum_{k=1}^{n} (X_k - \mu) \right) \overset{dist}{\to} \Phi$$

where Φ is the normal or Gaussian distribution with mean 0 and variance 1 (also called the standard normal distribution). ∎

B.3.4   Stationarity and Ergodicity
Strict Stationarity
A random process $X_n, n \geq 0$, is said to be *strictly stationary* or *stationary* if for all $k \geq 1$, and indices $n_1, n_2, \ldots, n_k$, and m,

$$\left(X_{n_1}, X_{n_2}, \ldots, X_{n_k} \right) \overset{dist}{=} \left(X_{n_1+m}, X_{n_2+m}, \ldots, X_{n_k+m} \right)$$

where $\stackrel{dist}{=}$ denotes equality in (joint) distribution. Thus if we take any subset of random variables of the process and then shift this subset in time by any amount, then the joint distribution of the constituent random variables is unchanged.

Stationary Increments

Consider a Poisson process $A(t), t \geq 0$; that is, $A(t)$ is the number of points in the interval $(0, t]$. Then obviously, $A(t)$ is nondecreasing and hence cannot be stationary. However, consider time points $t_1, t_2, \ldots, t_n$, and look at the *increments* of the process $A(t)$—that is $(A(t_2) - A(t_1), A(t_3) - A(t_2), \ldots, A(t_n) - A(t_{n-1}))$; this is the random vector of the number of arrivals over the intervals $(t_1, t_2], (t_2, t_3], \ldots, (t_{n-1}, t_n]$. Let us now shift all these intervals by some amount τ, yielding the increments $(A(t_2 + \tau) - A(t_1 + \tau), A(t_3 + \tau) - A(t_2 + \tau), \ldots, A(t_n + \tau) - A(t_{n-1} + \tau))$. If these two random vectors have the same distribution for any choice of n, $t_1, t_2, \ldots, t_n$, and τ, then we say that $A(t), t \geq 0$, has stationary increments. The Poisson process has stationary increments. The term also can, of course, be applied to discrete time processes. In fact, in addition, nonoverlapping increments of a Poisson process are also independent; that is, the random variables $(A(t_2) - A(t_1), A(t_3) - A(t_2), \ldots, A(t_n) - A(t_{n-1}))$ are mutually independent for any choice of $t_1 \leq t_2 \leq t_3 \leq \cdots \leq t_n$. Thus a Poisson process is said to have *stationary and independent increments*.

Ergodicity and the Ergodic Theorem

Consider a random process $X_n, n \geq 0$. Let us ask a question about this process whose answer does not depend on whether we ask the question about $X_n, n \geq 0$, or about $X_{n+k}, n \geq 0$, for any $k \geq 1$. Each such question yields an event on which the answer is yes, and its complement on which the answer is no. For example, consider the question, does $X_n, n \geq 0$, converge? Then the event $\{\omega: X_n(\omega) \text{ converges}\}$ is the same as the event $\{\omega: X_{n+k}(\omega) \text{ converges}\}$ for any $k \geq 1$, because whether or not a sequence converges does not depend on any finite shift of the sequence (i.e., does not depend on the time origin from which we start looking at the sequence). Such events are called *invariant* events. We say that $X_n, n \geq 0$, is *ergodic* if each such event has probability 0 or 1. This is the formal definition. Let us understand the concept by examining some implications.

Consider the following experiment. There are two biased coins, with heads probabilities p_1 and p_2, $p_1 \neq p_2$. A coin is chosen at random (with equal probability) and then tossed repeatedly. For $n \geq 0$, let $X_n = 1$ if the outcome is heads on the nth toss; otherwise, let $X_n = 0$. Examine the event $A = \{w: \lim_{n \to \infty} \frac{1}{n} \sum_{k=0}^{n-1} X_k = p_1\}$. This is an invariant event, but by the strong law of large numbers, $\Pr(A) = 0.5$, because the average value of the process converges almost surely to p_1 conditional on the first coin being chosen, and this happens with probability 0.5. Hence we do not have an ergodic process. Basically, the process evolves according to two different probability laws depending on how it starts.

An important consequence of $X_n, n \geq 0$, being an ergodic process is that time averages are the same as expectations. Note that this does not happen in the coin tossing example. Observe that $E(X_n) = 0.5 \times (1 \times p_1 + 0 \times (1 - p_1)) + 0.5 \times (1 \times p_2 + 0 \times (1 - p_2)) = 0.5(p_1 + p_2)$, whereas the time average converges to p_1 or p_2.

The following is a generalization of the strong law of large numbers to stationary and ergodic processes.

Theorem B.5   Birkhoff's Strong Ergodic Theorem

$X_n, n \geq 0$, is a stationary and ergodic process, and $f(\cdot)$ is a function that maps realizations of the process (i.e., $X_n(\omega), n \geq 0$) to $\mathbb{R}$, such that $E(|f(X_n, n \geq 0)|) < \infty$. Then, with probability 1,

$$\lim_{m \to \infty} \frac{1}{m} \sum_{k=0}^{m-1} f(X_{n+k}, n \geq 0) = E(f(X_n, n \geq 0))$$

∎

Remark: Note that the notation $f(X_{n+k}, n \geq 0)$ means "f evaluated at $X_{n+k}(\omega)$, $n \geq 0$—that is, a left shift of $X_n(\omega), n \geq 0$ by k steps." Thus, for each m the term $\frac{1}{m} \sum_{k=0}^{m-1} f(X_{n+k}, n \geq 0)$ is a random variable. The theorem states that this sequence of random variables converges with probability 1 to a constant.

As a simple application, take $f(X_n, n \geq 0) = X_0$; thus $f(X_{n+k}, n \geq 0) = X_k$. Then for a stationary and ergodic $X_n, n \geq 0$, the strong ergodic theorem states that

$$\lim_{m \to \infty} \frac{1}{m} \sum_{k=0}^{m-1} X_k = E(X_0)$$

Here, because $X_n, n \geq 0$, is stationary, $E(X_0)$ is the expectation of any of the random variables in the process. We have thus the result that is popularly stated as "time averages converge to the ensemble average," for a stationary and ergodic process.
Remark: If $X_n, n \geq 0$, is a sequence of i.i.d. random variables, then it is clearly stationary and can also be shown to be an ergodic process. Thus the strong ergodic theorem implies the strong law of large numbers (i.e., Theorem B.3).

All the foregoing discussion can also be written for a continuous time process, $X(t), t \geq 0$.

B.4   Notes on the Literature

Hoffman's book on real analysis [140] provides the rigorous theory, but a lot of intuition is also developed in the discussion. The book [232] by Papoulis, is a popular engineering textbook on probability, random variables, and random processes. Bremaud's book [41] is a rigorous but highly accessible book on probability theory.

APPENDIX C

Convex Optimization

Linear and nonlinear optimization is used in several places in this book (e.g., maximum utility formulation of the bandwidth-sharing problem in Chapter 7, optimal power control with an average power constraint in Chapter 8, and optimal routing in Chapter 14). We review standard material on convex nonlinear and linear optimization in this appendix. The material follows the notation and approach in the text by Bazaraa et al. [23].

C.1 Convexity
Definition C.1

A set $\mathcal{X} \subset \mathbb{R}^n$ is said to be *convex* if for any $\mathbf{x}_1, \mathbf{x}_2 \in \mathcal{X}$ it holds that $\lambda \mathbf{x}_1 + (1 - \lambda)\mathbf{x}_2 \in \mathcal{X}$, for every $\lambda \in [0, 1]$. ∎

Thus a set of real vectors is convex if the entire line segment joining any pair of elements of the set lies entirely within the set.

Definition C.2

$\mathcal{X}$ is a convex set in $\mathbb{R}^n$. A function $f : \mathcal{X} \to \mathbb{R}$ is said to be *convex* (respectively *concave*) if for $\mathbf{x}_1, \mathbf{x}_2 \in \mathcal{X}$, we have $f(\lambda \mathbf{x}_1 + (1 - \lambda)\mathbf{x}_2) \leq$ (respectively $\geq$) $\lambda f(\mathbf{x}_1) + (1 - \lambda)f(\mathbf{x}_2)$, for every $\lambda \in [0, 1]$. The function is said to be *strictly convex* or *strictly concave* if the inequality is strict for distinct $\mathbf{x}_1$ and $\mathbf{x}_2$ and $\lambda \in (0, 1)$. ∎

Observe that if $f : \mathcal{X} \to \mathbb{R}$ is convex over $\mathcal{X}$ then $-f$ is concave over $\mathcal{X}$. The following is a useful characterization of a real-valued convex function on $\mathbb{R}$.

Theorem C.1

A function $f : \mathbb{R} \to \mathbb{R}$ is convex if and only if for every $a < b < c$, the following holds:

$$\frac{f(b) - f(a)}{b - a} \leq \frac{f(c) - f(a)}{c - a}$$

∎

Remark: This theorem states that, for $a < b < c$, the slope of the chord joining $(a, f(a))$ and $(b, f(b))$ is no greater than the slope of the chord joining $(a, f(a))$ and $(c, f(c))$.

In constrained optimization problems, the set of $\mathbf{x} \in \mathbb{R}^n$ over which a real-valued function must be maximized or minimized is often specified in terms of inequality constraints on given functions. The following exercise illustrates a typical situation and relates to the previous two definitions.

Exercise C.1

a. $\mathcal{X}$ is a convex set in $\mathbb{R}^n$. Suppose that, for $1 \le i \le m$, the functions $g_i : \mathcal{X} \to \mathbb{R}$ are convex over $\mathcal{X}$, and b_i are given real numbers. Show that the set $\{\mathbf{x} \in \mathbb{R}^n : g_i(\mathbf{x}) \le b_i, 1 \le i \le m\}$ is a convex set.

b. Hence show that if $\mathbf{A}$ is an $m \times n$ matrix with real elements and if $\mathbf{b} \in \mathbb{R}^m$, then $\{\mathbf{x} \in \mathbb{R}^n : \mathbf{A}\mathbf{x} \le \mathbf{b}\}$ is a convex set.

A set of the form shown in part (b) of Exercise C.1 is called a *convex polyhedral set*. Each inequality defined by a row of matrix $\mathbf{A}$ yields a *half space* in $\mathbb{R}^n$, and the set defined by $\mathbf{A}\mathbf{x} \le \mathbf{b}$ is the intersection of these half spaces.

C.2   Local and Global Optima
Definition C.3

For the problem of minimizing a function $f : \mathcal{X} \to \mathbb{R}$, any $\mathbf{x} \in \mathcal{X}$ is said to be a *feasible* solution, and an element $\mathbf{x}^* \in \mathcal{X}$ is said to be a *global optimal solution*, or a *solution*, if $f(\mathbf{x}^*) \le f(\mathbf{x})$ for all $\mathbf{x} \in \mathcal{X}$. An element $\hat{\mathbf{x}} \in \mathcal{X}$ is said to be locally optimal if for some $\epsilon > 0$, $f(\hat{\mathbf{x}}) \le f(\mathbf{x})$, for all $\mathbf{x} \in \{\mathbf{x} \in \mathcal{X} : \|\mathbf{x} - \hat{\mathbf{x}}\| < \epsilon\}$. ∎

Observe that a global optimal solution need not be unique and must be locally optimal. A corresponding definition, obviously, holds for the maximization of f over $\mathcal{X}$.

Theorem C.2

Given $f : \mathcal{X} \to \mathbb{R}$ such that $\mathcal{X}$ is a convex set and f is a convex function over $\mathcal{X}$, a local minimum of f over $\mathcal{X}$ is also a global minimum. If f is strictly convex over $\mathcal{X}$, then a local optimum is the unique global optimum. ∎

If $f : \mathcal{X} \to \mathbb{R}$ is a concave function over the convex set $\mathcal{X}$, then the problem of maximizing f over X is equivalent to the problem of minimizing the convex function $-f$ over $\mathcal{X}$. Hence a theorem corresponding to Theorem C.2 clearly holds for the problem of maximizing a concave function over a convex set.

C.3   The Karush–Kuhn–Tucker Conditions

We now turn to an important *sufficient* condition for the optimality of a point $\mathbf{x}^* \in \mathcal{X} \subset \mathbb{R}^n$ for the following *primal* problem.

Primal Problem

$$\min f(\mathbf{x})$$

subject to

$$g_i(\mathbf{x}) \leq 0, \qquad 1 \leq i \leq m$$

$$\mathbf{x} \in \mathbb{R}^n \tag{C.1}$$

We limit our presentation here to the special situation in which $f : \mathbb{R}^n \to \mathbb{R}$ and, for $1 \leq i \leq m$, $g_i : \mathbb{R}^n \to \mathbb{R}$, *are all convex and differentiable functions* over $\mathbb{R}^n$. It follows from Exercise C.1 that this is a problem of minimizing a convex function over a convex constraint set.

Theorem C.3   Karush–Kuhn–Tucker

Given a feasible $\mathbf{x}^* \in \mathbb{R}^n$, if there exists $\boldsymbol{\lambda} \in \mathbb{R}^m$, with $\boldsymbol{\lambda} \geq 0$, such that

$$\nabla f(\mathbf{x}^*) + \sum_{i=1}^{m} \lambda_i \nabla g_i(\mathbf{x}^*) = 0 \tag{C.2}$$

and

$$\sum_{i=1}^{m} \lambda_i g_i(\mathbf{x}^*) = 0 \tag{C.3}$$

then $\mathbf{x}^*$ is a global optimal solution for the primal problem.   ∎

The sufficient conditions stated in Theorem C.3 are called the Karush–Kuhn–Tucker (KKT) conditions. If $\mathbf{x}^*$ satisfies the KKT conditions, then it is called a *KKT point*. The elements of the vector λ are called *Lagrange multipliers*, or *dual variables*; the latter term becomes clear later in this section when we discuss duality. There is one dual variable for every constraint $g_i(\mathbf{x}) \leq 0$.

The condition stated in Equation C.3 is called a *complementary slackness* condition; notice that it implies that if the primal constraint $g_i(\mathbf{x}) \leq 0$ is met with a strict inequality at $\mathbf{x}^*$ (i.e., there is a *slack* in the ith constraint) then (because the vector λ is nonnegative) the corresponding $\lambda_i = 0$. If at point $\mathbf{x}^*$ a constraint is met with equality, then that constraint is said to be *binding* or *active* at that point. Thus, because of the complementary slackness condition, we see that in the condition of Equation C.2 only the active constraints will appear.

Example C.1

$$\min \ (x_1 - 5)^2 + (x_2 - 5)^2$$

subject to

$$x_1^2 + x_2^2 - 5 \leq 0$$

$$\frac{1}{2}x_1 + x_2 - 2 \leq 0$$

$$-x_1 \leq 0$$

$$-x_2 \leq 0$$

$$\mathbf{x} \in \mathbb{R}^2$$

Identifying with the standard form of primal problem shown in Equation C.1, we see that $n = 2, m = 4, f(x_1, x_2) = (x_1 - 5)^2 + (x_2 - 5)^2, g_1(x_1, x_2) = x_1^2 + x_2^2 - 5$, $g_2(x_1, x_2) = \frac{1}{2}x_1 + x_2 - 2, g_3(x_1, x_2) = -x_1$, and $g_4(x_1, x_2) = -x_2$. Figure C.1 shows contours of constant value for $f(\mathbf{x})$; these are circles centered at $(5, 5)$. The figure also shows the curves with equations $g_i(\mathbf{x}) = 0, 1 \leq i \leq 4$.

We have $\nabla f(\mathbf{x}) = \begin{pmatrix} 2(x_1 - 5) \\ 2(x_2 - 5) \end{pmatrix}, \nabla g_1(\mathbf{x}) = \begin{pmatrix} 2x_1 \\ 2x_2 \end{pmatrix}, \nabla g_2(\mathbf{x}) = \begin{pmatrix} \frac{1}{2} \\ 1 \end{pmatrix}, \nabla g_3(\mathbf{x}) =$ $\begin{pmatrix} -1 \\ 0 \end{pmatrix}, \nabla g_4(\mathbf{x}) = \begin{pmatrix} 0 \\ -1 \end{pmatrix}$. Consider $\mathbf{x}^* = \begin{pmatrix} 2 \\ 1 \end{pmatrix}$. At this point the first and second

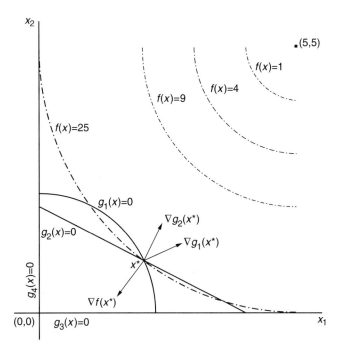

Figure C.1   Geometry of Example C.1. The quarter arcs of circles centered at (5, 5) are portions of contours of constant value of the objective function; that is, they have equations $(x_1 - 5)^2 + (x_2 - 5)^2 = h$, for various values of $h \geq 0$; shown are these curves for $h = 1, 4, 9, 25$.

constraints are binding. We have $\nabla f(\mathbf{x}^*) = \begin{pmatrix} -6 \\ -8 \end{pmatrix}$, $\nabla g_1(\mathbf{x}^*) = \begin{pmatrix} 4 \\ 2 \end{pmatrix}$, $\nabla g_2(\mathbf{x}^*) =$

$\begin{pmatrix} 1 \\ 2 \\ 1 \end{pmatrix}$. Take $\lambda = \begin{pmatrix} \frac{2}{3} \\ \frac{20}{3} \\ 0 \\ 0 \end{pmatrix}$. It can be checked that

$$\nabla f(\mathbf{x}^*) + \lambda_1 \nabla g_1(\mathbf{x}^*) + \lambda_2 \nabla g_2(\mathbf{x}^*) = 0$$

Also, $\sum_{i=1}^{4} \lambda_i g_i(\mathbf{x}^*) = 0$, because $\lambda_3 = \lambda_4 = 0$.

Thus $\mathbf{x}^* = \begin{pmatrix} 2 \\ 1 \end{pmatrix}$ is a KKT point, and hence, by Theorem C.3, is a global optimal solution. The optimum value is 25; in Figure C.1 notice that the contour $f(\mathbf{x}) = 25$ passes through $\mathbf{x}^*$. Because $f(\mathbf{x})$ is strictly convex, $\mathbf{x}^*$ is the unique global optimum. ∎

Remarks C.1

a. Figure C.1 can be used to obtain some intuition into the KKT condition. The derivative condition basically states that at $\mathbf{x}^*$ there is no direction in which the objective will improve while maintaining feasibility. To improve the objective function we must move along a direction d from $\mathbf{x}^*$ such that $\nabla f(\mathbf{x}^*)^T d < 0$; but any such direction leads to a violation of the binding constraints. The derivative condition states that this happens when the negative of $\nabla f(\mathbf{x}^*)$ lies in the *cone* generated by the gradients of the binding constraints. It follows that $\mathbf{x}^*$ is a local optimum. Because $f(\mathbf{x})$ is convex, we also have a global optimum.

b. In general, the KKT conditions are *not necessary*; that is, a point $\mathbf{x} \in \mathbb{R}^n$ can be optimal for the primal problem, and yet it may not satisfy the KKT conditions. Following is an example where this happens.

Example C.2

$$\min (x_1 - 5)^2 + (x_2 - 5)^2$$

subject to

$$(x_1 - 1)^2 + (x_2 - 1)^2 \le 0$$

$$\mathbf{x} \ge 0$$

In this example we have $f(\mathbf{x}) = (x_1 - 5)^2 + (x_2 - 5)^2$, $g_1(\mathbf{x}) = (x_1 - 1)^2 + (x_2 - 1)^2$, $g_2(\mathbf{x}) = -x_1$, and $g_3(\mathbf{x}) = -x_2$. There is only one $\mathbf{x}$ that satisfies the constraints— namely, $\mathbf{x}^* = \begin{pmatrix} 1 \\ 1 \end{pmatrix}$. At this point the binding constraint is $g_1(\mathbf{x}) \le 0$. But $\nabla g_1(\mathbf{x}^*) = \begin{pmatrix} 0 \\ 0 \end{pmatrix}$, whereas $\nabla f(\mathbf{x}^*) = \begin{pmatrix} -8 \\ -8 \end{pmatrix}$. Hence the KKT conditions do not hold at the optimum point in this problem. ∎

The following theorem provides two simple conditions under which the KKT conditions are necessary and sufficient for Primal Problem C.1. Recall that we have limited our discussion to convex and differentiable $f(\mathbf{x})$ and $g_i(\mathbf{x}), 1 \le i \le m$.

Theorem C.4   Necessity and Sufficiency of KKT Conditions

a. If $\mathbf{x}^*$ is such that there exists an $\mathbf{x} \in \mathbb{R}^n$ with $g_i(\mathbf{x}) < 0$ if i is active at $\mathbf{x}^*$, then $\mathbf{x}^*$ is optimal if and only if the KKT conditions hold at $\mathbf{x}^*$.

b. *Linear constraints:* If the constraints are linear (i.e., there are vectors $\mathbf{a}_i \in \mathbb{R}^n$ such that $g_i(\mathbf{x}) = \mathbf{a}_i^T \mathbf{x}$, for $1 \le i \le m$), then $\mathbf{x}^*$ is optimal if and only if the KKT conditions hold at $\mathbf{x}^*$.   ■

Observe that the hypothesis of part (a) of Theorem C.4 does not hold in Example C.2, nor are the constraints in Example C.2 linear as required in part (b) of Theorem C.4.

C.4   Linear Programming

We now use Theorem C.4 to discuss important special case of Primal Problem C.1, in which the objective function and the constraints are both linear. These are called *linear programs*.

Example C.3   A Linear Program (LP)

Given that $\mathbf{b} \in \mathbb{R}^n$, $\mathbf{A}$ is an $m \times n$ matrix with real elements, and $\mathbf{c} \in \mathbb{R}^m$, consider the following optimization problem.

Primal LP

$$\min \mathbf{b}^T \mathbf{x}$$

subject to

$$\mathbf{A}\mathbf{x} \ge \mathbf{c}$$

$$\mathbf{x} \ge 0$$

Identifying terms with the standard primal problem, we have $f(\mathbf{x}) = \mathbf{b}^T \mathbf{x}$, and $g_i(\mathbf{x}) = -\mathbf{A}_{i,\cdot}\mathbf{x} + c_i, 1 \le i \le m$. To represent the nonnegativity constraints, define $p_j(\mathbf{x}) = -x_j, 1 \le j \le n$. Hence we find that $\nabla f(\mathbf{x}) = \mathbf{b}$, $\nabla g_i(\mathbf{x}) = -\mathbf{A}_{i,\cdot}^T$, and

$\nabla p_j(\mathbf{x}) = -e_j$. With $\lambda \in \mathbb{R}^m$, $\boldsymbol{\nu} \in \mathbb{R}^n$, both nonnegative vectors, the KKT conditions at a point x become

$$\mathbf{b} - \sum_{i=1}^{m} \lambda_i \mathbf{A}_{i,\cdot}^T - \sum_{j=1}^{n} \nu_j e_j = 0, \qquad \text{(C.4)}$$

$$\lambda^T (\mathbf{A}\mathbf{x} - \mathbf{c}) = 0$$

and

$$\boldsymbol{\nu}^T \mathbf{x} = 0$$

Thus, we can see that $x_j > 0$ implies that $\nu_j = 0$ (because both of these vectors are nonnegative), and hence, looking at Equation C.4, we see that $x_j > 0$ implies that $b_j - \lambda^T \mathbf{A}_{\cdot,j} = 0$; or, equivalently,

$$(\mathbf{b}^T - \lambda^T \mathbf{A})\mathbf{x} = 0$$

From the foregoing and part (b) of Theorem C.4, it follows that the KKT conditions for the LP become the following: $\mathbf{x}^*$ is optimal for the LP *if and only if* there exists $\lambda \in \mathbb{R}^m$, $\lambda \geq 0$, such that

$$\lambda^T (\mathbf{A}\mathbf{x}^* - \mathbf{c}) = 0 \text{ and } (\mathbf{b}^T - \lambda^T \mathbf{A})\mathbf{x}^* = 0$$

These are called *complementary slackness* conditions. We say more about them after discussing duality. ∎

C.5   Duality

Consider the primal problem of Equation C.1, with m general constraints $g_i(\mathbf{x}) \leq 0$, $1 \leq i \leq m$, and n nonnegativity constraints $\mathbf{x} \geq 0$—that is, with a total of $(m + n)$ constraints, as follows:

$$\min f(\mathbf{x})$$

subject to

$$g_i(\mathbf{x}) \leq 0, \quad 1 \leq i \leq m$$

$$\mathbf{x} \geq 0 \qquad\qquad\qquad\qquad\qquad\qquad \text{(C.5)}$$

For the general primal problem of Equation C.5, define for $\lambda \in \mathbb{R}^m, \lambda \geq 0$,

$$\Theta(\lambda) = \inf\{\mathbf{x} \geq 0 : f(\mathbf{x}) + \sum_{i=1}^m \lambda_i g_i(\mathbf{x})\} \qquad\qquad \text{(C.6)}$$

$\Theta(\lambda)$ is called the *Lagrangian dual function* and is obtained by relaxing the constraints $g_i(\mathbf{x}) \leq 0, 1 \leq i \leq m$.

Exercise C.2

   a. Show that $\Theta(\lambda)$ is a concave function on $\lambda \in \mathbb{R}^m, \lambda \geq 0$.

   b. Show that if $\lambda \geq 0$ and if $\mathbf{x}$ satisfies the constraints of the primal problem, then $\Theta(\lambda) \leq f(\mathbf{x})$.

Now let us consider the following optimization problem.

Dual Problem

$$\max_{\lambda \geq 0} \Theta(\lambda) \qquad\qquad\qquad\qquad \text{(C.7)}$$

It is easily seen from Exercise C.2 that the solution to the dual problem will lower bound the solution to the primal problem. Under convexity, which we have been assuming throughout this section, more is true, as is stated by the following theorem. Again, we are stating the special case, which suffices for our purposes.

Theorem C.5   Strong Duality
Let the primal and dual problems be as defined in Equations C.5 and C.7, respectively. Suppose there exists $\mathbf{x} \geq 0$ such that $g_i(\mathbf{x}) < 0, 1 \leq i \leq m$. Then the primal and dual problems have the same optimum values. If the optimum value is finite, and if $\mathbf{x}^*$ and λ^* are solutions to the primal problem and the dual problem, then $\sum_{i=1}^m \lambda_i^* g_i(\mathbf{x}^*) = 0$. ∎

Example C.4   (Example C.1 Continued)

Let us relax the $g_i(\mathbf{x}) \leq 0$ constraints. This yields

$$\Theta(\lambda) = \inf\left\{\mathbf{x} \geq 0:\ ((x_1 - 5)^2 + (x_2 - 5)^2) + \lambda_1(x_1^2 + x_2^2 - 5) + \lambda_2\left(\frac{1}{2}x_1 + x_2 - 2\right)\right\}$$

Differentiating with respect to x_1 and x_2, we get the following equations.

$$2(x_1 - 5) + 2\lambda_1 x_1 + \frac{\lambda_2}{2} = 0$$

$$2(x_2 - 5) + 2\lambda_1 x_2 + \lambda_2 = 0$$

Thus

$$x_1 = \frac{20 - \lambda_2}{4(1 + \lambda_1)}; \quad x_2 = \frac{10 - \lambda_2}{2(1 + \lambda_1)}$$

Substituting x_1 and x_2 in $\Theta(\lambda)$, we get

$$\Theta(\lambda) = \frac{720\lambda_1 + 88\lambda_2 - 32\lambda_1\lambda_2 - 80\lambda_1^2 - 5\lambda_2^2}{16(1 + \lambda_1)}$$

It can be shown that $\Theta(\lambda)$ is concave in λ, and differentiating we get the equations

$$720 - 120\lambda_2 - 160\lambda_1 - 80\lambda_1^2 + 5\lambda_2^2 = 0$$

$$88 - 32\lambda_1 - 10\lambda_2 = 0$$

Solving these equations we get the optimal solution $\lambda_1^* = \frac{2}{3}$ and $\lambda_2^* = \frac{20}{3}$. Finally, substituting these we get the optimal solution $x_1^* = 2$ and $x_2^* = 1$. The optimal value $f(\mathbf{x}^*) = 25$. ∎

Example C.5 (Example C.3 Continued)

Relaxing the constraints $\mathbf{Ax} \geq \mathbf{c}$, we obtain

$$\Theta(\boldsymbol{\lambda}) = \inf\{\mathbf{x} \geq 0 \colon \mathbf{b}^T\mathbf{x} + \boldsymbol{\lambda}^T(\mathbf{c} - \mathbf{Ax})\}$$

$$= \inf\{\mathbf{x} \geq 0 \colon ((\mathbf{b}^T - \boldsymbol{\lambda}^T\mathbf{A})\mathbf{x} + \boldsymbol{\lambda}^T\mathbf{c}\}$$

$$= \begin{cases} \boldsymbol{\lambda}^T\mathbf{c} & \text{for } \mathbf{b}^T - \boldsymbol{\lambda}^T\mathbf{A} \geq 0 \\ -\infty & \text{otherwise} \end{cases}$$

where the last expression is obtained because if $b_j - \boldsymbol{\lambda}^T\mathbf{A}_{.j} < 0$ for some $j, 1 \leq j \leq n$, then we can make $x_k = 0, k \neq j$, and take x_j to ∞.

Hence the dual problem is as follows.

Dual LP

$$\max \boldsymbol{\lambda}^T\mathbf{c}$$

subject to

$$\boldsymbol{\lambda}^T\mathbf{A} \leq \mathbf{b}^T$$

$$\boldsymbol{\lambda} \geq 0$$

Thus we see that the dual problem of a linear program is also a linear program. Notice that there is one dual variable for each primal linear constraint (there are m of each), and one primal variable for each dual linear constraint (there are n of each). ∎

Remarks C.2

a. The term *complementary slackness* can now be better understood in the context of the primal and dual LPs. Recall that the KKT conditions for the LP became the following: $\mathbf{x}^*$ is optimal for the LP if and only if there exists $\boldsymbol{\lambda} \in \mathbb{R}^m, \boldsymbol{\lambda} \geq 0$, such that

$$\boldsymbol{\lambda}^T(\mathbf{Ax}^* - \mathbf{c}) = 0 \text{ and } (\mathbf{b}^T - \boldsymbol{\lambda}^T\mathbf{A})\mathbf{x}^* = 0$$

Thus if there is a slack in a primal constraint, then the corresponding dual variable must be 0, and if there is a slack in a dual constraint, then the corresponding primal variable must be 0.

b. From Theorem C.5 we can conclude that if the optimum value of the linear program is finite, then this value is equal to $(\boldsymbol{\lambda}^*)^T \mathbf{c}$. ∎

More generally, the primal LP can contain equality constraints. The following are the primal and dual LPs in that case. In addition to the earlier definitions, H is an $l \times n$ matrix, and $\mathbf{d} \in \mathbb{R}^l$.

Primal LP

$$\min \mathbf{b}^T \mathbf{x}$$

subject to

$$\mathbf{A}\mathbf{x} \geq \mathbf{c}$$

$$\mathbf{H}\mathbf{x} = \mathbf{d}$$

$$\mathbf{x} \geq 0$$

Dual LP

$$\max \boldsymbol{\lambda}^T \mathbf{c} + \boldsymbol{\mu}^T \mathbf{d}$$

subject to

$$\boldsymbol{\lambda}^T \mathbf{A} + \boldsymbol{\mu}^T \mathbf{H} \leq \mathbf{b}^T$$

$$\boldsymbol{\lambda} \geq 0 \quad \boldsymbol{\mu} \text{ unrestricted}$$

Notice that in the dual LP the dual variables corresponding to the equality constraints in the primal LP (i.e., $\boldsymbol{\mu}$) are unrestricted; that is, they are not constrained to be nonnegative.

The complementary slackness conditions for this general LP become as follows: $\mathbf{x}^*$ is optimal for the primal LP if and only if there exists $\boldsymbol{\lambda} \in \mathbb{R}^m$, $\boldsymbol{\lambda} \geq 0$, and $\boldsymbol{\mu} \in \mathbb{R}^l$ such that

$$\boldsymbol{\lambda}^T(\mathbf{Ax}^* - \mathbf{c}) = 0 \quad \text{and} \quad (\mathbf{b}^T - (\boldsymbol{\lambda}^T\mathbf{A} + \boldsymbol{\mu}^T\mathbf{H}))\mathbf{x}^* = 0$$

The following theorem states the complete relationship between the primal and dual LPs.

Theorem C.6

Given the primal and dual LPs as shown earlier,

a. For any feasible solution $\mathbf{x}$ of the primal LP and any feasible solution $(\boldsymbol{\lambda}^T, \boldsymbol{\mu}^T)$ of the dual LP, $\mathbf{b}^T\mathbf{x} \geq \boldsymbol{\lambda}^T\mathbf{c} + \boldsymbol{\mu}^T\mathbf{d}$; that is, the dual LP objective value lower bounds the primal LP objective value.

b. If the primal LP is infeasible, then the dual LP is unbounded, and if the dual LP is infeasible, then the primal LP is unbounded.

c. If the primal LP and the dual LP are both feasible, then they both have solutions and have the same optimal objective value.

d. When the optimal solutions exist, they must satisfy the complementary slackness conditions. ∎

C.6 Sensitivity of the Optimal Solution

Consider the primal problem in Equation C.5 and suppose that each of the constraints (other than the ones that require nonnegativity) are of the form

$$g_i(\mathbf{x}) \leq c_i$$

where $c_i, 1 \leq i \leq m$, are given constants. The functions $g_i(\mathbf{x})$ are convex and differentiable. Suppose we are interested in examining the sensitivity of the optimal value to perturbations in c_i. For the linear program, it is clear from Remark C.2 that this sensitivity is provided by the optimal dual variables $\boldsymbol{\lambda}^*$; that is, a small perturbation in c_i results in a proportional perturbation in the optimal value, and the proportionality factor is just λ_i^*. The same fact can be shown to be true under certain more general situations even if the constraints are nonlinear. Here we provide an example for the linear case.

Example C.6   The LP Knapsack

As an illustration of many of the concepts related to linear optimization, including the sensitivity of the optimal solution, consider the following so-called *LP knapsack* problem. There is a container of volume V that must be packed with some resources (say, rations for a camping trip). There are n items, indexed by $j \in \{1, 2, \ldots, n\}$ (e.g., flour, sugar, potatoes, candy, etc.). The available amount of the jth item is q_j. Item j has a per-unit value η_j (e.g., flour could have more value as a ration than candy) and a per-unit volume v_j. The problem is to maximize the total value of what is carried, subject to the constraints on the volume of the container and on what is available. Let us assume that $\sum_{j=1}^{n} v_j q_j > V$; otherwise, the value is maximized by taking all the available rations.

Let η denote the vector of per-unit values, let v denote the vector of per-unit volumes, and let q denote the vector of available quantities. The variables are the amounts of the various items that are carried, denoted by the n-vector x. Let I denote the $n \times n$ identity matrix.

Primal Knapsack LP

$$\max \eta^T x$$

subject to

$$\begin{bmatrix} v^T \\ I \end{bmatrix} x \leq \begin{bmatrix} V \\ q \end{bmatrix}$$

$$x \geq 0$$

There is one dual variable for the total volume constraint. Let us denote this by y. There is one dual variable for each of the quantity constraints. Denote this vector of dual variables by z. The dual LP can then be written as follows.

Dual Knapsack LP

$$\min \begin{bmatrix} y & z^T \end{bmatrix} \begin{bmatrix} V \\ q \end{bmatrix}$$

subject to

$$[y \quad z^T]\begin{bmatrix} v^T \\ I \end{bmatrix} \geq \eta^T$$

$$y \geq 0$$

$$z \geq 0$$

Consider the ratio $\frac{\eta_j}{v_j}$; this is the ratio of the per-unit value of item j to its per-unit volume. Intuitively, it will be better to carry items that have the most value for the space they take up in the container. Let us reindex the items so that $\frac{\eta_1}{v_1} \geq \frac{\eta_2}{v_2} \geq \cdots \geq \frac{\eta_{l-1}}{v_{l-1}} \geq \frac{\eta_l}{v_l} \geq \cdots \geq \frac{\eta_n}{v_n}$. The index l is such that $\sum_{j=1}^{l-1} v_j q_j \leq V$, but $\sum_{j=1}^{l} v_j q_j > V$. By our assumption that $\sum_{j=1}^{n} v_j q_j > V$, such an l exists and $1 \leq l \leq n$; $l = 1$ means that the available quantity of the first item itself is so much that it can fill up the container.

With the reindexing and the definition of l, the primal solution of the problem has the following form.

$$x_j = q_j \qquad\qquad 1 \leq j \leq l-1$$
$$x_l = \frac{(V - \sum_{j=1}^{l-1} v_j q_j)}{v_l}$$
$$x_j = 0 \qquad\qquad l \leq j \leq n$$

The dual solution has the following form:

$$y = \frac{\eta_l}{v_l}$$
$$z_j = \eta_j - \frac{\eta_l}{v_l} v_j \quad 1 \leq j < l$$
$$z_j = 0 \qquad\qquad l \leq j \leq n$$

Note that $z_j \geq 0$ by virtue of the reindexing of the items. The solution has a very simple intuitive form. It asks us to rank the items by decreasing order of the per-unit value to per-unit volume ratios and then fill up the container from the top of the list until the container is full.

Let us check the complementary slackness conditions. The primal volume constraint is binding, and hence the dual variable y can be positive, which it is. The quantity constraints for items $j, 1 \leq j \leq l - 1$, are binding, and hence

$z_j, 1 \leq j \leq l-1$, can be positive. For items $j, l \leq j \leq n$, the quantity constraints are not binding, and hence the dual variables $z_j, l \leq j \leq n$, are all 0.

It is interesting to study the sensitivity of the solution via the dual variables. Suppose we wish to ask how much the value of the primal problem improves if we were to increase the size of the container by a small amount ϵ. This will allow us to take an additional amount $\frac{\epsilon}{v_l}$ of item l, and the additional value will be $\frac{\eta_l}{v_l}\epsilon$. This is the interpretation of the optimal value of the dual variable y. If we increased any of the item quantities q_j, what is the sensitivity? Increasing the quantities of the items $j, l \leq j \leq n$, provides no improvement in the value; this is the interpretation of $z_j = 0, l \leq j \leq n$. If we increase q_j, for some $j, 1 \leq j \leq l-1$, by, say, δ, then we can take more of this item and reduce some amount of item l, thus increasing the total value of the solution. Let us verify this. The amount by which the value increases because we add δ more of item j is $\eta_j\delta$; the amount by which we must reduce the amount we take of item l is $\frac{v_j\delta}{v_l}$, and this reduces the value of the solution by $\frac{v_j\delta}{v_l}\eta_l$. The net change in value is $\eta_j\delta - \frac{v_j\delta}{v_l}\eta_l = z_j\delta$, which is positive.   ∎

APPENDIX D

Discrete Event Random Processes

D.1 Markov Chains and Some Renewal Theory

D.1.1 Discrete Time Markov Chains

We have a stochastic process $\{X_n, n \in \{0, 1, 2, \ldots\}\}$ that takes values in a countable *state space* $\mathcal{S}$.

Definition D.1

$\{X_n\}$ is called a discrete time Markov chain (DTMC) if, for all $n \in \{0, 1, 2, \ldots\}$, and $j, i_0, i_1, \ldots, i_{n-1}, i \in \mathcal{S}$,

$$\Pr(X_{n+1} = j \mid X_0 = i_0, X_1 = i_1, \ldots, X_{n-1} = i_{n-1}, X_n = i) = \Pr(X_{n+1} = j \mid X_n = i)$$

■

In words, the future (i.e., the state at time $n + 1$) is independent of the past (i.e., the states from time 0 to time $n - 1$), given the present (i.e., the state at time n). It is important to note that the future and the past are not *unconditionally* independent.

Definition D.2

Given a DTMC $\{X_n\}$, we define the following.

a. For $n \in \{0, 1, 2, \ldots\}$, and $i, j \in \mathcal{S}$,

$$p_{i,j}(n) := \Pr(X_{n+1} = j \mid X_n = i)$$

is called a *transition probability* of the DTMC at time n.

b. If $p_{i,j}(n) = p_{i,j}$, for every n, and for every $i, j \in \mathcal{S}$, then the DTMC is said to be *time homogeneous*.

c. In the time homogeneous case, the $|\mathcal{S}| \times |\mathcal{S}|$ matrix $\mathbf{P}$ with elements $p_{i,j}$ is called the *transition probability* matrix of the DTMC. ■

We will deal only with time homogeneous DTMCs, and hence in the following discussion the term DTMC can be read as "time homogeneous DTMC."

Definition D.3

A square matrix **P** with elements $p_{i,j}$ is a *stochastic matrix* if

a. For all i, j, $p_{i,j} \geq 0$, and

b. For all i, $\sum_j p_{i,j} = 1$.   ∎

Clearly, the transition probability matrix of a DTMC is a stochastic matrix.

Consider two independent DTMCs $X_n, n \geq 0$, on $\mathcal{S}$, and $Y_n, n \geq 0$, on $\mathcal{T}$, with transition probability matrices P and Q, respectively. Let $|\mathcal{S}| = K$ and $|\mathcal{T}| = L$. Hence, P is a $K \times K$ matrix and Q is an $L \times L$ matrix. Then $Z_n = (X_n, Y_n)$ is a DTMC on $\mathcal{S} \times \mathcal{T}$, with transition probabilities

$$\Pr(Z_{n+1} = (s_2, t_2) | Z_n = (s_1, t_1)) = p_{s_1, s_2} q_{t_1, t_2}$$

Thus the transition probability matrix has the form

$$
\begin{bmatrix}
Pq_{1,1} & Pq_{1,2} & \cdots & Pq_{1,L} \\
Pq_{2,1} & Pq_{2,2} & \cdots & Pq_{2,L} \\
\vdots & \vdots & \ddots & \vdots \\
Pq_{L,1} & Pq_{L,2} & \cdots & Pq_{L,L}
\end{bmatrix}
$$

which is a $KL \times KL$ matrix and is called the *Kronecker product* (also *outer product*) of the matrices P and Q. Several mutually independent DTMCs $X_n^{(i)}, 1 \leq i \leq m$, $n \geq 0$, can be composed in this way to obtain a vector DTMC $Z_n = (X_n^{(1)}, \ldots, X_n^{(m)})$, and the transition probability matrix of Z_n can be obtained by recursively applying the Kronecker product to the transition probability matrices of the component DTMCs.

Returning to a generic DTMC, $X_n, n \geq 0$, let us denote $p_{i,j}^{(n)} := \Pr(X_n = j | X_0 = i)$; that is, $p_{i,j}^{(n)}$ is the *n*-step transition probability of a DTMC. Denote by $\mathbf{P}^{(n)}$ the *n*-step transition probability matrix; clearly it is also a stochastic matrix. The *n*-step transition probabilities are completely determined by the one-step transition probabilities, as stated in the following theorem.

Theorem D.1

For a DTMC with transition probability matrix $\mathbf{P}$, for every n,

$$\mathbf{P}^{(n)} = \mathbf{P}^n$$ ∎

Given a DTMC $\{X_n\}$ and $i, j \in \mathcal{S}$, define, for $n \geq 1$,

$$f_{i,j}^{(n)} := \Pr(X_1 \neq j, X_2 \neq j, \dots, X_{n-1} \neq j, X_n = j | X_0 = i)$$

That is, $f_{i,j}^{(n)}$ is the probability that the DTMC *first hits*, or *first visits*, state j at time n, when it starts in state i at time 0. In this definition we allow $j = i$. We also define $f_{i,i}^{(0)} = 1$, and, for $j \neq i$, $f_{i,j}^{(0)} = 0$. Further define

$$f_{i,j} := \sum_{n=1}^{\infty} f_{i,j}^{(n)}$$

That is, $f_{i,j}$ is the probability that the DTMC visits the state j in *finite time* if it starts in state i. To see that $f_{i,j}$ is a probability, note that it can also be written as $f_{i,j} = \Pr(\cup_{n=1}^{\infty}\{X_1 \neq j, X_2 \neq j, \dots, X_{n-1} \neq j, X_n = j\} | X_0 = i)$, and the events in the unions are disjoint. In particular, $f_{i,i}$ is the probability of return to i in finite time.

Definition D.4

Given a DTMC $\{X_n\}$, we classify its states as follows. Given $j \in \mathcal{S}$,

a. j is *absorbing* if $p_{j,j} = 1$.

b. j is *transient* if $f_{j,j} < 1$.

c. j is *recurrent* if $f_{j,j} = 1$.

d. If j is recurrent, then it is *positive* if $\sum_{n=1}^{\infty} n f_{j,j}^{(n)} < \infty$; otherwise it is said to be *null*. ∎

For a transient state j it can be shown that

$$\sum_{n=1}^{\infty} p_{j,j}^{(n)} < \infty$$

and hence that $\lim_{n \to \infty} p_{j,j}^{(n)} = 0$; that is, eventually the probability of the DTMC being in state j goes to 0. It can be shown that a transient state is visited only finitely often by the DTMC. A recurrent state, after it is entered, is visited infinitely often. If the recurrent state is, in addition, positive, then the mean time between visits is finite. A positive recurrent state is also called positive.

Let us define $p_{i,j}^{(0)} = 0$ for $j \neq i$, and $p_{i,i}^{(0)} = 1$.

Definition D.5

Given a DTMC $\{X_n\}$, we say that

a. The state j can be reached from state i if for some $n \geq 0$ it holds that $p_{i,j}^{(n)} > 0$. Denote this by $i \to j$.

b. i and j *communicate* if $i \to j$ and $j \to i$. Denote this by $i \leftrightarrow j$.   ∎

By the special cases defined just before Definition D.5, it follows that, for all i, $i \leftrightarrow i$. It can then easily be checked that the binary relation $\leftrightarrow$ is an *equivalence*. The equivalence classes induced on S by $\leftrightarrow$ are called *communicating classes*; that is, $S = \cup_i C_i$ where the C_i are disjoint communicating classes.

Definition D.6

Given a DTMC $\{X_n\}$ on the state space S,

a. A communicating class C is said to be *closed* if for all $i \in C$ and $j \notin C$ we have $p_{i,j} = 0$; otherwise the class is said to be *open*.

b. The DTMC is said to be *irreducible* if for all $i, j \in S$, $i \leftrightarrow j$; that is, the entire state space S is one communicating class.   ∎

Theorem D.2

Given a DTMC $\{X_n\}$ on the state space S, and the induced communicating classes, for each communicating class exactly one of the following holds:

a. All the states in the class are transient.

b. All the states in the class are null recurrent.

c. All the states in the class are positive (recurrent).   ∎

Thus to show that an irreducible DTMC is positive recurrent, it is sufficient to show that one of its states is positive.

Theorem D.3

Given a DTMC $\{X_n\}$ on the state space $\mathcal{S}$, and the induced communicating classes, the following hold.

a. An open communicating class is transient.

b. A closed finite communicating class is positive recurrent.   ∎

Theorem D.3 leaves open the question of recurrence or transience of infinite closed classes. The following is an important criterion. By thinking of the restriction of a DTMC to any of its closed classes, we can confine ourselves to irreducible DTMCs.

Theorem D.4

An irreducible DTMC is positive recurrent if and only if there exists a positive probability distribution π on $\mathcal{S}$ such that (with the distribution viewed as a vector of probabilities) $\pi = \pi P$. Such a probability vector if it exists is unique.   ∎

If π is a probability vector such that $\pi = \pi P$, and if $\Pr(X_0 = i) = \pi_i$, then, for all $n \geq 1$, $\Pr(X_n = i) = \pi_i$. Hence such a π is also called an *invariant* probability vector for the DTMC. It can be shown that if π is invariant and if $\Pr(X_0 = i) = \pi_i$, then the DMTC so obtained is a *stationary* random process. When the DTMC is positive recurrent, Theorem D.4 states that an invariant probability vector exists and is unique.

What are the consequences of positive recurrence of an irreducible DTMC, X_k? The simplest fact is that, for each $j \in \mathcal{S}$,

$$\lim_{n \to \infty} \frac{1}{n} \sum_{k=0}^{(n-1)} I_{\{X_k = j\}} = \pi_j$$

where π is the stationary probability vector guaranteed by Theorem D.4. This states that the long-run fraction of steps during which the DTMC is in state j is π_j. But what about convergence of the distribution itself; does $\Pr(X_k = j | X_0 = i)$ itself converge? Consider a DTMC on $\mathcal{S} = \{0, 1\}$, with $p_{0,1} = 1 = p_{1,0}$. This DTMC is clearly positive recurrent, with $\pi_0 = 0.5 = \pi_1$, but X_k alternates between the two states and the distribution does not converge to π. For such convergence we require the condition of *aperiodicity*.

Definition D.7

Given a DTMC X_k, with transition probability matrix **P**, the *period* of a state $j \in S$ is given by

$$d_j := gcd\{n : f_{j,j}^{(n)} > 0\}$$

If $d_j = 1$, then the state j is called *aperiodic*.   ∎

Thus the DTMC returns to j in multiples of its period. It can be shown that

$$gcd\{n : f_{j,j}^{(n)} > 0\} = gcd\{n : p_{j,j}^{(n)} > 0\}$$

Evidently if j is such that $p_{j,j} > 0$, then j is aperiodic. The period is also a class property.

Theorem D.5

All the states in a communicating class of a DTMC have the same period.   ∎

Thus an irreducible DTMC may also be aperiodic. Finally, the following result states how the distribution of the DTMC can converge to a limit.

Theorem D.6

For an irreducible positive DTMC,

a. For all $j \in S$, $\lim_{n \to \infty} p_{j,j}^{(nd)} = d\pi_j$.

b. If the DTMC is aperiodic, then for all $i, j \in S$, $\lim_{n \to \infty} p_{i,j}^{(n)} = \pi_j$.   ∎

Thus, in the irreducible, aperiodic, positive case the distribution of the DTMC converges to the stationary distribution, irrespective of the initial distribution. We can thus say that the DTMC converges to a *steady state* distribution.

In analyzing queueing models, one of the first questions one asks is whether the system is *stable*. When a queueing system is modeled by a Markov chain, one of the criteria for stability is that the Markov chain characterizing the queueing system be positive recurrent. This at least ensures that there is a steady state distribution of the associated processes. Theorem D.4 suggests a method for determining whether a given irreducible DTMC is positive recurrent: One looks for a positive solution for the system of equations $\pi = \pi P, \sum_{i \in S} \pi_i = 1$. Although this can be done in many simple cases, in general this approach is intractable. Fortunately, there is another approach based on the technique of drift analysis of a suitable *Lyapunov function*. The following are the main theorems that provide sufficient conditions for an irreducible DTMC to be recurrent, transient, or positive. The state space S is countable and is viewed as $S = \{1, 2, 3, \ldots\}$. In the following three theorems the function $f(\cdot)$ is often called a Lyapunov function.

Theorem D.7

An irreducible DTMC $X_n, n \geq 0$, is *recurrent* if there exists a nonnegative function $f(j), j \in S$, such that $f(j) \to \infty$ as $j \to \infty$, and a *finite* set $A \subset S$, such that, for all $i \notin A$,

$$\mathsf{E}\big(f(X_{n+1})|X_n = i\big) - f(i) \leq 0$$

∎

Theorem D.8

An irreducible DTMC $X_n, n \geq 0$, is *transient* if there exists a nonnegative function $f(j), j \in S$, and a set $A \subset S$ such that, for all $i \notin A$,

$$\mathsf{E}\big(f(X_{n+1})|X_n = i\big) - f(i) \leq 0$$

and, there exists a $j \notin A$ such that for all $i \in A$

$$f(j) < f(i)$$

∎

Theorem D.9

An irreducible DTMC $X_n, n \geq 0$, is *positive recurrent* if there exists a nonnegative function $f(j), j \in S$, and a *finite* set $A \subset S$, such that, for all $i \notin A$,

$$\mathsf{E}\big(f(X_{n+1})|X_n = i\big) - f(i) \leq -\epsilon$$

for some $\epsilon > 0$, and for all $i \in A$

$$\mathsf{E}\big(f(X_{n+1})|X_n = i\big) < B$$

for some finite number B.

∎

Exercise D.1

Consider a Markov chain X_n, on $\mathcal{S} = \{0, 1, 2, \ldots\}$, with $p_{i,(i+1)} = p$, for all $i \geq 0$, $p_{0,0} = 1 - p$, and $p_{i,(i-1)} = 1 - p$, for all $i \geq 1$. This process is called a *random walk* on the nonnegative integers.

   a. Use Theorem D.8 to show that X_n is transient for $(1 - p) < p$. Hint: use $f(i) = \alpha^i$, with $0 < \alpha < 1$.

   b. Use Theorem D.7 to show that X_n is recurrent for $p \leq (1 - p)$. Hint: use $f(i) = i$.

   c. Use Theorem D.9 to show that X_n is positive recurrent for $p < (1 - p)$. Hint: use $f(i) = i$.

It may appear, intuitively, that if $E(X_{n+1}|X_n = i) - i \geq 0$ for all i larger than some finite value then the Markov chain should not be positive recurrent. However, this is not the case, and there are counterexamples. Basically, in these examples, even though the mean drift of the process is positive, the chain can return to small values of state *in a single transition* from any state i no matter how large. The following theorem basically eliminates this possibility in order to provide a "converse" to Theorem D.9.

Theorem D.10

An irreducible DTMC $X_n, n \geq 0$, on $i \in \{0, 1, 2, \ldots\}$, is *not* positive recurrent if there exist finite values $K > 0$ and $B > 0$, such that, for all $i \geq 0$,

$$E(X_{n+1} \mid X_n = i) < \infty$$

for all $i \geq K$,

$$E(X_{n+1} \mid X_n = i) - i \geq 0$$

and, for all $i \geq K$,

$$E\big((X_n - X_{n+1})^+ \mid X_n = i\big) \leq B$$

∎

Remark: In the context of Theorems D.7, D.8, and D.9, Theorem D.10 is stated for $f(j) = j$. In this theorem, the final requirement states that for large i, the mean *downward* drift must be bounded for states $i \geq K$ (notice that $(X_n - X_{n+1})^+$ is nonzero only if $X_{n+1} < X_n$). This result is very useful in establishing the instability of discrete time Markov queueing models with a finite number of servers, or bounded service capacity. Such queues evolve over so-called slots. In each slot there can be arrivals, and a certain maximum number of customers can be served. If the mean arrival rate per slot is more than the maximum service rate, then the mean drift of the queue length will be positive. Obviously, however, the number of customers cannot decrease in a single slot by more than the maximum number that can be served in a slot, and hence this maximum number of possible departures in a slot serves as the bound B. Hence, in such systems the condition of the arrival rate being greater than or equal to the maximum number of possible services in a slot (the second condition in Theorem D.10) immediately establishes the lack of positive recurrence (i.e., instability).

Exercise D.2

Use Theorem D.10 to show that the random walk in Exercise D.1 is null recurrent for $p = (1 - p)$.

D.1.2 Continuous Time Markov Chains

We have a continuous time stochastic process $\{X(t), t \geq 0\}$ that takes values in the discrete state space $\mathcal{S}$.

Definition D.8

$\{X(t)\}$ is a continuous time Markov chain (CTMC) on $\mathcal{S}$ if for all $t, s \geq 0$, for all $j \in \mathcal{S}$,

$$\Pr(X(t+s) = j | X(u), u \leq t) = \Pr(X(t+s) = j | X(t))$$

■

We assume time homogeneity and write

$$p_{i,j}(t) := \Pr(X(t+s) = j | X(s) = i)$$

which will be the elements of the transition probability matrix over time t, denoted by $\mathbf{P}(t)$. Unlike a DTMC, a CTMC has no basic "one-step" transition probability matrix that determines the collection of transition probability matrices $\mathbf{P}(t), t \geq 0$.

For all $t \geq 0$, define $W(t) = \inf\{s > 0 : X(t+s) \neq X(t)\}$; that is, $W(t)$ is the time after which the process leaves the state it is in at time t. The Markov property itself leads to the following important result.

Theorem D.11

For a CTMC $\{X(t)\}$, for all $i \in \mathcal{S}$, and all $t \geq 0$, and $u \geq 0$,

$$\Pr(W(t) > u \mid X(t) = i) = e^{-a_i u}$$

for some constant $a_i \geq 0$.   ■

Thus the time during which the CTMC stays in a state is exponentially distributed with a parameter that depends only on the state. Because of the *memoryless property* of the exponential distribution, the fact that the sojourn times in states are exponentially distributed is quite intuitive because the state process is Markov; otherwise, we would need to keep track of how long the process has spent in a state in order to determine the probability of future events.

A state $i \in \mathcal{S}$ is called *absorbing* if $a_i = 0$. We assume that for all states $a_i < \infty$. Under this assumption a CTMC is a *pure jump process*; the process enters a state, spends a positive amount of time in that state, and then moves on to another state (unless the state entered is absorbing). With this picture in mind, define $T_0 = 0, T_1, T_2, \ldots$ to be the successive jump instants of a CTMC, and let $X_n = X(T_n)$. The sequence $T_n, n \geq 0$, is called a sequence of *embedded instants*, and the state sequence $X_n, n \geq 0$, is called the *jump chain*, or an *embedded process*. The following is an important characterization of the jump instants and the jump process.

Theorem D.12

Given a CTMC $\{X(t)\}$, with jump instants $T_n, n \geq 0$, and jump chain $X_n, n \geq 0$, for $i_0, i_1, \ldots, i_{n-1}, i, j \in \mathcal{S}, t_0, t_1, \ldots, t_n, u \geq 0$,

$$\Pr\{X_{n+1} = j, T_{n+1} - T_n > u \mid X_0 = i_0, \ldots, X_{n-1} = i_{n-1},$$

$$X_n = i, T_0 = t_0, \ldots, T_n = t_n\} = p_{i,j} e^{-a_i u}$$

where $p_{i,j} \geq 0, \sum_{j \in \mathcal{S}} p_{i,j} = 1$, and if $a_i > 0$, then $p_{i,i} = 0$.   ■

This result states that, given the entire state process until the jump T_n (including the state, i, entered at this jump), the time spent in this state (i.e., $T_{n+1} - T_n$) and the state entered at the next jump (i.e., X_{n+1}) are independent, with the time spent in the state (i.e., i) being exponential with the same parameter as that of the unconditional sojourn time in the state.

We can conclude from this result that the embedded process is a DTMC with transition probabilities $p_{i,j}$, and further that the time that the process spends in a state is independent of the past and is exponentially distributed, with a parameter determined only by the state. Define **P** to be the transition probability matrix of the embedded DTMC. Thus a CTMC can be constructed as follows. First generate the DTMC using the transition probability matrix **P**, and then generate the sequence of state sojourn times—say, $W_0, W_1, W_2, \ldots$—using the parameters a_i. The jump times $T_n, n \geq 0$, are obtained by concatenating the sojourn times (i.e., $T_0 = 0$, and, for $n \geq 1$, $T_n = \sum_{i=0}^{n-1} W_i$). Then define $X(t) = X_n$ if $t \in [T_n, T_{n+1})$, for $n \geq 0$. In general, this construction defines the process for all t only if $\sum_{i=0}^{\infty} W_i = \infty$. Such CTMCs are called *regular*. We assume that this is the case for the CTMCs that we are concerned with. Hence by the foregoing construction we can think of a (regular) CTMC in terms of the jump DTMC and the sequence of state sojourn times.

The following are two criteria for recognizing that a CTMC is regular.

Theorem D.13

$\{X(t)\}$ is a CTMC with embedded DTMC $\{X_n\}$. The sojourn time parameters are $a_i, i \in \mathcal{S}$.

a. If there exists ν such that $a_i \leq \nu$ for all i, then the CTMC is regular.

b. If $\{X_n\}$ is recurrent then $\{X(t)\}$ is regular.   ■

We now turn to the question of transience and recurrence of a CTMC. It can easily be seen that the binary relation $\leftrightarrow$, on $\mathcal{S}$, is the same for a CTMC and its embedded DTMC, and hence a CTMC is irreducible if and only if its embedded DTMC is irreducible. For a state $j \in \mathcal{S}$, let $X(0) = j$ and define $\tau_{j,j}$ to be the time until the process returns to state j after leaving it.

Definition D.9

The state j in a CTMC is said to be *recurrent* if $\Pr(\tau_{j,j} < \infty) = 1$; otherwise, j is *transient*. A recurrent state j is *positive* if $E(\tau_{j,j}) < \infty$; otherwise, it is *null*.   ■

Just as in the case of DTMCs, it can be shown that the states of an irreducible CTMC are either all transient, all positive, or all null. Correspondingly we say that an irreducible CTMC is transient, positive, or null.

It can be argued that j is recurrent in the CTMC $\{X(t)\}$ if and only if it is recurrent in its embedded DTMC $\{X_n\}$, and an irreducible CTMC is recurrent if and only if its embedded DTMC is recurrent. A similar result does *not* hold for positivity of states of the CTMC.

Given the transition probabilities (i.e., $p_{i,j}$) of the embedded DTMC $\{X_n\}$ and the state sojourn time parameters $a_i, i \in \mathcal{S}$, we define the $|\mathcal{S}| \times |\mathcal{S}|$ matrix $\mathbf{Q}$ as follows. For $i, j \in \mathcal{S}, i \neq j$, $q_{i,j} = a_i p_{i,j}$, and for $i \in \mathcal{S}$, $q_{i,i} = -a_i$. The off-diagonal terms in $\mathbf{Q}$ can be interpreted as the rate of leaving the state i to enter the state j conditional on being in the state i. Notice that the sum of each row of $\mathbf{Q}$ is 0. The following theorem provides an important criterion for the positivity of an irreducible regular CTMC.

Theorem D.14

An irreducible regular CTMC is positive if and only if there exists a positive probability vector $\boldsymbol{\pi}$ (i.e., $\pi_i, i \in \mathcal{S}$, is a probability distribution on $\mathcal{S}$) such that $\boldsymbol{\pi}\mathbf{Q} = 0$. When such a $\boldsymbol{\pi}$ exists it is unique.   ∎

For an irreducible regular CTMC, a probability vector $\boldsymbol{\pi}$ such that $\boldsymbol{\pi}\,\mathbf{Q} = 0$ is also a stationary probability vector; that is, if $\Pr(X(0) = i) = \pi_i$, then $\Pr(X(t) = i) = \pi_i$ for all t. It can also be shown that $\pi_j = \frac{\frac{1}{a_j}}{\mathsf{E}(\tau_{j,j})}$ (i.e., the fraction of time that the process stays in state j). Furthermore, unlike DTMCs, CTMCs have no notion of periodicity, and the following holds for an irreducible positive recurrent CTMC:

$$\lim_{t \to \infty} p_{i,j}(t) = \pi_j$$

where $\boldsymbol{\pi}$ is the stationary probability vector.

When $\boldsymbol{\pi}$ is the stationary probability vector, the set of linear equations $\boldsymbol{\pi}\,\mathbf{Q} = 0$ have an important interpretation. The jth equation is

$$\sum_{i \in \mathcal{S}, i \neq j} \pi_i q_{i,j} = \pi_j a_j$$

The right side of this equation is the *unconditional* rate of leaving the state j, and each term in the summation on the left side is the unconditional rate of leaving the state i to enter j (and, hence, the sum is the unconditional rate of leaving the state i).

For many simple Markov chain models, examining the solutions of the system of linear equations $\pi Q = 0$, and $\sum_{i \in S} \pi_i = 1$, is the standard way to obtain a condition for positive recurrence and the corresponding stationary probability distribution.

Example D.1

Customers arrive to a queue with infinite storage space in a Poisson process with rate λ. The customers' service requirements are i.i.d. and exponentially distributed with mean $\frac{1}{\mu}$. There is a single server that serves at the rate of one unit of work per unit time. This is the M/M/1 queueing model. Let $X(t)$ be the number of customers at time t. Because of the Poisson arrivals and the exponential service requirements, it is easily seen that $X(t)$ is a CTMC. When $X(t) = 0$, the next state change occurs on an arrival, and hence $a_0 = \lambda$. When $X(t) \geq 1$, the next state change occurs at the earliest of two instants: the current service completion whose residual time is exponentially distributed with mean $\frac{1}{\mu}$, or the next arrival whose residual arrival time is exponentially distributed with mean $\frac{1}{\lambda}$. Hence the time until the next event is exponentially distributed with mean $\frac{1}{\mu+\lambda}$. It follows that $a_i = \mu + \lambda$, for $i \geq 1$. Clearly $p_{0,1} = 1$. Furthermore, it can be seen that, for $i \geq 1$, $p_{i,(i-1)} = \frac{\mu}{\mu+\lambda}$, and $p_{i,(i+1)} = \frac{\lambda}{\mu+\lambda}$. Thus we find that the transition rate matrix Q has the following form: $q_{0,0} = -\lambda = -q_{0,1}$; for $i \geq 1$, $q_{i,(i-1)} = \mu$, $q_{i,(i+1)} = \lambda$, and $q_{i,i} = -(\mu+\lambda)$. It can now be checked that the system of equations $\pi Q = 0$ has a positive, summable solution if and only if $\lambda < \mu$ (the arrival rate is less than the maximum rate at which customers can be served), and furthermore, defining $\rho = \frac{\lambda}{\mu}$, $\pi_i = (1 - \rho)\rho^i$, for $i \in \{0, 1, 2, \ldots\}$, is the stationary distribution. The fraction of time that the system has i customers (counting anyone in service) is π_i; in particular, the queue is empty during a fraction $(1 - \rho)$ of the time. If $\Pr(X(0) = i) = (1 - \rho)\rho^i$, then the CTMC is a stationary process. ∎

D.1.3 Renewal Processes

Renewal processes have a very simple characterization. They arise as components of many discrete event processes and are fundamental to the study of such processes. In fact, even though we have discussed Markov chains before renewal processes, the study of the latter is more fundamental, and the theory of Markov chains derives many results from *renewal theory*—that is, the theory of renewal processes.

There is a sequence of mutually independent random variables $X_k, k \in \{1, 2, 3, \ldots\}$, such that $X_k, k \geq 2$ are i.i.d., and X_1 can have a possibly different distribution from the rest. Think of the $X_k, k \geq 1$, as the *lifetimes* of a component that fails and is repeatedly replaced (e.g., a light bulb in a particular socket in one's home). The *renewal instants*, $Z_k, k \geq 1$, are defined as

$$Z_k = \sum_{i=1}^{k} X_i$$

The number of renewals in the interval $(0, t]$ is called the *renewal process* and is denoted by $M(t)$. Note that $M(t)$ jumps at each instant that a renewal occurs and stays "flat" in between. Because the lifetime distributions can have a point mass at 0, there could be multiple jumps at the same instant (just as a new light bulb can fail the moment it is first switched on).

Example D.2

There are renewal processes embedded inside Markov chains. Suppose that $B(t)$ is an irreducible CTMC with $B(0) = i$, and consider visits to state j. Let X_1 be the time at which the CTMC first hits state j, and let $X_k, k \geq 2$, be the times between subsequent visits to j. Because $B(t)$ is a Markov chain, $X_k, k \geq 1$, are mutually independent random variables, and $X_k, k \geq 2$, are i.i.d., with distribution the same as that of $\tau_{j,j}$ defined earlier. The distribution of X_1 is possibly different from that of the other lifetimes. Notice that if the CTMC is positive recurrent, then $X_k, k \geq 2$, have a (common) finite mean. For this example, $M(t)$ is the number of times that the process visits state j in the interval $(0, t]$. ∎

D.1.4 Renewal Reward Processes

A useful class of models arises when we associate a *reward* with each renewal interval. Consider a renewal process with lifetimes $X_k, k \geq 1$. Associated with the lifetime X_k is a reward R_k, such that the R_k, $k \geq 1$, are mutually independent. However, R_k can depend on X_k. For example, in the Markov chain example (Example D.2), R_k can be the time spent in the state i during a renewal interval; the longer the time between successive visits to j, the more time the Markov chain is likely to spend in i. Note from this example that the reward R_k could accrue over the interval X_k, rather than all at once. Because of the Markov property,

$R_k, k \geq 1$, are clearly mutually independent, but R_k and X_k are not independent. Furthermore, $R_k, k \geq 2$, are also i.i.d.

In this renewal reward framework, let us define $C(t)$ to be the total reward accrued until time t. With reference to Example D.2, $C(t)$ would be the time spent in the state i until t. Then we may be interested in the reward rate (i.e., in $\lim_{t\to\infty} \frac{C(t)}{t}$). So, in the example, $\lim_{t\to\infty} \frac{C(t)}{t}$ would be the fraction of time that $B(t)$ spends in the state i.

Theorem D.15   Renewal Reward Theorem

For $E(|R_k|) < \infty$ and $E(X_k) < \infty$ the following hold:

a.

$$\lim_{t\to\infty} \frac{C(t)}{t} = \frac{E(R_2)}{E(X_2)}$$

where the convergence is in the almost sure sense.

b.

$$\lim_{t\to\infty} \frac{E(C(t))}{t} = \frac{E(R_2)}{E(X_2)}$$

∎

Remarks D.1

a. The reason for the subscript 2 appearing in the limit on the right side is that $X_k, k \geq 2$, and $R_k, k \geq 2$, are i.i.d., and the values in the first cycle do not matter in the limit.

b. The form of the limit is important. In particular, the limit is *not* $E\left(\frac{R_2}{X_2}\right)$, and this is a common source of error. For example, suppose that $(X_k, R_k), k \geq 1$, are i.i.d.; we then denote the generic cycle length and reward pair by (X, R). Suppose that $(X, R) = (1, 10)$ with probability 0.5, and $(X, R) = (10, 1000)$ with probability 0.5. So, on the average, in half the intervals the reward rate is 10, and in half the intervals the reward rate is 100. One might want to say that the average reward rate is $0.5 \times 10 + 0.5 \times 100 = 55$; this would be the answer one would get if the formula $E\left(\frac{R}{X}\right)$ were used. Yet the theorem declares that the answer is $\frac{E(R)}{E(X)} = \frac{0.5 \times 10 + 0.5 \times 1000}{0.5 \times 1 + 0.5 \times 10} = 91.82$. The reason is that we seek a time average reward rate (not a cycle average), and over time more of the time is occupied by the longer intervals, which provide a larger reward rate. To see how this works out, observe that in a large time period T the average number of intervals will be roughly $\frac{T}{E(X)} = \frac{T}{5.5}$. Of these, a fraction 0.5 are intervals of length 1, and the rest are intervals of length 10. Up to time T, the time occupied by intervals of length 1 is $0.5 \times \frac{T}{5.5} \times 1$, and the time occupied by intervals of length 10 is $0.5 \times \frac{T}{5.5} \times 10$. Hence up to T the fraction of time occupied by intervals of length 1 is $0.5 \times \frac{T}{5.5} \times 1 \times \frac{1}{T} = \frac{0.5}{5.5}$, and the fraction of time occupied by intervals of length 10 is $0.5 \times \frac{T}{5.5} \times 10 \times \frac{1}{T} = \frac{0.5 \times 10}{5.5}$. Hence the time average reward rate is $\frac{0.5}{5.5} \times \frac{10}{1} + \frac{0.5 \times 10}{5.5} \times \frac{1000}{10} = \frac{E(R)}{E(X)} = 91.82$. The underlying phenomenon is called *length biasing*, and the fraction of time that is occupied with intervals of length x (i.e., $\frac{xp(x)}{E(X)}$, where $p(x)$ is the probability that an interval is of length x) is called the *length-biased distribution* of X.

c. The second part of the theorem is only a convergence of a deterministic sequence of numbers and does not follow simply by taking expectation in the first part, because convergence almost surely does not in general imply convergence of expectations. ∎

D.1.5   Markov Renewal Processes

Let $X_n, n \geq 0$, be a random sequence taking values in $\mathcal{S}$, and let $T_0 \leq T_1 \leq T_2 \ldots$ be a nondecreasing sequence of random times.

Definition D.10

The random sequence $(X_n, T_n), n \geq 0$, is a Markov renewal process (MRP) if for $i_0, i_1, \ldots, i_{n-1}, i, j \in \mathcal{S}, t_0 \leq t_1 \leq \ldots \leq t_n, u \geq 0$,

$$\Pr\{X_{n+1} = j, T_{n+1} - T_n \leq u | X_0 = i_0, \ldots, X_{n-1} = i_{n-1}, X_n = i, T_0 = t_0, \ldots, T_n = t_n\}$$

$$= \Pr\{X_{n+1} = j, T_{n+1} - T_n \leq u | X_n = i\}$$ ∎

Thus given X_n, the random vector $(X_{n+1}, T_{n+1} - T_n)$ is independent of anything else in the past. Note that this property holds for the embedded DTMC and the jump times of a CTMC (see Theorem D.12). In a CTMC, however, $X_{(n+1)}$ and $(T_{n+1} - T_n)$ are independent given X_n. Hence an MRP is a generalization. Furthermore, in an MRP the state sojourn times $(T_{n+1} - T_n), n \geq 0$, need not be exponential.

We define $p_{i,j} = \lim_{u \to \infty} \Pr\{X_{n+1} = j, T_{n+1} - T_n \le u | X_n = i\}$, assuming that the limit does not depend on n. Then $X_n, n \ge 0$, is a DTMC on S with transition probabilities $p_{i,j}, i, j, \in S$.

We also define, for $i, j, \in S$,

$$H_{i,j}(u) = \Pr((T_{n+1} - T_n) \le u | X_n = i, X_{n+1} = j)$$

Thus $H_{i,j}(u)$ is the distribution of the sojourn time in a state given this state and the state entered at the end of this sojourn. Then the following theorem can be proved.

Theorem D.16

$$\Pr((T_1 - T_0) \le u_1, (T_2 - T_1) \le u_2, \ldots, (T_n - T_{n-1}) \le u_n | X_0, X_1, \ldots, X_n)$$
$$= \Pi_{i=1}^{n} H_{X_{i-1}, X_i}(u_i) \qquad \blacksquare$$

This result asserts that the sequence of sojourn times are independent given the sequence of states at the end points. Thus we can construct a sample path of an MRP by first generating a sample of the embedded DTMC and then filling in the sojourn times in the states by using the distributions $H_{i,j}(u)$.

The distribution of the sojourn time in state i is given by

$$H_i(u) = \sum_{j \in S} p_{i,j} H_{i,j}(u)$$

and the mean sojourn time in state i, denoted by σ_i, is given by

$$\sigma_i = \sum_{j \in S} p_{i,j} \sigma_{i,j}$$

where $\sigma_{i,j}$ is the mean of the distribution $H_{i,j}(u)$.

Consider now a reward $R_k \ge 0$ associated with the interval (T_{k-1}, T_k), for $k \ge 1$, such that R_k is independent of anything else given (X_{k-1}, X_k) and $(T_k - T_{k-1})$ (the states at the end points of and the length of the interval (T_{k-1}, T_k)). Let r_j be the expected reward in an interval that begins in the state j. Let $C(t)$ be the cumulative reward until time t. Suppose that $X_k, k \ge 0$, is a positive recurrent DTMC on S, with stationary probability $\pi_j, j \in S$. Then, as in the renewal reward theorem (Theorem D.15), under the condition that $\sum_{j \in S} \pi_j \sigma_j < \infty$, it can be shown that

$$\lim_{t \to \infty} \frac{C(t)}{t} = \frac{\sum_{j \in S} \pi_j r_j}{\sum_{j \in S} \pi_j \sigma_j}$$

where the convergence is in the almost sure sense.

D.1.6   The Excess Distribution

Given a nonnegative random variable X with distribution $F(x)$ and with finite mean $EX = \int_0^\infty (1 - F(u)) du$, define a distribution $F_e(\cdot)$ as follows:

$$F_e(y) := \frac{1}{EX} \int_0^y (1 - F(u)) du$$

Clearly, $F_e(\cdot)$ has the properties of a probability distribution function; in particular, it is nondecreasing in its argument, and $\lim_{y \to \infty} F_e(y) = 1$. The term *excess* distribution, or *excess life* distribution, comes from the following fact. Consider a renewal process with i.i.d. lifetimes $X_k, k \geq 1$, with common distribution $F(\cdot)$, and with finite mean lifetime. Let $Y(t)$ be the *residual life* or *excess life* at time t; that is, at t, $Y(t)$ is the time until the first renewal in (t, ∞). Consider

$$\lim_{t \to \infty} \frac{1}{t} \int_0^t I_{\{Y(u) \leq y\}} du$$

or

$$\lim_{t \to \infty} \frac{1}{t} \int_0^t \Pr(Y(u) \leq y)\, du$$

The first expression is the long-run fraction of time that the excess life is less than or equal to y, and the second expression is the time average probability of the excess life being less than or equal to y. It can be shown, using Theorem D.15, that in each case these limits exist and are equal to $F_e(y)$. The first expression converges to this limit in the almost sure sense; the second converges as an ordinary limit of real numbers. Thus $F_e(\cdot)$ can be interpreted as the residual life distribution seen by a random observer of the renewal process.

D.1.7   Phase Type Distributions

Consider a CTMC $X(t)$ on the state space $\{1, 2, \ldots, M, a\}$ such that the set of states $\{1, 2, \ldots, M\}$ is transient, and a is an absorbing state. The transition rate matrix of $X(t)$ is thus of the form

$$\begin{bmatrix} \mathbf{Q} & \mathbf{q} \\ \mathbf{0} & 0 \end{bmatrix}$$

where $\mathbf{Q}$ is an $M \times M$ matrix, $\mathbf{q}$ is a column vector of size M, and $\mathbf{0}$ is a row vector of size M. There is a probability vector $\boldsymbol{\alpha}$ of size M (thus, $0 \le \alpha_i \le 1, \sum_{i=1}^{M} \alpha_i = 1$). The Markov chain starts in state j, $1 \le j \le M$, with probability α_j, and then evolves to absorption in a. The distribution of this time until absorption is said to be *phase type* with parameters $(\boldsymbol{\alpha}, \mathbf{Q})$. During its evolution from its initial state, the $X(t)$ enters the various states in $\{1, 2, \ldots, M\}$; when it is in state j, it is said to be in the *phase j*.

Example D.3

a. If $\boldsymbol{\alpha} = (1, 0, 0, \ldots, 0)$, $Q = \begin{bmatrix} -\mu & \mu & 0 & \cdots & 0 & 0 \\ 0 & -\mu & \mu & 0 & \cdots & 0 \\ 0 & 0 & \ddots & \ddots & 0 & \vdots \\ \vdots & \vdots & & \ddots & \ddots \\ 0 & 0 & 0 & 0 & -\mu & \mu \\ 0 & 0 & 0 & 0 & 0 & -\mu \end{bmatrix}$, and

$\mathbf{q} = (0, 0, \ldots, \mu)^T$, then the phase type distribution is just an Erlang distribution of order M, with each stage being exponentially distributed with mean $\frac{1}{\mu}$.

b. For an arbitrary α, and $Q = \begin{bmatrix} -\mu_1 & & 0 & \cdots & 0 & 0 \\ 0 & -\mu_2 & 0 & 0 & \cdots & 0 \\ 0 & 0 & \ddots & 0 & 0 & \vdots \\ \vdots & \vdots & & \ddots & 0 & 0 \\ 0 & 0 & 0 & 0 & -\mu_{M-1} & 0 \\ 0 & 0 & 0 & 0 & 0 & -\mu_M \end{bmatrix}$, and

$\mathbf{q} = (\mu_1, \mu_2, \ldots, \mu_M)^T$, then the phase type distribution is a mixture of M exponential distributions and hence is a *hyperexponential* distribution. ■

One reason that phase type distributions are important is that they can be used to approximate arbitrarily closely (in the sense of convergence in distribution) any given distribution. Although this fact may not always be useful for numerical approximation (because the number of phases required for a good approximation can become very large), it is very useful for theoretical purposes. We can often prove results using phase type distributions because they have a simple structure, and then we can prove that the result holds for any distribution by considering a sequence of phase type distributions converging to the general distribution. As an example of where this approach is used, see Remark D.2 later in this Appendix.

Overflow Process of the M/M/c/c System

Consider the M/G/c/c queue discussed later in Section D.2.1, and take the special case in which the holding times are exponentially distributed; this is the M/M/c/c queue. Now consider an instant at which an arrival occurs and cannot be assigned a server. At this instant the number of customers in the system is c. The residual holding times of the customers in service are i.i.d. and exponentially distributed with means $\frac{1}{\mu}$, and the future of the arrival process is independent of anything else because it is Poisson. Hence the sequence of times at which customers are denied service forms a renewal process. Furthermore, observe that the distribution of these times is phase type with

$$
\boldsymbol{\alpha} = (0,0,\ldots,0,1), \; Q =
\begin{bmatrix}
-\lambda & \lambda & 0 & \cdots & & 0 & 0 \\
\mu & -(\lambda+\mu) & \lambda & 0 & \cdots & & 0 \\
0 & \mu & \ddots & \ddots & & 0 & \vdots \\
\vdots & \vdots & \ddots & \ddots & & \ddots & 0 \\
0 & 0 & 0 & \mu & -(\lambda+\mu) & & \lambda \\
0 & 0 & 0 & 0 & & \mu & -(\lambda+\mu)
\end{bmatrix}, \text{ and}
$$

$$
\mathbf{q} = (0,0,\ldots,\lambda)^T.
$$

D.1.8   Poisson Arrivals See Time Averages (PASTA)

In a discrete event model, we may need to ask a question about a process $X(t), t \geq 0$, if it is observed only at a sequence of random time points $t_k, k \geq 0$. How does the answer to such a question relate to the process observed over all time? For example, consider a stable D/D/1 queue. In such a system, customers arrive periodically at intervals of length a and require a service time to $b < a$. Let $X(t)$ be the number of customers in the system at time t. Note that $X(t) \in \{0, 1\}$. Clearly, the following holds.

$$\lim_{t \to \infty} \frac{1}{t} \int_0^t X(u)du = \frac{b}{a}$$

That is, the average number of customers *over all time* is $\frac{b}{a}$. Let $t_k = ka$, for $k \geq 0$, and let us look at

$$\lim_{n \to \infty} \frac{1}{n} \sum_{k=0}^{n-1} X(t_{k-})$$

where $X(t_{k-})$ means $X(\cdot)$ observed just before t_k. Thus this is the average number of customers *seen by arrivals*. This is clearly 0, because arrivals always find the system empty (in a stable D/D/1 queue the preceding arrival always leaves before the next arrival comes). However, when the instants at which the process is observed form a Poisson process, and under an additional independence assumption, the time averages observed at these Poisson points are the same as the averages over all time. We now state this formally.

Let $X(t), t \geq 0$, be a random process. Let B denote a set in the state space of $X(t)$. Let $A(t)$ be a Poisson process of rate λ; $A(t)$ denotes the number of Poisson arrivals in $(0, t]$, and $t_k, k \geq 1$, denotes the points of the Poisson process. Define

$$V^B(t) = \frac{1}{t} \int_0^t I_{\{X(u) \in B\}} du$$

That is, $V^B(t)$ is the fraction of time over $[0, t]$ that the process $X(\cdot)$ is in the set B. Also, define

$$V_A^B(t) = \frac{1}{A(t)} \sum_{k=1}^{A(t)} I_{\{X(t_{k-}) \in B\}}$$

That is, $V_A^B(t)$ is the fraction of arrivals over $(0, t]$ that see the process in the set B (not counting the arriving customer). We are interested in relating the limit of $V_A^B(t)$ (i.e., the fraction of customers that find $X(\cdot)$ in the set B) to the limit of $V^B(t)$ (i.e., the fraction of time that the process $X(\cdot)$ spends in the set B). An independence assumption is required.

Lack of Anticipation Assumption: For all $t \geq 0$, $A(t+u) - A(t), u \geq 0$, is independent of $X(s), 0 \leq s \leq t$; that is, for all $t \geq 0$, the future arrivals (more precisely, the future increments of the arrival process) are independent of the past of the process $X(\cdot)$.

Remark: When a system receives external arrivals as independent Poisson processes, the foregoing independence assumption clearly holds. This follows because the Poisson process has independent increments. $X(s), 0 \leq s \leq t$, depends on the arrival process up to t, which is independent of the increments of the arrival process after t (see Section B.3.4.)

Theorem D.17 PASTA

Under the lack of anticipation assumption, $V_B(t) \overset{w.p. \ 1}{\to} \bar{V}_B$ if and only if $V_A^B(t) \overset{w.p. \ 1}{\to} \bar{V}_B$; that is, the time average and the arrival average converge with probability 1, and to the same values. ∎

As an application, consider the M/G/c/c model that is used for modeling telephone calls using a trunk group of c trunks. The arrival rate is λ, the mean holding time is $\frac{1}{\mu}$, and $\rho := \lambda\mu$. If $N(t)$ is the number of trunks occupied at time t, then Markov chain analysis (see Section D.2.1) yields, for $0 \leq n \leq c$,

$$\frac{1}{t} \int_0^t I_{\{N(u)=n\}} du \overset{w.p. \ 1}{\to} \frac{\frac{\rho^n}{n!}}{\sum_{j=0}^c \frac{\rho^j}{j!}} =: \pi_n$$

Because the arrival process is Poisson and is independent of the future evolution of the $N(u)$ process, we can conclude from PASTA (i.e., Theorem D.17) that

$$\frac{1}{A(t)} \sum_{k=1}^{A(t)} I_{\{N(t_{k-})=c\}} \overset{w.p. \ 1}{\to} \pi_c$$

The left side is the number of arrivals that find all c trunks occupied and hence is the probability of call blocking.

Bernoulli Arrivals See Time Averages

We often need to work with queueing processes that evolve at discrete times, $t_k = kT, k = 0, 1, 2, \ldots$. Let X_k denote a discrete time queue embedded at the instants t_k. There is a Bernoulli arrival process of rate p; that is, at times t_{k+}

an arrival can occur with probability p. This can also be called a geometric arrival process, because the interarrival times are geometrically distributed (the interarrival time is i with probability $(1 - p)^{i-1}p$, $i = 1, 2, 3, \ldots$). The mean interarrival time is $\frac{1}{p}$, and the arrival rate is p (arrivals per time interval). As can be expected, a result similar to PASTA holds for Bernoulli or geometric arrival processes and can be called GASTA: geometric arrivals see time averages; that is, under the lack of anticipation assumption, the fraction of arrival instants at which the process X_k is in a set of states (not counting the arrival) is the fraction of instants, t_k, at which the process is in that set of states.

D.1.9   Level Crossing Analysis

To obtain performance measures for a positive recurrent Markov chain, we usually must derive the stationary probability distribution. This can be quite difficult to do, in general, and hence it is useful to have ancillary equations that can provide some information about the stationary distribution. One way to obtain such equations is via the idea of analyzing level crossing rates.

Consider a random process $X(t)$ on $[0, \infty)$. Assume that $X(t)$ has piecewise continuous sample paths. View $X(t)$ as evolving over time, and, for $x \geq 0$, define $U_x(t)$ to be the number of times the process crosses the "level" x from below (the up-crossing rate) in the interval $(0, t]$, and define $D_x(t)$ to be the number of times the process crosses level x from above (the down-crossing rate). Obviously, the following inequality holds:

$$|U_x(t) - D_x(t)| \leq 1$$

Now, dividing by t and letting $t \to \infty$, it is clear that if either of the limits $\lim_{t \to \infty} \frac{1}{t} U_x(t)$ or $\lim_{t \to \infty} \frac{1}{t} D_x(t)$ exists, then the other must also exist and the limits must be equal. In other words, the up-crossing rate and the down-crossing rate of any level must be equal. This, along with the fact that the limits can be written in terms of the stationary distribution, is the basic idea of level crossing analysis. We illustrate the utility of this approach by deriving Equation 1.1 in Chapter 1.

Example D.4   Derivation of Equation 1.1

Recall that Figure 1.2 shows a sample path of the multiplexer buffer level with bit-dropping at the multiplexer. When accepting an arrival that would cause the buffer level to exceed the level B, bits are dropped from the arriving and existing packets so that the buffer level just hits B. Immediately after this instant, the buffer begins to decrease at the rate of C bits per second. Suppose that the stationary distribution of the buffer process exists, and let $f(x), 0 < x \leq B$, denote the density function, and z the point mass at 0. Note that because the process does not spend any positive time at B there is no point mass at B, but $f(B) > 0$ because the process spends positive time in the neighborhood of B. We use level crossing analysis to obtain expressions for $f(\cdot)$ and z.

Consider any level $x, 0 < x < B$. In steady state, the rate of up-crossing of the level x is equal to the rate of down-crossing. It can be shown (see [43]) that the steady state down-crossing rate of the level x is given by $Cf(x)$. Here is a rough argument. Take a long period of time T. The amount of time in $(0, T)$ that the process spends in the interval $(x, x+dx)$ is approximately $f(x)dx \times T$. Every time that the process enters this interval it spends an amount of time $\frac{dx}{C}$ in the interval. Hence the number of times in $(0, T)$ that the process crosses the level x in the downward direction is approximately $\frac{f(x)dx \times T}{\frac{dx}{C}} = Cf(x)T$. Dividing this by T gives the required down-crossing rate. An up-crossing occurs only on an arrival and then occurs if at the arrival the buffer level is below x and if the arriving data is sufficient to cause the buffer to cross over x. Hence the up-crossing rate is given by

$$\lambda \int_0^x f(u)e^{-\mu(x-u)}du + \lambda z e^{-\mu x}$$

In this equation we have used PASTA (see Section D.1.8) to conclude that a Poisson arrival "sees" the steady state distribution of the buffer occupancy. The first term corresponds to an arrival finding the buffer level in the interval $(u, u + du)$, with probability $f(u)du$, and requiring a packet length of at least $(x - u)$ to cause an up-crossing of the level x; for an exponentially distributed packet length (with mean $\frac{1}{\mu}$) this probability is $e^{-\mu(x-u)}$. Integration from 0 to x takes care of all the cases that can cause an up-crossing of the level x. The second term accounts for arrivals that see an empty buffer. By PASTA, arrivals see an empty buffer with probability z, and such arrivals must bring a packet larger than x to cause an up-crossing of the level x. Equating the up-crossing rate to the down-crossing rate obtained earlier, we obtain the following equation. For every $x, 0 < x < B$,

$$\lambda \int_0^x f(u)e^{-\mu(x-u)}du + \lambda z e^{-\mu x} = Cf(x)$$

This equation can be solved using the Laplace transform to yield, for $0 < x \leq B$,

$$f(x) = \frac{\lambda}{C} z e^{\left(\mu - \frac{\lambda}{C}\right)x}$$

Now z can be obtained by using the normalization condition:

$$z + \int_0^B f(x)dx = 1$$

Using this, and defining $\rho := \frac{\lambda}{\mu C}$, we obtain

$$f(x) = \frac{\mu(1-\rho)\rho}{1 - \rho e^{-\mu(1-\rho)B}} \; e^{-\mu(1-\rho)x}$$

and

$$z = \frac{(1-\rho)}{1 - \rho e^{-\mu(1-\rho)B}}$$

which, with the subscript *bit-dropping*, are shown as Equations 1.1 and 1.2 in the text in Chapter 1.   ∎

D.2   Some Important Queueing Models
D.2.1   The M/G/c/c Queue

Consider a queueing system in which there are c servers, each of which serves at unit rate. Customers arrive in a Poisson process of rate $\lambda, 0 < \lambda < \infty$. The service requirements are i.i.d. and are generally distributed, with distribution $F(\cdot)$, with finite mean $\frac{1}{\mu}$, $0 < \mu < \infty$. Each arriving customer is assigned to a free server if one exists; otherwise, the arriving customer is denied admission and it goes away, never to be heard from again. Because a customer "holds" a dedicated server for the entire duration of its service, the service requirements are also called *holding times*. Let $X(t)$ denote the number of customers in the queue at time t. Define $\rho = \frac{\lambda}{\mu}$. Observe that ρ is the average number of new arrivals during the holding time of a customer.

Exercise D.3
Show that if the holding time distributions are exponential, then $X(t)$ is a positive recurrent CTMC on the state space $\{0, 1, \ldots, c\}$. The stationary distribution is given by

$$\pi_n = \frac{\frac{\rho^n}{n!}}{\sum_{j=0}^{c} \frac{\rho^j}{j!}} \tag{D.1}$$

When the holding times are not exponentially distributed, then $X(t)$ is not a Markov chain. When $X(t) = n$, let $Y_i(t), 1 \leq i \leq n$, denote the residual service requirements of the customers in the system. It is then easily seen that the process $(X(t), Y_1(t), \ldots, Y_{X(t)}(t))$ is a Markov process. To see this, observe that no more information is required to "evolve" the process; the arrivals come in an independent Poisson process, and each of the residual service times decreases at the rate 1. It can be shown that the stationary joint distribution of the process is given by

$$\Pr(X(t) = n, Y_1 \leq y_1, \ldots, Y_n \leq y_n) = \pi_n \, \Pi_{i=1}^{n} F_e(y_i) \tag{D.2}$$

where π_n is as displayed in Equation D.1, and $F_e(\cdot)$ is the excess distribution of the holding time distribution (see Section D.1.6).

D.2.2   The Processor Sharing Queue

Consider a queue with an infinite amount of waiting space. Customers arrive in a Possion process with rate λ. The service requirements are i.i.d. and are generally distributed with common distribution $F(\cdot)$, with finite mean. As long as there is work to be done, the server reduces the total amount of unfinished work at the rate of 1 unit per second. When there are n customers in the system, the unfinished work on the ith customer decreases at rate $\frac{1}{n}$. This is called the *M/G/1 processor sharing* (PS) queueing model; the G in the notation refers to the generally distributed service requirements. Let $X(t)$ denote the number of customers at time t.

Exercise D.4
If the service requirements are exponentially distributed with mean $\frac{1}{\mu}$, then show that $X(t)$ is a CTMC, which is positive recurrent if and only

> if $\frac{\lambda}{\mu} < 1$. In that case, defining $\rho := \frac{\lambda}{\mu}$, the stationary distribution of $X(t)$ is given by
>
> $$\pi_n = (1 - \rho)\rho^n$$

In general, when the service requirements are not exponentially distributed, $X(t)$ is not a Markov chain. When $X(t) = n$, let $Y_i(t), 1 \leq i \leq n$, denote the residual service requirements of the customers in the system. It is then easily seen that the process $(X(t), Y_1(t), \ldots, Y_{X(t)}(t))$ is a Markov process. To see this, observe that no more information is required to evolve the process; the arrivals come in an independent Poisson process, and each of the residual service times decreases at the rate $\frac{1}{n}$. Furthermore, if $\rho < 1$, it can be shown that the stationary joint distribution of the process is given by

$$\Pr(X(t) = n, Y_1 \leq y_1, \ldots, Y_n \leq y_n) = (1 - \rho)\rho^n \; \Pi_{i=1}^n F_e(y_i) \qquad \text{(D.3)}$$

where $F_e(\cdot)$ is the excess distribution of the service time distribution (see Section D.1.6).

Remarks D.2

a. From the form of the stationary distribution displayed in Equation D.3, we conclude that the residual service times are independent, conditional on the number of customers in the system.

b. Furthermore, the stationary residual service time distribution is just the excess life distribution of the service time distribution.

c. This important result can be established in at least two ways. The brute force (but very instructive) way is to write the partial differential equations for the stationary probability distribution of the process, and show that it has the displayed product form solution. An elegant approach is via time reversal arguments. We begin by considering the special case of the problem with a phase type service time distribution. The distribution in Equation D.3 can be specialized to this case. A time reversal of the queue is proposed, and then Lemma D.1 (presented later in this Appendix) is used to show that the proposed distribution is indeed the stationary distribution. The proof is completed by invoking the fact that general distributions can be approximated arbitrarily closely by phase type distributions.

d. Finally, a very important conclusion from Equation D.3 is that the stationary distribution of the number of customers in an M/G/1 PS queue (i.e., the marginal distribution of $X(t)$) is the same as that in an M/M/1 queue and, hence, is *insensitive* to the distribution of the service time (except through its mean). ∎

Sojourn Times

Because the stationary distribution of the number of customers in an M/G/1 PS queue is $\pi_n = (1 - \rho)\rho^n$, the time average number of customers is $E(X) = \frac{\rho}{1-\rho}$. Hence by Little's Theorem (Theorem 5.1) the mean sojourn time is

$$E(W) = \frac{E(S)}{1 - \rho}$$

where S is the service time random variable. Again, we see that the mean sojourn time in the M/G/1 PS queue is insensitive to the actual distribution of the service time (except through its mean). Furthermore, an important result is that the mean conditional sojourn time of a customer with service requirement s is given by

$$E(W|S = s) = \frac{s}{1 - \rho}$$

D.2.3　Symmetric Queues

Let us consider a single-station queue constructed as follows (see Figure D.1). Customers of class c, $c \in \mathcal{C}$, arrive in independent Poisson processes, of rates λ_c. Customers of class c have a phase type service requirement with parameter $(\boldsymbol{\alpha}_c, \mathbf{Q}_c)$, where, without loss of generality, there are M phases for each customer.

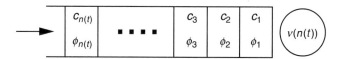

Figure D.1　A state of the queue with customers of several classes that have phase type service requirements. The state of the system keeps track of the position of each customer. The customer in position _i_ has class _c_i_ and is in service phase _ϕ_i_. When there are _n_ customers in the system, the service rate applied is _v_(_n_).

The mean service requirement of class c customers is $1/\mu_c$. Define

$$\rho = \sum_{c \in C} \frac{\lambda_c}{\mu_c}$$

When there are n customers in the queue, the service rate applied is $v(n)$ (for example, in an M/M/k queue with service rate v, $v(n) = v \min\{k, n\}$). There are given functions $\gamma(m, j)$ and $\delta(m, j)$ that govern the positions taken by arriving customers and the service effort obtained by customers in various positions. An arriving customer finding n customers in this system joins in position l, $1 \leq l \leq n+1$, with probability $\gamma(n+1, l)$ (it is thus required that $\sum_{l=1}^{n+1} \gamma(n+1, l) = 1$). The initial service phase of a class c customer is j with probability $\alpha_c(i)$. When there are n customers in the queue, a fraction $\delta(n, l)$ of the service effort (i.e., of $v(n)$) is applied to the customer at position l (it is thus required that $\sum_{l=1}^{n} \delta(n, l) = 1$).

Definition D.11

A queueing station, as described above, is said to be a *symmetric queue* if the functions $\gamma(\cdot, \cdot)$ and $\delta(\cdot, \cdot)$ are such that $\delta(n, l) = \gamma(n, l)$. ∎

Example D.5

The following are some examples of symmetric queues.

a. *The M/PH/1 queue with the last-come-first-served preemptive resume (LCFS-PR) discipline:* In such a queueing station there is a constant-rate server $v(n) = v$, $\gamma(n, 1) = 1$, and $\gamma(n, j) = 0, j > 1$; $\delta(n, 1) = 1$, and $\delta(n, j) = 0, j > 1$. The latter requirements on $\gamma(\cdot, \cdot)$ and $\delta(\cdot, \cdot)$ simply state that any arrival joins the first position, and all the service effort is applied to the first position. The preemptive resume property means that when a customer in service is "pushed back" by a new arrival, it retains its service phase, and when it returns to the service position, service *resumes* from the same phase.

b. *The M/PH/1 processor sharing queue (see also Section D.2.2):* There is a constant rate server $v(n) = v$. An arrival when there are n customers joins any of the $n + 1$ possible positions with equal probability (i.e., $\gamma(n, l) = \frac{1}{n}$, $1 \leq l \leq n$). The server applies an equal service effort to each customer in the system (i.e., $\delta(n, l) = \frac{1}{n}, 1 \leq l \leq n$). The customers can also be conceptually viewed as standing side-to-side, breadthwise in the buffer (rather than back-to-back, depthwise in the buffer); the server stands in front and services them equally.

c. *The M/PH/∞ queue:* There is one server per customer, and more servers can be assigned (and the extra ones released) as the number of customers increases or decreases. If each server has service rate v, we see that $v(n) = nv$. Also, it is clear that $\gamma(n, l) = \frac{1}{n}$, and $\delta(n, l) = \frac{1}{n}$. ∎

Theorem D.18

The stationary distribution of the number of customers in a symmetric queue is given by

$$\pi(x) = G \frac{\rho^{|x|}}{v(1)v(2) \ldots v(|x|)}$$

where $|x|$ denotes the number of customers in the state x, and G is the normalization constant. ∎

Remark: The key observation from this theorem is that the stationary distribution of the number of customers in a symmetric queue depends on the service requirement distributions of the various classes only through their means, $\frac{1}{\mu_c}, c \in \mathcal{C}$, and is thus said to be insensitive to the service requirement distributions.

D.3   Reversibility of Markov Chains, and Jackson Queueing Networks

D.3.1   Reversibility: Criteria and Consequences

Definition D.12

Given a process $\{X(t), \ t \in (-\infty, \infty)\}$, and given $\tau \in \mathbb{R}$, a *reversed process* $\{\tilde{X}(t), \ t \in (-\infty, \infty)\}$ is defined as $\tilde{X}(t) = X(\tau - t)$ (i.e., $X(t)$ time reversed and shifted by τ). ∎

Thinking in terms of sample paths, time reversal of a process is like playing a movie of the sample path backward.

Suppose that $\{X(t), \ t \in (-\infty, \infty)\}$ is a stationary regular CTMC, with state space S. Let $\mathbf{P}(t)$ denote the transition probability matrices for each $t \geq 0$. Let the stationary probability measure be $\pi_i, i \in S$, and let $\boldsymbol{\pi}$ denote the stationary probability vector. Consider the following probability of an event in the past given the present and the future:

$$\Pr(X(t - \tau) = j | X(s), s \geq t, \ X(t) = i)$$

$$= \frac{\Pr(X(t-\tau)=j)\ \Pr(X(s), s \geq t, X(t)=i | X(t-\tau)=j)}{\Pr(X(s), s \geq t, X(t)=i)}$$

$$= \frac{\Pr(X(t-\tau)=j)\ P_{j,i}(\tau)\Pr(X(s), s \geq 0 | X(0)=i)}{\Pr(X(t)=i)\ \Pr(X(s), s \geq 0 | X(0)=i)}$$

$$= \frac{\pi_j}{\pi_i}\ P_{j.i}(\tau)$$

where the second equality uses the Markov property (in forward time). Thus a time reversed stationary CTMC is again a CTMC with time homogeneous transition probabilities, which we denote by $\tilde{\mathbf{P}}(t), t \geq 0$, where

$$\tilde{p}_{i,j}(t) = \frac{\pi_j}{\pi_i}\ p_{j,i}(t)$$

and let $\mathbf{Q}$ denote the transition rate matrix. Because $\dot{\mathbf{P}}(0) = \mathbf{Q}$, differentiating the foregoing equation at $t = 0$, we can relate the transition rate matrix $\tilde{\mathbf{Q}}$ of the time reversed CTMC to $\mathbf{Q}$, the transition rate matrix of the original CTMC. We obtain

$$\tilde{q}_{i,j} = \frac{\pi_j}{\pi_i}\ q_{j,i}$$

Now observe that

$$\sum_{i \in S} \pi_i \tilde{q}_{i,j} = \sum_{i \in S} \pi_i \frac{\pi_j}{\pi_i}\ q_{j,i}$$

$$= 0$$

Thus we have $\pi \tilde{\mathbf{Q}} = 0$, and hence π is also the stationary measure for the time reversed CTMC $\tilde{X}(t)$. In addition, we have

$$\sum_{j \in S: j \neq i} \tilde{q}_{i,j} = \sum_{j \in S: j \neq i} \frac{\pi_j q_{j,i}}{\pi_i}$$

$$= \frac{-\pi_i\ q_{i,i}}{\pi_i} = \frac{\pi_i a_i}{\pi_i} = a_i$$

That is, the conditional rate of leaving i in the time reversed CTMC is a_i; equivalently, the mean time the reversed process spends in the state i, each time it enters i, is the same in the forward and reversed CTMCs. This is as expected because on reversing time, the sample path of a process does not change the amounts of time it spends in each state between jumps.

Proposition D.1

If $\{X(t), t \in \mathbb{R}\}$ is stationary, then a time reversed process $\{\tilde{X}(t), t \in \mathbb{R}\}$ is also stationary. ∎

Definition D.13

A process $\{X(t), t \in \mathbb{R}\}$ is said to be *reversible* if, for all $\tau \in \mathbb{R}$, $\{\tilde{X}(t) = X(\tau - t), t \in \mathbb{R}\}$ has the same probability law as $\{X(t), t \in \mathbb{R}\}$. ∎

Remark: Thus any probabilistic question asked about a reversible process has the same answer as the same question asked about *any* time reversal of the process.

Proposition D.2

A reversible process is stationary (and consequently any time reversal of a reversible process is stationary). ∎

The following result provides a condition for the reversibility of a stationary regular CTMC in terms of its stationary probability distribution and the transition rate matrix, Q.

Theorem D.19

A regular CTMC is reversible if and only if there exists a probability distribution π on S such that, for all i, and $j \in S$,

$$\pi_i q_{i,j} = \pi_j q_{j,i}$$

Such a π will also be the stationary probability vector of the CTMC. ∎

Remarks D.3

a. $\sum_{i \in S} \pi_i q_{i,j} = 0$, for all $j \in S$, are the well-known *full balance equations*, which are used to obtain the stationary probability vector π. In Theorem D.19, the equations $\pi_i q_{i,j} = \pi_j q_{j,i}$ are called *detailed balance equations*; it is easily seen, by summing them, that they imply the full balance equations. These equations are sometimes also called *local balance* equations.

b. It follows that if a CTMC $X(t)$ is reversible, then for two states $i, j \in S$, $q_{i,j}$ and $q_{j,i}$ are either both zero or both positive.

c. Consider an irreducible CTMC whose transition rate matrix is such that $q_{i,j} > 0$ if and only if $q_{j,i} > 0$ (see remark (b)). In the transition rate diagram of this CTMC, consider each pair of nodes between which there are directed edges, and replace each such pair of directed edges with an undirected edge. Let us suppose that the resulting graph is a *tree*; so, for example, a tree would be obtained from the transition diagram shown in Figure D.2. Now consider any pair of states i and j, such that $q_{i,j} > 0$ and $q_{j,i} > 0$. Let A and A^c be the sets into which the state space, S, is partitioned by the removal of the arcs between the states i and j in the transition graph of the CTMC; again, see Figure D.2. Because the transition rate diagram is a tree, there is only one edge connecting A and A^c. For a stationary CTMC, the rate of leaving A to enter A^c is equal to the rate of leaving A^c to enter A:

$$\pi_i q_{i,j} = \pi_j q_{j,i}$$

This allows us to conclude, from Theorem D.19, that *a CTMC whose transition rate diagram is a tree is reversible*. In particular the Markov chain of a birth-and-death process is reversible, because its transition graph is only

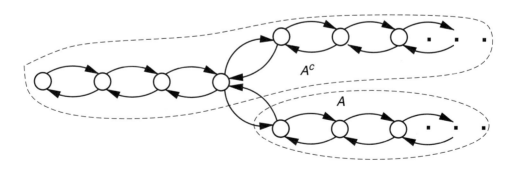

Figure D.2   An example of a tree transition rate structure for a CTMC.

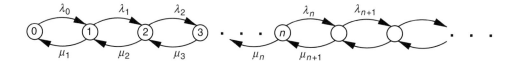

Figure D.3   The transition structure of a birth-and-death CTMC.

a chain of states (see Figure D.3). Thus, for example, the queue-length process of a stationary M/M/c queue is reversible.   ∎

The condition in Theorem D.19 requires us to first obtain, or guess, the stationary probability vector π. The following is a criterion for the reversibility of a CTMC that is only in terms of the transition rates of the CTMC.

Theorem D.20   Kolmogorov's Criterion
Let $X(t)$ be an irreducible regular positive CTMC. $X(t)$ is reversible iff for any finite sequence of states $i_i, i_2 \dots, i_m$ in S,

$$q_{i_1,i_2} \times q_{i_2,i_3} \times q_{i_3,i_4} \times \cdots \times q_{i_m,i_1} = q_{i_1,i_m} \times q_{i_m,i_{m-1}} \times q_{i_{m-1},i_{m-2}} \times \cdots \times q_{i_2,i_1}$$

∎

For a CTMC that satisfies the condition in Remark D.3(b), consider the undirected graph as constructed in Remark D.3(c). Note that Theorem D.20 requires one to check Kolmogorov's condition along every cycle in this undirected graph. When the graph is planar, however, then the verification of Kolmogorov's criterion is simplified, as the following theorem states.

Theorem D.21
A positive Markov process $X(t)$ with a planar (undirected) transition graph is reversible if the Kolmogorov criterion holds for every minimal closed path in the graphs (i.e., one that encircles a connected domain).   ∎

Consider a regular CTMC $X(t)$ on the state space S, and let $E \subset S$. By the restriction of $X(t)$ to E we mean the CTMC obtained by retaining only the transitions within the set E. Hence if $\mathbf{Q}$ is the transition rate matrix of the original

CTMC, then the restriction of $X(t)$ to E has the transition rate matrix Q_E, of size $|E| \times |E|$, where $[Q_E]_{i,j} = [Q]_{i,j}$ whenever $i, j \in E$.

Theorem D.22

If a reversible CTMC with stationary distribution π is restricted to the state space $E \subset S$, then the restricted Markov chain is reversible and has stationary distribution $\frac{\pi_j}{\sum_{i \in E} \pi_i}$ for all $j \in E$—that is, the restriction of the probability distribution π to the set E.   ■

Proof: If $i, j \in E$, then, because the original CTMC is reversible, by Theorem D.19, $\pi_i q_{i,j} = \pi_j q_{j,i}$. Because π is the stationary measure, $\sum_{i \in E} \pi_i > 0$. Hence we see that $\frac{\pi_j}{\sum_{i \in E} \pi_i}, j \in E$, is a probability measure on E and satisfies the detailed balance condition. Hence, by Theorem D.19, the truncated Markov chain is reversible and $\frac{\pi_j}{\sum_{i \in E} \pi_i}, j \in E$, is its stationary measure.   ■

Example D.6

We illustrate Theorem D.22 with an example (see Figure D.4). Two Poisson processes, with rates λ_1 and λ_2, arrive to separate buffers, each of infinite size. Each process of arrivals has its own server, and the service times are exponential, with means $\frac{1}{\mu_1}$ and $\frac{1}{\mu_2}$. Define $\rho_i = \frac{\lambda_i}{\mu_i}, i = 1, 2$, and assume that $\rho_i < 1, i = 1, 2$. The joint process of queue lengths $(X_1(t), X_2(t))$ is a CTMC. These are clearly two independent M/M/1 queue processes, and the stationary distribution is given by

$$\pi_{n_1, n_2} = (1 - \rho_1)\rho_i^{n_1}(1 - \rho_2)\rho_2^{n_2}$$

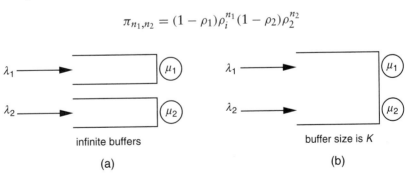

infinite buffers

(a)

buffer size is K

(b)

Figure D.4   (a) Two arrival processes with separate servers and separate infinite-size buffers; (b) the same arrival processes with the same servers, except that they share a buffer of size K.

It is easily checked that the process $(X_1(t), X_2(t))$ is reversible (for example, check the detailed balance condition). Now suppose that the same arrival processes, each with its own server, share a common buffer of size K (see Figure D.4(b)). Notice that this is simply the restriction of the original CTMC to the set $E = \{(n_1, n_2) \geq 0 : n_1 + n_2 \leq K\}$. Hence, applying Theorem D.22, we find that the restricted CTMC is reversible and has stationary distribution:

$$v_{n_1, n_2} = \frac{(1 - \rho_1)\rho^{n_1}(1 - \rho_2)\rho^{n_2}}{\sum_{(k_1, k_2) \in E} \pi_{k_1, k_2}}$$

∎

Birth-and-Death Processes: Poisson Departures

Consider a stationary birth-and-death process, $X(t)$, with constant arrival rate λ, and death rate μ_j when in state j (see Figure D.3). Denote by $\boldsymbol{\mu}(= (\mu_1, \mu_2, \ldots,))$ the vector of all the death rates, and call this a $(\lambda, \boldsymbol{\mu})$ process. The queue-length process of an M/M/c is an example of such a birth-and-death process with $\mu_j = j\mu$, where $\frac{1}{\mu}$ is the mean service time of any customer.

Being a birth-and-death process, $X(t)$ is reversible. Hence a time reversal of $X(t)$—say, $\tilde{X}(t)$—is also a $(\lambda, \boldsymbol{\mu})$ process (because, by the definition of reversibility, $X(t)$ and $\tilde{X}(t)$ have identical probability laws). Hence the process of arrivals (upward jumps) in $\tilde{X}(t)$ is a Poisson process of rate λ. Now, consider any departure in $X(t)$ (this is a downward jump in the process); when time is reversed, this departure, in forward time, will correspond to an upward jump in reversed time, and hence to an arrival in the time reversed process $\tilde{X}(t)$. It follows that the point process of departures in $X(t)$ is the time reversal of the point process of arrivals in $\tilde{X}(t)$. Hence the departure process in $X(t)$ is the time reversal of a Poisson point process of rate λ, which can easily be argued to be a Poisson point process (of rate λ). Thus we see that the departure process of a "stable" M/M/c queue— and, in particular, an M/M/1 queue—is a Poisson process of rate equal to the arrival rate.

More can be concluded from such arguments. Let us denote by $\Lambda(t)$ the arrival process in $X(t)$, and by $\Omega(t)$ the departure process. Similarly define $\tilde{\Lambda}(t)$ and $\tilde{\Omega}(t)$. In a Poisson process, past arrivals are independent of future arrivals, and the past of $\tilde{X}(t)$ depends only on past arrivals. We can conclude that, for all t,

$$\left\{\left(\tilde{\Lambda}(t + \tau) - \tilde{\Lambda}(t)\right), \tau \geq 0\right\} \amalg \{\tilde{X}(s), s \leq t\}$$

That is, in reversed time, the past of the queue-length process is independent of the future of the arrival process. Now using this information in the forward time process we can conclude that the past of the departure process is independent of the future of the queue-length process:

$$\{\Omega(u),\ u \le t\} \amalg \{X(s),\ s \ge t\}$$

We have thus established the following result.

Theorem D.23

The departure process, $\Omega(t)$, of a stationary (λ, μ) process, $X(t)$, is Poisson at rate λ. Furthermore, for all t, $\{\Omega(u), u \le t\}$ is independent of $\{X(s), s \ge t\}$.   ■

Queues in Tandem

Consider the situation shown in Figure D.5. Assume that $\lambda < \min\{\mu_1, \mu_2\}$. Denote by $X_i(t), i = 1, 2$, the two queue-length processes. Because $\lambda < \mu_1$, $X_1(t)$ is a positive recurrent CTMC, with stationary distribution $(1 - \rho_1)\rho_1^{n_1}$, where $\rho_1 = \lambda/\mu_1$. Let us consider the stationary version of $X_1(t)$. Then, by Theorem D.23, its departure process, $\Omega_1(t)$, is Poisson with rate λ, and furthermore $X_1(t)$ is independent of the past of $\Omega_1(t)$. Because $X_2(t)$ depends only on the past of the departures of $X_1(t)$, and because the service times at the two queues are independent, we see that $X_1(t)$ is independent of $X_2(t)$. Furthermore, because the arrival process into the second queue is a Poisson process, we see that

$$\lim_{t \to \infty} \Pr(X_2(t) = n_2) = (1 - \rho_2)\rho_2^{n_2}$$

Figure D.5   A Poisson arrival process into two queues in tandem, each with an exponential service time.

where $\rho_2 = \lambda/\mu_2 < 1$. Hence

$$\Pr(X_1(t) = n_1, X_2(t) = n_2) = \Pr(X_1(t) = n_1) \Pr(X_2(t) = n_2)$$

$$= (1 - \rho_1)\rho_1^{n_1} \Pr(X_2(t) = n_2)$$

$$\overset{t \to \infty}{\to} (1 - \rho_1)\rho_1^{n_1}(1 - \rho_2)\rho_2^{n_2}$$

Thus the stationary joint distribution of $(X_1(t), X_2(t))$ has a product form:

$$\pi_{n_1,n_2} = (1 - \rho_1)\rho_1^{n_1}(1 - \rho_2)\rho_2^{n_2}$$

Furthermore, the departure process of the tandem of queues is Poisson at rate λ. It is important to note that if $X_1(t)$ is not its stationary version, then it is not true that $X_1(t)$ and $X_2(t)$ are independent. The asymptotic result, however, still holds.

D.3.2   Jackson Networks

In Section D.3.1 we saw that a simple "network" of two M/M/1 queues in tandem, under suitable stability conditions, has a *product form* stationary distribution. This result can be generalized significantly. The simplest such generalization is to the class of Jackson networks.

Consider a queueing network consisting of J stations, indexed by $j = 1, 2, \ldots, J$, and let 0 denote the outside world (i.e., the source and sink of customers). Suppose that the vector of queue-length processes $\mathbf{X}(t) = (X_1(t), \ldots, X_J(t))$ is a CTMC. This would be the case, for example, if the arrival processes are independent Poisson processes, the service times at each queue are exponentially distributed, and on leaving a queue a customer probabilistically chooses to move to another queue or leave the system, the choice being made independent of anything else except its present location. Denote a state of $\mathbf{X}(t)$ by the vector $\mathbf{n} = (n_1, n_2, \ldots, n_J)$. Let the state space be denoted by S.

Let $q_{\mathbf{m},\mathbf{n}}$ denote the transition rate of the CTMC from state $\mathbf{m}$ to state $\mathbf{n}$, and let $\pi_{\mathbf{n}}, \mathbf{n} \in S$, denote the stationary probability distribution (when one exists). Let $\pi_k^{(j)} = \sum_{\{\mathbf{n}:n_j=k\}} \pi_{\mathbf{n}}$ denote the *marginal* stationary distribution of queue j.

Let $\mathbf{e}_i = (0, 0, \ldots, 1, 0, \ldots, 0)$, with the 1 in position i denoting a unit vector of length J.

Transitions in the CTMC $\mathbf{X}(t)$ occur only when there is an arrival into the system or there is a service completion. The possible transitions out of state $\mathbf{n}$ are denoted by the following:

$$S_{ij}(\mathbf{n}) = \mathbf{n} - \mathbf{e}_i + \mathbf{e}_j \text{ assuming } n_i > 0$$

$$A_i(\mathbf{n}) = \mathbf{n} + \mathbf{e}_i$$

$$D_i(\mathbf{n}) = \mathbf{n} - \mathbf{e}_i \text{ assuming } n_i > 0$$

In general, let us denote the transition rates out of state $\mathbf{n}$ by

$$q_{\mathbf{n},S_{ij}(\mathbf{n})} = \mu_{i,j}(\mathbf{n})$$

$$q_{\mathbf{n},A_i(\mathbf{n})} = \lambda_i(\mathbf{n})$$

$$q_{\mathbf{n},D_i(\mathbf{n})} = \mu_{i,0}(\mathbf{n})$$

It follows that the total service rate at station i when the system state is $\mathbf{n}$ is $\mu_i(\mathbf{n}) = \sum_{j=0}^{J} \mu_{ij}(\mathbf{n})$.

A *Jackson network* is the following special case of the foregoing general Markov queueing network.

$$\lambda_i(\mathbf{n}) = \lambda_i$$

$$\mu_{i,0}(\mathbf{n}) = \mu_i(n_i)r_{i,0}$$

$$\mu_{i,j}(\mathbf{n}) = \mu_i(n_i)r_{i,j}$$

where $\sum_{j=0}^{J} r_{ij} = 1$; $0 \leq r_{i,j} \leq 1$. These requirements are equivalent to the following:

- The arrival processes into the queues are independent Poisson processes with rates λ_j, $1 \leq j \leq J$. Let $\Lambda = \sum_{j=1}^{J} \lambda_j$.

- When the system state is $\mathbf{n}$, the service rate at queue j is $\mu_j(n_j)$, and hence it is allowed to depend on the number of customers at queue j, but not on the other queues.

- When a customer finishes service at queue i it moves to queue j with probability $r_{i,j}$ or leaves the network with probability $r_{i,0}$. Note that the routing probability depends on the current position of the customer, but not on the path that the customer followed to arrive at this queue. For this reason this is called the *Markov routing* model.

Exercise D.5

  a. Show that a network of M/M/c queues with Markov routing is a Jackson network.

  b. Let Q_k denote the sequence of queues visited by a given customer. Show that Q_k is a discrete time Markov chain.

Given a Jackson network, define the $J \times J$ matrix $\mathbf{R}$ by letting its elements be $r_{i,j}, 1 \le i, j \le J$. Two cases arise.

Open Jackson Networks

When $\Lambda > 0$, let $\mathbf{v}$ be a column vector of size J, with elements $v_j = \frac{\lambda_j}{\Lambda}$; that is, v_j is the fraction of external arrivals that enter queue j. Furthermore, define the vector $\mathbf{u}$ of size J by $u_j = r_{j,0}$. Then consider the following $(J+1) \times (J+1)$ matrix:

$$\tilde{\mathbf{R}} = \begin{bmatrix} 0 & \mathbf{v}^T \\ \mathbf{u} & \mathbf{R} \end{bmatrix}$$

Let us assume that this matrix is irreducible. This means that a customer entering the system and encountering finite delay at each queue will eventually leave the system in finite time. Such a queueing network is called an *open Jackson network*. It can be checked that the CTMC $\mathbf{X}(t)$ is also irreducible. Suppose, in addition, that $\mathbf{X}(t)$ is positive recurrent. Then the throughput of each component queue j is well defined, and let us denote it by γ_j; this is the rate at which customers pass through queue j, in steady state. Then by equating the flow of customers out of queue j and into queue j, we obtain

$$\gamma_j = \lambda_j + \sum_{i=1}^{J} \gamma_i r_{i,j}$$

or

$$\gamma = \lambda + \gamma R$$

where γ and λ are the vectors of throughputs and arrival rates. Equivalently,

$$\gamma(I - R) = \lambda$$

where, as usual, I is the identity matrix. Because $\tilde{R}$ has been assumed to be irreducible, $(I - R)$ can be shown to be nonsingular. To see this, note that it is sufficient to argue that $(\sum_{k=1}^{\infty} R^k) \cdot 1 < \infty$; but this is true because the jth element of this vector is the mean number of queues visited by a customer arriving into queue j before eventually leaving the system, and this is finite. It follows that we can solve for the throughputs, to obtain

$$\gamma = \lambda(I - R)^{-1} \tag{D.4}$$

and this solution is unique. Define

$$g_i^{-1} = 1 + \sum_{k=1}^{\infty} \frac{(\gamma_i)^k}{\mu_i(1) \dots \mu_i(k)}$$

The following theorem establishes the condition for the stability of an open Jackson network and also shows that the stationary joint distribution of the queue lengths has a product form.

Theorem D.24   Jackson
Let γ be the unique solution of $\gamma = \lambda + \gamma R$. The process $X(t)$, $t \geq 0$, is positive recurrent iff $g_j > 0$ (i.e., $(g_j^{-1} < \infty)$, or all j). The stationary distribution is of the form

$$\pi_{\mathbf{n}} = \prod_{j=1}^{J} \pi_{n_j}^{(j)}$$

where

$$\pi_0^{(j)} = g_j$$

$$\pi_k^{(j)} = g_j \frac{(\gamma_j)^k}{\mu_j(1) \dots \mu_j(k)}$$

Remarks D.4

a. To apply Theorem D.24, we proceed as follows. First calculate the steady state throughputs from the flow balance equations $\gamma = \lambda + \gamma R$. Now consider each queue in isolation, with a Poisson arrival process of rate γ_j, and with service rate $\mu_j(n)$, when there are n customers in this isolated queue. Check that this isolated queue is stable (i.e., positive recurrent). Note that the queue-length process of each isolated queue is a (γ_j, μ_j) process. If each such isolated queue is stable, then the joint queue-length process of the open Jackson network is positive recurrent, and the joint stationary distribution is the product of the stationary distributions of the queues that were examined in isolation.

b. It is important to note that the actual aggregate arrival process into a queue of an open Jackson network is *not* Poisson if customers can return to the queue immediately or after visiting other queues (i.e., the queue is part of a customer *feedback* loop). In *feedforward* open Jackson networks, however, the input processes into queues are Poisson. The tandem of M/M/1 queues examined in Section D.3.1 is a simple example of a feedforward open Jackson network.

c. Although the stationary distribution of an open Jackson network has the product form, if the network starts from some arbitrary initial state, then the queues are not independent at any finite time. See also the example of two M/M/1 queues in tandem at the end of Section D.3.1.

Proof of Jackson's theorem: One way to prove Theorem D.24 is to simply check that the proposed probability distribution satisfies the detailed balance equations. It is also possible to identify certain *partial balance* equations, the satisfaction of which is easier to check, and which imply the detailed balance equations. Here we provide the proof based on the following lemma.

Lemma D.1

Consider an irreducible regular CTMC $X(t)$, on the state space S, with transition rate matrix $\mathbf{Q}$. Suppose there exists a transition rate matrix $\tilde{\mathbf{Q}}$ and a probability distribution π on S, such that, for all $x, y \in S$,

$$\pi_x \tilde{q}_{x,y} = \pi_y q_{y,x}$$

Then

i. π is the stationary measure of the CTMC $X(t)$.

ii. $\tilde{\mathbf{Q}}$ is the rate matrix of the time reversal of the stationary version of the CTMC $X(t)$.

Proof of lemma: This is a simple exercise of checking the required conclusions. We now return to the proof of Jackson's theorem. With Lemma D.1 in mind, the idea is to produce a new open Jackson network whose transition rate matrix $\tilde{\mathbf{Q}}$ satisfies the hypothesis of the lemma, along with the proposed product form distribution. To this end, consider an open Jackson network whose external arrival rates and routing probabilities are related to the original Jackson network as follows:

$$\tilde{\lambda}_i = \gamma_i r_{i,0}$$

$$\tilde{r}_{i,0} = \frac{\lambda_i}{\gamma_i}$$

$$\tilde{r}_{i,j} = \frac{\gamma_j r_{j,i}}{\gamma_i}$$

Note that these definitions also specify the routing probabilities in the proposed network. It can be checked that the queue throughputs in this network and the original one remain the same. The service rates in the new network are the same as in the original network (i.e., $\mu_j(n_j)$).

Let $\tilde{\mathbf{Q}}$ denote the transition rate matrix of this Jackson network. Now we can verify that for the proposed stationary distribution, the following holds for all $x, y \in S$,

$$\pi_x \tilde{q}_{x,y} = \pi_y q_{y,x}$$

For example, consider $\mathbf{x}$ such that $x_i > 0$. Let us verify that

$$\pi_{\mathbf{x}} \tilde{q}_{\mathbf{x}, S_{i,j}(\mathbf{x})} = \pi_{S_{i,j}(\mathbf{x})} q_{S_{i,j}(\mathbf{x}), \mathbf{x}}$$

From the product form of $\pi_{\mathbf{x}}$, we then see that we need

$$\pi_{\mathbf{x}} \mu_i(x_i) \tilde{r}_{i,j} = \pi_{\mathbf{x}} \frac{\mu_i(x_i)}{\gamma_i} \frac{\gamma_j}{\mu_j(x_j + 1)} \mu_j(x_j + 1) r_{j,i}$$

but these are equal by the way we have defined $\tilde{r}_{i,j}$. Hence, applying the lemma, we conclude that the proposed π is the stationary distribution of the original open Jackson network. ∎

The lemma, however, yields more information. We also conclude that the time reversal of the original open Jackson network is also an open Jackson network. The arrival processes in the time reversed network are independent Poisson processes. These are time reversals of the departure processes of the original open Jackson network. Hence the departure processes in an open Jackson network are also independent Poisson processes. Similarly, arguing as we did for Theorem D.23, we can also conclude that the state of an open Jackson network, at any time, is independent of past departures. We thus can state the following theorem.

Theorem D.25

The departure processes from the various queues of an open Jackson network are independent Poisson processes. The departure from queue j is Poisson with rate $\gamma_j r_{j,0}$. ∎

It is important to note, however, that, in general, an open Jackson network need not be reversible.

Exercise D.6

Classify each of the Jackson networks in Figure D.6 as being reversible or irreversible.

Closed Jackson Networks

In closed Jackson networks, customers do not arrive to or depart from the system. The routing matrix $\mathbf{R}$ is a stochastic matrix. Let us assume that it is irreducible.

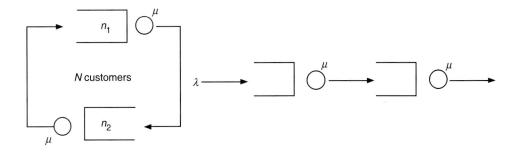

Figure D.6   Figures for Exercise D.6.

There are M customers trapped forever in the network. The CTMC $\mathbf{X}(t)$ is irreducible and has a finite state space, and hence is positive recurrent. Let γ_j be the throughput of queue j. By balancing customer flows we obtain, for all $j, 1 \le j \le J$,

$$\gamma_j = \sum_{i=1}^{J} \gamma_i r_{i,j}$$

Writing this in vector notation we have

$$\gamma = \gamma \mathbf{R}$$

But $\mathbf{R}$ is an irreducible stochastic matrix; hence there is a positive solution unique up to a multiplicative constant.

Remark: A solution of $\gamma = \gamma \mathbf{R}$ can be interpreted as the *relative throughputs* of the queues; that is, $\frac{\gamma_j}{\gamma_i}$ is the mean number of times that a customer visits queue j for each visit to queue i. This interpretation easily follows from the viewpoint that the sequence of queues visited by a customer is a Markov chain on $\{1, 2, \ldots, J\}$ with transition probability matrix $\mathbf{R}$.

Exercise D.7
Consider a closed Jackson network with J nodes and with servers having rates μ_j, $1 \le j \le J$. The routing matrix $\mathbf{R}$ is irreducible, and there is exactly one customer in the system, whose service requirement is exponentially

distributed with unit mean at every queue. Let γ be a solution of $\gamma = \gamma R$. Show that the throughput of queue i is

$$\frac{1}{\sum_{j=1}^{J} \left(\frac{\gamma_j}{\gamma_i}\right) \cdot \frac{1}{\mu_j}}$$

The following theorem provides the form of the stationary distribution of the joint queue-length process of a closed Jackson network.

Theorem D.26

The process $\mathbf{X}(t)$ is a positive recurrent CTMC with stationary distribution:

$$\pi_{\mathbf{n}} = G_M \prod_{j=1}^{J} \frac{(\gamma_j)^{n_j}}{\mu_j(1)\ldots\mu_j(n_j)}$$

where γ is any solution of $\gamma = \gamma R$ and G_M is obtained by normalizing π over $\{\mathbf{n}: \sum_{j=1}^{J} n_j = M\}$.

Proof: Take any solution of $\gamma = \gamma R$ and construct a closed Jackson network over the same queues, with the same number of customers and the same service rates, but with the following routing probabilities.

$$\tilde{r}_{i,j} = \frac{\gamma_j r_{j,i}}{\gamma_i}$$

Then check that the proposed stationary distribution and the transition rates of the joint queue-length process of this Jackson network satisfy the hypotheses of Lemma D.1. ∎

From the foregoing proof we also infer the following theorem.

Theorem D.27

A time reversal of a closed Jackson network is also a closed Jackson network. ∎

D.4   Notes on the Literature

This appendix brings together classical results in Markov chains, renewal processes, and queueing theory, as a ready reference when reading the main chapters of the book. There are many excellent books on these topics. Our treatment of Markov chains and renewal theory is based on the textbook by Wolff [300]. A first course on probabilistic modeling and queueing theory is provided by Mitrani [216]. A more sophisticated treatment is available in the books by Fayolle et al. [97] and by Bremaud [42]. The Lyapunov drift criteria for the stability analysis of Markov chains are available in these two books. The two volumes by Kleinrock ([173] and [174]) provide a compendium of results on a vast variety of single-station queueing models. The topic of reversibility and queueing networks is covered in the monograph by Kelly [161], and an insightful treatment is provided by Walrand in [294].

APPENDIX E

Complexity Theory

We discuss briefly a few terms and concepts commonly used in complexity theory. Our discussion follows [7].

Problem

A *problem* is a statement of a requirement pertaining to a generic model. For example, a shortest path problem expresses the requirement of finding a shortest path between a source node and other nodes in a directed graph, with given link lengths. Note that the elements of a problem are specified in general terms; for example, in the shortest path problem statement, the elements are the graph, the source and destination nodes, and the link lengths.

Instance

When a problem's parameters are explicitly specified, an *instance* of a problem is generated. For example, in the shortest path problem, if the graph $\mathcal{G} = (\mathcal{N}, \mathcal{L})$ and the source and destination nodes and the link lengths are specified, we have an instance of the shortest path problem.

Size of an Instance

To get an idea of the running time of an algorithm, we need to consider the complexity of a problem instance. The more complex, or "larger," the problem instance, the greater is the running time of the algorithm. The size of a data item whose value is x is taken to be $\log_2 x$, because the number of bits required in a binary representation of x is the smallest integer greater than $\log_2 x$. For convenience, we write $\log_2 x$ as $\log x$. The size of a problem would then be defined as the number of bits required to store the values of the parameters of the problem.

Note that the size of a problem can depend on how the problem is stated. For example, in a network problem, the topology is often specified by a node–link incidence matrix. Suppose a network has N nodes and L links and that each link has a cost metric that is $\leq \delta$ and a capacity that is $\leq C$. A representation would use one pointer each to store every node and arc, and one data element each for every link cost and the link capacity. Because there are N nodes, the space required to provide the identity of each node is at most $\log N$. Similarly, the space required to store the cost value corresponding to each link is at most $\log \delta$. This gives a problem size s, expressed as

$$s = N \log N + L \log L + L \log C + L \log \delta$$

The running time of an algorithm could, in principle, be expressed as a function of s. But it is more convenient to express it as a function of the network parameters N, L, $\log C$, and $\log \delta$.

Complexity

A time complexity function of an algorithm for a problem specifies the largest amount of time needed to solve any problem instance of a given size. Because the *largest* time needed to solve any problem instance of a given size is provided by the time complexity function, it is also called the *worst case complexity*, or simply the *complexity* of the algorithm. Accordingly, *polynomial complexity* means that the worst case running time of an algorithm is a polynomial function of the problem size.

Optimization Problem

We are concerned with several optimization problems in this book. In such problems, the objective is to maximize or minimize a specified objective function. The *traveling salesman problem* (TSP) provides an example. Suppose that we are given a directed graph $\mathcal{G} = (\mathcal{N}, \mathcal{L})$ and an integer link length $\delta_{i \to j}$ associated with each link $i \to j \in \mathcal{L}$. A *tour* is defined as a directed cycle that visits each node in the network exactly once. The TSP asks us to determine a tour W with the smallest possible value of tour length $\sum_{i \to j \in W} \delta_{i \to j}$. We refer to this problem as TSP-O, where O is for "optimization."

Decision Problem

Formally, the theory of complexity applies to decision problems. A decision problem is one that is stated such that its answer is of the form yes or no. For example, an instance of a decision problem associated with the TSP is as follows: Given a directed graph $\mathcal{G} = (\mathcal{N}, \mathcal{L})$, an integer link length $\delta_{i \to j}$ associated with each link $i \to j \in \mathcal{L}$, and an integer k^*, does the network contain a tour W satisfying the condition $\sum_{i \to j \in W} \delta_{i \to j} \leq k^*$? We refer to this problem as TSP-D, where D is for "decision." We also note that there could be other decision problems associated with TSP-O. For example, given a directed graph $\mathcal{G} = (\mathcal{N}, \mathcal{L})$, an integer link length

$\delta_{i \to j}$ associated with each link $i \to j \in \mathcal{L}$, and an integer k^*, does every tour W in $\mathcal{G}$ satisfy the condition $\sum_{i \to j \in W} \delta_{i \to j} > k^*$?

It can be seen that if we have a polynomial time algorithm for solving TSP-O, then we also have a polynomial time algorithm for answering TSP-D. It can also be shown that if a polynomial time algorithm for TSP-D is available, then we can use it several times to obtain a polynomial time algorithm for solving TSP-O. This actually shows that the optimization and decision versions of the traveling salesman problem are *polynomially equivalent*.

It has also been noted that for most optimization problems arising in practice, *some* decision version of the problem is polynomially equivalent to the optimization version. This allows the use of complexity theory in assessing the complexity of optimization problems, even though, formally, complexity theory addresses decision problems only.

Yes Instance and No Instance

Consider an instance of a given decision problem. If the answer to the instance is yes, then the instance is referred to as a *yes instance* of the problem. If the answer to the instance is no, then the instance is referred to as a *no instance* of the problem.

Polynomial Reduction

A problem P_1 polynomially reduces to another problem P_2 if some polynomial time algorithm A_1 that solves P_1 uses an algorithm A_2 for solving P_2 at *unit cost*. Thus, algorithm A_1 uses the algorithm A_2 as a subroutine. We note that A_1 is polynomial time only when it is *assumed* that A_2 runs in unit time.

Polynomial reduction is useful because of the following property: If a problem P_1 polynomially reduces to some problem P_2, and if some polynomial time algorithm solves P_2, then some polynomial time algorithm solves P_1. The proof relies on the fact that a composition of two polynomial functions yields another polynomial function.

Polynomial Transformation

Polynomial transformation is a special case of polynomial reduction. A problem P_1 *polynomially transforms* to another problem P_2 if for every instance I_1 of problem P_1, we can construct in polynomial time (i.e., polynomial in terms of the size of I_1) an instance I_2 of problem P_2 such that I_1 is a yes instance of P_1 if and only if I_2 is a yes instance of P_2.

In polynomial reduction, too, there is an instance I_2 of problem P_2 that is implicitly associated with the reduction. This is the instance solved by the subroutine in A_1. When talking about polynomial reduction, we did not consider I_2 explicitly. We were concerned only with whether A_1 is polynomial time, assuming that the subroutine A_2 takes unit time to run. Here, in addition, we examine the corresponding instance I_2 and relate it to I_1 of P_1.

If P_1 polynomially transforms to P_2, then P_2 is at least as hard as P_1. This can be said because given an algorithm for solving P_2, we can always construct an algorithm for solving P_1 that runs in "comparable" time. For example, if the algorithm for solving P_2 runs in polynomial time, then the corresponding algorithm for P_1, which uses the algorithm for P_2 as a subroutine, also runs in polynomial time. Thus, if a polynomial time algorithm for P_2 is available, we can be sure that a polynomial time algorithm for P_1 is available. This already indicates that P_1 cannot be harder than P_2.

Class $\mathcal{P}$

Decision problem P_1 is said to belong to class $\mathcal{P}$ (for "polynomial") if a polynomial time algorithm solves P_1. This means that given any instance of P_1, we can determine in polynomial time whether the instance is a yes instance or a no instance.

Class $\mathcal{NP}$, Certificate, Certificate-Checking Algorithm

The class $\mathcal{NP}$ (for "nondeterministic polynomial") uses the concept of *verification*. Roughly speaking, decision problem P_1 is said to be in class $\mathcal{NP}$ if, for every yes instance of P_1, there is a polynomial-length verification that the instance is a yes instance.

For example, consider the problem TSP-D. Suppose that we are given a yes instance and also a tour W of length at most k^* for verifying whether this is a yes instance. Now the time taken to *actually verify* is polynomial, because this can be done in $O(n)$ time: We need to check whether the given tour W really passes through every node exactly once and whether its length is really at most k^*. After this verification is done, we are sure that the given instance is a yes instance. It has been *proved* to be a yes instance by the given tour.

On the contrary, suppose that we are given a no instance and a tour W of length greater than k^*. Here, even after verifying that W visits every node exactly once and that its length is greater than k^*, we cannot conclude that the given instance is a no instance. All that we have checked is that the given tour is really a tour and that it costs more than k^*. To conclude that the instance is a no instance, we would need to check *every* possible tour!

This shows a peculiar asymmetry in the problem TSP-D. Given an appropriate tour, we have a *short* proof that the instance is a yes instance; but there is no short proof that the instance is a no instance. We note here that the issue of how to obtain this appropriate tour is different. We may have to *guess* this tour, which accounts for the word "nondeterministic" in the name $\mathcal{NP}$.

The given tour that proves that an instance is a yes instance is called a *certificate* of the yes instance. The algorithm that verifies that the given tour visits every node exactly once and that its length is at most k^* is called a *certificate-checking algorithm.*

Formally, we say that a decision problem P is in the class $\mathcal{NP}$ if some certificate-checking algorithm $\mathcal{AL}$ and polynomial $p(\cdot)$ satisfy the following:

- Every yes instance I of P has a certificate $CR(I)$.

- The algorithm $\mathcal{AL}$ can verify the correctness of $CR(I)$ in at most $p(|I|)$ steps.

$\mathcal{P} \subset \mathcal{NP}$

From the definition of the class $\mathcal{NP}$, it can be seen that $\mathcal{P} \subset \mathcal{NP}$. Consider a problem $P \in \mathcal{P}$ and an algorithm $\mathcal{AL}$ that solves P. Then we can take an instance itself to be its own certificate and take the algorithm $\mathcal{AL}$ as the certificate-checking algorithm. Then every yes instance I has a certificate, and $\mathcal{AL}$ verifies the correctness of the certificate in polynomial time because $\mathcal{AL}$ is a polynomial time algorithm. Hence $\mathcal{P} \subset \mathcal{NP}$. It is very widely believed that $\mathcal{P}$ is a *strict* subset of $\mathcal{NP}$, but this has not yet been proved.

NP-Complete

A decision problem P is said to be NP-complete if

- $P \in \mathcal{NP}$

- All other problems in the class $\mathcal{NP}$ polynomially transform to P.

Recalling the property of polynomial reduction (and, therefore, of polynomial transformation) stated before, we can say the following. If there is an efficient (i.e., polynomial time) algorithm for some NP-complete problem, then there is an efficient algorithm for *every* problem in the class $\mathcal{NP}$. This would actually show that $\mathcal{NP} \subset \mathcal{P}$, and we would therefore have $\mathcal{P} = \mathcal{NP}$. However, an efficient algorithm for an NP-complete problem has never been found, and most researchers believe it will never be found.

From the definition, it is clear that every NP-complete problem polynomially transforms to every other NP-complete problem. Therefore, all NP-complete problems can be expected to be comparable in terms of computational difficulty.

Proving the existence of the first NP-complete problem was a remarkable achievement. However, after one NP-complete problem was discovered, it was much easier to show that some other problems are NP-complete. From the definition, two criteria must be satisfied to show that a problem P is NP-complete. The first criterion (i.e., whether $P \in \mathcal{NP}$) is usually somewhat simpler to establish. The second criterion is established by showing that a *known* NP-complete problem—say, P_1—polynomially transforms to P. Because P_1 is NP-complete, every other problem in $\mathcal{NP}$ polynomially transforms to P_1. Moreover, because P_1 polynomially transforms to P, every problem in $\mathcal{NP}$ polynomially transforms to P. Identifying the known NP-complete problem P_1 that would polynomially transform to the given problem P can be a challenging task.

NP-Hard

A decision problem P is said to be NP-hard if every problem in NP polynomially transforms to P. This is nothing but the second criterion in the definition of NP-completeness. Evidently, every NP-complete problem is also NP-hard, but the converse is not true. From the definition, it is clear that a problem P is NP-hard when the second criterion in the definition of NP-completeness is satisfied, but not the first—that is, when we do not know whether $P \in \mathcal{NP}$. To prove that P is NP-hard, we proceed as before by taking a known NP-complete problem P_1 and polynomially transforming it to P.

Bibliography

[1] S. P. Abraham and A. Kumar. Max-min fair rate control of ABR traffic with nonzero MCRs. In *Proceedings of IEEE GLOBECOM*. IEEE, 1997.

[2] S. P. Abraham and A. Kumar. A stochastic approximation approach for max-min fair, adaptive rate control of ABR sessions with MCRs. In *Proceedings of IEEE INFOCOM*. IEEE, 1998.

[3] S. P. Abraham and A. Kumar. A new approach for asynchronous distributed rate control of elastic sessions in integrated packet networks. *IEEE/ACM Transactions on Networking*, 9(1):13–30, February 2001.

[4] N. Abramson. The ALOHA system – another alternative for computer communications. In *Proceedings of Fall Joint Computing Conference*, pages 281–285, 1970.

[5] R. Agrawal, R. L. Cruz, C. Okino, and R. Rajan. Performance bounds for flow control protocols. *IEEE/ACM Transactions on Networking*, 7(3):310–323, June 1999.

[6] H. Ahmadi and W. E. Denzel. A survey of modern high-performance switching techniques. *IEEE Journal on Selected Areas in Communications*, 7(7):1091–1103, 1989.

[7] R. K. Ahuja, T. L. Magnanti, and J. B. Orlin. *Network Flows*. Prentice Hall, NJ, 1993.

[8] J. M. Akinpelu. The overload peformance of engineered networks with nonhierarchical and hierarchical routing. *AT&T Bell Laboratories Technical Journal*, 63(7):1261–1281, 1983.

[9] I. Akyildiz, W. Su, Y. Sankarasubramanian, and E. Cayirci. Wireless sensor networks: A survey. *Computer Networks*, 38:393–422, 2002.

[10] E. Altman, K. Avrachenkov, and C. Barakat. A stochastic model of TCP/IP with stationary random losses. In *Proceedings of ACM SIGCOMM*, 2000.

[11] E. Altman, K. Avrachenkov, and C. Barakat. TCP network calculus: The case of large delay-bandwidth product. In *Proceedings of IEEE INFOCOM*, 2002.

[12] T. Anderson, S. Owicki, J. Saxe, and C. Thacker. High speed switch scheduling for local area networks. *ACM Transactions on Computer Systems*, 11(4):319–352, 1993.

[13] D. Anick, D. Mitra, and M. M. Sondhi. Stochastic theory of a data-handling system. *The Bell System Technical Journal*, 61(8):1871–1894, October 1982.

[14] F. Anjum and L. Tassiulas. On the behavior of different TCP algorithms over a wireless channel with correlated packet losses. In *Proceedings of ACM SIGMETRICS*. ACM, 1999.

[15] G. Apostolopoulos, D. Williams, S. Kamat, R. Guérin, A. Orda, and D. Przygienda. QoS routing mechanisms and OSPF extensions. Internet RFC 2676, August 1999.

[16] S. Athuraliya, D. Lapsley, and S. Low. An enhanced random early marking algorithm for internet flow control. In *Proceedings of IEEE INFOCOM*. IEEE, 2000.

[17] ATM-Forum. Traffic management specification version 4.0. Technical Report af-tm-0056.000, April 1996.

[18] D. Awduche, L. Berger, D. Gan, T. Li, V. Srinivasan, and G. Swallow. RSVP-TE: Extensions to RSVP for LSP tunnels. Internet Draft draft-ietf-mpls-rsvp-lsp-tunnel-09.txt, December 2001.

[19] F. Baboescu and G. Varghese. Scalable packet classification. In *Proceedings of ACM SIGCOMM*, pages 199–210, 2001.

[20] M. L. Bailey, B. Gopal, M. Pagels, L. L. Peterson, and P. Sarkar. PATHFINDER: A pattern based packet classifier. In *Proceedings of First Symposium on Operating System Design and Implementation*, pages 115–123, 1994.

[21] C. Barakat and E. Altman. Performance of short TCP transfers. In *Proceedings of Networking*, Paris, May 2000.

[22] F. Baskett and A. J. Smith. Interference in multiprocessor computer systems with interleaved memory. *Communications of the ACM*, 19(6):327–334, 1976.

[23] M. S. Bazaraa, H. D. Sherali, and C. M. Shetty. *Nonlinear Programming— Theory and Algorithms*. Series in Discrete Mathematics and Optimization. John Wiley, second edition, 1993.

[24] J. C. Bellamy. *Digital Telephony*. John Wiley and Sons, third edition, 2002.

[25] R. Bellman and K. L. Cooke. *Differential-Difference Equations*. Academic Press, New York, 1963.

[26] V. E. Benes. *Mathematical Theory of Connecting Networks and Telephone Traffic*. Academic Press, New York, 1965.

[27] L. Benmohamed and S. M. Meerkov. Feedback control of congestion in packet switching networks: The case of a single congested node. *IEEE/ACM Transactions on Networking*, 1(6):693–708, December 1993.

[28] J. C. R. Bennett and H. Zhang. WF^{2}Q: Worst-case fair weighted fair queuing. In *Proceedings of IEEE INFOCOM*, pages 120–128. IEEE, March 1996.

[29] J. C. R. Bennett, C. Partridge, and N. Shectman. Packet reordering is not pathological network behaviour. *IEEE/ACM Transactions on Networking*, 7(6):789–798, December 1999.

[30] J. Beran. *Statistics of Long-Memory Processes*. Chapman and Hall, New York, 1994.

[31] A. W. Berger and Y. Kogan. Dimensioning bandwidth for elastic traffic in high-speed data networks. *IEEE/ACM Transactions on Networking*, 8(5):643–654, 2000.

[32] R. A. Berry and R. G. Gallager. Communication over fading channels with delay-constraints. *IEEE Transactions on Information Theory*, 48(5):1135–1149, May 2002.

[33] D. Bertsekas and R. Gallager. *Data Networks*. Prentice-Hall of India, second edition, July 1992.

[34] V. Bharghavan, A. Demers, S. Shenker, and L. Zhang. MACAW: A media access protocol for wireless LANs. In *Proceedings of ACM SIGCOMM*, pages 212–225, 1994.

[35] G. Bianchi. Performance analysis of the IEEE 802.11 distributed coordination function. *IEEE Journal on Selected Areas in Communications*, 18(3):535–547, March 2000.

[36] A. Birman. Computing approximate blocking probabilities for a class of all optical networks. *IEEE Journal on Selected Areas in Communications*, 14(5):852–857, 1996.

[37] J. C. Bolot and A. U. Shankar. Analysis of a fluid approximation to flow control dynamics. In *Proceedings of IEEE INFOCOM*, pages 2398–2407. IEEE, 1992.

[38] T. Bonald and L. Massoulie. Impact of fairness on Internet performance. In *Proceedings of ACM SIGMETRICS*, 2001.

[39] T. Bonald and J. W. Roberts. Performance of bandwidth sharing mechanisms for service differentiation in the Internet. In *Proceedings of the ITC Specialist Seminar on IP Traffic*, 2000.

[40] R. R. Boorstyn, A. Burchard, J. Liebeherr, and C. Oottamakorn. Statistical service assurances for traffic scheduling algorithms. *IEEE Journal on Selected Areas in Communications*, 18(12):2651–2664, December 2000.

[41] P. Bremaud. *An Introduction to Probabilistic Modelling.* Springer Verlag, 1988.

[42] P. Bremaud. *Markov Chains: Gibbs Fields, Monte Carlo Simulation, and Queues.* Springer, 1999.

[43] P. H. Brill and M. J. M. Posner. Level crossings in point processes applied to queues: Single-server case. *Operations Research*, 25(4):662–674, July-August 1977.

[44] P. Brown. Resource sharing of TCP connections with different round trip times. In *Proceedings of IEEE INFOCOM*, April 2000.

[45] T. Bu and D. Towsley. Fixed point approximations for TCP behaviour in an AQM network. In *Proceedings of ACM SIGMETRICS*, 2001.

[46] R. Cahn. *Wide Area Network Design: Concepts and Tools for Optimization.* Morgan Kaufmann, 1998.

[47] F. Cali, M. Conti, and E. Gregori. IEEE 802.11 protocol: Design and performance evaluation of an adaptive backoff mechanism. *IEEE Journal on Selected Areas in Communications*, 18(9):1774–1780, September 2000.

[48] N. Cardwell, S. Savage, and T. Anderson. Modeling TCP latency. In *Proceedings of IEEE INFOCOM*, 2000.

[49] C. Casetti and M. Meo. A new approach to model the stationary behaviour of TCP connections. In *Proceedings of IEEE INFOCOM*, April 2000.

[50] Y. Chait, C. V. Hollot, V. Misra, D. Towsley, H. Zhand, and J. C. S. Lui. Providing throughput differentiation for TCP flows using adaptive two-color marking and two-level AQM. In *Proceedings of IEEE INFOCOM*, pages 1–8, 2002.

[51] P. S. Chanda, A. Kumar, and A. A. Kherani. An approximate calculation of max-min fair throughputs for non-persistent elastic sessions. In *Proceedings of IEEE GLOBECOM (Internet Performance Symposium)*. IEEE, 2001.

[52] C. S. Chang. Stability, queue length, and delay of deterministic and stochastic queueing networks. *IEEE Transactions on Automatic Control*, 39(5):913–931, May 1994.

[53] C. S. Chang. *Performance Guarantees in Communication Networks.* Telecommunications Networks and Computer Systems. Springer, 2000.

[54] A. Charny and K. K. Ramakrishnan. Time scale analysis of explicit rate allocation in ATM networks. In *Proceedings of IEEE INFOCOM*, pages 1182–1189, 1996.

[55] A. Charny, D. D. Clark, and R. Jain. Congestion control with explicit rate indication. In *Proceedings of IEEE ICC*, 1995.

[56] A. Charny, P. Krishna, N. Patel, and R. Simcoe. Algorithms for providing bandwidth and delay guarantees in input crossbar with speedup. *IEEE Journal on Selected Areas in Communications*, 17(6):1057–1066, 1999.

[57] R. J. Cherian. On active queue management for providing assured service to TCP controlled elastic flows. Master's thesis, Dept. of Electrical Communications Engg., Indian Institute of Science, Bangalore, India, January 2001.

[58] G. Cheung and S. McCanne. Optimal routing table design for IP address lookups under memory constraints. In *Proceedings of IEEE INFOCOM*, pages 1437–1444, 1999.

[59] P. Chevillat, J. Jelitto, A. N. Barreto, and H. L. Truong. A dynamic link adaptation algorithm for IEEE 802.11a wireless LANs. In *Proceedings of IEEE ICC*, 2003.

[60] L. Chisvin and R. J. Duckworth. Content addressable and associative memory. *IEEE Computer*, 22(7):51–64, 1989.

[61] D. M. Chiu and R. Jain. Analysis of the increase and decrease algorithms for congestion avoidance in computer networks. *Computer Networks and ISDN Systems*, 17(1):1–14, 1989.

[62] F. M. Chiussi and A. Francini. A distributed scheduling architecture for scalable packet switches. *IEEE Journal on Selected Areas in Communications*, 18(12), 2000.

[63] J. Choe and N. B. Shroff. A central-limit-theorem based approach for analysing queue behavior in high-speed networks. *IEEE/ACM Transactions on Networking*, 6(5):659–671, October 1998.

[64] G. L. Choudhury, D. M. Lucantoni, and W. Whitt. Squeezing the most out of ATM. Technical Report BL041334-930222-02TM, AT&T Bell Laboratories, 1993.

[65] S. T. Chuang, A. Goel, N. McKeown, and B. Prabhakar. Matching output queueing with combined input output queued switch. *IEEE Journal on Selected Areas in Communications*, 17(1030–1039), 1999.

[66] D. E. Comer and D. L. Stevens. *Internetworking with TCP/IP*, volume III. Prentice Hall of India, second edition, 1995.

[67] D. R. Cox. *Renewal Theory*. Methuen & Co. Ltd., 1962.

[68] D. R. Cox. Long-range dependence: A review. In *Proceedings of the 50th Anniversary Conference, Iowa State Statistical Laboratory*, pages 55–74, 1984.

[69] M. E. Crovella and A. Bestavros. Self-similarity in World Wide Web traffic: Evidence and possible causes. *IEEE/ACM Transactions on Networking*, 5:835–846, 1997.

[70] R. L. Cruz. A calculus for network delay, part I: Network elements in isolation. *IEEE Transactions on Information Theory*, 37(1):114–131, January 1991.

[71] R. L. Cruz. A calculus for network delay, part II: Network analysis. *IEEE Transactions on Information Theory*, 37(1):132–141, January 1991.

[72] J. Dai and B. Prabhakar. The throughput of data switches with and without speedup. In *Proceedings of IEEE INFOCOM*, pages 556–564, 2000.

[73] N. G. De Bruijn. *Asymptotic Methods in Analysis*. Dover Publications, New York, 1981.

[74] A. Demers, S. Keshav, and S. Shenker. Analysis and simulation of a fair queueing algorithm. *Internetworking Research and Experience*, 1:3–26, 1990.

[75] H. De Neve and P. Van Mieghem. TAMCRA: A tunable accuracy multiple constraints routing algorithm. *Computer Communications*, 23:667–679, 2000.

[76] M. P. Desai and D. Manjunath. On the connectivity in finite ad hoc networks. *IEEE Communications Letters*, 10(6):437–490, 2002.

[77] M. P. Desai, R. Gupta, A. Karandikar, K. Saxena, and V. Samant. Reconfigurable state machine based IP lookup engine for high speed router. *IEEE Journal on Selected Areas in Communications*, 21(4):501–512, 2003.

[78] M. De Prycker. *Asynchronous Transfer Mode: Solution for Broadband ISDN*. Ellis Norwood, third edition, 1995.

[79] G. de Veciana, G. Kesidis, and J. Walrand. Resource management in wide-area ATM networks using effective bandwidths. *IEEE Journal on Selected Areas in Communications*, 13(6):1081–1090, August 1995.

[80] G. de Veciana, T. J. Lee, and T. Konstantopoulos. Stability and performance analysis of networks supporting services with rate control – Could the Internet be unstable? In *Proceedings of IEEE INFOCOM*, April 1999.

[81] D. Dias and J. R. Jump. Analysis and simulation of buffered delta networks. *IEEE Transactions on Communications*, 30(8):273–282, 1981.

[82] M. Degermark, A. Bridnik, S. Carlsson, and S. Pink. Small forwarding tables for fast routing lookups. In *Proceedings of ACM SIGCOMM*, pages 3–13, 1997.

[83] A. Diwan and J. Kuri. Asymptotic performance limits for routing and rate allocation in rate-based multi-class networks. In *Proceedings of IEEE GLOBECOM*, 2002.

[84] A. Diwan, R. A. Guerin, and K. N. Sivarajan. Performance analysis speeded up high-speed packet switches. *Journal of High Speed Networks*, 10(3):161–186, 2001.

[85] W. Doeringer, G. Karjoth, and M. Nassehi. Routing on longest matching prefixes. *IEEE/ACM Transactions on Networking*, 4(1):86–97, 1996.

[86] B. T. Doshi. Deterministic rule based traffic descriptors for broadband ISDN: Worst case behaviour and connection acceptance control. In J. Labetoulle and J. W. Roberts, editors, *Proceedings of the International Teletraffic Congress (ITC)*, pages 591–600. Elsevier, 1994.

[87] B. A. Douglass and A. Y. Oruc. On self routing in Clos connection networks. *IEEE Transactions on Communications*, 41(1):121–124, 1993.

[88] P. Druschel, M. B. Abbott, M. A. Pagels, and L. L. Peterson. Network subsystem design. *IEEE Network*, 7(4):8–17, 1993.

[89] N. G. Duffield and N. O'Connell. Large deviations and overflow probabilities for general single server queues, with applications. *Math. Proceedings of the Cambridge Philosophical Society*, 118:363–375, 1995.

[90] A. Elwalid. Adaptive rate-based congestion control for high-speed wide area network. In *Proceedings of IEEE ICC*, pages 48–53. IEEE, 1995.

[91] A. I. Elwalid and D. Mitra. Effective bandwidth of general Markovian traffic sources and admission control of high speed networks. *IEEE/ACM Transactions on Networking*, 1(3):329–343, June 1993.

[92] A. Elwalid, D. Heyman, T. V. Lakshman, D. Mitra, and A. Weiss. Fundamental bounds and approximations for ATM multiplexers with applications to video teleconferencing. *IEEE Journal on Selected Areas in Communications*, 13(6):1004–1016, August 1995.

[93] A. Elwalid, D. Mitra, and R. H. Wentworth. A new approach for allocating buffers and bandwidth to heterogeneous, regulated traffic in an ATM node. *IEEE Journal on Selected Areas in Communications*, 13(6):1115–1127, August 1995.

[94] A. Erramilli, O. Narayan, and W. Willinger. Experimental queueing analysis with long-range dependent packet traffic. *IEEE/ACM Transactions on Networking*, 4(2):209–223, April 1996.

[95] D. E. Everitt and N. W. MacFayden. Analysis of multicellular mobile radio telephone systems with loss. *British Telecom Journal*, 1(2):37–45, 1983.

[96] G. Fayolle, E. Gelenbe, and J. Labetoulle. Stability and optimal control of the packet switching broadcast channel. *Journal of the ACM*, 24(3):375–386, 1977.

[97] G. Fayolle, V. A. Malyshev, and M. V. Menshikov. *Topics in the Constructive Theory of Countable Markov Chains*. Cambridge University Press, 1995.

[98] A. Feldmann and S. Muthukrishnan. Tradeoffs for packet classification. In *Proceedings of IEEE INFOCOM*, pages 397–413, 2000.

[99] W. Feller. *An Introduction to Probability Theory and Its Applications*, volume I. Wiley Eastern Limited, 1968.

[100] K. W. Fendick. Evolution of controls for the available bit rate service. *IEEE Communications Magazine*, pages 35–39, November 1996.

[101] V. Firoiu and M. Borden. A study of active queue management for congestion control. In *Proceedings of IEEE INFOCOM*, 2000.

[102] V. Firoiu, I. Yeom, and X. Zhang. A framework for practical performance evaluation and traffic engineering in IP networks. In *Proceedings of IEEE International Conference on Telecommunications*, 2001.

[103] S. Floyd and V. Jacobson. Random early detection gateways for congestion avoidance. *IEEE/ACM Transactions on Networking*, 1(4):397–413, August 1993.

[104] S. Floyd and V. Paxson. Difficulties in simulating the Internet. *IEEE Transactions on Networking*, 9(4):392–403, August 2001.

[105] L. R. Ford and D. R. Fulkerson. *Flows in Networks*. Princeton University Press, Princeton, NJ, 1962.

[106] B. Fortz and M. Thorup. Internet traffic engineering by optimizing OSPF weights. In *Proceedings of IEEE INFOCOM*, 2000.

[107] B. Fortz, J. Rexford, and M. Thorup. Traffic engineering with traditional IP routing protocols. *IEEE Communication Magazine*, October 2002.

[108] S. B. Fredj, T. Bonald, A. Proutiere, G. Regnie, and J. W. Roberts. Statistical bandwidth sharing: A study of congestion at flow level. In *Proceedings of ACM SIGCOMM*. ACM, 2001.

[109] R. Freeman. *Telecommunication System Engineering*. John Wiley and Sons, third edition, 1996.

[110] S. W. Fuhrman. Performance of a packet switch with a crossbar architecture. *IEEE Transactions on Communications*, 41(3):486–491, 1993.

[111] V. Fuller, T. Li, J. Yu, and K. Varadhan. Classless interdomain routing (CIDR): An address assignment and aggregation strategy. Internet RFC 1519, September 1993.

[112] C. Fulton, S. Q. Li, and C. S. Lim. UT: ABR feedback control with tracking. In *Proceedings of IEEE INFOCOM*, pages 806–815, 1997.

[113] Y. Ganjali, A. Keshavarzian, and D. Shah. Input queued switches: Cell switching vs. packet switching. In *Proceedings of IEEE INFOCOM*, 2003.

[114] L. Georgiadis, R. Guerin, V. Peris, and R. Rajan. Efficient support of delay and rate guarantees in an internet. In *Proceedings of ACM SIGCOMM*, pages 106–116. ACM, August 1996.

[115] L. Georgiadis, R. Guerin, V. Peris, and K. N. Sivarajan. Efficient network QoS provisioning based on per node traffic shaping. *IEEE/ACM Transactions on Networking*, 4(4):482–501, August 1996.

[116] M. Gerla, R. Bagrodia, L. Zhang, K. Tang, and L. Wang. TCP over wireless multi-hop protocols: Simulations and experiments. In *Proceedings of IEEE ICC*, pages 1089–1094. IEEE, 1999.

[117] D. Ghosal, T. V. Lakshman, and Y. Huang. Parallel architectures for processing high speed network signalling protocols. *IEEE/ACM Transactions on Networking*, 3(6):716–728, 1995.

[118] R. J. Gibbens and P. J. Hunt. Effective bandwidths for the multi-type UAS channel. *Queueing Systems*, 9:17–28, 1991.

[119] A. Girard. *Routing and Dimensioning in Circuit Switched Networks*. Addison-Wesley, 1990.

[120] P. W. Glynn and W. Whitt. Logarithmic asymptotics for steady state tail probabilities in a single server queue. *Studies in Applied Probability*, 1994.

[121] A. J. Goldsmith and S. G. Chua. Variable-rate variable-power MQAM for fading channels. *IEEE Transactions on Communications*, 45(10):1218–1230, October 1997.

[122] A. J. Goldsmith and P. P. Varaiya. Capacity of fading channels with channel side information. *IEEE Transaction on Information Theory*, 43(6):1986–1992, November 1997.

[123] S. J. Golestani. A self clocked fair queueing scheme for broadband applications. In *Proceedings of IEEE INFOCOM*, pages 636–646. IEEE, June 1994.

[124] D. J. Goodman. *Wireless Personal Communications Systems*. Addison-Wesley, 1997.

[125] M. Goyal, A. Kumar, and V. Sharma. Power constrained and delay optimal policies for scheduling transmission over a fading channel. In *Proceedings of IEEE INFOCOM*. IEEE, April 2003.

[126] P. Goyal, H. M. Vin, and H. Cheng. Start-time fair queueing: A scheduling algorithm for integrated services packet switching networks. *IEEE/ACM Transactions on Networking*, 5(5):690–704, October 1997.

[127] I. S. Gradshteyn and I. M. Ryzhik. *Table of Integrals, Series, and Products*. Academic Press, 1980.

[128] R. L. Graham, D. E. Knuth, and O. Patashnik. *Concrete Mathematics*. Addison-Wesley, second edition, 1998.

[129] R. Guérin and A. Orda. QoS routing in networks with inaccurate information: Theory and algorithms. *IEEE/ACM Transactions on Networking*, 7(3):350–364, June 1999.

[130] R. Guérin, H. Ahmadi, and M. Naghshineh. Equivalent capacity and its application to bandwidth allocation in high-speed networks. *IEEE Journal on Selected Areas in Communications*, 9(7):968–981, 1991.

[131] P. Gupta and P. R. Kumar. Critical power for asymptotic connectivity in wireless networks. In W. M. McEneaney, G. Yin, and Q. Zhang, editors, *Stochastic Analysis, Control, Optimization and Applications: A Volume in Honor of W. H. Fleming*. Birkhauser, Boston, 1998.

[132] P. Gupta and P. R. Kumar. The capacity of wireless networks. *IEEE Transactions on Information Theory*, 46(2):388–404, 2000.

[133] P. Gupta and N. McKeown. Packet classification on multiple fields. In *Proceedings of ACM SIGCOMM*, pages 147–160, 1999.

[134] P. Gupta, S. Lin, and N. McKeown. Routing lookups in hardware at memory access speeds. In *Proceedings of IEEE INFOCOM*, pages 1240–1247, 1998.

[135] R. Händel, M. N. Huber, and S. Schroder. *ATM Networks: Concepts, Protools, Applications*. Elecronic Systems Engineering. Addison-Wesley Publishing Company, second edition, 1994.

[136] M. Hassan, A. Nayandoro, and M. Atiquzzaman. Internet telephony: Service, technical challenges, and products. *IEEE Communications Magazine*, 38(4):96–103, April 2000.

[137] D. Heath, S. Resnick, and G. Samorodnitsky. Heavy tails and long range dependence in ON/OFF processes and associated fluid models. *Mathematics of Operations Research*, 23:145–165, 1998.

[138] D. P. Heyman, T. V. Lakshman, and A. L. Neidhardt. A new method for analyzing feedback-based protocols with applications to engineering Web traffic over the Internet. *Performance Evaluation Review*, 25(1), 1997.

[139] M. G. Hluchyj and M. J. Karol. Queueing in high-performance packet switching. *IEEE Journal on Selected Areas in Communications*, 6(9):1587–1608, 1988.

[140] K. Hoffman. *Analysis in Euclidean Space*. Prentice Hall, 1975.

[141] T. Holliday, A. Goldsmith, and P. Glynn. Wireless link adaptation policies: QoS for deadline constrained traffic with imperfect channel estimates. In *Proceedings of IEEE ICC*, 2002.

[142] C. Hollot, V. Misra, D. Towsley, and W. B. Gong. A control theoretic analysis of RED. In *Proceedings of IEEE INFOCOM*, April 2001.

[143] J. Y. Hui. *Switching and Teletraffic Theory for Integrated Broadband Networks*. Kluwer Academic Publishers, 1990.

[144] P. Hurley, J. Y. Le Boudec, and P. Thiran. A note on the fairness of additive increase and multiplicative decrease. In *Proceedings of the International Teletraffic Congress (ITC)*, 1999.

[145] F. K. Hwang. *The Mathematical Theory of Non Blocking Switching Networks*. World Scientific Press, 2000.

[146] R.-H. Hwang, J. F. Kurose, and D. Towsley. On call processing delay in high speed networks. *IEEE/ACM Transactions on Networking*, 3(6):628–639, 1995.

[147] IEEE Standard 802.3. *802.3: Carrier Sense Multiple Access with Collision Detection*. IEEE, New York.

[148] S. Iyer, R. R. Kompella, and A. Shelat. ClassiPI: An architecture for flexible packet clasification. *IEEE Network*, 15(2):33–41, 2001.

[149] L. Jacob and A. Kumar. Saturation throughput of an input queueing ATM switch with multiclass bursty traffic. *IEEE Transactions on Communications*, 43(2/3/4):757–761, 1995.

[150] L. Jacob and A. Kumar. Establishing the region of stability for an input queuing cell switch. *IEE Proceedings–Communications*, 148(6):343–347, 2001.

[151] R. Jain. Congestion control and traffic management in ATM networks: Recent advances and a survey. *Computer Networks and ISDN Systems*, January 1995.

[152] B. N. Jain and A. K. Agrawala. *Open Systems Interconnection*. Elsevier, 1990.

[153] Y. C. Jenq. Performance analysis of a packet switch based on single-buffered banyan network. *IEEE Journal on Selected Areas in Communications*, 1(6):1014–1021, 1983.

[154] N. Jindal and A. Goldsmith. Capacity and optimal power allocation for fading broadcast channels with minimum rates. In *Proceedings of IEEE GLOBECOM*, November 2001.

[155] S. Kalyanaraman, R. Jain, S. Fahmy, R. Goyal, and B. Vandalore. The ERICA switch algorithm for ABR traffic management in ATM networks. *IEEE/ACM Transactions on Networking*, 8(1):87–98, February 2000.

[156] K. Kar, M. Kodialam, and T.V. Lakshman. Minimum interference routing of bandwidth guaranteed tunnels with MPLS traffic engineering applications. *IEEE Journal on Selected Areas in Communication*, 18(12):2566–2579, December 2000.

[157] P. Karn. A new channel access method for packet radio. In *Proceedings of Ninth Computer Networking Conference*, pages 134–140, 1990.

[158] M. Karol, M. Hluchyj, and S. Morgan. Input versus output queueing on a space-division packet switch. *IEEE Transactions on Communications*, 35(12):1347–1356, 1987.

[159] J. S. Kaufman. Blocking in a shared resource environment. *IEEE Transactions on Communications*, 29(10):1474–1481, 1981.

[160] V. Kawadia and P. R. Kumar. A cautionary perspective on cross layer design. Unpublished manuscript, 2003.

[161] F. P. Kelly. *Reversibility and Stochastic Networks*. John Wiley, 1979.

[162] F. P. Kelly. Stochastic models of computer communication systems. *Journal of the Royal Statistical Society: Series B*, 47(3):379–395, 1985.

[163] F. P. Kelly. Blocking probabilities in large circuit switched networks. *Advances in Applied Probability*, 18:473–505, 1986.

[164] F. P. Kelly. Effective bandwidths at multi-class queues. *Queueing Systems*, 9:5–16, 1991.

[165] F. P. Kelly. Loss networks. *Annals of Applied Probability*, 1(3):319–378, 1991.

[166] F. P. Kelly, A. Maulloo, and D. Tan. Rate control for communication networks: Shadow price proportional fairness and stability. *Journal of the Operations Research Society*, 49:237–252, 1998.

[167] A. Kershenbaum. *Telecommunication Network Design Algorithms*. McGraw Hill, 1993.

[168] S. Keshav. *An Engineering Approach to Computer Networking*. Addison-Wesley, 1997.

[169] G. Kesidis, J. Walrand, and C. S. Chang. Effective bandwidths for multiclass Markov fluids and other ATM sources. *IEEE/ACM Transactions on Networking*, 1(4):424–428, August 1993.

[170] A. Kherani and A. Kumar. Stochastic models for throughput analysis of randomly arriving elastic flows in the Internet. In *Proceedings of IEEE INFOCOM*. IEEE, 2002.

[171] A. Kherani and A. Kumar. Closed loop analysis of the buffer with TCP controlled HTTP-like transfers. In *Proceedings of IEEE INFOCOM*. IEEE, 2003.

[172] A. Kherani and A. Kumar. The lightening effect of adaptive window control. *IEEE Communications Letters*, 2003.

[173] L. Kleinrock. *Queueing Systems*, volume 1. John Wiley, 1975.

[174] L. Kleinrock. *Queueing Systems*, volume 2. John Wiley, 1976.

[175] L. Kleinrock and F. Tobagi. Packet switching in radio channels: Part I: Carrier sense multiple access and their throughput delay characteristics. *IEEE Transactions on Communications*, 23(12):1400–1412, 1975.

[176] E. W. Knightly and N. B. Shroff. Admission control for statistical QoS: Theory and practice. *IEEE Network Magazine*, pages 20–29, March/April 1999.

[177] D. E. Knuth. *The Art of Computer Programming: Sorting and Searching*. Addison-Wesley, 1973.

[178] A. Kolarov and G. Ramamurthy. A control-theoretic approach to the design of an explicit rate controller for ABR service. *IEEE/ACM Transactions on Networking*, 7(5):741–753, October 1999.

[179] T. J. Kostas, M. S. Borella, I. Sidhu, G. M. Schuster, J. Grabiec, and J. Mahler. Real-time voice over packet-switched networks. *IEEE Network Magazine*, pages 18–27, 1998.

[180] A. Kulshreshtha and K. N. Sivarajan. Maximum packing channel assignment in cellular networks. *IEEE Transactions on Vehicular Technology*, 48(3):858–872, 1999.

[181] A. Kumar. Comparative performance analysis of versions of TCP in a local area network with a lossy link. *IEEE Transactions on Networking*, 6(4):485–498, August 1998.

[182] A. Kumar and J. M. Holtzman. Performance analysis of versions of TCP in a local network with a mobile radio link. *SADHANA: Indian Academy of Sciences Proceedings in Engineering Sciences*, 23(1):113–129, February 1998. Also a WINLAB Technical Report, Rutgers University, 1996.

[183] A. Kumar and A. Karnik. Performance analysis of wireless ad-hoc networks. In *The Handbook of Ad-Hoc Wireless Networks*. CRC Press, 2003.

[184] A. Kumar, K. V. S. Hari, R. Shobhanjali, and S. Sharma. Long range dependence in the aggregate flow of TCP controlled elastic sessions: An investigation via the processor sharing model. In *Proceedings of the National Conference on Communications (NCC)*, New Delhi, 2000.

[185] D. Kumar, J. Kuri, and A. Kumar. Routing guaranteed bandwidth virtual paths with simultaneous maximization of additional flows. In *Proceedings of IEEE ICC*, 2003.

[186] J. W. Kurose and K. W. Ross. *Computer Networking—A Top-Down Approach Featuring the Internet*. Addison-Wesley, 2001.

[187] T. V. Lakshman and U. Madhow. The performance of TCP/IP for networks with high bandwidth-delay products and random loss. *IEEE/ACM Transactions on Networking*, 5(3):336–350, June 1997.

[188] T. V. Lakshman and D. Stiliadis. High speed packet forwarding using efficient multidimensional range matching. In *Proceedings of ACM SIGCOMM*, pages 203–214, 1998.

[189] S. S. Lam. A carrier sense multiple access protocol for local networks. *Computer Networks*, 4(1):21–32, 1980.

[190] D. Lapsley and S. Low. An optimization approach to ABR control. In *Proceedings of IEEE ICC*, June 1998.

[191] J. Y. Le Boudec and P. Thiran. *Network Calculus*. Lecture Notes in Computer Science. Springer, 2001.

[192] T. T. Lee. Non blocking copy networks for multicast packet switching. *IEEE Journal on Selected Areas in Communications*, 6(9):1455–1467, 1988.

[193] T. J. Lee and G. de Veciana. A decentralized framework to achieve max-min fair bandwidth allocation for ATM networks. In *Proceedings of IEEE GLOBECOM*, 1998.

[194] E. A. Lee and D. G. Messerschmitt. *Digital Communication*. Kluwer Academic Publishers, February 1988.

[195] W. E. Leland, M. S. Taqqu, W. Willinger, and D. V. Wilson. On the self-similar nature of Ethernet traffic (extended version). *IEEE/ACM Transactions on Networking*, 2(1):1–15, February 1994.

[196] J. Liebherr, S. Patek, and E. Yilmaz. Tradeoffs in designing networks with end-to-end statistical QoS guarantees. In *Proceedings of the International Workshop on QoS (IWQoS)*, 2000.

[197] N. Likhanov and R. Mazumdar. Cell loss asymptotics for buffers fed with a large number of independent stationary sources. *Journal of Applied Probability*, 36(1):86–96, 1999.

[198] S.-Q. Li. Nonuniform traffic analysis on a nonblocking space-division packet switch. *IEEE Transactions on Communications*, 38(7):1085–1096, 1990.

[199] S.-Q. Li. Performance of a non-blocking space-division packet wth correlated input traffic. *IEEE Transactions on Communications*, 40(1):97–108, 1992.

[200] F. LoPresti, Z. Zhang, D. Towsley, and J. Kurose. Source time scale and optimal buffer/bandwidth trade-off for regulated traffic in an ATM node. In *Proceedings of IEEE INFOCOM*. IEEE, April 1997.

[201] S. H. Low and D. E. Lapsley. Optimization flow control—I: Basic algorithm and convergence. *IEEE/ACM Transactions on Networking*, 7(6):861–874, December 1999.

[202] G. Malkin. RIP Version 2. Internet RFC 2453, November 1998.

[203] S. Mangold, S. Choi, P. May, O. Klein, G. Hiertz, and L. Stibor. IEEE 802.11e wireless LAN for quality of service. In *Proceedings of European Wireless*, February 2002.

[204] D. Manjunath and B. Sikdar. Variable length packet switches: Delay analysis of crossbar switches under Poisson and self similar traffic. In *Proceedings of IEEE INFOCOM*, pages 1055–1054, 2000.

[205] D. Manjunath and B. Sikdar. Input queued switches for variable length packets: Analysis for Poisson and self similar traffic. *Computer Communications: Special Issue on Terabit Switching*, 25(6):590–610, 2002.

[206] L. Massoulie and J. W. Roberts. Bandwidth sharing: Objectives and algorithms. In *Proceedings of IEEE INFOCOM*, 1999.

[207] A. J. McAuley and P. Francis. Fast routing table lookup using CAMs. In *Proceedings of IEEE INFOCOM*, pages 1382–1391, 1993.

[208] S. McCanne and V. Jacobson. The BSD packet filter: A new architecture of user level packet capture. In *Proceedings of Winter Usenix Conference*, pages 259–266, 1994.

[209] N. McKeown. The iSLIP scheduling algorithm for input queued switches. *IEEE/ACM Transactions on Networking*, 7(2):188–201, 1999.

[210] N. McKeown, V. Anatharam, and J. Walrand. Achieving 100% throughput in an input queued switch. In *Proceedings of IEEE INFOCOM*, pages 296–302, 1996.

[211] N. McKeown, A. Mekkittikul, V. Anantharam, and J. Walrand. Achieving 100% throughput in an input queued switch. *IEEE Transactions on Communications*, 47(8):1260–1267, 1999.

[212] R. Metcalfe and D. Boggs. Ethernet: Distributed packet switching for local computer networks. *Communications of the ACM*, 19(7):395–404, 1976.

[213] P. P. Mishra, D. Sanghi, and S. K. Tripathi. TCP flow control in lossy network: Analysis and enhancement. In S. V. Raghavan, G. von Bochmann, and G. Pujolle, editors, *Computer Networks, Architectures and Applications (C-13)*, pages 181–192. Elsevier Science, 1993.

[214] V. Misra, W. B. Gong, and D. Towsley. Stochastic differential equation modeling and analysis of TCP-windowsize behavior. In *Proceedings of PERFORMANCE*, October 1999.

[215] V. Misra, W. B. Gong, and D. Towsley. Fluid-based analysis of a network of AQM routers supporting TCP flows with an application to RED. In *Proceedings of ACM SIGCOMM*, 2000.

[216] I. Mitrani. *Probabilistic Modelling*. Cambridge University Press, 1997.

[217] J. Mo and J. Walrand. Fair end-to-end window-based congestion control. *IEEE/ACM Transactions on Networking*, 8(5), October 2000.

[218] J. C. Mogul, R. F. Rashid, and M. J. Acetta. The packet filter: An efficient mechanism for user level network code. In *Proceedings of Eleventh ACM Symposium on Operating System Principles*, pages 39–51, 1987.

[219] M. L. Molle, K. Sohraby, and A. N. Venetsanopoulos. Space-time models for asynchronous CSMA protocols for local area networks. *IEEE Journal on Selected Areas in Communications*, 5(6):956–968, 1987.

[220] J. Moy. OSPF Version 2. Internet RFC 2328, April 1998.

[221] R. Nelson and L. Kleinrock. Maximum probability of successful transmission in a random planar packet radio network. In *Proceedings of IEEE INFOCOM*, pages 365–360, 1983.

[222] R. Nelson and L. Kleinrock. The spatial capacity of a slotted ALOHA multihop packet radio network with capture. *IEEE Transactions on Communications*, 32(6):684–694, June 1984.

[223] P. Newman. Traffic management for ATM local area networks. *IEEE Communications Magazine*, August 1994.

[224] S. Nilsson and G. Karlsson. IP-address lookup using LC-tries. *IEEE Journal on Selected Areas in Communications*, 17(6):1083–1092, 1999.

[225] G. Nong, J. K. Muppala, and M. Hamdi. Analysis of nonblocking ATM switches with multiple input queues. *IEEE/ACM Transactions on Networking*, 7(1):60–74, 1999.

[226] I. Norros. A storage model with self-similar input. *Queueing Systems*, 16:387–396, 1994.

[227] H. Ohsaki, M. Murata, H. Suzuki, C. Ikeda, and H. Miyahara. Rate-based congestion control for ATM networks. *ACM Computer Communication Review*, 1995.

[228] Y. Oie, M. Murata, K. Kubota, and H. Miyahara. Effect of speedup in nonblocking packet switch. In *Proceedings of IEEE ICC*, pages 410–414, 1989.

[229] A. Orda. Routing with end-to-end QoS guarantees in broadband networks. *IEEE/ACM Transactions on Networking*, 10(1):365–374, June 1999.

[230] J. Padhye, V. Firoiu, D. Towsley, and J. W. Kurose. Modeling TCP throughput: A simple model and its empirical validation. In *Proceedings of ACM SIGCOMM*, 1998.

[231] P. Panchapakesan and D. Manjunath. On the transmission range in dense ad hoc radio networks. In *Proceedings of Signal Processing and Communications Conference (SPCOMM)*, 2001.

[232] A. Papoulis. *Probability, Random Variables, and Stochastic Processes*. McGraw Hill, 1984.

[233] A. Parekh and R. G. Gallager. A generalised processor sharing approach to flow control in integrated services networks: The single–node case. *IEEE/ACM Transactions on Networking*, 1(3):344–357, June 1993.

[234] J. H. Patel. Performance of processor-memory interconnections for multiprocessors. *IEEE Transactions on Computers*, 30(10):771–780, 1981.

[235] M. D. Penrose. On k-connectivity for a geometric random graph. *Random Structures and Algorithms*, 15(2):145–164, 1999.

[236] C. Perkins. *Ad Hoc Networking*. Addison-Wesley, 2001.

[237] L. Peterson and B. Davie. *Computer Networks: A Systems Approach*. Morgan Kaufman, third edition, 2003.

[238] J. G. Proakis. *Digital Communications*. Electrical Engineering. McGraw Hill International Editions, 1995.

[239] D. Qiao and S. Choi. Goodput enhancement of IEEE 802.11a wireless LAN via link adaptation. In *Proceedings IEEE ICC*. IEEE, 2001.

[240] K. K. Ramakrishnan. Performance considerations in designing network interfaces. *IEEE Journal on Selected Areas in Communications*, 11(2):203–219, 1993.

[241] K. K. Ramakrishnan and S. Floyd. A proposal to add explicit congestion notification (ECN) to IP. Internet Draft, March 1999.

[242] S. Ramanathan and E. L. Lloyd. Scheduling algorithms for multi-hop radio networks. *IEEE/ACM Transactions on Networking*, 1(2):166–177, 1993.

[243] R. Ramaswami and K. N. Sivarajan. *Optical Networks: A Practical Perspective*. Morgan Kaufmann Publishers, second edition, 2001.

[244] T. S. Rappaport. *Wireless Communications: Principles and Practice*. Prentice Hall PTR, 1996.

[245] P. A. Raymond. Performance analysis of cellular networks. *IEEE Transactions on Communications*, 39(12):1787–1793, 1991.

[246] K. M. Rege. Equivalent bandwidth and related admission criteria for ATM systems: A performance study. Technical Report BL045370F-980930-01TM, AT&T Bell Laboratories, 1998.

[247] M. Reisslein, K. W. Ross, and S. Rajgopal. Guaranteeing statistical QoS to regulated traffic: The multiple node case. In *Proceedings of IEEE CDC*. IEEE, 1998.

[248] Y. Rekhter and P. Gross. Application of the Border Gateway Protocol in the Internet. Internet RFC 1772, March 1995.

[249] Y. Rekhter and T. Li. Border Gateway Protocol 4. Internet RFC 1771, March 1995.

[250] R. F. Rey, editor. *Engineering and Operations in the Bell System*. AT&T Bell Laboratories, 1983.

[251] J. W. Roberts. A service system with heterogeneous user requirements. In G. Pujolle, editor, *Performance of Data Communications Systems and Their Applications*, pages 423–431. North-Holland, 1981.

[252] J. W. Roberts and L. Massoulie. Bandwidth sharing and admission control for elastic traffic. In *Proceedings of the ITC Specialist Seminar*, October 1998.

[253] R. Rom and M. Sidi. *Multiple Access Protocols: Performance and Analysis*. Springer-Verlag, 1990.

[254] M. T. Rose. *The Simple Book*. Prentice Hall, 1991.

[255] E. Rosen, A. Viswanathan, and R. Callon. Multiprotocol label switching architecture. Internet RFC 3031, January 2001.

[256] K. Ross. *Multiservice Loss Models for Broadband Telecommunication Networks*. Springer-Verlag, 1995.

[257] M. A. Ruiz-Sanchez, E. W. Biersack, and W. Dabbous. Survey and taxonomy of IP address lookup algorithms. *IEEE Network*, 15(2):8–23, 2001.

[258] M. Schwartz. *Broadband Integrated Networks*. Prentice Hall PTR, 1996.

[259] N. Shah. Understanding network processors. Master's thesis, Department of Electrical Engineering and Computer Science, University of California, Berkeley, 2001.

[260] V. Sharma and P. Purkayastha. Performance analysis of TCP connections with RED control and exogenous traffic. In *Proceedings of IEEE INFOCOM*, 2001.

[261] J. F. Shoch, Y. K. Dalal, D. D. Reddel, and R. C. Crane. Evolution of the Ethernet local computer network. *IEEE Computer Magazine*, 18(8):10–27, 1982.

[262] K. Y. Siu and H. Y. Tzeng. Intelligent congestion control for ABR service in ATM networks. *ACM Computer Communication Review*, 25(5):81–106, 1995.

[263] K. Sklower. A tree-based routing table for Berkeley Unix. Technical report, University of California, Berkeley, 1991.

[264] J. L. Sobrinho. Algebra and algorithms for QoS path computation and hop-by-hop routing in the Internet. In *Proceedings of IEEE INFOCOM*, 2001.

[265] V. Srinivasan and G. Varghese. Fast address lookups using controlled prefix expansion. *ACM Transactions on Computer Systems*, 17(1):1–40, 1999.

[266] V. Srinivasan, S. Suri, and G. Varghese. Packet classification using tuple-space search. In *Proceedings of ACM SIGCOMM*, pages 147–160, 1999.

[267] V. Srinivasan, G. Varghese, S. Suri, and M. Waldvogel. Fast and scalable layer four switching. In *Proceedings of ACM SIGCOMM*, pages 191–202, 1998.

[268] K. Sriram and D. M. Lucantoni. Traffic smoothing effects of bit dropping in a packet voice multiplexer. *IEEE Transactions on Communications*, 37(7):703–712, July 1989.

[269] K. Sriram, G. A. Mariano, D. O. Bowker, and W. J. Giguere. An integrated access terminal for wideband packet networking: Design and performance overview. In *Proceedings of the International Switching Symposium*, volume 6, pages 17–24, June 1990.

[270] W. Stallings. *ISDN and Broadband ISDN with Frame Relay and ATM*. PHI, 1995.

[271] W. Stallings. *SNMP, SNPv2, SNMPv3 and RMON 1 and RMON 2*. Addison-Wesley, 1999.

[272] D. Stephens and H. Zhang. Implementing a distributed packet fair queueing in a scalable switch architecture. In *Proceedings of IEEE INFOCOM*, pages 282–290, 1998.

[273] W. R. Stevens. *TCP/IP Illustrated*. Addison-Wesley, 2000.

[274] D. Stiliadis and A. Varma. Rate proportional servers: A design methodology for fair queueing algorithms. *IEEE/ACM Transactions on Networking*, 6(2):164–174, April 1998.

[275] D. Stiliadis and A. Varma. Efficient fair queueing algorithms for packet-switched networks. *IEEE/ACM Transactions on Networking*, 6(2):175–185, April 1998.

[276] I. Stoica and H. Zhang. Exact emumlation of an output queueing switch by a combined input output queueing switch. In *Proceedings IEEE/IFIP International Workshop on Quality of Service (IWQoS)*, pages 218–224, 1998.

[277] A. L. Stolyar. Asymptotic behavior of the stationary distribution for a closed queueing system. *Problems of Information Transmission*, 25:321–331, 1989.

[278] G. L. Stuber. *Principles of Mobile Communication*. Kluwer, second edition, 2001.

[279] M. Subramanian. *Network Management: Principles and Practice*. Addison-Wesley, 2000.

[280] S. Suri, M. Waldvogel, and P. R. Warkhede. Profile-based routing: A new framework for MPLS traffic engineering. In *Lecture Notes in Computer Science, no. 2156*, 2001.

[281] A. S. Tanenbaum. *Computer Networks*. Prentice Hall, 2003.

[282] L. Tassiulas and A. Ephremides. Stability properties of constrained queueing systems and scheduling policies for maximum throughput in multihop radio networks. *IEEE Transactions on Automatic Control*, 37(12):1936–1948, 1992.

[283] I. E. Telatar and R. G. Gallager. Combining queueing theory with information theory for multiaccess. *IEEE Journal on Selected Areas in Communications*, 13(6):963–969, August 1995.

[284] F. A. Tobagi. Fast packet switch architectures for broadband integrated services digital networks. *Proceedings of the IEEE*, 78(1):133–167, 1990.

[285] F. A. Tobagi and V. B. Hunt. Performance analysis of carrier sense multiple access with collision detection. *Computer Networks*, 1(5):245–259, 1980.

[286] T. Tripathi and K. N. Sivarajan. Computing approximate blocking probabilities in wavelength routed all optical networks with limited range wavelength conversion. *IEEE Journal on Selected Areas in Communications*, 18(10):2123–2129, 2000.

[287] B. Tsybakov and N. D. Georganas. Overflow and loss probabilities in a finite ATM buffer fed by self-similar traffic. *Queueing Systems*, 32(1-3):233–256, 1999.

[288] J. Turner and N. Yamanaka. Architectural choices in large scale ATM switches. Technical Report WUCS-97-21, Computer Science and Engineering Department, Washington University, St. Louis MO, 1997.

[289] J. G. van Bosse. *Signaling in Telecommunication Networks*. John Wiley and Sons, New York, 1998.

[290] A. Varma and D. Stiliadis. Hardware implementation of fair queueing algorithms for asynchronous transfer networks. *IEEE Communications Magazine*, pages 54–68, December 1997.

[291] R. Venkatesh. Performance analysis of aggregates of non-persistent TCP flows in very high speed networks. Master's thesis, Dept. of ECE, Indian Insitute of Science, Bangalore, January 2002.

[292] M. Vojnovic, J. Y. Le Boudec, and C. Boutremans. Global fairness of additive increase and multiplicative decrease with heterogenous round-trip times. In *Proceedings of IEEE INFOCOM*. IEEE, 2000.

[293] M. Waldvogel, G. Varghese, J. Turner, and B. Plattner. Scalable high-speed prefix matching. *ACM Transactions on Computer Systems*, 19(4), 2001.

[294] J. W. Walrand. *An Introduction to Queueing Networks*. Prentice Hall, 1988.

[295] J. Walrand and P. Varaiya. *High-Performance Communication Networks*. Morgan Kaufman, second edition, 1999.

[296] R. J. Walsh and C. M. Ozveren. The GIGASwitch control processor. *IEEE Network*, 9(1):36–43, 1995.

[297] Y. Wang, Z. Wang, and L. Zhang. Internet traffic engineering without full mesh overlaying. In *Proceedings of IEEE INFOCOM*, 2001.

[298] D. West. *Introduction to Graph Theory*. Prentice Hall of India Private Limited, 1996.

[299] W. Willinger, M. S. Taqqu, W. E. Leland, and D. V. Wilson. Self-similarity in high-speed packet traffic: Analysis and modeling of Ethernet traffic measurements. *Statistical Science*, 10(1):67–85, 1995.

[300] R. W. Wolff. *Stochastic Modelling and the Theory of Queues*. Prentice-Hall, Englewood Cliffs, New Jersey, 1989.

[301] T. C. Y. Woo. A modular approach to packet classification: Algorithms and results. In *Proceedings of IEEE INFOCOM*, pages 1213–1222, 2000.

[302] L. L. Xie and P. R. Kumar. A network information theory for wireless communication: Scaling laws and optimal operation. To appear in IEEE Transactions on Information Theory, 2003.

[303] S. Xu and T. Saadawi. Does the IEEE 802.11 MAC protocol work well in multihop wireless ad hoc networks? *IEEE Communications Magazine*, pages 130–137, June 2001.

[304] S. Xu and T. Saadawi. Revealing the problems with 802.11 medium access control protocol in multi-hop wireless ad hoc networks. *Computer Networks*, 38:531–548, 2002.

[305] Y.-S. Yeh, M. G. Hluchyj, and A. S. Acampora. The knockout switch: A simple, modular architecture for high-performance packet switching. *IEEE Journal on Selected Areas in Communications*, 5(8):1274–1283, 1987.

[306] W. H. Yuen, H. N. Lee, and T. D. Andersen. A simple and effective cross layer networking system for mobile ad hoc networks. In *Proceedings of PIMRC*, September 2002.

[307] E. W. Zegura. Evaluating blocking probability in generalised connectors. *IEEE/ACM Transactions on Networking*, 3(4), 1995.

[308] H. Zhang. Service disciplines for guaranteed performance service in packet switching networks. *IEEE Proceedings*, 83(10):1374–1395, October 1995.

[309] D. Zhang and M. Wasserman. Transmission schemes for time-varying wireless channels with partial state observations. In *Proceedings of IEEE INFOCOM*, June 2002.

[310] M. Zorzi and R. R. Rao. Effect of correlated errors on TCP. In *Proceedings of the Conference on Information Sciences and Systems (CISS)*, March 1997.

[311] M. Zorzi, A. Chockalingam, and R. R. Rao. Throughput analysis of TCP on channels with memory. *IEEE Journal on Selected Areas in Communications*, 18(7):1289–1300, July 2000.

[312] M. Zorzi, R. R. Rao, and L. B. Milstein. On the accuracy of a first-order Markov model for data transmission on fading channels. In *Proceedings of ICUPC*, November 1995.

Index